Inverse Methods in Electromagnetic Imaging

Part 1

This Advanced Research Workshop was co-sponsored by:
NATO, Scientific Affairs Division
DFVLR, Institute for Radio Frequency Technology
U.S.-ARO, European Branch Office
SIEMENS A.G., Medical Division
UIC/EECS, Communications Laboratory

NATO ASI Series

Advanced Science Institutes Series

A series presenting the results of activities sponsored by the NATO Science Committee, which aims at the dissemination of advanced scientific and technological knowledge, with a view to strengthening links between scientific communities.

The series is published by an international board of publishers in conjunction with the NATO Scientific Affairs Division

A	Life Sciences	Plenum Publishing Corporation
B	Physics	London and New York
C	Mathematical and Physical Sciences	D. Reidel Publishing Company Dordrecht, Boston and Lancaster
D	Behavioural and Social Sciences	Martinus Nijhoff Publishers
E	Engineering and Materials Sciences	The Hague, Boston and Lancaster
F	Computer and Systems Sciences	Springer-Verlag
G	Ecological Sciences	Berlin, Heidelberg, New York and Tokyo

Series C: Mathematical and Physical Sciences Vol. 143 - Part 1

Inverse Methods in Electromagnetic Imaging

Part 1

edited by

Wolfgang-M. Boerner
Communications Laboratory, EECS Department,
University of Illinois at Chicago, U.S.A.

Hans Brand
High Frequency Engineering Laboratories,
Universität Erlangen-Nürnberg, F.R.G.

Leonard A. Cram
Thorn EMI, Electronics Ltd.,
U.K.

Dag T. Gjessing
ESTP, Royal Norwegian Council for Industrial
and Scientific Research, Kjeller, Norway

Arthur K. Jordan
Space Sciences Division,
Naval Research Laboratory, U.S.A.

Wolfgang Keydel
Institüt für HF-Technik,
DFVLR, F.R.G.

Günther Schwierz
Siemens Medical Division,
Siemens AG, Erlangen, F.R.G.

Martin Vogel
Institüt für HF-Technik,
DFVLR, F.R.G.

D. Reidel Publishing Company

Dordrecht / Boston / Lancaster

Published in cooperation with NATO Scientific Affairs Division

Proceedings of the NATO Advanced Research Workshop on
Inverse Methods in Electromagnetic Imaging
Bad Windsheim, Franconia, F.R.G.
18-24 September 1983

Library of Congress Cataloging in Publication Data

NATO Advanced Research Workshop on Inverse Methods in Electromagnetic Imaging
(1983: Bad Windsheim, Germany)
Inverse methods in electromagnetic imaging.

(NATO ASI series. Series C, Mathematical and physical sciences; vol. 143)
"Proceedings of the NATO Advanced Research Workshop on Inverse Methods in Electromagnetic Imaging, Bad Windsheim, Franconia, F.R.G., September 18–24, 1983"–T.p. verso.
"Published in cooperation with NATO Scientific Affairs Division."
Bibliography: p.
Includes index.
1. Imaging systems–Congresses. 2. Electromagnetic waves–Congresses.
I. Boerner, Wolfgang M. II. North Atlantic Treaty Organization, Scientific Affairs Divison. III. Title. IV. Series: NATO ASI series. Series C, Mathematical and physical sciences; no. 143.
TK8315.N36 1983 621.36'7 84-22843
ISBN 90-277-1890-3 (set)
ISBN 90-277-1885-7 (v. 1)
ISBN 90-277-1888-1 (v. 2)

Published by D. Reidel Publishing Company
P.O. Box 17, 3300 AA Dordrecht, Holland

Sold and distributed in the U.S.A. and Canada
by Kluwer Academic Publishers,
190 Old Derby Street, Hingham, MA 02043, U.S.A.

In all other countries, sold and distributed
by Kluwer Academic Publishers Group,
P.O. Box 322, 3300 AH Dordrecht, Holland

D. Reidel Publishing Company is a member of the Kluwer Academic Publishers Group

---

Printed in The Netherlands.

TABLE OF CONTENTS (Part 1) v

TOPIC II - NUMERICAL INVERSION METHODS
(Papers are cross-referenced with the **Final Technical Program Outline.** Given are session and sequence of presentation, i.e., OS.3)

CONTENTS OF PART 2

TOPIC V - HOLOGRAPHIC AND TOMOGRAPHIC IMAGING AND RELATED PHASE PROBLEMS

(Papers are cross-referenced with the **Final Technical Program Outline.** Given are session and sequence of presentation, i.e., OS.3)

ADDRESS OF THE HOST NATION

It is an honour and pleasure for me to welcome the experts of inverse methods in electromagnetic imaging in Bad Windsheim, which is located in Franconia, the historical heart of Germany. If more than 100 experts from 11 countries representing the key persons on this important and fundamental science area meet for a workshop, then the scientific community will expect relevant results. I hope this little town, with its quiet and easy-going atmosphere, will provide the environment for fruitful discussions and for stimulating scientific interaction. The inverse methods in electromagnetic imaging are fundamental tools for remote sensing and they should find application in several military and civilian areas. However, they are in wide parts not yet developed completely. A lot of scientific and technical problems are still to be solved and I want to express my hope that the results of this workshop will contribute to the further advancement of the state of the art in this scientific field. Therefore, I request that we teach, learn and work together in order to bring the different inverse methods in electromagnetic imaging a step further into the direction of operational applicability.

A more general text and source book on this subject matter does not exist. It seems an impossible task for one man to write a suitable book because of the great variety of subjects and the wide range of applications; even if one or a group of experts could have produced a book, it would certainly have removed them from active research for several years. Therefore, in preparing the Proceedings of this Workshop, several review papers have been included which together with more specific papers should provide a good overview on the subject matter. Thus, may the resulting Proceedings serve as a source and reference book where none existed before.

Dr. Wolfgang Keydel, Director
Institute for Radio Frequency Technology
German Aerospace Research Establishment
DFVLR-Oberpfaffenhofen, FR Germany

IN MEMORIAM
EDWARD M. KENNAUGH
PROFESSOR EMERITUS
1922 - 1983

The Electrical Engineering profession and the electromagnetics discipline lost a great luminary and teacher with the death of Professor Edward M. Kennaugh at age 60 on March 11, 1983. Beyond the borders of the campus of his alma mater and that of the U.S.A he became known and highly admired, specifically through his contributions to NATO Advanced Study Institutes and AGARD Workshops.

Coming to The Ohio State University in 1946 he stayed as a graduate student under Professor Victor Rumsey, and labored for 35 yrs. on electromagnetic scattering theory, bringing international fame to himself and honor to the ElectroScience Laboratory in which he worked. His fundamental theory of the polarizing properties of radar scatterers, accomplished in the early 1950's while he was still a student, remains today the definitive work on this subject. His introduction in 1958 of the "Impuse Response" concept for three dimensional scattering obstacles brought new importance to the time-domain viewpoint in scattering theory and opened avenues of research which continue to branch even today. The large body of literature which has been built upon this one concept is overwhelming witness to the depth of Professor Kennaugh's insight into the scattering process. For twenty years, he and his students labored to refine the theory and contribute to this literature while proceeding to develop basic ideas in other directions. In the early 1960's he laid the foundations of antenna scattering theory and called attention to the importance of antennas in the modification and control of radar scattering. This work was in advance of its time and came to be widely appreciated only with the advent of low cross section aircraft. In the middle 1960's he inspired and fostered the development of modal descriptions of scattering by obstacles of arbitrary shape. In this fundamental work he displayed his genius for seeing common things in an uncommon way by bringing together the familiar ideas of power factor and eigenvectors, a union which continues to bear fruit even today. In the early 1970's, Professor Kennaugh took on the directorship of the ElectroScience Laboratory and remained at that post for three years while continuing his researches in scattering theory. His "retirement" in 1977 was one in name and not in fact, for he proceeded in the six years given to him since then to develop the so-called "K-pulse" concept which he formally introduced in 1981 and which he was carrying forward at the time of his death. This concept crowns Professor Kennaugh's years of research in time-domain methods and, with its singular originator now gone, its development remains a formidable challenge to present and future investigators.

PREFACE

This NATO Advanced Research Workshop on "Inverse Methods in Electromagnetic Imaging" has come at a time of greatly increased interest in using electromagnetic imaging methods as tools for remote sensing, material testing and medical diagnosis. Spurred by a combination of recent advances in sensor and device technology spanning the entire m-to-sub-mm wavelengths, the infra-red, optical, ultraviolet and x-ray regions of the electromagnetic spectrum, the rapid development of radar polarimetry including the deployment of active/passive sensors on spacecraft platforms, it has become evident that these advanced imaging techniques can only then become useful when supported by well-developed inverse scattering theories for proper consistent and efficient data interpretation. However, these data interpretative techniques, which are urgently needed in several military and civilian areas, are in many aspects not yet completely developed. Thus, the main purpose of this NATO-ARW-IMEI-1983 was to bring together internationally renowned key experts of this new scientific discipline of Inverse Methods in Electromagnetic Imaging and have them prepare well-written overviews, treatises on new areas, papers on special topics; have them deliberate on still unresovled problems in working discussion groups, and then report back on their important findings to the entire NATO Scientific Community in dealing with these exciting current developments.

The scientific and technical aspects of the organization, the structuring and grouping of seventy-five papers into the seven topics of the Proceedings are presented in the lead article on "Inverse Methods in Electromagnetic Imaging - An Overview of and Introduction to the Workshop Papers with a State-of-the-Art-Review", prepared in collaboration with Dr. Arthur K. Jordan.

All participants of this NATO-ARW-IMEI-1983 were actively and constructively engaged in the Working Discussion Group Program. All invitees were provided with clear and rather detailed instructions, forwarded with the pre-workshop information package, explaining why these working discussion group activities are most essential to the success of any NATO-ASI and/or ARW. Originally, we had planned five separate groups, which increased by two, and which met separately each day for about two-to-three hours. The well-written informative and stimulating reports, published in Part 2, Topic VI, of the Proceedings, give witness to the fact that although we were pressed for time, the adopted plan worked and was very successful in spite of these unavoidable time limitations and the relatively extensive working discussion group material that had to be absorbed, filtered, and put to use in formulating recommendations for the NATO Scientific Affairs Division, ASI/ARW Selection Panel Committees. We thank all discussion group participants, and particularly, commend the hard work done by the coordinators, moderators, advisors and especially the reporters for their tireless work.

After inspection of the headings and the memorial address you will observe that our workshop has been dedicated to the late Professor Edward Morton Kennaugh, who departed from us on March 11, 1983, at the age of 60. This dedication was made because of his extensive and numerous contributions to the field of inverse methods in electromagnetic imaging for which reason he had originally been chosen as one of the main speakers and as moderator and advisor for the working discussion group programs. On this occasion we are happy to report that Ed's dear wife, Mrs. Mary Kennaugh, was present during the dedication address at the Sunday evening, September 18, 1983 Opening Session and that she was able to join us for the duration of the workshop; and so we know that Professor Kennaugh's invigorating spirit and humor will be still alive and carried among us and with us into the future.

Finally, it is instructive to know that NATO Advanced Research Workshops are primarily sponsored with contributions from the United Kingdom. In retrospect, we would like to honor these financial commitments of its Ministry of Defense to the programmes of the NATO Scientific Affairs Division. In this NATO-ARW-IMEI-1983 the British contributions were doubled by the active participation of some of its finest engineering scientists as was so strongly reflected in the working discussion group activities. Our British colleagues, especially Mr. Len Cram, Prof. Alan P. Anderson, Dr. Roderick Logan, Prof. John O. Thomas, Dr. Ed Pike, together with Dr. Dag T. Gjessing, Dr. Tilo Kester, Dr. Wolfgang Keydel and Dr. Martin Vogel, punctually at 7:00 every morning sat down with us for breakfast to preview the day's scheduled events. Recalling the active participation of the latter German colleagues, it is also gratefully recorded here that they made possible the badly needed financial "fill-in" support from their Institute budget within DFVLR, Oberpfaffenhofen, to keep the workshop budget afloat. Here we also wish to acknowledge the financial contributions made by the US Army Research Office, European Branch, Edison House, London, for their assistance, and for having Dr. Karl Steinbach of MERADCOM, Ft. Belvoir, present it. Many of us were totally or partly supported by our home base institutes and we thank the Department Heads, Deans, Division Directors, and the Graduate Research Colleges for their willingness to support us. We thank them all for their contributions and the NATO Scientific Affairs Division for awarding this prestigious ARW-Award No. SA.5.2.04(0009/82)191. And last but not least, the financial sacrifices of every single participant are acknowledged, since without such individual cost sharing, only a trickle of all the key experts and their accompanients would have entered the appended list of participants, and without their presence, this NATO-ARW-IMEI-1983 just would not have taken place.

W-M. Boerner
Workshop Director

NATO-ARW-IMEI-1983
Group of Participants
present during break of
FINAL WORKING GROUP SESSION
KuK Hotel Residenz, Bad Windsheim
Friday, Sept. 23, 1983, 15:30
- Exit to the Kurpark -
"Auf der Brücke"

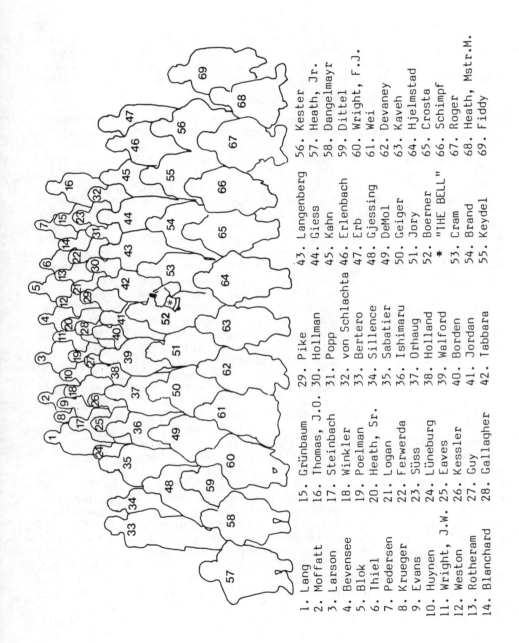

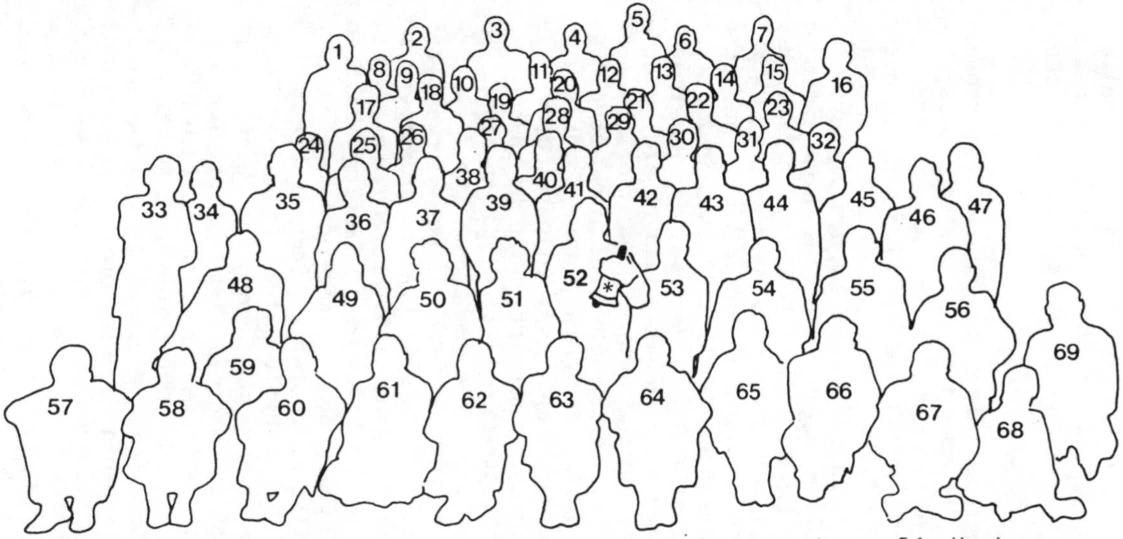

1. Lang
2. Moffatt
3. Larson
4. Bevensee
5. Blok
6. Thiel
7. Pedersen
8. Krueger
9. Evans
10. Huynen
11. Wright, J.W.
12. Weston
13. Rotheram
14. Blanchard
15. Grünbaum
16. Thomas, J.O.
17. Steinbach
18. Winkler
19. Poelman
20. Heath, Sr.
21. Logan
22. Ferwerda
23. Süss
24. Lüneburg
25. Eaves
26. Kessler
27. Guy
28. Gallagher
29. Pike
30. Hollman
31. Popp
32. von Schlachta
33. Bertero
34. Sillence
35. Sabatier
36. Ishimaru
37. Orhaug
38. Holland
39. Walford
40. Borden
41. Jordan
42. Tabbara
43. Langenberg
44. Giess
45. Kahn
46. Erlenbach
47. Erb
48. Gjessing
49. DeMol
50. Geiger
51. Jory
52. Boerner

* "THE BELL"

53. Cram
54. Brand
55. Keydel
56. Kester
57. Heath, Jr.
58. Dangelmayr
59. Dittel
60. Wright, F.J.
61. Wei
62. Devaney
63. Kaveh
64. Hjelmstad
65. Crosta
66. Schimpf
67. Roger
68. Heath, Mstr.M.
69. Fiddy

DIRECTOR'S FOREWORD

This foreword deals exclusively with the planning, organization and execution of the Workshop's scientific, as well as cultural programs. An informal discussion about some of the organizational details, may recapture the spirit of this NATO Advanced Research Workshop and will reveal how successful it was and how its accomplishments may be appreciated.

From the very outset, a rather ambitious program with a densely packed schedule was prepared. In order to reach our set goals, together with Prof. Hans Brand and Dr. Martin Vogel, we approached (in April 1981) the NATO Scientific Affairs Division and requested funding for a combined Advanced Study Institute plus Advanced Research Workshop of two weeks' duration on the subject matter. After some minor alterations and the expansion of the organizational committee to include two experienced NATO ASI/ARW organizers, Mr. Leonard A. Cram and Dr. Dag T. Gjessing, an Advanced Research Workshop of six days' duration was approved in April 1982. We found the reduction in working days expedient in that it enabled a larger number of key experts, specifically from industrial and governmental laboratories to participate; experts who would otherwise have not been able to attend. Although the original date for the staging of the Workshop of September 12-25, 1982 could have been feasible logistically, in order to allow sufficient time for planning and preparation of Workshop arrangements and, in particular, for the authors to prepare print-ready manuscripts well in advance of the Workshop, a final date of September 18-24, 1983 was set. In selecting a suitable Workshop site we, the members of the first Organizing Committee — originating from FR Germany, were looking for a modern Congress Center located off the main traffic crossroads in the historical heart of a cultural region within Germany where relevant contributions to the Arts and Sciences of Imaging had been accomplished. The choice was an easy one. I received my secondary education in Ansbach, the political center of Central Franconia, and, Professor Brand is currently a Professor in Erlangen, Franconia's largest University, we consequently chose the quiet, idyllic Franconian spa, Bad Windsheim, a medieval Reichsstadt placed in the heart of the famed Franconian Arts Center of the Fifteenth/Sixteenth Century. In this lovely Franconian region with Nürnberg to the East, Rothenburg to the Southwest, Würzburg to the Northwest and Bamberg to the Northeast, renowned contributors to the arts and sciences of imaging, such as Tilman Riemenschneider, Albrecht Dürer, Veit Stoss, Adam Kraft, Matthias Grünewald, Martin Behaim and Peter Henlein, lived and strongly interacted with one another during those turbulent years of the reformation. It was this region of Germany, which at that time, was the cultural heart of the Roman Empire of Germanic Nations North of the Alps; where the cultural and industrial crossroads of Europe, "South ↔ North", "Southwest ↔ Northeast", "Northwest ↔ Southeast", met, bringing great wealth, industrial and scholarly craftsmanship to this part of Franconia. This tradition of Franconia being a cultural and scientific center of the arts and sciences in imaging has regained new impetus and value today with the famed Siemens Zentral Laboratorium and Siemens Medical Division having their research centers in Erlangen, where exciting R&D&T advances in zeugmatography, nuclear magnetic resonance imaging are accomplished. Therefore, it was not a mere coincidence that the NATO-ARW-IMEI-1983 should take place in this historical heart of Germany. In Bad Windsheim, about forty km straight west from Erlangen, we found the ideal workshop center in the Kur und Kongress Hotel Residenz, and all those who actively participated in that Workshop will agree with us that it was a choice location. The center is embedded in the lovely Kurpark and is close to that dreamy, restful, pretty township, remote from modern life's hustle and well removed from rustling tourist traps.

It was the intent of the Workshop Organizing Committee to provide a forum for internationally renowned key experts to expand the interaction and to enlarge the scope of their activities in pursuit of promoting this rapidly growing new engineering sciences discipline of "Inverse Methods in Electromagnetic Imaging", which covers mathematical, metrological, numerical, signal processing, statistical, and also technical apparatus design topics in the engineering, physical and aeronomic sciences. As a consequence, the list of interested key experts became excessively large and, because the NATO-Scientific Affairs Division requested the total number of active participants not to exceed sixty, by too many, a careful choice in the selection of specific topic areas and of invitees had to be made so that a meaningful program and harmonious execution of the adopted approach could be guaranteed. We note here that by early September 1983, the number of key experts requesting late acceptance for participation rose steadily and reached twice as many as we were able to accommodate with the room capacity provided by the KuK Hotel Residenz. We regret that so many other valuable contributions to this, otherwise, well functioning and highly productive workshop did not materialize. The final total of accepted invited scientific participants was one hundred twenty-five, and that of accompanients was twenty-four, with approximately twenty no-shows due to illness or other sudden priority commitments. It was the determined position of the Organizing Committee to limit the total number of participants to that not exceeding the maximum available hotel capacity, and not to allow any participants to board in neighboring hotels. Thus, we achieved the assembly of all participants under one roof so as to allow optimum direct interaction and the ultimate use of a day's working hours (from 7:00 to 22:00 hours). The Workshop Director requested early (6:00) daily wake-up calls to be executed by a hotel bellboy for all participants, which indeed caused some rumbling complaints during the first two mornings of the NATO-ARW-IMEI-1983 conference, without which, however, we would otherwise not have successfully completed a long week of hard work and inspiring interaction.

The Workshop Organizing Committee had, at one time, considered the possibility of reducing the number of papers selected for oral presentation to approximately twenty and to collect the remaining papers in a Poster Session. The Workshop Director dropped this plan and every invited paper was presented and is included in the Workshop Proceedings. We note that invited papers which could not be presented due to illness or other priority commitments are also formally included in the Workshop Proceedings, except for those not cleared for publication by their administrations. In total some seventy-five papers were presented and, as a consequence, the duration for oral paper presentation and subsequent open discussions was regretfully limited. Yet, it was a pleasure to witness that the speakers, with the exception of some inexperienced junior members, obeyed the time limitations strictly. We also wish to mention that there were no simultaneous lectures scheduled nor presented; and that the language of paper presentations was exclusively English. Because all papers were of direct input to the working discussion group activities, the main purpose of oral paper presentation was to declare it an open forum for identifying unresolved questions which resulted from the research to be discussed in more detail in the Proceedings papers. Therefore, speakers were strictly advised to make optimum use of the alloted time, i.e. for a normal paper with twenty minutes time alloted for presentation, the first three minutes were to identify the specific new research under consideration, to summarize succinctly the research results of their Proceedings' paper during the next ten minutes; to use the next three to four minutes for clearly amplifying and stating the particular research problems that are unresolved; and the remainder was intended for questions from the audience. This procedure certainly required a great deal of concentration on the part of all participants, as well as strict adherence to the set session schedules which was admirably executed by all session chairmen. The Workshop Director permitted himself the liberty of using the KuK-Congress-Center-bell, whenever "Call to Order" was required, to get us back to the pre-set time schedule. In retrospect, everyone will agree that this approach not only worked, but, as so many stated in their enthusiastic, positive thank you notes, it also contributed to the optimum information exchange, stimulation of new ideas, and the creation of new interactions among participants who only knew of each other by names on publications, and will now desire to meet again in the near future. In concluding, the Organizing Committee wishes to thank all speakers and session chairmen for the demonstrated harmonious collaboration, for the excellence of paper content, as well as paper presentation and open-minded question handling.

In enriching the overall program, cultural events were scheduled for the evenings on a daily basis in addition to the two main cultural scientific events of Wednesday, September 21 and Saturday,

September 24, 1983. Furthermore, deliberately on late call, a rather functional ladies program was intelligently improvised with the dear assistance of Ms. Ursula Allmendinger, Verkehrsamt, Bad Windsheim, by Mrs. Barbara Kester, Mrs. Mary Jordan and Mrs. Eileen Boerner, whose inspiration made the cultural ladies tours, sight-seeing, walking and biking tours a memorable event for the accompanients and also some scientists. The two cultural/scientific tour events were mainly planned, organized and executed by Prof. Dr.-Ing. Hans Brand and his pleasant institute staff members, Mrs. Margarete Geiger, Dipl-Ing. Siegfried Osterrieder and Dipl-Ing. Roman Glöckler who deserve our highest praise and admiration. The cultural evening programs were arranged with the assistance of Mr. Rolph K. Erlenbach, the Hotel Manager, Mr. Hans-Dieter Erb, the Congress Center Manager, and Ms. Ursula Allmendinger who went out of their way to help, advise and assist us in any way they could. Therefore, we wish to use this ocassion to express, on behalf of all participants, our admiration and gratitude for the thorough and sincere assistance we received during our stay in Bad Windsheim, not only from its friendly citizens during our walks and shopping sprees through their inviting township but, in particular, to the staff and management of the KuK Hotel Residenz, specifically, the head receptionist, Mrs. Rossmann and her friendly assistants, the most impressive, capable master of the Restaurant "die Brücke", Herr Dick, with his well selected team of friendly, attentive waiters, and not to forget our friend "Bruno". It was a pleasure to be their guests, because the excellent service we received and the peaceful atmosphere of the restaurant enabled us to work hard, yet at the same time, eat well and relax. Then, after "DAY'S END", some of us disappeared downstairs into the hotel lounge to enjoy the relaxed atmosphere, created by lively Fräulein Kleine. The KuK Hotel Residenz with its recreational facilities and the adjacent Kurpark certainly provided an ideal stimulating, yet pleasant environment for such a work-loaded Advanced Research Workshop, lying as it does in the historical heart of the modern German culture, Franconia.

In the following, some of the important highlights of our scientific/cultural events will be summarized with the intention of concluding the description of our NATO-ARW-IMEI-1983 which was a successful, well-rounded scientific as well as cultural experience.

Monday Evening Event (September 19, 1983)

We recall with pleasure the leisurely walk from the KuK Hotel Center through the Kurpark and downtown Bad Windsheim to the City Hall, where we were given a reception planned by the First Mayor Mr. Otmar Schaller, with the Second Mayor, Mr. Josef Heinrich present to receive us in the Historische Rathaussaal. For those U.S. citizens of German descent it was an historical event in that Francis Daniel Pastorius (1651-1719), the son of a former Bad Windsheim Mayor Pastorius (1675-1691), was raised and received his outstanding early education in this city, passed his inaugral disputation (doctorate of law) at the age of 25 in Altdorf bei Nürnberg, the famed medieval Franconian University which in the 18th century was moved to Erlangen, preparing him to become the co-founder of the first official German Settlement in the United States of America on October 6, 1683 in Germantown, PA. It is noteworthy to mention that it was Francis Daniel Pastorius the well travelled, highly respected lawyer and scholar, who formulated the first decree in Northern America banning slavery and racial segregation, in 1697, which later on became one of Abraham Lincoln's most admired documents on the subject matter. It was a pleasant evening and made everyone relax after having survived the first day of our "Arbeitszusammentreffen" which, indeed, was tough for many who arrived the day before from North America.

Tuesday Evening Event (September 20, 1983)

A total surprise was our pleasant reception in the Franconian Open Air Museum, where on a perfect "Indian-Summer" evening, we viewed original small villages and farms from the Fifteenth to Eighteenth Century in Franconia which were brought from the surrounding countryside, salvaged from complete destruction, and reassembled in this lovely location. The delicious carp dinner (the traditional Franconian fall festivity) was followed with the superb presentation by the "Bad Windsheimer Sänger", so masterfully conducted by Professor Horst Steinmetz, who, with his capable artists, introduced us to Franconian folk and art songs of the Fifteenth/Sixteenth Century including many enjoyable cultural originals created within the vicinity of Bad Windsheim. This lively event provided us with the pleasure to get to know many new and dear friends, to appreciate the deep humorous and relaxed character of the Franconians, and set the foundation for a harmonious working environment for the remainder of the Workshop.

Wednesday, Scientific/Cultural Tour (September 21, 1983)

By now, all of the participants of this NATO-ARW-IMEI-1983 were able to appreciate that not only are the scientific Workshop programs on schedule, but so are the cultural ones, and this according to a motto: "hard work deserves hard play"! This Wednesday Scientific/Cultural Tour in itself must have made it worth coming to Franconia, and our praise, admiration and thanks go equally to the Siemens Medical Division, Siemens AG., and the High Frequency Engineering Laboratories, UEN, both in the famed Hugenottenstadt, Erlangen. While Dr. Manfred Pfeiler, Dr. Werner J. Haas. and Dr. Günter Schwierz provided the financial support for the elegant Workshop portfolio, the bus tour, the scientific visits and most of the contributions towards the overhead of other cultural events; Prof. Dr.-Ing. Hans Brand and his institute staff planned these tour programs. Dipl-Ing. Roman Glöckler, who was responsible for organizing the tours, was unable to attend due to illness and other priority commitments. We all commend the pleasant, dear, intelligent and polite manner in which Dipl-Ing. Siegfried Osterrieder substituted and executed the tour programs so efficiently.

After the late morning lectures on Tomographic Imaging and the excellent overview of Dr. Günter Schwierz on the subject matter, introducing us to one of the most exciting fields of "Inverse Methods in Electromagnetic Imaging", namely Zeugmotography or Nuclear Magnetic Resonance Imaging, we were introduced to the development of this exciting new medical imaging technology which sooner or later may replace hazardous, detrimental x-ray tomographic methods. However, the ultimate deployment of these not only very demanding advanced imaging techniques, side-by-side, will also require a very expansive high technology base plus expertise in "Inverse Methods in Electromagnetic Imaging", at a cost level of about three to five million dollars per final imaging unit which includes very large, super-cooled magnets (5m x 3m x 3m), very sensitive RF receivers, fifth generation parallel processing computers equal to or probably larger than the CRAY-type, and many other extensive software supporting programs. To the participants of this workshop it was very interesting to be informed that as large and extensive a global enterprise as the Siemens, AG. is, the small and disadvantaged business component suppliers are now from almost all NATO-member countries, due to their highly selective technological expertise, providing another reason for our close future collaboration and the further expansion of the ASI/ARW programmes of the NATO-Scientific Affairs Division. It was also interesting to note that a great demand on improved "Inverse Methods in Electromagnetic Imaging" is still required to make Magnetic Resonance Imaging become of real practical use in everyday medical diagnostic imaging. We express our fond gratitude to all of the scientists and engineers who guided us and replied so knowledgeably and enthusiastically to our questions all pertaining to and demonstrating the rapid advancement of the new engineering sciences discipline of "Inverse Methods in Electromagnetic Imaging" of which Erlangen now has become one of the most important R&D&T centers, certainly another valid reason to have our NATO-ARW-IMEI-1983 take place adjacent to its centers of research activities in this new exciting engineering sciences discipline.

The subsequent visit of the "Germanische National Museum", where concurrently, the "500 Anniversary Exhibition Martin Luther (1483-1546)" and the "450 Anniversary Exhibition Veit Stoss (1440-1533)" were taking place, and the late evening "Franconian Smorgasbord at the Heiliggeist Spital" were well organized and culturally inspiring events. At the museum we split up into several groups, all visiting the Veit Stoss Exhibition, and unfortunately, missing the Martin Luther Anniversary, which took place in that Section of the Museum where Dürer's, Behaim's, Kepler's, Henlein's and all the other great artist's and scientists' of the Fifteenth/Sixteenth Century contributions to Imaging of Nature, the Human Mind, and the Extraterrestrial Bodies are collected. Yet, there is so much to be comprehended on the subject matter; and, our guided tours through the Veit Stoss Exhibition certainly were gratifying in that the interaction of all of the contemporaries of this culmination point of medieval Trans-European culture was so vividly demonstrated to us. So, it is interesting to note, that Albrecht Dürer, the painter; Veit Stoss, the painter/sculptor; and Tilman Riemenschneider, the sculptor/artist/statesman had produced combined works of art in Imaging, not only of the lively body, but also of the human mind, certainly the most daring and demanding subfield of expertise in the general sciences of inverse methods. In concluding this exciting visit to the Germanische National Museum in Nürnberg, it is worthwhile mentioning that Albrecht Dürer set some formal foundations of tomographic imaging in his treatise on stereographic projections. Thus, all those participants and others who are going to revisit Nürnberg, don't miss the Germanische National Museum and the Dürer Haus.

This exciting day was concluded with an elegant, neatly prepared, Franconian Smorgasbord Buffet at the Heiliggeist-Spital, located on a bridge across the river Pegnitz, which provided another event of socializing and of finding new friends and scientific interaction. By suprise, the traditional "Altstadt Fest: Die Pegnitz brennt!" took place with fireworks right below the restaurant on the river, an historic event which dates back to the middle ages. By this time, many of those who wanted to depart and disappear in tourist traps, hung on to us, with the understanding and conviction that cultural/scientific offerings of a NATO-ARW, such as ours, could in no way be surpassed or substituted by individually arranged sidestepping tours, i.e. we were successful in aiding a flock of approximately one-hundred-and-thirty individualistically minded professors and engineering scientists to actively live together and interact with one another until the very end of the Workshop.

Friday Evening Workshop Dinner (September 23, 1983)

After the closing down of the Scientific Paper Presentations and completing the Working Discussion Group Reports, it was a delight to sit down, relax and get prepared for the final event, the introduction to the Life of Tilman Riemenschneider, Artist/Scholar/-Statesman. Before our invited dinner speaker, Mr. William McKee Wisehart, was given access to the podium, Dr. Dag T. Gjessing, Dr. Tilo Kester and Mr. Leonard A. Cram discharged the Workshop Director from his duty and made him replace the KuK Congress Center (school) bell by a tiny mountain bell (with Saxionian pitch: very high in tone, clear and not too loud). With pleasure he accepted this relief and handed over the loud instrument which was used for "Call to Order", to the next NATO-ARW Director, Dr. Robert C. Worrest, of the Oregon State University in Corvallis, who was thus given the maddening instrument to punctually commence with his workshop on Monday morning, September 26, 1983 (who knows whether he needed the bell for some thirty or less participants). Here, I also wish to thank my friend, Prof. Walter K. Kahn, for participating in the Workshop, and for making himself feel at home again in the beloved homeland of his honorable ancestors who had contributed so much to its wealth of knowledge, scholarship, and industry. Also, to say thank you to all the participants for some lively days of hard scientific work and friendly, accomodating and intelligent interaction.

The post-dinner speech of Bill McKee Wisehart was by no means a standard one, and with intelligent, sensitive elaborations, he introduced us, most of whom were not capable of speaking the German Language, nor familiar with its way of life, into the Franconian sphere of life at the turn of the Fifteenth/Sixteenth Century and then, step-by-step, made us appreciate the artistic and scholarly achievements of one of mankind's greatest craftsman, Tilman Riemenschneider. Bill had done a marvelous job and no Professional Scholar of the Historical Fine Arts could easily have replaced him. It is important to note that with his fine selection of slides he was able to trace the connections and interactions of Tilman Riemenschneider with Veit Stoss and Albrecht Dürer and to explain why Tilman's contribution became almost forgotten. In the early 1500's Bad Windsheim also sponsored and displayed one of Riemenschneider's Works of Art, which in reproduction has been recently reinstalled in its original shrine in the Stadtkirche.

In the final presentation of the evening, Prof. Brand and Dipl-Ing. Osterrieder were kind enough to provide us with a preview of the Workshop Concluding Scientific/Cultural Tour planned for Saturday which was centered on the artistic productions of the sculptor, Tilman Riemenschneider. Every intermediate stop and the highlights of that program were clearly identified. Both the presentations of Bill McKee Wisehart and of Prof. Brand, with Dipl-Ing. Osterrieder complimented one another very well, and got us all excited about participating in the Final Workshop Close-Down Tour Activities. Equal thanks are extended to all three.

<u>Saturday Cultural/Scientific Close-Down Tour (September 24, 1983)</u>
As requested by the KuK Hotel Residenz management, by 9:00 AM all participants of this NATO-ARW-IMEI-1983 had vacated the hotel. Approximately sixty participants were transported by bus, others followed in their rental/own cars and departed Bad Windsheim, guided across the Frankenhöhe to Rothenburg by Bill, who, himself a U.S. citizen and fine scholar, carefully explained all the specific differences in village structure, farming, forresting, etc. that could be observed during our tour, which was highly appreciated by those not acquainted with Franconia and/or Central Europe.

In Rothenburg, a totally rebuilt city, everyone was on his own to view this "idol" of a medieval township: city walls, narrow "Gassen", medieval wood-framed housefronts, old cathedrals and also a Judaic cemetery rebuilt next to the house from which the famed scholar of the Talmud, and temporary citizen of Rothenburg, Meier Ben Baruch (1215-1293), fought against the unsuccessful attempt by

Emperor Rudolf to have the Jews pay taxes, resulting in his imprisonment (1286) and death in the Emperial prison at Worms. The main event though, was to view one of Tilman Riemenschneider's masterpieces, the Heiligblutaltar in the St. Jakob's Kirehe, Rothenburg ob der Tauber. After viewing the many memorabilia of the middle ages, most of us must have been grateful that the Workshop took place in Bad Windsheim and not in Rothenburg; otherwise, many of us would have simply gone astray in this quaint and inviting tourist trap which has so many exciting views to offer.

Bill guided us during the next part of the bus tour to Creglingen, where we admired the Marienaltar, which by many occidental and oriental fine arts historians is considered to be the culmination and absolute perfection in wood carving, demonstrating the grandeur of that Franconian phase of high cultural, but also high technological, activities (e.g. the colour woodprinting process invented by Albrecht Dürer, the first pocket-watch instrumented by Peter Henlein, the first globe created by Martin Behaim, etc.)

In leaving Creglingen, we soon departed from the deep-rutted Tauber valley and ended up near the Käpelle above Würzburg for lunch under open blue skies to enjoy the view of the River Main, and the famed baroque spires of Würzburg, being overpowered by the mighty fortress, Marienburg, encroaching the river and the city on the south side. We mention here that the 450 Anniversary Exhibition of Tilman Riemenschneider (1460-1531) took place in the Mainfränkische Museum in the Marienburg in, 1981, setting the prerogatives of the Workshop Director in selecting this cultural/scientific program. After viewing the baroque Käpelle or the Roccocco Residenz, with the famed wall paintings and frescoes by the Venetian Master, Giovanni Batista Tiepolo (1696-1770), and its famous fountain, "Der Künstler Brunnen", depicting the contributions of three of Mainfranken's world-famed artists: Walther von der Vogelweide (1170-1230), Matthias Grünewald (1465-1535), and Tilman Riemenschneider, the tour bus returned along the main valley road and then via the Autobahn to the Nürnberg-Airport and to Nürnberg Hauptbahnhof, where the Workshop came to a fulfilled end. The remaining participants dispersed for a relaxing weekend either in Nürnberg, in nearby München and/or the Alps and we hope that they all returned home safely with the understanding that there is a good reason why we need to have a North Atlantic Treaty Organization and that its Scientific Affairs Division serves the noble purpose of reminding us that there is a lot to be protected in any one of our many NATO member countries, rich in history, talent, and free societies.

In retrospect, it had been the intention, that next to accomplishing outstanding technical and scientific workshop results, to introduce the participants of this NATO-ARW-IMEI-1983 to the rich cultural and scientific life of one of the most productive periods of German and European history. Although born in Austral-Asia, of Saxonian descent, the Workshop Director made Franconia his

"Wahlheimat" (motherland of choice) and prefers to speak its dialect rather than use the official instrument of German linguistic interaction "Hochdeutsch". The reason for this laudable action is his early realization that this part of Germany had discovered a lot of basic truths, produced honest statesmenship and generated great artistic values which still are not fully comprehended, but had so much to offer to make real day-to-day life more livable and easier to appreciate for all of mankind. Hence, the Workshop Director personally hopes that all participants acquired deep respect for this dreamy, well-tempered stronghold of German medieval culture which had so much to offer toward obtaining a more realistic view, down-to-earth pure image of our terrestrial and celestial environment.

In preparing such an extensive NATO-ARW, enormous secretarial assistance, organizational ability and know-how was required. The prime source of such contributions came from within the Communications Laboratory, Department of Electrical Engineering & Computer Science, Univeristy of Illinois at Chicago, and all our thanks go directly to Richard and Deborah Foster, without whose talents and dedication the pre-workshop and post-workshop processing of all relating material, particularly producing the Workshop Proceedings could not have been accomplished. Here, we would like to thank Drs. Sujeet K. Chaudhuri, Koroda Umashankar and Anthony J. Devaney for their assistance in proof reading the manuscripts.

On this occasion, I would like to express my very special, warm and hearty thanks to Dr. Mario Di Lullo and Dr. Craig Sinclair, the Directors of the NATO-Scientific Affairs Division, ASI/ARW Programmes, within the NATO Headquarters in Brussels; together with Dr. Tilo Kester and his capable, dear wife, Barbara Kester, of the NATO Publications Coordination Office, for their generous advice and many discussion hours so freely offered to me during the preplanning and organization phase of this NATO-ARW-IMEI-1983.

Last, but not least, my understanding wife, Eileen Annette, deserves my deep appreciation for her patient tolerance of the many long nights and lost weekends of extra work during the entire pre to post workshop engagement.

Wolfgang-M. Boerner
The Workshop Director

NATO-ARW-IMEI-1983
Friday, Sept. 23, 1983, 18:30
KuK Hotel Residenz, Bad Windsheim, FR Germany
The Ladies Group

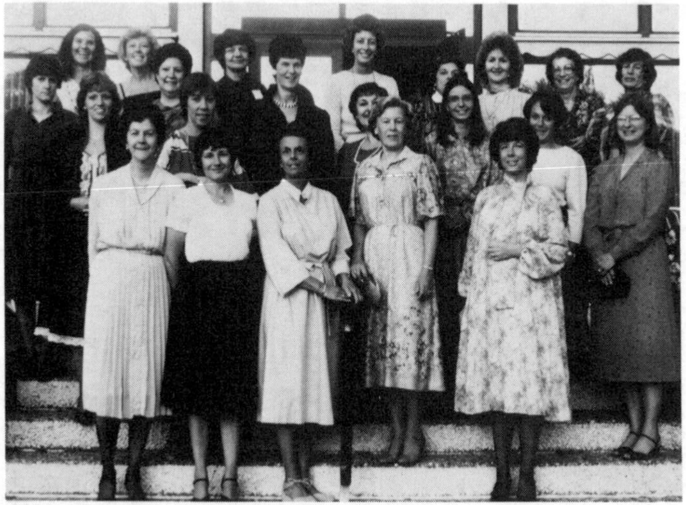

1. Barbara Kester
2. Joan Rotheram
3. Vila Eaves
4. Virginia Jory
5. Jackie Weston
6. Mary Jordan
7. Susan Moffatt
8. Mae Bevensee
9. Mary Kennaugh
10. Aartji Block
11. Dianne Logan
12. Cindy Manson
13. Phylene Raven
14. Carli Poelman
15. Dixie Carpenter
16. Rie Fewerda
17. Ruth Wright
18. Etty Huynen
19. Christine de Mol
20. Avril Guy
21. Maureen Anderson
22. Eileen Boerner

Photographed by: Bruce Z. Hollmann
Portrayed by : Richard W. Foster
Reproduced by : John Darby
Available from : Communications Lab/EECS Dept., M/C 154
University of Illinois at Chicago
P.O. Box 4348, Chicago, IL 60680 USA

ORGANIZATIONAL COMMITTEE

PLANNING COMMITTEE

DIRECTOR: Professor Wolfgang-M. Boerner
CL-EECS, University of Illinois at Chicago
Chicago, IL 60680 USA

CO-DIRECTORS:

Professor Hans Brand
Universität Erlangen-Nürnberg
Erlangen, FR GERMANY

Professor Dag T. Gjessing
ESTP, Royal Norwegian Council for Industrial and Scientific Research, Kjeller, NORWAY

Dr. Günter Schwierz
SIEMENS A.G., Medical Division
Erlangen, FR GERMANY

Mr. Leonard A. Cram
THORN EMI, Radar Electronics
Wells/Somerset, UK

Dr. Wolfgang Keydel
DFVLR, Institute for Radio Frequency Technology
Oberpfaffenhofen, FR GERMANY

Dr. Martin Vogel
DFVLR, Institute for Radio Frequency Technology
Oberpfaffenhofen, FR GERMANY

NATO-ADVISORS

Dr. Mario Di Lullo
ARW/ASI Programmes, Scientific Affairs Division
NATO HQ, Brussels BELGIUM

Professor Anton de Hoop
Delft University of Technology
Delft, THE NETHERLANDS

Dr. Tilo Kester
NATO-Publications Coordination Office
Overyse, BELGIUM

Dr. Craig Sinclair
ARW Programme, Scientific Affairs Division
NATO HQ, Brussels, BELGIUM

TECHNICAL COMMITTEE AND EDITORIAL BOARD

Professor Wolfgang-M. Boerner
Professor Hans Brand
Dr. Günter Schwierz
Dr. Arthur K. Jordan
SSD, Naval Res. Lab., USA
Mr. Leonard A. Cram
Dr. Wolfgang Keydel
Professor Dag T. Gjessing
Dr. Martin Vogel

COORDINATORS OF SCIENTIFIC/CULTURAL TOURS

Dipl.-Ing. Roman Glöckler
Dipl.-Ing. Siegfried Osterrieder
UEN, Erlangen, FR GERMANY
Professor Hans Brand
Dr. Günter Schwierz
Professor Wolfgang-M. Boerner

HOSTESSES OF THE LADIES/CULTURAL PROGRAMMES

Frau Margarete Geiger
UEN, Erlangen, FR GERMANY
Mrs. Eileen A. Boerner
Northbrook, IL, USA
Frau Ursula Allmendinger
Verkehrsamt, Bad Windsheim, FRG
Frau Barbara Kester
Overyse, BELGIUM

WORKSHOP PROGRAM AND PROCEEDINGS DEVELOPMENT AND PREPARATION

Mr. Richard W. Foster
Mrs. Deborah A. Foster
CL-EECS, University of Illinois at Chicago, Chicago IL, USA

0.1 (OS.3)

TRANSIENT CURRENT DENSITY WAVEFORMS ON A PERFECTLY CONDUCTING SPHERE

E.M. Kennaugh
D.L. Moffatt

The Ohio State University
Department of Electrical Engineering
2015 Neil Avenue
Columbus, Ohio 43212

FOREWORD

The transient current densities induced on a conducting spherical scatterer by an incident plane wave with impulsive time dependence, while in existence for some time at the ElectroScience Laboratory, have never been formally published. The time-dependent current density waveforms are an elegant example of how the impulse response concept can illuminate and simplify relatively complex relationships. Much of the time domain research by Prof. E.M. Kennaugh, together with his students and colleagues, has appeared in the literature. This paper, while it is not his final work, serves to further complete a report of his unique contributions. I am indebted to Prof. W-M Boerner for the opportunity to include this paper in the Workshop Proceedings.

ABSTRACT

The time-dependent surface current density waveforms at various points on the surface of a perfectly conducting sphere illuminated by a plane electromagnetic wave are presented. The incident plane wave has an impulsive or shock-type time dependence. Current density waveforms in the E-plane (strong creeping wave) and H-plane on both the "lit" and shadowed sides of the sphere are shown. On the illuminated side of the sphere, removal of the Kirchhoff or physical optics approximation for the surface current density permits a detailed examination

W.-M. Boerner et al. (eds.), Inverse Methods in Electromagnetic Imaging – Part 1, 1–31.

of the deficiencies of this approximation for short times (high frequencies). From these results the form of a first order correction to the Kirchhoff current is suggested. On the shadowed side of the sphere, the changing character of the current waveform is noted.

INTRODUCTION

Development of the impulse response concept in 3-dimensional electromagnetic scattering problems[1,2] was materially aided by a Fourier synthesis procedure whereby harmonically related complex scattering data were used to synthesize approximate far-zone canonical response waveforms[2,3,4,5]. The same procedure permits one to generate the time-dependent surface current density waveforms on a scatterer or radiator. A comparison of Fourier synthesis and space-time integral equation calculations has been made[6]. Both procedures have an inherent resolution limitation dictated either by a finite summation (Fourier synthesis) or a Gaussian pulse excitation (space-time integral equation), but it is not correct to suggest that an infinite summation[7] or equivalently an impulsive excitation must be used. However, some care must be exercised in interpreting singularities and/or jump discontinuities particularly if they occur simultaneously.

Impetus for this paper stems from a depolarization correction to physical optics derived by Bennett[8] and utilized by Boerner[9]. We show that even for an object where no depolarization occurs (the conducting sphere), the form of a basic correction to the Kirchhoff current can be deduced. It is suggested that similar first order corrections will improve the Kirchhoff approximation for a number of different object geometries.

Surface Current Density Waveforms

Mie series calculations of the tangential magnetic field and radial electric field at the surface of a conducting sphere of radius a immersed in an incident plane wave have been made for sphere circumferences in wavelengths of 0.2 (0.2)20.0. Locations on the sphere surface in both the E-plane and H-plane for theta angles of 0 (15) 180 degrees were used with 0 degrees corresponding to the specular point. The real, time-dependent surface current density waveforms synthesized from these data are shown in Figures 1, 2 and 3. Figure 1 combines the E-plane (positive) and H-plane (negative) waveforms on the illuminated side of the sphere. The abscissa scale is in units of transit time for the sphere diameter. On the illuminated side of the sphere, the weighted quasi-impulses at the origin of each waveform illustrate the resolution obtained (approximately o.75a). On the illuminated side of the sphere, the H-plane waveforms have a sign reversal. The abscissa scales are in units of transit time for the sphere diameter. On the illuminated side of the sphere, the current density is primarily impulsive; details for short times are best seen after the Kirchhoff approximation for the current is removed. There is clear evidence of the creeping wave contributions. The surface current density waveforms on the shadowed side of the sphere are shown in Figure 2 (E-plane) and Figure 3 (H-plane). Note that the ordinate scales for H-plane are not constant. On the shadowed side of the sphere note that the character of the current density at the onset of the waveform is still apparently impulsive near the shadow boundary (theta less than 120 degrees) in the H-plane but not in the E-plane. There is a strong creeping wave contribution in the E-plane and some evidence of a very weak creeping wave contribution in the H-plane.

In Figure 4 the surface current density waveforms on the illuminated side of the sphere are shown after the Kirchhoff approximation to the current density has been removed. The H-plane waveforms are dashed. It is very evident that as the observation point progresses from the specualr point to the shadow boundary, the Kirchhoff approximation becomes progressively too large in the E-plane and too small in the H-plane. At a given angle, the magnitude of the errors in the E-plane and H-plane are approximately equal. This would seem to imply that on a cut at 45 degrees from the E-plane and H-plane the Kirchhoff approximation to the current may be quite good even near the shadow boundary. In the E-plane at theta equal 60 degrees and 75 degrees one can also clearly discern two distinct creeping wave contributions, both launched from the shadow boundary and delayed from the waveform onset by different travel times on the sphere surface.

Correction

Based on the results shown for the conducting sphere, the following formulas for correcting the physical optics estimates for the induced currents are suggested for smooth (no edge) scatterers. In the E-plane modify the physical optics currents by the multiplicative factor

$$[1 - K \sin\theta] \quad . \tag{1}$$

In the H-plane modify the physical optics currents by the multiplicative factor

$$[1 + K \frac{\sin\theta}{\cos\theta}] \;, \qquad (2)$$

where θ is the angle between the outward normal to the surface of scatterer and the backscatter direction. The H-plane correction becomes infinite at the shadow boundary (θ=90°) but when multiplied by the physical optics current estimate ($\cos\theta$ factor) yields a finite current estimate. The specular error (exact minus physical optics) before correction is shown in Figure 5. Figures 6 and 7 show the effect of the correction term for values of the constant K of 0.2 and 0.1, respectively. For the simple form of the correction suggested, a value K near 0.1 appears optimum. Other forms for the correction, possibly exponential, may suggest themselves to the reader. Our purpose here, however, was to show that simple corrections are possible.

Conclusions

It has been demonstrated that relatively simple formulas can be developed for correcting the physical optics estimates of the current densities induced on a conducting spherical scatterer. It has also been suggested that similar simple corrections could be made for other "smooth" scatterers. It is noted that the corrections indicated here are not related to depolarization properties of the scatterer.

The surface current density waveforms on the shadowed side of the sphere have been included here for the possible benefit of other researchers. The physical optics estimate has long been a favorite direct scattering solution for inverse scattering because the estimate can be generally related to a target cross sectional area function on the illuminated side of the scatterer. A similar relationship (general) for the shadowed side of the scatterer would greatly improve imaging techniques based on the physical optics estimate.

REFERENCES

[1] E.M. Kennaugh and R.L. Cosgriff, "The Use of Impulse Responses in Electromagnetic Scattering Problems," 1958 IRE Nat'l Conv. Rec., Pt. 1, pp. 72-77.

[2] E.M. Kennaugh and D.L. Moffatt, "Transient and Impulse Response Approximations," Proc. IEEE, Vol. 53, No. 8, pp. 893-901, August, 1965.

[3] E.M. Kennaugh, "The Scattering of Short Electromagnetic Pulses by a Conducting Sphere," Proc. IRE (Correspondence) Vol. 49, p. 380, January, 1961.

[4] D.L. Moffatt and E.M. Kennaugh, "The Axial Echo Area of a Perfectly Conducting Prolate Spheroid," IEEE Trans. on Antennas & Propagation, Vol. AP-13, pp. 401-409, May, 1965.

[5] E.M. Kennaugh and D.L. Moffatt, "On the Axial Echo Area of the Cone Sphere Shape," Proc. IRE (Correspondence) Vol. 50, p. 199, February, 1962; see also K.M. Siegel, et. al., "Radar Cross Section of a Cone Sphere," Proc. IRE (Correspondence) Vol. 51, pp. 231-232, January, 1963.

[6] C.L. Bennett, "The Numerical Solution of Transient Electromagnetic Scattering Problems," SCRC-RP-77-55, Sperry Research Center, August, 1977.

[7] E.M. Kennaugh and D.L. Moffatt, Comments on "Impulse Response of a Conducting Sphere Based on Singularity Expansion Method," Proc. IEEE, Vol. 70, No. 3, pp. 294-295, March, 1982.

[8] C.L. Bennett, A.M. Auckanthaler, R.S. Smith and J.D. DeLorenzo, "Space Time Integral Equation Approach to the Large Body Scattering Problem," Sperry Rand Research Center, Sudbury, Mass., May 1973. (AD 763794) (RADC-CR-73-70)

[9] S.K. Chaudhuri and W-M. Boerner, "A Monostatic Inverse Scattering Model Based on Polarization Utilization," Applied Physics, Vol. 11, pp. 337-350, 1976.

ADDITIONAL REFERENCES

A relatively complete list of publications and reports of The Ohio State University's ElectroScience Laboratory on the application of time domain concepts to electromagnetic scattering and radiation problems is given here.

Publications

[1] E.M. Kennaugh and R.L. Cosgriff, "The Use of Impulse Responses in Electromagnetic Scattering Problems," 1958 National Convention IRE Record, Pt. 1, pp. 72-77.

[2] E.M. Kennaugh, "The Scattering of Short Electromagnetic Pulses by a Conducting Sphere," Proc. IRE (Correspondence), Vol. 49, 1961, p. 380.

[3] E.M. Kennaugh and D.L. Moffatt, "On the Axial Echo Area of the Cone Sphere Shape," Proc. IRE (Correspondence) Vol. 50, 1962, p. 199. Also see Proc. IRE (Correspondence) Vol. 51, 1963, p. 232.

[4] D.L. Moffatt and E.M. Kennaugh, "The Axial Echo Area of a Perfectly Conducting Prolate Spheroid," Proc. IEEE (Correspondence), Vol. 52, 1964, pp. 1252-1253.

[5] D.L. Moffatt and E.M. Kennaugh, "The Axial Echo Area of a Perfectly Conducting Prolate Spheroid," IEEE Trans. on Antennas and Propagation, Vol. AP-13, 1965, pp. 401-409.

[6] E.M. Kennaugh and D.L. Moffatt, "Transient and Impulse Response Approximations," Proc. IEEE, Vol. 53, 1965, pp. 893-901.

[7] D.L. Moffatt, "The Echo Area of a Perfectly Conducting Prolate Spheroid," IEEE Trans. on Antennas and Propagation, Vol. AP-17, No. 3, 1969, pp. 299-307.

[8] D.L. Moffatt, "Impulse Response Waveforms of a Perfectly Conducting Right Circular Cylinder," Proc. of IEEE, Vol. 57, No. 5, 1969, pp. 816-817.

[9] D.L. Moffatt, R.H. Paul and R.A. Voss, "The Echo Area of a Perfectly Conducting Prolate Spheroid," IEEE Trans. on Antennas and Propagation, AP-21, No. 2, p. 231, March 1973.

[10] J.D. Young, D.E. Svoboda and W.D. Burnside, "A Comparison of Time- and Frequency-Domain Measurement Techniques in Antenna Theory," IEEE Trans., Vol. AP-21, No. 4, 1973, pp. 581-583.

[11] D.L. Moffatt and R.K. Mains, "Detection and Discrimination of Radar Targets," IEEE Trans. on Antennas and Propagation, Vol. AP-23, No. 3, 1975, pp. 78-83.

[12] J.D. Young, "Radar Imaging from Ramp Response Signatures," IEEE Trans. on Antennas and Propagation, Vol. AP-24, No. 3, May 1976.

[13] D.L. Moffatt and R.J. Puskar, "A Subsurface Electromagnetic Pulse Radar," Geophysics, Vol. 41, No. 3, 1976, pp. 506-518.

[14] C.W. Chuang and D.L. Moffatt, "Natural Resonances Via Prony's Method and Target Discrimination," IEEE Trans. Aerospace Electronic Systems, Vol. 12, No. 5, 1976.

[15] K.A. Shubert, J.D. Young and D.L. Moffatt, "Synthetic Radar Imagery," IEEE Transactions on Antennas and Propagation, Vol. AP-25, No. 4, 1977.

[16] D.L. Moffatt and K.A. Shubert, "Natural Resonances Via Rational Approximants," IEEE Transactions on Antennas and Propagation, Vol. AP-25, No. 5, 1977.

[17] L.C. Chan, D.L. Moffatt and L. Peters, Jr., "A Characterization of Subsurface Radar Targets," Proceedings of the IEEE, Vol. 67, No. 7, 1979.

[18] L.C. Chan, D.L. Moffatt and L. Peters, Jr., "Estimation of the Complex Natural Resonances from a Class of Subsurface Targets," in Acoustic, Electromagnetic and Elastic Wave Scattering - Focus on the T-Matrix Approach. Edited by V.K. Varadan, Pergamon Press, 1980, pp. 463-482.

[19] J.D. Young, "Approximate Image Reconstruction from Transient Signature," Part 9 of Acoustic, Electromagnetic and Elastic Scattering - Focus on the T-Matrix Approach, (edited by Varadan and Varadan), Pergamon Press, 1980.

[20] E.M. Kennaugh, "Opening Remarks," IEEE Transactions on Antennas and Propagation, Vol. AP-29, No. 2, 1981.

[21] D.L. Moffatt, J.D. Young, A.A. Ksienski, H.C. Lin and C.M. Rhoads, "Transient Response Characteristics in Identification and Imaging," IEEE Trans. on Antennas and Propagation, Vol. AP-29, No. 2, 1981. Invited.

[22] E.M. Kennaugh, "The K-Pulse Concept," IEEE Trans. on Antennas and Propagation, Vol. AP-29, No. 2, 1981.

[23] D.L. Moffatt, "Ramp Response Radar Imagery Spectral Content," IEEE Trans. on Antennas and Propagation, Vol. AP-29, No. 2, March, 1981.

[24] L.C. Chan, D.L. Moffatt and L. Peters, Jr., "Improved Performance of a Subsurface Radar Target Identification System Through Antenna Design," IEEE Trans. on Antennas and Propagation, Vol. AP-29, No. 2, 1981. Invited.

[25] E.M. Kennaugh, "Polarization Dependence of RCS - A Geometrical Interpretation," IEEE Trans. on Antennas and Propagation, Vol. AP-29, No. 2, 1981.

[26] L.C. Chan, D.L. Moffatt and L. Peters, Jr., "Subsurface Radar Target Imaging Estimates," IEEE Trans. on Antennas and Propagation, Vol. AP-29, No. 2, March, 1981.

[27] J.D. Young and J.L. Volakis, "Phase Linearization of a Broad-Band Antenna Response in Time Domain," IEEE Trans. Antennas and Propagation, Vol. AP-30, No. 2, March, 1982.

[28] D.L. Moffatt and C.M. Rhoads, "Radar Identification of Naval Vessels," IEEE Transactions on Aerospace and Electronic Systems, Vol. AES-18, No. 2, March, 1982.

[29] T.W. Johnson and D.L. Moffatt, "Electromagnetic Scattering by an Open Circular Waveguide," Radio Science, Vol. 17, No. 6, pp. 1547-1536, November-December, 1982.

[30] E.M. Kennaugh and D.L. Moffatt, Comments on "Impulse Response of a Conducting Sphere Based on Singularity Expansion Method," Proceedings of the IEEE, Vol. 70, No. 3, March, 1982.

[31] J.L. Volakis and L. Peters, Jr., "Improved Identification of Underground Targets Using Video Pulse Radars by Elimination of Undesired Natural Resonances," IEEE Trans. Antenna Propagation, Vol. AP-31, No. 2, 1983.

[32] D.L. Moffatt, C-Y. Lai and T.C. Lee, "Time Domain Electromagnetic Scattering by Open Ended Circular Waveguide and Related Structures," To be published in WAVE MOTION.

Reports

MONOSTATIC AND BISTATIC MEASUREMENT OF SCATTERING SHAPE AND SYNTHESIS OF SCATTERING SHAPES
Air Force Cambridge Research Center
Contract No. AF 19(604)-6157
Project No. 1073

J.W. Eberle and R.W. St. Clair, "Echo Area of Combinations of Cones, Spheroids, and Hemispheres as a Function of Bistatic Angle and Target Aspect," 1073-1, 30 June 1960. (AD 242472) (AFCRC-TN-60-795)

R. Tsu, "The Theory and Application of the Scattering Matrix for Electromagnetic Waves," 1073-2, 1 August 1960. (AD 243689) (AFCRC-TN-60-950)

R. Tsu, "The Evaluation of Incomplete Normalization Integrals and Derivatives with Respect to the Order of Associated Legendre Polynomials," 1073-3, 1 April 1960. (AD 238953) (AFCRC-TN-60-376)

"Problems in Electromagnetic Scattering Analysis," Final Report, 1073-4, 1 January 1961. (AD 255839) (AFCRL-193)

RESEARCH ON RADAR CAMOUFLAGE AND ANTENNA ECHOS
Aeronautical Systems Division
Contract No. AF 33(616)-8039
Project No. 1223

D.L. Moffatt and S.A. Redick, "Memorandum on a Camouflaged Model of the B-47 Aircraft," 1223-1, 1 August 1961. (AD 326301)

D.L. Moffatt, "Selective Camouflage of Spheres and Cylinders," 1223-2, 15 October 1961. (AD 327636)

R.J. Garbacz, "Electromagnetic Scattering by Radially Inhomogeneous Spheres," 1223-3, 9 January 1962. (AD 275370)

D.L. Moffatt, "Low Radar Cross Sections, The Cone-Sphere," 1223-5, 1 June 1962. (AD 283338)

Annual Summary Engineering Report, 1223-6, 1 January 1962. (AD 328318)

R.J. Garbacz, "A Memorandum on a Measurement Technique for Low Cross Sections," 1223-7, 1 May 1962. (AD 276102)

R.J. Garbacz, "The Determination of Antenna Parameters by Scattering Cross-Section Measurements. I. Antenna Impedance," 1223-8, 30 September 1962. (AD 286760)

R.J. Garbacz, "The Determination of Antenna Parameters by Scattering Cross-Section Measurements. II. Antenna Gain," 1223-9, 30 November 1962. (AD 297953)

R.J. Garbacz, "The Determination of Antenna Parameters by Scattering Cross-Section Measurements. III. Antenna Scattering Cross Section," 1223-10, 30 November 1962. (AD 295031)

E.M. Kennaugh and D.L. Moffatt, "A Memorandum of Antenna Camouflage," 1223-11, 31 December 1963. (AD 347113)

D.L. Moffatt, "Determination of Antenna Scattering Properties from Model Measurements," 1223-12, 1 January 1964. (AD 441216)

Annual Summary Engineering Report, 1223-13, 31 December 1962. (AD 335006)

D.L. Moffatt, "Camouflage of the Parabolic Reflector Antenna," 1223-14, 1 April 1963.

D.L. Moffatt and R.H. Ott, "Echo Patterns of Spheres with Radial Stubs," 1223-15, 10 May 1963. (AD 337610)

E.M. Kennaugh and R.H. Ott, "Fields in the Focal Region of a Parabolic Receiving Antenna," 1223-16, 31 August 1963. (AD 420437)

R.B. Green, "The General Theory of Antenna Scattering," 1223-17, 31 August 1963. (AD 429186)

S.J. Skarote, "A Parabolic Reflector Receiving Antenna Utilizing Heterodyne Conversion and Control at the Focus," 1223-18, 1 January 1964. (AD 351271)

Final Engineering Report, 1223-19, 1 January 1964. (AD 348098)

RESEARCH ON RADAR ECHO REDUCTION STUDIES
Air Force Avionics Laboratory
Wright-Patterson Air Force Base, Ohio
Contract No. AF 33(615)-1318
Project No. 1774

E.M. Kennaugh and D.L. Moffatt, "The Axial Echo Area of a Perfectly Conducting Prolate Spheroid," 1774-1, 15 June 1964 (AD 446398)

J.H. Richmond, "Narrow-Band Window Design for Antenna Camouflage," 1774-2, 1 September 1964. (AD 354599)

R.H. Ott, "Camouflage of the Parabolic Reflector Type Antenna: II. Use of Tuned Surfaces," 1774-3, 1 September 1964. (AD 354979)

R.H. Ott, "Design Curves for Series RLC Tuned Surfaces," 1774-4, 22 January 1965. (AD 457198)

R.E. Ryan, Jr., "Camouflage of the Parabolic Reflector Type Antenna: III. The Concave Fresnel Reflector," 1774-5, 26 Janaury 1965. (AD 358349)

R.H. Ott and S.A. Redick, "Camouflage of the Parabolic Reflector Type Antenna: IV. The OSU Tuned Reflector," 1774-6, 5 January 1965. (AD 358348)

D.L. Moffatt, R.H. Ott and C.E. Ryan, Jr., "Internal Memorandum, A Memorandum on Test Results of a Camouflaged Parabolic Reflector Type Antenna," 1774-7, 22 April 1965.

Annual Summary Engineering Report, 1774-8, 31 December 1964 (AD 358054)

J.H. Richmond and R.H. Ott, "A Linear-Equation Solution for Scattering by a Periodic Plane Array on Thin Conducting Plates," 1774-9, 1 May 1965.

R.H. Ott and S.A. Redick, "Internal Memorandum, Camouflage of the Parabolic Reflector Type Antenna: V. Monostatic Echo Area Measurements on the OSU Tuned Reflector," 1774-10, 5 May 1965.

SCATTERING OF ELECTROMAGNETIC ENERGY FROM
HIGHLY CONDUCTING BODIES
Air Force Cambridge Research Laboratories
Office of Aerospace Research
Contract No. F19628-67-C-0239
Project No. 2415

D.L. Moffatt, "Interpretation and Application of Transient and Impulse Response Approximations in Electromagnetic Scattering Problems," 2415-1, 27 March 1968. (AD 668124) (AFCRL-67-0690)

R. Schaffer, "Transient Currents on a Perfectly Conducting Cylinder Illuminated by Unit-Step and Impulsive Plane Waves," 2415-2, 3 May 1968. (AD 668536) (AFCRL-67-0691)

C.E. Ryan, Jr., "Diffraction Analysis of Scattering by a Cube with Application to the Time Response Waveforms," 2415-3, 13 March 1970. (AD 702874) (AFCRL-69-0490)

D.B. Hodge, "The Calculation of the Spheroidal Wave Equation Eigenvalues and Eigenfunctions," 2415-4, 22 September 1969. (AD 694711) (AFCRL-69-0359)

D.B. Hodge, "Spectral and Transient Response of a Circular Disk to Plane Electromagnetic Waves," 2415-5, 28 May 1970. (AD 707493) (AFCRL-70-0237)

D.L. Moffatt, Final Report: "Scattering of Electromagnetic Energy From Highly Conducting Bodies," 2415-6, 7 September 1970. (AD 875533) (AFCRL-70-0238)

RESEARCH IN THE RADAR REFLECTING PROPERTIES OF DISCRETE
TARGET IN THE PRESENCE OF AN INTERFERING MEDIUM
Air Force Cambridge Research Laboratories
Office of Aerospace Research
United States Air Force
Contract No. F44620-67-C-0095

Li-Jen Du, "Rayleigh Scattering from Leaves," 2467-1, January 1969. (AD 682335) (AFCRL-68-0454)

D.A. Hill and D.L. Moffatt, "The Transient Fields of Dipole Antennas in the Presence of a Dielectric Half-Space," 2467-2, 13 March 1970. (AD 707106) (AFCRL-70-0016)

J.D. Young, D.L. Moffatt and E.M. Kennaugh, "Time-Domain Radar Signature Measurment Techniques," 2467-3, 30 July 1969. (AFCRL-69-0202)

D.L. Moffatt, "Characteristic Waveforms of Metallic Targets in the Presence of Background Signal," 2467-4, 13 March 1970. (AD 870168) (AFCRL-69-0527)

Final Summary Report, 2467-5, 13 March 1970. (AD 870169) (AFCRL-69-0402)

ELECTROMAGNETIC SCATTERING RESEARCH
Air Force Systems Command
Laurence G. Hanscom Field
Contract No. F19628-70-C-0125
Project No. 2971

D.A. Hill, "Electromagnetic Scattering Concepts Applied to the Detection of Targets Near the Ground," 2971-1, 4 September 1970. (AD 875889) (AFCRL-70-0250)

D.L. Moffatt, Final Report: "Time Domain Electromagentic Scattering from Highly Conducting Objects," 2971-2, May 1971. (AD 885883) (AFCRL-71-0319)

ELECTROMAGNETIC PULSE SOUNDING FOR GEOLOGICAL SURVEYING WITH APPLICATION IN ROCK MECHANICS AND RAPID EXCAVATION PROGRAM
Bureau of Mines, Department of Interior (ARPA)
Contract No. H0230009
Project No. 3408-A1

D.L. Moffatt, L. Peters, Jr., and R.J. Puskar, Semiannual Report: "Electromagnetic Pulse Sounding for Geological Surveying with Application in Rock Mechanics and the Rapid Excavation Program," 3408-1, April 1973.

D.L. Moffatt, R.J. Puskar and L. Peters, Jr., Final Report: "Electromagnetic Pulse Sounding for Geological Surveying with Application in Rock Mechanics and the Rapid Excavation Program," 3408-2, September 1973. (AD 772065)

INVESTIGATION OF THE MULTI-FREQUENCY RADAR REFLECTIVITY OF RADAR TARGET
Dept. of the Air Force, Hanscom AFB, MA
Contract No. F19628-72-C-0203
Project No. 3424

R.K. Mains and D.L. Moffatt, "Complex Natural Resonances of an Object in Detection and Discrimination," 3424-1, June 1974. (AFCRL-TR-74-0282)

S.C. Lee, "Control of Electromagnetic Scattering by Antenna Impedance Loading," 3424-2, July 1974. (AFCRL-TR-74-0426)

C.W. Chuang and D.L. Moffatt, "Complex Natural Resonances of Radar Targets Via Prony's Method," 3424-3, April 1975. (AFCRL-TR-75-0203)

J. Aas, "Control of Electromagnetic Scattering from Wing Profiles by Impedance Loading," 3424-4, August 1973. (AFCRL-TR-75-0463)

D.L. Moffatt, R.C. Rudduck, C.W. Chuang and J.A. Aas, "Continuation of the Investigation of Multi-Frequency Radar Reflectivity and Radar Target Identification," 3424-5, July 1975. (AFCRL-TR-75-0417)

C.W. Chuang, "Radiation of Backscatter by Impedance Loading," 3424-6, October 1976. (RADC-TR-76-376) (AD/A-035 509)

C.W. Chuang, D.L. Moffatt, L. Peters, Jr. and K.A. Shubert, "Continuation of the Investigation of the Multi-Frequency Radar Reflectivity and Radar Target Identification," 3424-7, Final Report, March 1977. (RADC-TR-77-115)

THREE-D IMAGES FOR SPECTRAL SIGNATURE DATA
Rome Air Development Center
Contract No. F30602-74-C-0170

J. D. Young, "Three-D Images for Spectral Signature Data," 3919-1, August 1974.

J.D. Young, Quarterly Report: "Three-D Images for Spectral Signature Data," 3919-2, November 1974.

K.A. Shubert and D.L. Moffatt, "Three-D Images for Spectral Signature Data," 3919-3, March 1975.

K.A. Shubert and D.L. Moffatt, "Three-D Images for Spectral Signature Data," 3919-4, May 1975.

K.A. Shubert and D.L. Moffatt, "Three-D Images for Spectral Signature Data," 3919-5, August 1975.

K.A. Shubert, D.L. Moffatt, J.D. Young and W.E. Cory, "Multiple Frequency Scattering Data," 3919-6, October 1975.

K.A. Shubert and J.D. Young and D.L. Moffatt, "Target Imaging from Multiple Frequency Radar Data," 3919-7, Final Report, November 1975.

S.C. Lee, "Control of Electromagnetic Scattering by Antenna Impedance Loading," 3424-2, July 1974. (AFCRL-TR-74-0426)

C.W. Chuang and D.L. Moffatt, "Complex Natural Resonances of Radar Targets Via Prony's Method," 3424-3, April 1975. (AFCRL-TR-75-0203)

J. Aas, "Control of Electromagnetic Scattering from Wing Profiles by Impedance Loading," 3424-4, August 1973. (AFCRL-TR-75-0463)

D.L. Moffatt, R.C. Rudduck, C.W. Chuang and J.A. Aas, "Continuation of the Investigation of Multi-Frequency Radar Reflectivity and Radar Target Identification," 3424-5, July 1975. (AFCRL-TR-75-0417)

C.W. Chuang, "Radiation of Backscatter by Impedance Loading," 3424-6, October 1976. (RADC-TR-76-376) (AD/A-035 509)

C.W. Chuang, D.L. Moffatt, L. Peters, Jr. and K.A. Shubert, "Continuation of the Investigation of the Multi-Frequency Radar Reflectivity and Radar Target Identification," 3424-7, Final Report, March 1977. (RADC-TR-77-115)

THREE-D IMAGES FOR SPECTRAL SIGNATURE DATA
Rome Air Development Center
Contract No. F30602-74-C-0170

J. D. Young, "Three-D Images for Spectral Signature Data," 3919-1, August 1974.

J.D. Young, Quarterly Report: "Three-D Images for Spectral Signature Data," 3919-2, November 1974.

K.A. Shubert and D.L. Moffatt, "Three-D Images for Spectral Signature Data," 3919-3, March 1975.

K.A. Shubert and D.L. Moffatt, "Three-D Images for Spectral Signature Data," 3919-4, May 1975.

K.A. Shubert and D.L. Moffatt, "Three-D Images for Spectral Signature Data," 3919-5, August 1975.

K.A. Shubert, D.L. Moffatt, J.D. Young and W.E. Cory, "Multiple Frequency Scattering Data," 3919-6, October 1975.

K.A. Shubert and J.D. Young and D.L. Moffatt, "Target Imaging from Multiple Frequency Radar Data," 3919-7, Final Report, November 1975.

DEVELOPMENT OF A SUBTERRANEAN RADAR FOR TUNNEL LOCATION
U.S. Army Mobility Equipment Research
& Development Command
Ft. Belvoir, VA
Contract DAAG53-76-C-0179
Project 4460

G.A. Burrell and B.A. Munk, "The Array Scanning Method and Applying it to Determine the Impedance of Linear Antennas in a Lossy Half Space," 4460-1, October 1976.

G.A. Burrrell, L. Peters, Jr. and A.J. Terzuoli, Jr., "The Propagation of Electromagnetic Video Pulses with Application to Subsurface Radar for Tunnel Application," 4460-2, December 1976.

L. Peters, Jr., G.A Burrell and H.B. Tran, "A Scattering Model for Detection of Tunnels Using Video Pulse Systems," 4460-3, February 1977.

C.A. Tribuzi, "An Antenna for Use in an Underground (HFW) Radar System," 4460-4, November 1977.

D.O. Stapp, "Method for Gray Scale Mapping of Underground Obstacles Using Video Pulse Radar Return," 784460-5, December 1978.

CHARACTERIZATION OF SUBSURFACE ELECTROMAGNETIC SOUNDINGS
National Sciene Foundation
Grant ENG76-04-04344
Project 784490

D.L. Moffatt, "Characterization of Subsurface Electromagnetic Soundings," 784490-1, May 1977.

D.L. Moffatt and L.C. Chan, "Characterization of Subsurface Electromagnetic Soundings," Final Report, 784490-2.

RADAR IDENTIFICATION OF NAVAL VESSELS
Office of Naval Research
Arlington, VA
Contract N00014-76-C-1079
Project 784558

D.L. Moffatt and C.M. Rhoads, "Radar Identification of Naval Vessels," 784558-1, December 1977. (AD/A050 047) (AD/A071 921)

D.L. Moffatt and C.M. Rhoads, "Radar Identification of Naval Vessels," Final Report, 784558-2, April 1979.

A STUDY OF RADAR TARGET IDENTIFICATION BY THE MULTIPLE-FREQUENCY RESONANCE AND TIME DOMAIN POLE METHOD
Deputy for Electronic Technology (RADC/ETER)
Hanscom AFB, MA 01731
Contract F19628-77-C-0125
Project 784677

K.A. Shubert and D.L. Moffatt, "Swept Frequency Scattering Measurements of Aircraft," 784677-1, January 1979. (AD/A071 474) (RADC-TR-79-110)

D.L. Moffatt, K.A. Shubert and E.M. Kennaugh, "Radar Target Identification," Final Report, 784677-2. (AD/A070 793) (RADC-TR-79-118)

BASIC RESEARCH IN 3-DIMENSIONAL IMAGING FOR TRANSIENT RADAR SCATTERING SIGNATURES
Ballistic Missile Defense Systems Command
Contract DASG60-77-C-0133
Project 784785

J.D. Young, R.A. Day, F.R. Gross and E.K. Walton, "Basic Research in the Three-Dimensional Imaging from Transient Radar Scattering Signatures," Annual Report, 784785-1, July 1978.

R.A. Day, "Automated Imaging of Cone-Like Targets from Transient Radar Signature Data," 784785-2, March 1979.

F. Gross, "Application of Physical Optics Inverse Diffraction to the Identification of Cones from Limited Scattering Data," 784785-4, June 1979.

J.D. Young, "Basic Research in Three-Dimensional Imaging from Transient Radar Scattering Signatures," Final Report, 784785-5. (AD/A079 423)

JOINT SERVICES ELECTRONICS PROGRAM
Dept. of the Navy
Office of Naval Research
Arlington, Virginia 22217
Contract N00014-78-C0049
Project 710816

"Joint Services Electronics Program," First Annual Report, Annual Report, 710816-1, December 1978. (AD/A064 791)

D.B. Hodge, "The Calculation of Far Field Scatter by a Circular Metallic Disk," 710816-2, Feburary 1979. (AD/A068 208)

D.P. Mithouar and D.B. Hodge, "Electromagnetic Scattering by a Metallic Disk," 710816-3, September 1979. (AD/A080 441)

I.L. Volakis, "Improved Identification of Underground Targets Using Video-Pulse Radars by Elimination of Undesired Natural Resonances," 710816-4, October 1979. (AD/A078 328)

"Joint Services Electronics Program," Second Annual Report, 710816-6, December 1979. (AD/A087 061)

T.W. Johnson and D.L. Moffatt, "Electromagnetic Scattering by Open Circular Waveguides," 701816-9, December 1980.

"Joint Services Electronics Program," Third Annual Report, 710816-10, December 1980.

"Joint Services Electronics Program," Fourth Annual Report, 710816-11, December 1981.

"Joint Services Electronics Program," Fifth Annual Report, 710816-12, December 1982.

RESEARCH IN HIGH RESOLUTION RADAR TARGET POLARIMETRY

Polarization Studies for Echo Characteristics
Rome Air Development Center
Contract No. AF 28(099)-90
Project No. 389

389-1 to 389-15, and 398-17 to 389-24
Quarterly Progress & Final Reports, "EFFECT OF TYPE OF POLARIZATION ON ECHO CHARACTERISTICS, Sept., 1949 to Oct., 1954

Landing System Antennas and Echo Enhancement Problems
Rome Air Development Center
Griffiss Air Force Base, New York
Contract No. AF 30(635)-2811
Project No. 612

612-1 to 612-16
Quarterly Progress and Interim & Final Reports, "POLARIZATION DEPENDENCE OF RADAR ECHOES", January, 1955 to January, 1957

List of Reports on Polarimetric Properties of Radar Targets is in "Proceedings of the Second Workshop on Polarimetric Radar Technology", Conducted at the U.S. Army Missile Command, DRSMI-RN, Redstone Arsenal, AL 35898, 3-5 May, 1983, Vol. 1, pp. 178-180. A commemorative collection of unpublished notes and reports in 4 volumes will be made available by the ElectroScience Lab., June 1984.

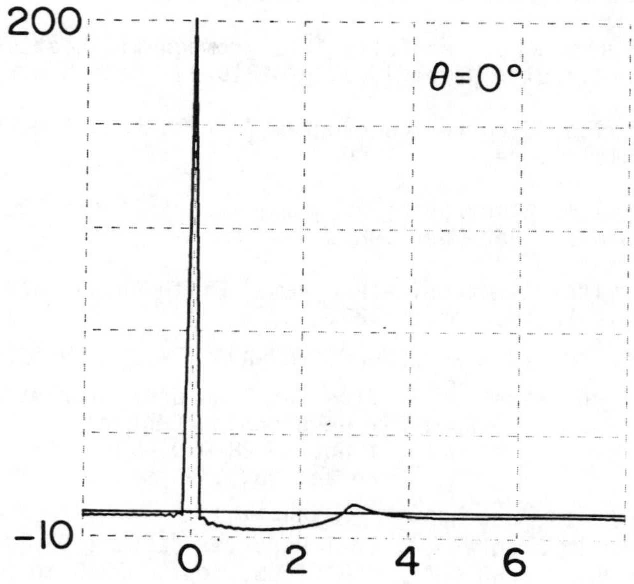

a. θ = 0° (specular)

Figure 1. E-plane (positive) and H-plane (negative) surface current density waveforms induced by an impulsive plane wave on a perfectly conducting sphere. Illuminated side of sphere.

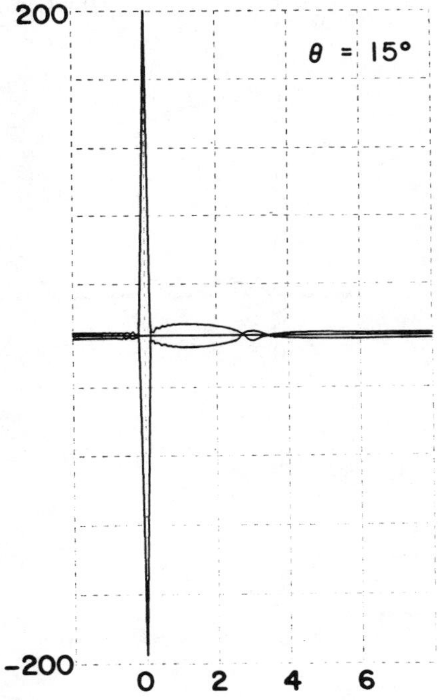

b. θ = 15°

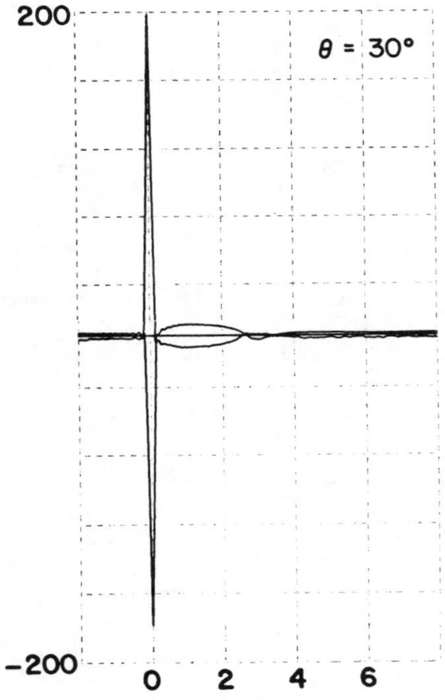

c. θ = 30°

Figure 1. E-plane (positive) and H-plane (negative) surface current density waveforms induced by an impulsive plane wave on a perfectly conducting sphere. Illuminated side of sphere.

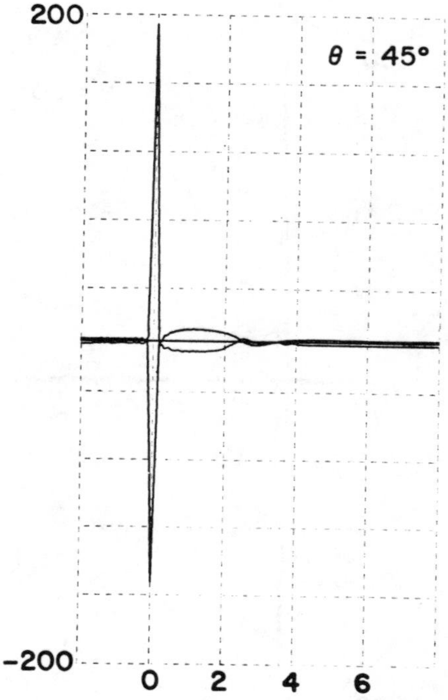

d. θ = 45°

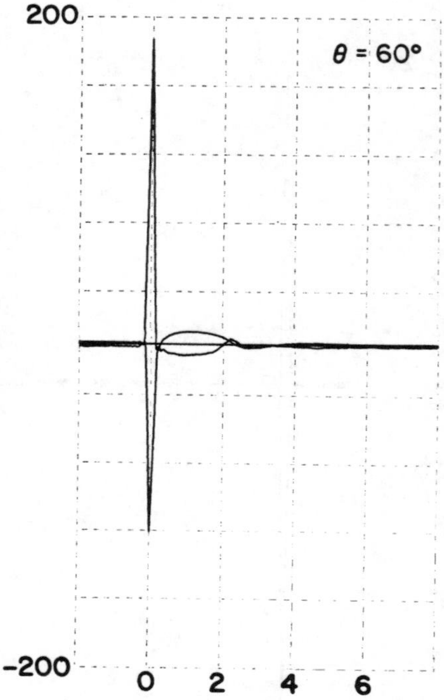

e. θ = 60°

Figure 1. E-plane (positive) and H-plane (negative) surface current density waveforms induced by an impulsive plane wave on a perfectly conducting sphere. Illuminated side of sphere.

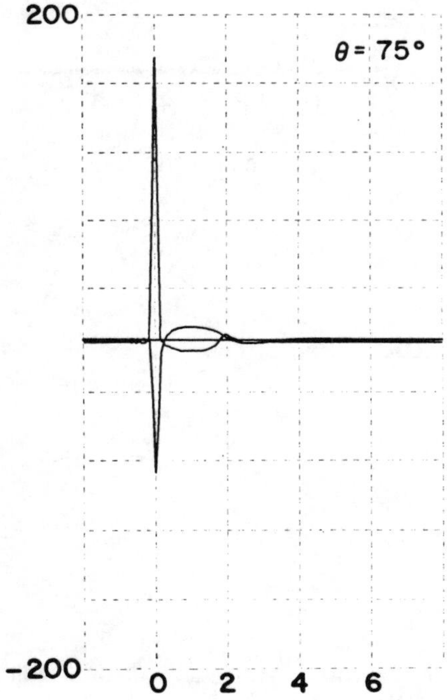

f. θ = 75°

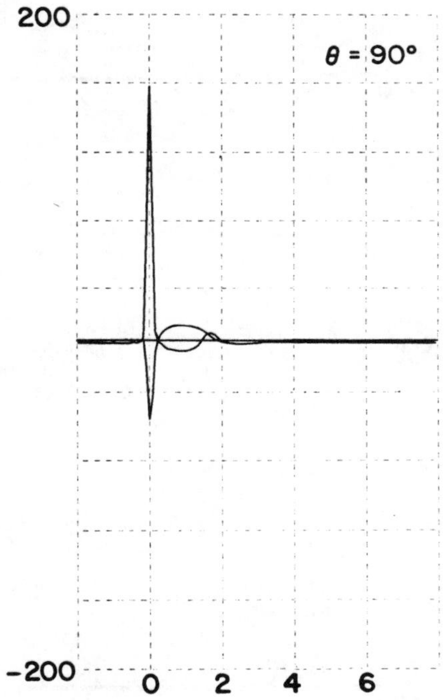

g. θ = 90°

Figure 1. E-plane (positive) and H-plane (negative) surface current density waveforms induced by an impulsive plane wave on a perfectly conducting sphere. Illuminated side of sphere.

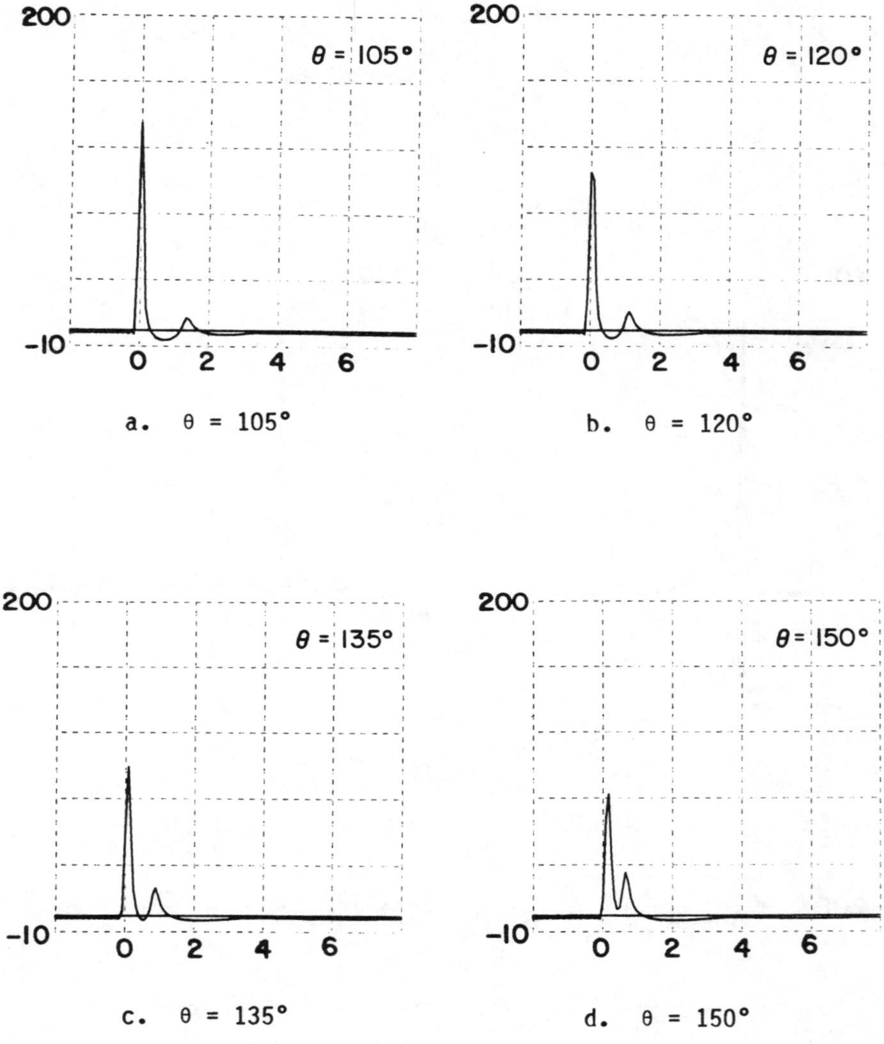

a. θ = 105°

b. θ = 120°

c. θ = 135°

d. θ = 150°

Figure 2. E-plane surface current density waveforms induced by an impulsive plane wave on a perfectly conducting sphere. Shadowed side of sphere.

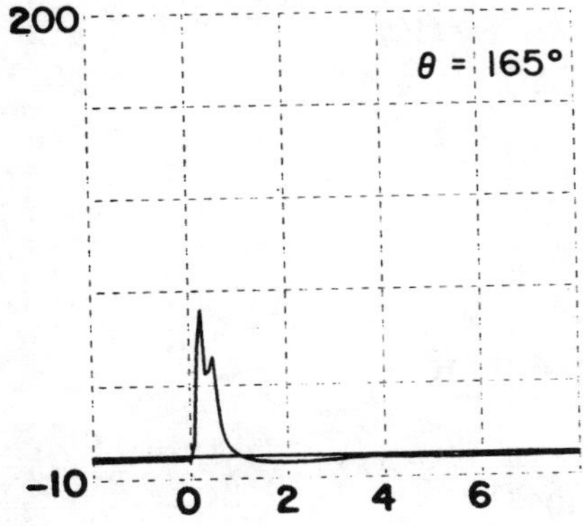

e. θ = 165°

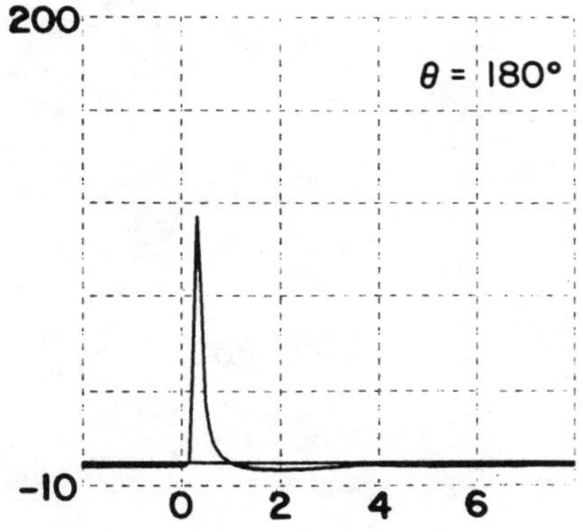

f. θ = 180°

Figure 2. E-plane surface current density waveforms induced by an impulsive plane wave on a perfectly conducting sphere. Shadowed side of sphere.

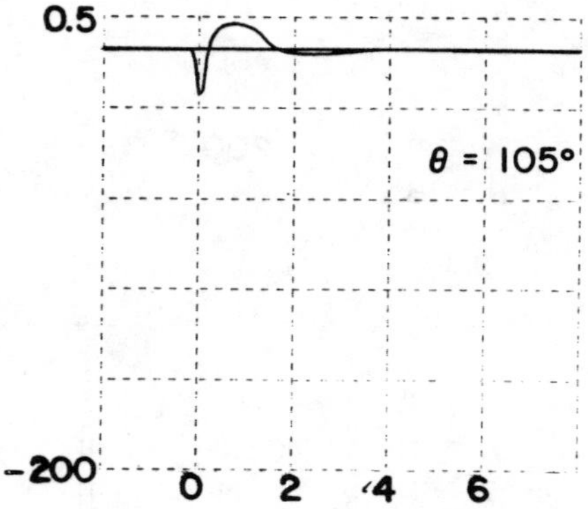

a. θ = 105°

Figure 3. H-plane surface current density waveforms induced by an impulsive plane wave on a perfectly conducting sphere. Shadowed side of sphere.

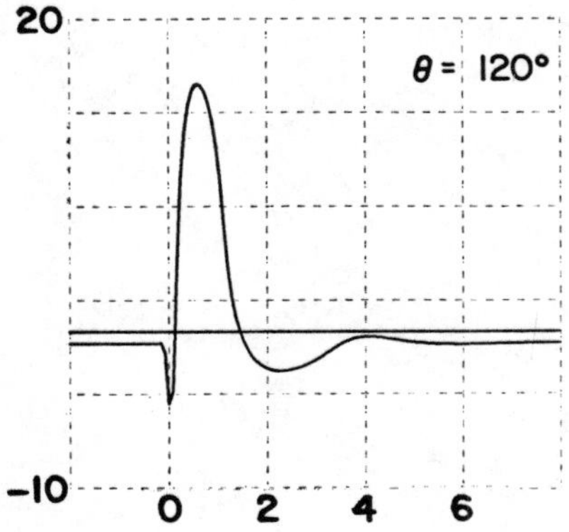

b. θ = 120°

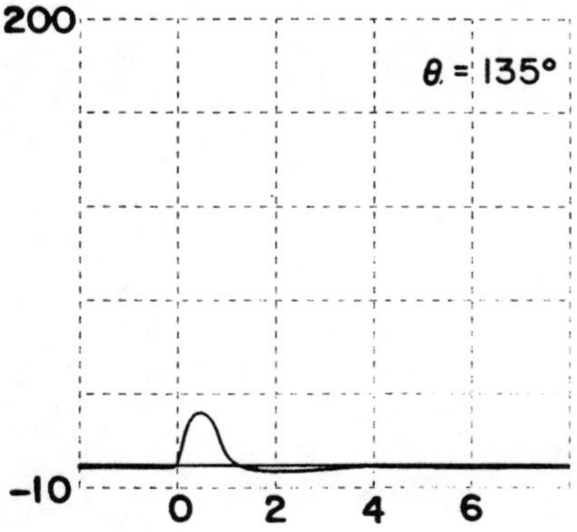

c. θ = 135°

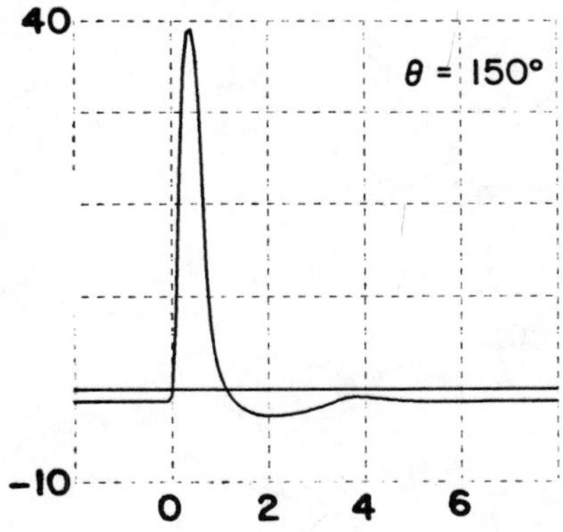

d. θ = 150°

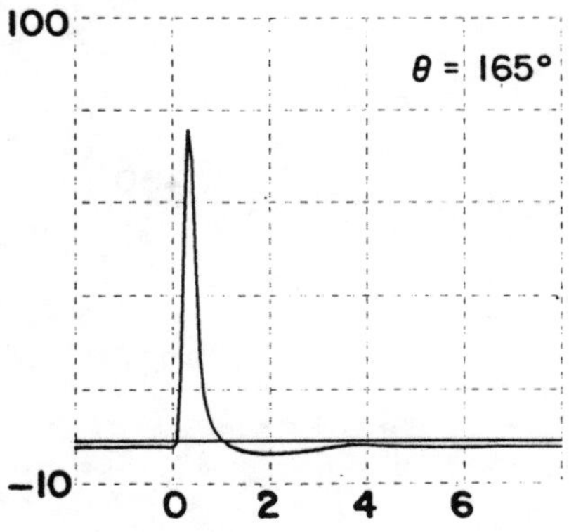

f. θ = 165°

Figure 3. H-plane surface current density waveforms induced by an impulsive plane wave on a perfectly conducting sphere. Shadowed side of sphere.

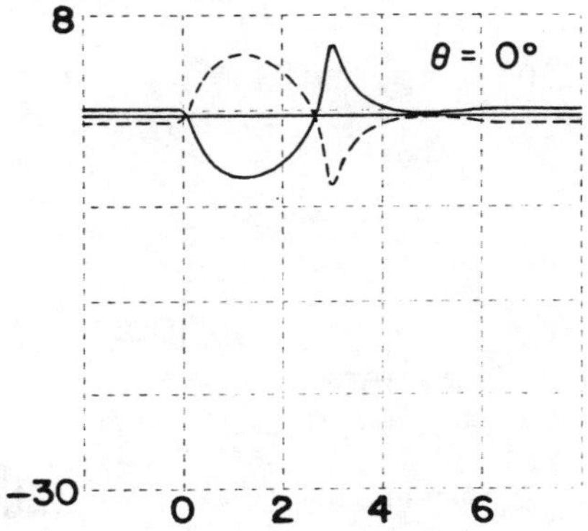

a. θ = 0°

Figure 4. E-plane (solid) and H-plane (dashed) surface current density waveforms induced by an impulsive plane wave on a perfectly conducting sphere after the physical optics approximation for the surface current density has been removed. Illuminated side of sphere.

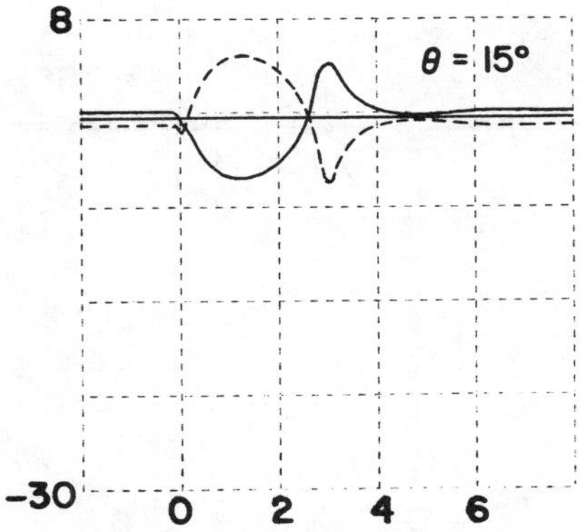

b. θ = 15°

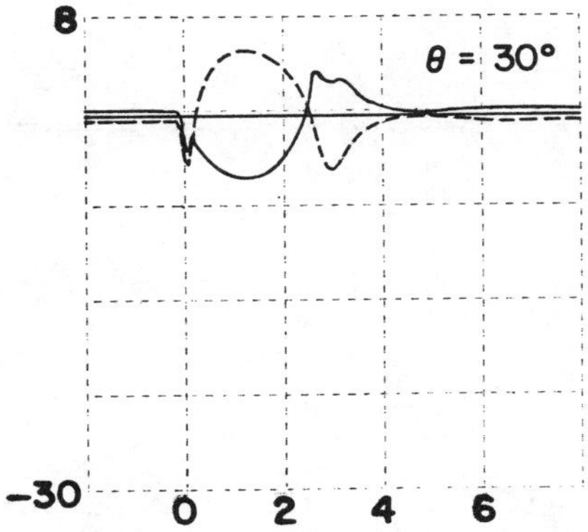

c. θ = 30°

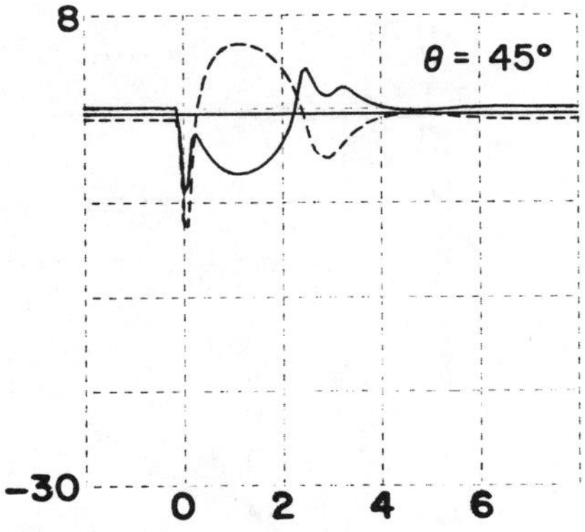

d. θ = 45°

Figure 4. E-plane (solid) and H-plane (dashed) surface current density waveforms induced by an impulsive plane wave on a perfectly conducting sphere after the physical optics approximation for the surface currrent density has been removed. Illuminated side of sphere.

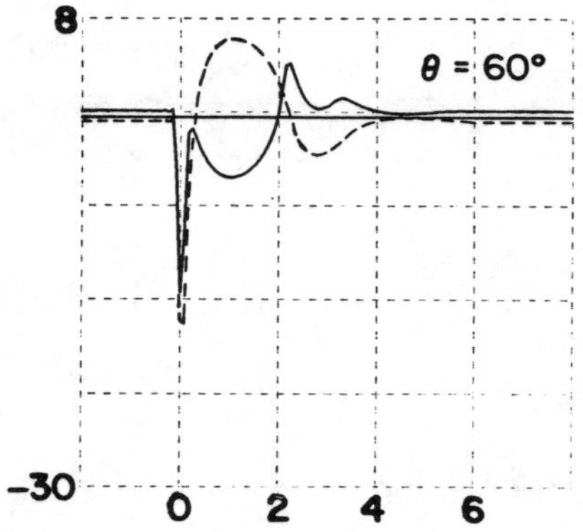

e. θ = 60°

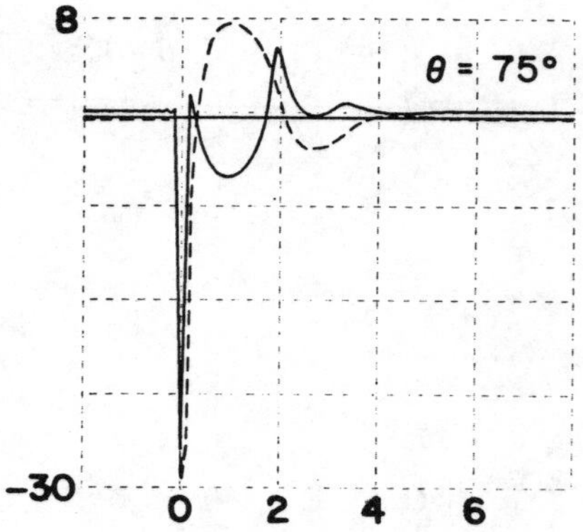

f. θ = 75°

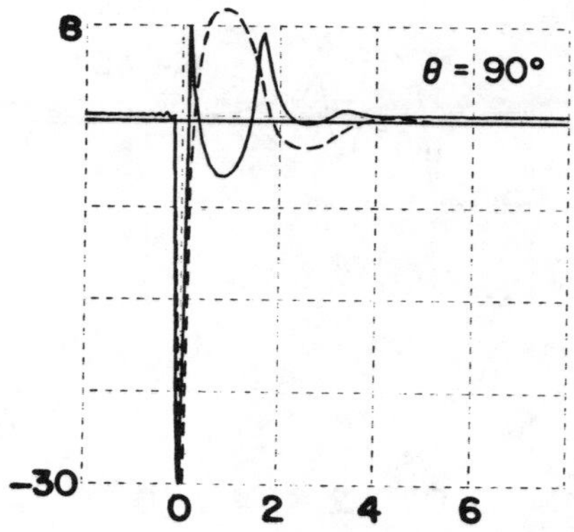

g. θ = 90°

Figure 4. E-plane (solid) and H-plane (dashed) surface current density waveforms induced by an impulsive plane wave on a perfectly conducting sphere after the physical optics approximation for the surface current density has been removed. Illuminated side of sphere.

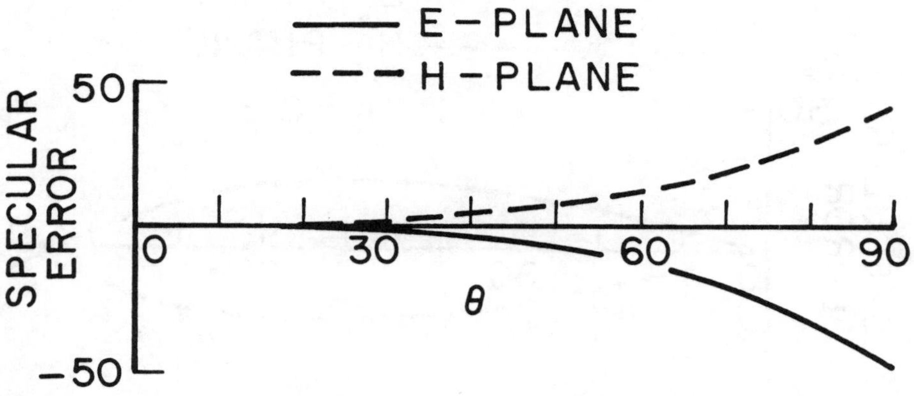

Figure 5. E-plane (solid) and H-plane (dashed) error for specular term (exact minus physical optics) for surface current density induced by an impulsive plane wave on a perfectly conducting sphere.

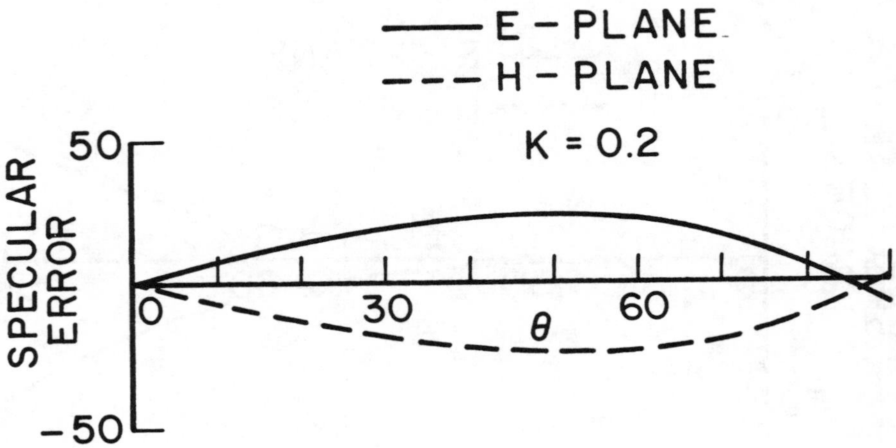

Figure 6. E-plane (solid) and H-plane (dashed) error for specular term (exact minus corrected physical optics, value of K constant is 0.2) for surface current density induced by an impulsive plane wave on a perfectly conducting sphere.

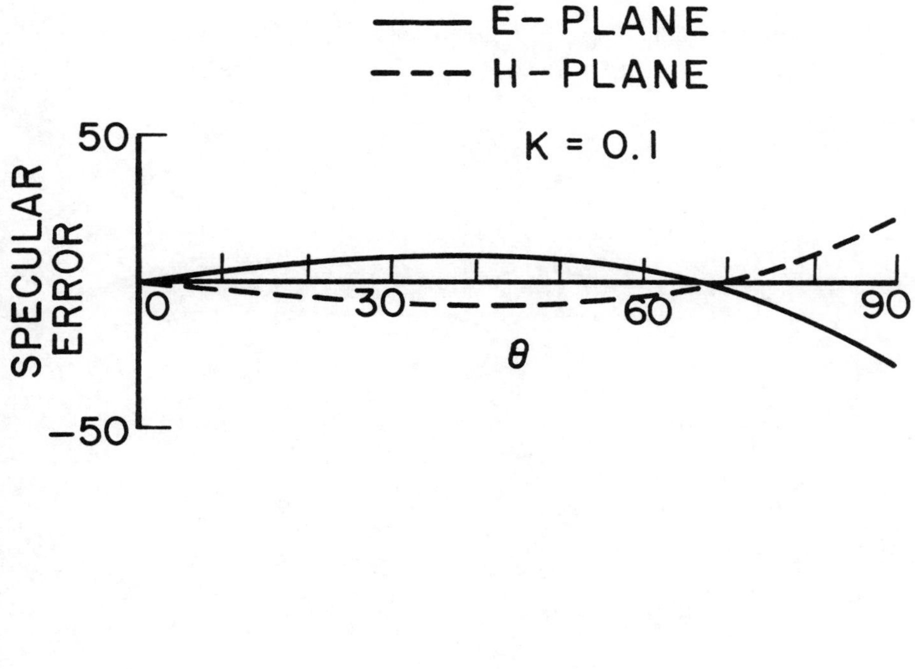

Figure 7. E-plane (solid) and H-plane (dashed) error for specular term (exact minus corrected physical optics, value of K constant is 0.2) for surface current density induced by an impulsive plane wave on a perfectly conducting sphere.

0.2 (OS.4)

INVERSE METHODS IN ELECTROMAGNETIC IMAGING

Arthur K. Jordan

Space Science Division,
Naval Research Laboratory,
Washington, DC 20375

Wolfgang-M. Boerner

Communications Laboratory
EECS Department
University of IL at Chicago
Chicago, IL 60680

Abstract

Imaging can be described as the generation of the image field I from an object field O by means of appropriate electromagnetic and computational systems L, so that $I = L \cdot O$. In remote sensing, non-destructive evaluation and similar applications, the object field O is not known explicitly and must be estimated by an inverse method $O = L^{-1} \cdot I$ subject to some constraint $\| I - O \| < \varepsilon$, which results in inherent numerical instabilities. These problems in general are "ill-posed" and are treated by a variety of methods: constraints on the solution for O, use of *a priori* knowledge of O, relaxation of the error limit ε on image quality, addition of more data in the image I, etc. Using this rationale, the seventy-five papers comprising these Proceedings are summarized and organized into five Topics: Topic I was concerned with fundamental inversion theories; Topic II with numerical instabilities; Topic III with the utilization of polarization information to improve high resolution imaging; Topic IV with the effects of environmental noise and clutter on image quality; and Topic V with holographic and tomographic imaging and their associated phase problems. This General Review concludes with a summary of recent relevant advances.

1. Introduction

In general, images are representations of objects that are indirectly or remotely sensed. Since indirect sensing is used, inverse methods are necessary to reconstruct images of these objects. This NATO Advanced Research Workshop was concerned with imaging by electromagnetic waves in two or three space dimensions and with the mathematical and experimental implementation of inverse methods for

W.-M. Boerner et al. (eds.), Inverse Methods in Electromagnetic Imaging – Part 1, 33–42.

imaging systems.

The complete proceedings of this NATO Advanced Research Workshop on Inverse Methods in Electromagnetic Imaging are collected in two parts. A perusal of the Table of Contents reveals the diversity of topics presented, ranging from abstract mathematical discussions to practical designs and experimental results. For such a rapidly developing research area, it is necessary to classify the progress reported here approximately according to topics. This arrangement will enable the reader to become acquainted with the current research trends and to appreciate the fundamental unity of these inverse methods. It has not been possible to include many important contributions that were not presented at this ARW. However, most of the papers in these Proceedings contain extensive references which serve to extend the reader's understanding. This general review section also summarizes recent related monographs that contain reviews with extensive references; recent relevant NATO Advanced Study Institutes and Workshops Proceedings are also listed.

Since the participants in the ARW presented a wide-ranging variety of papers, it will be worthwhile to give a general background and description of imaging and why inverse methods are essential to any practical imaging system. It is convenient to organize this overview by first considering a mathematical representation of the imaging process which serves to highlight the roles of inverse theory, image quality and image reconstruction techniques. This problem definition also serves to organize these Proceedings into six topics with their corresponding Working Groups. The Proceedings are then summarized with one or two key sentences abstracted from each paper (identified by the authors' names). We conclude with a brief review of references on imaging and inverse methods, on progress in related sciences and technologies, and also related NATO Advanced Study Institutes.

2. Problem Definition

For the purposes of this discussion, imaging can be described as the generation of the image field, I, from an object field, O, by means of appropriate electromagnetic and computational systems L:

$$I = L \cdot O$$

The image quality is measured (this subjective process is, of course, difficult to quantify) by the "closeness" of I to O in some appropriate mathematical representation, i.e.,

$$\| I - O \| < \varepsilon$$

In remote sensing, nondestructive evaluation, medical imaging, and similar applications, the object field O is not known explicitly and must be estimated by indirect or inverse methods, L^{-1}, from data contained in the image fields I. This procedure is conveniently written as:

$$O = L^{-1} \cdot I \quad ,$$

although we may not be able to define L^{-1} mathematically. Inverse problems of this type have inherent numerical instabilities, so that small variations in the image data, I, can cause large errors in the reconstructed object, O. These problems are called "ill-posed" and are treated by a variety of methods: constraints on the solution for O, use of *a priori* knowledge of O, relaxation of the error limit ε on image quality, addition of more data in the image I, etc. Using this rationale we have organized (somewhat arbitrarily) the papers comprising these Proceedings into five topics, roughly corresponding to the five Working Groups, and these papers are briefly summarized in the next section.

3. Synopsis of Proceedings of the Workshop

3.0 The memorial paper by Kennaugh & Moffatt also contains extensive lists of references of the unpublished research reports by the late Prof. Edward M. Kennaugh and his coworkers on transient electromagnetic problems, those on the polarization properties of radar targets are now available in four volumes (0).

3.1 The papers of Topic 1 were concerned with the fundamental inversion theories required to obtain the object from the image. In the first paper Sabatier, provides an analysis of the common structure of inverse methods and demonstrates this by examples of three important aspects of inverse methods:

a. the importance of *a priori* information,

b. the necessary use of geometrical methods in function spaces,

c. the interest in these methods from the algorithmic points of view.

Basic topological considerations are used by Güttinger to classify singularities in waves and rays in order to study the formation of patterns. Three-dimensional, time-dependent inverse problems are considered by Moses, who obtains solutions for the current distributions that are not unique, but conditions are given that make them so. Langenberg et al consider two types of transient methods: broadband signals, such as the singularity expension method, and synthetic aperture methods with band-limited data in K-space. Roger describes a numerical method (Newton-Kantorovich iterative algorithm) that demonstrates a characteristic instability of inverse problems. Krueger et al use time-domain methods to reconstruct both the permittivity and conductivity profiles of a one-dimensional dissipative medium: four numerical examples are given.

Jordan et al extend inverse scattering theory to include profiles that are almost-periodic functions; applications to modulated inhomogeneous and non-deterministic media are indicated. Several time-domain methods are compared by Blok & Tijhuis: direct-approach which involves a step-by-step determination of the unknown profile, iterative approach for successive approximations to the unknown constitutive coefficient, optimization approach by minimization of the mean square error. Approaches to electromagnetic inverse problem are given by Eftimiu by using analogies to the quantum mechanical inverse problem. The Abel integral is applied to a variety of remote sensing problems by Süss; a combined analytic-numeric inversion algorithm has relatively high stability. The importance of group representations, i.e., symmetry analysis, in scattering problems, direct and inverse, is reviewed by Wacker. Borden has used partial scattering data to obtain the shape of a convex body from a variational solution to an inverse scattering problem. Ferwerda et al presented a method of profile reconstruction based on the Gelfand-Levitan-Marchenko approach; bound states were accounted for.

3.2 The papers of Topic II were concerned with the numerical instabilities inherent in inverse problems. One cause of numerical instability is the lack of sufficient data; this was discussed by Weston for large perturbations so a nonlinear iteration approach is used. Grünbaum considers the problem of recovering a real function, e.g. the density, from bandlimited data, the projections; extension to the hyperbolic case enables the use of prolate spheroidal wave functions. Crosta analyzes the modeling procedures in direct and inverse problems. Singular function expansions are used by Bertero et al in two papers to obtain stable solutions and to define resolution limits. Conjugate gradient methods are used by Sarkar to obtain stable results as long as the number of effective bits in the signal is correctly chosen. Jory uses approximation schema for the Moore-Penrose generalized inverse to obtain regularization operators. Several maximum entropy methods are reviewed by Bevensee and fundamental limitations on the resolution on various radiating sources are discussed. Complex electromagnetic problems are treated by Taflove & Umashankar using a combined finite-difference time-domain method; accurate results have been obtained for three-dimensional structures as large as 20 wavelengths. Geometrical and quasi-geometrical imaging are considered by Danglmayr with special attention to the caustics associated with topological singularities of the distance function. Dittel has investigated three algorithms for the deconvolution of radiometry data using *a priori* information. Time-domain inverse scattering methods have been reviewed by Tabbara. Complex natural resonances of an object have been extracted by Moffatt & Lee using rational function approximants; these can be used to correct and supplement poles obtained by a K-pulse technique.

3.3 The papers of Topic III were concerned with the appropriate utilization of polarization information to improve high resolution target imaging. The tools used for handling polarization problems are reviewed by Deschamps. Suggestions are made to account for symmetries and their use in inverse problems. Utilization of polarization information by radar systems is reviewed by Poelman. Thiel, briefly, notes the transformation of the Stokes' vectors. After a review of fundamentals, Grüner proposes a correlation measuring method, using polarization switching for use in microwave radiometry. Three-dimensional radar images are constructed by polarization-dependent methods by Heath; efficient Fourier or Radon transforms are used. The phenomenological theory of radar targets is reviewed by Davidovitz & Boerner and the optimum polarization case is extended to asymmetric polarization scattering matrices. The graphical display of polarization parameters was discussed by Raven and polarization nulls of simulated, complex reflectors were displayed. Meteorological information is obtained from polarimetric measurements by Nespor & Boerner. The related problem of polarization glint has been considered from the view point of the measuring system by Stock. The use of polarization signatures for target angle tracking systems has been considered in Carpenter, whereas, it is shown in Manson & Boerner how high resolution scattering matrix pulse compression radar data can be used for reconstructing the three-dimensional target image downrange using Kennaugh's target characteristic operator and Huynen's Mueller matrix decomposition concept. Agreement between experiment and simulation provides support for the single-photon coherence zone concept of Wei et al. The criteria for reducing errors in recovering polarized scattering parameters are provided by Blanchard & Newton. Five conceptual issues relating to polarization processing are discussed by Vannicola & Lis.

3.4 The papers of Topic IV were concerned with the effects of environmental noise and clutter on image quality and methods for improvement of imagery. A fundamental approach to electromagnetic imaging is radiative transfer theory. Ishimaru has reviewed this theory for arbitrary polarized electromagnetic waves; limitations due to multiple scattering effects are also discussed. Image quality also depends upon the perception of the data by the observer; Huynen reviews the methods of data presentation of objects based upon fields and powers. Radar imagery is reviewed by Orhaug from the viewpoint of image quality, image encoding, and computer analysis. Controlled experiments to assess and develop inverse processing algorithms have been described by Cram. Optimum methods of detection and identification of surface chemicals (e.g. paints) and target shapes are discussed by Gjessing et al. Inverse methods are used by Rotheram & Macklin to apply linear demodulation and speckle removal to synthetic aperture radar images. Rough surface properties have been obtained by inversion of scattering data by Fung. SAR imagery was also discussed by Larson et al with respect to fine resolution imagery of stationary scenes and imaging of a moving surface. Radio exploration of glaciers was reviewed by Walford with respect to the inverse problems for polar and temperate glaciers. Results from fast millimeter wave imaging for traffic studies were reviewed by Brand. Low-frequency imaging algorithms were presented by Chaudhuri, using ellipsoidal models. Dielectric inhomogeneities have been reconstructed by Chaloupka, using the Born approximation. Experimental results from a millimeter-wave imaging system were presented by Detlefsen.

3.5 The papers of Topic V were concerned with holographic and tomographic imaging and the phase problems that are critical for successful imaging. Lohmann reviewed the basic holographic principle, pointing out that information can be lost to phase-, color-, and polarization blindness. Medical tomographic imaging was reviewed by Schwierz from algebraic techniques to iterative and integral transform methods. Using circuit theory analogies, Kahn considers inverse methods using resonance conditions, transformation invariants, and physical symmetry. Anderson & Adams have produced satisfactory images using fast convenient FFT-based algorithms. Devaney has reviewed the theoretical foundations of computed tomography using diffracting wavefields (diffraction tomography) and contrasted them with the foundations of conventional computed tomography. Born and Rytov approximations were presented as special cases of a general view of diffraction tomography by Kaveh & Soumekh. An approach to extend scalar propagation tomography to the electromagnetic vector case is presented in James, Yang & Boerner. Techniques to increase the number of degrees-of-freedom in image processing were reviewed by Szu. The iterative method for image restoration has been reviewed by Gori & Wabnitz and the discrete and continuous cases contrasted. Cases where mono-pulse techniques can be applied to synthetic aperture radar imagery were proposed by Gough. Methods for probing dielectric radomes were reviewed by Tricoles. Methods for constructing and retrieving authenticity features were reviewed by Baltes & Huiser. Methods for improving target recognition from microwave imagery were presented by Gniss & Magura. The inverse problem of far-field-to-near-field transformation in spherical coordinates was considered by Hess incorporating many concepts discussed in previous papers.

3.6 Five Working Discussion Groups evaluated the state-of-the-art of each of these Topics and made specific recommendations for future work. Their reports, which are collected at the end of Part 2, provide a concise summary of the present status and the important future emphasis of inverse methods applied to electromagnetic imaging. In addition, Working Group VI considered the methods of enhancing the interactions between workers and institutions concerned with this multi-disciplinary technology.

4. Reviews of Related Progress

There have been several recent, comprehensive reviews with voluminous references to various aspects of imaging and inverse methods. Several of these are: Colin (1), which has reviews of the state-of-the-art of inverse scattering circa 1972; a survey of the literature up to 1978 has been presented by Boerner (2); Boerner, Jordan & Kay (3), which reviews progress in inverse methods in electromagnetics up to 1981; Baltes (4,5), which reviews progress in inverse methods in optics up to 1978 and 1981; Sabatier (6) reviews applied inverse problems and electromagnetic inverse scattering (12).

Several important topics were not discussed or were only briefly mentioned in this lead article. Zuev & Naats (7) review inverse problems of lidar (laser radar) sensing; Wilde & Barrett (8) review the mathematical basis for digital image processing and Devaney (9) reviews related optical inverse methods. Mathematical methods for inverse problems of remote sensing are treated by Twomey (10). A comparison of phase retrieval methods, which topic was considered by Working Group V, has been provided by Fienup (11).

Nonlinear processes and their effects on imaging merit considerable attention. Sabatier (12), Scott et al (13), Dodd et al (14), among others, discuss these problems and their applications to solitons. Remote sensing of the environment uses digital imaging techniques; Gjessing (15), Bernstein (16), Lillesand (17), Ulaby & Moore (18), and Rosenfeld (19) have made extensive reviews. Laser beam propagation was only briefly considered in this overview; Ishimaru (20), Zuev (21) and Hinkley (22) provide further information. Artificial intelligence, robot vision and related topics inherently depend upon inverse methods for imaging; Winston (23) has reported on the progress in this rapidly developing field. Problems of perception and psycho-physics are important for the subjective interpretation of image quality; Geissler (24) has reported on recent developments. Imaging methods for radiology have been extensively reviewed by Barrett and Swindell (25).

NATO Advanced Study Institutes and Research Workshops have provided a continuing forum for fundamental results and latest developments. Several recent and relevant conferences are:

Physics and Engineering of Medical Imaging (26)

Industrial Robotic Vision (27)

Vision & Image Understanding (28)

Pictorial Data Analysis (29)

Diagnostic Imaging in Medicine (30)

Atmospheric Effects on Radar Target Identification and Imaging (31)

Surveillance of Environmental Pollution and Resources by Electromagnetic Waves (32)

Theoretical Methods for Determining the Interaction of Electromagnetic Waves with Structures (33)

Pattern Recognition Theory and Applications (34)

Remote Sensing Applications in Marine Science and Technology (35)

Optical Metrology (36)

The Application of Laser Light Scattering to the Study of Biological Motion (37)

Nonlinear Phenomena at Phase Transitions and Instabilities (38)

In conclusion, this NATO Advanced Research Workshop has demonstrated both the basic unity of inverse methods and the diversity of their applications to imaging. The Proceedings of this Workshop represent, however, only one sampling of the development of this active area of science and technology. These Proceedings should provide workers with sufficient reference materials and suggestions for new research topics to further their continuing development.

5. References

0. Kennaugh, E.M., Contributions to the Polarization Properties of Radar Targets, Commemorative Collection of Unpublished Research Reports in Four Volumes, ElectroScience Lab., Dept. of Electrical Engineering, Ohio State University, Columbus, OH 43212, June 1984.

1. Colin, L., ed., Mathematics of Profile Inversion, NASA TMX-62, 150; Aug., 1972.

2. Boerner, W.-M., Detailed State of the Art Review, Communication Lab., Univ. of Ill. (Chicago), Report 78-3; Oct., 1978.

3. Boerner, W.-M., Jordan, A. K., & Kay, I., Special Issue on Inverse Methods in Electromagnetics, IEEE Trans. Ant. & Prop. AP-29; March, 1981.

4. Baltes, H. P., Inverse Source Problems in Optics, Springer Verlag; 1978.

5. Baltes, H. P., Inverse Scattering Problems in Optics, Springer Verlag; 1981.

6. Sabatier, P. C., Applied Inverse Problems, Springer Verlag; 1978.

7. Zuev, V. E., & Naats, I. E., Inverse Problems of Lidar Sensing of the Atmosphere, Springer Verlag; 1983.

8. Wilde, C. O., & Barrett, E., Image Science Mathematics, Western Periodicals Co., 1977.

9. Devaney, A. J., Inverse Optics, SPIE Proceedings, vol. 413; 1983.

10. Twomey, S., Introduction to the Mathematics of Inversion in Remote Sensing and Indirect Measurements, Elsevier Scientific Publ. Co.; 1977.

11. Fienup, J. R., "Phase retrieval algorithms: a comparison", Appl. Optics 21, 2758; 1982.

12. Sabatier. P. C., "Theoretical considerations for inverse scattering", Radio Science 18. 1-18; 1983.

13. Scott, A. C. et al, "The soliton: a new concept in applied science", Proc. IEEE 61, 1443-1483; 1973.

14. Dodd, R. K. et al, Solitons and Nonlinear Wave Equations, Academic Press; 1982.

15. Gjessing, Dag T., Adaptive Radar in Remote Sensing, Ann Arbor Publishers; 1981.

16. Bernstein, R., ed., Digital Image Processing for Remote Sensing, J. Wiley Sons; 1978.

17. Lillesand, T. M., Remote Sensing and Image Interpretation, J. Wiley Sons; 1979.

18. Ulaby, F. T. & Moore, R. K., Microwave Remote Sensing: Active and Passive, Addison-Wesley Publishing Co., 1981.

19. Rosenfeld, A., Digital Picture Processing, Academic Press; 1982.

20. Ishimaru, A., Wave Propagation and Scattering in Media, Vols. 1 & 2, Academic Press; 1978.

21. Zuev, V. E., Laser Beam Propagation in the Atmosphere, Consultants Bureau; 1982.

22. Hinkley, E. D., ed., Laser Monitoring of the Atmosphere, Springer Verlag; 1976.

23. Winston, P. H., Artificial Intelligence, Addison-Wesley; 1984.

24. Geissler, H.-G., Modern Issues in Perception, North-Holland; 1983.

25. Barrett, H.H. & Swindell, W., Radiological Imaging, Vols. 1&2, Academic Press; 1981.

26. Physics & Engineering of Medical Imaging, NATO Advanced Study Institute, 23 Sept.-5 Oct. 1984; Maratea, Italy (R. Guzzardi).

27. Industrial Robotic Vision, NATO Advanced Study Institute, 6-17 August 1984; Leuven, Belgium (A. Oosterlinck).

28. Vision and Image Understanding, NATO Advanced Study Institute, 1-12 July, 1984; Erice, Italy (A. Borsellino).

29. Haralick, R.M., ed., Pictorial Data Analysis, Proc. of ASI, Springer Verlag; 1983.

30. Reba, R.C., Goodenough, D.J., Diagnostic Imaging in Medicine, Proc. of ASI, M. Nijhoff Publishers; 1983.

31. Jeske, H., ed., Atmospheric Effects on Radar Target Identification and Imaging, Proc. of ASI, D. Reidel Publishing Co.; 1976.

32. Lund, T., ed., Surveillance of Environmental Pollution and Resources by Electromagnetic Waves, Proc. of ASI, D. Reidel Publishing Co.; 1979.

33. Skwirzynski, J.F., ed., Theoretical Methods for Determining the Interaction of Electromagnetic Waves with Structures, Proc. of ASI, M. Nijhoff Publishers; 1981.

34. Kittler, J., Fu, K., Pau, L., eds., Pattern Recognition Theory and Applications, Proc. of ASI, D. Reidel Publishing Co.; 1982.

35. Cracknell, A.P., ed., Remote Sensing Applications in Marine Science and Technology, Proc. of ASI, Reidel Publishing Co.; 1983.

36. Optical Metrology, NATO Advanced Study Institute, 16-27 July, 1984, Viana do Castelo, Portugal.

37. Earnshaw, J.C., ed., The Application of Laser Light Scattering to the Study of Biological Motion, NATO Advanced Study Institute, Plenum Publishing Co.; 1983.

38. Riste, T., ed., Nonlinear Phenomena at Phase Transitions and Instabilities, NATO Advanced Study Institute, Plenum Press; 1982.

I.1 (TS.1)

CRITICAL ANALYSIS OF THE MATHEMATICAL METHODS USED IN ELECTROMAGNETIC INVERSE THEORIES: A QUEST FOR NEW ROUTES IN THE SPACE OF PARAMETERS

Pierre C. SABATIER

Département de physique mathématique
Université des sciences et Techniques du Languedoc
34060 MONTPELLIER CEDEX - FRANCE

It has been remarked by Lanczos that no mathematical trick can remedy a lack of information. This point is illustrated in two kinds of problem belonging to Inverse Scattering Theory. The first one is the one dimensional inverse potential scattering, as it is used for instance in electron-density profile reconstructions). The smoothness of the surface is discussed as a possible source of uncertainties. By using a generalized study of the Schrödinger equation with rational reflection coefficients, one produces a simple method to take it into account. In turn, this method suggests a new way to explore the space of parameters, by which paths are followed through geometrical transforms. This way enables one to study possible bifurcations of solutions and exact ambiguities of the one dimensional impedance inversion, which may yield several different physical impedances for one set of reflection and transmission coefficients. It is proved that the previous results showing these ambiguities are not isolated in the parameters space but are only points on continuous paths of equivalent solutions. This non uniqueness of the one-dimensional Inverse Scattering Theory for impedances illustrates in an illuminating way the limitations of most mathematical methods currently used in inverse theories : for obvious reasons related with the existence of algorithms, they are local methods in the space of solutions, and therefore they cannot deal with this kind of problems.

W.-M. Boerner et al. (eds.), Inverse Methods in Electromagnetic Imaging - Part 1, 43–64.

INTRODUCTION

The word "critical study" is not used here for giving a title to a more or less censorious review of other methods, with some "good" remarks on their relevance and their usefulness. We do not feel competent or interested in this kind of work, although it may prove to be very useful. Rather, we would like to show that the common structure of all current methods, ie the obvious fact that they are local methods in the spaces of parameters and datas, is neither a necessary or a particularly good feature of inverse methods to be used in electromagnetic inverse problems. Indeed, the usual analyses only try to locate inside the space $\mathcal{C}$ of parameters an element that "reasonably" corresponds to the given "data" in the space $\mathcal{E}$ of results. To be more specific, let us assume that it has been possible to define distances in $\mathcal{C}$ and $\mathcal{E}$ and that the physical problem has been translated into mathematical words by defining a mapping $\mathcal{M}$ of $\mathcal{C}$ into $\mathcal{E}$. One must keep in mind that $\mathcal{E}$ has to contain both $\mathcal{M}(\mathcal{C})$ and all data with all possible errors. Now, the word "reasonably" can imply several meanings. If $\mathcal{M}$ is a bicontinuous bijection between $\mathcal{C}$ and $\mathcal{E}$ ("well-posed problem"), certainly one can call "reasonable" the true solution of the problem. However, suppose that $\mathcal{M}^{-1}$ is not Holder-continuous, and suppose that small errors in $\mathcal{E}$ give rise to large shifts inside $\mathcal{C}$, which may get at very unlikely parameters. Then using the exact solution may be a not very reasonable way of interpreting datas. In slightly more general cases, $\mathcal{M}(\mathcal{C})$ does not cover $\mathcal{E}$, but there is still a continuous mapping of $\mathcal{M}(\mathcal{C})$ back into $\mathcal{C}$. Then it is sound to define a "quasisolution" corresponding to an arbitrary element $e \in \mathcal{E}$ as an element c of $\mathcal{C}$ such that $\mathcal{M}(c)$ is close to e. This "quasisolution" is "reasonable" if choosing $\mathcal{M}(c)$ can be done by a continuous mapping of $\mathcal{E}$ into $\mathcal{M}(\mathcal{C})$. e.g. the orthogonal projection if $\mathcal{E}$ is a Hilbert space and $\mathcal{M}(\mathcal{C})$ a convex compact set. In other cases, the quasisolution may be not reasonable and the only acceptable analyses define "approximate solutions" for which a trade-off is achieved between the "quality of the fit" and the "likelihood of the parameter". No matter how this likelihood is defined (distance to a reference point in $\mathcal{C}$, best estimator in given statistics, maximum entropy, singular values analyses, etc), the structure of this analysis is necessarily "local" in the metric space $\mathcal{C}$. Furthermore, the "best" algorithms are those which are confined in the smallest subsets of $\mathcal{C}$.

As we already said, we are only interested by the general features of the mathematical methods which introduce local algorithms. Their most obvious quality is their flexibility. Most of these methods can be adapted to many kinds of inverse problems, with almost the same algorithms. However, they have also two defects. First, if there are several branches of solutions, these methods are unable to discover more than one of them.

Next, they are often sensitive to "a priori" assumptions, which may be clear cut, but which are in most cases hidden in the very structure of the algorithms.

The only way to implement a creative criticism is by giving an example of what one might wish to obtain. Hence, we would like to give here an example of a "global" method for exploring the parameter space, and connecting the paths we follow in $\mathcal{C}$ with the corresponding ones in $\mathcal{E}$. The word "global" is of course understood as an opposite word to "local", but it is also reminiscent of the way we explore our natural world, the surface of the Earth. In the naval History, there has been hardly any approach of a port by an algorithmic like approach, sort of spiralling around and down into the harbour. Captains prefer knowing the harbour coordinates and following a cross-coordinated route. Admittedly, a similar method cannot exist in most inverse problems, and its main defect is a complete lack of flexibility. But we can present it in some of the best known problems, and in particular in the scalar one-dimensional inverse scattering problem for waves governed by the Schrödinger equation or by the telegraph equation. This problem is met in several geophysical situations (ionospheric soundings, electromagnetic soundings of the Earth, elastic sounding of layered media by parallel plane waves, etc.). The paths we follow in $\mathcal{C}$ and $\mathcal{E}$ are geometrical paths along which the parameter undergoes "Darboux transforms" which preserve the wave equation structure, whereas the reflection and transmission coefficients are simply multiplied by phase factors. Following a given path, we always get at what we call an "undressed parameter", which is "trivial" in some way, and from which one can construct the desired parameter by following backwards the same path.

In section 1, we recall the elements of the one-dimensional scattering, and show how the geometrical transforms can span the spaces $\mathcal{C}$ and $\mathcal{E}$. In this section we restrict $\mathcal{C}$ and $\mathcal{E}$ to the sets of parameters and data that satisfy the so-called "Deift-Trubowitz" conditions, for which $\mathcal{M}$ is a bicontinuous bijection. Admittedly, even with this assumption, the net of our paths does not completely cover $\mathcal{C}$ and $\mathcal{E}$, since one only gets at coefficients that are rational fractions of k. But the subspaces which are thus generated are everywhere dense in $\mathcal{C}$ and $\mathcal{E}$, since any reflection or transmission coefficient can be arbitrarily approximated by a rational fraction (e.g. a Padé approximant) and $\mathcal{M}^{-1}$ is continuous. Then we show that this "global" method gives us a simple way to study a practical problem : in plasma soundings, we study profile inversion uncertainties related with uncertainties of the plasma surface position.

In section 2, giving up the Deift-Trubowitz conditions enables us to demonstrate ambiguities in $\mathcal{C}$. An "ambiguity" shows up when two or more parameters exactly correspond to the same data. In the one-dimensional Schrödinger inverse problem some examples of ambiguities were recently given by Abraham, de Facio, Moses(15),

Brownstein[16], Moses[17], but they can be objected on physical grounds. These objections break down in the one-dimensional inverse problems where the physical parameter is the impedance or an equivalent parameter. If this parameter is twice differentiable, the problem is equivalent to the Schrödinger problem. But physics only guarantees that the impedance is piecewise continuous and ambiguities are allowed. In this paper, we like better giving methods for constructing ambiguities rather than showing a few examples like in previous studies of the subject. It is also shown for the first time that ambiguities are not isolated but follow paths in the parameter space, where they "travel" through Darboux transforms. Two paths can cross each other at a point where they stop, because the two equivalent parameters become equal (bifurcation point). A path can also "stop" when a new transform yields a "singular" potential, that goes out of $\mathcal{C}$. It is important to notice that deriving these results is done by means of a new transform, which generalizes the Darboux one, and fits the requirements of piecewise continuous impedances. This transform can also be used to undress (of their background) impedances that are made of layered jump discontinuities and of a background that can be represented by rational coefficients.

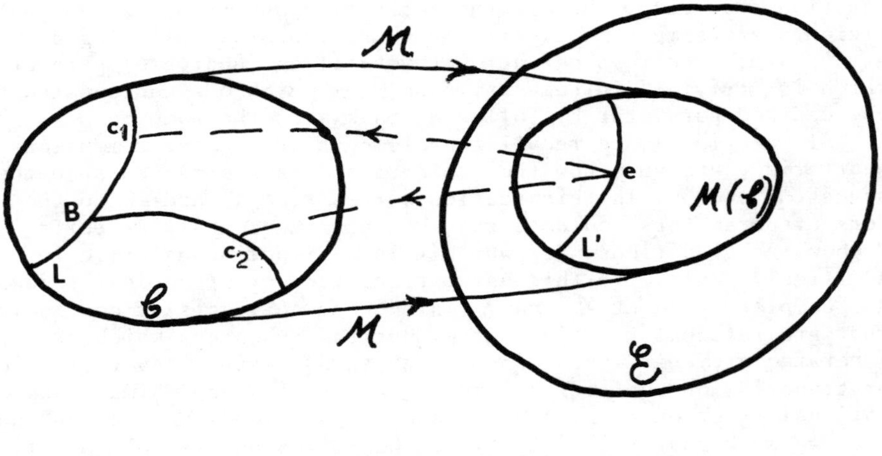

This figure shows the sets $\mathcal{C}$, $\mathcal{E}$, the mapping $\mathcal{M}$, and an example of ambiguity : 2 parameters c_1 and c_2 correspond to the same element e of $\mathcal{M}(\mathcal{C})$. When e follows in $\mathcal{M}(\mathcal{C})$ the line L' whose elements differ from each other by a multiplicative phase factor only, the corresponding parameters c_1 and c_2 follow in $\mathcal{C}$ the graph L, which shows at B a bifurcation point, and shows also three end-points at the boundaries of $\mathcal{C}$. Is it necessary to emphasize the fact that local methods are not able to deal with this kind of problems ?

In a concluding section, we survey some other aspects of a global approach, and some open questions.

SECTION 1. ONE DIMENSIONAL SCATTERING AND RATIONAL COEFFICIENTS

We first recall the mathematical problem, next we briefly survey known results for rational reflection coefficients and their connection with the surface position. Then we present our own methods, which are partly founded on the Darhoux method or on the "Wronskian method" used by Calogero and Degasperis in their analysis of non linear partial differential equation, and we briefly discuss the surface problem in this framework.

1.1. The mathematical problem

In its simplest form, the one-dimensional inverse scattering problem reduces to the Schrödinger equation :

$$[-\frac{d^2}{dx^2} + V(x)]\varphi(k,x) = k^2\varphi(k,x) \qquad (x \in \mathbb{R}) \qquad (1)$$

where we shall definitely assume that V belongs to the class C of functions defined by

$$\int_{-\infty}^{+\infty} dx[1 + |x|]\ |V(x) \qquad \infty \qquad (2)$$

For real k, it is possible to define the (unique) solution of (1) asymptotic to exp[ikx] for $x \to +\infty$, say, $f_+(k,x)$, and that asymptotic to exp [-ikx] for $x \to -\infty$, say $f_-(k,x)$. They are called the Jost solutions. Their Fourier transforms are traveling waves, which enable us to identify the physical meaning of coefficients in the formula :

$$T(k)f_-(k,x) = f_+(-k,x) + R_+(k)f_+(k,x) \qquad (3)$$

T(k) is the transmission coefficient, $R_+(k)$ is the reflection coefficient from right to left. The direct scattering problem ($V \to R$) and the inverse one ($R \to V$) have been the object of a huge body of studies. With the only assumption (2), Faddeyev[(2)] obtained a wide set of results. He had used for that only usual techniques of inverse scattering (Chadan-Sabatier[(3)]). With the help of higher mathematics, Deift and Trubowitz[(4)] were able to complete this study. However, a pedagogical presentation was needed, at least for the direct part and the reconstruction part of the problem, that are most useful in so many problems. Such a presentation obviously needs minimizing the use of high mathematics and the length of proofs. The first goal was already attained by Atkinson[5] and we shall use the general lines of his lecture. We think that the proofs we have introduced[6] and which are reproduced here achieve the second goal. Setting $F_+(k,x) = \exp[-ikx] f_+(k,x)$ (resp. $F_- = \exp[+] f_-$), one readily derives from (1) the Volterra equations :

$$F_+(k,x) = 1 - \int_x^\infty dy\, L_+(x,y) F_+(k,y) V(y) \tag{4a}$$

$$F_-(k,x) = 1 - \int_{-\infty}^x dy\, L_-(x,y) F_-(k,y) V(y) \tag{4b}$$

where $L_+(x,y) = L_-(y,x) = -k^{-1} \sin[k(y-x) \exp ik(y-x)]$ (5)

In order to continue $F_\pm(k,x)$ into the upper half plane, all authors have used the following bound, which holds for any $y \geq x$, and where $\sigma(k) = |\mathrm{Im}k| - \mathrm{Im}k$:

$$\exp[-(y-x)\sigma(k)]\, |L_+(x,y)| \leq 2\inf\{|k|^{-1}, y - x\} \tag{6}$$

This is used in two steps, with first keeping $2|k|^{-1}$ only at the right hand side of (6) (we call it the first choice in (6)), then using $(y-x)$ instead. However, this last bound is not convenient, and led to some errors. Our improvement was to replace $(y-x)$ in the right hand side of (6) by $[1+|x|\theta(-x)][1+|y|\theta(y)]$, where θ is the Heaviside function (for $L_-(y,x)$, and $x \geq y$, it suffices to permute x and y). This way of proceeding is called below the second choice. Hence the kernel of (4a) (resp. 4b) can be absolutely majorized by either of two separable kernels of the general form $\alpha(x)\beta(y)$. For instance, with the second choice, and with corresponding labels $\pm$ put at the same level, we have :

$$\begin{aligned} \alpha_\pm(x) &= \exp[\mp x\sigma(k) \{1+|x|\theta(\mp x)\} \\ \beta_\pm(x) &= 2\exp[\pm x\sigma(k)]\, |V(x)|\{1+|x|\theta(\pm x)\} \end{aligned} \tag{7}$$

The same formulas hold for the first choice, with k^{-1} instead of $[1+|x|\theta]$. The interest of such separable kernels is that the corresponding Volterra equation can be exactly solved and its solution expansion yields uniform absolute bounds for corresponding terms in the solution expansion to (4a) or (4b). Since each term obviously is an entire function of k, we readily prove that $F_\pm(k,x)$ is holomorphic in the half plane $\mathrm{Im}k \geq 0$ and bounded as follows :

$$\begin{aligned} |F_+(k,x)-1| &\leq \alpha_+(x)\int_x^\infty du\; \beta_+(u)\exp[2\int_x^u dt|V(t)|\gamma(t)] \\ |F_-(k,x)-1| &\leq \alpha_-(x)\int_-^x du\; \beta_-(u)\exp[2\int_u^x dt|V(t)|\gamma(t)] \end{aligned} \tag{8}$$

where $\gamma(t)$ is equal to $|k|^{-1}$ for our "first choice", to $(1+|t|)$ for our "second choice". By the way, one sees here the interest of assumption (2). It follows that $|F_+(k,x)-1|$ is $O(k^{-1})$ as $|k| \to \infty$, $\mathrm{Im}k \geqslant 0$. Hence we can define, at least in L_2, the following Fourier transforms :

$$K_\pm(x,y) = (2\pi)^{-1}\int_{-\infty}^{+\infty} dk \exp[\mp ik(y-x)]\{F_\pm(k,x)-1\} \tag{9}$$

By iterating (4) once, and inserting the result into (9), one easily sees that the integral in (9) uniformly converges for $y \neq x$. By closing the contour in the upper-half k-plane, we deduce from (9) that $K_+(x,y)$ vanishes for $x > y$ and that so does $K_-(x,y)$ for $x < y$. Hence, the inverse Fourier transform yields, at least in the L_2 sense, the transformation formulas :

$$F_+(k,x)-1 = \int_x^\infty dy \exp[ik(y-x)]K_+(x,y) \tag{10a}$$

$$F_-(k,x)-1 = \int_{-\infty}^x dy \exp[-ik(y-x)]K_-(x,y) \tag{10b}$$

By substituting (4) into (9), then using (10) to eliminate $F_\pm$ in favor of $K_\pm$, and performing the Fourier transforms, we obtain the integral equations :

$$K_+(x,y) = \frac{1}{2}\int_{\frac{1}{2}(x+y)}^{\infty} ds\{V(s)+2\int_0^{\frac{1}{2}(y-x)} drV(s-r)K_+(s-r,s+r)\} \tag{11a}$$

$$K_-(x,y) = \frac{1}{2}\int_{-\infty}^{\frac{1}{2}(x+y)} ds\{V(s)+2\int_{\frac{1}{2}(y-x)}^{0} dr\, V(s-r)K_-(s-r,s+r)\} \tag{11b}$$

It follows from (11) that $K_+(x,x)$ and $K_-(x,x)$ are trivially related to V. In fact, the formulas (10) and (11) show that the transformation kernels $K_\pm$ **are the key to** the direct problem. They can be constructed from V by integrating (11), and, in turn, they readily yield V and F through (10), ie the parameter <u>and</u> the solutions of (1). We now go to the inverse problem, which can proceed either from $R_+(k)$ or from the reflection coefficient from left to right, $R_-(k)$, **here defined by**

$$f_-(-k,x) + R_-(k)f_-(k,x) = T(k)f_+(k,x) \tag{12}$$

Some algebraic derivations of wronskians yield the properties :

$$T(k) = 2ik[\Delta(k)]^{-1} \tag{13a}$$
$$R_\pm(k) = \pm[\Delta(k)]^{-1} W[f_\pm(-k,.),f_\mp(k,.)] \tag{13b}$$
$$\text{where } \Delta(k) = W\, f_-(k,.),f_+(k,.) \tag{13c}$$
$$\text{and } R_+(k)T(-k) + R_-(-k)T(k) = 0 \tag{14a}$$
$$R_+(k)R_+(-k)+T(k)T(-k) = R_-(k)R_-(-k)+T(k)T(-k) = 1 \tag{14b}$$

Because of the reality of V(x), we have in addition for real k :

$$R_{\pm}(-k) = [R_{\pm}(k)] \qquad T(-k) = [T(k)] \tag{15}$$

The so called "Jost solution" $\Delta(k)$ is holomorphic in Imk $\geqslant$ 0 and therefore so is T(k), except may be at the zeros of $\Delta(k)$. Now it can be proved that these zeros correspond to bound states ("trapped modes") and we can exclude them in a study intended for applications in electromagnetic scattering. Assuming this is done, we can give a method of reconstruction of V(x) from R(k). Indeed, the analytic and asymptotic properties of $F_{\pm}(k,.)$ in Imk $\geqslant$ 0 enable us to write down the Cauchy formula

$$F_{\pm}(k,x)-1 = \lim_{\varepsilon\to 0^+} \frac{1}{2\pi i} \int_{-\infty}^{+\infty} dk' \frac{F_{\pm}(k',x)-1}{k'-k-i\varepsilon} \tag{16}$$

which can be transformed by making x → -x and by inserting (3) or (12), to yield

$$f_{\pm}(k,x) \exp[\mp ikx -1 = \lim_{\varepsilon\to 0^+}\{$$

$$\frac{1}{2\pi i} \int_{-\infty}^{+\infty} dk' \frac{1-T(k')f_{\mp}(k',x)\exp[\pm ik'x]}{k'+k+i\varepsilon}$$

$$+ \frac{1}{2\pi i} \int_{-\infty}^{+\infty} dk' \frac{R\ (k')f_{\pm}(k',x)\exp[\pm ik'x]}{k'+k+i\varepsilon}\} \tag{17}$$

Using (13 a,c), the asymptotic behavior of $F_{\pm}$, and the holomorphy assumption for T(k) in Imk $\geqslant$ 0, we can show that the first term in the right-hand side of (17) vanishes. Hence we obtain an equation that is convenient for the desired reconstruction :

$$F_{\pm}(k,x)-1 = \frac{1}{2\pi i} \lim_{\varepsilon\to 0^+} \int_{-\infty}^{+\infty} dk' \frac{R_{\pm}(k')F_{\pm}(k',x)\exp[\pm 2ik'x]}{k'+k+i\varepsilon} \tag{18}$$

The interest of equations like (18) for reconstruction problems has been shown by Karlsson[7]. In fact, (18) is equivalent to the Marchenko equation, which is readily obtained by Fourier-transforming (18) :

$$K_{\pm}(x,y)+M_{\pm}(x+y)+\int_{-\infty}^{+\infty} dz\theta(z\mp x)M(y+z)K_{\pm}(x,z) \tag{19}$$

where

$$M_{\pm}(x) = (2\pi)^{-1} \int_{-\infty}^{+\infty} dk\ R_{\pm}(k)\exp[ikx] \tag{20}$$

The characterization problem in the class defined by (2) is still open. In a slightly narrower class, Deift and Trubowitz[4] gave existence and uniqueness theorems. However, we are interested here in a still narrower class. Thus we shall not recall these complicated results but only refer to the original paper when necessary.

1.2. On "rational" reflection coefficients

Reflection coefficients that are rational fractions of k already appear in the very first important paper on the inverse scattering problem on the line, an unpublished report by Kay[8], who recognized the closed form of potentials corresponding to a coefficient R(k) holomorphic everywhere but a finite number of poles in the lower half-plane. Kay[9,10] gave then a complete study of these "rational" reflection coefficients for potentials vanishing identically after a finite distance on one side of the real axis ("cut off potentials"), whereas applications of reflectionless potentials were sought in electromagnetic propagation[11]. From that time on, and although the method given by Kay could have been used for the general case, almost all authors only considered cut off potentials (say $V(x) = 0$ for $x < 0$). (See for instance Reilly & Jordan[12], Pechenick and Cohen[13], and the references listed therein).Cut potentials are studied because they naturally appear in plasma sounding. Exact calculations with two or three poles in the reflection coefficient were performed, either with Kay's method or with a method beginning by a Laplace transforming the Marchenko equation, or with one using a linear differential operator to cancel R(x) in the Marchenko equation and derive instead a linear differential equation. Concerning full range potentials, let us cite the papers of Moses[14] and of Abraham, de Facio, Moses[15], in which some assumptions of §(1.1) are given up in order to get singular examples. Studies of non linear partial differential equations (NLPDE connected with the inverse scattering problem (ISP) involved the studies of N-soliton solutions, for which R(k) has a finite number of simple poles on the upper half imaginary axis. They gave rise, among others, to a large number of papers on rational solutions of NLPDE (rational functions of x and t). However, our knowledge of reflection coefficients that exhibit poles in the lower half plane, the most interesting in scattering studies, was hardly improved by them. Hence we have undertaken a systematic study of rational reflection coefficients in the ISP. Here we shall only present some results of interest for electromagnetic inverse scattering. To derive them, we introduce two new methods. The first one is a generalization of Kay's method. The second one, founded on the wronskian method, is completely different and yields a sequential algorithm for solving the ISP with rational reflection coefficients. Throughout these methods[6] new distinction between poles has been suggested to us by our current interest in isospectral transformations and has proved to be essential <u>also</u> in the inverse scattering problem.

1.3. The methods

We start with the following assumptions :

(a) $T(k)$ is a rational fraction, with poles located only in the lower half-plane, goes to 1 for $|k| \to \infty$, vanishes on the real axis at $k = 0$ only, where it has a simple pole, and satisfies (15).

(b) $R_\pm(k)$ are rational fractions, go to zero as $O(|k|^{-2})$ for $|k| \to \infty$, go to -1 for $|k| \to 0$, and satisfy the formulas (14) and (15).

(c) Either $R_+(k)$ or $R_-(k)$ is given, consistently with Assumptions (a) and (b).

Thanks to the assumptions (a) and (b), it is possible to check the Deift and Trubowitz conditions that are sufficient to guarantee the existence and uniqueness of a real potential in the class defined by (2), with reflection coefficients $R_\pm(k)$, transmission coefficient $T(k)$, and for which all the formulas in (1.1) do hold. These formulas involve products like $R(k)R(-k)$, which are invariant under inserting into $R(k)$ a "phase factor" $P(k)/P(-k)$. This remark, and a few algebraic manipulations, show that the only forms consistent with no bound state are

$$T(k) = \frac{\prod_{i=1}^{q}(\kappa_i+k)}{\prod_{j=1}^{q}(\lambda_j-k)} = \frac{N(k)}{D(k)} \tag{21}$$

$$R_-(k) = \frac{P(k)}{\prod_{j=1}^{q}(\lambda_j-k)} \prod_{i=q+1}^{r} \frac{\kappa_i-k}{\kappa_i+k} \prod_{\ell=p+1}^{s} \frac{\lambda_\ell-k}{\lambda_\ell+k} \tag{22}$$

$$R_+(k) = \frac{P(-k)}{\prod_{j=1}^{q}(\lambda_j-k)} \prod_{i=2}^{r} \frac{\kappa_i+k}{\kappa_i-k} \prod_{\ell=p+1}^{s} \frac{\lambda_\ell+k}{\lambda_\ell-k} \tag{23}$$

where the κ_i's and λ_ℓ's are complex numbers such that $\kappa_1 = 0$, $\mathrm{Im}\,\kappa_i > 0$ $(i > 1)$, $\mathrm{Im}\,\lambda_i < 0$, and $p \leqslant q \leqslant r,s$. The polynomial $P(k)$ degree is at most $q-1$. Some factors may disappear if these forms are algebraically reduced. In any case, the poles of $R_-(k)$ in $\mathrm{Im}k > 0$ belong to two sets, $\mathcal{P}_s^-$ and $\mathcal{P}_I^-$, with $\mathcal{P}_s^- \subset \{-\lambda_\ell | \ell \in (p+1,q)\}$, $\mathcal{P}_I^- \subseteq \{-\lambda_\ell | \ell \in (q+1,s)\}$. The poles of $R_+(k)$ in $\mathrm{Im}k > 0$ belong to $\mathcal{P}_s^+$ and $\mathcal{P}^+$, with $\mathcal{P}_s^+ \subset \{\kappa_i | i \in (2,q)\}$, $\mathcal{P}_s \subset \{\kappa_i | i \in (q+1,r)\}$. Together with $T(k)$, the sets labelled s correspond to the spectral properties of our problem. Those labelled I do not depend on T and correspond to isospectral transformations of $R_{\pm}$. The whole sets $\mathcal{P}^{+}(= \mathcal{P}_{I}^{+} \cup \mathcal{P}_{s}^{-})$ and (similarly) $\mathcal{P}^{-}$ contribute in the derivation of V from R±. Equations (18) are very convenient for solving this problem. In the simple case when all poles are simple, with residues R_{+}^{i} and R_{-}^{j}, we readily derive for x > 0

$$F_{+}(k,x)-1 = \sum_{\kappa_i \in \mathcal{P}^{+}} \frac{R_{+}^{i} F_{+}(\kappa_i,x)\exp[2i\kappa_i x]}{k+\kappa_i} \tag{24}$$

and for x < 0

$$F_{-}(k,x)-1 = \sum_{-\lambda_j \in \mathcal{P}^{-}} \frac{R_{-}^{i} F_{-}(-\lambda_j,x)\exp[2i\lambda_j x]}{k-\lambda_i} \tag{25}$$

We can identify $K_{\pm}(x,y)$ for instance by comparing (24-25) and (10a-10b) :

$$K_{+}(x,y) = -i \sum_{K_i \in \mathcal{P}^{+}} R_{+}^{i} F_{+}(\kappa_i,x)\exp[i\kappa_i(x+y)] \tag{26a}$$

$$K_{-}(x,y) = -i \sum_{-\lambda_j \in \mathcal{P}^{-}} R_{-}^{i} F_{-}(-\lambda_j,x)\exp[i\lambda_j(x+y)] \tag{26b}$$

Let k run through $\mathcal{P}_{+}$ in (24), through $\mathcal{P}_{-}$ in (25). We hereby obtained two linear systems whose solution yields everything in our problem through (26a) and (26b). It is interesting to notice that R(k) being a rational fraction depends on the position of the origin. Changing this position would insert in R(k) a factor exp[-2ika]. Besides, the two linear systems are independent, so that taking care of the phase isospectral factor enables us to modify almost independently the two sides ($x \gtrless 0$) of the potential. Hence, this way of processing is convenient for a study of the smoothing of the "surface" of V(x) at x = 0. However, there is a better method, that fully makes use of the isospectral transformations properties. This method(6) could be based either on the Darboux theorem or on the generalized Wronskian formula. Here we use the Darboux theorem. We start with a potential V_1 that already corresponds to rational coefficients. Hence (24) and (25) hold. They mean that $F_{\pm}(k,x)$ and the relations (3) or (12) can be continued everywhere in the k-plane except at isolated poles. Let us now consider a solution u_o of

$$u_o'' + (\lambda_o^2 - V_1)u_o = 0 \tag{27a}$$

where, say, Im $\lambda_o > 0$. Let f be a solution of

$$f'' + (k^2 - V_1)f = 0 \tag{27b}$$

Following Darboux, we easily check that the function ψ defined by

$$\psi = f' - f\, u_o'/u_o \tag{28}$$

is a solution of the equation

$$\psi'' + (k^2 - V_2)\psi = 0 \tag{29}$$

with the modified potential

$$V_2 = V_1 - 2\frac{d}{dx}(u_o'/u_o) \tag{30}$$

Let us now derive the corresponding reflection and transmission coefficients, setting

$$u_o = \alpha\, f_-^{(1)}(\lambda_o,x) + \beta\, f_-^{(1)}(-\lambda_o,x) \tag{31}$$

where the upper label $^{(1)}$ refers to V_1. A study of (3) and (12) shows the possible cases :

(1) $\beta \neq 0$, any α but $\alpha = \beta\, R_-(\lambda_o)$

In this case, u_o'/u_o is asymptotic to $i\lambda_o$ for $x \to -\infty$, and to $-i\lambda_o$ for $x \to +\infty$. Starting from the solution $f \equiv f_-^{(1)}(k,x)$ of (27b), we can write down the formula analogous to (3) and show by identification that

$$T_2(k) = -\frac{\lambda_o+k}{\lambda_o-k}\, T_1(k) \quad ; \quad R_2^{\pm}(k) = \frac{\lambda_o+k}{\lambda_o-k}\, R_1^{\pm}(k) \tag{32}$$

(2) $\beta \neq 0$, $\alpha = \beta\, R_-(\lambda_o)$. Then

$$T_2(k) = T_1(k) \quad ; \quad R_2^{\pm}(k) = \frac{\lambda_o-k}{\lambda_o+k}\, R_1^{\pm}(k) \tag{33}$$

(3) $\beta = 0$, λ_o not a bound state. Then

$$T_2(k) = T_1(k) \quad ; \quad R_2^{\pm}(k) = \frac{\lambda_o+k}{\lambda_o-k}\, R_1^{\pm}(k) \tag{34}$$

(4) $\beta = 0$, λ_o bound state. Then

$$T_2(k) = -\frac{\lambda_o-k}{\lambda_o+k}\, T_1(k) \quad ; \quad R_2^{\pm}(k) = \frac{\lambda_o-k}{\lambda_o+k}\, R_1^{\pm}(k) \tag{35}$$

Only the cases (2) and (3) correspond to isospectral transformations. They enable one to insert into R(k) a phase factor with a pole located either in the upper half plane or in the lower half plane. This possibility of sequentially inserting the poles of R(k) can be used to write down a new algorithm. In this algorithm[19], the coefficients $R_\pm(k)$ are written as products of "undressed" coefficients $R_\pm^o(k)$ without poles in the ± half k-plane respectively, times finitely many phase factors. These "routes" in $\mathcal{G}$ correspond to "routes" in $\mathcal{C}$ starting at "undressed" $V_\pm^{(0)}$

which (see (24),(25)), resp. vanish on $x \gtrless 0$, and one travels along these routes by Darboux transforms.

As an application, we start from a coefficient $R_+(k)$ which corresponds to a potential cut-off at $x = 0$, and we represent uncertainties on a data almost equal to $R_+(k)$ by introducing an isospectral extra factor $(i\mu-k)/(i\mu+k)$, with $\mu >> 0$, so as to appraise the uncertainties on the surface position due to data errors. For a more specific example, let $R_-(k)$ have only two poles in Im $k < 0$.

$$R_-(k) = (a^2+b^2)[(k+ib)^2-a^2]^{-1} \tag{36}$$

where a, b, > 0, $b^2 > a^2$. The corresponding other coefficients are

$$T(k) = k(k+i\kappa)[(k+ib)^2-a^2]^{-1} \tag{37}$$

$$R_+(k) = -\frac{k+i\kappa}{k-i\kappa}\,\frac{a^2+b^2}{(k+ib)^2-a^2} \tag{38}$$

where $\kappa = [2(b^2-a^2)]^{1/2}$. The corresponding potential is

$$V(x) = \begin{cases} 0 & \text{for } x < 0 \\ 8\kappa^2 \dfrac{c^2 \exp[-2\kappa x]}{\{1-c^2\exp[-2\kappa x]\}^2} & \text{for } x > 0 \end{cases} \tag{39}$$

$$\text{where } c^2 = [c(\kappa)]^2 = [(b+\kappa)^2+a^2]^{-1}(a^2+b^2) \tag{40}$$

Assume now the uncertainties on $R_-(k)$ are such that the isospectral phase factor $(i\mu-k)/(i\mu+k)$ in front of $R_+(k)$ can represent them, with $\mu \geqslant \mu_o$. Then the uncertainty on $V(x)$ is seen in the formula

$$V_\mu(x) = \begin{cases} 8\mu^2 d^2 \exp[2\mu x]\{1-d^2\exp[2\mu x]\}^{-2} & (x < 0) \\ -8\kappa^2 c^2 \exp[-2\kappa x]\dfrac{\kappa^2-\mu^2}{\{(\kappa+\mu)+(\kappa-\mu)c^2\exp[-2\kappa x]\}^2} & (x > 0) \end{cases} \tag{41}$$

where $d^2 = c^2(\mu)$.
It clearly shows that for $\mu/\kappa >> 1$, the uncertainty for $x > 0$ is much weaker than the one for $x < 0$. Set for instance $a^2 = 1$, $b^2 = 1{,}125$, so that $\kappa = 0{,}5$, $c^2 \sim 0{,}6$, $\mu = 10$, $d^2 \sim 0{,}016$. Then $V(x)$ has values between 0,1 and 1 for x between 0 and 2. The perturbation due to the phase factor lies for $x > 0$ all along $V(x)$ with a relative value not larger than 1/200, whereas for $x < 0$, it is approximately $13 \exp[20x]$, which has values larger than 0.2 for $x > -0.2$. Hence these uncertainties on $R_+(k)$ result into an uncertainty on the position $x = 0$ rather than an uncertainty on the shape of $V(x)$.

SECTION 2.

Let us now deal with exact ambiguities. We know that the inverse problem on the line is well-posed if V is sought in the set L_2' where $\int_{-\infty}^{+\infty} dx(1+|x|^2)|V(x)|$ is finite and if R(k) satisfies the Deift-Trubowitz conditions :
(a) T(k), $R_+(k)$, $R_-(k)$, satisfy on the real axis (14) and (15)
(b) T(k) is holomorphic in Imk > 0, continuous down to the real axis
(c) $T(k)-1 = 0(|k|^{-1})(\text{Imk} \geqslant 0)$; $R_\pm(k) = 0(|k|^{-1})(k \in \mathbb{R})$
(d1) $|T(k)| > 0$ (Imk $\geqslant 0$, $k \neq 0$) .
(d2) either $|T(k)| > 0$ (Imk $\geqslant$ 0) or there exists T and $\rho_\pm$ such that $T(k)-kT = 0(k)$, (Imk $\geqslant$ 0) and $1+R_\pm(k)-k\rho_\pm = 0(k)$, ($k \in \mathbb{R}$)
(e) $M_\pm(x)$ is absolutely continuous, and there exists c(a) finite such that $\int_{-\infty}^{+\infty} dx\ \theta[\pm(x-a)]|M_\pm'(x)|(1+x^2) \leqslant c(a)$ for any real a.
In order to show exact ambiguities in potential reconstructions, one must violate at least one of these conditions. This has been done by Abraham, de Facio and Moses[15] whose example involves a δ-function in V, and violates conditions (d) and (e). In particular, T(k) vanishes at k = 0, but $R_+(0)$ is equal to +1 (this feature corresponding to a Jost solution singular at k = 0). Other examples were given by Brownstein[16] and by Moses[17]. None of these authors described how they discovered their examples. It is certainly not a systematic way of exploring the subject. Here we shall try to give one of the possible ways which can be applied in the cases of interest arising in the one-dimensional inverse scattering problem. Thus we study the equation

$$A^{-1}\frac{d}{dx}A\frac{d}{dx}p + k^2p = 0 \tag{42}$$

where A may be the admittance, p the potential, k the frequency. This equation arises in various electromagnetic and geophysical problems (see for example references 18 and 19). It must be solved in such a way that p and $A\frac{dp}{dx}$ are continuous. A must be non negative but is not necessarily continuous. We studied elsewhere[19] the case where A shows an arbitrary number of simple discontinuities (jumps). So as to get a chance of discovering a simple way towards ambiguities, we assume here that A is twice differentiable everywhere except at the point x = 0, where A and $\frac{dA}{dx}$ may show simple discontinuities (jumps). Let V be the function defined at each point $x \neq 0$ by

$$V(x) = [A(x)]^{-1/2}\frac{d^2}{dx^2}[A(x)]^{1/2} \tag{43}$$

We assume that V belongs to L_2'. This guarantees the existence of the "Jost solutions" $p_\pm(k,x)$ of (42), that go respectively to $[A(x)]^{-1/2}\exp[\pm ikx]$ as x goes to $\pm\infty$. The reflection and transmission coefficients are defined by the relations

$$\tilde{T}(k)p_-(k,x) = p_+(-k,x)+\tilde{R}_+(k)p_+(k,x) \quad (44)$$
$$p_-(-k,x)+\tilde{R}_-(k)p_-(k,x) = \tilde{T}(k)p_+(k,x) \quad (45)$$

It is easy to see that $\tilde{T}(k)$ and $\tilde{R}_\pm(k)$ satisfy (14) and (15). Now, both for $x > 0$ and for $x < 0$, the equation (42) is equivalent to the Schrödinger equation (1), with $V(x)$ as in (43). The function $p_\pm(k,x)$ are related (19) to the ordinary Jost solutions $f_\pm(k,x)$ of (1) on both sides of $x = 0$. We can write down for instance

$$p_-(k,x) = u_\pm[A(x)]^{-1/2}f_+(-k,x)+v_\pm[A(x)]^{-1/2}f_+(k,x) \quad (46)$$

where the coefficients u_+ and v_+ hold for $x > 0$, u_- and v_- for $x < 0$. They are trivially related with $R_+, T, \tilde{R}_+, \tilde{T}$. Writing down the continuity relations at $x = 0$ we obtain :

$$u_+ = [\tilde{T}(k)]^{-1} = [T(k)]^{-1}[\alpha(k)+\beta(k)R_+(k)] \quad (47)$$

$$v_+ = [\tilde{T}(k)]^{-1}\tilde{R}_+(k) = [T(k)]^{-1}[\beta(-k)+\alpha(-k)R_+(k)] \quad (48)$$

where

$$\alpha(k) = t^{-1}\{1+s(ik)^{-1}f_+(k)f_+(-k)+r(2ik)^{-1}[f_+'(k)f_+(-k)+f_+'(-k)f_+(k)]\} \quad (49)$$

$$\beta(k) = (ikt)^{-1}\{s[f_+(k)]^2+r\, f_+(k)f_+'(k)\} \quad (50)$$

$$f_\pm(k,0) \equiv f_\pm(k) \quad (51)$$

$$t^{-1} = \frac{1}{2}\{[A(0^+)/A(0^-)]^{1/2} + [A(0^-)/A(0^+)]^{1/2}\} \quad (52)$$

$$r/t = \frac{1}{2}\{[A(0^+)/A(0^-)]^{1/2} - [A(0^-)/A(0^+)]^{1/2}\} \quad (53)$$

$$s/t = -\frac{1}{4}[A(0^+)A(0^-)]^{-1/2}[A'(0^+)-A'(0^-)] \quad (54)$$

We can obtain in the same way the relations

$$T(k)/\tilde{T}(k) = \bar{\alpha}(-k)-\bar{\beta}(k)R_-(k) \quad (55)$$

$$\tilde{R}_-(k)T(k)/\tilde{T}(k) = -\bar{\beta}(-k)+\bar{\alpha}(k)R_-(k) \quad (56)$$

where

$$t\,\bar{\alpha}(k) = 1-(ik)^{-1}sf_-(k)f_-(-k)-(2ik)^{-1}r[f_-'(k)f_-(-k)+f_-'(-k)f_-(k)] \quad (57)$$

$$-ikt\,\bar{\beta}(k) = s[f_-(k)]^2 + r\, f_-(k)f_-'(k) \quad (58)$$

We can work either on $R_+(k)$ and $T(k)$, using the system (47)-(48), or $R_-(k)$ and $T(k)$, using the system (57)-(58). Once R_+ and T

are fixed, R_- is fixed by (14) (or conversely), so that the two approaches are equivalent. In the following, we choose working on $R_+(k)$, and drop everywhere the index +. We look for two different admittances A_1 and A_2 such that $\tilde{R}_1(k) = \tilde{R}_2(k)$ (with $\tilde{R}$ standing for $\tilde{R}_+$). Our working assumption is that the reflection and transmission coefficients are rational fractions of k. Now it follows from the algebraic proofs recalled in (1.3) that giving $\tilde{R}(k)$ determines $\tilde{T}(k)$ up to the "bound state factor", (not written in (21)), which we shall ignore in this first study, since it is related to unphysical zeros of the admittance[20]. Hence we are led also to impose $\tilde{T}_1(k) = \tilde{T}_2(k)$. Thus it is natural to take as an additional working assumption that $T_1(k) = T_2(k), s_1=s_2$. With these three equalities, it follows from (47) and (48) that $R_1(k)$ and $R_2(k)$ satisfy the system

$$\left.\begin{aligned} &\beta_1(k)R_1(k)-\beta_2(k)R_2(k)+\alpha_1(k)-\alpha_2(k) = 0 \\ &\alpha_1(-k)R_1(k)-\alpha_2(-k)R_2(k)+\beta_1(-k)-\beta_2(-k) = 0 \end{aligned}\right\} \tag{59}$$

We shall now reduce our study to the case where the impedance is continuous, ie $t = 1$, $r = 0$. Provided $\alpha_1(k) \neq \alpha_2(k)$, and $\beta_1(k) \neq \beta_2(k)$, the system (59) is equivalent to the system

$$R_1(k) = -\psi(k)/m(k) \tag{60}$$

$$R_2(k) = -\psi(-k)/m(k) \tag{61}$$

where

$$\psi(k)=sf_2(k)f_1(-k)k^{-1}[f_2(k)f_1(-k)-f_2(-k)f_1(k)]+i[f_1(k)f_1(-k)-f_2(k)f_2(-k)] \tag{62}$$

and

$$m(k)=sf_1(k)f_2(k)k^{-1}[f_2(k)f_1(-k)-f_2(-k)f_1(k)]+i\{[f_1(k)]^2-[f_2(k)]^2\} \tag{63}$$

Hence $R_1(k)/R_2(k)$ is a "phase factor". The first factor in the right-hand side of (23) is the same one for $R_1(k)$ and $R_2(k)$, a result which is fortunately consistent with our working assumption $T_1(k) = T_2(k)$. Hence one goes from V_1 to V_2 by means of one or several Darboux transforms of the form (33) or (34). Referring to (23), let us define $R_o(k)$ as a common factor of $R_1(k)$ and $R_2(k)$ which has no pole in Imk > 0 and also (so as to guarantee its uniquess), no zero in Imk > 0. Setting

$$R_1(k) = R_o(k)\psi_1(k)/\psi_1(-k) \quad R_2(k) = R_o(k)\psi_2(k)/\psi_2(-k) \tag{64}$$

we see that the Darboux transform which are associated with these two factors give rise to an ambiguity if two conditions are

fulfilled :
(a) there exists a rational function of k^2, $C(k^2)$, such that

$$\psi_1(k)\psi_2(-k) = C(k^2)\psi(k) \qquad (65)$$

(b) $$R_o(k) = -\psi_1(-k)\psi_2(-k)[C(k^2)]^{-1}[m(k)]^{-1} \qquad (66)$$

Indeed, the condition (65) guarantees that $R_1(k)/R_2(k) = \psi(k)/\psi(-k)$, and the condition (66) guarantees that $R_1(k)+R_2(k)$ is consistent with this condition and (60)-(61). The functions $\psi_j(k)$ are polynomials of the form $\prod_{i=1}^{n_j} (\mu_i-k)$, and the (n_1+n_2) parameters μ_i completely describe the problems corresponding to R_1 and R_2. Using the results of §(1.3) enables one to construct $f_1(k)$ and $f_2(k)$ and hence $\psi(k)$ and $m(k)$ in terms of the μ_i's and of the numbers $R_o(\mu_i)$. The parity condition (65) imposes then constraints on the μ_i's and $R_o(\mu_i$'s). $R_o(k)$ can then be constructed from (66) and must satisfy the consistency condition on its zeros and poles location in the lower half plane. We believe that these constraints can in general be fulfilled for whatever choice of the degrees of ψ_1 and ψ_2 (ie n_1 and n_2). Our guess is supported on one hand by a counting of the number of (non linear) equations, constraints, and variables, on the other hand by the existence of a solution in the very simplest and most constrained case where ψ_1 is of degree 0 and ψ_2 of degree 1. Let us now give this example :

<u>Example</u>

Let $\psi_1(k) = 1$, $\psi_2(k) = i\gamma s + k$, with $s > 0$, $0 < \gamma < 2(\sqrt{2}-1)$. The method yields two "equivalent potentials", say $-2s\delta(x)-8\theta(x)\gamma^2 s^2[1+(\gamma-1)e^{-2\gamma sx}]^{-2}(\gamma-1)e^{-2\gamma sx}$ and the symmetrical one, with the same reflection <u>and</u> the same transmission coefficients :

$$\tilde{T}(k) = k(k+i\gamma s)(k-i\gamma s)^{-1}(k+is)^{-1} \qquad (67)$$

$$\tilde{R}(k) = is(k+i\gamma s)(k-i\gamma s)^{-1}(k+is)^{-1} \qquad (68)$$

These potentials have a bound state so that the corresponding impedance has a zero and is not physical. The method is consistent because the "background" potential (ie without the δ function) has no bound state.

In the case of continuous admittances, the formula (43) defines $V(x)$ as a generalized function which contains at $x = 0$ the distribution $-2s\delta(x)$. It is easy to see that the Schrödinger equation (1), with the thus defined function $V(x)$, and the equation (42) are equivalent. They yield the same reflection and transmission coefficients, and the Jost solutions of (42) are those of the Schrödinger equation multiplied by $[A(x)]^{-1/2}$. The Darboux transforms apply without modification to the Schrödinger equation with V(x) involving δ-functions. Let us call "equivalent" two parameters V which correspond to the same scattering coefficients $\tilde{R}_{\pm}$ and $\tilde{T}$. If a given Darboux transform is applied to two "equivalent" parameters, the corresponding scattering coefficients are multiplied by the same phase factors, and hence the two new parameters V are equivalent. It follows that each pair of equivalent parameters V generates a pair of continuous paths of equivalent parameters in the space of parameters. One can do it as well by starting from the example cited above as by starting from the pairs (or multiplets) introduced by Abraham, de Facio, Moses[14,15] or Brownstein[16]. Some non-physical features are kept in the transformations (e.g. $R_{\pm}(0) \neq -1$). Others may be introduced (e.g. zeros of the admittances or double poles of the potential on real x). In fact, there may be several points of special interest on two paths of equivalent parameters. Look for instance on the example given above. We may find it interesting to remove the bound state by using the Darboux transform (35). This is easy but the result is the potential $2s\,\delta(x)$ for whatever potential V_1 or V_2 we were starting from. Hence the ambiguity disappears beyond this point in our travel along the paths. In other words, this is a multiple point of the transform (see (32)). We can also find it interesting to introduce arbitrary phase factors, for instance factors $(i\beta s-k)^{-1}(i\beta s+k)$, which keep V real. Then we see that for a range of values of β, the Darboux transform introduces singularities (double poles on the real axis coming from the zeros of f_-). In the space of scattering data, the reflection coefficients violate the constraint $|R_{\pm}(k)| < 1$ on the real axis. This constraint actually restricts the domain of allowed pole positions in the complex k-plane, as noticed by several authors[13]. Hence the paths of equivalent parameters may also stop at the boundary of the set of allowed parameters.

In the general case (piecewise continuous impedance), the equation (42) is no longer equivalent to the Schrödinger equation with a potential. Nevertheless, since we have still more freedom in the choice of parameters (see (49),(50)), it is clear that ambiguities still exist. The problem is to know whether a generalized transform still enables us (or not) to construct paths in the parameter space that correspond to multiplying by phase factors in the spaces of scattering coefficients. We are able to exhibit such a transform :

Let A be a piecewise continuous function, asymptotically constant at $x = \pm\infty$, bounded away from zero, and which has only a finite set of (jump) discontinuities. A is the "parameter" in the equation (42). Let $p_k \equiv p(k,x)$ be a continuous solution of (42), whose definition is achieved by imposing $Ap'_k \equiv A \frac{d}{dx} p(k,x)$ absolutely continuous, and (arbitrary) convenient asymptotic conditions. Let $p_\mu \equiv p(\mu,x)$ be an absolutely continuous solution of the equation

$$A^{-1} \frac{d}{dx} A \frac{dp_\mu}{dx} + \mu^2 p_\mu = 0 \tag{69}$$

with Ap'_μ absolutely continuous, and convenient asymptotic conditions. Now, we claim
(1) that the function A^T defined by

$$(A^T)^{1/2} = -A^{1/2} \, p'_\mu / p_\mu \tag{70}$$

is piecewise continuous, with a finite set of jump discontinuities, and is asymptotically constant at $\pm\infty$.
(2) that the function

$$p_k^T = p_k - p'_k \, p_\mu / p'_\mu \tag{71}$$

is a continuous solution of the equation

$$(A^T)^{-1} \frac{d}{dx} A^T \frac{d}{dt} p_k + k^2 p_k = 0 \tag{72}$$

such that $A^T \frac{d}{dx} p_k$ is absolutely continuous. The proof of these results is easy. We only give to the reader an intermediate result:

$$\frac{d}{dx} p_k^T = Ap_\mu (Ap'_\mu)^{-2} [k^2 p_k Ap'_\mu - \mu^2 \, p_\mu Ap'_k] \tag{73}$$

Now, choosing the asymptotic conditions like in (31) to (35), enables us to transform parameters in such a way that the corresponding scattering coefficients are multiplied by phase factors like in (32), (33), (34), and (35). In other words, this generalized transform exactly works like the Darboux transform. In particular, if two admittances are equivalent, the generalized transform (70) enables us to construct two paths of equivalent admittances.

Needless to say, all the other applications of these transforms are also generalized. In particular, suppose an admittance is made of jump discontinuities, with homogeneous and constant parts between discontinuities (e.g. Goupillaud or Claerbout models in geophysics). Such a problem is well-known and can be dealt with by matrix methods. Suppose now the parts between discontinuities are no longer homogeneous constant but are such that their effect can be represented[19] by a rational phase factor in the scattering coefficients. Then it is possible to use the generalized transform to undress the problem, reducing it to the simplest one.

Conclusions

So as to conclude our "critical study", it is interesting now to care for the physical "teachings" of a global approach to electromagnetic inverse problems.

(a) the first point which has been made shows the importance of the (generally concealed) a priori assumptions. We saw in section 1 that choosing the gap position of a truncated impedance, and allowing or not some diffuseness of this surface may have large consequences on the whole reconstruction. We saw in section 2 that allowing surface diffuseness can lead to ambiguities in the reconstruction, may-be (hopefully) discarded by convenient physical requirements. Other authors have discussed for example the importance of the no absorption assumption, either when it is related to the potential [13] or to the spectral problem itself[20]. But this assumption usually is not forgotten.
(b) the second point is the necessary use of special geometrical methods for constructing all solutions of the inverse problem. Of course, geometrical methods in function spaces are well-known: they enable us to define most quasisolutions or approximate solutions. But here we sought the exact solutions that correspond to results which are rational fractions of k (and any result can be arbitrarily approached by a rational fraction of k). And remarkably enough, the only way to explore completely the function space containing these solutions is by using Lie-Backlund transforms, more specially the Darboux ones which are described in § 1. The result is very elegant.
(c) last but not least, the third point is the particular interest of these methods from the algorithmic point of view. We have not given a detailed study of this problem in the present paper, but we did it elsewhere[19]. Of course the reader can imagine that using a cross-coordinated route in the parameter space is the shortest way to go from one point to another one. The value of this guess for a large number N of poles has been proved[19] : inversion algorithms based on Darboux transforms involve $O(N^2)$ operations only, in contrast with $O(N^3)$ for other algorithms.

Connected with these physical points there is still a number of open problems. Let us say honestly that the first one is how the scattering coefficients, with their zeros and poles, can be constructed from the experimental data. More theoretical open questions are connected with the ambiguities : is the number of their paths finite ?, and, if so, is it possible to construct all of them - what are the minimum requirements on A such that there is no ambiguity - and what is the influence of absorption ?

References

(1) Calogero F. and Degasperis A. Spectral transform and solitons I. North-Holland 1982.

(2) Faddeyev L.D. Properties of the S-matrix of the one-dimensional Schrödinger equation, Trudy Mat. Inst. Steklov 73, 1964 pp. 314-333 ; AMST 2, 65, 139-166.

(3) Chadan K., and Sabatier P.C. Inverse Problems in Quantum Scattering Theory, Springer-Verlag, New York 1977.

(4) Deift P., and Trubowitz E. Inverse Scattering on the Line Comm. Pure Appl. Math. 32, 121-251 (1979).

(5) Atkinson D. Marchenko in one dimension. Internal Rpt Inst. for Theoretical Physics Groningen, Netherlands (1979).

(6) Sabatier P.C. Rational reflection coefficients and Inverse Scattering on the line, to be published in Nuovo Cimento.

(7) Karlsson B. Inverse method for off-shell continuation of the scattering amplitude in quantum mechanics - in "Applied Inverse Problems" Lecture Notes in Physics n° 85, Springer-Verlag Berlin, Heidelberg, New York 1978.

(8) Kay I. The Inverse Scattering Problem. Research Report n° EM-74 New York University. Inst. Math. Sciences 1955.

(9) Kay I. On the determination of the free electron distribution of an ionized gas. Research Report n° EM-141 New York University Inst. Math. Sciences 1959.

(10) Kay I. The Inverse Scattering Problem when the reflection coefficient is a rational function. Comm. Pure and Appl. Math. 13, 371-393 (1960).

(11) Kay I. & Moses H.E. Reflectionless transmission through dielectrics and scattering potentials. Journ. Appl. Phys. 27 1503-1508 (1956).

(12) Pechenick K.R. & Cohen J.M. Exact solutions to the valley problem in inverse scattering. Journ. Math. Phys. 24 406-409 (1983).

(13) Reilly M.H. & Jordan A.K. The applicability of an inverse method for reconstruction of electron-density profiles IEEE Trans on Ant. and Prop. AP-29 245-252 (1982).

(14) Moses H.E. An example of the effect of rescaling of the reflection coefficient on the scattering potential for the one-dimensional equation Stud. Appl. Math. 60, 177-181 (1979).

(15) Abraham P.B., De Facio B., & Moses H.E. Two distinct local potentials with no bound states can have the same scattering operator : A non uniqueness in inverse spectral transformations. Phys. Rev. Lett. 46, 1657-59 (1981).

(16) Brownstein K.R. Non uniqueness of the inverse-scattering problem and the presence of c = 0 bound states. Phys. Rev. D. 25, 2704-2705 (1982).

(17) Moses H. Example of two distinct potentials without point eigenvalues which have the same scattering operator with the reflection coefficient c(0) = -1, Phys. Rev. A 27, 2220-2221 (1983).

(18) Sabatier P.C. Theoretical considerations for inverse scattering Radio Science 18, 1-18 (1983).

(19) Sabatier P.C. Rational reflection coefficients in one dimensional inverse scattering and applications Proceedings of Tulsa Conference on Scattering Theory, to be published by S.I.A.M.

(20) See for example Riska D.O. On the electro-dynamic inversion problem for vertically layered earth Commentationes Physico Mathematicae 53 1-24 (1982).

I.2 (SR.2)

TOPOLOGICAL APPROACH TO INVERSE SCATTERING IN REMOTE SENSING*

Werner Güttinger & Francis J. Wright

Institute for Information Sciences
University of Tübingen
Köstlinstr. 6, 7400 Tübingen, FRG

The topological problem underlying inverse scattering is analyzed by determining the caustic singularities that an unknown surface generically impresses on a sensing wavefield. Imposing the principle of structural stability (i.e., qualitative insensitivity to slight perturbations) on the inversion process shows the following: The dominant singularities that generically occur in recorded signals, travel-time curves, surface contour maps and Fresnel-zone topographies can, together with the associated diffraction patterns, be classified into a few topological normal forms described by catastrophe polynomials. As the source-receiver positions vary, the observed patterns change their morphologies according to universal bifurcation sets. These provide a new quasi-geometric processing methodology for surface reconstruction.

1. INTRODUCTION

This paper is designed to review the present state of our knowledge concerning the application of singularity and bifurcation theory to the inverse scattering problem in remote sensing and to sketch some new developments and techniques which promise to become important tools for future research. The central idea underlying this geometrical approach is the following. In order that the reconstruction of surfaces and subsurface structures from backscattered waves be physically feasible, i.e., repeatable, the scattering process underlying remote sensing has to be structurally stable, i.e., qualitatively insensitive to slight perturbations of the sensing wave system. Otherwise today's experiment would not reproduce yesterday's result. Imposing this principle of structural

* Research supported in part by the Stiftung Volkswagenwerk, FRG

W.-M. Boerner et al. (eds.), Inverse Methods in Electromagnetic Imaging – Part 1, 65–76.

stability [19] on the inversion process has the following consequences [10]. First, it permits us to classify the geometric singularities, that an unknown surface or structure generically impresses on a sensing wavefield, into a few universal topological normal forms described by catastrophe polynomials. These geometrical singularities produce the dominant analytic singularities and typical configurations that can generically occur in recorded signals, traveltime curves, contour maps, and in the associated diffraction patterns. As the source-receiver positions vary, the patterns change their morphologies according to typical universal bifurcation sets (caustics). These, and the resulting universal power laws for the frequency dependent diffraction amplitudes, permit a reconstruction of 3D profiles in a genuine zero-offset survey. Second, the topological singularity and bifurcation concepts deriving from structural stability provide a unifying framework for all sensing techniques currently in use. The resulting conceptually simple processing methodology yields directly the desired end result of interpretation and comes much closer to the interpreter's intuitive qualitative approach than wave-equation based methods [17]. Moreover, the topological singularities provide an explanation for the similarity and universality of the high-intensity patterns encountered in surface sensing [5], [6], [13], seismology [16], ocean acoustics [14], electromagnetic sensing [4], [7], ultrasound tomography, phonon spectroscopy [2], and so forth. The same singularities also govern the spontaneous formation of qualitatively similar spatio-temporal structures in systems of various genesis [11], [12], [18], [19].

2. WAVEFIELD SINGULARITIES

Suppose a point source at $\underline{x}_o$ in space emits a spherical wave pulse that propagates through a layered medium made up of partially reflecting surfaces. The scattered wave $\psi(\underline{x}_o,t)$ that is received back at the source -- the echo -- exhibits as a function of time t, a number of strong peaks which vary with the source position $\underline{x}_o$. In this echogram (a seismogram in a geological survey) caustic structures are discernible, e.g., those shown in Figure 1 where the intensity $|\psi|^2$ of the echo is plotted as a function of the (vertical) time, or depth, and of the (horizontal) source position. Known as bright spots in seismology, these caustic events and the diffraction patterns around them are the singularities which the medium impresses on the sensing wavefield. They are at the root of the interpreter's intuitive geometric, i.e., qualitative, approach to remote sensing. Indeed, an interpreter readily points out that the layered medium which produces the echogram of Fig. 1 is the one shown in Fig. 2.

In discussing the effects of caustics we confine ourselves to a single reflecting surface. Since source and receiver are at

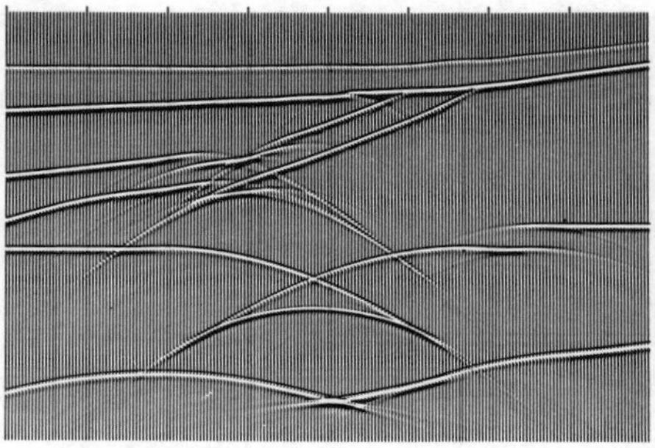

Fig. 1

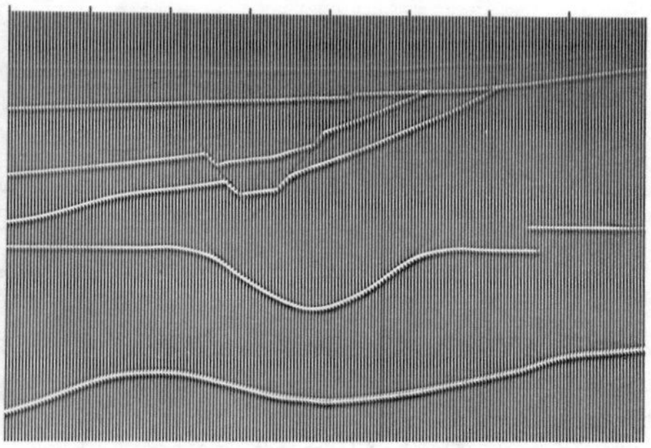

Fig. 2

the same place, only the geometrical optics specular reflection points of the surface, whose distance vectors to the source are normal to the surface, contribute to the echo. Then, as an observer at $\underline{x}_o$ moves along a line on an observation surface Σ above the surface, the reflecting point moves in the same direction for a surface with low curvature. But when the line is above the lowest curvature center of a concave part of the surface S, the reflecting point suddenly moves in the opposite direction. This reversal in direction occurs when $\underline{x}_o$ on Σ crosses the surface's evolute E (the set of loci of the surface's curvature centers). The evolute is the envelope of the surface normals where neighboring rays, normal to S, touch and focusing occurs (Fig. 3(a)). When the source-receiver -- the shotpoint for short -- crosses E, then the number of rays going through $\underline{x}_o$ suddenly changes by two and so does the number of specularly reflecting points. This implies that the number of arrival peaks in the echo also changes by two. Therefore, the reversals in the propagation direction of the specularly reflecting points, as seen from $\underline{x}_o$, are recorded as cusps in the echo profile (Fig. 3(b)). Indeed, while there is but one echo when $\underline{x}_o$ is at A, there are three when $\underline{x}_o$ is at B. Fig. 3(b) is revealed in Fig. 1.

In three-dimensional space the observation surface Σ intersects the evolute sheet generically in smooth

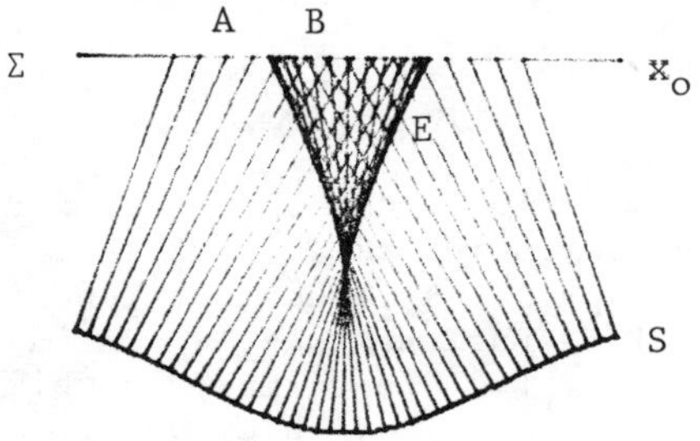

(a)

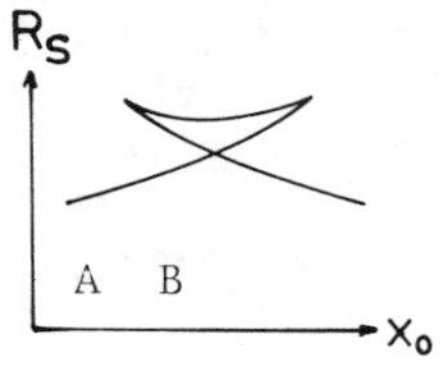

(b)

Fig. 3

curves L, called fold (A_2) lines, and in isolated cusp (A_3) points $\hat{C}$ (Fig. 4). As the height of Σ varies, the cusp points evolve into lines called ribs. If the rib is curved upwards (or downwards) and Σ is nearly tangent to a point on the rib, then the observation surface intersects the evolute in two beak-to-beak cusp singularities (or in a lip). This is shown in Fig. 4 for the winged evolute sheets E of a saddle-surface S. Let $R=R(\underline{x}_o,\underline{x})=|\underline{x}_o-\underline{x}|$ be the distance from $\underline{x}_o$ to a point $\underline{x}$ on S and $T=2R/c$ the two-way travel-time (with speed c). The signals that are received back at the source after reflections by the specular points $\underline{x}=\underline{x}_s(\underline{x}_o)$ on S have travel-times $T_s=T(\underline{x}_o,\underline{x}_s(\underline{x}_o))$ or distances $R_s=R(\underline{x}_o,\underline{x}_s(\underline{x}_o))$. These form a multisheeted 3-dimensional hypersurface in 4-dimensional $(T_s,\underline{x}_o)$-space. For example in case of a plane observation surface Σ, with $\underline{x}_o=(x_o,y_o)$, the saddle-surface S of Fig. 4 gives rise to the travel-time surface on top of Fig. 5 with two dimensional sections shown below.

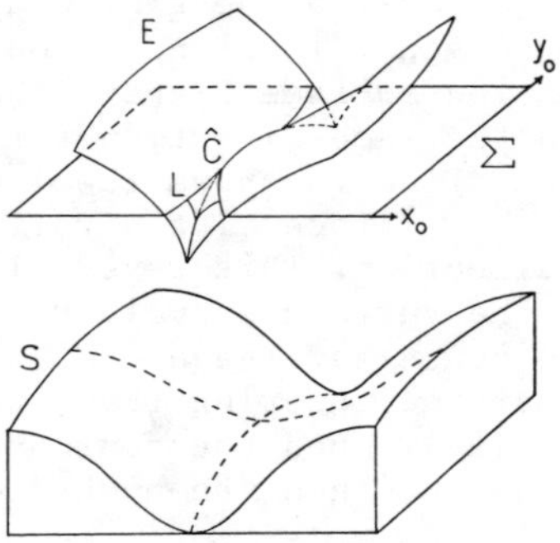

Fig. 4

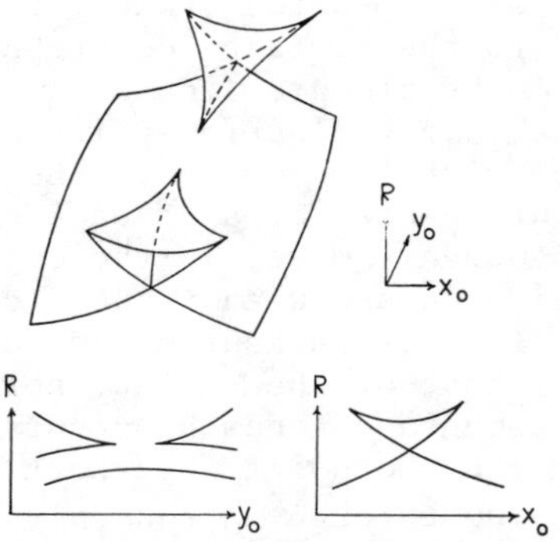

Fig. 5

A rib is smooth except for isolated points where either of the following singularities can occur: (i) the rib itself has a cusp point, called a swallowtail point (A_4) because near it the evolute has the shape shown in Fig. 8. (ii) a rib may touch a fold or two other ribs; in this case two evolute sheets have a contact in a common curvature center and are either hyperbolic or elliptic umbilics (D_4^+ or D_4^-) shown in Figs. 9 and 10, respectively, and the surface has an umbilical point. Since the observed singularities in echo recordings derive from the focal surfaces which constitute evolutes, and since still more complex caustic morphologies will arise if a surface possesses edges [1], [9], or if source and receiver are different [10], the question arises how to classify the singularities a surface impresses on a sensing wavefield or ray family. To understand this, we need catastrophe theory.

3. STRUCTURAL STABILITY OF REMOTE SENSING

Let the point $\underline{x}$ on the smooth surface S be parametrized by surface coordinates, $\underline{x}=\underline{x}(x,y)$. The distance vector $\underline{R}_s=\underline{x}_o-\underline{x}_s$ from a specular reflection point $\underline{x}=\underline{x}_s(\underline{x}_o)$ to the source is normal to S, $\underline{R}_s=R_s(\underline{x}_o)\underline{n}(\underline{x}_s)$ where $R_s=|\underline{x}_o-\underline{x}_s|$ (cf., Fig. 6 for the basic scattering geometry). The distance R has an extremum R_s at $\underline{x}_s$: $\underline{t}\cdot\nabla_{\underline{x}}R=0$ for vectors $\underline{t}$ tangent to S, or

$$\nabla R = 0 \qquad (3.1)$$

with $R=R(\underline{x}_o,\underline{x}(x,y))$ and $\nabla=(\partial/\partial x,\partial/\partial y)$. By definition, a solution $\underline{x}=\underline{x}_s(\underline{x}_o)$ of (3.1) is a stationary point, or a "topological singularity" of R. The type of singularity R possesses at $\underline{x}_s$ depends on the Hessian determinant

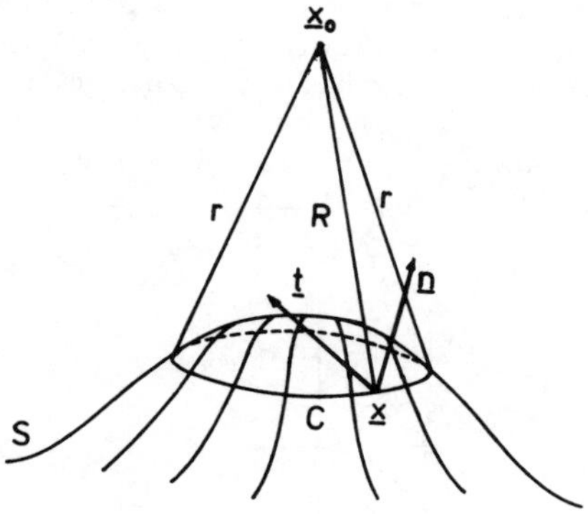

Fig. 6

$$H(R)=\begin{vmatrix} R_{xx} & R_{xy} \\ R_{xy} & R_{yy} \end{vmatrix} \qquad (3.2)$$

evaluated at $\underline{x}_s$. If $H\neq 0$, $\underline{x}_s$ is said to be nondegenerate. In this case, R can be represented by a stable Morse quadratic form. If $H=0$, $\underline{x}_s=\underline{x}_{sc}$ is called a degenerate specular point. In that case $\underline{x}_o=\underline{x}_c$ lies in a curvature center $\underline{x}_c$ on the surface's evolute E. The equation for the evolute follows by eliminating x and y from (3.1) and H=0, leaving one equation for $\underline{x}_o=x_c$ alone.

At this point we require that the scattering process underlying remote sensing be structurally stable, i.e., that the observed images preserve their quality under slight perturbations of the system caused, e.g., by slight deformations of the surface or by small variations of the source. Then Thom's theorems [19], [3] assert that -- in an appropriate curvilinear surface coordinate system denoted again by (x,y) and in a curvilinear source coordinate system $\underline{x}_o-\underline{x}_c=(u,v,w)$ -- the structurally stable distance function R can take on but five "catastrophe" polynomial normal forms near a degenerate specular point $\underline{x}_{sc}$ inside a smooth part of S, viz.,

$$R = R_o+P(x,y;\underline{x}_o)+Dy^2/2 \qquad (3.3)$$

where $R_o=|\underline{x}_c-\underline{x}_{sc}|$ and $D=\kappa_1-\kappa_2$ where κ_i are the surface's principal curvatures in $\underline{x}_{sc}$. If $D\neq 0$, P is one of the cuspoid polynomials A_2 (fold): $P=x^3+ux$; A_3(cusp): $P=x^4+ux^2+vx$; A_4 (swallowtail):

$P=x^5+ux^3+vx^2+wx$. If D=0, $P=xy^2 \pm x^3+wy^2+uy+vx$ represents the hyperbolic (D_4^+) resp. elliptic (D_4^-) umbilic polynomials. The bifurcation set of R is just the evolute E which, therefore, is classified into the five types mentioned above. They are shown in Figs. 7 (A_3), 8 (A_4), 9(D_4^+) and 10 (D_4^-). As a consequence, there are but five structurally stable and generic travel-time singularities observable in an echogram. All others are concatenations of these. If source and receiver are at different positions, there are 14 genuine singularities [10], but the above five describe completely the observed high intensity focal surfaces. Their structural stability can be attributed to the fact that because of the high intensity accumulated in the caustics these are insensitive to small perturbations.

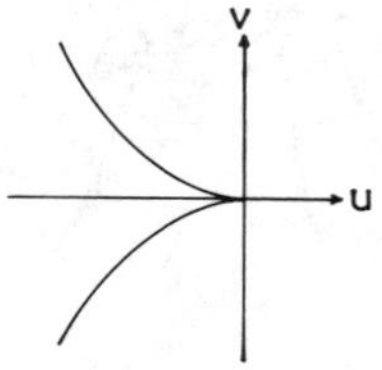

Fig. 7

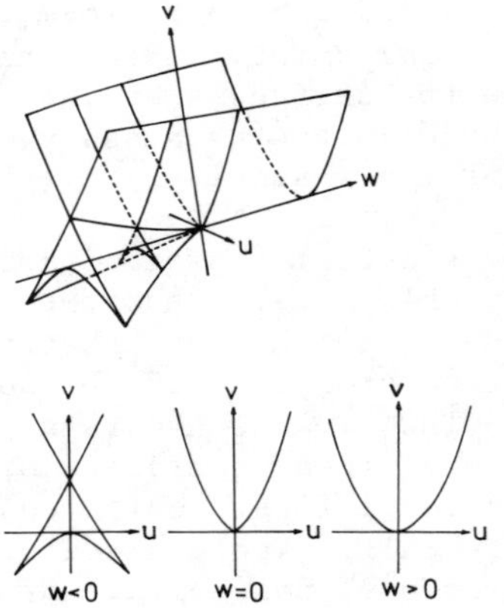

Fig. 8

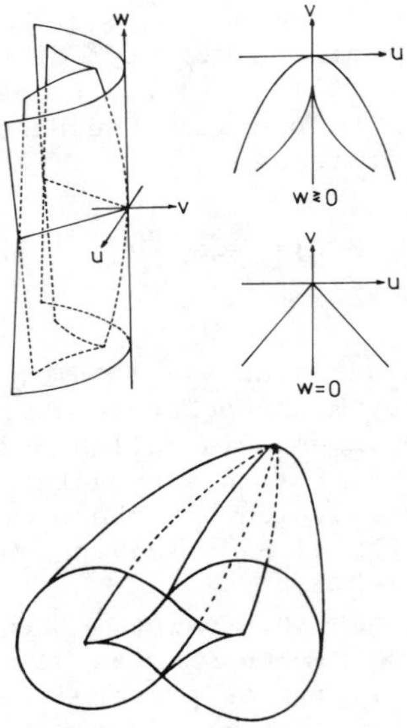

Fig. 9

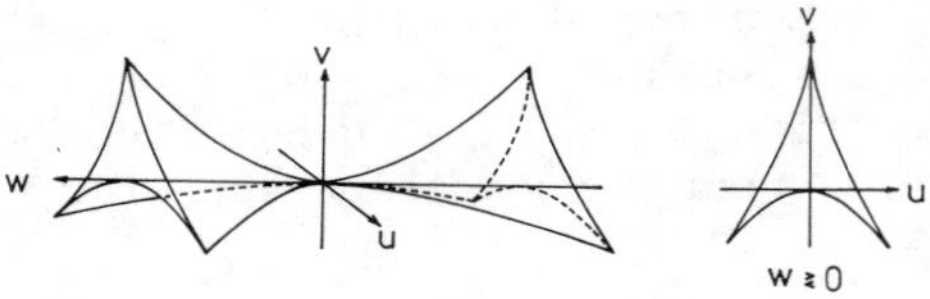

Fig. 10

For A_3, Equ. (3.1) yields with (3.3) the equation $4x^3+2ux^2+v=0$, i.e., the "overhanging cliff" of Fig. 11 where $\rho=x_o=-v$, X=x and u=h are identified with the source position, surface point and height h of the observation line. Projecting the cliff's edges onto the (x_o,h)-plane gives a cusp which is precizely the evolute E of Fig. 3. A geometric interpretation of Fig. 11 in another context is given in Sec. 7.

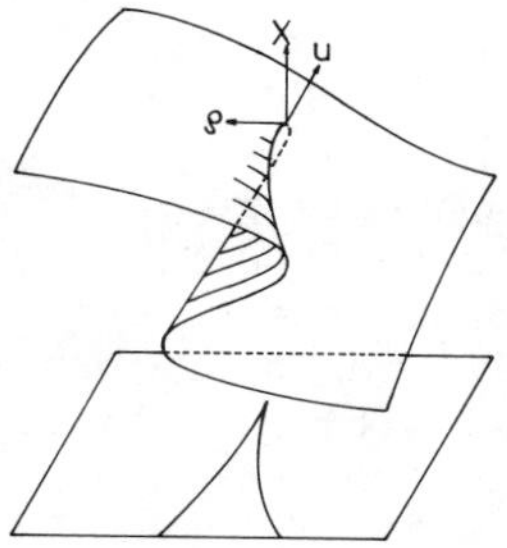

Fig. 11

4. TRAVEL-TIMES, CONTOUR GEOMETRY AND SURFACE RECONSTRUCTION

Structural stability implies that, as the source position varies, the travel-times change their morphologies according to universal bifurcation sets. The travel-times recorded in the echo as a function of the source position are obtained by eliminating x and y from the equations (3.1) and (3.3). One example is shown in Fig. 5. A bifurcation sequence of travel-times coming from an umbilical surface point of elliptic type when $\underline{x}_o$=u varies on a line above the curvature center, is shown in Fig. 12 where $\rho_s=R_s-R_o$. The spherical wavefronts of constant radius r=R, centered on $\underline{x}_o$, cut the surface in a series of contours $C(r,\underline{x}_o)$ (cf., Fig. 6). An observer moving with $\underline{x}_o$ sees the Fresnel-zone contour topography changing according to universal bifurcation sets. A typical topographic change is shown in Fig. 13 for an umbilical surface point of hyperbolic type. The contours follow by setting $R-R_o=r=const$ in (3.3).

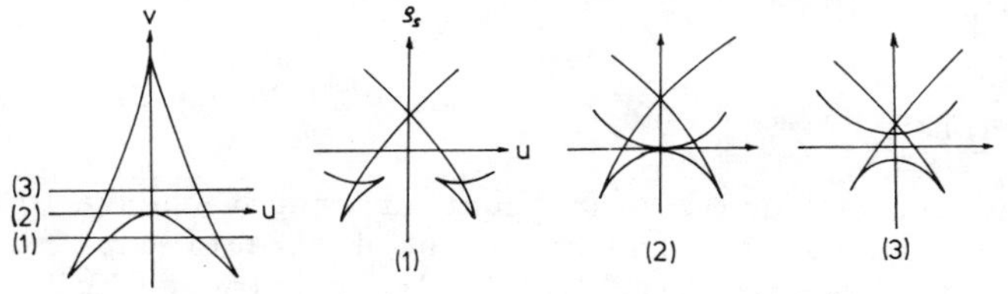

Fig. 12

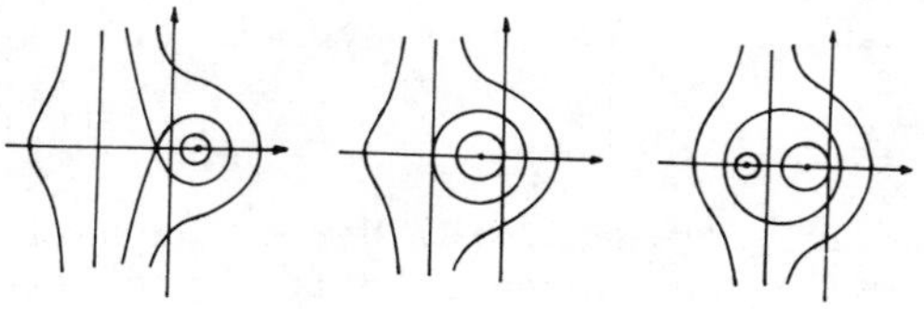

Fig. 13

As the source-receiver $\underline{x}_o$ varies above the surface, each point of S eventually becomes specular. In virtue of the specularity condition (3.1), the surface normal at $\underline{x}_s$ is given by $\underline{n}(\underline{x}_s(\underline{x}_o))=\nabla_{\underline{x}_o}R_s(\underline{x}_o)$. Then we obtain from $\underline{x}_o-\underline{x}_s=R_s\underline{n}$ the equation for the surface

$$\underline{x} = \underline{x}_o - R_s(\underline{x}_o)\nabla_{\underline{x}_o} R_s(\underline{x}_o) \tag{4.1}$$

Since $|n|=1$, only two components of $\nabla_{\underline{x}_o}R_s$ are needed and it suffices therefore, to vary $\underline{x}_o$ on a surface Σ, say, on a plane. Inferring R_s from the arrivals or peaks the echo ψ possesses at times $t=T_s$ $=2R_s/c$, the surface profile can be reconstructed directly from (4.1). For an input signal F(t) with basic source pulse frequency ω_o (cf., Sec. 5), the echogram near a degenerate singularity $\underline{x}_s=\underline{x}_{sc}$ has the form

$$\psi(t,\underline{x}_c) \propto \omega_o^{\alpha}\{AF(t-T_{sc})+B\bar{F}(t-T_{sc})\} \tag{4.2}$$

where $\alpha=\{1/6,\ 1/4,\ 3/10,\ 1/3\}$ for the singularities $\{A_2, A_3, A_4, D_4^{\pm}\}$ and $\bar{F}$ is the Hilbert transform of F. Varying ω_o and measuring the rate of change of the received power permits one to identify and to classify the structural details of the reflecting surface near travel-time reversals. Typical travel-time recordings for a surface with a fault and two domes are shown on top of Fig. 14 (D. Lang [10]) for a source-receiver moving along horizontal survey lines above the lines 1, 6, 14 and 20. In Fig. 14a, b and c the height of the observation plane is 1.5 times, 2 times and 5 times the height of the domes, respectively. Travel-times are plotted downwards. In the travel-time recording of Fig. 14a one sees four swallowtail points coming from the fault and corresponding to a rib lying entirely in the observation plane. In Fig. 14b the typical swallowtail bifurcation sets have developed into cusp lines. In addition, the slopes of the domes give rise to new swallowtail points. Fig. 14c shows the most complex travel-time recordings pertaining to the surface in which the swallowtails have fully developed.

5. DIFFRACTION PATTERNS

In classifying travel-times and surface contours the language of ray theory has been used so far. The wave-type diffraction patterns around the caustics follow from Kirchhoff's diffraction formula [10] for the backscattered wave ψ in the shortwave limit,

$$\psi(t,\underline{x}_o)=(1/8\pi^2c)\int_0^{\infty} dr G(r,\underline{x}_o)\dot{F}(t-2r/c) \tag{5.1}$$

with the incident spherical wave $F(t-R/c)/4\pi R$. Here, the surface structure function G -- the wavefront sweep velocity or scatter-

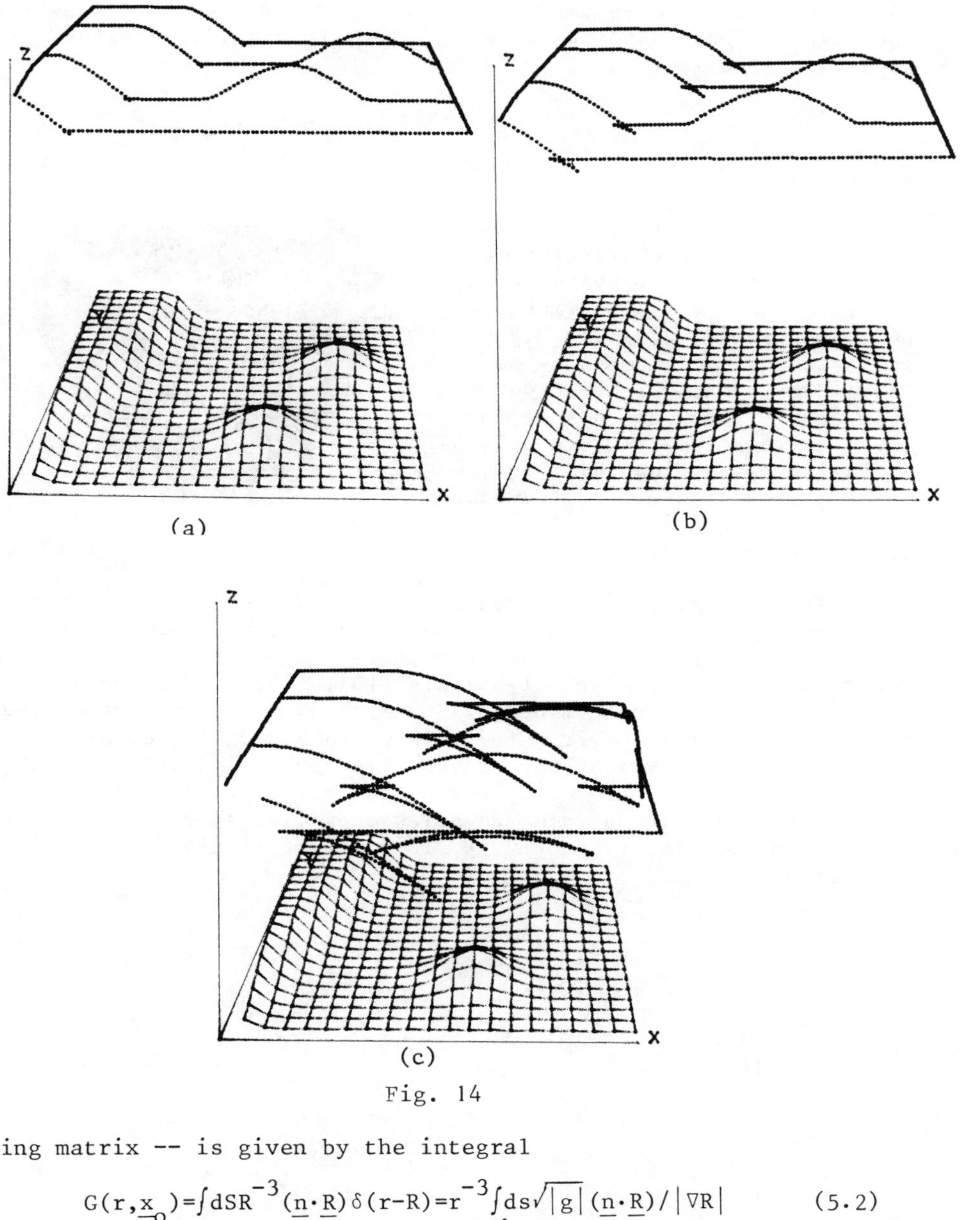

Fig. 14

ing matrix -- is given by the integral

$$G(r,\underline{x}_o)=\int_S dSR^{-3}(\underline{n}\cdot\underline{R})\,\delta(r-R)=r^{-3}\int_{C'} ds\sqrt{|g|}\,(\underline{n}\cdot\underline{R})/|\nabla R| \qquad (5.2)$$

where δ is Dirac's function and the contour C' is the projection of C (Fig. 6) onto the (x,y)-plane. From (5.2) the main contributions to G (and ψ) are seen to come from those values of r for which the integrand is infinite, i.e., from the geometrical optics specular reflection points given by $\nabla R=0$, Equ. (3.1). The Fourier transform of G is

$$\tilde{G}(k,\underline{x}_o) = \int_S dS(\partial R^{-1}/\partial n)\exp(ikR) \tag{5.3}$$

where $k=\omega/c$. Hence, $\tilde{\psi}(\omega,\underline{x}_o)=-i\omega\tilde{G}(2\omega/c,\underline{x}_o)\tilde{F}(\omega)/8\pi^2 c$. The generic diffraction patterns associated with the travel-time singularities follow by substituting (3.3) into (5.3). Asymptotic evaluation of (5.3) gives high-intensity diffraction patterns of Airy, Pearcey and higher-order type [10]. The ω_o-dependence in (4.2) follows from (5.3) and (3.3) by scaling. A typical diffraction pattern around an elliptic umbilic caustic is shown in Fig. 15 (cf., also Fig. 1).

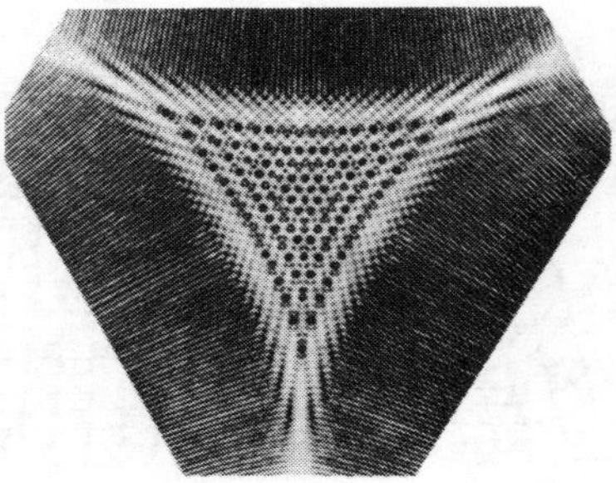

Fig. 15

If the surface possesses edges and faults [1], the evolutes are amputated by the surface's shadow boundaries (Fig. 16). The effects produced by such discontinuities can be classified by six "constraint" catastrophe polynomials [10], [20]. For example, a fault (Fig. 16) is determined by $P=x^3+vx^2+ux$ $(x>0)$ and the diffraction pattern following from (5.3) is of combined Fresnel-Airy type: The intensity near the travel-time reversal decreases much faster than the one produced by a slope. Figs. 17 (a) and (b) show the travel-time curves when $\underline{x}_o$ varies on a line above and below the point B in Fig. 16, respectively. This effect is also present in Fig. 1.

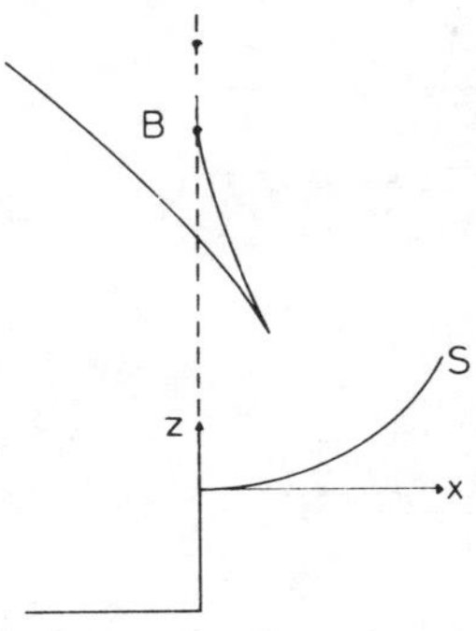

Fig. 16

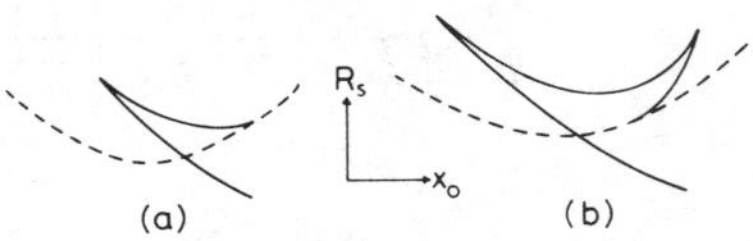

Fig. 17

7. CONCLUSIONS

The above classification of singularities in wavefields and ray families scattered back by a surface can be generalized to deal with laterally inhomogeneous media by using Maslov's representation [8], [10]. Incorporating these topological concepts into S-matrix techniques of inversion offers new and practicable tools for all the sensing processes presently in use in a wide variety of physical and technical systems, ranging from ocean acoustics [14] to phonon spectroscopy [2]. An important thing to note about the universal singularities and bifurcations we have described is that they also determine the spontaneous formation of qualitatively similar spatio-temporal patterns in systems of various genesis exhibiting similar critical behavior. Comparison of, e.g., Fig. 11 with familiar phase transition diagrams, or identifying the singularities observed in flow fields [15] with the above travel-time bifurcations, strongly indicates that structure formation and structure recognition are governed by the same topological principles [11]. The reason for these analogies is that the geometry underlying pattern formation and pattern recognition is subjected to the general principle of structural stability. As a consequence, structure formation and recognition are just the two sides of the same topological coin. This fact raises fascinating questions that promise a great challenge for future research.

REFERENCES

[1] Achenbach, J.D. and Norris, A.: 1981, J. Acoust. Soc. Am. 70, pp. 165-171.

[2] Armbruster, D., Dangelmayr, G., and Güttinger, W.: Nonlinear Phonon Focusing, Proc. 4th Int. Conference on Phonon Scattering in Condensed Matter (Stuttgart 1983). Armbruster, D. and Dangelmayr, G.: Topological Singularities and Phonon Focusing, Z. Phys. B (1983), in print.

[3] Arnold, V.: 1978, Russ. Math. Surveys 33, pp. 99-116.

[4] Baltes, H.P. (ed.): Inverse Scattering Problems in Optics (Springer 1980)

[5] Berry, M.V.: 1972, J. Phys. A 5, pp. 272-291.

[6] Bleistein, N. and Cohen, J.K.: A survey of recent progress on inverse scattering problems, US Naval Res. Rep. MS-R-7806 (1978).

[7] Boerner, W.M., Polarization utilization in electromagnetic inverse scattering, in Baltes, Ref. [4].

[8] Chapman, C.H. and Drummond, R.: 1982, Bull. Seism. Soc. Am. 72, pp. 227-317.

[9] Dangelmayr, G.: Singularities in quasigeometrical imaging, this volume.

[10] Dangelmayr, G. and Güttinger, W.: 1982, Geophys. J. R. Astr. Soc. 71, pp. 79-126; Güttinger, W. and Lang, D.: in preparation (1983).

[11] Güttinger, W. and Eikemeier, H. (Eds.): Structural Stability in Physics (Springer 1979); Güttinger, W.: Geometrical principles underlying synergetics, in preparation (1983).

[12] Haken, H.: Synergetics (Springer 1980)

[13] Hilterman, F.J.: 1975, Geophysics 35, pp. 1020-1037 and Geophysics 40, pp. 745-762 (1975).

[14] Keller, J.B. and Papadakis, J.S. (Eds.): Wave propagation and underwater acoustics (Springer 1977).

[15] Nye, J.F.: Structural Stability in evolving flow fields, in Güttinger, W. and Eikemeier, H., Ref. 11 .

[16] Ristow, D.: Three-dimensional finite-difference migration, Ph.D. thesis, Utrecht 1980.

[17] Schneider, W.A.: 1978, Geophysics 43, pp. 49-76.

[18] Stewart, I.: 1981, Physica 2D, pp. 245-305.

[19] Thom, R.: Structural Stability and Morphogenesis (Benjamin 1975)

[20] Wright, F.J., Güttinger, W. and Dangelmayr, G.: in preparation (1983).

I.3 (IS.3)

A SOLUTION OF THE TIME-DEPENDENT INVERSE SOURCE PROBLEM FOR THREE-DIMENSIONAL ELECTROMAGNETIC WAVE PROPAGATION

Harry E. Moses

University of Lowell, Center for Atmospheric Research

The usual form of the inverse source problem for electromagnetic theory is given in a time-independent version in which the radiation pattern is prescribed and the sources are required to be contained in a sphere of prescribed radius. The time-dependent inverse source problem, which we discuss in the present paper, has currents which are switched on for a prescribed interval of time and the final time-dependent field is prescribed. One seeks the currents which will give rise to the fields. Thus the time-dependent inverse source problem seeks to provide sources for pulsed fields and is complementary to the time-independent problem. Our approach makes use of eigenfunctions of the curl operator, whose properties are reviewed. It also represents an extension of our earlier work. The currents which give rise to the prescribed fields are not unique but conditions are given which make them so.

1.0 INTRODUCTION. STATEMENT OF THE PROBLEM

Let us consider Maxwell's equations with time-dependent sources. If, for simplicity, we take units in which the velocity of light c is unity, Maxwell's equations are

$$\begin{aligned} \underset{\sim}{\nabla} \times \underset{\sim}{H}(\underset{\sim}{x};t) &= 4\pi \underset{\sim}{j}(\underset{\sim}{x};t) + \frac{\partial}{\partial t} E(\underset{\sim}{x},t), \\ \underset{\sim}{\nabla} \times \underset{\sim}{E}(\underset{\sim}{x};t) &= -\frac{\partial}{\partial t} \underset{\sim}{H}(\underset{\sim}{x};t), \\ \underset{\sim}{\nabla}\cdot\underset{\sim}{E}(\underset{\sim}{x};t) &= 4\pi\rho(\underset{\sim}{x};t) \\ \underset{\sim}{\nabla}\cdot\underset{\sim}{H}(\underset{\sim}{x};t) &= 0. \end{aligned} \tag{1}$$

W.-M. Boerner et al. (eds.), Inverse Methods in Electromagnetic Imaging - Part 1, 77–85.

For our purpose the direct time-dependent source problem for pulsed electromagnetic waves is to prescribe the current $j(x;t)$ and density $\rho(x;t)$ as functions of x and t such that they vanish outside a finite time interval. Before the sources are turned on we prescribe, for the simplicity of discussion, that the field be zero. We want to find the field after the currents and sources are turned on and then off. This problem is a classical one for Maxwell's equations. After the currents and sources are turned off, the solution of Maxwell's equations represents a pulse of the electromagnetic field.

In the inverse source problem, as we treat it for the purpose of the present discussion, we prescribe the time interval within which the sources and currents are on. We also prescribe the pulse of radiation which we want to have after we have turned the source off.

Usually the inverse source problem is described in a time-independent version. One takes Maxwell's equations of Eq. (1) and separates out the time variable by taking a Fourier transformation with respect to time. The time-independent inverse problem is to find sources and currents within a finite domain in space which gives rise to a prescribed radiation pattern. Results are summarized in Ref. 1. The time-independent approach to the inverse source problem was first given by the present author in Ref. 2. The present paper represents an extension of Ref. 2. It is seen that the time-dependent problem is complementary to the time-independent problem.

The time-dependent problem has many applications. For example for use in radar systems or for ionosondes which probe the ionosphere one wishes to shape and direct pulses of radiation. What current distributions in the antennas can be used for this purpose? A second application is decoying. One can make a source which simulates the emitted electromagnetic signature of an aircraft, for example, and thus can be used to confuse detectors used by an enemy. A closely related problem is the source detection problem which is concerned with determining the position, size and shape of a source of electromagnetic radiation. The inverse source problem, as we shall formulate it, is valid in regions where the space is homogeneous, even when the homogeneous space is enclosed by or abuts an inhomogeneity. For example, one can use our method - or its extensions - to find the currents within a waveguide which will give rise to prescribed pulsed fields within the waveguide before the pulse hits the boundary of the guide. Such a shaping of the initial radiation is useful in design problems.

2.0 EIGENFUNCTIONS OF THE CURL OPERATOR AND GENERALIZATION OF THE HELMHOLTZ THEOREM

In our original treatment of the inverse source problem for Maxwell's equations (Ref. 2), we recast Maxwell's equations into the form of Dirac's equation and expanded the fields in terms of the eigenfunctions of the Dirac-like Hamiltonian. In this way the transverse components of the fields and the currents which gave rise to them were separated from the longitudinal field and its sources. In the present paper we shall use eigenfunctions of the curl operator to carry out this separation and find a simple differential equation for the time amplitudes. The eigenfunctions of the curl operator were introduced by us in Ref. 3 to provide a rotationally invariant generalization of the Helmholtz decomposition theorem in the infinite domain. [It might be mentioned that a still further generalization which gives analogues to the Helmholtz theorem for tensor fields is discussed in Ref. 4.] The eigenfunctions were intended to simplify equations in which the curl operator appears, such as Maxwell's equations and the equations of fluid mechanics.

Let us consider an arbitrary unit vector $\underset{\sim}{\eta}$ and a discrete variable $\lambda = \pm 1,\ 0$. For each value of $\underset{\sim}{\eta}$ we define a vector $\underset{\sim}{Q}_\lambda(\underset{\sim}{\eta})$ by

$$\underset{\sim}{Q}_0(\underset{\sim}{\eta}) = -\underset{\sim}{\eta},$$
$$\underset{\sim}{Q}_\lambda(\underset{\sim}{\eta}) = -\frac{\lambda}{\sqrt{2}}\left[\frac{\eta_1(\eta_1+i\lambda\eta_2)}{1+\eta_3} - 1,\ \frac{\eta_2(\eta_1+i\lambda\eta_2)}{1+\eta_3} - i\lambda,\ \eta_1 + i\lambda\eta_2\right],$$
$$(\lambda = \pm 1). \tag{2}$$

The vectors $\underset{\sim}{Q}_\lambda(\underset{\sim}{\eta})$ satisfy the following orthogonality and completeness relations:

$$\underset{\sim}{Q}^*_\lambda(\underset{\sim}{\eta})\cdot\underset{\sim}{Q}_\mu(\underset{\sim}{\eta}) = \delta_{\lambda\mu},\quad \sum_\lambda Q^*_{i\lambda}(\underset{\sim}{\eta})Q_{j\lambda}(\underset{\sim}{\eta}) = \delta_{ij}. \tag{3}$$

Also

$$\underset{\sim}{\eta}\cdot\underset{\sim}{Q}_\lambda(\underset{\sim}{\eta}) = -\delta_{\lambda 0},\quad \underset{\sim}{\eta}\times\underset{\sim}{Q}_\lambda(\underset{\sim}{\eta}) = -i\lambda\underset{\sim}{Q}_\lambda(\underset{\sim}{\eta}). \tag{4}$$

The eigenfunctions of the curl operator $\underset{\sim}{\chi}(\underset{\sim}{x}|\underset{\sim}{p},\lambda)$ are defined by

$$\underset{\sim}{\chi}(\underset{\sim}{x}|\underset{\sim}{p},\lambda) = \frac{1}{(2\pi)^{3/2}}\, e^{i\underset{\sim}{p}\cdot\underset{\sim}{x}}\, \underset{\sim}{Q}_\lambda\left(\frac{\underset{\sim}{p}}{p}\right),\quad p = |\underset{\sim}{p}|. \tag{5}$$

The following orthogonality and completeness relations are satisfied:

$$\int \underset{\sim}{\chi}^*(\underset{\sim}{x}|\underset{\sim}{p},\lambda)\cdot\underset{\sim}{\chi}(\underset{\sim}{x}|\underset{\sim}{p}',\lambda')\, d\underset{\sim}{x} = \delta(\underset{\sim}{p}-\underset{\sim}{p}')\,\delta_{\lambda,\lambda'},$$
$$\sum_\lambda \int \chi^*_i(\underset{\sim}{x}\ \underset{\sim}{p},\lambda)\,\chi_j(\underset{\sim}{x}'\ \underset{\sim}{p},\lambda)\, d\underset{\sim}{p} = \delta(\underset{\sim}{x}-\underset{\sim}{x}')\,\delta_{ij}. \tag{6}$$

Moreover, the eigenfunctions <u>are</u> eigenfunctions of the curl operator with eigenvalue λp, since

$$\underset{\sim}{\nabla}\times\underset{\sim}{\chi}(\underset{\sim}{x}|\underset{\sim}{p},\lambda) = \lambda p\underset{\sim}{\chi}(\underset{\sim}{x}|\underset{\sim}{p},\lambda). \tag{7}$$

Also

$$\nabla \cdot \chi(x|p,\lambda) = -ip \frac{e^{ip\cdot x}}{(2\pi)^{3/2}} \delta_{\lambda,0}. \tag{8}$$

We can now generalize the Helmholtz theorem.

Any vector $v(x)$ can be written in the form

$$v(x) = \sum_{\lambda} v_{\lambda}(x), \text{ where } v_{\lambda}(x) = \int \chi(x|p,\lambda)\, V(p,\lambda)\, dp, \tag{9}$$

where

$$V(p,\lambda) = \int \chi^{*}(x|p,\lambda)\cdot v(x)\, dx. \tag{10}$$

Moreover,

$$\nabla \cdot v_{\lambda}(x) = 0 \text{ for } \lambda = \pm 1, \; \nabla \times v_{0}(x) = 0. \tag{11}$$

We have thereby sharpened the Helmholtz theorem by showing that there are two, rather than one, transverse components of a vector. As shown in Ref. 3 the decomposition is rotationally invariant.

The second part of the Helmholtz theorem states that the rotational part of a vector can be expressed as the curl of a vector and the irrotational part as the gradient of a vector. We now discuss our sharpening of this part of the theorem. We can introduce vector potentials $A_{\lambda}(x)$ defined by

$$\nabla \times A_{\lambda}(x) = v_{\lambda}(x), \lambda = \pm 1, \tag{12}$$

and a scalar potential V(s) defined by

$$\nabla V(x) = v_{0}(x). \tag{12a}$$

We shall give the general form of the vector potentials $A_{\lambda}(x)$ and the scalar potential $V(x)$ which we regard as solutions of Eqs. (12) and (12a) respectively. To find the vector potentials we write the vector potentials in the form of an expansion in terms of the eigenfunctions of the curl operator, as in Eq. (9)

$$A_{\lambda}(x) = \sum_{\lambda'} \int \chi(x|p,\lambda')B(p,\lambda')\, dp. \tag{13}$$

On substituting into Eq. (12), and using the linear independence of the eigenfunctions, we obtain

$$B(p,\lambda) = \frac{\lambda V(p,\lambda)}{p}, \; B(p,-\lambda) = 0, \; B(p,0) \text{ arbitrary}. \tag{14}$$

Thus the vector potentials $A_{\lambda}(x)$ can be written as

$$A_{\lambda}(x) = \int \chi(x|p,\lambda)\, B(p,\lambda)\, dp + W(x),$$

where

$$W(x) = \int \chi(x|p,0)\ B(p,0)\ dp. \tag{15}$$

Thus each of the two vector potentials (one for $\lambda = +1$ and one for $\lambda = -1$) consists of the sum of an essential, minimal part which depends uniquely on $v_\lambda(x)$ and an arbitrary vector $W(x)$ which has the form

$$W(x) = \nabla Q(x),\quad Q(x) = \frac{i}{(2\pi)^{3/2}} \int e^{ip\cdot x}\, \frac{B(p,0)}{p}\, dp. \tag{15a}$$

Thus $W(x)$ is the gradient of an arbitrary function $Q(x)$.

The function $W(x)$ is the gauge. Our procedure has isolated the essential parts of the vector potential and has removed the mystery of the gauge.

We now want to find the function $V(x)$ from $v_0(x)$. From Eq. (9) and the form of $\chi(x|p,0)$ we have

$$V(x) = -\frac{i}{(2\pi)^{3/2}} \int e^{ip\cdot x}\, \frac{V(p,0)}{p}\, dp. \tag{16}$$

The reality property is important. For $v(x)$ to be real

$$V(p,\lambda) = -\frac{p_1 - i\lambda p_2}{p_1 + i\lambda p_2}\, V^*(-p,\lambda). \tag{17}$$

3.0 APPLICATION OF CURL EIGENFUNCTIONS TO MAXWELL'S EQUATIONS

We rewrite Maxwell's equations in the Bateman form. Thus we introduce the complex vector.

$$\psi(x;t) = E(x;t) - iH(x;t). \tag{18}$$

Maxwell's Equations Eq. (1) become

$$\nabla \times \psi(x;t) = -i\frac{\partial}{\partial t}\psi(x;t) - 4\pi i j(x;t),$$
$$\nabla\cdot\psi(x;t) = 4\pi\rho(x;t). \tag{19}$$

The wave function $\psi(x;t)$ is generally complex. However, the current $j(x;t)$ and charge density $\rho(x;t)$ are real.

We use the eigenfunctions of curl operator and write

$$\psi(x;t) = \sum_\lambda \int \chi(x|p,\lambda)\ \psi(p,\lambda;t)\ dp, \tag{20}$$

$$4\pi j(x;t) = \sum_\lambda \int \chi(x|p,\lambda)\ \gamma(p,\lambda;t)\ dp, \tag{21}$$

$$4\pi\rho(x;t) = \frac{1}{(2\pi)^{3/2}} \int e^{ip\cdot x}\ r(p;t)\ dp. \tag{22}$$

On substituting into Eq. (19) and using the properties of the eigenfunctions Eqs. (7) and (8)

$$\psi(\underset{\sim}{p},0;t) = \frac{i}{p} r(\underset{\sim}{p};t),$$
$$-\frac{\partial}{\partial t}\psi(\underset{\sim}{p},0;t) = \gamma(\underset{\sim}{p},0;t), \tag{23}$$

which relate the longitudinal components of the electromagnetic field to the charge distribution and the longitudinal component of the current. The transverse part of the electromagnetic field is related to the transverse part of the current in the following way:

$$\frac{\partial}{\partial t}\psi(\underset{\sim}{p},\lambda;t) - ip\lambda\psi(\underset{\sim}{p},\lambda;t) = -\gamma(\underset{\sim}{p},\lambda;t), \quad (\lambda = \underline{+}1). \tag{24}$$

Our use of the eigenfunctions of the curl operator has enabled us to split the equations into a longitudinal part and two uncoupled transverse parts in a rotationally invariant way.

The longitudinal field is given essentially by $\psi(\underset{\sim}{p},0;t)$ does not concern us, since it is only the radiation field which is of interest. Using the first of Eq. (23) we can set $\rho(\underset{\sim}{x};t) \equiv 0$ and thus cause the longitudinal field to vanish. The second of Eq. (23) also pertains only to the longitudinal field and is essentially the equation of continuity. This equation also can be disregarded if the charge density is identically zero.

We are then left only with the two ordinary differential uncoupled equations for $\psi(\underset{\sim}{p},\underline{+}1;t)$ which are readily integrated. We take the fields before the currents are turned on to be identically zero and choose the origin of the time variable to be such that the currents are turned on at $t = -T$ and turned off again at $t = T$. Then the solution of Eq. (24) is

$$\psi(\underset{\sim}{p},\lambda;t) = -e^{i\lambda pt}\int_{-T}^{t} e^{-i\lambda pt'}\gamma(\underset{\sim}{p},\lambda;t')\,dt'. \tag{23}$$

(We recollect that $\gamma(\underset{\sim}{p},\lambda;t) \equiv 0$ for $t < -T$.)

Thus as our initial and final conditions we have

$$\underset{\sim}{\psi}(\underset{\sim}{x};t) \equiv 0 \text{ for } t < -T,$$
$$\underset{\sim}{\psi}(\underset{\sim}{x};t) \equiv \underset{\sim}{\psi}_{+}(\underset{\sim}{x};t) \text{ for } t > T, \tag{24}$$

where

$$\underset{\sim}{\psi}_{+}(\underset{\sim}{x};t) = \sum_{\lambda=\underline{+}1}\int \underset{\sim}{\chi}(\underset{\sim}{x}|\underset{\sim}{p},\lambda)\, e^{i\lambda p(t-T)}\,\psi(\underset{\sim}{p},\lambda;T)\,d\underset{\sim}{p}. \tag{25}$$

It is clear that $\psi_{+}(\underset{\sim}{x};t)$ is a solution of Maxwell's equations without currents and thus represents the final radiation field.

Before we proceed further it is useful to give a physical interpretation to the discrete variable $\lambda = \underline{+}1$.

It is readily seen that the general solution of Maxwell's equations without sources is of the form Eq. (25) in which $\psi(\underset{\sim}{p},\lambda;T)$ is replaced by $\psi(\underset{\sim}{p},\lambda)$ where this latter function is an arbitrary function of $\underset{\sim}{p},\lambda$. Moreover, it is convenient to take

T = 0. Let us consider the case in which

$$\psi(\underset{\sim}{p},1) = \delta(p_1)\ \delta(p_2)\ \delta(p_3-k). \quad \text{Also } \psi(\underset{\sim}{p},-1) \equiv 0. \tag{26}$$

Then

$$E_1(\underset{\sim}{x};t) = H_2(\underset{\sim}{x};t) = \frac{1}{4(\pi)^{3/2}} \cos k(x_3+t),$$

$$E_2(\underset{\sim}{x};t) = -\ H_1(\underset{\sim}{x};t) = -\ \frac{1}{4(\pi)^{3/2}} \sin k(x_3+t),$$

$$E_3(\underset{\sim}{x};t) = H_3(\underset{\sim}{x};t) \equiv 0. \tag{26a}$$

It is clear that this solution of Maxwell's equations is a positively polarized circularly polarized wave moving with wave number k in the negative z-direction. Hence $\lambda = 1$ corresponds to positive circular polarization.

Likewise the choice of $\psi(p,-1) = \delta(p_1)\delta(p_2)\delta(p_3-k)$, $\psi(\underset{\sim}{p},+1) \equiv 0$ gives rise to a negative circularly polarized wave of wave number k moving in the positive z-direction. Then $\lambda = -1$ means negative circular polarization, and therefore decomposition of $\psi(\underset{\sim}{x};t)$ into eigenvectors of curl operator is equivalent to expressing the electromagnetic field as a sum of circularly polarized radiation of various frequencies moving in various directions.

4.0 THE INVERSE PROBLEM

We shall now treat the inverse source problem for Maxwell's equations.

We are given $\underset{\sim}{\psi}_+(\underset{\sim}{x};t)$ and we are required to find a current $\underset{\sim}{j}(\underset{\sim}{x};t)$ of the form

$$\underset{\sim}{j}(\underset{\sim}{x};t) = \underset{\sim}{j}_e(\underset{\sim}{x})\ h_e(t) + \underset{\sim}{j}_o(\underset{\sim}{x})\ h_o(t) \tag{27}$$

where $h_e(t)$ is a given real even function and $h_o(t)$ is a given real odd function of t. We shall show how <u>real</u> transverse vectors $\underset{\sim}{j}_{e,o}(\underset{\sim}{x})$ can be obtained uniquely from the <u>complex</u> vector $\underset{\sim}{\psi}_+(\underset{\sim}{x};t)$.

Knowing $\psi+(x;t)$ is equivalent to knowing $\psi(\underset{\sim}{p},\pm1;T)$. Let us define $F(\underset{\sim}{p},\lambda;k)$ for $\lambda = \pm1$ by

$$F(\underset{\sim}{p},\lambda;k) = \frac{1}{\sqrt{2\pi}} \int_{-T}^{+T} e^{-ikt}\ \gamma(\underset{\sim}{p},\lambda;t)\ dt$$

$$= \frac{4\pi}{\sqrt{2\pi}} \int_{-T}^{+T} e^{-ikt}\ dt \int \underset{\sim}{\chi}^*\ (\underset{\sim}{x}|\underset{\sim}{p},\lambda)\cdot\underset{\sim}{j}(\underset{\sim}{x};t)\ d\underset{\sim}{x}. \tag{28}$$

Then we define

$$G(\underset{\sim}{p},\lambda) \equiv F(\underset{\sim}{p},\lambda;\lambda p) = -\frac{1}{\sqrt{2\pi}} e^{-i\lambda pT} \psi(\underset{\sim}{p},\lambda;T) \tag{29}$$

It is our intent to find $\underset{\sim}{j}_{e,o}(\underset{\sim}{x})$ of Eq. (27) from $G(\underset{\sim}{p},\lambda)$.

We also define,

$$F_e(\underset{\sim}{p},\lambda) = \frac{1}{2g_e(p)} \left[G(\underset{\sim}{p},\lambda) - \frac{p_1-i\lambda p_2}{p_1+i\lambda p_2} G^*(-\underset{\sim}{p},\lambda)\right],$$

$$F_o(\underset{\sim}{p},\lambda) = \frac{\lambda}{2g_o(p)} \left[G(\underset{\sim}{p},\lambda) + \frac{p_1-i\lambda p_2}{p_1+i\lambda p_2} G^*(-\underset{\sim}{p},\lambda)\right]. \tag{30}$$

where

$$g_{e,o}(k) = \frac{1}{\sqrt{2\pi}} \int_{-T}^{+T} h_{e,o}(t)\, e^{-ikt}\, dk. \tag{31}$$

Note that $g_e(k)$ is a <u>real</u>, even function of k and $g_o(k)$ is an <u>imaginary</u>, odd function of k.

Then

$$4\pi \underset{\sim}{j}_e(\underset{\sim}{x}) = \sum_\lambda \int \underset{\sim}{\chi}(\underset{\sim}{x}|\underset{\sim}{p},\lambda)\, F_e(\underset{\sim}{p},\lambda)\, d\underset{\sim}{p},$$

$$4\pi \underset{\sim}{j}_o(\underset{\sim}{x}) = \sum_\lambda \int \underset{\sim}{\chi}(\underset{\sim}{x}|\underset{\sim}{p},\lambda)\, F_o(\underset{\sim}{p},\lambda)\, d\underset{\sim}{p}. \tag{32}$$

The fact that

$$F_{e,o}(\underset{\sim}{p},\lambda) = -\frac{p_1-i\lambda p_2}{p_1+i\lambda p_2} F^*_{e,o}(-\underset{\sim}{p},\lambda) \tag{33}$$

assures us that $\underset{\sim}{j}_{e,o}(\underset{\sim}{x})$ are real vectors [see Eq. (17)].

Let us consider the special case

$$h_e(t) = \delta(t),\quad h_o(t) = \delta'(t). \tag{34}$$

Moreover, we let $T \to +0$ and define

$$\underset{\sim}{\psi}(\underset{\sim}{x}) = \psi_+(\underset{\sim}{x};0_+) \equiv \underset{\sim}{\psi}(\underset{\sim}{x};0_+). \tag{35}$$

That is, we prescribe the radiation fields immediately after the current is turned off. Of course, the given vector $\underset{\sim}{\psi}(\underset{\sim}{x})$ must be a purely transverse field. We obtain

$$4\pi \underset{\sim}{j}_e(\underset{\sim}{x}) = -\mathrm{Re}\, \underset{\sim}{\psi}(\underset{\sim}{x}),\quad 4\pi \underset{\sim}{j}_o(\underset{\sim}{x}) = \frac{1}{4\pi} \mathrm{Im}\, \underset{\sim}{\nabla} \times \int \frac{1}{|\underset{\sim}{x}-\underset{\sim}{x}'|} \underset{\sim}{\psi}(\underset{\sim}{x}')\, d\underset{\sim}{x}'. \tag{36}$$

However, since

$$\mathrm{Re}\, \underset{\sim}{\psi}(\underset{\sim}{x}) = \underset{\sim}{E}(\underset{\sim}{x};0_+),\quad \mathrm{Im}\, \underset{\sim}{\psi}(\underset{\sim}{x}) = -\underset{\sim}{H}(\underset{\sim}{x};0_+), \tag{37}$$

Eq. (36) becomes

$$4\pi \underset{\sim}{j}_e(\underset{\sim}{x}) = -\underset{\sim}{E}(\underset{\sim}{x};0_+),\quad 4\pi \underset{\sim}{j}_o(\underset{\sim}{x}) = -\frac{1}{4\pi} \underset{\sim}{\nabla} \times \int \frac{1}{|\underset{\sim}{x}-\underset{\sim}{x}'|} \underset{\sim}{H}(\underset{\sim}{x}';0_+)\, d\underset{\sim}{x}'. \tag{38}$$

From Eq. (38) it is clear that the current is a transverse vector as indeed are all currents as constructed by our inverse source method.

REFERENCES

1. B.J. Hoenders, Inverse Source Problems in Optics, ed. H.P. Baltes, Springer, New York, p. 41ff (1978).

2. H.E. Moses, Phys. Rev., 113, 1970 (1959).

3. H.E. Moses, SIAM J. App. Math., 21, 114 (1971).

4. H.E. Moses and A.F. Quesada, Arch. Rat. Mech., 50, 194 (1973).

ACKNOWLEDGEMENT

The work in this paper was done under the National Science Foundation Grant MCS-8024640.

I.4 (IS.2)

TRANSIENT METHODS IN ELECTROMAGNETIC IMAGING

K.J. Langenberg, G. Bollig, M. Fischer, D. Brück

Gesamthochschule Kassel
Fachgebiet Theoretische Elektrotechnik, FB 16
Wilhelmshoeher Allee 71
D-3500 Kassel, F.R.G.

The theoretical background of pulsed electromagnetic imaging is reviewed. We differentiate between stationary methods relying on the information contained in the broadband transient signal scattered into a prescribed spatial direction, like the Singularity Expansion Method, and synthetic aperture methods, which generally need data within a subset of $\underline{K}$-space, where $\underline{K}$ is a spatial Fourier variable comprising both aperture, i.e. spatial, and frequency information. Different support of these variables yields different imaging algorithms, which can either be derived by explicit inversion of Kirchhoff's integral under low or high frequency assumptions or by the heuristic backward wave propagation argument. We outline the general potential as well as the defects and limitations of these procedures.

INTRODUCTION

Generally, imaging methods try to extract scatterer specific information out of the field scattered by a certain obstacle illuminated by an incident wave, which may either be chosen as an electromagnetic, acoustic or elastodynamic one according to applications in radar detection, remote sensing, nondestructive testing of materials, geophysics or medical diagnostics. Here, we concentrate on mathematically based methods referred to as inverse scattering or identification methods. In order to outline the common theoretical background of some presently used synthetic aperture algorithms, and the feasability of a proposed stationary procedure, we ignore the vector- or tensor-nature of certain above-mentioned applications and use a scalar notation throughout the paper; polarization effects and methods utilizing them as developed by

W.-M. Boerner et al. (eds.), Inverse Methods in Electromagnetic Imaging - Part 1, 87–110.

Boerner et al. {1,2} are therefore excluded. We emphasize that part of our work is strongly related to that of Devaney and Porter et al. {3,4,5,6}.

Producing scattered wave fields and extracting information out of them can be achieved by several arrangements (Fig. 1), which can be separated into "stationary" (fixed source point Q

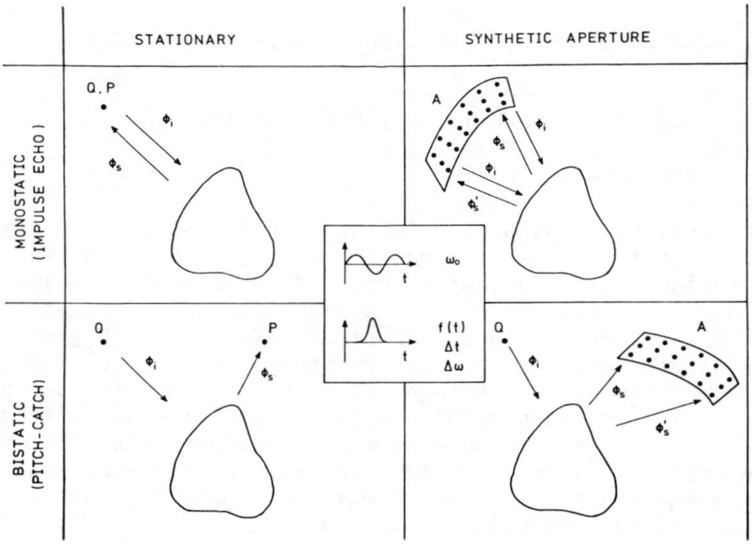

Fig. 1: Data recording arrangements (Φ_i incident field, Φ_s scattered field)

and observation point P) and "synthetic aperture" arrangements (moving observation point P within a prescribed aperture A). A further distinction can be made into "monostatic" (coinciding Q and P) and "bistatic" (spatially separated Q and P). If the application is nondestructive testing of materials the corresponding expressions are "impulse echo" and "pitch-catch". It is clear that a stationary arrangement requires a broadband incident field with a prescribed impulse structure f(t) of duration Δt or bandwidth $\Delta\omega$ in order to contain enough relevant information. In contrast, if sufficient information is collected within a synthetic aperture, a time harmonic excitation of circular frequency ω_0 might suffice to reconstruct a scattering surface described by two variables. We will make it evident that this depends on the general structure of the scatterer (flat or voluminous) and on resolution requirements combined with the accessible aperture range.

Mathematical algorithms related to stationary arrangements rely on general system identification procedures {7,8} or pattern recognition techniques {9}. Relatively simple ideas have been applied to nondestructive testing or underwater acoustics and distinguish between different scattering centres estimating travel times between single distinctive scattered pulses {10,11} or estimating their spectral counterpart in terms of resonance distances {12,13}. A further more sophisticated procedure includes the parameters "resonance width" and is therefore based on a complex frequency plane. Originally developed by Baum {14} the mathematical background of this so-called Singularity Expansion Method is presently not far from being settled {15,16} whereas practical applications to experimental data are still few {17,18,19} and lack reliability {20} on behalf of the strong ill-posedness of the resonance extraction; Dudley {21} states that presently available algorithms {22,23,24} match data and not resonances.

Reconstruction of the geometrical boundary of scattering objects is strongly related to the identification of the equivalent sources induced on the scatterer by the incident field giving rise to the scattered field via Huygens' principle. If the scattering process is considered as a linear time-invariant filter, we know that a time harmonic excitation exp $(j\omega t)$ can be regarded as an eigenfunction with the eigenvalue "frequency response" $G(\underline{\hat{r}},\underline{\hat{r}}_0,\omega)$ depending on the direction of observation $\underline{\hat{r}}$ and the direction $\underline{\hat{r}}_0$ of the incident field ("^" denotes unit vectors). In the far-field G is essentially determined by the scattering amplitude or field pattern $H(\underline{\hat{r}},\underline{\hat{r}}_0,\omega)$ via

$$G(\underline{\hat{r}},\underline{\hat{r}}_0,\omega) = H(\underline{\hat{r}},\underline{\hat{r}}_0,\omega)\,\frac{e^{jkr}}{r} \qquad (1)$$

where r is the distance of the observation point to some coordinate origin, and k denotes the wave number of the circular frequency ω. As outlined in {5}, the scattered field is uniquely determined by $H(\underline{\hat{r}},\underline{\hat{r}}_0,\omega)$; therefore, the scattering amplitude is the maximum amount of information available to an observer in the far-field. On the other hand, $H(\underline{\hat{r}},\underline{\hat{r}}_0,\omega)$ is related to the above-mentioned equivalent sources (under additional assumptions) by a spatial Fourier transform, where the Fourier transform variable $\underline{K}$ is essentially composed of the product $k\underline{\hat{r}}$. Hence, for example, a time harmonic (bistatic) synthetic aperture experiment yields data on an aperture subset A of the spherical surface $\underline{K} = k\underline{\hat{r}}$ = const., the so-called Ewald-sphere (Fig. 2); it is clear then that, even for a 4π - aperture, this experiment is not sufficient to allow per-

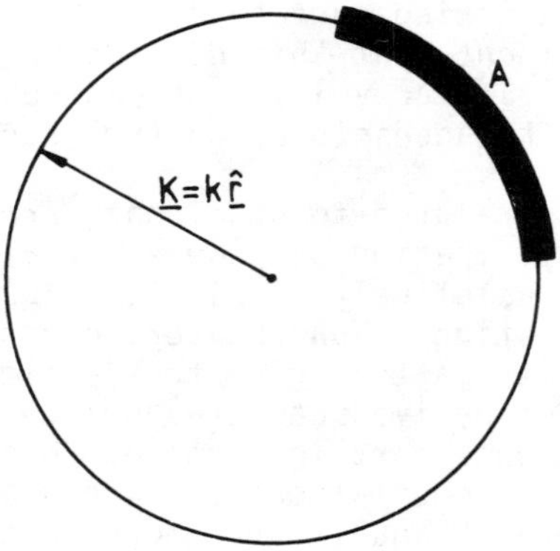

Fig. 2: The Ewald-sphere in K-data space

formance of the inverse spatial Fourier transform to yield the (equivalent) sources confined to the scattering surface. In fact, the Ewald surface has to be swept throughout the K-data space, i.e. broadband or transient data seem to be necessary to solve the imaging problem.

Within the visible range of electromagnetic waves our eye evolved a solution to the imaging problem during the evolutionary process in terms of "pattern matching" {25}, and we consider this as a key word in inverse scattering: our mathematical reconstruction scheme should be matched to the scattering procedure, i.e. we are looking for something like a spatial matched filter.

BROADBAND SYNTHETIC APERTURE ALGORITHMS

Basic Approaches

Contour imaging of scattering objects is mainly based on Kirchhoff's formulation of Huygens' principle (Fig. 3). Enclose the scatterer with surface S_c and volume V_c inside the surface S_M^i surrounded by another surface S_M^a, consider single and double layer sources $\partial\Phi/\partial n$ and Φ as known on $S_M^i \cup S_M^a$, then the (scalar) field Φ in the volume $V_M^a - V_M^i$ between the surfaces S_M^i and S_M^a is given by Kirchhoff's integral with the free space Green's function G

$$\Phi = \int_{S_M^a \cup S_M^i} \left(G \frac{\partial \Phi}{\partial n} - \Phi \frac{\partial G}{\partial n} \right) dS \qquad (2)$$

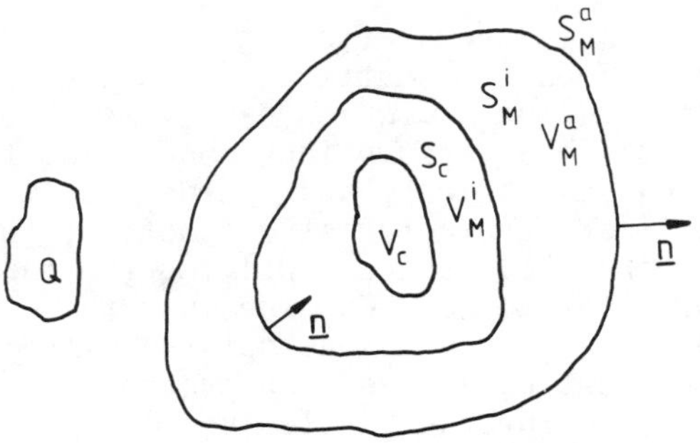

Fig. 3: Inverse scattering with Kirchhoff's integral

Extending S_M^a to infinity, the corresponding integral yields Φ_i the incident field produced by the source distribution Q. Therefore

$$\Phi = \Phi_i + \int_{S_M^i} \left(G \frac{\partial \Phi}{\partial n} - \Phi \frac{\partial G}{\partial n}\right) dS \tag{3}$$

As Bojarski already pointed out {26}, application of (2) or (3) for inverse scattering purposes, i.e. for points inside S_M^i where the scatterer lives, is useless, because the value zero and no information about the scatterer, contained in the total field Φ, is obtained. Otherwise spoken, according to (3), integration of known (measured) values of $\partial\Phi/\partial n$ and Φ on a surface S_M^i surrounding the scatterer yields nothing but the (negative) incident field inside S_M^i. Two alternative approaches seem appropriate:

1. Backward wave propagation

Induced by (3), define a quantity Θ_h ("effectal field") according to

$$\Theta_h = \int_{S_M^i} \left(G^* \frac{\partial \Phi}{\partial n} - \Phi \frac{\partial G^*}{\partial n}\right) dS \tag{4}$$

with the complex conjugate G^* of the Green's function defining incoming waves. This expression is nonzero inside V_M^i, and one can either hope that it propagates backward to the equivalent sources on the scattering surface where the data $\partial\Phi/\partial n$ and Φ originated from, or one might use it as a starting point for further processing. This backward wave propagation approach is dealt with under the headlines "Generalized Holography" and "Exact Inverse Scattering" {26,27}.

2. Explicit inversion

Constrain the surface S_M^i to the scattering surface itself; then the integral in (3) defines the scattered field due to the equivalent sources $\partial\Phi/\partial n$ and Φ on S_c, which can be measured throughout $\underline{K}$-space. Under additional assumptions concerning the sources (low or high frequency assumptions) and introducing the support function of the scatterer (characteristic function) the integral can be extended over the whole space; applying the far-field approximation it turns out to be a Fourier integral making an explicit inversion possible. Acronymes associated with this approach and defining (highly identical) inversion procedures are POFFIS (Physical Optics Far-Field Inverse Scattering) {28,29} and IBA (Inverse Born Algorithm) {5,30}. They can be both explicitly interpreted as transient methods thus rediscovering the basic idea of SAR (Synthetic Aperture Radar) {31}.

Generalized Holography and Exact Inverse Scattering

It is worthwhile to notice that the backward wave propagation argument formulated in terms of the holographic reconstruction or effectal field Θ_h is a time harmonic approach thus reducing the amount of data to be recorded considerably, and it comprises no far-field approximation. The trade-off is the drawback of being restricted to specific geometries, i.e. scatterers with planar surfaces more or less parallel to an also planar recording surface, giving rise to the utilization of a special Green's function being zero-valued on that surface {32}; equ. (4) then reads

$$\Theta_h = - \int_{S_M^i} \Phi \frac{\partial G^*}{\partial n} dS \tag{5}$$

Denoting the kernel $\partial G^*/\partial n$ by Γ^*, the spatial Fourier transform of (5) with respect to the measurement surface variables x and y of a cartesian coordinate system, indicated by a tilde, yields

$$\tilde{\Theta} = - \tilde{\Phi}\, \tilde{\Gamma}^* \tag{6}$$

where {33} (under the assumption kr >> 1)

$$\tilde{\Gamma} = e^{-jz\sqrt{k^2-\tilde{x}^2-\tilde{y}^2}} \tag{7}$$

can be interpreted as a spatial reconstruction filter into the depth z from the measurement surface. On the other hand, applying Kirchhoff's integral (3) to a planar scatterer in that depth, oriented parallel to S_M^i, and assuming a Neumann boundary condition ($\partial\Phi/\partial n = 0$ on S_c), we obtain for the (Fourier transformed) data

$$\tilde{\Phi} = - \tilde{\Phi}_s \tilde{\Gamma} \tag{8}$$

where Φ_s denote the sources, i.e. $\Phi_s = \Phi$ on S_c. Inserting (8) into (6) yields

$$\tilde{\theta}_h = \tilde{\Phi}_s \tilde{\Gamma}\tilde{\Gamma}^* \tag{9}$$

Hence, for the above special scattering arrangement, we have a transformation of the holographic reconstruction (its Fourier transform) into the Fourier transformed equivalent sources through the spatial filter $\tilde{\Gamma}\tilde{\Gamma}^*$, which, due to (7), is approximately (for kz sufficiently large) an identity transformation, as long as the spatial spectrum of the sources falls into the range {-k,k}, which is the case for (sufficiently extended) scatterers not inclined with respect to S_M^i {34}. Therefore, θ_h is actually a reconstruction of the (lateral) dimensions of such scatterers, and due to the relationship "reconstruction filter Γ^* equal to complex conjugate of scattering filter Γ", it represents a spatial matched filter with all the consequences for the noise behavior of such filters. In the field of nondestructive testing of materials this algorithm has been termed Rayleigh-Sommerfeld-reconstruction {34} and is used with good success as long as its limitations are well observed {35}. It should be noted that also axial resolution is obtained if a full 4π-aperture is available and the complete expression (4) is used (provided, $\underline{\hat{r}}_o$ is not in the plane of the scatterer); the procedure has then recently been termed BIS (Basic Imaging System) {36}; numerical experiments show that this is true, in some respect, even for multidimensional scatterers (like a sphere for instance) {37}. The loss of axial resolution is always associated with a reduced aperture and the only non ill-posed remedy is the utilization of broadband or transient excitation. A further illustration and a mathematical proof of that need (apart from our introductory remarks) can be accomplished by a closer look to an extension of Generalized Holography termed Exact Inverse Scattering.

According to Fig. 4 one might consider $\tilde{\theta}_h$ as pseudo-data to fill up the Ewald-sphere for further processing. Porter {27} as well as Bojarski {26} pursued this idea deriving an integral equation of the first kind for the (equivalent) sources with θ_h as inhomogeneity. Unfortunately, due to the nonuniqueness of the solution of this equation {6}, the set of eigenfunctions is incomplete {37}, and the eigenvalues decay rapidly {36,37}, which has two consequences: θ_h is already the "best" possible solution, and, to ensure completeness of the eigenfunctions, one has to integrate over frequency confirming the above statement that transient imaging is appropriate (at least for single experiments).

Another heuristic extension to increase the resolution of the reconstruction θ_h is the use of monostatic data in (4).

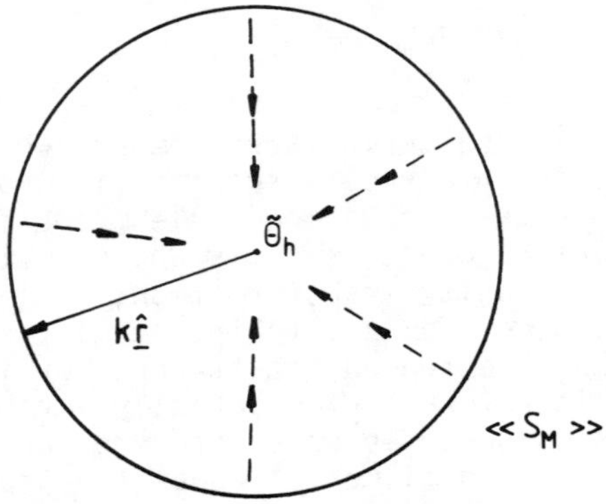

Fig. 4: Holographic reconstruction as pseudo-data in $\underline{K}$-space; $<<S_M>>$ denotes the measurement surface in $\underline{K}$-space

Transient "inverse scattering"

For monostatic as well as bistatic excitation the surface integral in (3) accounting for the scattered field can be converted into a volume integral {38,39} after introducing a Dirichlet or Neumann boundary condition (soft or rigid scatterer) on the support function (characteristic function $\gamma(\underline{r})$), assuming far-field distances, and, after all, applying a low frequency principle of equivalent sources of constant magnitude, and phase proportional to the incident field, on the whole surface of the scatterer. This integral turns out to be a Fourier transform relationship making the explicit inversion of Kirchhoff's integral possible.

An alternative approach (yet yielding quite similar results) starts with the low frequency Born approximation for a weak inhomogeneity in an otherwise constant material {49,5}. In acoustic notation, consider a homogeneous fluid with constant compressibility κ_o and density ρ_o where a scattering body with surface S_c and spatially varying and frequency dependent $\kappa(\underline{r},k)$ and $\rho(\underline{r},k)$ is embedded. The total pressure field $\Phi(\underline{r},k)$ is then given by the solution of {40}

$$(\Delta+k^2)\Phi(\underline{r},k) = \nabla \cdot \gamma_\rho(\underline{r},k)\nabla\Phi(\underline{r},k) - k^2\gamma_\kappa(\underline{r},k)\Phi(\underline{r},k) \tag{10}$$

with

$$\gamma_\rho(\underline{r},k) = \frac{\rho(\underline{r},k) - \rho_o}{\rho_o} \quad , \quad \gamma_\kappa(\underline{r},k) = \frac{\kappa(\underline{r},k) - \kappa_o}{\kappa_o} \tag{11}$$

The right hand side of (10) can be interpreted as a source term $-q(\underline{r},k)$ giving rise to an iterative solution in terms of

$$\Phi_{n+1}(\underline{r},k) = \Phi_i(\underline{r},\underline{r}_o,k) + \int_{V_c} d^3\underline{r}' \frac{e^{jk|\underline{r}-\underline{r}'|}}{4\pi|\underline{r}-\underline{r}'|} q(\underline{r}',k)\Phi_n(\underline{r}',k) \qquad n=0,1,2,.. \tag{12}$$

where the incident field $\Phi_i(\underline{r}',\underline{r}_o,k)$ from a source point $\underline{r}_o$ is inserted as zero-order approximation for the total field $\Phi_o(\underline{r}',k)$ under the integral, which might be sufficiently valid for a weak scatterer. Assuming a point source at $\underline{r}_o$, introducing the far-field approximation

$$\frac{e^{jk|\underline{r} - \underline{r}'|}}{4\pi|\underline{r} - \underline{r}'|} \cong \frac{e^{jk(r - \hat{\underline{r}} \cdot \underline{r}')}}{4\pi r}$$

and manipulating the differential operators in q yields for the first-order scattered field $\Phi_s = \Phi_1 - \Phi_i$

$$\Phi_s(\underline{r},k) = k^2 \frac{e^{jk(r+r_o)}}{(4\pi)^2 r r_o} \int d^3\underline{r}' e^{-jk(\hat{\underline{r}}+\hat{\underline{r}}_o)\cdot\underline{r}'} \gamma(\underline{r}')p(\underline{r}',k) \tag{13}$$

with the so-called object function

$$p(\underline{r},k) = -\{\gamma_\rho(\underline{r},k) + \gamma_\kappa(\underline{r},k)\} - \frac{1}{2}k^2\Delta\gamma_\rho(\underline{r},k) \tag{14}$$

The integral (13) is a spatial Fourier transform with the transform variable $\underline{K} = k(\hat{\underline{r}} + \hat{\underline{r}}_o)$; therefore, inversion leads to an expression ("$*$" denoting convolution)

$$\tilde{\gamma}(\underline{K}) \star \tilde{p}(\underline{K},k) = (4\pi)^2 r r_o e^{-jk(r+r_o)} \frac{\Phi_s(\underline{r},k)}{k^2} \tag{15}$$

which can be experimentally determined within the volume of the "double" or limiting Ewald sphere $|\underline{K}| < 2k$. Depending on the nature and feasability of the experiment (compare Fig. 1) the subsequent alternatives can be pursued:

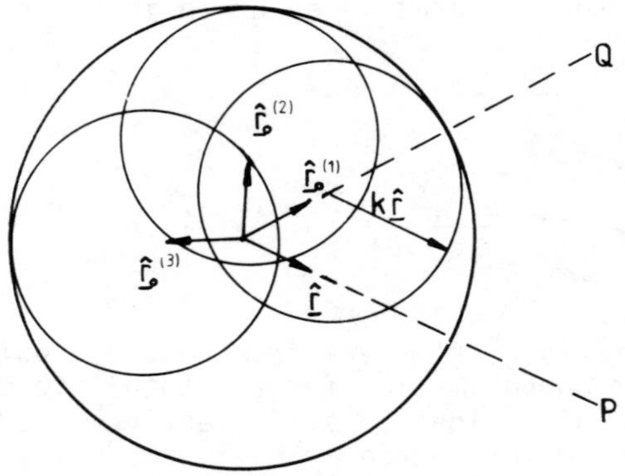

Fig. 5: Bistatic time harmonic multiple experiment

1. Bistatic time harmonic multiple experiment
The limiting Ewald sphere can be filled by a manifold of Ewald sphere surfaces with radii k and subsequent midpoints $k\hat{\underline{r}}_o^{(i)}$ (i = 1,2,3,...) (Fig. 5), indicating a new synthetic aperture bistatic time harmonic experiment for each i {5}. As outlined in {41} this means a considerable experimental effort leading to huge data storage problems. Nevertheless, for optical purposes this time harmonic approach has proved to be useful, especially since recent computational improvements have been reported {41,42} closing the gap to tomographic reconstruction.
2. Monostatic broadband single experiment (IBA, POFFIS)
As indicated by Fig. 6 monostatic broadband experiments yielding $\underline{K} = 2k\hat{\underline{r}}$ are also appropriate to fill the limiting Ewald sphere. The resulting algorithm, as extensively applied for nondestructive testing of materials, has been termed IBA (Inverse Born Algorithm) {30}; yet, the inversion requires an explicit frequency independence of the object function.

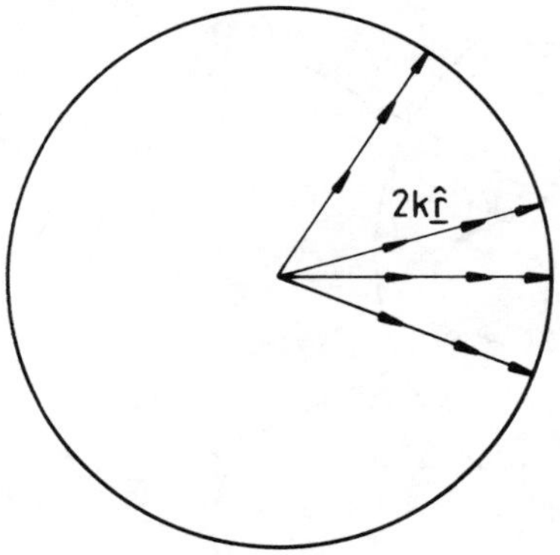

Fig. 6: Monostatic broadband single experiment (IBA, POFFIS)

It has recently been emphasized {43} that the monostatic broadband version of (15) is very similar to the POFFIS-identity {28, 29}, even though the latter approach is a high frequency one. The deeper reason for this coincidence is that both POFFIS and IBA concentrate on the scattering surface, the first procedure by definition and the latter one by assuming an undisturbed incident field inside the scatterer.

3. Bistatic broadband single experiment (FIFFIS)

According to Fig. 7, fixed $\underline{r}_0$ combined with varying k and $\hat{\underline{r}}$ leads to a subset of the data space, and the resulting algorithm has been termed FIFFIS (Frequency Independent Far-Field Scattering) {44}, because, once again, the object function has to be frequency independent. As IBA, FIFFIS is derived under the low frequency assumption, and it is not quite clear if there exists, like for IBA and POFFIS, a formally correct high frequency counterpart, because corresponding bistatic POFFIS identities reveal intrinsic complications {45,46}. Nevertheless, it is somehow heuristically justified for high frequencies by assuming a priori given sources {47} with (frequency independent!) spatially varying magnitudes and a phase related to the incident wave. It might then be interpreted as a broadband Fourier-type (i.e. far-field) holography, which has been used successfully in microwave imaging {48,49}; a broadband nearfield counterpart based on Rayleigh-Sommerfeld holography has also been reported {50}.

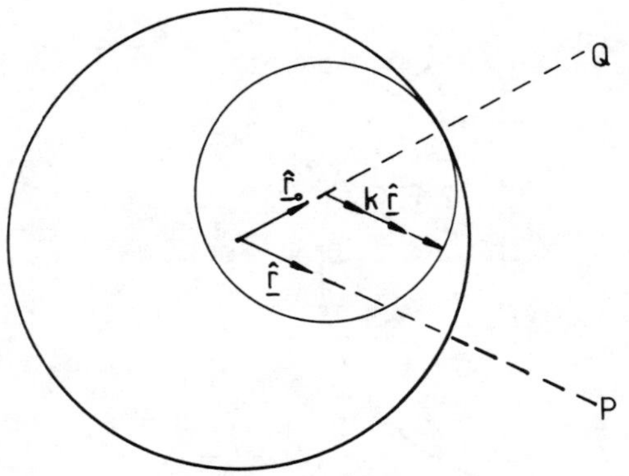

Fig. 7: Bistatic broadband single experiment (FIFFIS)

Transient interpretation of broadband "inverse scattering" (SAR)

The interpretation of IBA and POFFIS in the time domain {29,30} reveals very nicely their intrinsic properties and identifies intuitively their possible pitfalls. Let $\underline{K} = 2k\hat{\underline{r}}$ and rewrite the Fourier transformed equ. (15) in terms of spherical coordinates in $\underline{K}$-space $d^3\underline{K} = K^2 dK d\Omega$

$$\gamma(\underline{r}')p(\underline{r}') = \frac{32r^2}{c} \int d\Omega$$

$$\cdot \frac{1}{2\pi} \int_{-\infty}^{+\infty} d\omega \Phi_s(\hat{\underline{r}},\frac{\omega}{c})\, e^{-2j\frac{r}{c}\omega}\, e^{2j\frac{\hat{\underline{r}}\cdot\underline{r}'}{c}\omega}\, U(\omega) \tag{16}$$

where $U(\omega)$ denotes the unit step-function. The ω-integral is an inverse Fourier integral for the time $t = 0$ and therefore

$$\gamma(\underline{r}')p(\underline{r}') = \frac{32r^2}{c} \int d\Omega \Phi_s(\hat{\underline{r}},t) * U(t)\Big|_{t = 2\frac{r}{c} - 2\frac{\hat{\underline{r}}\cdot\underline{r}'}{c}} \tag{17}$$

with

$$U(t) = \frac{1}{2}\,\delta(t) + \frac{1}{2\pi}\,\mathrm{Pf}\,\frac{1}{jt} \tag{18}$$

and $\Phi_s(\hat{\underline{r}},t)$ the scattered time domain impulse response.

Hence

$$\gamma(\underline{r}')p(\underline{r}') = \frac{16r^2}{c} \int d\Omega\Phi_s(\hat{\underline{r}}, 2\frac{r}{c} - 2\frac{\hat{\underline{r}}\cdot\underline{r}'}{c}) + j\frac{16r^2}{c} \int d\Omega H\{\Phi_s(\hat{\underline{r}},t)\}\Big|_{t = 2\frac{r}{c} - 2\frac{\hat{\underline{r}}\cdot\underline{r}'}{c}} \tag{19}$$

where H is the Hilbert transform. We state that the broadband monostatic reconstruction under the Born assumption is composed of a real and imaginary part, whereas the derivation of its time domain PO counterpart yields only the real part of (19) {29}. We suspect that this is due to the fact that the POFFIS identity is strictly valid only for *Physical Optics* data, because numerical experiments have shown {51,52} that the additional imaginary part of (19) accounts for the deviation of exact scattering data from Physical Optics data emphasizing the scattering centres of the body (edges, corners etc.) whose singularities are not accounted for by the smooth PO equivalent sources.

The above fact of differing real life data yields a pitfall of POFFIS and IBA even for 4π-apertures and infinite bandwith as illustrated by Fig. 8.

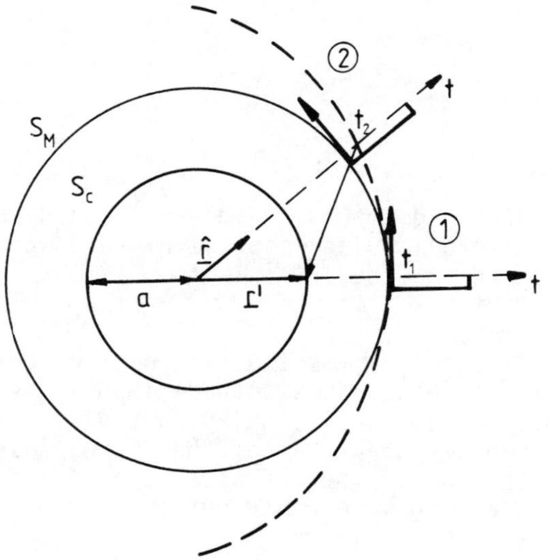

Fig. 8: Time domain illustration of IBA and POFFIS

Consider a spherical scatterer of radius a with a perfectly hard boundary surrounded by the measurement surface S_M, and two monostatic experimental points displaying the (infinite broadband) backscattered impulse response as function of time, which is composed of the specular reflection peak and the constant PO part ending abruptly at the time corresponding to the shadow boundary. According to the real part of (19), choice of a reconstruction point $\underline{r}'$ means integration of the transient data amplitudes along a surface indicated by the dashed line; we see that one specular and additional PO data points are included, where the amplitude at t_2 corresponds to the PO contribution from our reference point thus being correctly accounted for. The integration procedure then

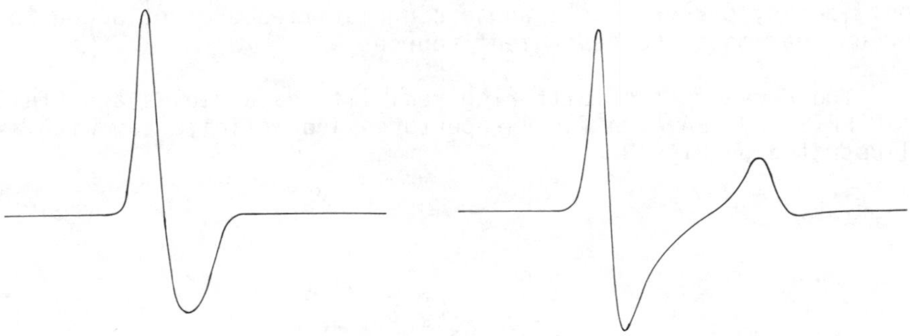

Fig. 9: Left: PO far-field impulse response of hard sphere. Right: Far-field impulse response of hard sphere (both Gaussian smoothed)

exactly yields the characteristic function of the sphere. Fig. 9 now compares the exact (Gaussian smoothed) impulse response with the PO counterpart, and obviously the PO part of the exact response decays more rapidly than that of the PO response yielding an indentation of the top of the characteristic function (Fig. 10). Additionally, the creeping wave contribution to the exact impulse response shows up as a ghost image.

The POFFIS identity itself is independent on the boundary condition "soft" or "hard"; nevertheless, the soft impulse response differs considerably from the hard one (Fig. 11) since its PO part is much less pronounced compared to the specular part.

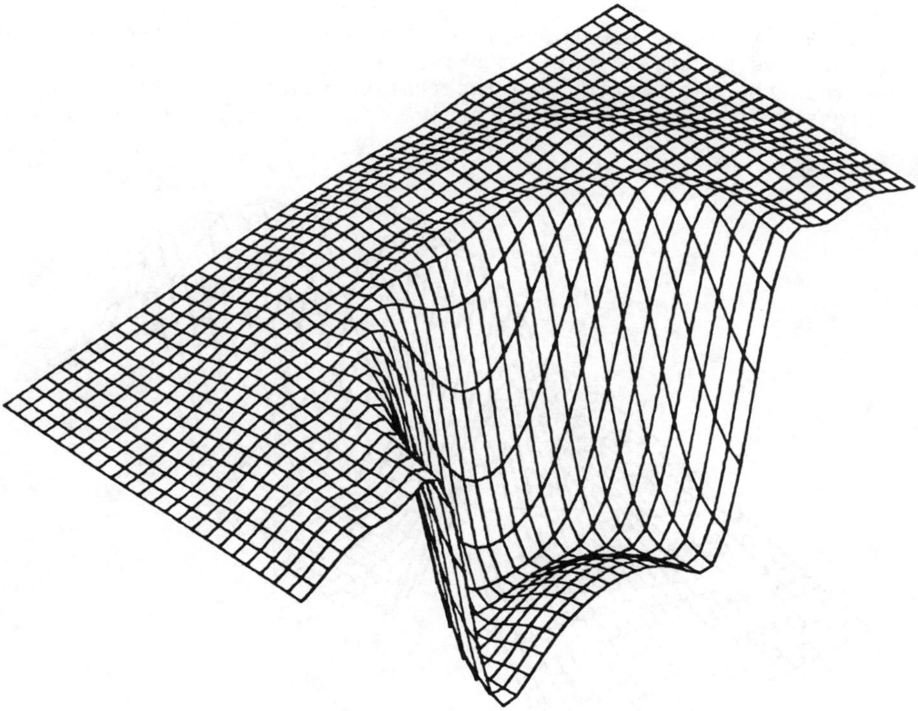

Fig. 10: POFFIS reconstruction of hard sphere

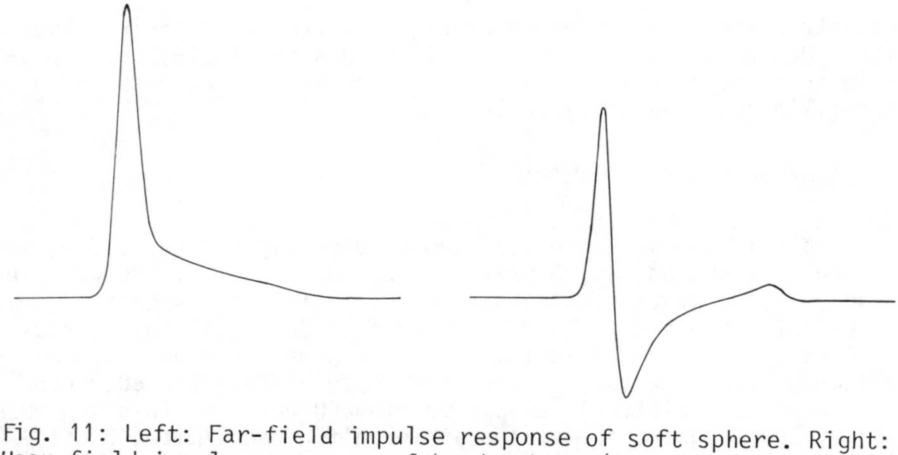

Fig. 11: Left: Far-field impulse response of soft sphere. Right: Near-field impulse response of hard sphere (both Gaussian smoothed)

This results in an overemphasis of the scattering surface when reconstructed by this algorithm (Fig. 12). Therefore, this fact underlines the engineering feeling that specular reflection contributions to the impulse response give rise to a contour or surface

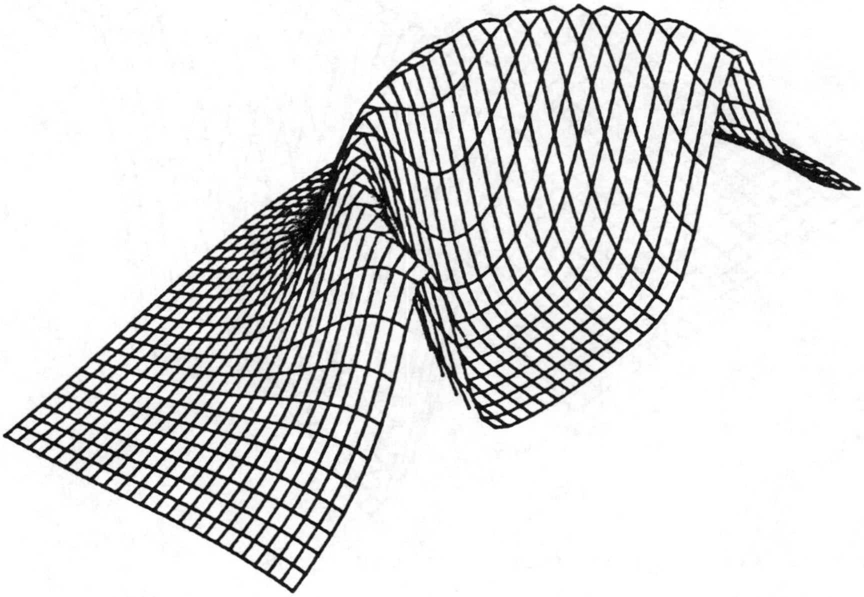

Fig. 12: POFFIS reconstruction of soft sphere

reconstruction when summed up properly according to their travel times. Hence, a heuristic near-field extension of (19) seems possible in terms of (expecially when one looks at a corresponding near-field response as given in Fig. 12).

$$\gamma_{surface}\ (\underline{r}') \stackrel{\sim}{=} \int d\Omega \Phi_s(\hat{\underline{r}}, \frac{2}{c}|\underline{r} - \underline{r}'|) \tag{20}$$

Explicit surface detection can be directly accounted for, if a singular function, which peaks on the surface, is introduced instead of the characteristic function {53}. This is essentially equivalent in multiplying the right hand side of (15) by k, thus suppressing low frequencies; in the time domain the data to be collected according to (19) are then to be differentiated, suppressing again the relatively slowly varying PO portion. This suppression of low frequencies is also achieved for bandlimited excitations, say amplitude modulated sine bursts or chirp signals, i.e.

real life data. We therefore prefer to use the singular function formulation for experimental purposes {43,44}.

Unprocessed reconstruction with bandlimited signals introduces unwanted image side lobes {51}; a hint for further processing is given by our "Real life data POFFIS identity" (19): in the context of analytic or radar signals the time domain data and their Hilbert transform can be combined to the computation of envelopes, correlations etc. thus improving resolution, and, if extended to the near-field version, rediscovering the synthetic aperture radar (SAR), which recently found its counterpart SAFT (Synthetic Aperture Focussing Technique) {54} for nondestructive testing of materials.

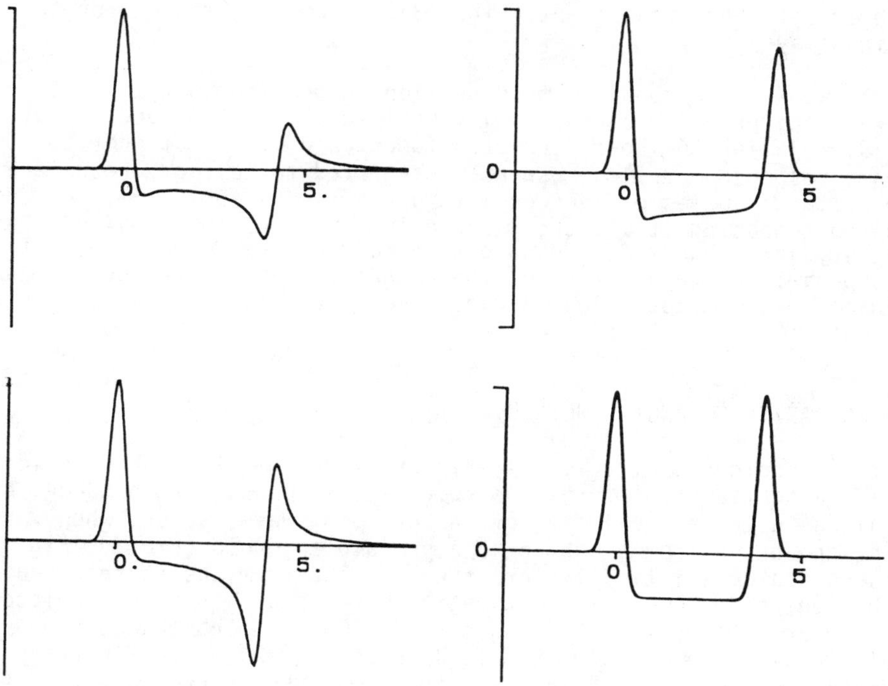

Fig. 13: Far-field backscattered Gaussian smoothed impulse response of hard elliptical scatterers for nose-on incidence (above: axis ratio 5:1, below: axis ratio 100:1). Left: infinitely long cylinder with elliptical cross-section (two-dimensional problem). Right: prolate spheroid (three-dimensional problem)

Certain severe drawbacks of POFFIS or IBA reconstructions are readily derived via simulations {43,44,51} and well understood through their time domain interpretation: a limited spatial aperture, for instance, misoriented with respect to a planar part of the scatterer, may lead to a lack of specular information; then, high amplitude corner or edge diffracted pulses are interpreted as such by the algorithm and placed at the proper position. When applied to nondestructive testing this missing information could be dangerous because the image of a long crack might look like a point scatterer, namely the crack tip.

When simulations are performed in testing POFFIS and IBA another problem has to be carefully considered: the time domain algorithm is essentially independent of the dimension of the scatterer but the data are not (Fig. 13); therefore, misinterpretations may arise.

We want to conclude this section in pointing out that other transient procedures essentially utilize this k^{-2}-dependence of (15) relating the so-called area function of the scatterer to its ramp response (impulse response integrated twice) for prescribed monostatic angles of incidence ("look angles") {55,56,57} instead of compensating it via the $\underline{K}$-space integration. These methods need a very large bandwidth because they rely on the PO portion of the response; on the other hand, the amount of data is reduced because three equally spaced look angles have proved to be generally sufficient.

SINGULARITY EXPANSION METHOD

If, according to Fig. 1, a stationary arrangement is used to identify a scatterer, information about its size is, for instance, contained in the arrival time of the creeping wave, which, when varying the shape from a sphere to a prolate spheroid (Fig. 9, Fig. 13), turns characteristically into the tip echo from the farther end of the spheroid. Time delay estimation methods are then appropriate to separate single scattering centres. On the other hand, timely distinguishable wave fronts yield resonances when Fourier transformed to a spectral domain and these may then serve as a tool for identification (Fig. 14). Unfortunately, both time and frequency results depend upon the look angle. An identification procedure which overcomes this problem is derived in terms of the Singularity Expansion Method. Consider for example the Mie series for the scattered far-field of a rigid sphere in the Laplace (complex frequency) domain ($k \rightarrow js/c$)

$$\Phi_s(\underline{r},s) = \frac{e^{-\frac{s}{c}r}}{\frac{s}{c}} \sum_{n=0}^{\infty} (-1)^n (2n+1) \frac{j_n'(j\frac{s}{c}a)}{h_n^{(1)\prime}(j\frac{s}{c}a)} P_n(\cos\Theta) \tag{21}$$

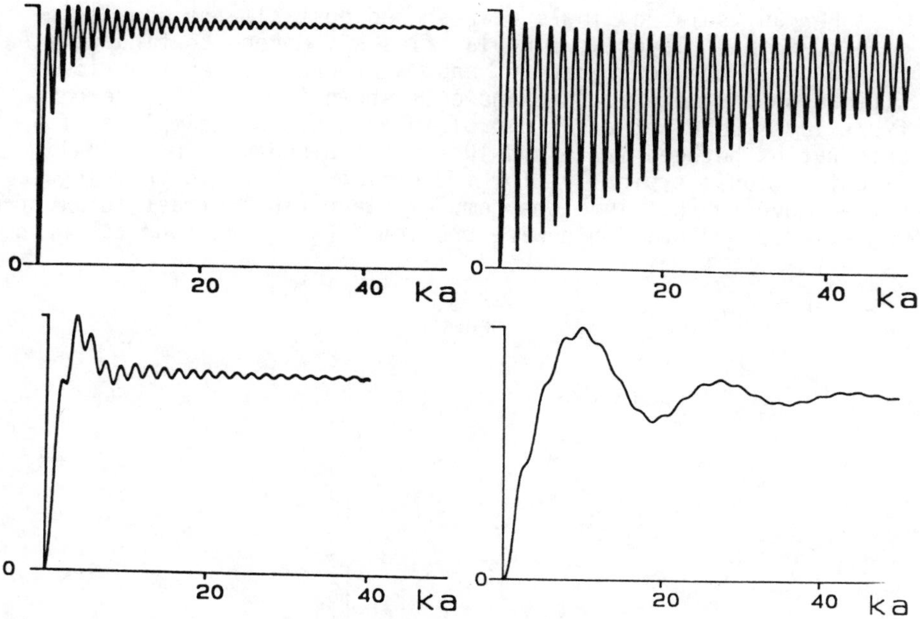

Fig. 14: Far-field backscattered frequency response (magnitude) for nose-on incidence of hard prolate spheroid (above; left: axis ratio 1:1, i.e. sphere; right: axis ratio 5:1) and hard oblate spheroid (below; left: axis ratio 2:1; right: axis ratio 5:1); a denotes major axis

where Θ denotes the polar angle of a spherical coordinate system; j_n, $h_n^{(1)}$ are spherical Bessel and Hankel functions and P_n Legendre polynomials. A singularity representation of (21) is obtained applying the Mittag-Leffler theorem to the corresponding induced surface distribution and integrating over the surface {20} ending up with

$$\Phi_s(\underline{r},s) = js\frac{a^2}{c}j_o'(j\frac{s}{c}a)\,\frac{e^{-\frac{s}{c}(r-a)}}{r} + \tag{22}$$

$$jsa\sum_{n=0}^{\infty}(-1)^n(2n+1)j_n'(j\frac{s}{c}a)P_n(\cos\Theta)\,\frac{e^{-\frac{s}{c}(r-a)}}{r}\,\cdot$$

$$\sum_{l=1}^{n+1}\frac{e^{-\frac{s_{nl}}{c}a}}{\left[n(n+1)+\frac{s_{nl}^2}{c^2}a^2\right]h_n^{(1)}(j\frac{s_{nl}}{c}a)}\left(\frac{1}{s-s_{nl}}+\frac{1}{s_{nl}}\right)$$

For an isolated single star-shaped obstacle the double-indexed, and thus layered, singularities s_{nl} depend continuously on the shape of the scatterer {16} and may therefore serve as identification parameters. An example is shown in Fig. 15, where the first layer poles of a rigid prolate spheroid are displayed for axis ratios between 1.2:1 and 10:1 as obtained via spheroidal eigenfunctions; typically, for increasing axis ratio the singularities move further into the complex plane (in contrast to electromagnetic spheroidal "thin-wire problems" {58}) and tend to arrange

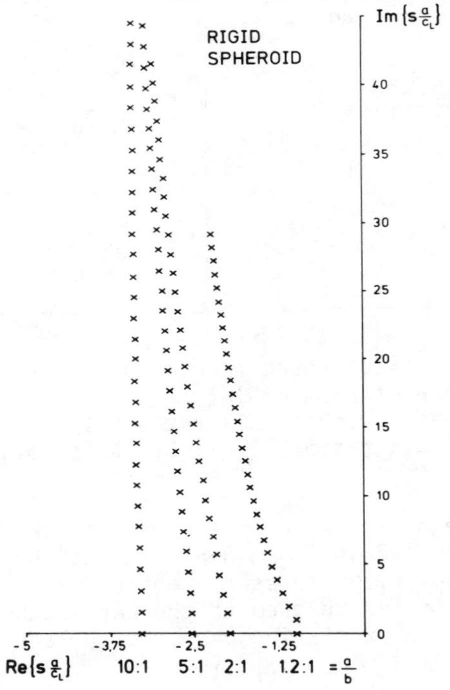

Fig. 15: First layer singularities of prolate rigid spheroid

themselves along a straight line parallel to the imaginary axis, thus approaching the high-frequency ray theoretical result. Additional investigations of that kind using T-matrix methods are contained in {59,60}, and the "elastodynamic" spheroid is addressed in {20}. Even though promising at this point that SEM might be used for imaging and identification purposed, the pole extraction from experimental data has not yet been satisfactorily solved {20, 21}. The reason for this may be illustrated the following way {61}: consider the (simulated) identity between timely separated (by a distance T) high-frequency wave fronts and resonances

$$\sum_{n=-\infty}^{+\infty} \delta(t - nT) = \frac{1}{T} \sum_{n=-\infty}^{+\infty} e^{jn\frac{2\pi}{T} t} \tag{23}$$

where δ denotes the δ-distribution. Assuming 20 resonances the result is shown in Fig. 16, but, if the resonances are corrupted by Gaussian noise with spectral power N according to $2\pi n(1+\sqrt{N}/n)/T$ and $N = 5\cdot 10^{-3}$, Fig. 17 illustrates the ill-posedness of resonance extraction out of wave fronts, like creeping waves for instance.

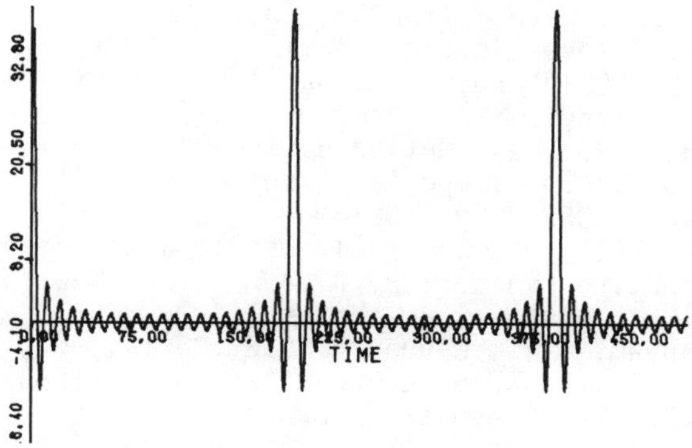

Fig. 16: Wave fronts composed of 20 resonances

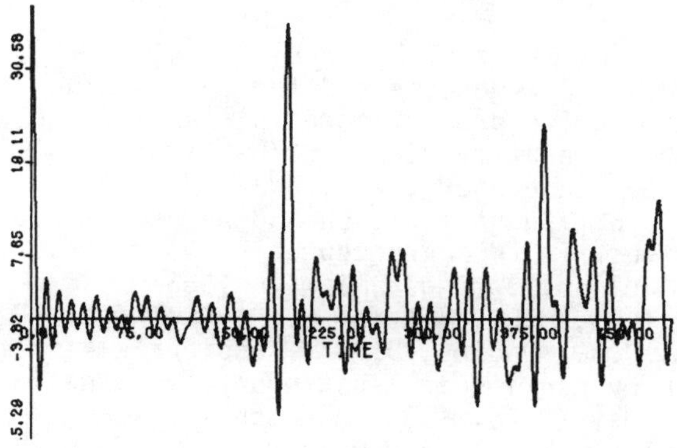

Fig. 17: Wave fronts composed of 20 noisy resonances

REFERENCES

{ 1} Boerner, W.-M.: Polarization Utilization in Electromagnetic Inverse Scattering. In: Inverse Scattering Problems in Optics. Ed.: H.P. Baltes, Springer, New York 1980

{ 2} Boerner, W.-M.: El-Arini, M.B., Chan, C.-Y., Mastoris,P.M.: IEEE Trans. AP-29 (1981)

{ 3} Porter, R.P., Devaney, A.J.: J. Opt. Soc. Am. 72 (1982) 327

{ 4} Porter, R.P., Devaney, A.J.: J. Opt. Soc. Am. 72 (1982) 1707

{ 5} Devaney, A.J.: Inverse source and scattering problems in ultrasonics. Proc. Int. Symp. IEEE Ultrasonics, San Diego/Ca 1982

{ 6} Devaney, A.J., Sherman, G.C.: IEEE Trans. AP-30 (1982) 1034

{ 7} Dudley, D.G.: Wave Motion 5 (1983)

{ 8} Ljung, L., Glover, K.: Automatica 17 (1981) 71

{ 9} Rose, J.L., Goldberg, B.G.: Basic Physics in Diagnostic Ultrasound. J. Wiley & Sons, New York 1979

{10} Tittmann, B.R.: Wave Motion 5 (1983)

{11} Carter, G.C. (Ed.): Special Issue on Time Delay Estimation. IEEE Trans. ASSP - 29 (1981)

{12} Adler, L., Fitting, D.W.: Ultrasonic Spectral Analysis for Nondestructive Evaluation. Plenum Press, New York 1980

{13} Überall, H., Moser, P.J., Murphy, J.D., Nagl, A., Igiri, G., Subrahmanyam, J.V., Gaunaurd, G.C., Brill, D., Delsanto, P.P., Alemar, J.D., Rosario, E.: Wave Motion 5 (1983)

{14} Baum, C.E.: The Singularity Expansion Method. In: Transient Electromagnetic Fields. Ed.: L.B. Felsen, Top. Appl. Phys. 10, Springer, Berlin 1976

{15} Pearson, L.W., Marin, L. (Eds.): Special Issue on the Singularity Expansion Method. Electromagnetics 1 (1981)

{16} Ramm, A.G.: J. Math. Anal. Applic. 86 (1982) 562

{17} Chan, L.C., Moffatt, D.L., Peters, L.: Proc. IEEE 67 (1979) 991

{18} Volakis, J.L., Peters, L.: IEEE Trans. AP - 31 (1983) 334

{19} Lytle, R.J.: A remote method for determining the thicknesses and constitutive properties of planar layered media. Lawrence Livermore Laboratory UCRL - 51789 (1975)

{20} Bollig, G., Langenberg, K.J.: Wave Motion 5 (1983)

{21} Dudley, D.G.: Identification and parametric modeling of transient waves. In: New Procedures in Nondestructive Testing. Ed.: P. Höller, Springer, Berlin 1983

{22} van Blaricum, M.L., Mittra, R.: IEEE Trans. AP - 23 (1975) 777

{23} Goodman, D.M., Dudley, D.G.: Transient electromagnetic identification by non-linear least squares. US National Radio Science Meeting, May 23-26, 1983, Houston/Tx

{24} Sarkar, T.K., Nebat, J., Weiner, D.D., Jain, V.K.: IEEE Trans. AP - 28 (1980) 928

{25} Lorenz, K.: Die Rückseite des Spiegels. Piper Verlag, München 1973

{26} Bojarski, N.N.: Radio Science 16 (1981) 1025

{27} Porter, R.P.: J. Opt. Soc. Am. 60 (1970) 1051
{28} Bojarski, N.N.: IEEE Trans. AP - 30 (1982) 980
{29} Bleistein, N.: J. Acoust. Soc. Am. 60 (1976) 1249
{30} Rose, J.H., Richardson, J.M.: J. Nondestr. Eval. 3 (1983)
{31} Skolnik, M.I.: Radar Handbook. McGraw Hill, New York 1970
{32} Sommerfeld, A.: Optik. Akademische Verlagsgesellschaft, Leipzig 1964
{33} Collier, R.J., Burckhardt, Ch.B., Lin, L.H.: Optical Holography. Academic Press, New York 1971
{34} Berger, M., Brück, D., Fischer, M., Langenberg, K.J., Oberst, J., Schmitz, V.: J. Nondestructive Evaluation 2 (1982) 85
{35} Müller, W., Schmitz, V., Schäfer, G., Langenberg, K.J., Hoppstädter, K., Gräber, B.: Materialprüfung 25 (1983) 9
{36} Porter, R.P., Devaney, A.J.: J. Opt. Soc. Am. 72 (1982) 1707
{37} Fischer, M., Langenberg, K.J.: Limitations and defects of Exact Inverse Scattering. Submitted to Electromagnetics for publication (1983)
{38} Bojarski, N.N.: IEEE Trans. AP - 30 (1982) 775
{39} Bojarski, N.N.: J. Acoust. Soc. Am. 73 (1983) 733
{40} Morse, P.M., Ingard, K.U.: Theoretical Acoustics. McGraw Hill, New York 1968
{41} Devaney, A.J.: Inverse Source and Scattering Problems in Optics. In: Optics in Four Dimensions. Eds.: M.A. Machado, L.M. Narducci, Am. Inst. Phys., New York 1981
{42} Devaney, A.J.: Optics Letters 7 (1982) 111
{43} Langenberg, K.J., Bollig, G., Brück, D., Fischer, M.: Remarks on recent developments in inverse scattering theory. In: Orientation and Localization in Engineering and Biology. Ed.: D. Varjú, Springer, Berlin 1983
{44} Langenberg, K.J., Brück, D., Fischer, M.: Inverse scattering algorithms. In: New Procedures in Nondestructive Testing. Ed.: P. Höller, Springer, Berlin 1983
{45} Rosenbaum-Raz, S.: IEEE Trans. AP - 24 (1976) 66
{46} Meckelburg, H.J.: Electronics Letters 18 (1982)
{47} Detlefsen, J.: Kleinheubacher Berichte 25 (1982) 297
{48} Chan, C.K., Farbat, N.H.: IEEE Trans. AP - 29 (1981) 312
{49} Paolini, F.J., Duffy, M.J.: IEEE Trans. AP - 31 (1983) 389
{50} Karg, R.: Archiv Elektronik Übertrag.techn. 31 (1977) 150
{51} Brück, D.: Simulationsrechnungen zur Fehlerrekonstruktion mit dem POFFIS-Algorithmus. Diploma Thesis, Universität des Saarlandes, Saarbrücken, FRG, 1982
{52} Rose, J.H.: private communication of the Ames Research Lab, Iowa State University, Ames/Iowa
{53} Cohen, J.K., Bleistein, N.: Wave Motion 1 (1979) 153
{54} Ganapathy, S., Wu, W.S., Schmult, B.: Ultrasonics 20 (1982) 249
{55} Kennaugh, E.M., Moffatt, D.L.: Proc. IEEE 53 (1965) 893
{56} Moffatt, D.L., Young, J.D., Ksienski, A.A., Lin, H.C., Rhoads, C.M.: IEEE Trans. AP - 29 (1981) 192

{57} Boerner, W.M., Ho, C.M., Foo, B.Y.: IEEE Trans. AP - 29 (1981) 336

{58} Marin, L.: IEEE Trans. AP - 22 (1974) 266

{59} Boström, A.: J. Math. Phys. 23 (1982) 1444

{60} Kristensson, G.: Natural frequencies of circular discs. Report of the Institute of Theoretical Physics, Chalmers University, Göteborg, Sweden, March 1983

{61} Felsen, L.B.: Progressive wave and oscillartory formulations of propagation and scattering. Plenary Session of the Joint AP-S/URSI Symposium, Houston/Tx, 1983

I.5 (IS.4)

THEORETICAL STUDY AND NUMERICAL RESOLUTION OF INVERSE PROBLEMS VIA THE FUNCTIONAL DERIVATIVES

A. Roger

Laboratoire d'Optique Electromagnétique, ERA 597 CNRS, Faculté des Sciences et Techniques, Centre de Saint-Jérôme, 13397 MARSEILLE CEDEX 13, France

ABSTRACT : A rigorous method of inverse scattering is described, which enables one to solve numerically many practical inverse problems and throws light upon the fundamental instability of these problems. The algorithm is described on the example of a finite two dimensional rough surface, and several profile reconstructions are presented.

1. INTRODUCTION

This paper presents a method which could be called "numerical inverse diffraction". It is a rigorous method, meaning that is it limited only by purely numerical phenomenons. Also, it is a method of inverse diffraction, since it works directly in the framework of functional spaces as opposed to methods of parameter optimization which are limited to the search of a finite number of a-priori chosen parameters. Our algorithm has permitted us to solve successfully practical inverse problems associated with layered media, cylinders and gratings. The method relies upon linearization and iteration. Its implementation includes the following two steps :

First, one must explicitely find the linear operator which links small variations of the object profile to the corresponding small variations of the scattering pattern ; of course it is a first order relation. This operator is the "functional derivative" and it may be exhibited with the help of a so called "adjoint direct problem". Not only the functional derivative is a powerful numerical tool, but also it provides a way of analyzing the information contained in the data via the study of its singular

W.-M. Boerner et al. (eds.), Inverse Methods in Electromagnetic Imaging - Part 1, 111–120.

values. On the other hand, the compactness of these operators explains why the associated inverse problems are unstable and thus require the use of regularization techniques.

Second, one must use the functional derivative to implement an iterative algorithm capable to retrieve the characteristics of the scatterer. Following optimization methods, one may minimize a "cost function" with the help of conjugate gradient algorithms. One may also construct a Newton type algorithm, often much faster than the conjugate gradient.

The main features of these methods will be described on a particular example, i.e. the scattering by a finite rough surface.

2. THE FINITE ROUGH SURFACE : DIRECT AND INVERSE PROBLEMS

Fig. 1A shows the physical device : x,y,z being cartesian coordinates, we study the scattering by a perfectly conducting plane situated at y = 0 which presents a local z-independent deformation. This is a model for a finite rough surface S, with equation $y = f(x)$ ($f(x) = 0$ outside $[a,b]$).

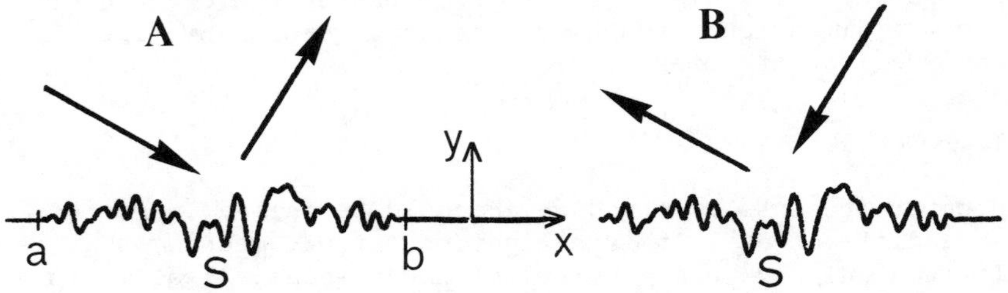

Fig. 1 : The diffraction by a finite rough surface ; fig. 1A shows the initial direct problem, fig. 1B an associated adjoint direct problem.

Let an electromagnetic plane wave, with wavevector $\underline{k}$ ($k = |\underline{k}|$) parallel to the xy plane, be incident upon S with an angle of incidence θ. The electric field $\underline{u}$ is supposed to be parallel to Oz. Under these conditions, the component u of $\underline{u}$ (which we call "the field" in the rest of the paper) along Oz verifies :

$$\Delta u + k^2 u = 0 \quad \text{above } S \tag{1}$$

$$u = 0 \qquad \text{on and below } S$$

$$u = \exp(i\alpha x - i\beta y) - \exp(i\alpha x + i\beta y) + \int_{-\infty}^{+\infty} B(\alpha')\exp(i\alpha' x + i\beta' y)\,d\alpha' \quad (y > \sup f(x))$$

where α, β, β' are defined by :

$$\alpha = k \sin\theta \qquad \beta = k \cos\theta$$

$$\beta' = (k^2 - \alpha'^2)^{1/2} \quad \text{if} \quad |\alpha'| \leqslant k$$

$$\beta' = i(\alpha'^2 - k^2)^{1/2} \quad \text{if} \quad |\alpha'| > k .$$

The direct problem of diffraction may be defined as : knowing $f(x)$, calculate $B(\alpha')$. It has been recently studied and solved [1], and now we have at our disposal a computer code capable to calculate the diffracted amplitude $B(\alpha')$ rigorously, at least in the limit of accuracy of the computer. The numerical resolution is achieved via the use of the Green function of the free space, which leads to an integral equation.

In the following of the paper, the following theorem will be needed : let v and w be two functions of x and y, defined and bounded for $f(x) \leqslant y \leqslant A$ (with $A > \sup f(x)$) and which verify the Helmholtz equation (1). Then, under some conditions of regularity on v and w, we have that :

$$\int_{S_0} \left(v \frac{\partial w}{\partial n} - w \frac{\partial v}{\partial n}\right) dS = -4i\pi \int_{-\infty}^{+\infty} \beta'(\alpha')[V^+(\alpha')W^-(-\alpha') - V^-(\alpha')W^+(-\alpha')]d\alpha' \quad (2)$$

where S_0 is any sufficiently regular surface above S, and V^+, V^-, W^+, W^- are the coefficients of the plane wave expansions of v and w (valid only for $y \geqslant A > \sup f(x)$) :

$$v(x,y) = \int_{+\infty}^{-\infty} [V^-(\alpha')\exp(i\alpha'x - i\beta'y) + V^+(\alpha')\exp(i\alpha'x + i\beta'y)]d\alpha' \quad (3)$$

$$w(x,y) = \int_{+\infty}^{-\infty} [W^-(\alpha')\exp(i\alpha'x - i\beta'y) + W^+(\alpha')\exp(i\alpha'x + i\beta'y)]d\alpha' \quad (4)$$

The demonstration is given in [2]. The reciprocity theorem is a straighforward consequence of this lemma : let $u^*(x,y)$ be the field corresponding to the angle of incidence θ^* ($\alpha^* = \sin\theta^*$, $\beta^* = \cos\theta^*$). The coefficients of the plane wave expansions of u and u^* similar to (3) and (4) are :

$$U^-(\alpha') = \delta(\alpha' - \alpha) \qquad U^{*-}(\alpha') = \delta(\alpha' + \alpha^*)$$

$$U^+(\alpha') \overset{\text{def}}{=} B(\alpha') - \delta(\alpha' - \alpha)$$

$$U^{*+}(\alpha') \overset{\text{def}}{=} B^*(\alpha') - \delta(\alpha' + \alpha^*).$$

Now, let us apply the theorem given above, with $v = u$ and $w = u^*$. Since u and u^* are zero on S, the left hand member of (2) is also zero. The evaluation of the right hand member gives the reciprocity theorem :

$$\beta^* B(-\alpha^*) = \beta B^*(-\alpha). \tag{5}$$

3. CALCULATION OF THE FUNCTIONAL DERIVATIVE

Clearly, the connection between f and B is complicated : it is non linear and, moreover, it cannot be exhibited in closed form since B may be computed only by solving integral or differential equations. So, as a first step, we will restrict our attention to small variations δf and δB. To first order in δf, there exists a linear relation between these two quantities, and we will exhibit this relation.

Let $u'(x,y)$ be the total field created by the incident plane wave $\exp i(\alpha x-\beta y)$ in presence of the diffracting surface S' ; in order to simplify the demonstration, we suppose for a moment that S' is everywhere above S (i.e. $\delta f(x) \geqslant 0\ \forall x$). Let $B'(\alpha')$ be the corresponding diffracted amplitude ; apply the lemma of section 2, with $v = u'$, $w = u^*$ and $S_0 = S'$; the right hand member is :

$$-4i\pi \int_{-\infty}^{+\infty} \beta'(\alpha')[U'^{+}(\alpha')U^{*-}(-\alpha') - U'^{-}(\alpha')U^{*+}(-\alpha')]d\alpha' =$$

$$= -4i\pi(\beta^* B'(-\alpha^*) - \beta B^*(\alpha)) = -4i\pi\beta^*(B'(-\alpha^*) - B(-\alpha^*)) =$$

$$= -4i\pi\beta^*\delta B(-\alpha^*)$$

where the reciprocity relation (5) has been used. So, it is enough to evaluate the left hand member in order to obtain the expression of δB ; u' is zero on S' and we have that :

$$-4i\pi\beta^*\delta B(-\alpha^*) = -\int_{S'} u^* \frac{\partial u'}{\partial n}\, dS' .$$

This relation is rigorous. Now, to first order in δf :

$$\int_{S'} u^* \frac{\partial u'}{\partial n'}\, dS' \simeq \int_{S} \frac{\partial u^*}{\partial y} . \partial f . \frac{\partial u'}{\partial n'}\, dS \simeq \int_{-\infty}^{+\infty} \frac{\partial u}{\partial n}\frac{\partial u^*}{\partial n}\, \delta f(x)dx$$

$$\delta B(-\alpha^*) = \frac{1}{4i\pi\beta^*}\int_{-\infty}^{+\infty} \frac{\partial u}{\partial n}\frac{\partial u^*}{\partial n}\, \partial f(x)dx . \tag{6}$$

For sake of simplicity, it has been supposed that $\delta f(x) > 0$, but the result is the same for any sufficiently regular δf : the demonstration may be achieved from the same lemma with a well chosen surface of integration.

The quantity $\frac{\partial u}{\partial n}\frac{\partial u^*}{\partial n}$ is often called the "functional derivative" of B with respect to f, and is denoted by $\frac{\delta B}{\delta f}$. We have obtained a very compact and elegant expression for $\delta B/\delta f$: it involves only the normal derivatives of the fields u (initial direct problem, Fig. 1A) and u* (adjoint direct problem, Fig. 1B). This is a general feature of all the diffraction problems we have studied : the functional derivative of the diffraction pattern with respect to the profile of the scatterer may always be expressed very simply in terms of quantities related either to the initial direct problem or to an adjoint direct problem. When the source, in the initial problem, is a plane wave and the diffraction pattern is observed at infinity, then the adjoint problem corresponds to the same scatterer illuminated by a plane wave propagating in the direction of observation (Fig. 1B). Functional derivatives have already been calculated for 2-D and 3-D scatterers of any conductivity [3]. In the case of diffraction gratings, their expressions can be found in [4] [5].

The functional derivative of B with respect to f contains all the necessary information on the "local" relationship between B and f : in particular, it yields very simply the partial derivative of B with respect to any parameter linked with the profile. Suppose for instance that the profile f is restricted to a class of functions depending upon m parameters $p_1, p_2, \dots p_m$ and described by the known function F :

$$f(x) = F(x, p_1, p_2, \dots p_m) .$$

A variation δf thus takes the form :

$$\delta f(x) = \sum_{i=1}^{m} \frac{\partial F}{\partial p_i} \partial p_i$$

and the corresponding variations δB is :

$$\delta B(-\alpha^*) = \sum_{i=1}^{m} \delta p_i \int_{-\infty}^{+\infty} \frac{\partial u}{\partial n} \frac{\partial u^*}{\partial n} \frac{\partial F}{\partial p_i} dx$$

which means that :

$$\frac{\partial B}{\partial p_i} = \int_{-\infty}^{+\infty} \frac{\partial u}{\partial n} \frac{\partial u^*}{\partial n} \frac{\partial F}{\partial p_i} dx .$$

The functional derivative is simply the generalisation of the concept of gradient to the case of infinitely dimensional spaces. Eq.(6) has a straightforward simple consequence : when f = 0, $\partial u/\partial n$ and $\partial u^*/\partial n$ are analytically known since S reduces to a plane ; moreover, $B(\alpha')$ is zero when f = 0, and thus δB and δf may be confused with B and f. Eq.(6) becomes :

$$B(\alpha') = \frac{i\beta}{\pi} \int_a^b \exp[i(\alpha - \alpha')x] f(x) dx$$

which shows that the diffracted amplitude is simply the Fourier transform of the profile. Of course, this relation holds only for shallow surfaces !

4. CONJUGATE GRADIENT AND NEWTON METHODS

Eq.(6) is a linear relation, valid only to first order approximation in δf. However, it may be employed to construct rigorous numerical algorithms capable to solve inverse scattering problems. In order to show this fact, this subsection deals with the following particular problem : given a diffracted amplitude $B_s(\alpha')$ $(\alpha_1 \leqslant \alpha' \leqslant \alpha_2)$, find the corresponding profile $f_s(x)$ $(a \leqslant x \leqslant b)$.

4.1. Newton-Kantorovitch algorithm

Let us start from a profile $f^{(0)}$ which, for example, may be a rough estimate of the solution f_s. The corresponding amplitude $B^{(0)}$ may be easily calculated with the help of the computer code written for the direct problem. We define :

$$\delta B_s^{(0)} = B_s - B^{(0)}$$

and let $\delta f^{(0)}$ be the solution of the integral equation :

$$\delta B_s^{(0)}(\alpha') = \int_a^b \frac{\partial u}{\partial n}(x)\, \frac{\partial u^*}{\partial n}(x,\alpha')\, \delta f^{(0)}(x)\, dx \qquad (7)$$

where $\partial u/\partial n$, $\partial u^*/\partial n$ are relative to the known profile $f^{(0)}$ and thus are easily calculable with the help of the direct code. Since the actual profile perturbation $f - f^{(0)}$ and $\delta B_s^{(0)}$ verify approximatively eq.(6) - only approximatively because (6) is exact to first order in δf - $\delta f^{(0)}$ is expected to be approximatively equal to $f - f^{(0)}$. Thus the function $f^{(1)}$ defined as :

$$f^{(1)} = f^{(0)} + \delta f^{(0)}$$

should be a better approximation of f_s than $f^{(0)}$. Starting then from $f^{(1)}$, the same calculation can be made and yields a new approximation $f^{(2)}$, and so on. An iterative process is so generated, which, when it converges, yields a numerically rigorous solution to the inverse scattering problem.

This algorithm is the analogue in terms of functions of the well-known Newton Raphson algorithm often used to find the zeros of a real function. It is called the "Newton Kantorovitch" algorithm, because L.Y. Kantorovitch first exhibited conditions of convergence [6]. The N.K. algorithm has been successfully applied by the author to inverse scattering problems associated with layered media [7], gratings [8] [9] and cylinders [10].

It is worth noting that the resolution of the integral equation of the first kind (7) is not a simple problem, because the solution $\delta f^{(0)}$ is highly sensitive to rounding or experimental errors on $\delta B_s^{(0)}$. It is necessary to make use of so-called "regularization techniques". This "instability" is highly significant and has a clear physical meaning.

4.2. Conjugate gradient algorithm

This type of algorithm is often used in Numerical Analysis. In the framework of optimization theory, the inverse problem is formulated as the minimization of a given real functional J(f) called the "cost function". The choice of J expresses the desired features of the required solutions. In the particular problem defined above, one may choose :

$$J(f) = \int_{\alpha_1}^{\alpha_2} |B(\alpha') - B_s(\alpha')|^2 \, d\alpha' \tag{8}$$

the first point is to calculate the "gradient" of J, i.e. the functional derivatife ∂J/∂f. The calculation is straighforward from eqs.(6) (8) :

$$\frac{\partial J}{\partial f} = 2 \operatorname{Re} \{ \int_{\alpha_1}^{\alpha_2} [\overline{B}(\alpha') - \overline{B}_s(\alpha')] \frac{\partial u}{\partial n}(x) \frac{\partial u^*}{\partial n}(-\alpha', x) \, d\alpha' \}$$

Then J may be easily minimized using a conjugate gradient method, such as the Fletcher-Powell algorithm. It is not the purpose of the present paper to describe this kind of methods precisely ; for more details see for instance [11]. The main advantage of this method is to be simpler to carry out than the N.K. algorithm, and to require less memory storage. However, its convergence is much slowier. Apparently no regularization should be needed. In fact, numerical computations have shown that the cost function (8) often leads to oscillatory reconstructed profiles, and this is not astonishing : the instability of the inverse problem is a fundamental feature and cannot be avoided so easily! It is necessary to re-introduce a kind of regularization by choosing another type of cost function, such as :

$$J(f) = \int_{\alpha_1}^{\alpha_2} |B(\alpha') - B_s(\alpha')|^2 \, d\alpha' + r \int_a^b \left(\frac{df}{dx}\right)^2 dx \,.$$

In this context, the physical significance of the regularizing term is not as clear as in the Newton algorithm.

5. INSTABILITY

As is now well known, the resolution of integral equations of the first kind as (7) constitutes an ill-posed problem in Hadamard's sense ; i.e. an arbitrarily small perturbation on B_s and thus on $\delta B_s^{(0)}$ may correspond to an arbitrarily large perturbation on $\delta f^{(0)}$.

Consider for instance the following variation :

$$\delta f^{(0)}(x) = A \sin \omega x \qquad a \leqslant x \leqslant b$$

it is easily seen that the corresponding variation $\delta B_s^{(0)}$ tends towards zero when ω tends towards infinite, and thus the property holds for sufficiently large values of ω and A. Note that in this example, $\delta f^{(0)}$ oscillates very quickly ; in other words, high frequency oscillations of the profile f have little influence on the scattering pattern B, and the device acts as a low-pass filter. This property is particularly clear for shallow surfaces : in this case, the relation between B and f is simply a Fourier transform, and the inverse problem reduces to the reconstruction of f knowing its Fourier transform on a finite bandwith. For more details on this classical problem, see for instance [12]. The restoration of the stability may be achieved with the help of regularization methods [13]. The problem of solving (7) is replaced by a variational problem where physical constraints are incorporated, and the solution is computed taking into account for instance experimental errors on the data.

6. A FEW RESULTS

Figure 2 shows the reconstruction of a locally deformed surface. The scattered amplitude is known for $-60° < \theta < 60°$, the angle of incidence is 30°, and $\lambda = 1$; line 1 is the true profile, line 2 the initial estimation, line 3 the reconstructed profile.

Figure 3 presents the reconstruction of the profile of a manufactured grating with 1800 gr/mm. The efficiency curve in the -1 order has been measured in the range $0.485\ \mu < \lambda < 0.600\ \mu$ and the algorithm described above has been used, in conjunction with well-checked rigorous integral theories of gratings [14] full line is the reconstructed profile from the efficiency in TM polarization (magnetic field parallel to the grooves) and the dotted one corresponds to the efficiency curve in T.E. polarization (electric field parallel to the grooves). Both of them are in very good agreement with the profile found by the manufacturer with electron microscopy.

7. CONCLUSION

We have described a method capable of solving various inverse problems numerically, which provides a theoretical explanation of the fundamental instability of the inverse problems. This method has proven to be successful in many problems of inverse scattering in electromagnetics. Further developpements seem to be straighforward, especially in the case of three dimensional

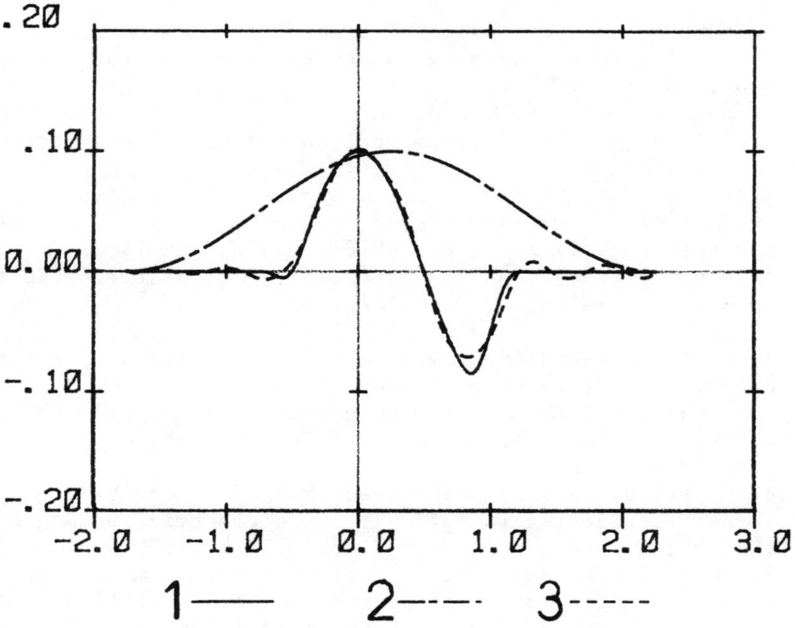

Fig. 2 : Reconstruction of the profile of a locally deformed surface from its scattered amplitude.

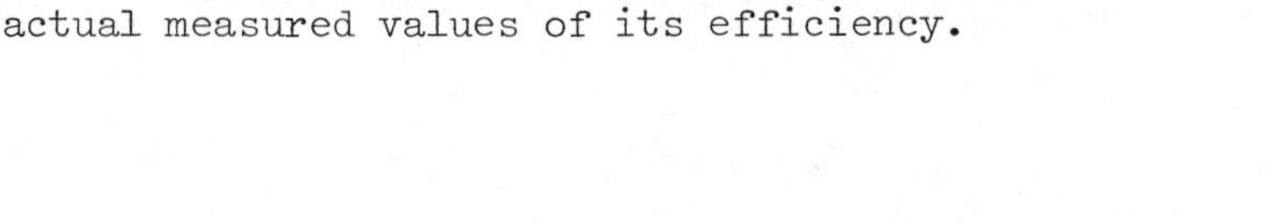

Fig. 3 : Reconstruction of the groove profile of a grating from actual measured values of its efficiency.

scatterers where the required functional derivatives have already been calculated [3]. So, the algorithm described here is general and convenient, and it could be of great help for the numerical solution of many inverse problems.

References

[1] D. Maystre, "Electromagnetic scattering from perfectly conducting rough surfaces in the resonance region", to be published in IEEE Trans. Antennas Propagat.

[2] D. Maystre, O. Mata Mendez and A. Roger, "A new electromagnetic theory for scattering from shallow rough surfaces", submitted to Opt. Acta.

[3] A. Roger, "Reciprocity theorem applied to the computation of functional derivatives of the scattering matrix", Electromagnetics, vol. 2, n° 1, pp. 69-83, Jan.-March 1982.

[4] A. Roger, "Generalised reciprocity relations for perfectly conducting gratings", Opt. Acta, vol. 29, n° 10, pp. 1427-1439, 1982.

[5] A. Roger, "Generalised reciprocity for gratings of finite conductivity", to be published in Opt. Acta.

[6] L.Y. Kantorovitch, "On Newton's method for functional equations", Dokl. Akad. Nauk. SSSR, vol. 59, pp. 1237-1240, Jan. 1948.

[7] A. Roger, D. Maystre and M. Cadilhac, "On a problem of inverse scattering in optics : the dielectric inhomogeneous medium", J. Opt., vol. 9, pp. 83-90, Feb. 1978.

[8] A. Roger and D. Maystre, "Inverse scattering method in electromagnetic optics : application to diffraction gratings", J. Opt. Soc. Amer., vol. 70, pp. 1483-1495, dec. 1980.

[9] A. Roger and M. Breidne, "Grating profile reconstruction by an inverse scattering method", Opt. Commun., vol. 35, pp. 299-301, Dec. 1980.

[10] A. Roger, "Newton Kantorovitch Algorithm applied to an Electromagnetic Inverse Problem", IEEE Trans. Antennas Propagat., vol. 29, n° 2, pp. 232-238, March 1981.

[11] R. Fletcher and J.D. Powell, "A rapidly convergent descent method for minimization", The Computer Journal, vol. 6, pp. 163-168, 1963.

[12] M. Bertero, C. De Mol, and G. Viano, "Restoration of optical objects using regularization", Opt. Lett., vol. 3, pp. 51-53, 1978.

[13] A.N. Tikhonov and V.Y. Arsenin, Solutions of Ill-Posed Problems. New York : Winston-Wiley, 1977.

[14] R. Petit, Ed., Electromagnetic Theory of Gratings, New York : Springer, 1980.

I.6 (IS.5)

DISSIPATIVE INVERSE PROBLEMS IN THE TIME DOMAIN

J.P. Corones*, M.E. Davison*, R.J. Krueger**

*Applied Mathematical Sciences, Ames Laboratory, Iowa State University, Ames, IA 50011 USA

**Dept. of Mathematics, University of Nebraska, Lincoln, NE 68588 USA

ABSTRACT. This work focuses on the time domain electromagnetic inverse problem for a one dimensional inhomogeneous dissipative medium. In practice, reflection data from a normally incident plane wave is not sufficient for reconstructing both the permittivity and conductivity profiles. However, if one of these profiles is known a-priori, then the other can be reconstructed. A straightforward technique for doing this is shown. It is based on a solution of the direct scattering problem which is derived from a wave splitting approach coupled with an invariant imbedding analysis of the scattering matrix. The technique is summarized in an integro-differential equation for the kernel of the reflection operator. Numerical examples are given which study the degradation of the reconstruction when the assumed profile differs from the true profile.

1. PROBLEM FORMULATION

A new approach to time domain inverse problems is presented in this paper. It is based upon a representation of the solution of the corresponding direct scattering problem. Thus, the direct problem will be formulated first.

Consider a one dimensional inhomogeneous dissipative medium occupying the region $0 < z < L$ with a homogeneous nondissipative medium outside of that region. A normally incident electromagnetic plane wave produces a field $E(z,t)$ in the medium such that

$$E_{zz} - c^{-2}(z)E_{tt} - b(z)E_t = 0 \qquad (1)$$

W.-M. Boerner et al. (eds.), Inverse Methods in Electromagnetic Imaging - Part 1, 121–130.

where

$$c^{-2}(z) = \varepsilon(z)\mu_0$$

$$b(z) = \sigma(z)\mu_0$$

and ε, σ, μ_0 are the permittivity, conductivity and (constant) magnetic permeability respectively. To simplify the analysis assume that c is continuous at $z = 0$ and $z = L$ so that $c(z) = c(0)$ for $z < 0$ and $c(z) = c(L)$ for $z > L$ and observe that $b(z) = 0$ for $z < 0$ or $z > L$. Finally, within the slab $0 < z < L$ assume c is C^1 and b is C^0. This model of a finite medium is used only for purposes of motivation. A semi-infinite model can also be considered, as will be obvious as the analysis proceeds.

A right moving incident plane wave in the region $z < 0$,

$$E^i(z,t) = f(t - z/c(0)),$$

produces reflected and transmitted fields,

$$E^r(z,t) = g(t+z/c(0)),$$

$$E^t(z,t) = h(t-z/c(L))$$

in the regions $z < 0$ and $z > L$, respectively. These fields are related by scattering operators for the slab

$$E^r(z,t) = [\tilde{R}^+(0,L)E^i(z,\cdot)](t)$$

$$E^t(z,t) = [\tilde{T}^+(0,L)E^i(z,\cdot)](t)$$

where the reflection operator is given by

$$[\tilde{R}^+(0,L)f](t) = \int_{-\infty}^{t} R^+(0,L,t-s)f(s)ds \qquad (2)$$

and the transmission operator by

$$[T^+(0,L)f](t) = [\frac{c(L)}{c(0)}]^{\frac{1}{2}} \exp[-\tfrac{1}{2}\int_0^L b(s)c(s)ds]f(t-\Delta t(0,L)) + \int_{-\infty}^{t-\Delta t(0,L)} T^+(0,L,t-s)f(s)ds, \qquad (3)$$

and

$$\Delta t(0,L) = \int_0^L \frac{1}{c(s)} \, ds. \tag{4}$$

These expressions follow from [6] or [7]. The kernels $R^+(0,L,t)$ and $T^+(0,L,t)$ are dependent on the properties of the slab, but are independent of $E^i(z,t)$. The endpoints $z = 0$ and $z = L$ are explicitly displayed because in the analysis that follows, scattering operators for subregions of the original medium will be considered.

If a left moving plane wave in the region $z > L$ impinges on the slab, then a second pair of scattering operators, $\tilde{R}^-$, $\tilde{T}^-$ is used to determine the right going reflected and left going transmitted waves. Notice that because Eq. (1) is linear, it follows by superposition that the operators $\tilde{R}^\pm$, $\tilde{T}^\pm$ can be used to describe the outcome of any scattering experiment performed on the slab.

The definitions of the scattering operators presumes a concept of right and left moving waves. This gives no difficulty because the medium is homogeneous outside of the slab. Since the notion of a scattering operator is to be extended to subregions of the original slab, it is also necessary to extend the idea of right and left going waves to the case in which the properties of the medium are nonconstant. This is done via a splitting method.

2. SPLITTING METHOD FOR A HOMOGENEOUS MEDIUM

In order to motivate the general notion of a splitting, consider a homogeneous, nondissipative medium. The field satisfies the wave equation

$$E_{zz} - c_0^{-2} E_{tt} = 0 \tag{5}$$

where c_0 is constant. Thus, E can be split into a sum of right and left moving waves,

$$\begin{aligned} E(z,t) &= F(z-c_0 t) + G(z+c_0 t) \\ &= E^+(z,t) + E^-(z,t). \end{aligned} \tag{6}$$

Now consider the problem of rewriting Eq. (5) as a system

involving $E^{\pm}$. This is straightforward, but a general mechanism for doing it will be displayed. First, rewrite Eq. (5) as

$$\frac{\partial}{\partial z}\begin{pmatrix} E \\ E_z \end{pmatrix} = \begin{pmatrix} 0 & 1 \\ (\partial_t/c_0)^2 & 0 \end{pmatrix}\begin{pmatrix} E \\ E_z \end{pmatrix} \equiv D\begin{pmatrix} E \\ E_z \end{pmatrix} \qquad (7)$$

and notice that Eq. (6) implies

$$\begin{pmatrix} E^+ \\ E^- \end{pmatrix} = \tfrac{1}{2}\begin{pmatrix} 1 & -c_0\partial_t^{-1} \\ 1 & c_0\partial_t^{-1} \end{pmatrix}\begin{pmatrix} E \\ E_z \end{pmatrix} \equiv T\begin{pmatrix} E \\ E_z \end{pmatrix} \qquad (8)$$

Combining (7) and (8) it follows that

$$\begin{pmatrix} E^+ \\ E^- \end{pmatrix} = [-T\frac{\partial T^{-1}}{\partial z} + TDT^{-1}]\begin{pmatrix} E^+ \\ E^- \end{pmatrix} .$$

Since T^{-1} is independent of z, this becomes

$$\begin{pmatrix} E^+ \\ E^- \end{pmatrix} = \begin{pmatrix} -\partial_t/c_0 & 0 \\ 0 & \partial_t/c_0 \end{pmatrix}\begin{pmatrix} E^+ \\ E^- \end{pmatrix}$$

Notice that in this case, T diagonalizes D.

3. SPLITTING METHOD FOR A NONHOMOGENEOUS DISSIPATIVE MEDIUM

Now consider Eq. (1) instead of Eq. (5). This can be written as

$$\begin{pmatrix} E \\ E_z \end{pmatrix} = \begin{pmatrix} 0 & 1 \\ (\partial_t/c)^2+b\partial_t & 0 \end{pmatrix}\begin{pmatrix} E \\ E_z \end{pmatrix} \equiv D\begin{pmatrix} E \\ E_z \end{pmatrix} . \qquad (9)$$

Using Eq. (8) as a guide, choose a splitting of E(z,t) according to

$$\begin{pmatrix} E^+ \\ E^- \end{pmatrix} = \frac{1}{2}\begin{pmatrix} 1 & -c(z)\partial_t^{-1} \\ 1 & c(z)\partial_t^{-1} \end{pmatrix}\begin{pmatrix} E \\ E_z \end{pmatrix} \equiv T\begin{pmatrix} E \\ E_z \end{pmatrix} \qquad (10)$$

Notice that this reduces to (8) in the region $z < 0$. Combining (9) and (10) it follows that

$$\begin{pmatrix} E^+ \\ E^- \end{pmatrix} = \begin{pmatrix} -(\partial_t/c)-\tfrac{1}{2}(bc-c'/c) & -\tfrac{1}{2}(bc+c'/c) \\ \tfrac{1}{2}(bc-c'/c) & (\partial_t/c)+\tfrac{1}{2}(bc+c'/c) \end{pmatrix} \begin{pmatrix} E^+ \\ E^- \end{pmatrix}$$

$$= \begin{pmatrix} \tilde{\alpha}(z) & \tilde{\beta}(z) \\ \tilde{\gamma}(z) & \tilde{\delta}(z) \end{pmatrix} \begin{pmatrix} E^+ \\ E^- \end{pmatrix} \tag{11}$$

4. SCATTERING MATRIX EQUATIONS

For arbitrary x and y consider the portion of the medium in the interval $x \leq z \leq y$, where $0 \leq x, y \leq L$. Assume that outside of this subinterval the medium is replaced with a nondissipative, homogeneous medium with $c(z) = c(x)$ for $z < x$, $c(z) = c(y)$ for $z > y$. Waves incident on this subinterval are described by $E^+(x,t)$ and $E^-(y,t)$ and the resulting scattered waves by $E^-(x,t)$ and $E^+(y,t)$. Hence, the scattering matrix $S(x,y)$ for the region $x \leq z \leq y$ takes the form

$$\begin{pmatrix} E^+(y,\cdot) \\ E^-(x,\cdot) \end{pmatrix} = \begin{pmatrix} \tilde{T}^+(x,y) & \tilde{R}^-(x,y) \\ \tilde{R}^+(x,y) & \tilde{T}^-(x,y) \end{pmatrix} \begin{pmatrix} E^+(x,\cdot) \\ E^-(y,\cdot) \end{pmatrix}$$

$$= \tilde{S}(x,y) \begin{pmatrix} E^+(x,\cdot) \\ E^-(y,\cdot) \end{pmatrix} \tag{12}$$

The elements of $\tilde{S}(x,y)$ are the scattering operators for the region $x \leq z \leq y$. They are defined as in (2) and (3) with x,y replacing 0,L.

A formal procedure for deriving differential equations for the scattering operators is now given. Evaluate (12) at t and differentiate with respect to x, and use (11) evaluated at x to obtain

$$\frac{\partial \tilde{S}(x,y)}{\partial x} = -\begin{pmatrix} \tilde{T}^+(x,y) & 0 \\ \tilde{R}^+(x,y) & I \end{pmatrix} \begin{pmatrix} \tilde{\alpha}(x) & \tilde{\beta}(x) \\ -\tilde{\gamma}(x) & -\tilde{\delta}(x) \end{pmatrix} \begin{pmatrix} \tilde{I} & \tilde{0} \\ \tilde{R}^+(x,y) & \tilde{T}^-(x,y) \end{pmatrix}$$

In particular, the equation for $\tilde{R}^+(x,y)$ is

$$\frac{\partial \tilde{R}^+(x,y)}{\partial x} = \tilde{\gamma}(x) + \tilde{\delta}(x)R^+ - \tilde{R}^+\tilde{\alpha}(x) - \tilde{R}^+\tilde{\beta}(x)R^+ \tag{13}$$

An alternate method for obtaining (13) is given in [5]. It uses an invariant imbedding [1] approach to determine the behavior of the scattering operators by appending a portion of the medium of width Δx to the left hand edge of the region $x \leq z \leq y$ and computing the change in the scattering matrix; i.e.,

$$\lim_{\Delta x \to 0} - \frac{1}{\Delta x} [\tilde{S}(x-\Delta x,y) - \tilde{S}(x,y)].$$

This approach conveys the idea that the scattering matrix $\tilde{S}(0,L)$ for the physical medium is the *-product [8] of scattering matrices for thin, contiguous subregions which can be thought of as comprising the original medium.

Equation (13) can be rewritten in terms of the kernel, $R^+(x,y,t)$, of the convolution operator $\tilde{R}^+(x,y)$. This yields

$$\frac{\partial R^+(x,y,t)}{\partial x} - \frac{2}{c}\frac{\partial R^+(x,y,t)}{\partial t} = bcR^+ + \tfrac{1}{2}(bc+c'/c)\int_0^t R^+(x,y,t-s)R^+(x,y,s)ds$$

$$0 < x < y, \quad t > 0 \tag{14}$$

where b,c and c' are evaluated at z=x. From this same calculation, it also follows that

$$R^+(x,y,0^+) = -\tfrac{1}{4}(bc^2-c')\Big|_{z=x} \tag{15}$$

$$R^+(y,y,t) = 0, \quad t > 0 \tag{16}$$

Equations for the remaining scattering kernels, $R^-(x,y,t)$, $T^{\pm}(x,y,t)$, can be found in a similar way. They are not pertinent to the present discussion and consequently are not written here. The interested reader can readily obtain them from [2] or [5].

Notice that Eq. (14) and (15) are not dependent on the finite thickness of the medium. Consequently, these same equations apply to a semi-infinite ($L=\infty$) medium. It follows from causality that

$$R^+(0,\infty,t) = R^+(0,y,t) \text{ for } 0 < t < 2\Delta t(0,y).$$

5. SOLUTION PROCEDURE

Both the direct and inverse scattering problem can be solved via Eq. (14) - (16). Figure 1 illustrates the space-time region in which these equations are to be considered.

In the direct problem, $c(z)$ and $b(z)$ are given for $0 < z < L$ and $R^+(0,L,t)$ is to be found for $t > 0$. Thus, set $y = L$ in Eq. (14) - (16). The integro-differential equation (14)

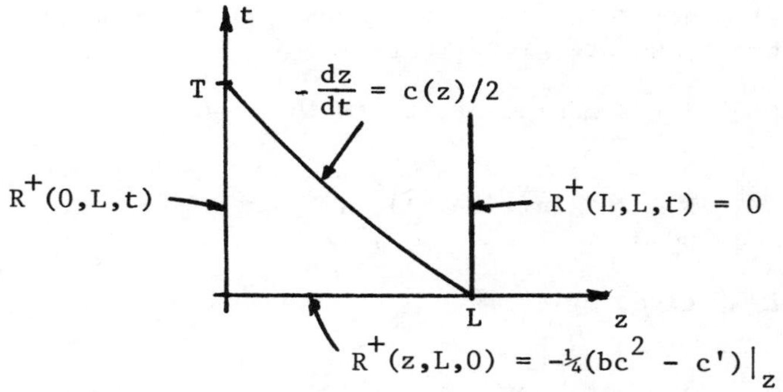

Fig. 1. Initial and boundary values for $R^+(z,L,t)$.

then propagates the initial data (15) and the boundary data (16) to the line z = 0, thus solving the direct problem.

In the inverse problem $R^+(0,L,t)$ is known for $0 < t < T$ where T is one round trip travel time through the medium. In the situation considered here, either c(z) or b(z) is also assumed to be known, and the problem is to determine the remaining function. To do this, Eq. (14) is used to propagate the boundary data $R^+(0,L,t)$ onto the line t = 0 at which point the initial condition (15) and known function b(z) (or c(z)) are used to determine c(z) (or b(z)).

Details regarding the numerical implementation of these procedures are given in [3].

6. EXAMPLES

The following examples were generated using the inversion procedure described above. In these examples, it is assumed that the relative permittivity ε_r or conductivity σ is only known approximately. Using an approximation σ_A of the true σ in the inversion procedure results in a reconstructed permittivity ε_r^* which differs from the true permittivity ε_r, with the accuracy of the reconstruction generally decreasing as the accuracy of σ_A decreases. Roughly speaking, a trade-off is occurring between permittivity gradients and energy dissipation.

If σ_A is set equal to the true σ in the inversion algorithm, then ε_r^* is found to be indistinguishable from the true ε_r in the examples that follow. In fact, the plots of $\varepsilon_{r,TRUE}$ shown below were obtained by setting $\sigma_A = \sigma_{TRUE}$ and carrying out the inversion (and conversly for the plot of σ_{TRUE}).

Some preliminary studies on the influence of noisy data in the inversion process are given in [4].

Example 1: Consider a medium of length 10 m. with

$$\varepsilon_r(z) = \begin{cases} 50 & , z < .25 \\ 75 + 25\sin(2(z-.5)), & .25 < z < .75 \\ 100 & , .z > .75, \end{cases}$$

and σ (in mho/m) given by

$$\sigma(z) = \begin{cases} .02 & , z < .25 \\ .02 - .039\ (z-.25) & , .25 < z < .75 \\ .0005 & , .z > .75. \end{cases}$$

Figure 2 shows ε_r^* for various approximations σ_A of σ. (The data $R^+(0,L,t)$ is the same for all cases.) Notice that over- and underestimating σ in the thin region of relatively high conductivity has considerable influence on ε_r^*. This in turn affects the apparent depth of the medium.

Example 2: Again assume L = 10 m. Let ε_r be given by

$$\varepsilon_r(z) = 50 + kz(10-z), \quad 0 < z < 10 \tag{17}$$

with k=6 and let σ be constant, $\sigma \equiv 0.01$ mho/m. Figure 3 shows ε_r^* for several σ_A's. Observe that even a 2% error in the approximation of σ can produce significant errors in the permittivity profile.

Using the same scattering data, reconstructed conductivity profiles are shown in Fig. 4. The assumed permittivity profiles are given by Eq. (17) for various values of k.

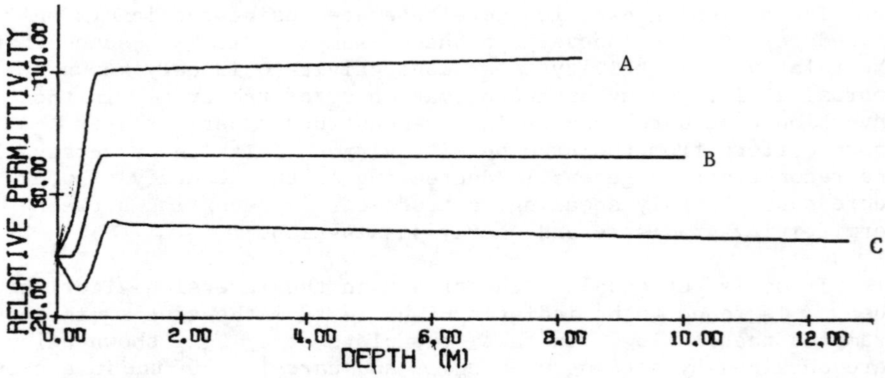

Fig. 2. Reconstructed permittivity profiles. A. $\sigma_A = 0.01$ in the thin region. B. $\sigma_A = \sigma_{TRUE}$ C. $\sigma_A = 0.03$ in the thin region.

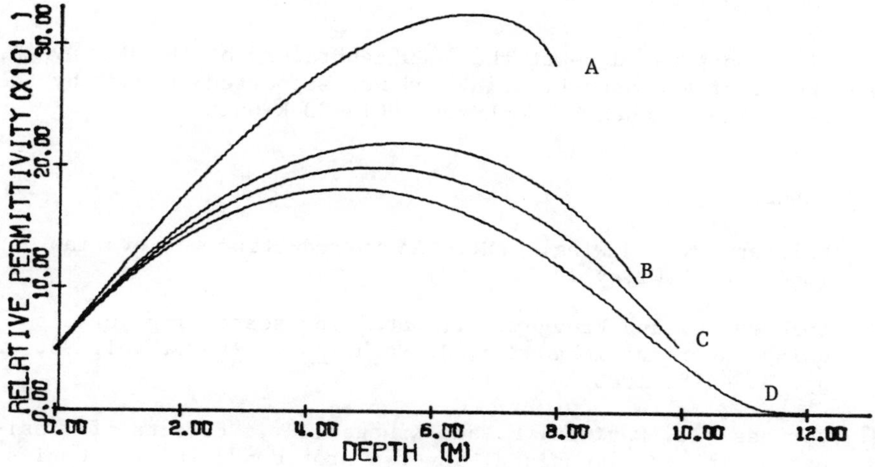

Fig. 3. Reconstructed permittivity profiles for example 2.
A. $\sigma_A = 0.009$ B. $\sigma_A = 0.0098$ C. $\sigma_A = \sigma_{TRUE} = 0.01$
D. $\sigma_A = 0.0102$.

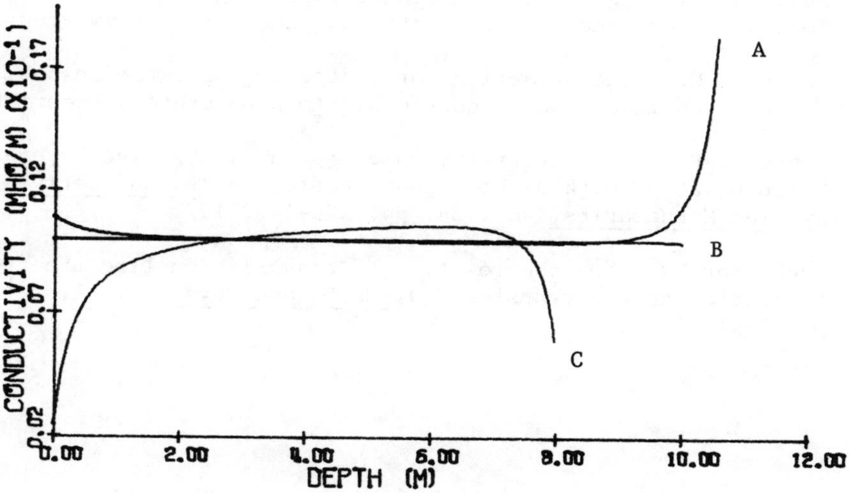

Fig. 4. Reconstructed conductivity profiles for example 2.
A. $k = 5.5$ B. $k = k_{TRUE}$ C. $k = 10.0$.

ACKNOWLEDGEMENTS

This work was done at the Ames Laboratory of the U.S.D.O.E. and the U. of Nebraska-Lincoln, and was supported in part by the Office of Naval Research Contract N0014-83-K-0038.

REFERENCES

1. Bellman, R. and Wing, G.M., "An introduction to invariant imbedding," Wiley 1975.

2. Corones, J. and Krueger, R., Obtaining scattering kernels using invariant imbedding, J. Math. Anal. Appl., Vol. 95, pp. 393-415. 1983.

3. Corones, J., Davison, M. and Krueger, R., "Effects of dissipation in one-dimensional inverse problems", Inverse Optics, A.J. Devaney, Editor, Proc. SPIE 413, pp. 107-114. 1983.

4. Corones, J., Davison, M. and Krueger, R., Direct and inverse scattering in the time domain via invariant imbedding equations, J. Acoustical Soc. Am., to appear.

5. Davison, M., A general approach to splitting and invariant imbedding for linear wave equations. Ames Laboratory Preprint.

6. Davison, M., Bremmer series for scattering operators associated with linear wave equations. Ames Laboratory Preprint.

7. Krueger, R. J., An inverse problem for a dissipative hyperbolic equation with discontinuous coefficients, Quarterly of Applied Mathematics, Vol. 34, pp. 129-147. 1976.

8. Redheffer, R., "On the relation of transmission-line theory to scattering and transfer," J. Math. and Phys. Vol 41, pp. 1-41. 1962.

I.7 (IM.3)

APPLICATIONS OF ALMOST-PERIODIC FUNCTIONS TO INVERSE SCATTERING THEORY

A.K. Jordan*, S. Ahn* and D.L. Jaggard**
*Naval Research Laboratory
Washington, DC 20375

**University of Pennsylvania
Philadelphia, PA 19104

ABSTRACT. Inverse scattering theory is used to reconstruct dielectric profiles that are almost-periodic functions. If the scattering data can be represented as a rational function of the wave number, then it is possible to obtain solutions to the Gelfand-Levitan integral equation of inverse scattering theory that will be, in general, almost-periodic functions. The dielectric profile function will also be an almost-periodic function, for which the number and strengths of the almost-periodic "tones" can be related to the pole-zero configuration of the scattering data. An example is considered where the complex reflection coefficient r(k) is represented as a rational function of the wave number k with four poles and two zeros.

1. INTRODUCTION

Inverse scattering theory includes the many different analytical methods that can be used to interpret electromagnetic and optical wave scattering data. Despite their diversity of applications, these methods are related by their basic theoretical structures. The mathematical methods used are typically rigorous and "exact" for simple and idealized physical models; these theories are useful for providing basic understanding of physical phenomena. It is possible to obtain knowledge about a class of complicated systems by extending an "exact" inversion method to include systems that can be represented by almost-periodic functions. These functions have been used in direct scattering theories for inhomogeneous dielectric media, integrated optical multilayers, and heterogeneous dielectric media, that exhibit stochastic properties.

W.-M. Boerner et al. (eds.), Inverse Methods in Electromagnetic Imaging – Part 1, 131–140.

As Sabatier [1,2] notes, the word "exact" refers to the fact that the inversion methods to be used are, in principle, not approximations. A one-dimensional physical model is used here, so that the inverse problem is properly termed "profile inversion"; this idealization will allow us to focus on the mathematical structure of the inverse scattering problem and its solution.

The direct scattering problem can be studied formally as a transformation or mapping M between a set C of theoretical parameters and a set E of experimental variables [1,2]. The inverse scattering problem can be studied in terms of the mapping M^{-1} between E and C; this study usually begins by considering a subset E' and its mapping into a subset C'.

In the physical model of Fig. 1, we have assumed that the plane-polarized electromagnetic field is normally incident on a semi-infinite inhomogeneous region whose permittivity $\varepsilon(k,x)$ is assumed to obey the dispersion relation

$$\varepsilon(k,x) = \begin{cases} \varepsilon_o\left(1 - \frac{1}{k^2}q(x)\right), x\geq 0 \\ \varepsilon_o \end{cases}$$

where ε_o is the permittivity of free space.
Here the subset E' consists of the reflection coefficient r(k) which is an analytic function of the wavenumber k; the subset C' consists of q(x), the profile function.

The inverse problem is to determine q(x) from r(k); the mapping M^{-1} is the "exact" method developed by the Soviet mathematicians Gel'fand, Levitan and Marchenko, [8,9] circa 1950, for use in quantum mechanical problems. Subsequent work extended the availability of these results and expanded the theory to include the electromagnetic problem. Of particular interest is the treatment by Kay [10] in which several examples display the usefulness of the Gel'fand-Levitan-Marchenko theory. Much of the initial work [10,11,12] emphasized exact, closed-form solutions to the problem when the scattered data were represented by rational functions. Numerical and approximate [13,15] solutions provided additional understanding of this problem.

2. EXTENSION OF C' TO INCLUDE ALMOST-PERIODIC FUNCTIONS

We now explore the feasibility of extending the subset C' to include "almost periodic functions". For our purpose, we will consider almost periodic functions as defined by Bohr [3] and Besicovitch [4], that is, as functions which possess a discrete or line spectrum. This implies that such a function f(x) can be written in a generalized Fourier series

$$f(x) = \sum_n a_n e^{ik_n x}, \tag{1}$$

where k_n is the wavenumber or spatial frequency of the nth harmonic and a_n is its amplitude. (Often we will refer to k_n as the tone frequency and a_n as the tone strength.) This equation can be inverted to produce the harmonic amplitudes

$$a_n = \lim_{X\to\infty} \int_{-X}^{X} f(x) e^{-ik_n x} dx \tag{2}$$

These functions have been applied to electromagnetic wave propagation in inhomogeneous regions by Jaggard [5,6,7], where the resemblance to random wave propagation is apparent. Furthermore, almost-periodic functions form a large class of examples of entire functions of exponential type [14]. Clearly, (1) represents a periodic function if only one tone is present or if all of the tone frequencies are commensurable. In this case (1) and (2) become the usual expressions for the Fourier series and Fourier inversion formula, respectively. Although almost periodic functions do not possess a period, as noted above, they do possess an almost period or translation number τ such that

$$|f(x+\tau) - f(x)| < \sigma , \tag{3}$$

where $\tau = \tau(\sigma)$. This relation provides an alternative definition to the more restrictive one expressed by eq. (1). Another property of interest is the Parseval relation expressed as

$$\lim_{X\to\infty} \frac{1}{2X} \int_{-X}^{X} |f(x)|^2 dx = \sum_n |a_n|^2 . \tag{4}$$

These and other characteristics can be shown from the defining equations (1) and (2) and can be found in the literature [3,4].

If the potential function q(x) is almost periodic, an approximate theory can be developed [4,5,6] for the reflection coefficient. For example, the three-tone case described by

$$q(x) = \begin{cases} -\frac{\kappa^2\eta}{4}\left[\frac{m}{2}\cos\{(\kappa+\Delta)x\} + \cos\{\kappa x\} + \frac{m}{2}\cos\{(\kappa-\Delta)x\}\right], & 0\leq x\leq \ell \\ 0 , \quad x < 0 , \; x > \ell \end{cases} \tag{5}$$

This function represents a modulated periodic structure with carrier wavenumber κ and side-band wavenumbers $\kappa \pm \Delta$; clearly Δ is the tone spacing. Here η is the relative amplitude of the carrier and m is the modulation index. By assuming an appropriate form for the electric field amplitude,

$$u(k,x)=F_o(x)e^{i\frac{\kappa}{2}x} + B_-(x)e^{-i(\frac{\kappa}{2}-\Delta)x} + B_o(x)e^{-i\frac{\kappa}{2}x} + B_+(x)e^{-i(\frac{\kappa}{2}+\Delta)x} \tag{6}$$

in the vicinity of the primary Bragg resonance defined by

$$k = \kappa/2 \tag{7}$$

it is possible to obtain a Riccati differential equation for the reflection matrix $\underline{\underline{R}}$:

$$\underline{\underline{R}}' = i\underline{\chi} + i\underline{\underline{\delta_F}} \cdot \underline{\underline{R}} + i\underline{\underline{R}} \cdot \underline{\underline{\delta_B}} + i(\underline{\underline{R}}\underline{\chi})^{\dagger}\underline{\underline{R}}, \tag{8}$$

where the dagger indicates Hermitean transpose and the initial condition is

$$\underline{\underline{R}}\ (0) = [\ 0 \quad 0 \quad 0\ ]. \tag{9}$$

This formulation is amenable to numerical solution and can be used for any number of tones. A result for the three tone case corresponding to (5) is shown in Fig. 2. Here the scalar reflection is given by

$$\tilde{R} = \sqrt{\underline{\underline{R}}\ \underline{\underline{R}}^{\dagger}} \tag{10}$$

and is plotted for a single value of normalized coupling

$$\chi\ell = \sqrt{(\chi_o\ell)^2 + (\chi_+ + \chi_-)^2\ell^2} = 2 \tag{11}$$

as a function of normalized frequency $\tilde{f} = 2\Delta\kappa/\chi$. Note particularly the large peak at the primary Bragg resonance defined by eq. (7), or $\tilde{f} = 0$, and the secondary Bragg resonances at

$$k = (\kappa \pm \Delta)/2 \quad . \tag{12}$$

The tone spacings are $\Delta = 0, 2\chi_c, 4\chi_c, 6\chi_c, 8\chi_c$ (top to bottom) where χ_c is some critical coupling and the loss is $L\ell$ with m = 1.

2.1 Example: Widely Spaced Tones

For extremely widely spaced tones ($\Delta >> \kappa$) in which the tones act independently, the results of scalar coupled mode theory can be used in the vicinity of each Bragg resonance. For example, if an N tone medium is defined by

$$q(x) = \begin{cases} -\frac{1}{4}\sum_{n=1}^{N} \eta_i \kappa_i^2 \cos\kappa_i x, & 0 \leq x \leq \ell \\ 0, & x<0,\ x>\ell \end{cases} \tag{13}$$

where $|\kappa_i - \kappa_j| \ll \kappa_i$ or κ_j; $i,j = 1,2,\ldots N$, then the scalar reflection $\tilde{R}_i$ in the vicinity of $k = \kappa_i/2$ is given by first order coupled mode theory as

$$\tilde{R}_i \simeq \frac{i\kappa_i \ell}{(\chi_i^2 - \delta_i^2)\ell \coth\sqrt{(\chi_i^2 - \delta_i^2)}\ell - i\delta_i \ell}, \tag{14}$$

where

$$\chi_i = \eta_i \kappa_i/8 \tag{15}$$

$$\delta_i = k - \kappa_i/2 \ . \tag{16}$$

The peak value of these primary resonances is

$$|\tilde{R}_i|^{pri}_{max} = \tanh(\chi_i \ell) \ , \tag{17}$$

which can be inverted to get all the $\chi_i \ell$. The placement of the peak gives the values of the κ_i.

The criterion for the applicability of this method is that the tones are separated sufficiently so that they act independently. This occurs when the bandgaps due to each periodicity are spaced in frequency so that they do not coalesce. This condition is described by the relative balance of the periodicity amplitudes with the wavenumbers as expressed by

$$|\eta_i|^2 \ll 16|\kappa_i - \kappa_j|; i,j = 1,2,\ldots N, i\neq j \ . \tag{18}$$

2.2 Example: Closely Spaced Tones

If the tones are closely spaced so that condition (18) does not hold, then the simplified expression (14) for the reflection coefficient cannot be used. The traditional approach of iterating between the direct problem, via the matrix Riccati equation, and the experimental data leads to formidable algebra and a problem of physical interpretation. Alternatively, the inverse problem for closely spaced tones can be approached by observing that the graph of the scalar reflection $\tilde{R}$ for reflection from an almost-periodic medium with three closely spaced tones (Fig. 2, second plot, $\Delta=2\chi_c L\ell = 0.4$) can be approximated by a reflection coefficient that is a rational function of the wave number k,

$$r(k) = r_o \frac{(k-\mu_1)(k-\mu_2)\ldots}{(k-k_1)(k-k_2)\ldots}, \tag{19}$$

where k_1, k_2,...are the poles and μ_1, μ_2 ,...are the zeros of r(k) in the complex k-plane and r_o is a normalization constant.

Scattering data that can be represented by rational reflection coefficients belong to a class of inverse problems solvable by "exact" methods that demonstrate the mathematical structure of inverse scattering theory.

3. "EXACT" METHODS FOR INVERSE PROBLEMS

A physical description of the inverse scattering procedure [10] is obtained by using the time-dependent differential equation

$$\frac{\partial^2}{\partial x^2} U(x,t) - \frac{\partial^2}{\partial t^2} U(x,t) = q(x)\, U(x,t). \tag{20}$$

The incident plane-wave impulse, $U_{inc} = \delta\,(x - t)$, produces a reflected transient

$$R(x+t) = \frac{1}{2\pi} \int_{-\infty}^{\infty} r(k) e^{-ik(x+t)} dk - i \sum_{\nu} r_\nu e^{-ik_\nu (x+t)}, \tag{21}$$

where the r_ν in the second term are the residues at the poles, k_ν if any, of r(k) on the positive imaginary k-axis. It is possible to relate the wave amplitude, U(x,t), in the inhomogeneous region, $x \geq 0$, with the wave amplitude in the free space region, by the transformation

$$U(x,t) = U_o(x,t) + \int_{-\infty}^{x} K(x,z) U_o(z,t) dz, \tag{22}$$

where

$$U_o(x,t) = \delta(x-t) + R(x+t). \tag{23}$$

Substituting the expression for the field in the free-space, $U_o(x,t)$, gives the integral equation

$$R(x+t) + K(x,t) + \int_{-t}^{x} K(x,z) R(z+t) dz = 0, \tag{24}$$

which is Kay's version of the Gel'fand-Levitan-Marchenko integral equation and is to be solved for the function K(x,t). Conditions on K(x,t) are found by substituting the expression (22) for U(x,t) in the differential equation (20); the function $K(x,t)$ will satisfy the same differential equation as $U(x,t)$, subject to the conditions

$$K(x, -x) = 0 \tag{25}$$

and

$$\frac{d}{dx} K(x,x) = \frac{1}{2} q(x), \; x \geq 0. \tag{26}$$

From condition (26) we see that the inverse scattering problem is solved for $q(x)$ if we can solve the integral equation (24). Several exact and approximate methods of solving (24) have been reviewed by Jordan [12]; the most important for obtaining closed-form solutions useful for engineering applications is that of Kay [10], where $r(k)$ is a rational function of the wavenumber k, as represented by eq. (17). Examples of these solutions have been given by Jordan et al. [11,12] for several pole-zero configurations.

The band-pass characteristics of the almost-periodic reflection coefficient in the vicinity of the central Bragg resonance suggests that poles on the unit circle ("Butterworth poles") could be used as a first approximation. The number of minimum of $|r(k)|^2$ will be determined by the number of zeros while the characteristics of the peaks are determined by the pole locations. The energy-conservation condition

$$|r(k)|^2 \leq 1 \tag{27}$$

determines the "allowed regions" for the zeros and poles of $r(k)$ in the k-plane.

We note that the $R(x)$ defined by eq. (22) will be, in general, an almost-periodic function as defined by Bohr in equation (3), when $r(k)$ is a rational function of k. Thus that $K(x,y)$ in the Gelfand-Levitan integral equation (24), and hence $q(x)$ will also be an almost-periodic function, due to the multiplication theorem for almost-periodic functions [3].

This method for solving the integral eq. (22) will be demonstrated by using our example for closely-spaced tones. The scalar reflection coefficient can be approximated by a rational reflection coefficient

$$\tilde{R} \simeq r(k) = r_B(k) \cdot r_\delta(k), \tag{28}$$

where

$$r_B(k) = \frac{-k_1 k_2 k_3 k_4}{(k-k_1)(k-k_2)(k-k_3)(k-k_4)},$$

and

$$r_\delta(k) = \frac{(k-\mu_1)(k-\mu_2)}{\mu_1\mu_2}$$

The pole-zero configuration is chosen so that

$$k_2 = -\bar{k}_1 = c_1 - ic_2,$$
$$k_4 = -\bar{k}_3 = c_3 - ic_4,$$
$$\mu_2 = -\bar{\mu}_1 = c_3 - i\delta c_4, 0 < \delta \leq 1.$$

The fourth-order Butterworth poles [11] are given by

$$c_1 = \cos(3\pi/8),\ c_2 = \sin(3\pi/8),$$
$$c_3 = \cos(\pi/8),\ \ c_4 = \sin(\pi/8).$$

The reflection coefficient $|r(k)|$ is shown in Fig. 3. The solution of the Gelfand-Levitan integral equation proceeds as outlined in [11,12]; it is interesting to note that the characteristic function R(x), eq. (21), is related to the corresponding $R_B(x)$, for the Butterworth approximation with no zeros, by

$$R(x) = R_B(x) + \frac{1}{c_3^2+\delta^2 c_4^2}\left[R''_B(x) - 2\delta\ c_4 R'_B(x)\right] \tag{29}$$

The function K(x,y) has the form

$$K(x,y) \doteq \sum_{n=1}^{4}\left[C_n(x)e^{if_n y} + D_n(x)e^{-if_n y}\right], \tag{30}$$

where the f_n are functions of $R(x)|_{x=0}$ and its derivatives and C_n, D_n are found from the boundary conditions on K(x,y) and its derivatives. The graph of q(x) found from eq. (26) is shown in Fig. 4.

4. DISCUSSION

It has been demonstrated that when the scattering data are represented as a rational function of the wave number, then it is possible to obtain solutions to the Gelfand-Levitan integral equation of inverse scattering theory that are almost-periodic functions. The profile of the equivalent dielectric constant for inhomogeneous media can be reconstructed from the scattering data. The use of zeros in addition to poles in the analytic representation of the data provides additional degrees of freedom for the reconstruction procedure. This method may be useful to analyze heterogeneous systems, such as polydisperse distributions of dielectric particles.

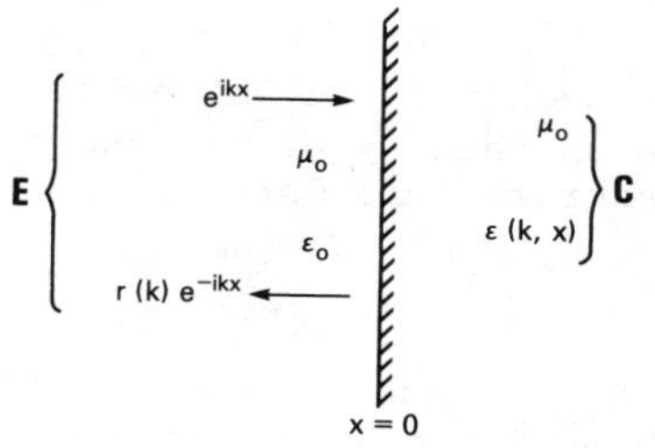

Fig. 1. Idealized physical model.

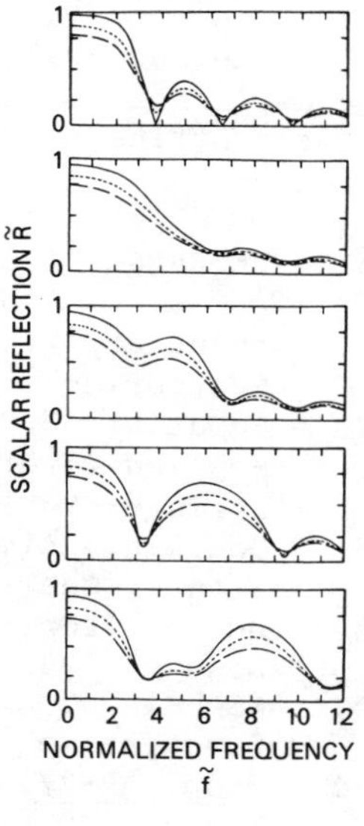

Fig. 2. Reflection coefficients $\tilde{R}$ for three-tone medium (from {7}).
—— $L\ell = 0$
---- $L\ell = 0.2$
— — $L\ell = 0.4$

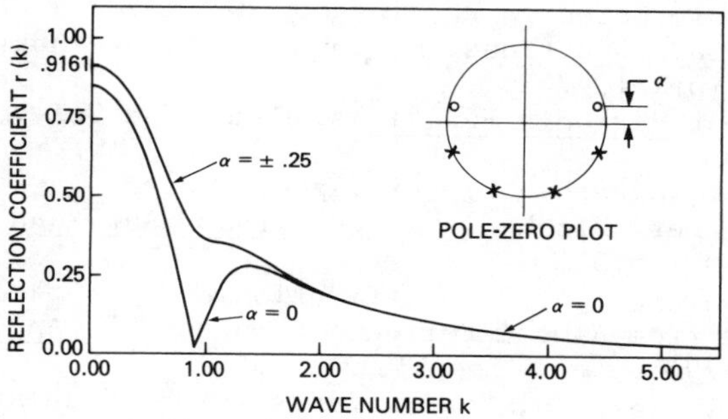

Fig. 3. Reflection coefficient $|r(k)|$ for example of eq. (28).

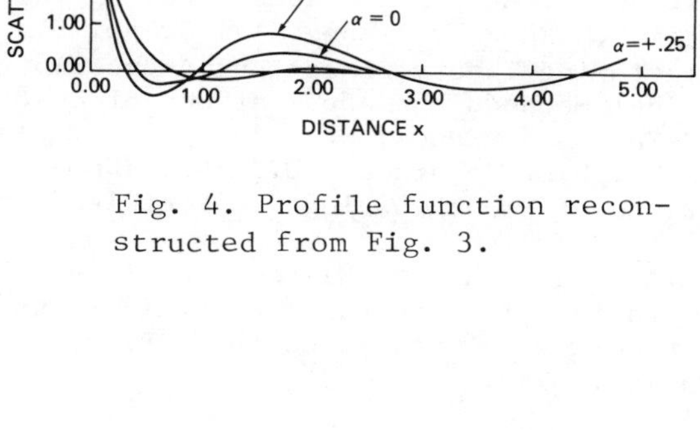

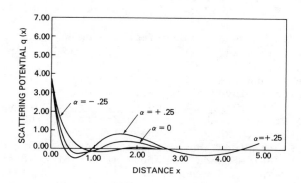

Fig. 4. Profile function reconstructed from Fig. 3.

5. ACKNOWLEDGEMENTS

The authors acknowledge the support of the University of Pennsylvania through the Committee on Faculty Grants and Awards and of the Office of Naval Research under the 6.1 Core Research Program.

6. REFERENCES

1. Sabatier, P.C., "Theoretical Considerations for Inverse Scattering", Radio Science 18, pp. 1-18 (1983)
2. Sabatier, P.C., "Introduction to Applied Inverse Problems",in Applied Inverse Problems", ed. by P.C. Sabatier, Springer-Verlag, Lecture Notes in Physics No. 85 (1978).
3. Bohr, H., Almost Periodic Functions, H. Cohn and F. Steinhardt, Trans., Chelsea, New York (1974).
4. Besicovitch, A., Almost Periodic Functions, Cambridge Univ., London (1932).
5. Michelson, A. and D.L. Jaggard, "Electromagnetic Wave Propagation in Almost Periodic Media", IEEE Trans. Ant. and Prop. AP-27, pp. 34-40 (1979).
6. Jaggard, D.L. and A. Michelson, "The Reflection of Electromagnetic Waves from Almost Periodic Structures", Appl. Phys. 19, 405-412 (1979).
7. Jaggard, D.L., and G.T. Warhola, "Characteristics of Modulated Periodic Media: I-Passive Structures", Radio Sci. 16, pp. 467-474 (1981); "II-Active Structures", pp. 475-480 (1981).
8. Gel'fand, I.M. and B.M. Levitan, "On the Determination of a Differential Equation by Its Spectral Function", Am. Math.Soc. Trans. 1, pp. 253-304 (1955).
9. Marchenko, V.A., "Reconstruction of the Potential Energy from the Phase of Scattered Waves", Dokl, Akad. Nauk. SSR 104, pp. 695-698 (1955).
10. Kay, I., "The Inverse Scattering Problem When the Reflection Coefficient is a Rational Function", Comm. Pure Appl. Math. 13, pp, 371-393 (1960).
11. Jordan, A.K. and H.N. Kritikos, "An Application of One-Dimensional Inverse Scattering Theory for Inhomogeneous Regions", IEEE Trans. Ant. and Prop. AP-21, pp. 209-211 (1973)
12. Jordan, A.K. and S. Ahn, "Inverse Scattering Theory and Profile Reconstruction", IEE Proc. (London) 126, pp. 945-950 (1979) or "Inverse Scattering Theory; Exact and Approximate Solutions", in Mathematical Methods ana Applications of Scattering Theory, ed. by J. DeSanto, A. Saenz, W. Zachary, Springer-Verlag (1980).
13. Kritikos, H.N., D.L. Jagard and D.B. Ge, "Numerical Reconstructions of Smooth Dielectric Profiles", Proc. IEEE 70, pp. 295-297 (1982).
14. Ross, G., M.A. Fiddy, and M. Nieto-Vesperinas, "The Inverse Scattering Problems in Structural Determinations", in Inverse Scattering Problems in Optics, ed. by H.P. Baltes, Springer-Verlag, 1980.

I.8 (TS.3)

ONE DIMENSIONAL TIME-DOMAIN INVERSE SCATTERING APPLICABLE TO ELECTROMAGNETIC IMAGING

HANS BLOK and ANTON G. TIJHUIS

Delft University of Technology
Department of Electrical Engineering
P.O. Box 5031, 2600 GA Delft
The Netherlands

Abstract

Various inverse-scattering methods applied to the electromagnetic imaging of the permittivity or the conductivity profile of an inaccessible layered medium are reviewed. Special emphasis is placed upon the time-domain analysis. The inaccessible region is illuminated by a pulsed plane wave at normal incidence. The problem of estimating the profile from this set of data is formulated in terms of a source-type integral equation for the electric-field strength in the layered medium, e.g. a slab. In the direct-scattering problem, space-time discretization leads to a step-by-step updating procedure; alternatively the Singularity Expansion Method is applied. For the solution of the inverse-scattering problem various methods will be discussed: a direct method, an iterative method and methods based on the iterative minimization of an appropriatedly chosen integrated square error.

1. INTRODUCTION

In this paper, we consider the scattering by a one-dimensional, inhomogeneous, lossy dielectric slab (Fig. 1.) of thickness d, embedded in vacuum. Normally incident on the slab is a given, pulsed, E-polarized electromagnetic wave E^i of finite duration. The case of oblique incidence is a straightforward extension. The electric field strength can be written as $\underline{E}(\underline{r},t) = E(z,t)\underline{i}_y$ and the magnetic field strength as $\underline{H}(\underline{r},t) = H(z,t)\underline{i}_x$. The investigation into this configuration is subdivided into two parts: (i) solution of the direct-scattering problem, i.e. the computation of the total electromagnetic field for given constitutive para-

W.-M. Boerner et al. (eds.), Inverse Methods in Electromagnetic Imaging - Part 1, 141–156.

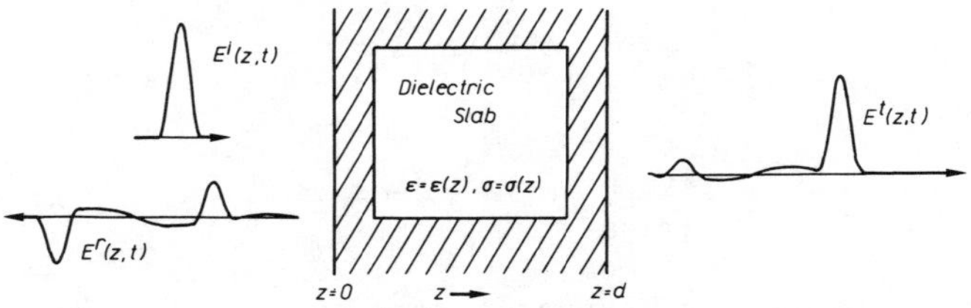

Fig. 1. Reflected and transmitted field corresponding to a pulse incident on an inhomogeneous, lossy dielectric slab.

meters; (ii) solution of the inverse-scattering problem, i.e. the identification of the slab from known values of the incident and the scattered electromagnetic field. In Fig. 2 we show an outline of the various methods that we have applied to solve these problems. Hereafter, the time-domain case, as well as the application of the Singularity Expansion Method (S.E.M.), will be discussed briefly. Whenever possible, we will also show some illustrative numerical results. The first part of this paper is devoted to the direct-scattering problem. In Section 2, we show how, for the time-domain case, the total field can be computed by marching-on--in-time [1, 2]. When the reflected field is known, e.g. in an inverse-scattering solution problem, an alternative scheme is proposed. There the total field can be computed by marching--on-in-space. In section 3, we describe a different way of solving the time--domain problem. In this so-called Singularity Expansion Method [3, 4, 5, 6, 7] the field quantities are expressed in terms of the natural modes of the slab, each corresponding to a complex natural frequency. These natural modes are dependent on the slab characteristics only. To this end, the time--domain electric-field strength is written as a Laplace contour integral. If the contour is deformed properly in the complex plane, only residual con-

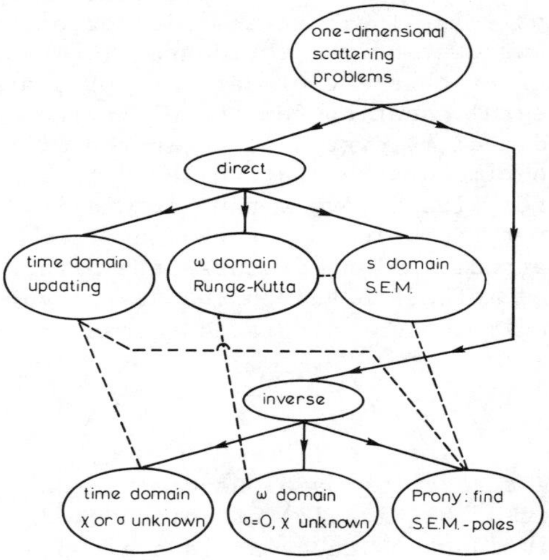

Fig. 2. Outline of the methods that were applied to solve the one-dimensional direct or inverse-scattering problem.

tributions need to be taken into account. At each point in space, the total electric field can then from an initial instant be written as a series of damped exponentials with complex arguments being exactly the natural frequencies.

The second part of the paper is devoted to the inverse-scattering problem. In Section 4, we discuss the reconstruction of either the permittivity or the conductivity profile from the known reflected field at the front of the slab. This problem can be posed in various ways: (i) the classical time-domain approach where the scattered data are used to construct the kernel and forcing term for a *Gelfand-Levitan type of integral equation*. The solution of this integral equation is then used to generate the permittivity and conductivity profiles [8, 9, 10]. This approach will not be discussed here. For an extensive review the reader is referred to [11].

(ii) An alternative approach in the time-domain where the direct- and inverse-scattering problem is formulated in terms of a *source-type integral equation*. This integral equation is subsequently discretized in the space-time domain. For the *direct--scattering problem* this leads to an infinite system of algebraic equations from which the total field can be solved. By virtue of the choice of equally large steps in space and time, these equations can be solved in a *step-by-step updating procedure* (marching on in time or, when the reflected field is known, marching on in space). For the solution of the inverse-scattering problem various methods will be discussed. (a) The *direct approach* which involves a step-by-step determination of the unknown profile by accuratedly tracing the wave front of the field [12, 13, 14].

(b) The *iterative approach* which basically consists of alternately solving equations for an approximate direct-scattering problem and an approximate inverse-scattering problem yielding successive approximations for the electric field in the slab and the unknown constitutive coefficient [15, 16].

(c) The *optimization approach* where the conductivity or susceptibility profiles are reconstructed by iterative minimization of an appropriatedly chosen integrated square error. In this review two different optimization schemes will be discussed: the *separate-optimization scheme*, where the square error depends on both the approximated profile and the approximated field and the *direct--optimization scheme*, where the square error is introduced as a function of the unknown profile only [17, 18].

Finally, in Section 5, we describe an alternative way of *identifying* the layered structure, namely the decomposition of the field into a set of damped exponentials with a Prony-type algorithm. According to the SEM-theory, the corresponding natural frequencies depend on the scatterer geometry only and can therefore be used for identification purposes [19].

2. DIRECT SCATTERING: TIME-DOMAIN UPDATING AND SIDE DATING

Consider the direct-scattering problem for the case of an incident field $E^i(z,t)$ with arbitrary time-dependence. With the time-domain Green's function technique, the following integral-equation for $E(z,t)$ can be derived

$$E(z,t) = E^i(z,t)-(Z_0/2)\int_0^d [\sigma(z')+\varepsilon_0\chi(z')\partial_t] E(z' t-|z-z'|/c_0)dz' \quad \text{when } 0<z<d \text{ and } 0<t<\infty. \tag{1}$$

The reflected field in front of the slab can be written as $E^r(z,t)=R(t+z/c_0)=E(z,t)-E^i(z,t)$ when $-\infty<z<0$. When this reflected field is known, e.g. in an inverse-scattering problem, an alternative integral equation can be derived

$$E(z,t) = E^i(z,t)+R(t+z/c_0)-(Z_0/2)\int_0^z [\sigma(z')+\varepsilon_0\chi(z')\partial_t] [E(z',t-(z-z')/c_0)-E(z',t+(z-z')/c_0)]dz'. \tag{2}$$

Note that in the integral on the right-hand side of (1) besides the unknown field $E(z,t)$ at instant t, only field values at pre-

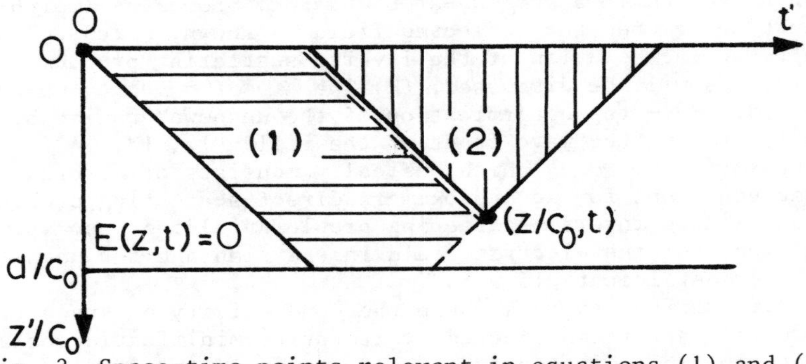

Fig. 3. Space-time points relevant in equations (1) and (2).

vious instants occur, while the corresponding integral in (2) contains only field values at previous space points. The space-time points that are relevant in eqs. (1) and (2) are shown in Fig. 3. The reflected field at the front of the slab is given by

$$R(t) = -(Z_0/2)\int_0^d [\sigma(z')+\varepsilon_0\chi(z')\partial_t]E(z',t-z'/c_0)dz'. \tag{3}$$

In order to implement the integral equations (1) and (2) numerically, they are discretized in the space-time domain. When the time step Δt and the space step Δz satisfy $\Delta z=c_0\Delta t=\Delta=d/N$, this discretization results for eq. (1) into

$$E(l,m) = E^i(l,m) - \tfrac{1}{4} \sum_{n=0}^{N} \alpha_n \chi^{(n)} \left[\tfrac{3}{2}E(n,m') - 2E(n,m'-2) + \tfrac{1}{2}E(n,m'-4)\right] - \frac{1}{2N} \sum_{n=0}^{N} \alpha_n \sigma(n)\, E(n,m'), \quad (4)$$

where $m'=m-|l-n|$, $l=0,1,2,\ldots,N$ and $m=0,1,2,\ldots,\infty$. In (4) we have $E(l,m)\triangleq E(l\Delta z, m\Delta t)$, $E^i(l,m)\triangleq E^i(l\Delta z, m\Delta t)$, $\chi(l)\triangleq\chi(l\Delta z)$, $\sigma(l)\triangleq\sigma(l\Delta z)$, $\alpha_n=1$ for $0<l<N$, while $\alpha_0=\alpha_N=\frac{1}{2}$. In this discretization, the spatial integral is approximated by a repeated trapezoidal rule and the time derivative by a three-point backward interpolation formula with a time-interval equal to the double time step. For given E^i, χ and σ, (4) represents an infinite system of algebraic equations from which the total electric field E can be solved. This special discretization allows a step-by-step updating procedure. The choice of the double time step in (4) allows the use of convenient recurrence relations similar to the ones derived in [2]. The increase in error in each iteration step is $O(\Delta^2)$. That means that for each finite time interval the maximum error is $O(N\Delta^2)=O(\Delta)$. This implies that stability can always be achieved by increasing N, i.e. by choosing Δ small enough.
In the case of eq. (2), this discretization allows a similar step-by-step side-dating procedure. The resulting schemes are often referred to as *marching on in time* and *marching on in space*, respectively. For some representative numerical results we refer to Fig. 1, Fig. 4 and Fig. 5.

SPACE-TIME FIELD DIAGRAM

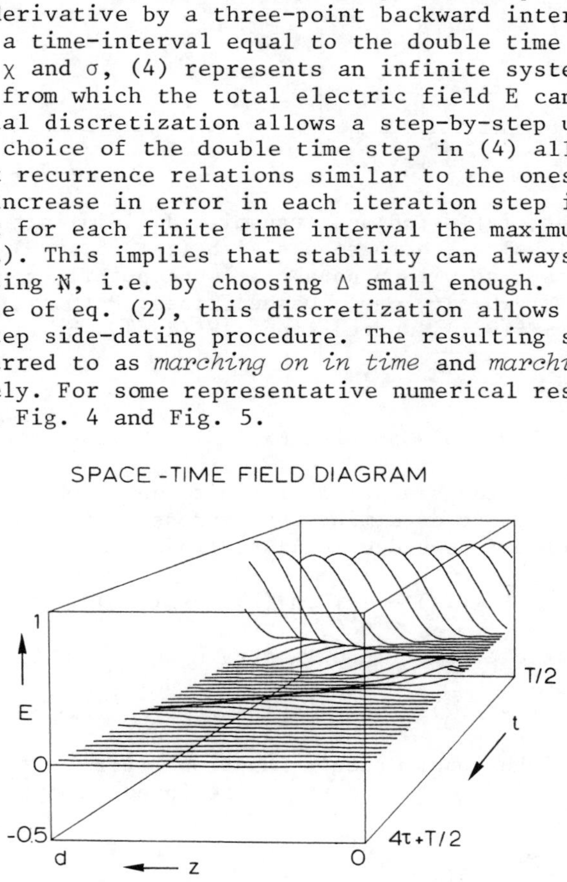

Fig. 4. Three-dimensional plot of field in a homogeneous slab for 4 runs of a pulse in the slab (χ=1.25, σ=0).

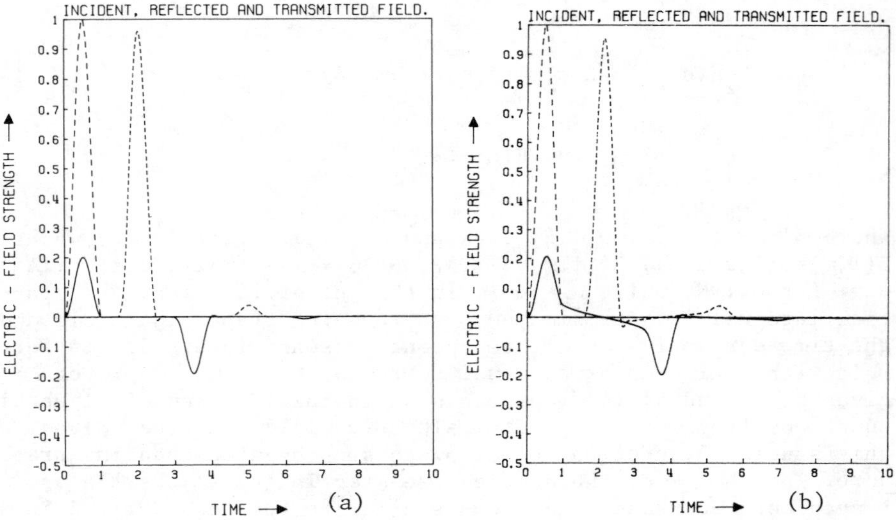

Fig. 5. Incident, reflected and transmitted field in the case of (a) a lossless, homogeneous slab (χ=1.25, σ=0) and (b) a lossless slab with a parabolic susceptibility profile ($\chi=2-4(z/d-0.5)^2$, σ=0). Dashed lines: $E^i(0,t)$, solid lines: $-E^r(0,t)$, dotted lines: $E^t(d,t)$. The time variable is c_0t/d.

3. DIRECT SCATTERING, SINGULARITY EXPANSION METHOD

If the incident field is a pulse of finite duration T, the electric field can also be determined by expressing the field quantities in terms of a (one-sided) Laplace transform

$$\{E,H\}(z,t)\leftrightarrow\{\tilde{E},\tilde{H}\}(z,s) = \int_{0^-}^{\infty} \{E,H\}(z,t)\exp(-st)dt. \tag{5}$$

For the incident electric field we obtain

$$\tilde{E}^i(z,s) = \exp(-sz/c_0)F_T(s) = \exp(-sz/c_0)\int_{0^-}^{T} E^i(0,t)\exp(-st)dt.$$

Using the one-dimensional Green's function $G(z;z')=(c_0/2s)\exp(-s|z-z'|/c_0)$, one derives the integral equation in operator form as

$$L(s)\tilde{E}(z,s) = \tilde{E}^i(z,s), \tag{6}$$

in which

$$L(s)\tilde{E}(z,s) = \int_{-\infty}^{\infty} \{\delta(z-z')+s^2C(z,s)G(z;z')/c_0^2\}\tilde{E}(z',s)\, dz', \tag{7}$$

where $C(z,s) \triangleq \varepsilon_r(z) - 1 + \sigma(z)/\varepsilon_0 s$ denotes the *contrast function* of the slab. From (6), we formally have $\tilde{E}(z,s) = L^{-1}(s)\tilde{E}^i(z,s)$. The time-domain response is then found by the inverse Laplace integral

$$E(z,t) = (1/2\pi i) \int_{\beta-i\infty}^{\beta+i\infty} \tilde{E}(z,s)\exp(st)\, ds, \tag{8}$$

where in the Bromwich contour C_B, $\mathrm{Re}(\beta)>0$.
In the evaluation of (8), we encounter simple poles and entire functions. Then the following Mittag-Leffler expansion is valid

$$\tilde{E}(z,s) = \sum_m e_m(z)/(s-s_m) + \tilde{E}^e(z,s), \tag{9}$$

where $\tilde{E}^e(z,s)$ is an entire function and $e_m(z) \triangleq \mathrm{res}\ \tilde{E}(z,s)|_{s=s_m}$. The complex natural frequencies s_m occur in pairs. With each m corresponds a complex natural mode, defined by

$$L(s_m)E_m(z) = 0. \tag{10}$$

Following now the SEM procedure as outlined in [3, 4, 5, 6, 7] we can express $e_m(z)$ as

$$e_m(z) = D_m F_T(s_m) E_m(z), \tag{11}$$

in which the coupling coefficient D_m depends on the slab configuration only and defines the strength of the relevant natural mode. The residual contribution of the poles to the time-domain response is then given by

$$E_{poles}(z,t) = \sum_m D_m F_T(s_m) E_m(z)\exp(s_m t). \tag{12}$$

The natural frequencies and the corresponding natural modes are found as follows. Outside the slab we redefine the electric field as

$$E(z,s) = \begin{cases} \exp(-sz/c_0)/t(s) + r(s)\exp(sz/c_0)/t(s) \\ \qquad\qquad \text{when } -\infty<z<0, \\ \exp(-s(z-d)/c_0) \quad \text{when } d<z<\infty. \end{cases} \tag{13}$$

Since the boundary conditions require the continuity of $E(z,s)$ and $H(z,s)$ at both ends of the slab, the unknown coefficients $r(s)/(t(s)$ and $1/t(s)$ in (13) can be computed by direct numerical integration for the Laplace-transformed, source-free electromagnetic field equations for E and H (a Runge-Kutta type method). The natural frequencies $\{s_m\}$ are then found by searching the zero's of $1/t(s)$ with Muller's method. The natural modes $E_m(z) \triangleq \tilde{E}(z, s_m)$ are obtained in this proces, while the coupling coefficients can be determined by numerically performing a multi-

ple integration of a product containing factors $C(z,s_m)E_m(z)$. In the total time-domain response, the entire function in (9) represents the contribution of the closing contour either in the left half or in the right half of the s-plane. In order to analyze the role of the entire function an asymptotic solution is constructed using the WKB method. Applying Jordan's lemma, we then find that we can avoid computing contributions along the closing contour by taking into account only that part of the incident pulse that has emerged into the slab and accelerating the convergence using the obtained asymptotic solution. In this way, the well known problem of the uniqueness of the SEM-representation has been resolved. For a more detailed description of this procedure, the reader is referred to [7].
In the Figures 6-8, we consider the scattering by a slab with a parabolic permittivity profile and a constant conductivity. In Figs. 6b and 6c, the magnitudes of the corresponding coupling coefficients are plotted. The field distributions of some of the natural modes with σ=0 are given in Fig. 7. The incident, reflec-

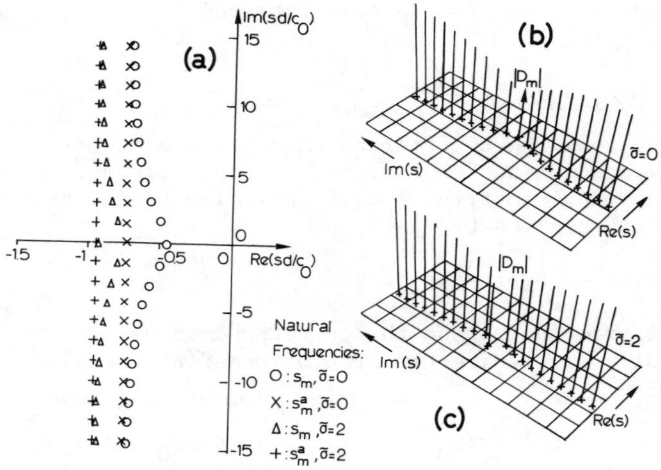

Fig. 6. Configuration-dependent parameters plotted in the complex s-plane for a dielectric slab with a parabolic permittivity profile $\varepsilon_r(z)=6.25-16(z/d-0.5)^2$ and constant conductivity $Z_0\sigma d=0$ and 2.
(a): Complex natural frequencies s_m with $|m|\leq 10$.
(b,c): Magnitudes of the corresponding coupling coefficients D_m.

ted and transmitted fields for a sine-squared pulse and a sinusoidal pulse train incident on a lossless, parabolic permittivity profile are shown in Fig. 8.

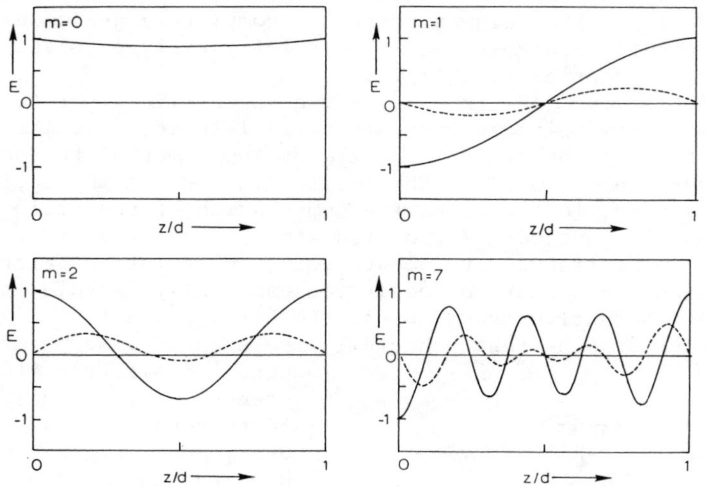

Fig. 7. Field distributions of the natural modes with m=0, 1, 2 and 7 for the lossless slab specified in Fig. 6. (——— = real part, ---- = imaginary part).

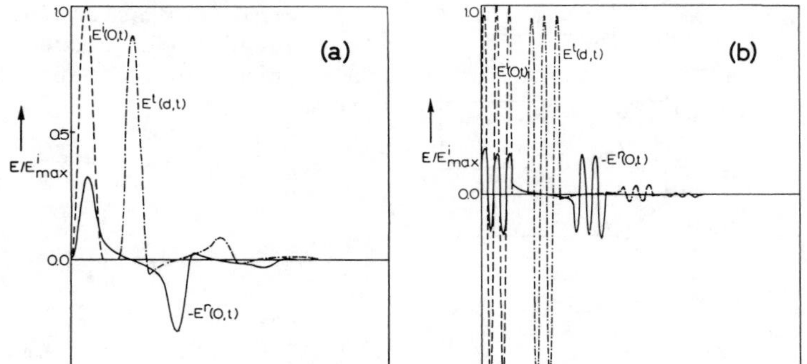

Fig. 8. (a,b): Incident, reflected and transmitted field for the lossless slab specified in Fig. 6 and for the incident pulses (a) $F(t)=\sin^2(\pi t/T)\mathrm{rect}(t-T/2;T)$; (b) $F(t)=\sin(5\pi t/T)\mathrm{rect}(t-T/2;T)$, with $c_0T/d=1.5$.

4. INVERSE SCATTERING: TIME DOMAIN

In the inverse scattering problem, we want to determine either the susceptibility profile $\chi(z)$ or the conductivity profile $\sigma(z)$

when the incident and reflected field at the front of the slab are known. Since the reconstruction is completely analogous for both constitutive parameters, we restrict ourselves to the case of unknown susceptibility $\chi(z)$.
In the *direct approach* [12, 13, 14], a space-time discretization of the source-type integral equation (1) is used. Calculating step by step, in the slab, the field at the front of the pulse, one can make an estimate of the local value of the susceptibility or conductivity. This enables the computation of the field in the next step. This method has the disadvantage that the front of the pulse has to be traced very accurately, the relative error in the computed field there being large, especially for dielectric slabs. An alternative method [16], the *iterative method*, is based on the consideration that especially those space-time points where the electric-field strength is relatively large and therefore allows an accurate numerical computation, should be taken into account. To this end, $\chi(z)$ is assumed to vary linearly within z-intervals $[kN_{cell}\Delta z, (k+1)N_{cell}\Delta z]$ with N_{cell} a positive integer. Then the discretized version of the integral expression for the reflected field reduces to a matrix equation of the form

$$\sum_{k=0}^{K} E_{mk}\chi_k = R_m - V_m, \qquad (14)$$

with $m=0,2,4,\ldots,\infty$.

In (14) we have $K \triangleq N/N_{cell}$, $\chi_k=\chi(kN_{cell})$, while E_{mk} is a weighted sum of the $E(l,n)$ with $|l-kN_{cell}|<N_{cell}$ and $n=m-l$, $m-l-2$ or $m-l-4$. R_m is given by the reflected field, while V_m depends on both σ and E.

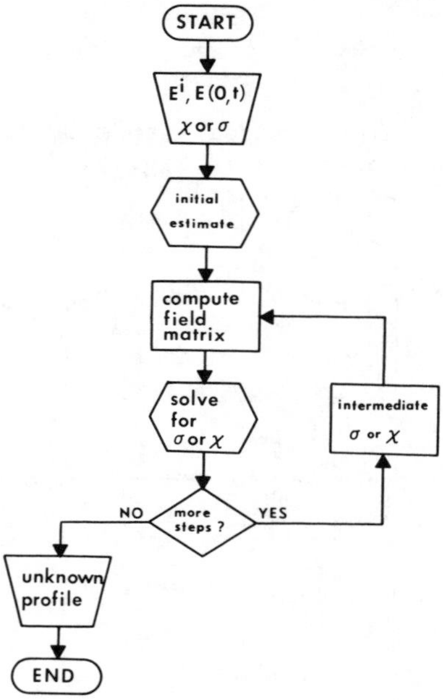

Fig. 9. Simplified block diagram of the iterative procedure for solving the time-domain in-verse-scattering problem.

Equation (14) constitutes an infinite system of linear equations. If the electric field E in the slab were known, χ_k could be determined as follows. The infinite system is reduced to a finite one by selecting those equations that cover the "first run" of the pulse through the slab:

$$T/2 \le m\Delta t \le \tau + T/2 + d/c_0 \tag{15}$$

In (15), $\tau = c_0^{-1}\int_0^d \varepsilon_r^{\frac{1}{2}}(z)dz$ is the time it takes for the front of the pulse to reach the end of the slab.

The remaining system contains considerably more equations than unknowns. So the problem is overdetermined. In order to overcome this difficulty, two methods are applied to construct K+1 equations, from which χ_k can be exactly solved by Gauss elimination. In the first method, the total time interval is subdivided into K+1 subintervals, each corresponding to an equation given by the sum of all equations with $t=m\Delta t$ in the subinterval.

The second method is a linear least squares inversion, minimizing the error

$$e = \sum_{\text{subsystem}} (R_m - \sum_{k=0}^{K} E_{mk}\chi_k)^2. \tag{16}$$

Since the electric field is known at the front of the slab only, the inverse problem is solved with the iterative procedure outlined in Fig. 9. Starting with an appropriate initial value of the unknown susceptibility, we obtain successive approximations on both the electric field in the slab and the susceptibility. In the first step of this procedure, the electric field is approximated by a field that results from the incident pulse in a slab with a suitably chosen, constant susceptibility. This field is computed with the method outlined in Section 2. The thus obtained results are converted into an equation of the form (14), solution of which yields the first approximation of χ_k. In subsequent steps, this process is repeated until a stable solution is reached.

We have solved this inverse-scattering problem numerically for various permittivity and conductivity profiles, using the time--domain response of a sine-squared incident pulse. The results indicate that, although correction procedures may be required, for low contrast, convergence is achieved within five iteration steps. Also, the method turns out to be stable with respect to variation of the initial guess and addition of random noise to the incident and reflected field. As an illustration, numerical results of the reconstruction of a susceptibility profile are shown in Figs. 10 and 11. For more details, the reader is referred to [15, 16].

Finally, we discuss an approach where the conductivity or susceptibility profiles are reconstructed by iterative minimization of an appropriatedly chosen integrated square error. Two different optimization schemes will be discussed: the *separate-optimization scheme* and the *direct optimization scheme*. First we will introduce the relevant integrated square errors. To this aim we rewrite the equations (1), (2) and (3) in the form $F_1(z,t)=0$, $F_2(z,t)=0$ and $F_3(t)=0$, respectively. Next we introduce the integrated square errors

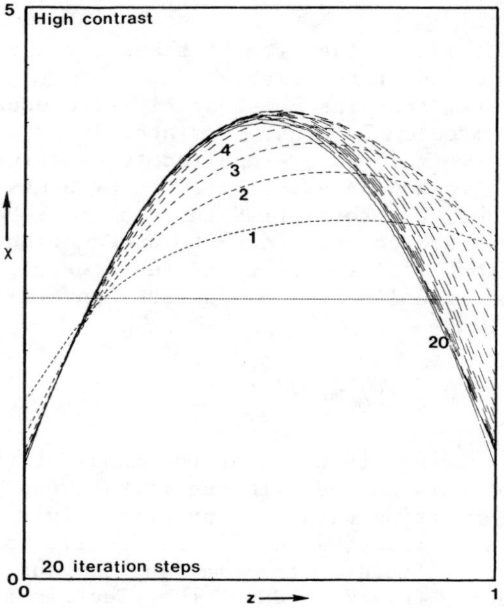

Fig. 10. Actual susceptibility profile $\chi(z)=4-12(z/d-0.5)^2$ (solid line) compared with the results obtained after 1,2,...,20 iteration steps for $\sigma(z)=0$, K=20, N=100, $c_0T_p/d=0.5$ and an initial estimate computed from the travel time (dashed lines; dash length increases with number of the iteration step).

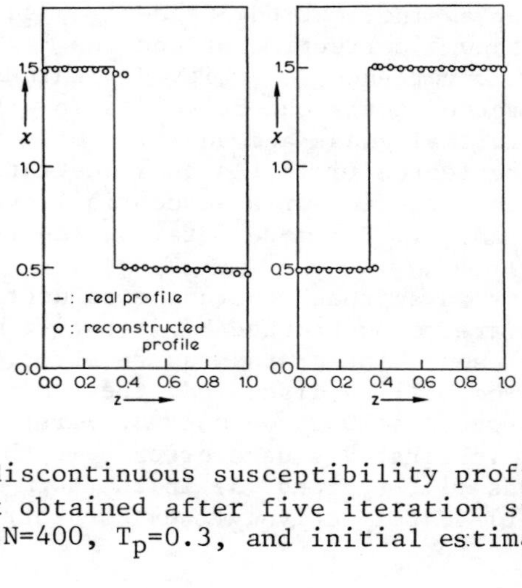

Fig. 11. Real discontinuous susceptibility profiles compared to result obtained after five iteration steps for $\sigma(z)=0$, K=40, N=400, $T_p=0.3$, and initial estimate $\chi=1.25$.

$$I_{1,2} \triangleq \int_0^\infty dt \int_0^d dz\ F_{1,2}^2(z,t),\quad I_3 \triangleq \int_0^\infty dt\ F_3^2(t), \tag{17}$$

where I_1 denotes the integrated square update error, I_2 the integrated square side-date error and I_3 the integrated square error in the approximated reflection factor. In both methods the electric field strength E is discretized as indicated in Section 2, while the susceptibility profile is, as indicated in Section 4, represented by a piecewise linear approximation.

In the *separate-optimization method* we minimize either $I(E,\chi,\sigma) = I_1+I_2$ or $I(E,\chi,\sigma)=I_1+cI_3$, where c is some chosen constant. In the initial step, the unknown profile has a chosen value and the unknown field is approximated by solving $F_1(E_1,\chi,\sigma)=0$ or $F_2(E_2,\chi,\sigma)=0$.

The profile is optimized by minimizing the integrated square error I, while the approximation of the field is not changed. Next, the field is improved by minimizing I without altering the profile. This process of separately improving profile and field is repeated until I has reached some minimum. Numerical experiments show that this method converges badly and only gives reasonable results for the determination of conductivity profiles for $\chi=0$.

In the *direct-optimization method*, the integrated square error is introduced as a function of the unknown profile only [16, 17]. The method works for both conductivity and susceptibility profiles. We will discuss the case of a χ-profile. Here we minimize either $I_2(E_\chi)$ or $I_3(E_\chi)$ where E_χ satisfies $F_1(E,\chi;z,t)\equiv 0\ \forall(z,t)$. It turns out that the minimization of $I_3(E_\chi)$ leads to a more efficient scheme. Two different minimization methods have been applied: the quasi-Newton method and the conjugate-gradient method [20]. In one version of the quasi-Newton method and in the conjugate-gradient method the first derivative of the integrated square error has to be determined analytically. It turns out that application of the conjugate-gradient method is more efficient than the quasi-Newton method. A numerical result is shown in Fig. 12.

As a conclusion it can be noted that the separate-optimization scheme can only be applied to the determination of conductivity profiles, provided that the susceptibility is equal to zero. The direct-optimization scheme is applicable for both conductivity and susceptibility profiles. In the numerical implementation it turns out that the conjugate-gradient method is more efficient than the quasi-Newton method. It is important to notice that for the determination of the profile less parameters are needed than for the determination of the field. This in contrast with the approach of Lesselier [17]. Finally it is remarked that the occurrence of noise on the measured fields may have a considerable effect on the retrieved profiles. There exist various ways to reduce this influence.

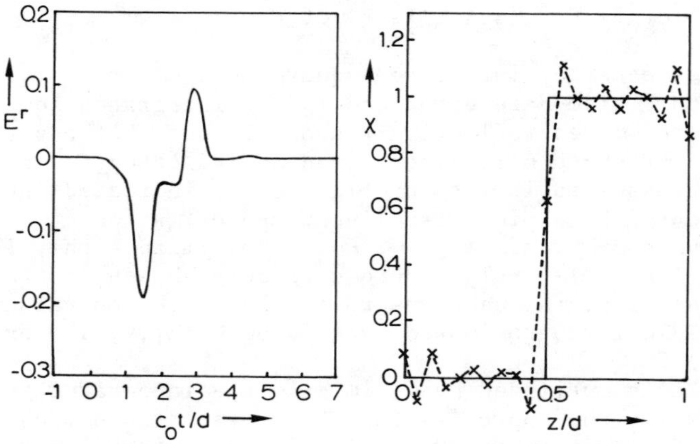

Fig. 12. Step-index susceptibility profile with linearly varying conductivity reconstructed by minimizing error I_3 with the conjugate-gradient method. 3% uncorrelated noise has been added on both incident and reflected field. $Z_0\sigma(z)/d=z/d$, K=20, N=100, $c_0T_p/d=1$.

5. INVERSE SCATTERING: PRONY ALGORITHM

As we discussed in Section 3, the Singularity Expansion Method shows that after the incident pulse has completely emerged into the slab, the reflected field is given by

$$E(z,t) = \sum_{m=-\infty}^{\infty} F_T(s_m)D_mE_m(0)\exp\{s_m(t+z/c_0)\}, \text{ for } z<0. \quad (18)$$

In (18), the constants s_m and $D_mE_m(0)$ depend on the configuration only, while $F_T(s_m)$ is given by the incident field. Therefore, if certain obstacle properties can be determined from the reflected field, there is also a one-to-one correspondence between the parameters s_m and $D_mE_m(0)$ and the unknown slab configuration. This correspondence can then be used to identify the slab uniquely.

Recently [19], methods have been described to determine the parameters s_m and $D_mE_m(0)$ numerically. Most of these methods are based on Prony's algorithm. This algorithm yields a characteristic equation, from which the natural frequencies s_m can be determined. The residual coefficients $D_mE_m(0)$ can subsequently be solved from a system of linear equations. It is noted that for the dielectric slab considered in this paper, the number of natural modes is infinite, whereas the Prony method was devised for a finite number of exponentials. Presently, techniques are being studied to circumvent this difficulty.

ACKNOWLEDGEMENTS

The authors with to thank the students C.P.L. Cames van Batenburg, R. Koster, B.L. Michielsen and R.M. van der Weiden for their contributions to parts of this research.

REFERENCES

[1] Bennett, C.L. and G.F. Ross, "Time-domain electromagnetics and its applications", Proc. IEEE, vol. 66, 1978, pp. 299-318.

[2] Bolomey, J.Ch., Ch. Durix and D. Lesselier, "Time domain integral equation approach for inhomogeneous and dispersive slab problems", IEEE Trans. Antennas Propagat., vol. AP-26, 1976, pp. 1598-1615.

[3] Felsen, L.B., ed., "Transient electromagnetic fields", Springer Verlag, Berlin, 1976, Chapter 3 (C.E. Baum).

[4] Uslenghi, P.L.E., ed., "Electromagnetic scattering", Academic Press, New York, 1978, Chapter 3 (C.E. Baum).

[5] Marin, L., "Natural-mode representations of transient scattered fields", IEEE Trans. Antennas Propagat., vol. AP-21, 1973, pp. 809-818.

[6] Tijhuis, A.G. and H. Blok, "SEM approach to the transient scattering by an inhomogeneous, lossy dielectric slab", Proceedings International URSI-Symposium 1980, Munich, Aug. 26-29, 1980, 221 B.

[7] Tijhuis, A.G. and H. Blok, "SEM approach to the transient scattering by an inhomogeneous, lossy dielectric slab; Part 1: The homogeneous case. Part 2: The inhomogeneous case." To be published in Wave Motion.

[8] Weston, V.H., "On the inverse problem for a hyperbolic dispersive partial differential equation", J. Math. Phys. 13, 1972, pp. 1952-1956.

[9] Krueger, R.J., "An inverse problem for a dissipative hyperbolic equation with discontinuous coefficients", Quart. Appl. Math. 34, 1976, pp. 129-147.

[10] Krueger, R.J., "Numerical aspects of a dissipative inverse problem", IEEE Trans. Antennas Propagat., vol. AP-29, 1981, pp. 253-261.

[11] Burridge, R., "The Gelfand-Levitan, the Marchenko, and the Gopinath-Sondhi integral equations of inverse scattering theory, regarded in the context of inverse impulse-response problems", Wave Motion 2, 1980, pp. 305-323.

[12] Bolomey, J.C., C. Durix and D. Lesselier, "Determination of conductivity profiles by time-domain reflectrometry", IEEE Trans. Antennas Propagat., vol. AP-27, 1979, pp. 244-248.

[13] Lesselier, D., "Determination of index profiles by time domain reflectrometry", J. Optics, vol. 9, 1978, pp. 349-358.

[14] Bojarski, N.N.,"One-dimensional direct and inverse scattering in causal space", Wave Motion, vol. 2, 1980, pp. 115-124.

[15] Tijhuis, A.G., "Iterative determination of permittivity and conductivity profiles of a dielectric slab in the time domain", Proceedings International URSI-Symposium, 1980, Munich, Aug. 26-29, 324 C.

[16] Tijhuis, A.G., "Iterative determination of permittivity and conductivity profiles of a dielectric slab in the time domain", IEEE Trans. Antennas Propagat., vol. AP-29, 1981, pp. 239-245.

[17] Lesselier, D.,"Optimization techniques and inverse problems: reconstruction of conductivity profiles in the time domain", IEEE Trans. Antennas Propagat., vol. AP-30, 1982, pp. 59-65.

[18] Cames van Batenburg, C., "One dimensional inverse scattering in the time domain using an error criterium", Report 1983-06, Laboratory of Electromagnetic Research, Delft University of Technology.

[19] Miller, E.K., "Natural mode methods in frequency and time domain analysis", Proceedings Nato advanced study institute on theoretical methods for determining the interaction of electromagnetic waves with structures, Norwich, July 23-August 4, 1979, Chapter H7.

[20] NAG, "E04-Minimizing or maximizing a function", Fortran library manual, Mark 9, vol. 3, Numerical Algorithms Group, 1982.

I.9 (IS.2)

INVERSE ELECTROMAGNETIC SCATTERING FOR RADIALLY INHOMOGENEOUS DIELECTRIC SPHERES

C. Eftimiu

McDonnell Douglas Research Laboratories
St. Louis, Missouri 63166

The scattering of a plane electromagnetic wave by a dielectric sphere with a radially distributed index of refraction is analyzed in terms of Debye potentials. The scattering amplitude, expanded in vector spherical harmonics, is given in terms of two sets of phase shifts. The inverse problem considered consists in reconstructing the index of refraction when one of the phase shifts is known as a function of frequency.

The approach is based on the use of Liouville transformations which cast the equations for the radial parts of the Debye potentials in the form of Schroedinger-like equations with frequency independent potentials. Given the frequency dependence of one of the phase shifts, one such potential can be found by solving the quantum mechanical inverse problem by using the Marchenko formalism. The index of refraction can be reconstructed from the potential by solving a nonlinear integro-differential equation; the existence and uniqueness of its solution is discussed.

1. INTRODUCTION

The electromagnetic inverse scattering problem consists in the reconstruction of a target for electromagnetic waves from scattering data. In general, the reconstruction concerns the electromagnetic characteristics of the target (dielectric constant, magnetic permeability, conductivity) the knowledge of which includes the spatial extension of the object from a set of measurable quantities directly related to the scattered field.

No attempt will be made here to examine this problem in its total generality. Rather, in order to reduce it to manageable proportions, partial **a priori** information regarding the target will be presumed available. Specifically, it will be assumed that the target has spherical shape, the conductivity is identically zero, its magnetic permeability is everywhere that of vacuum, and its dielectric constant is a continuous, twice differentiable function of the distance from its center, only. The only unknown of the inverse scattering problem will thus be the function $\epsilon(r)$. The dielectric sphere will also be assumed to have a finite extension, implying that there exists a radius r_o (not **a priori** known) such that $\epsilon(r) \equiv 1$ for $r \geq r_o$, whereas for $0 \leq r \leq r_o$, $\epsilon(r) \geq 1$.

W.-M. Boerner et al. (eds.), Inverse Methods in Electromagnetic Imaging - Part 1, 157–176.

The formulation of the inverse electromagnetic problem depends, of course, on the nature of the assumed set of given scattering data, and in this respect several choices are possible. To describe them, an analysis of the scattering process is necessary, and this is briefly done in the next section to make the exposition self-sufficient.

2. DIRECT ELECTROMAGNETIC SCATTERING

We begin by disposing of the complications arising from the vectorcharacter of the electromagnetic field, by reducing the electromagnetic scattering problem to two scalar problems. The procedure involves the introduction of the Debye potentials and, since it is well-known (1), will be described briefly. It follows from Maxwell's equations:

$$\nabla \times \vec{E} = -\frac{1}{c}\vec{B} = ik\vec{H} \qquad \nabla \cdot \vec{B} = 0$$

$$\nabla \times \vec{H} = \frac{1}{c}\vec{D} = -ik\epsilon\vec{E} \qquad \nabla \cdot \vec{D} = 0 \tag{2.1}$$

that the total (incident + scattered) electromagnetic field (the harmonic time dependence, in which $\omega = kc$, is suppressed):

$$\vec{E} = \vec{E}^{inc} + \vec{E}^{sc} = \vec{u}_x e^{ikz} + \vec{E}^{sc}$$

$$\vec{H} = \vec{H}^{inc} + \vec{H}^{sc} = \vec{u}_y e^{ikz} + \vec{H}^{sc} \tag{2.2}$$

satisfies the uncoupled equations:

$$\nabla \times (\nabla \times \vec{E}) = k^2\epsilon\vec{E}$$

$$\nabla \times \left(\frac{1}{\epsilon}\nabla \times \vec{H}\right) = k^2\vec{H}\,. \tag{2.3}$$

Addressing the second of Eqs. (2.3) first, and taking into account the transverse character of the magnetic field, one makes the Ansatz

$$\vec{H} = \vec{M} + \vec{N}\,, \tag{2.4}$$

where

$$\vec{M} = \nabla \times (n\vec{r}\chi) \tag{2.5}$$

introduces χ, the electric multipole Debye potential, and

$$\vec{N} = k^{-1}\nabla + [\nabla + (\vec{r}\psi)] \tag{2.6}$$

introduces ψ, the magnetic multipole Debye potential, with

$$n = \sqrt{\epsilon}\,. \tag{2.7}$$

The Debye potentials satisfy the equations:

$$\nabla^2\chi \times [k^2n^2 - U(r)]\chi = 0$$

$$\nabla^2\psi \times k^2n^2\psi = 0\,, \tag{2.8}$$

in which

$$U(r) \equiv n(r)\,\frac{d^2}{dr^2}\,\frac{1}{n(r)}\,. \tag{2.9}$$

Since the electric field can be obtained in terms of the magnetic field from the second of Maxwell's equations (2.1), the electromagnetic problem reduces to two scalar problems.

For the purpose of performing a phase-shift analysis of the scattering process, it is appropriate to introduce the vector spherical harmonics, defined as

$$\vec{\mu}_{\ell m}(\theta, \phi) = \nabla \times (\vec{r}Y_{\ell m})$$

$$\vec{\nu}_{\ell m}(\theta, \phi) = 2r\,\nabla Y_{\ell m}$$

$$\vec{\tau}_{\ell m}(\theta, \phi) = \ell(\ell + 1)\,\frac{\vec{r}}{r}\,Y_{\ell m}\,, \tag{2.10}$$

where $Y_{\ell m}$ are the standard spherical functions:

$$Y_{\ell m}(\theta, \phi) = \left[\frac{2\ell + 1}{4\pi}\,\frac{(\ell - |m|)!}{(\ell + |m|)!}\right]^{1/2} \cdot P_\ell^m(\cos\theta)\,e^{im\phi}\,. \tag{2.11}$$

One can check now that the second of Eqs. (2.3) **in vacuo** ($\epsilon \equiv 1$) possesses two sets of linearly independent solutions. One set, consisting of solutions regular in the origin (as it will be seen below, $\ell \geq 1$) is given by the functions

$$\vec{M}^{(1)}_{\ell m} = j_\ell(kr)\,\vec{\mu}_{\ell m}(\theta, \phi) \tag{2.12}$$

and

$$\vec{N}^{(1)}_{\ell m} = \frac{1}{k}\,\nabla \times \vec{M}^{(1)} = \frac{1}{kr}\,j_\ell(kr)\,\vec{\tau}_{\ell m} + \frac{1}{2kr}\,\frac{d}{dr}\,[rj_\ell(kr)]\,\vec{\nu}_{\ell m} \tag{2.13}$$

and another set, describing outgoing spherical waves, by

$$\vec{M}^{(2)}_{\ell m} = h^{(1)}_\ell(kr)\,\vec{\mu}_{\ell m}(\theta, \phi) \tag{2.14}$$

and

$$\vec{N}^{(2)}_{\ell m} = \frac{1}{k}\,\nabla \times \vec{M}^{(2)} = \frac{1}{kr}\,h^{(1)}_\ell(kr)\,\vec{\tau}_{\ell m} + \frac{1}{2kr}\,\frac{d}{dr}\,[rj_\ell(kr)]\,\vec{\nu}_{\ell m} \tag{2.15}$$

where j_ℓ and h_ℓ are the usual Riccati-Bessel and Riccati-Hankel functions, respectively. In terms of the first set, the incident electromagnetic field can be written as

$$\vec{E}^{inc} = \sqrt{\pi}\,\sum_{\ell=1}^{\infty} i^{\ell-1}\left[\frac{\ell(\ell + 1)}{2\ell + 1}\right]^{1/1}\left(\vec{M}^{(1)}_{\ell m} + \vec{N}^{(1)}_{\ell m} - \vec{M}^{(1)}_{\ell m} + \vec{N}^{(1)}_{\ell m}\right)$$

$$\vec{H}^{inc} = \sqrt{\pi}\,\sum_{\ell=1}^{\infty} i^{\ell+2}\left[\frac{\ell(\ell + 1)}{2\ell + 1}\right]^{1/1}\left(\vec{M}^{(1)}_{\ell m} + \vec{N}^{(1)}_{\ell m} + \vec{M}^{(1)}_{\ell m} - \vec{N}^{(1)}_{\ell m}\right), \tag{2.16}$$

The fact that only $\pm$ 1 values for m appear in the expansions (2.16) implies that $\ell \geq 1$, so that only such values for ℓ and m will appear in similar expansions for the scattered field as well. The latter will be **asymptotically** ($r \mapsto \infty$) described by the solutions from the second set:

$$\vec{E}^{sc} \underset{r \mapsto \infty}{\sim} \sum_{\ell=1}^{\infty} \sum_{m=\pm 1} i^{\ell-1} \left[\frac{\ell(\ell+1)}{\ell(\ell+1)} \right]^{1/2} \left(a_{\ell m} \vec{M}^{(2)}_{\ell m} + b_{\ell m} \vec{N}^{(2)}_{\ell m} \right)$$

$$\vec{H}^{sc} \underset{r \mapsto \infty}{\sim} \sum_{\ell=1}^{\infty} \sum_{m=\pm 1} i^{\ell+2} \left[\frac{\ell(\ell+1)}{\ell(\ell+1)} \right]^{1/2} \left(b_{\ell m} \vec{M}^{(2)}_{\ell m} + a_{\ell m} \vec{N}^{(2)}_{\ell m} \right) \tag{2.17}$$

An examination of symmetry properties of the fields, and of the vector spherical harmonics under the transformation $\phi \mapsto -\phi$, shows that

$$a_{\ell 1} = -a_{\ell-1}, \qquad b_{\ell 1} = b_{\ell-1}, \tag{2.18}$$

and hence, by introducing the phase shifts δ_ℓ and η_ℓ through the relations:

$$a_{\ell 1} = ie^{i\delta_\ell} \sin \delta_\ell, \qquad b_{\ell 1} = ie^{i\eta_\ell} \sin \eta_\ell, \tag{2.19}$$

one obtains for the asymptotic expansions of the scattered field the expressions:

$$\vec{E}^{sc} \underset{r \mapsto \infty}{\sim} \vec{\mathfrak{a}}(\theta, \phi) \frac{e^{ikr}}{r}$$

$$\vec{H}^{sc} \underset{r \mapsto \infty}{\sim} \frac{\vec{r}}{r} \times \vec{\mathfrak{a}}(\theta, \phi) \frac{e^{ikr}}{r}, \tag{2.20}$$

where $\vec{\mathfrak{a}}(\theta, \phi)$, the scattering amplitude, is

$$\vec{\mathfrak{a}}(\theta, \gamma) = -\frac{\sqrt{\pi}}{k} \sum_{\ell=1}^{\infty} \left[\frac{2\ell+1}{\ell(\ell+1)} \right]^{1/2} \left[i(\vec{\mu}_{\ell 1} - \vec{\mu}_{\ell-1}) e^{i\delta_\ell} \sin \delta_\ell \right.$$

$$\left. - \frac{1}{2} (\vec{\nu}_{\ell 1} + \vec{\nu}_{\ell-1}) e^{i\eta_\ell} \sin \eta_\ell \right]. \tag{2.21}$$

Hence, given the orthonormality properties of the vector harmonics, one obtains for the total scattering cross-section the well-known expression:

$$\sigma = \int d\Omega |\mathfrak{a}|^2 = \frac{2\pi}{k^2} \sum_{\ell=1}^{\infty} (2\ell+1) \left[\sin^2\delta_\ell + \sin^2\eta_\ell \right] \tag{2.22}$$

The direct electromagnetic scattering problem, consisting in finding the scattering amplitude (2.21), and hence the differential, $|\vec{\mathfrak{a}}|^2$, or total, σ, cross-sections, given the (spherically symmetric) scatterer, reduces thus to a determination of the two sets of phase-shifts, δ_ℓ and η_ℓ with $\ell = 1, 2, \ldots$, which are in general functions of the frequency ω of the incident field.

The asymptotic expressions for the total fields follow from Eqs. (2.16) and (2.17):

$$\vec{E} \underset{r \mapsto \infty}{\sim} \frac{\sqrt{\pi}}{kr} \sum_{\ell=1}^{\infty} \left[\frac{2\ell+1}{\ell(\ell+1)}\right]^{1/2} i^{\ell-1}\Big[(\vec{\mu}_{\ell 1} - \vec{\mu}_{\ell-1})\, e^{i\delta_\ell} \sin\left[kr - \frac{\ell\pi}{2} + \delta_\ell\right]$$

$$+ \frac{1}{2}(\vec{\nu}_{\ell 1} + \vec{\nu}_{\ell-1})\, e^{i\eta_\ell} \cos\left(kr - \frac{\ell\pi}{2} + \eta_\ell\right)\Big]$$

$$\vec{H} \underset{r \mapsto \infty}{\sim} \frac{\sqrt{\pi}}{ikr} \sum_{\ell=1}^{\infty} \left[\frac{2\ell+1}{\ell(\ell+1)}\right]^{1/2} i^{\ell-1}\Big[(\vec{\mu}_{\ell} + \vec{\mu}_{\ell-1})\, e^{i\eta_\ell} \sin\left(kr - \frac{\ell\pi}{2} + \eta_\ell\right)$$

$$+ \frac{1}{2}(\vec{\nu}_{\ell 1} + \vec{\nu}_{\ell-1})\, e^{i\delta_\ell} \cos\left(kr - \frac{\ell\pi}{2} + \delta_\ell\right)\Big] \tag{2.23}$$

The total fields having the above asymptotic expressions, and satisfying Eqs. (2.3) everywhere, can be written in the form:

$$\vec{E} = \frac{\sqrt{\pi}}{kn^2} \sum_{\ell=1}^{\infty} \left[\frac{2\ell+1}{\ell(\ell+1)}\right]^{1/2} i^{\ell-1} \left(\vec{M}^{(4)}_{\ell 1} + \vec{N}^{(4)}_{\ell 1} - \vec{M}^{(4)}_{\ell\text{-}1} + \vec{N}^{(4)}_{\ell\text{-}1}\right)$$

$$\vec{H} = \frac{\sqrt{\pi}}{ik} \sum_{\ell=1}^{\infty} \left[\frac{2\ell+1}{\ell(\ell+1)}\right]^{1/2} i^{\ell-1} \left(\vec{M}^{(3)}_{\ell 1} + \vec{N}^{(3)}_{\ell 1} + \vec{M}^{(3)}_{\ell\text{-}1} - \vec{N}^{(3)}_{\ell\text{-}1}\right) \tag{2.24}$$

where

$$\vec{M}^{(3)}_{\ell m} = F_\ell(r)\vec{\mu}_{\ell m}$$

$$\vec{N}^{(3)}_{\ell m} = \frac{1}{kr} G_\ell(r)\vec{\tau}_{\ell m} + \frac{1}{2kr}\frac{d}{dr}\left[rG_\ell(r)\right]\vec{\nu}_{\ell m} \tag{2.25}$$

and

$$\vec{M}^{(4)}_{\ell m} = \frac{1}{k}\nabla \times \vec{N}^{(3)}_{\ell m} = n^2 G_\ell(r)\vec{\mu}_{\ell m}$$

$$\vec{N}^{(4)}_{\ell m} = \frac{1}{k}\nabla \times \vec{M}^{(3)} = \frac{1}{kr} F_\ell(r)\vec{\tau}_{\ell m} + \frac{1}{2kr}\frac{d}{dr}\left[rF_\ell(r)\right]\vec{\nu}_{\ell m} \tag{2.26}$$

with the radial functions F_ℓ and G_ℓ **regular** (in the origin) solutions of the equations

$$\frac{d^2}{dr^2}\frac{rF_\ell(r)}{n(r)} + \left[k^2n^2(r) - \frac{\ell(\ell+1)}{r^2} - U(r)\right]\frac{rF_\ell(r)}{n(r)} = 0$$

$$\frac{d^2}{dr^2} rG_\ell(r) + \left[k^2n^2(r) - \frac{\ell(\ell+1)}{r^2}\right] rG_\ell(r) = 0, \tag{2.27}$$

and behaving asymptotically as:

$$rF_\ell(r) \underset{r \mapsto \infty}{\sim} e^{i\eta_\ell} \sin\left(kr - \frac{\ell\pi}{2} + \eta_\ell\right)$$

$$rG_\ell(r) \underset{r \mapsto \infty}{\sim} e^{i\delta_\ell} \cos\left(kr - \frac{\ell\pi}{2} + \delta_\ell\right). \tag{2.28}$$

We will refer to the total fields defined by the asymptotic conditions (2.28) as the **physical** fields. In addition to these solutions of Eqs. (2.3), we shall also construct auxiliary solutions playing an important role in the following section. One such set of solutions, called regular, are defined as solutions of the Eqs. (2.27):

$$\frac{d^2}{dr^2}\frac{u_\ell(r)}{n(r)} + \left[k^2n^2(r) - \frac{\ell(\ell+1)}{r^2} - U(r)\right]\frac{u_\ell(r)}{n(r)} = 0$$

$$\frac{d^2}{dr^2}v_\ell(r) + \left[k^2n^2(r) - \frac{\ell(\ell+1)}{r^2}\right]v_\ell(r) = 0 \tag{2.29}$$

through their behavior at the origin

$$u_\ell(r),\ v_\ell(r) \underset{r\to 0}{\sim} \frac{r^{\ell+1}}{(2\ell+1)!!} \tag{2.30}$$

which, up to a factor, is also the behavior at the origin of the function $rj_\ell(kr)$, a solution of either of the Eqs. (2.29) with $n \equiv 1$.

Another set of solutions, called **Jost solutions,** are defined as solutions of Eqs. (2.29):

$$\frac{d^2}{dr^2}\frac{U_\ell(k,r)}{n(r)} + \left[k^2n^2(r) - \frac{\ell(\ell+1)}{r^2} - U(r)\right]\frac{U_\ell(k,r)}{n(r)} = 0$$

$$\frac{d^2}{dr^2}V_\ell(k,r) + \left[k^2n^2(r) - \frac{\ell(\ell+1)}{r^2}\right]V_\ell(k,r) = 0 \tag{2.31}$$

with the asymptotic behavior:

$$U_\ell(k,r) \underset{r\to\infty}{\sim} e^{i(kr+\ell\pi/2)}$$

$$V_\ell(k,r) \underset{r\to\infty}{\sim} e^{i(kr+[\ell-1]\pi/2)} \tag{2.32}$$

which, up to constant factors, is also the asymptotic behavior of the function $rh_\ell^{(1)}(kr)$ another solution of the equations (2.31) with $n \equiv 1$. It is important to note that, if $k \neq 0$, the solution $U_\ell(-k,r)$, is independent of $U_\ell(k,r)$, as $V_\ell(-k,r)$ is independent of $V_\ell(k,r)$ since their Wronskians are non-vanishing:

$$W\left[\frac{U_\ell(k,r)}{n(r)}, \frac{U_\ell(-k,r)}{n(r)}\right] = \frac{1}{n^2}W\left[U_\ell(k,r), U_\ell(-k,r)\right] = 2ik\,(-1)^{\ell+1}$$

$$W\left[V_\ell(k,r), V_\ell(-k,r)\right] = 2ik\,(-1)^\ell \tag{2.33}$$

However, it follows from (2.31) and (2.32) that

$$U_\ell(-k,r) = (-1)^\ell\, U_\ell^*(k,r)$$

$$V_\ell(-k,r) = (-1)^{\ell-1}\, V_\ell^*(k,r)\,. \tag{2.34}$$

Consequently, the regular solutions, which are real, even functions of k, can be written in the form of the linear combinations:

$$u_\ell(r) = \frac{i}{2k^{\ell+1}}\left[U_\ell(k)\,U_\ell(-k,r) - (-1)^\ell U_\ell(-k)\,U_\ell(k,r)\right]$$

$$v_\ell(r) = \frac{i}{2k^{\ell+1}}\left[V_\ell(k)\,V_\ell(-k,r) - (-1)^\ell V_\ell(-k)\,V_\ell(k,r)\right] \tag{2.35}$$

in which the coefficients $U_\ell(k)$ and $V_\ell(k)$ are, by definition, the so-called Jost functions. From (2.35) and (2.33) it also follows that the Jost functions can be given explicitly as:

$$U_\ell(k) = (-k)^\ell W\left[\frac{U_\ell(k,r)}{n(r)}, \frac{u_\ell(r)}{n(r)}\right]$$

$$V_\ell(k) = -(-k)^\ell W\left[V_\ell(k,r), v_\ell(r)\right] \tag{2.36}$$

In view of Eqs. (2.34) and the reality of the regular solutions, Eqs. (2.36) imply that

$$U_\ell(-k) = U_\ell^*(k)$$

$$V_\ell(-k) = -V_\ell^*(k)\,. \tag{2.37}$$

From Eqs. (2.35) one can also obtain the asymptotic expressions of the regular solutions. These expressions in fact must be, up to constant factors, those of the physical solutions, as given by Eqs. (2.28) (the physical solutions are also regular and thus, up to multiplicative constants, equal to the regular solutions). To reproduce the behavior (2.28) from Eqs. (2.35), one must assume that

$$U_\ell(k) = |U_\ell(k)|e^{-i\eta_\ell}$$

$$V_\ell(k) = |V_\ell(k)|e^{-i\delta_\ell} \tag{2.38}$$

One obtains thus the asymptotic expressions

$$u_\ell(r) \underset{r\mapsto\infty}{\sim} \frac{|U_\ell(k)|}{k^{\ell+1}} \sin\left(kr - \frac{\ell\pi}{2} + \eta_\ell\right)$$

$$v_\ell(r) \underset{r\mapsto\infty}{\sim} \frac{|V_\ell(k)|}{k^{\ell}} \cos\left(kr - \frac{\ell\pi}{2} + \delta_\ell\right) \tag{2.39}$$

which establish the connection between the physical phase-shifts and the arguments of the complex Jost functions.

While listing the various sets of solutions of the Eqs. (2.27) it is also of interest to mention the lack of a special type of solutions (generally present in Quantum Mechanics), namely bound state solutions, defined as real, regular, square-integrable solutions corresponding to **negative** values of k^2 (purely imaginary k, with positive imaginary part). The absence of such solutions in the problem under consideration could be argued on physical, intuitive grounds, but a direct mathematical proof can be readily exhibited. Indeed considering first the second of the equations (2.27), assume that $w_\ell(r)$ is a square integrable solution of the equation:

$$\frac{d^2}{dr^2} w_\ell(r) + \left[k^2 n^2(r) - \frac{\ell(\ell+1)}{r^2}\right] w_\ell(r) = 0. \tag{2.40}$$

Multiplying this equation by $w_\ell(r)$ and integrating the first term by parts, one obtains

$$\int_0^\infty \left[\left(\frac{dw_\ell}{dr}\right)^2 + \frac{\ell(\ell+1)}{r^2} w_\ell^2\right] dr = k^2 \int_0^\infty n^2 w_\ell^2 \, dr\,. \tag{2.41}$$

The integrated part is discarded because, to be square integrable, the solution w must vanish at infinity, while in the origin it vanishes like $r^{\ell+1}$. It is clear now that Eq. (2.41) cannot be satisfied with negative values of k^2.

The same conclusion can be drawn for the first of Eqs. (2.27), but not as directly as for the second equation. In this case, it is convenient to first apply to the equation

$$\frac{d^2}{dr^2} w_\ell + \left[k^2 n^2 - \frac{\ell(\ell+1)}{r^2} - n \frac{d^2}{dr^2}\frac{1}{n}\right] w_\ell = 0 \tag{2.42}$$

a Liouville transformation, consisting of the transformation of variable:

$$r \mapsto s:\ s = \int_0^r n^2(t)dt \tag{2.43}$$

and simultaneously the transformation of function:

$$w_\ell \mapsto z_\ell:\ z_\ell(s) = m(s) w_\ell(r) \tag{2.44}$$

where $m(s) = n(r(s))$. Then

$$\frac{d^2 w_\ell(r)}{dr^2} = m^2(s)\left[m(s)\frac{d^2 z_\ell(s)}{ds^2} - z_\ell(s)\frac{d^2 m(s)}{ds^2}\right]$$

and

$$\frac{d^2}{dr^2}\frac{1}{n(r)} = -\,m^2(s)\frac{d^2 m(s)}{ds^2}$$

so that Eq. (2.42) becomes

$$\frac{d^2 z_\ell(s)}{ds^2} + \left[\frac{k^2}{m^2(s)} - \frac{\ell(\ell+1)}{m^4(s) r^2(s)}\right] z_\ell(s) = 0 \tag{2.45}$$

Because the transformation (2.43) is monotonic (indeed linear for $r > r_o$) and $s \sim r$ in the origin, the argument presented above can be applied to Eq. (2.43), with the general conclusion that no bound states are present in the context of our scattering problem.

3. INVERSE QUANTUM MECHANICAL SCATTERING

In the inverse electromagnetic scattering problem we are finally ready to begin discussing, the unknown is the function n(r). The input, obviously, must involve the vector scattering amplitude, which is given by the expression (2.21) in terms of two infinite sequences of phase-shifts, each of which is a function of k. The problem of retrieving the scattering amplitude from measurements, while important, will not be considered here. We assume that whatever information regarding the phase-shifts $\eta_\ell(k)$ and $\delta_\ell(k)$ needed to solve the inverse scattering problem is available.

The general approach adopted is that the inverse electromagnetic problem can be solved by reducing it to the quantum mechanical inverse scattering problem. The motivation for this approach[2] is based on the observation that either of the Eqs. (2.27) can be given the form of the reduced Schroedinger equation for radial functions

$$\frac{d^2}{dr^2}\psi_\ell(r) + \left(k^2 - \frac{\ell(\ell+1)}{r^2}\right)\psi_\ell(r) = V(r)\,\psi_\ell(r) \tag{3.1}$$

where, for the first of Eq. (2.27)

$$V(r) = k^2(1 - n^2(r)) + n(r)\frac{d^2}{dr^2}\frac{1}{n(r)}, \tag{3.1a}$$

while for the second,

$$V(r) = k^2(1 - n^2(r)) \tag{3.1b}$$

As far as the inverse scattering problem at fixed k is concerned, this is indeed all that is required for the reduction, since the techniques available for the reconstruction of a potential V(r) from one set of phase shifts values are directly applicable to the problem at hand. Moreover, since the potential to be reconstructed is known to have finite range, the reconstruction is theoretically unique. From a practical viewpoint, there still are questions relating to the use of only a finite number of phase shifts and to numerical stability which deserve more attention.

The inverse electromagnetic problem of the reconstruction of the function n(r) from **one** phase-shift, assumed known as a function of k, also reduces to the corresponding quantum mechanical inverse problem, but not as directly as the problem at fixed k. The main reason for the complications arising in this case resides in the k dependence of the potential appearing in the Schroedinger equation (3.1). However, this dependence can be readily disposed of by means of a Liouville transformation. Namely, if it is assumed that $\delta_\ell(k)$ is the phase shift available for the reconstruction problem, then we must investigate Eq. (3.1) with the potential (3.1b). In this case, the Liouville transformation of the equation for the physical radial solution (the asymptotic behavior of which is given by the second of Eqs. (2.28)), defined as:

$$r \mapsto \rho: \qquad \rho(r) = \int_0^r n(s)ds$$

$$\psi_\ell(r) \mapsto \bar{\psi}_\ell(\rho): \quad \bar{\psi}_\ell(\rho) = \sqrt{\nu(\rho)}\; u_\ell(r(\rho)) \tag{3.2}$$

where $\nu(\rho) = n(r(\rho))$, yields the Schroedinger-like equation

$$\frac{d^2}{d\rho^2}\bar{\psi}_\ell(\rho) + \left[k^2 - \frac{\ell(\ell+1)}{R^2(\rho)}\right]\bar{\psi}_\ell(\rho) = \bar{V}(\rho)\,\bar{\psi}_\ell(\rho) \tag{3.3}$$

with a potential independent of k:

$$\bar{V}(\rho) = [\nu(\rho)]^{-1/2}\frac{d^2}{d\rho^2}\,\nu(\rho)^{1/2} \tag{3.4}$$

and a "centrifugal potential" in which

$$R(\rho) = \nu(\rho)\,r(\rho)\,, \tag{3.5}$$

a function of ρ that can only vanish in the origin:

$$R(\rho) \underset{\rho \mapsto 0}{\sim} \rho \tag{3.6}$$

and which for $\rho \geq \rho_o$, where

$$\rho_o = \int_o^{r_o} n(s)ds\,, \tag{3.7}$$

becomes linear:

$$R(\rho) = \rho + r_o - \rho_o \cdot (\rho \geq \rho_o) \tag{3.8}$$

Eq. (3.3) can also be written as a Schroedinger equation:

$$\frac{d^2}{d\rho^2}\bar{\psi}_\ell(\rho) + \left[k^2 - \frac{\ell(\ell+1)}{\rho^2}\right]\bar{\psi}_\ell(\rho) = \bar{V}_\ell(\rho)\,\bar{\psi}_\ell(\rho) \tag{3.9}$$

with the ℓ-dependent potential:

$$\bar{V}_\ell(\rho) = \bar{V}(\rho) - \ell(\ell+1)\left[\frac{1}{\rho^2} - \frac{1}{R^2(\rho)}\right]. \tag{3.10}$$

Because of our assumptions on n(r) and the definition (3.5),

$$\lim_{\rho \mapsto 0} \rho\,\bar{V}_\ell(\rho) = -\,\ell(\ell+1)\,\frac{n'(o)}{n(o)}\,, \tag{3.11}$$

while asymptotically,

$$\lim_{\rho \mapsto \infty} \rho^3\,\bar{V}_\ell(\rho) = 2\ell(\ell+1)\,(\rho_o - r_o) \tag{3.12}$$

and hence both the integrals

$$\int_0^\infty \rho\,|\bar{V}_\ell(\rho)|\,d\rho \text{ and } \int_0^{\rho_o} \rho^2|\bar{V}_\ell(\rho)|\,d\rho \text{ exist.} \tag{3.13}$$

It follows then that a regular solution $\bar{v}_\ell(\rho)$ of the equation (3.9), obeying the boundary condition

$$\bar{v}_\ell(\rho) \underset{\rho \mapsto o}{\sim} \frac{\rho^{\ell+1}}{(2\ell+1)!!}, \tag{3.14}$$

exists and can be obtained by iteration as a solution of the Volterra integral equation

$$\bar{v}_\ell(\rho) = k^{-\ell}\rho j_\ell(k\rho) + \int_o^\rho g_\ell(k,\rho,\rho')\,\bar{V}_\ell(\rho')\,\bar{v}_\ell(\rho')d\rho'\,. \tag{3.15}$$

where

$$g_\ell(k,\rho,\rho') = \frac{i(-1)^\ell}{2}\,k\rho\rho'\Big[\,h_\ell^{(1)}(k\rho)\,h_\ell^{(1)}(-k\rho') - h_\ell^{(1)}(k\rho')\,h_\ell^{(1)}(-k\rho)\Big] \tag{3.16}$$

In our case, Eq. (3.15) only needs to be solved for $0 \le \rho \le \rho_o$, since for $\rho \ge \rho_o$, $\bar{v}_\ell(\rho)$is a linear combination of the functions $rj_\ell(kr)$ and $ry_\ell(kr)$ where $r = \rho - \rho_o + r_o$, the value of $R(\rho)$ or of $r(\rho)$ for such ρ. It is also known that, for fixed ρ, $\bar{v}_\ell(\rho)$ is an entire function of k.

Similarly, we can define a Jost solution of Eq. (3.9) through the asymptotic condition

$$\bar{V}_\ell k,\rho) \underset{\rho \mapsto \infty}{\sim} e^{i(k\rho + [\ell-1]\pi/2)} \tag{3.17}$$

Actually, $\bar{V}_\ell(k,\rho)$ is known explicitly for $\rho \ge \rho_o$ to be the function

$$\bar{V}_\ell^{II}(k,\rho) = e^{i\pi\ell}kr\;h_\ell^{(1)}(kr)\,e^{ik(\rho_o - r_o)} \tag{3.18}$$

where again r stands for $\rho - \rho_o + r_o$. The expression $\bar{V}_\ell^I(k,\rho)$ of $\bar{V}_\ell(k,\rho)$ for $0 < \rho \le \rho_o$ is then a solution of the Volterra equation:

$$\bar{V}_\ell^I(k,\rho) = \bar{V}_\ell^{II}(k,\rho_o) + w_\ell(k\rho) - w_\ell(k\rho_o)$$

$$- \int_\rho^{\rho_o} g_\ell(k,\rho,\rho')\bar{V}_\ell(\rho')\bar{V}_\ell^I(k,\rho')\,d\rho'\,, \tag{3.19}$$

in which

$$w_\ell(k\rho) = e^{i\pi\ell}k\rho\;h_\ell^{(1)}(k\rho)\,. \tag{3.20}$$

It can be shown now that, under the conditions (3.13), $\bar{V}_\ell^I(k,\rho)$and thus $\bar{V}_\ell(k,\rho)$ can be obtained by iterating Eq. (3.19) and that, for fixed $\rho > 0$, $k^\ell\bar{V}_\ell(k,\rho)$ thus constructed is an entire function of k of exponential type, vanishing at infinity in the upper complex k-plane. Another Jost solution, $\bar{V}_\ell(-k,\rho)$, can also be introduced, **mutatis mutanda,** and finally the Jost functions $\bar{V}_\ell(k)$and $\bar{V}_\ell(-k)$can be defined through the relation

$$\bar{v}_\ell(\rho) = \frac{i}{2k^{\ell+1}} [\bar{V}_\ell(k)\,\bar{V}_\ell(-k,\rho) - (-1)^\ell\,\bar{V}_\ell(-k)\,\bar{V}_\ell(k,\rho)] \tag{3.21}$$

or, explicitly, as

$$\bar{V}_\ell(k) = -(-k)^\ell W\,[\bar{V}_\ell(k,\rho), \bar{v}_\ell(\rho)] \tag{3.22}$$

From Eqs. (3.15), (3.17) and (3.21) with $\rho \mapsto \infty$ it follows that

$$\bar{V}_\ell(k) = i - k^{\ell+1} \int_0^\infty \rho'\, h_\ell^{(1)}(k\rho')\,\bar{V}_\ell(\rho')\,\bar{v}_\ell(\rho')d\rho' \tag{3.23}$$

and that

$$\bar{V}_\ell(-k) = i + (-k)^{\ell+1} \int_0^\infty \rho'\, h_\ell^{(1)*}(k\rho')\,\bar{V}_\ell(\rho')\,\bar{v}_\ell(\rho')d\rho' = -\bar{V}_\ell^{(1)*}(k) \tag{3.24}$$

Hence

$$\bar{V}_\ell(k) \underset{|k| \mapsto \infty}{\sim} i\,. \tag{3.25}$$

If then one sets

$$\bar{V}_\ell(k) = |\bar{V}_\ell(k)| e^{-i\bar{\delta}_\ell + i\pi/2},\ \bar{\delta}_\ell \text{ real,} \tag{3.26}$$

it follows that

$$e^{2i\bar{\delta}_\ell} \equiv \frac{\bar{V}_\ell(-k)}{\bar{V}_\ell(k)} \underset{|k| \mapsto \infty}{\sim} 1 \tag{3.27}$$

i.e., that

$$\bar{\delta}_\ell(k) \underset{|k| \mapsto \infty}{\sim} 0\,. \qquad (\text{mod } 2\pi) \tag{3.28}$$

The meaning of $\bar{\delta}_\ell(k)$ emerges from an examination of the asymptotic form of the regular solution, $\bar{v}_\ell(\rho)$. Indeed, from Eqs. (3.17), (3.21) and (3.26)

$$\begin{aligned} \bar{v}_\ell(\rho) &\underset{\rho \mapsto \infty}{\sim} \frac{|\bar{V}_\ell(k)|}{k^{\ell+1}} \sin(k\rho - \ell\pi/2 + \bar{\delta}_\ell) \\ &= \frac{\bar{V}_\ell(k)}{k^{\ell+1}} e^{i\bar{\delta}_\ell} \sin(k\rho - \ell\pi/2 + \bar{\delta}_\ell) \\ &= \frac{\bar{V}_\ell(k)}{2k^{\ell+1}} i^{\ell+1} [e^{-ik\rho} - (-1)^\ell e^{2i\bar{\delta}_\ell} e^{ik\rho}]\,, \end{aligned} \tag{3.29}$$

so that $\bar{\delta}_\ell$ are the phase-shifts of a quantum mechanical scattering by the potential $\bar{V}_\ell(\rho)$ described by a "physical" solution with the asymptotic behavior

$$\bar{\psi}_\ell(\rho) \underset{\rho \mapsto \infty}{\sim} e^{i\bar{\delta}_\ell} \sin(k\rho - \ell\pi/2 + \bar{\delta}_\ell)\,. \tag{3.30}$$

The behavior (3.28) of $\bar{\delta}_\ell(k)$ as $k \mapsto \infty$ is "physically" reasonable since, as the energy increases, the effect of the potential $\bar{V}_\ell(\rho)$, itself independent of energy, should correspondingly fade away.

However, this quantum mechanical process does not occur as such, so that it is "physical" only to the extent that it can be related to the truly physical process we are investigating, namely the scattering of electromagnetic waves by a sphere of index of refraction n(r). The phase-shift $\bar{\delta}_\ell$ is not one of the real phase-shifts δ_ℓ appearing in Eq. (2.28). Nevertheless, these phase-shifts are interrelated, since $\psi_\ell(\rho)$ and $\psi_\ell(r) = G_\ell(r)$ are interrelated through the Liouville transformation defined by Eqs. (3.2). It follows thus that

$$\delta_\ell(k) = \bar{\delta}_\ell(k) + k(\rho_o - r_o) - \frac{\pi}{2} \tag{3.31}$$

which implies that $\delta_\ell(k)$ increase linearly with k, a reasonable conclusion in view of the fact that the potential (3.1b) also increases with k (quadratically).

The relation (3.31) shows now that if $\delta_\ell(k)$ for some fixed ℓ is known as a function of k, then $\bar{\delta}_\ell(k)$ is also known since, taking into account Eq. (3.28),

$$\bar{\delta}_\ell(k) = \delta_\ell(k) - k \lim_{s \mapsto \infty} s^{-1}\delta_\ell(s) + \frac{\pi}{2} . \tag{3.32}$$

The electromagnetic inverse scattering problem of finding the index of refraction n(r), given a phase-shift is thus reduced to the quantum- mechanical inverse scattering problem of finding the potential $\bar{V}_\ell(\rho)$, given the phase-shift $\bar{\delta}_\ell(k)$. The latter has a well-established solution we will now briefly describe.

One begins by noting that, since both $w_\ell(k\rho)$ and $k\rho\, j_\ell(k\rho)$ satisfy Eq. (3.9) with an identically zero right-hand side,

$$\int_\rho^\infty w_\ell(k\rho')j_\ell(k'\rho')d\rho' = \frac{w_\ell(k\rho)\frac{d}{d\rho}[k'\rho j_\ell(k'\rho)] - k'\rho j_\ell(k'\rho)\frac{d}{d\rho}w_\ell(k\rho)}{k'^2 - (k + i\epsilon)^2} \tag{3.33}$$

Then one considers the integral

$$\mathfrak{F}_\ell(k, \rho) = \frac{(-1)^\ell}{\pi} \int_C dk' \bar{V}_\ell(k', \rho) \int_\rho^\infty d\rho' w_\ell(k\rho') k'\rho' j_\ell(k'\rho') , \tag{3.34}$$

where the closed contour of integration, C, consists of the real axis and a semi-circle of indefinitely large radius in the upper half-plane. The integral (3.34) is evaluated in two different ways.

First, because the integrand is an analytic function in the upper half-plane, except for a pole at $k + i\epsilon$, the theorem of residues can be employed. Since at this pole the numerator in (3.33) becomes a Wronskian, one obtains immediately the result:

$$\mathfrak{F}_\ell(k, \rho) = \bar{V}_\ell(k, \rho) . \tag{3.35}$$

Second, the contributions from the real axis and semi-circular portions of C are separated, viz.,

$$\mathfrak{F}_\ell(k,\rho) = \mathfrak{F}_\ell^{(1)}(k,\rho) + \mathfrak{F}_\ell^{(2)}(k,\rho) \tag{3.36}$$

and these contributions are separately evaluated. Because

$$k\rho j_\ell(k\rho) = -\frac{1}{2}\,[w_\ell(-k\rho) - (-1)^\ell w_\ell(k\rho)]\,, \tag{3.37}$$

the contribution of the real axis portion can be written as

$$\mathfrak{F}_\ell^{(1)}(k,\rho) = \frac{(-1)^{\ell+1}}{2\pi}\int_{-\infty}^{\infty} dk'[\bar{V}_\ell(-k,\rho) - (-1)^\ell\,\bar{V}_\ell(k',\rho)]$$

$$\times\int_\rho^\infty d\rho'\,w_\ell(k\rho')w_\ell(k'\rho')\,. \tag{3.38}$$

Using the relation between the Jost solution and the physical solution

$$\bar{\psi}_\ell(k,\rho) = \frac{k^{\ell+1}}{\bar{V}_\ell(k)}\,\bar{v}_\ell(\rho) = \frac{i}{2}\,[\bar{V}_\ell(-k,\rho) - (-1)^\ell e^{2i\bar{\delta}_\ell(k)}\,\bar{V}_\ell(k,\rho)]\,, \tag{3.39}$$

one obtains

$$\mathfrak{F}_\ell^{(1)}(k,\rho) = \frac{-1}{2\pi}\int_{-\infty}^{\infty} dk'\,[e^{2i\bar{\delta}_\ell(k')}-1]\bar{V}_\ell(k',\rho)\int_\rho^\infty d\rho'\,w_\ell(k\rho')w_\ell(k'\rho')$$

$$+\frac{(-1)^{\ell+1}}{\pi i}\int_{-\infty}^{\infty} dk'\,\bar{\psi}_\ell(k',\rho)\int_\rho^\infty d\rho'\,w_\ell(k\rho')\,w_\ell(k'\rho')\,. \tag{3.40}$$

Because the integrand of the term containing the physical solution $\bar{\psi}_\ell(k,\rho)$ is an entire function of k′ (as we have seen, there are no bound states in this problem), this integral equals its negative evaluated along the semi-circle portion of C. Thus it can be combined with $\mathfrak{F}_\ell^{(2)}$ and evaluated by using the asymptotic forms of the solutions for large k in Imk ≥ 0. The combined result is $w_\ell(k\rho)$. Therefore the integral equation

$$\bar{\psi}_\ell(k,\rho) = w_\ell(k\rho) - \frac{1}{2\pi}\int_{-\infty}^{\infty} dk'\,[e^{2i\delta_\ell(k')} - 1]\bar{\psi}_\ell(k',\rho)$$

$$\times\int_\rho^\infty d\rho' w_\ell(k\rho')w_\ell(k'\rho')\,, \tag{3.41}$$

is obtained which, in principle, solves the inverse scattering quantum mechanical problem: given $\bar{\delta}_\ell(k)$, one solves Eq. (3.41) for the Jost solution $\bar{V}_\ell(k,\rho)$ which, when substituted in Eq. (3.1), allows for immediate identification of the potential $\bar{V}_\ell(\rho)$.

However, Eq. (3.41) has the unpleasant feature of being singular, a circumstance which, when possible, is preferably avoided. In the present case, it can be avoided by recasting the solution according to the so-called Marchenko formulation. By introducing the function

$$A_\ell(\rho, \sigma) = \frac{(-1)^\ell}{\pi} \int_{-\infty}^{\infty} [\bar{\Psi}_\ell(k, \rho) - w_\ell(k, \rho)]\, k\sigma j_\ell(k\sigma)\, dk \tag{3.42}$$

and taking into account the bounds:

$$|\bar{V}_\ell(k, \rho) - w_\ell(k, \rho)| < C \left(\frac{1 + |k|\rho}{|k|\rho} \right)^\ell e^{-\rho \mathrm{Im} k}$$

$$|k\sigma j_\ell(k\sigma)| < e^{\sigma \mathrm{Im} k}, \tag{3.43}$$

one recognizes that

$$A_\ell(\rho, \sigma) = 0 \qquad \text{for } \rho > \sigma . \tag{3.44}$$

The integral representation

$$\bar{V}_\ell(k, \rho) = w_\ell(k\rho) + \int_\rho^\infty A_\ell(\rho, \sigma) w_\ell(k\sigma)\, d\sigma \tag{3.45}$$

is obtained by taking the Hankel transform of Eq. (3.42). Substitution of this representation in Eq. (3.41) yields the Marchenko equation:

$$A_\ell(\rho, \sigma) = B_\ell(\rho, \sigma) + \int_\rho^\infty d\tau\, A_\ell(\rho, \tau) B_\ell(\tau, \sigma), \tag{3.46}$$

where

$$B_\ell(\rho, \sigma) = \frac{1}{2\pi} \int_{-\infty}^{\infty} dk [e^{2i\bar{\delta}_\ell(k)} - 1] w_\ell(k\rho) w_\ell(k\sigma) . \tag{3.47}$$

Finally, one establishes the direct connection between $A_\ell(\rho, \sigma)$ and the potential $\bar{V}_\ell(\rho)$:

$$\bar{V}_\ell(\rho) = -2 \frac{d}{d\rho} A_\ell(\rho, \rho) . \tag{3.48}$$

Eqs. (3.46), (3.47) and (3.48) provide the solution to the quantum mechanical inverse scattering problem: given $\bar{\delta}_\ell(k)$, one evaluates the function $B_\ell(\rho, \sigma)$ by calculating the integral (3.47). Substitution of the result in Eq. (3.46) allows integration of the Marchenko equation, the solution of which provides the potential $\bar{V}_\ell(\rho)$.

4. RECONSTRUCTION OF n(r)

Once the quantum mechanical inversion is carried out and the potential $\bar{V}_\ell(\rho)$ is found, we may return to the electromagnetic inverse problem. The task now is, given $\bar{V}_\ell(\rho)$, find $\nu(\rho)$. In other words, we must solve Eq. (3.10) which, with (3.4), reads

$$[\nu(\rho)]^{-1/2} \frac{d^2}{d\rho^2} [\nu(\rho)]^{1/2} - \ell(\ell+1)[\rho^{-2} - R^{-2}(\rho)] = \bar{V}_\ell(\rho) \tag{4.1}$$

where, from Eqs. (3.5) and (3.2),

$$R(\rho) = \nu(\rho) \int_0^\rho \frac{dt}{\nu(t)} \tag{4.2}$$

Eq. (4.1), as an equation for $\nu(\rho)$, has a formidable aspect: strongly nonlinear, integro-differential, singular and, its solution should also be ℓ-independent. Realistically speaking, such an equation could be expected to have a solution, unique and ℓ-independent, only if a "miracle" happened.

However, the nature of the miracle one expects to occur in this case is closely related to the miracle discussed by Newton in his solution of the three dimensional quantum mechanical inverse scattering problem. The situation can be described as follows.

If one starts with a given, arbitrary function $\delta_\ell(k)$ there is no reason to expect the reconstructed potential $\bar{V}_\ell(\rho)$ to be such that Eq. (4.1) will have a solution. Only if $\delta_\ell(k)$ is a phase shift resulting from the actual scattering by a target with the assumed general properties of n(r) is it reasonable to attempt to reconstruct the underlying function n(r). It is not reasonable to attempt to **construct** a function n(r) for each $\delta_\ell(k)$ one can think of, but it does make sense to **reconstruct** the function n(r) which has yielded a given $\delta_\ell(k)$.

The implication of this attitude is that a solution of Eq. (4.1) should be sought only for that class $\mathcal{C}_\ell$ of functions $\bar{V}_\ell(\rho)$ that can be obtained from functions $\delta_\ell(k)$ representing physically performable scattering experiments on objects with the general properties outlined above. In particular, functions $\bar{V}_\ell(\rho)$ from this class must be such that, given $\delta_\ell(k)$ (and thus the limit $\rho_o - r_o = \lim_{s\to\infty} s^{-1}\delta_\ell(s)$), there exists a distance ρ_o such that, for $\rho \geq \rho_o$,

$$\bar{V}_\ell(\rho) = \ell(\ell+1)[\rho^{-2} - (\rho - \rho_o + r_o)^{-2}]. \tag{4.3}$$

Hence, as $\bar{V}_\ell(\rho)$ is given, the distance ρ_o beyond which $\nu(\rho)$ is known to be identically unity is readily identifed. Moreover, the functions in the class $\mathcal{C}_\ell$ must all be such that

$$\bar{V}_\ell(\rho) + \ell(\ell+1)(\rho^{-2} - R^{-2}(\rho)) \text{ is bounded for all } \rho \geq 0, \tag{4.4}$$

including the origin $\rho = 0$, which implies that only such solutions of Eq. (4.1) should be sought for which $R(\rho) > 0$ has the representation

$$R(\rho) = \begin{cases} \rho - \dfrac{[s\, V_\ell(s)]_{s=0}}{2\ell(\ell+1)}\rho^2 + \rho^3 S(\rho) & (0 \leq \rho \leq \rho_o) \\ \\ \rho - \rho_o + r_o & (\rho \geq \rho_o) \end{cases} \tag{4.5}$$

where $S(\rho)$ is a continuous, twice differentiable function, bounded for all $\rho \epsilon [0, \rho_o]$. The fact that such information is available for the function $R(\rho)$ suggests that it might be more convenient to consider this function rather than $\nu(\rho)$ as the unknown function in Eq. (4.1). This change is easily done because Eq. (3.50) can be immediately solved for $\nu(\rho)$ yielding

$$\nu(\rho) = \exp \int_\rho^{\rho_o} \frac{1 - R'(t)}{R(t)\,.}\, dt \tag{4.6}$$

Substitution of (4.6) into Eq. (4.1) yields a second order nonlinear differential equation for R which, in turn, can be written as a system of two first order nonlinear differential equations. Namely, if we introduce the function

$$P(\rho) = \frac{1 - R'(\rho)}{2R(\rho)}\,, \qquad (0 \leq \rho \leq \rho_o) \tag{4.7}$$

the following system results:

$$R' = 1 - 2PR$$

$$P' = P^2 - \ell(\ell+1)(\rho^{-2} - R^{-2}) - \bar{V}_\ell(\rho) \tag{4.8}$$

to be solved for $P(\rho)$ and $R(\rho)$ for $0 \leq \rho \leq \rho_o$ with the boundary conditions $P(\rho_o) = 0$, $R(\rho_o) = r_o$, say, where r_o may be assumed known, since both $\rho_o - r_o$ and ρ_o and have already been identified. The first of the Eqs. (4.8) can then be given the integral form:

$$R(\rho) = r_o \exp\left(2\int_\rho^{\rho_o} P(s)ds\right) - \int_\rho^{\rho_o} du \exp\left(2\int_\rho^{u} P(s)ds\right). \tag{4.9}$$

Similarly, the second of Eqs. (4.8) can be given the integral form:

$$P(\rho) = -\int_\rho^{\rho_o} ds\, P^2(s) + \int_\rho^{\rho_o} [\ell(\ell+1)[s^{-2} - R^{-2}(s)] + \bar{V}_\ell(s)]ds \tag{4.10}$$

The original equation (4.1) is thus reduced to a system of two coupled (non-linear) Volterra equations, (4.9) and (4.10). To prove the existence and uniqueness of a solution (R, P) of this system, all we need to show, in principle, is that there exists a domain D(s,R, P) for the variables s, R and P in which the integrands of Eqs. (4.9) and (4.10) are continuous as functions of s, and satisfy Lipschitz conditions as functions of R or P (which, of course, would be implied if the integrands had in D continuous partial derivatives of first order with respect to R and P). This can be easily seen to be the case, provided one excludes the origin $\rho = 0$.

In order to analyze the situation in the origin we must examine carefully the second integrand in Eq. (4.10). Since $R(\rho)$ must have the representation (4.5), the quantity

$$\alpha = \frac{[s\, V_\ell(s)]_{s=0}}{2\ell(\ell+1)} \tag{4.11}$$

must exist and be ℓ-independent. This implies that the class $\mathcal{C}_\ell$ of potentials $\bar{V}_\ell(\rho)$ corresponding to underlying functions n(r) consists of functions of the form

$$\bar{V}_\ell(\rho) = \begin{cases} \dfrac{2\ell(\ell+1)\,\alpha}{\rho} + \bar{\bar{V}}_\ell(\rho) & (0 \le \rho \le \rho_o) \\ \\ \ell(\ell+1)\,[(\rho - \rho_o + r_o)^{-2} - \rho^{-2}] & (\rho \ge \rho_o) \end{cases} \tag{4.12}$$

where $\bar{\bar{V}}_\ell(\rho)$ is integrable in $[0, \rho_o]$.

Likewise, from (4.5) and (4.7) it follows that $P(\rho)$ must have the representation:

$$P(\rho) = \alpha + \rho\, T(\rho) \tag{4.13}$$

where $T(\rho)$ is continuous in $[0, \rho_o]$, so that

$$P(0) = \alpha \tag{4.14}$$

Using (4.14) and $R(0) = 0$ as boundary conditions, Eqs. (4.8) can be given the integral form

$$R(\rho) = \int_o^\rho du \exp\left(-2\int_u^\rho P(s)\, ds\right) \tag{4.15}$$

$$P(\rho) = \alpha + \int_o^\rho P^2(s)\, ds - \int_o^\rho [\ell(\ell+1)\,[s^{-2} - R^{-2}(s)] + \bar{V}_\ell(s)]ds.$$

The first of these equations coincides with Eq. (4.9) provided

$$r_o = \int_o^{\rho_o} du \exp\left(-2\int_u^{\rho_o} P(s) ds\right) = \int_o^{\rho_o} \frac{du}{\nu(u)}$$

which is actually the definition of ρ_o. This equation also shows that $R(\rho)$ vanishes only at the origin (linearly), and that for all $\rho \epsilon (0, \rho_o]$,

$$\rho^{-1} R(\rho) = 1 - \alpha\rho + \rho^2 S(\rho) > 0\,. \tag{4.16}$$

Taking into account the representations (4.5), (4.12) and (4.13), Eq. (4.15) becomes a system of equations for $S(\rho)$ and $T(\rho)$:

$$S(\rho) = 2\rho^{-3} \int_0^\rho s^2 ds \, [\alpha^2 + \alpha s \, T(s) - T(s)] \exp\left[-2\alpha(\rho - s) - 2 \int_s^\rho t \, T(t) dt\right]$$

$$T(\rho) = \rho^{-1} \int_0^\rho \left\{ [\alpha + s\,T(s)]^2 + \ell(\ell+1)\left(\frac{[\alpha - s\,S(s)]^2}{[1 - \alpha s + s^2 S(s)]^2} + 2\,\frac{\alpha^2 + (1 - \alpha s)\,S(s)}{1 - \alpha s + s^2 S(s)} \right) - \overline{\overline{V}}_\ell(s) \right\} ds \tag{4.17}$$

The values at the origin of the functions $S(\rho)$ and $T(\rho)$ (and of their derivatives) are easily seen to be given in terms of the value of $\overline{\overline{V}}_\ell(\rho)$ (and its derivatives) at the origin:

$$S(o) = \frac{2}{3}\,\frac{\ell(\ell+1)(4\alpha^2 - 7) + 3\,\overline{\overline{V}}_\ell(o)}{3 + 4\,\ell(\ell+1)}$$

$$T(o) = \frac{3\,\alpha^2 + 7\ell(\ell+1) - 3\,\overline{\overline{V}}_\ell(o)}{3 + 4\ell(\ell+1)} \tag{4.18}$$

so that their existence requires corresponding existence conditions on the behavior of $\overline{\overline{V}}_\ell(\rho)$ at the origin. In any event, it is apparent now that the integrand in the second of Eq. (4.17) is no longer singular. The arguments leading now to a proof of existence and uniqueness of the solution are rather standard and will not be detailed here.

Clearly, once $R(\rho)$ is known, $\nu(\rho)$ follows via Eq. (4.6). Hence, the inverse Liouville transformation

$$r(\rho) = \int_0^\rho \frac{d\tau}{\nu(\tau)} \tag{4.19}$$

yields the function $r(\rho)$, the inverse of which, upon substitution in $\nu(\rho)$, finally produces $n(r)$.

To conclude, let us just mention that if the input of the electromagnetic scattering problem were an η_ℓ phase shift rather than a δ_ℓ, the reduced Schroedinger Eq. (3.1) with the potential (3.1a) also becomes a Schroedinger-like equation under the Liouville transformation (3.2), except that in this case, instead of the k-independent potential $\overline{V}(\rho)$ of Eq. (3.4) one obtains the potential

$$\overline{V}(\rho) = [\nu(\rho)]^{1/2} \frac{d^2}{d\rho^2} [\nu(\rho)]^{-1/2}, \tag{4.20}$$

also k-independent. The procedure outlined above can then be followed without essential modifications.

ACKNOWLEDGEMENTS

This work was performed under the McDonnell Douglas Independent Research and Development program.

REFERENCES

1. See e.g.: H. C. van de Hulst, *Light Scattering by Small Particles,* John Wiley and Sons, Inc., New York; R. Newton, *Scattering Theory of Waves and Particles,* McGraw-Hill Book Co., New York 1966; M. Kerker, *The Scattering of Light and other Electromagnetic Radiation,* Academic Press, New York 1969.

2. See e.g.: C. Eftimiu, *Direct and Inverse Scattering by a Sphere of Variable Index of Refraction,* J. Math. Phys. Vol 23, pp. 2140-2146, 1982, and the references given therein.

I.10 (SP.6)

APPLICATIONS OF THE ABEL TRANSFORM IN REMOTE SENSING AND RELATED PROBLEMS

H. Süss

Deutsche Forschungs- und Versuchsanstalt
für Luft- und Raumfahrt
Institut für Hochfrequenztechnik
8031 Oberpfaffenhofen, FRG

ABSTRACT

The Abel transform often applied in remote sensing on account of its analytical solution is used to determine density profiles from integrated line measurements if cylindrical or spherical symmetry is present. The purpose of this paper is to describe and to dispose the common physical aspects for three different fields of application in remote sensing with electromagnetic or elastic waves: atmospheric sounding, plasma- and geophysics. In a further application example - the sounding of the solar corona during the Helios occultation experiment - a combined analytical-numerical method is introduced and fully described. The physical and numerical applicability is demonstrated by inversion of simulated and measured electron content data sets.

1. INTRODUCTION TO THE MATHEMATICAL AND PHYSICAL BACKGROUND

In 1826 Niels Henrik Abel solved in one of his earliest works [1] the tautochrone problem, which arises in mathematical physics and was posed by Huygens: A material point moving under the influence of gravity along a curve s(y) in a vertical xy-plane takes the time $t(h = y_o)$ to move from the vertical height h to a fixed point O on the curve. The question is to find a curve s(y) such that $T = T(h)$ is a given function. As usual using the mechanical energy conservation law

potential energy: $mg\,(h - y)$
kinetic energy: $1/2\ mv^2$

(m: mass, g: acceleration of gravity) and the common expression for

W.-M. Boerner et al. (eds.), Inverse Methods in Electromagnetic Imaging – Part 1, 177–201.

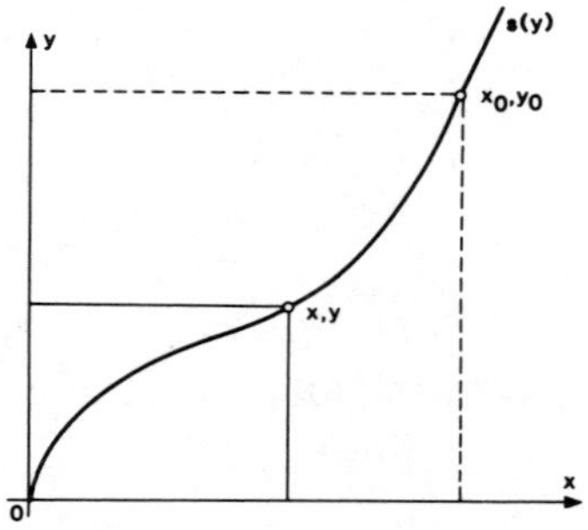

Figure 1. Representation of the coordinate frame for a material point moving along the curve s(y).

velocity v = ds(y)/dt the expression

$$dt = - \frac{ds(y)}{\sqrt{2g(h - y)}}$$

follows. Substitution of the relation ds/dy = s'(y) and integration from y = 0 to y = h leads to the following integral

$$T(h) = \int_0^h \frac{u(y)}{\sqrt{h - y}}\, dy, \qquad (1)$$

where $T(h) = (2g)^{-1/2}\, t(h)$ and s'(y) = u(y). The problem is to find the function u(y) when T(h) is given in such a way that equation (1) holds. The difficulty arises that T(h) depends not only on a single value but on all values of the unknown function u(y). The above equation is known as the Abelian equation.

The arguments for detailed derivation of this problem which is relatively easy to survey and often described concern the analogies between classical mechanics, geometrical optics and wave mechanics. The link is given by the formal correspondence between the Schrödinger equation, the iconal and the Hamilton-Jacobian equation (see [2]). They lead immediately from classical mechanics to other ranges of applications of the Abel transform e.g. to remote sensing with electromagnetic or seismic waves. Simplifying assumptions in mechanics namely the model of the material point, the frictionless gliding along the curve and the presence of the gravitation potential (only depending on the radial distance from the earth's center) are also necessary in the other scopes. In the field of remote sensing - the main point of this paper - it will be assumed that Fermat's principle is valid. This means that diffraction and interference effects can be neglected. Furthermore Fermat's principle includes that the propagation medium is reciprocal and the characteristical scale length is much greater than the vacuum wavelength of the sounding wave.

Equation (1) originates by a chain of simplifying assumptions of the generalized Abel transform. Now this course will be followed in the opposite direction. Equation (1) belongs to a class of integral equations of the more general type

$$T(h) = \int_{o}^{h} \frac{s(y)}{(h - y)^{\lambda}} dy, \qquad (2)$$

where λ takes a value between 0 and 1 (for the case $\lambda = 1/2$ the equations (1) and (2) get identity). The singular integral equation is one of the Volterra type of first the kind. For equation (2) exists the analytical solution

$$s(h) = \frac{\sin \pi \lambda}{\pi} \int_{o}^{h} \frac{T'(y)}{(h - y)^{1-\lambda}} dh. \qquad (3)$$

For $\lambda = 1/2$ the solution of the special case of equation (1) can easily be derived.

As a further generalization of equation (2) there exists the generalized Abel equation

$$T(h) = u(h) \int_{o}^{h} \frac{s(y)}{(h-y)^{\lambda}} dy + v(h) \int_{h}^{b} \frac{s(y)}{(h-y)^{\lambda}} dy, \qquad (4)$$

where the given functions u(h) and v(h) are defined on the interval [a, b] with derivatives satisfying Hölder's condition. A description of their extensive solutions which rarely arise in elastic theory, is given for review in [3, 4, 5].

Only for complete termination of the pursued course should the mathematical treatement of systems of generalized Abelian integral equations be mentioned here. Whereas mathematical solutions can be found in [6, 7], practical application in remote sensing is outside the author's scope.

After this short outline of the mathematical background the basics of the physical background will follow for an illustrative understanding of the application in remote sensing. The fundamental physical ideas are limited for demonstration only to a two dimensional medium (an extension to three dimensions works in the same way). Generally the problem can be formulated as (see Fig. 2): The unknown, inhomogenuous two-dimensional density distribution f(x,y)

- x and y are the coordinates of the reference frame - is detected by the measurement of the value of a particular line integral along the integration path l

$$g(p,\Theta) = \int_{l} f(x,y)\, dl. \tag{5}$$

For the fixed parameter Θ and by variation of the ray or impact parameter p the measurement quantity $g(p,\Theta)$ gives one projection onto a straight line parallel to p. Variation of the parameter Θ leads for practical applications to a finite set of projections, which will be used for the (density-)reconstruction of $f(x,y)$. The function $g(p,\Theta)$ is known as the Radon transform of $f(x,y)$. The solution of this inversion problem arising nowadays in computerized tomography [8], radio astronomy [9] and radar target shape estimation [10] was given by J. Radon [11].

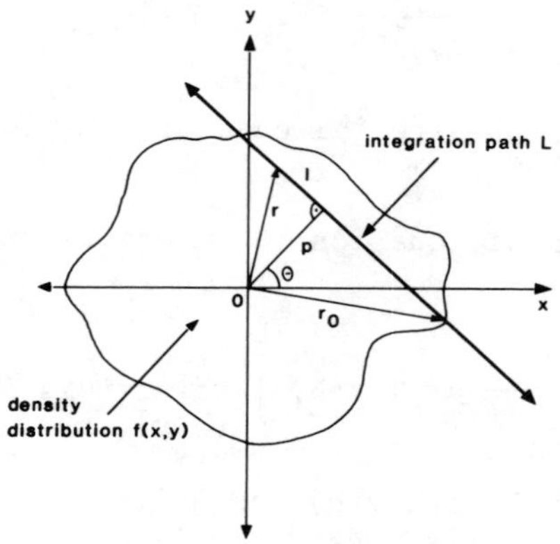

Figure 2. Geometrical representation of the line L (represented in dependence on the parameters p and Θ) passing the two-dimensional density distribution $f(x,y)$ to be probed indirectly.

In a special case which leads to the Abel transform the two-dimensional density distribution $f(x,y)$ takes on circular symmetry; therefore in equation (5) the function $g(p,\Theta)$ reduces to $g(p)$. Transformation of the line element dl into a cylindrical coordinate frame (using $r^2 = p^2 + l^2$ for the derivation of dl/dr) leads to the integral equation (for the case without ray bending)

$$g(p) = 2 \int_{p}^{r_o} \frac{rf(r)}{\sqrt{r^2 - p^2}}\, dr. \tag{6}$$

Performing the substitution of variables $h = r_0^2 - p^2$ and $y = r_0^2 - r^2$ the equation (6) is identical to the Abel equation (1). Only in this special case for rotational symmetry all projections are identical with one another by varying the angle Θ (see Fig. 2).

As the last point of this introduction it will be hinted to some useful relationships - mainly used in radio astronomy - between the Abel, Fourier and Hankel transforms [9]. Fig. 3 demonstrates that the Fourier transform of the Abel transform is a Hankel transform and that Abel transform in the function domain corresponds to the inverse Abel transform in the frequency domain. Furthermore in [9] is given a table with various and often used functions which are related by these transforms.

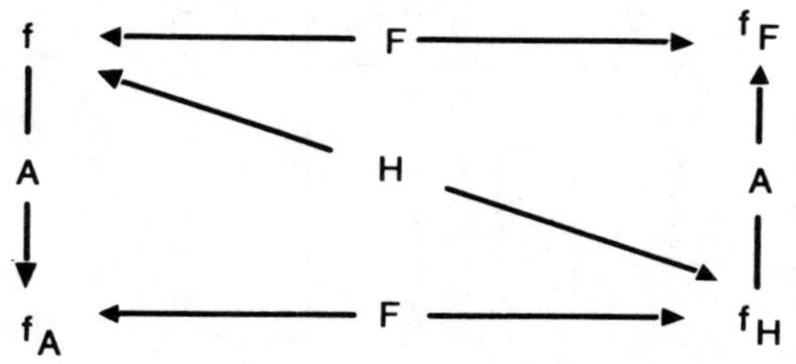

Figure 3. Schematic representation of the relationships between the Abel (A), Fourier (F) and Hankel (H) transforms [9]. f_A, f_F and f_H are the corresponding transforms of f.

2. APPLICATIONS IN VARIOUS FIELDS OF REMOTE SENSING

After some hints of possible applications in the introduction, three important areas of application of the Abel transform will be described in more detail in this section, namely the application in geo- and plasmaphysics and in atmospheric sounding. For an easy survey the meaning of the physical parameters of the integral equation are summarized in table 1.

The Abel transform found its first application in seismology in 1907, introduced by Herglotz and Wiechert [12]. It was used to determine from the measurement of travel times the velocity distribution of compressional (P) and transverse (S) waves which propagate inside the earth's mantle with different velocities. If an earthquake or an artificial explosion takes place at a point on or beneath the earth's surface the influence will travel to a distant point B (see Fig. 4). The travel time between A and B is related to the vertical distribution of the propagation speed which increases with depth. As a result the downward travelling ray paths are bent . Each ray is

field of application	integral equation	interpretation of the kernel K(p,r)	measured quantity T(p)	unknown quantity F(r)	upper integration boundary r_o
geophysics	$T(p) = \int_p^{r_o} K(p,r)\,dr = 2\int_p^{r_o} \frac{rF(r)\,dr}{\sqrt{\gamma^2 - p^2}}$	interaction of elastic waves with matter	travel time	density distribution inside the earth	earth's surface
plasma physics		interaction of electromagnetic waves with gaseous plasmas	integrated line intensity	distribution of particles in the plasma	plasma boundary
atmospheric sounding			ray bendig angle, phase or time delay	distribution of electrons in the atmosphere	effective atmospheric boundary

Table 1. Synoptical table of applications of the Abel integral equation in remote sensing (γ denotes the product of the refractive index of the propagation medium times the radial distance from the coordinate center).

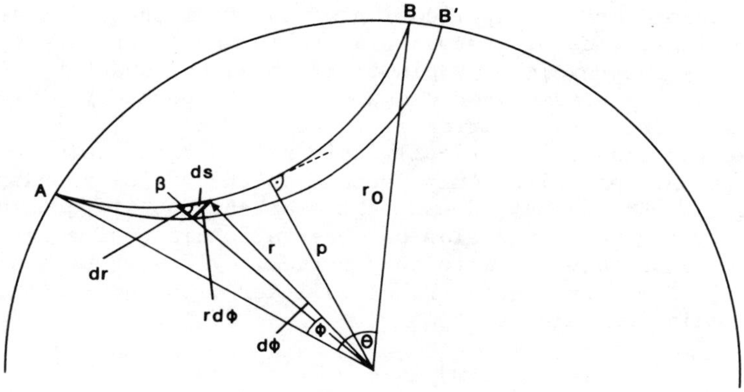

Figure 4. Geometrical notations used for the seismic problem.

characterized by the ray parameter p which is constant along the whole ray and for which the following relation holds (see Fig. 4)

$$p = \frac{r \sin \beta}{v} = \frac{r}{v} \frac{rd\phi}{ds} \tag{7}$$

(known as Bouger's rule).
Using a further relation also derived from Fig. 4

$$(ds)^2 = (dr)^2 + r^2 (d\phi)^2 \tag{8}$$

and the ray equation for the travel time

$$T(p) = \int_S \frac{1}{v} ds \tag{9}$$

($v = v(r)$: wave velocity of the elastic medium) after elimination of $d\phi$ and straightforward computation the integral equation

$$T(p) = 2 \int_p^{r_o} \frac{\eta^2}{\sqrt{\eta^2 - p^2}} \frac{dr}{r} \tag{10}$$

(where $\eta = r/v(r)$) can be obtained.

Analytical solutions of this kind of Abel integral equations can be found in [13, 14] where a lot of results are described too. The limits of this method are given by discontinuous changes of the density e.g. on the border between the earth's core and mantle where reflections of the waves occur.

The second area of application of the Abel transform deals with the diagnostics of gaseous plasmas in the laboratory and in space, whereby the physical origin of the measured quantity is the radiation of the plasma. For diagnostics of laboratory plasmas the measurement configuration is shown in Fig. 5. It will be assumed that the plasma is optically thin and that the emission and absorption coefficient only depends on the radial coordinate. Consequently the ray path doesn't have a bent but it forms a straight line along which the side-on observation is performed. The plasma tube can be divided into concentrical cylinders with different radii. For every concentrical cylinder with index k the relation of radiation balance holds

$$\begin{aligned} dI_k(r) &= \varepsilon_k(r)\ dx - \alpha_k(r)\ I_k(r)\ dx \\ &= i_k(r)\ dx \end{aligned} \tag{11}$$

(I_k, ε_k, α_k: intensity, emission and absorption coefficients of the k-th shell: dx infinitesimal length element). The measurement quantity, the integrated line intensity along the direction of observation can be expressed as

$$I(p) = \int_{-X}^{X} i(r)\ dx. \tag{12}$$

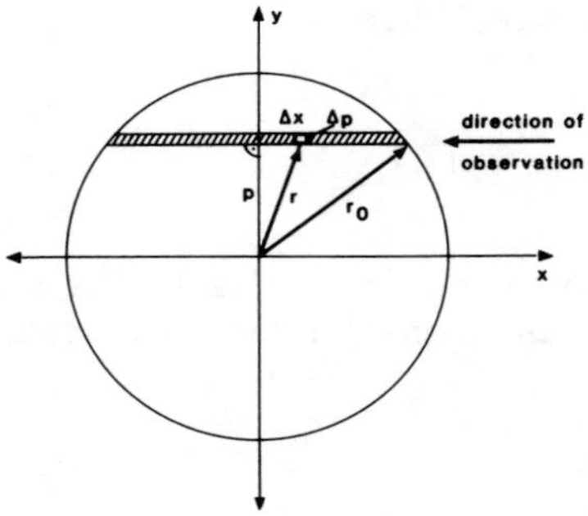

Figure 5. Sketch of the geometrical configuration for probing gaseous plasmas in cylindrical tubes.

Transforming this equation into cylindrical coordinates by using the relation $r^2 = x^2 + p^2$ for the calculation of dx the corresponding equation is yielded and nearly identical to equation (6). For this kind of remote probing of laboratory plasmas, which is widely used, a large number of numerical results are described in [15, 16].

Another related and currently used method is the determination of the electron density distribution of the inner solar corona (called

K-corona which extends to about 3 R_Θ (solar radii)) by measuring the polarization-brightness product of the white scattered light during a natural or artificial eclipse [17].The data were collected with a ground-based K-coronameter which records the product of polarization

$$P = \frac{I_T - I_R}{I_T + I_R}$$

(I_T, I_R: tangentially or radially emitted intensities) and the brightness which is the denominator of P.

The visibility of the K-corona can be derived from the Thomson scattering cross section which is related to the emitted power (scattered per steradian and per unit area) by

$$\sigma(r) = N(r) \cdot R_\Theta \cdot \frac{8\pi}{3} \left(\frac{e^2}{mc^2}\right)^2$$

(e,m: charge and mass of a free electron; c: velocity of light; N(r) radial dependent electron density distribution). Whereas in

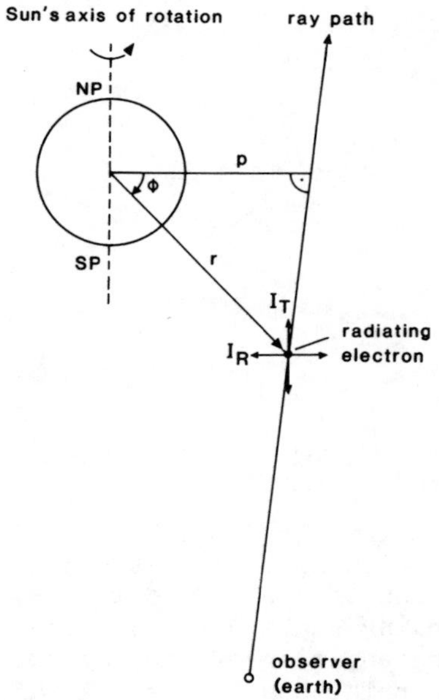

Figure 6. Sketch of the geometrical configuration for indirect probing the K-corona (NP, SP: north-, southpole of the sun).

equation (12) I(p) has to be replaced by the measured quantity pB, the unkown distribution i(r) is replaced by σ(r) or N(r). The connection to the Abel transform follows in the same way as in the previous example. Analytical results of this method and numerical computations can be found in [18, 19].

The last part of this chapter deals with the applications for studying planetary atmospheres by radio occultation methods. The orbit of a typical occultation experiment (e.g. the occultations of the Helios B spacecraft) is shown in Fig. 7 demonstrating remote probing of the solar corona. The principal observables of a radio occultation experiment are the amplitude (mainly used to determine absorption profiles) and the frequency of the signal. This last mentioned measured quantity forms the base for deriving the radial distribution of the refractive index. After elimination of observed frequency contributions resulting from spacecraft motion, rotation of the earth etc. the remaining frequency residual is the time rate of the change of the total phase caused by delay and refractive bending of the propagation medium.

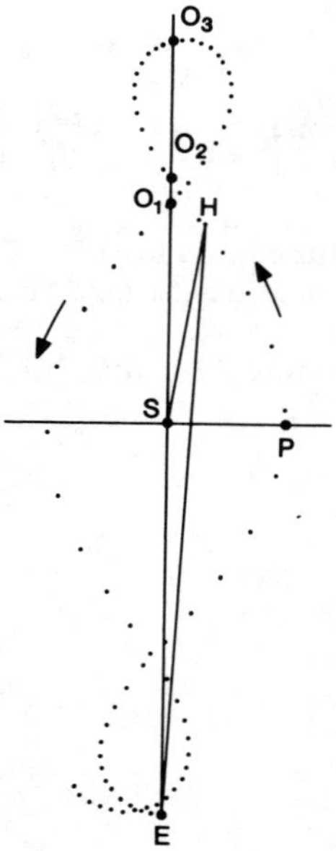

Figure 7. Geometry of the Helios B occultation experiment presented in a coordinate frame whose origin lies at the center of the sun and whose axis sun-earth remains fixed; O_1, O_2 and O_3 situations of occultations (E: earth, S: sun, H: Helios, P: perihelion).

Since the phase path plays a fundamental role in the next chapter, the relationships for determination of the refractive index profile from the observed refractive bending angle ψ is described here as an alternative example. The representation of the ray path geometry in Fig. 8 can be used to get the relations between the angle of incidence β, the bending angle ψ and the coordinates r and ϕ. They are linked by the following equations

$$\tan\beta = \frac{r d\phi}{dr}, \quad \text{and} \qquad d\psi = d\beta + d\phi, \tag{13}$$

whereby the last relation results from the fact that the sum of ψ,$-\beta$ and ϕ is equal to $\pi/2$. The expression for $d\beta$, which can be

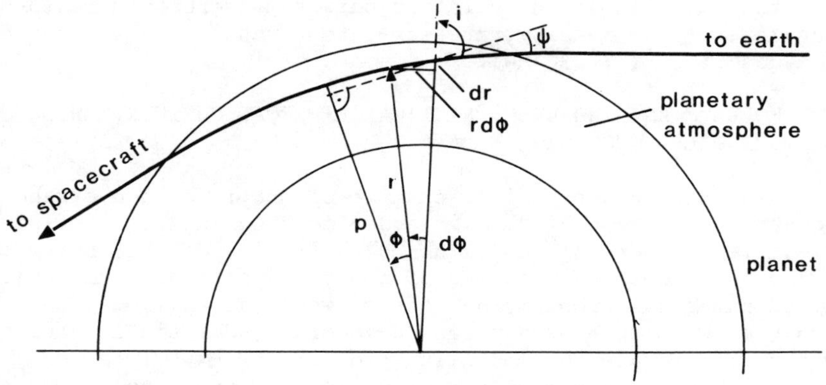

Figure 8. Refraction of a ray path in planetary atmosphere.

obtained by straightforward computation after differentiation of the first relation of equation (13), is given by

$$d\beta = \frac{-\,p\,(n + r\,\frac{dn}{dr})}{nr\,\sqrt{n^2r^2 - p^2}}\,dr.$$

An expression for $d\beta$ is deduced by the relation $\tan\beta = \sin\beta \cdot (1 - \sin^2\beta)^{-1/2}$ and by Bouger's rule (see equation (7)). For $d\phi$ the following holds

$$d\phi = \frac{p\,dr}{r\,\sqrt{n^2r^2 - p^2}} \quad .$$

Substitution of the expressions for $d\beta$ and $d\phi$ in the second relation of equation (12) leads to the integral equation for the bending angle

$$\psi(p) = -2 \int_p^{r_o} \frac{p\,dn/dr}{n\,\sqrt{n^2r^2 - p^2}}\,dr \qquad (14)$$

where r_o marks the effective boundary of the atmosphere (if no one exists: $r_o \to \infty$). These theories are well outlined in [20, 21] where also a lot of numerical exampels and results can be found inclusive of an exhaustive error analysis [22].

As a further possibility of probing atmosphere, which will only be hinted at here, is the radio probing the earth's atmosphere by a system of several **occultation** satellites. The experimental configuration as well as the relation between interesting meteorological parameters like density, pressure, temperature and water content are described in [23, 24].

3. PROBLEMS ARISING FROM PRACTICAL APPLICATIONS; INVERSION OF TIME DELAY OCCULTATION DATA

In this chapter a practical example of the inversion of Abel's equation will be performed in order to show in detail arising difficulties concerning numerical instabilities of the inversion algorithm and modelling problems of the propagation medium [25]. To avoid numerical problems an attempt was made to find a compromise for the solution by evaluating a combined analytical-numerical method; to reduce the modelling problem the experiment was started at a time where the best approximations of spherical symmetry of the propagation medium exist.

The physical background of this inversion problem is given by the experimental indirect probing of the solar corona measuring the time delay of radio signals(carrier frequency 2.115 GHz uplink/ 2.296 GHz downlink) during the occultation of the Helios B spacecraft [26]. The measured quantity is the difference between the phase- and group-delay time which is proportional to the electron content, the number of electrons integrated along the propagation path per unit area [27]. The main advantages of the employed measurement technique - developed at the Jet Propulsion Laboratory and called Differenced Range Versus Integrated Doppler (DRVID) - includes that all frequency independent errors (e.g. frequency drifts, inaccuracies in locus and velocity of the satellite) are left out whereas all frequency dependent errors (e.g. delay times in electronic systems on ground and on spacecraft, electron content of the terrestrial ionosphere) have to be taken into consideration *).
As shown in [25] the total electron content error takes a value of $1.3 \cdot 10^{16}$ $1/m^2$ (corresponding delay time 14.84 ns) compared with characteristic electron content variations of $9 \cdot 10^{17}$ to $2.7 \cdot 10^{19}$ $1/m^2$ (variations of delay time from 100 ns to 3 μs).

As briefly mentioned above for a clear interpretation of the electron content measurements the following properties of the propagation medium concerning the launch time and the orbit of the spacecraft have to be regarded:

*) The DRVID technique records the change of the electron content with respect to some unknown reference value at the initial time of a measurement set.

i) The spacecraft was launched so that during the period of data acquisition the solar activity had a minimum. During that time the density distribution of the solar corona nearly takes spherical symmetry above all relative to the ecliptic plane.

ii) The orbit lies nearly in the ecliptic plane, so that cylindrical symmetry is a good approximation.

iii) During the solar minimum structures of the charge density distribution keep for more than one solar rotation (period 27 days).

Further simplifying assumptions concerning the modelling of the propagation medium and the derivation of the integral equation are necessary:

iv) The propagation medium is fully ionized and quasineutral.

v) The magnetic field in the medium is so small that it can be neglected.

vi) The medium is so tenuous that the functional relation between signal frequency and refractive index can be linearized.

Then it is valid to write for the group and phase path (see also Fig. 9)

$$\rho_g(p) = \int_0^{L(t)} \frac{1}{n}\, ds, \qquad \rho_{ph}(p) = \int_0^{L(t)} n\, ds$$

($L(t)$: length of the signal path as a function of time). Using the linearization of the dispersion law, which relates the index of refraction with the electron density N by

$$n^2 = 1 - \frac{e^2}{\pi\, m}\,\frac{N}{f^2}$$

(signal frequncy $f \approx 2.2$ GHz, solar plasma frequency $f_p \approx 0.9$ MHz e.g. at 6 $R_\odot$; this means that $f_p/f \approx 0.4 \cdot 10^{-3}$) and subtracting the phase and group path, for the electron content holds

$$I(p) = \frac{\pi\, m}{e^2}\, f^2\, (\rho_{ph}(p) - \rho_g(p)) = \int_0^{L(t)} N(r,\phi,t)\, ds. \qquad (15)$$

The electron density in the solar plasma depends on the radial distance r, the azimuth angle ϕ and on time t. In order to get an analytical solution of equation (15) the following last additional assumptions of the coronal model have to be made:

vii) The medium remains stationary for the period of measurements.

viii) The electron density only depends on the radial direction.

After transformation of equation (15) into cylindrical coordinates the following generalized Abel equation (see eq. (4)) holds:

$$I(p) = \int_p^{r_o} \frac{rN(r)}{\sqrt{r^2 - p^2}} dr + \int_p^{r_1} \frac{rN(r)}{\sqrt{r^2 - p^2}} dr. \qquad (16)$$

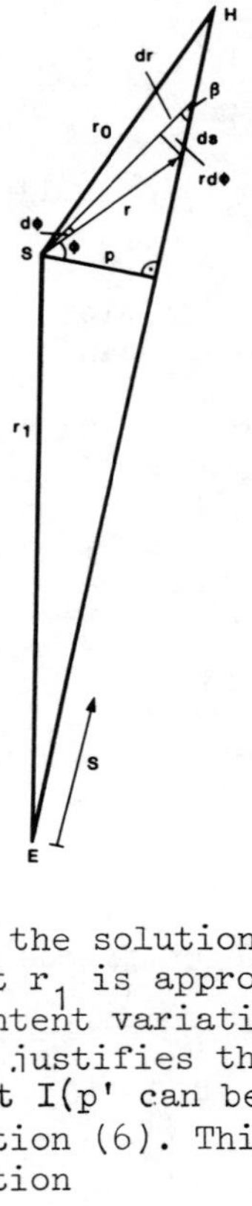

Figure 9. Occultation geometry during data acquisition.

The expense for the solution of this equation can be diminuished by assuming that r_1 is approximately equal to r_o. Estimation of the electron content variations in the region around the earth of about 10^{14} $1/m^2$ justifies this approximation. In equation (15) the electron content I(p' can be written as twice the first integral and looks like equation (6). This integral equation has the following analytical solution

$$N(r) = -\frac{1}{\pi} \frac{1}{r} \frac{d}{dr} \int_r^{r_o} \frac{p\, I(p)}{\sqrt{p^2 - r^2}} dp \qquad (17)$$

when these assumptions are fulfilled:

- I(p) has to be continuous in the integral [p, r_o]
- I(p) must be zero at p = r_o (includes the convergence of the integral)
- The derivation of I(p) may have a finite number of discontinuities.

Choosing a suitable solution, a combined analytical-numerical algorithm turned out to be numerically the most stable. Whereas a purely analytical solution is impossible because the measured quantity is a set of discrete values and not a continuous function the disadvantages of a purely numerical solution are:

- discretization errors caused by the transition from continuous to discrete formulation
- rounding errors caused by the finite number of places of the computer
- break-off errors caused by linearization (e.g. integration)
- measurement errors or noise with which every measured value is contaminated
- errors caused by amplification of noise by numerical differentiation in equation (18).

In order to avoid these error sources the inversion including the differentiation is performed analytically and the remaining part is done numerically. This method is outlined in the following part.

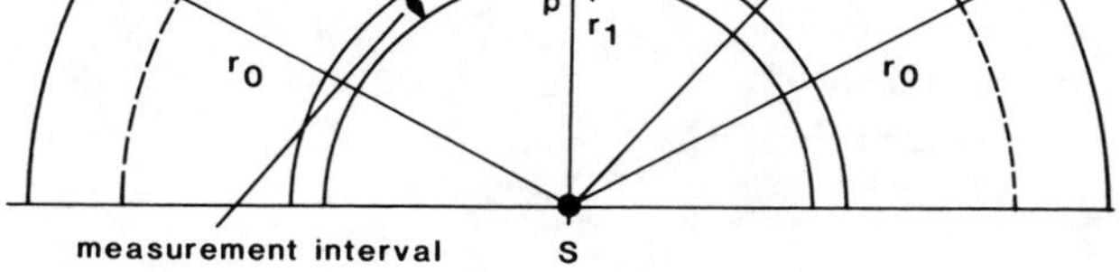

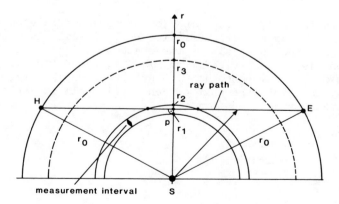

Figure 10. Occultation geometry at spherical symmetry of the density distribution.

For an analytic determination of the electron density N(r) the total ray path (see Fig. 10) is divided into three sections $[r_1,r_2]$, $[r_2,r_3]$ and $[r_3,r_o]$ which lead to a corresponding splitting of equation (17)

$$N(r) = -\frac{1}{\pi}\frac{1}{r}\frac{d}{dr} \tag{18}$$

$$\left\{ \int_r^{r_2} \frac{p\, I_1(p)}{\sqrt{p^2 - r^2}}\, dp + \int_{r_2}^{r_3} \frac{p\, I_2(p)}{\sqrt{p^2 - r^2}}\, dp + \int_{r_3}^{r_o} \frac{p\, I_3(p)}{\sqrt{p^2 - r^2}}\, dp \right\}$$

shell I	shell II	shell III
measured electron content values	electron content model	

where each shell corresponds to one section of the ray path in the order mentioned above. For the principal and most important point the intervall $[r_1, r_2]$ is divided into M-1 concentrical shells as shown in Fig. 11 (M: number of measured values). In order to make the analytical evaluation of the integral for shell I possible, in each partial interval $[r'_i, r'_{i+1}]$ the electron content is approximated by a straight line of the form

$$I_i(p) = a_i + b_i p \qquad \text{with} \qquad i = 1, 2, 3, \ldots, M$$

where the coefficients b_i are determined by

$$b_i = (I(r'_i) - I(r'_{i-1}))/(r'_i - r'_{i+1}).$$

(As shown in [25], a determination of the coefficients a_i is not necessary, because they neutralize by limit calculation between two neighboured intervals by the evaluation from table 2 to equation (19)). As derived from Fig. 11, where for example the courses of the ray paths 1, 2, 3 are shown explicitly, the entrance of a ray path into the next concentrical shell causes a further additional term of the form

$$\int_r^{r'_i} \frac{p(a_i + b_i p)}{\sqrt{p^2 - r^2}}\, dp$$

interval	expressions for density
$r_1' < r < r_2'$	$N_{11}(r) = \frac{-1}{\pi}\frac{1}{r}\frac{d}{dr}\left\{\int_r^{r_1'} \frac{p(a_1 + b_1 p)}{\sqrt{p^2 - r^2}} dp + \int_{r_1'}^{r_3} \frac{p(A_2 + C_2/p)}{\sqrt{p^2 - r^2}} dp\right\}$
$r_2' < r < r_3'$	$N_{12}(r) = \frac{-1}{\pi}\frac{1}{r}\frac{d}{dr}\left\{\int_r^{r_2'} \frac{p(a_2 + b_2 p)}{\sqrt{p^2 - r^2}} dp + \int_{r_2'}^{r_1'} \frac{p(a_1 + b_1 p)}{\sqrt{p^2 - r^2}} dp + \int_{r_1'}^{r_3} \frac{p(A_2 + C_2/p)}{\sqrt{p^2 - r^2}} dp\right\}$
$r_3' < r < r_4'$	$N_{13}(r) = -\frac{1}{\pi}\frac{1}{r}\frac{d}{dr}\left\{\int_r^{r_3'} \frac{p(a_3 + b_3 p)}{\sqrt{p^2 - r^2}} dp + \int_{r_3'}^{r_2'} \frac{p(a_2 + b_2 p)}{\sqrt{p^2 - r^2}} dp + \int_{r_2'}^{r_1'} \frac{p(a_1 + b_1 p)}{\sqrt{p^2 - r^2}} dp + \int_{r_1'}^{r_3} \frac{p(A_2 + C_2/p)}{\sqrt{p^2 - r^2}} dp\right\}$
$r_{M-1} < r < r_M$	$N_{1M}(r) = -\frac{1}{\pi}\frac{1}{r}\frac{d}{dr}\left\{\int_r^{r_{M-1}'} \frac{p(a_{M-1} + b_{M-1} p)}{\sqrt{p^2 - r^2}} dp + \dots + \int_{r_{i+1}'}^{r_i'} \frac{p(a_i + b_i p)}{\sqrt{p^2 - r^2}} dp + \dots + \int_{r_1'}^{r_3} \frac{p(A_2 + C_2/p)}{\sqrt{p^2 - r^2}} dp\right\}$
$r_j < r < r_{j+1}$	$N_{1j}(r) = -\frac{1}{\pi}\frac{1}{r}\frac{d}{dr}\left\{\int_r^{r_j'} \frac{p(a_j + b_j p)}{\sqrt{p^2 - r^2}} dp + \sum_{i=2}^{j} \int_{r_i'}^{r_{i-1}'} \frac{p(a_{i-1} + b_{i-1} p)}{\sqrt{p^2 - r^2}} dp + \int_{r_1'}^{r_3} \frac{p(A_2 + C_2/p)}{\sqrt{p^2 - r^2}} dp\right\}$ valid for $j \geq 2$

Table 2. Table of integrals for derivation of the recursive electron density formula

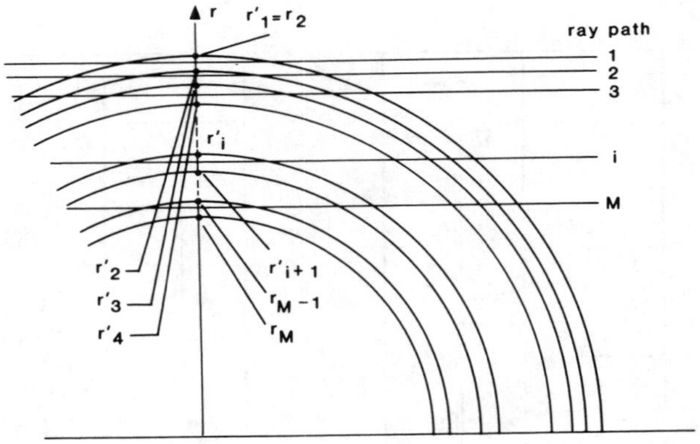

Figure 11. Model of concentrical shells of the measurement interval $[r_1, r_2]$ with passing ray paths.

to the preceding terms. The density expressions are summarized from the first until to the (M-1)-th shell in table 2. For shell II and III the following remarks have to be added (see [25]): first, estimations of the absolute values of the integral of shell III have shown, that it can be neglected compared with the integrals of shell I and II; second, the electron content of shell II is approximated by $I_2(p) = A_2 + C_2/p$ (expansion of equation (A2) into a Taylor series of second order) where the constants A_2 and C_2 are determined by continuity conditions at the boundaries. By straightforward computation the following recursive formula for the electron density distribution can be obtained from table 2

$$N(r_j) = \begin{cases} \dfrac{1}{\pi} \dfrac{A_2 + C_2\, r_3/r_1^2}{\sqrt{r_3^2 - r_1^2}} & \text{valid for } j = 1 \\[2ex] \dfrac{1}{\pi}\left[\dfrac{a_1 + b_1\, r'_1}{\sqrt{r_1'^2 - r_j^2}} + \dfrac{A_2 + C_2\, r_3/r_j'^2}{\sqrt{r_3'^2 - r_j^2}} - \dfrac{A_2 + C_2\, r'_1/r_j^2}{\sqrt{r_1'^2 - r_j'^2}} \right. \\ \left. \quad + b_1 \operatorname{arcosh} r'_1/r_j + \sum_{i=2}^{j-1} (b_i - b_{i-1}) \operatorname{arcosh} r'_{j-1}/r_j \right] & \text{valid for } j \geq 2. \end{cases} \tag{19}$$

In order to examine the inversion error of equation (19) the following steps of computations were carried out:

$$N_A(r) \xrightarrow[\text{integration}]{\text{analytical}} I_A(p) \xrightarrow[\text{algorithm}]{\text{inversion}} N(\mathbf{r})$$

This is a suited procedure for determination of the inversion error because the calculations of the single steps can be performed analytically using for $N_A(r)$ and $I_A(p)$ the relations (A1) and (A2) (see appendix). The advantage is that break-off and discrimination errors are avoided and only rounding and approximation errors are taken into account. From the computational results - shown in Fig. 12 - can be seen that the absolute error defined as $(N_A(r) - N(r))/N_A(r)$ takes a value of 1.8% and that the slopes differ by 1.2%. Furthermore can be observed the increasing numerical stability with increasing number of simulated (also valid for measured) electron content values.

In order to demonstrate the influence of noise to the inversion algorithm, the same simulated electron content data set was contaminated by pseudo noise with a variance of 0.04% (corresponds to an electron content variance of $3.7 \cdot 10^{16}$ $1/m^2$ which is related to an absolute value of 10^{20} $1/m^2$; model calculations with formula (A2)). The result of the inversion is shown in Fig. 13. To get a nu-

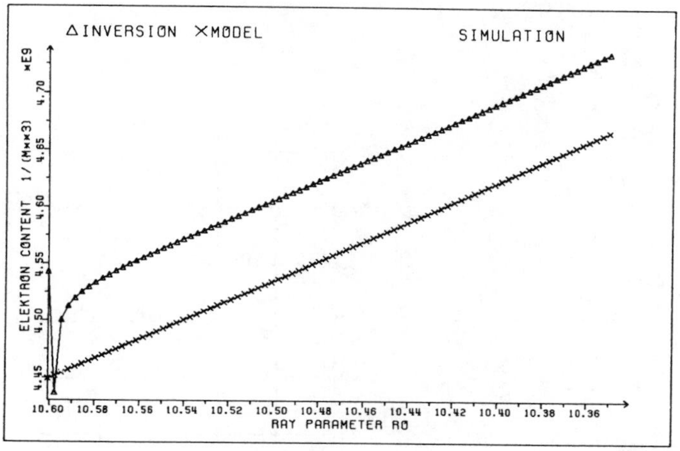

Figure 12: Illustration of the error of the combined analytical-numerical inversion algorithm.

merical quantity of the error magnification the ratios $\sigma_I/\bar{I}$ and $\sigma_N/\bar{N}$ are compared, where σ_I, σ_N and $\bar{I}$ and $\bar{N}$ are respectively

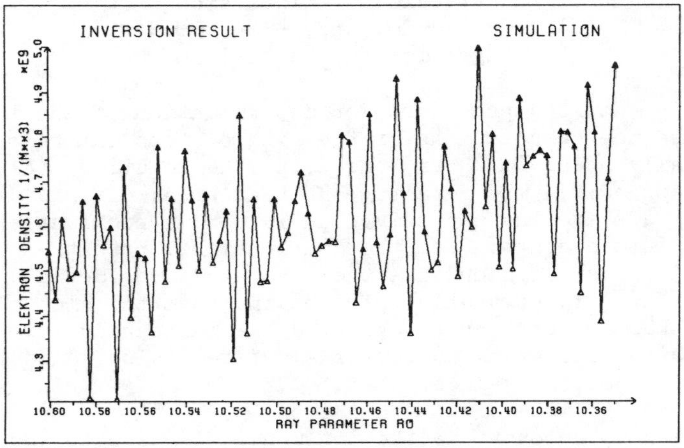

Figure 13. Illustration of the error magnification of a simulated electron content data set contaminated by (pseudo) noise.

the variances and means of the electron content and density. The quotient ρ_1 defined as

$$\rho_1 = \frac{\sigma_I/\bar{I}}{\sigma_N/\bar{N}}$$

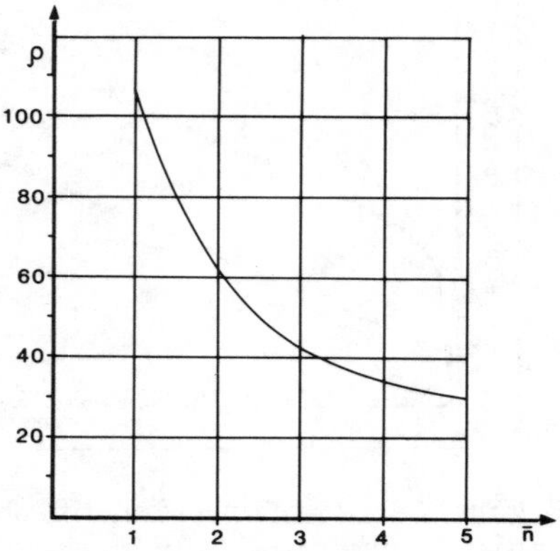

Figure 14. Illustration of the error magnification as a function of averaged electron content values.

has the value of about 100. Averaging over measured electron content values and calculating the corresponding factors to ρ_1, called ρ_2, ρ_3, ..., lead to a decrease of the error amplification. The result of successive averaging is demonstrated in Fig. 14 and shows a continuous decrease of ρ with an asymptotical approach to $\rho = 20$. The averaging process effects that spatial and temporal resolution decrease whereas the statistical resolution will increase because of noise supression. By the asymptotical character of the function ρ it follows (shown in Fig. 14) that the straight line shown in Fig. 13 cannot be reconstructed from noisy measurements.

From the large number of measured electron content data sets of the Helios occultation experiment one characteristical series is choosen in order to demonstrate the applicability of the developed inversion method. This electron content data set, acquired at an impact parameter between 10.6 and 10.35 $R_\odot$, shows during the entry phase of the occultation as indicated by the least-squares approximation, the accepted slope corresponding to the undisturbed coronal

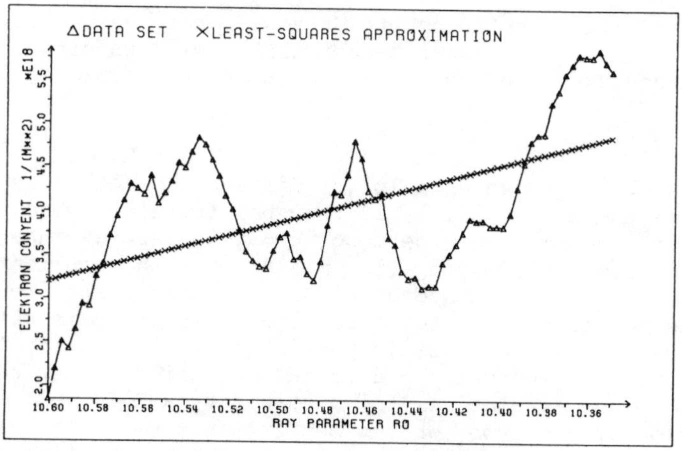

Figure 15. Measured electron content data set at a ray parameter of about 10.5 $R_\odot$.

plasma (see (A1) and (A2)). Furthermore the measured series proves that close to the sun immense plasma disturbances of solar origin pass the ray path. The result of the application of the inversion algorithm is presented in Fig. 16. The noisy character of the electron density distribution is obvious but on the other hand the oscillatory character as indicated in the measurements is significantly transformed into the density distribution. This shows that the physical information content gained by measurements will be

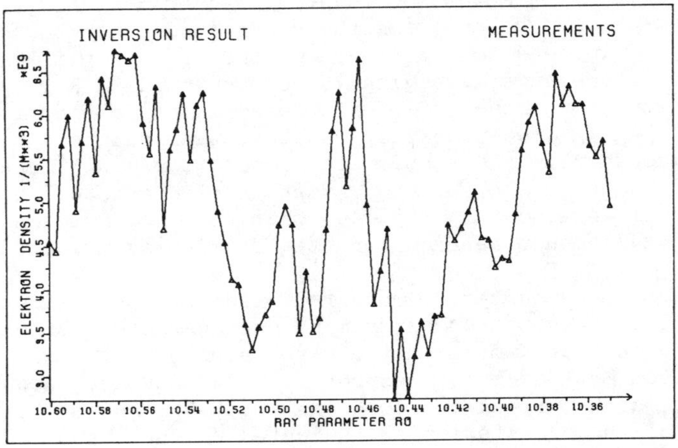

Figure 16. Electron density distribution calculated with the inversion algorithm of equation (19).

preserved. Furthermore has to be added that the density fluctuations of $3 \cdot 10^9\ 1/m^3$ can be correlated with in situ-electron density measurements of the Helios B spacecraft [25, 28].

5. CONCLUSION

This paper demonstrates a lot of different examples in remote sensing and the applicability of the Abel transform which is used when transport phenomena can be described with the ray theory in a rotational symmetric propagation medium. For the application in inhomogenuous media many constraints concerning experimental design as well as modelling have to be considered in order to get a solution whose results are physically realistic after a chain of simplifying assumptions. The developed combined analytical-numerical inversion algorithm has a relatively high stability and is well suited for the inversion of electron content measurements as well as for the inversion of other types of integrated line intensities.

APPENDIX

Stationary Model of Coronal Electron Density Distribution

The stationary, rotational symmetric electron density distribution of the solar corona in the ecliptic can be described by

$$N(r) = \frac{A}{r^6} + \frac{B}{r^2 + \varepsilon} \tag{A1}$$

where r is the dimensionless heliocentric distance in the units of solar radii (N given in electrons per cubicmeter). The coronal parameters A, B and ε determined by experimental data [29] as spatial and temporal averages have the following values and variances: $A = 1.0 \cdot 10^{14}\ 1/m^3$, $\sigma_A = 0.2 \cdot 10^{11}$; $B = 0.5 \cdot 10^{12}\ 1/m^3$, $\sigma_B = 0.2 \cdot 10^{14}$; ε: $0 < \varepsilon < 0.3$ (depends on the phase of the solar cycle; **solar minimum: $\varepsilon \approx 0$**).

Since the contribution of the first term of eq. (A1) to the electron density is effective below 4 R_Θ, for this inversion problem the density is approximated by the second term. Using only the quadratic term of N(r) to calculate an electron content model using the simplification of equation (16) the following expression is yielded

$$I(p) = \frac{2 \cdot B \cdot R_o}{p} \arccos \frac{p}{r_o} .$$

6. REFERENCES

[1] Abel, N.H., Résolution d'un Problème de Mécanique, J. Reine Angew. Math. 1, pp. 153 - 157, 1826.

[2] Goldstein, H., Klassische Mechanik, Frankfurt, 1963.

[3] Sakalyuk, K.D., Abel's generalized equation, Doklady Akad. Nauk. SSSR 131, pp. 332 - 335, 1960.

[4] Samko, S.G., The generalized Abel equation and fractional integration operator, Differential equations 4, pp. 157-166, 1968.

[5] Wolfersdorf, L.V., Zur Lösung der verallgemeinerten Abelschen Integralgleichung mit konstanten Koeffizienten, Z. Angew. Math. Mech. 49, pp. 759 - 761, 1969.

[6] Lowengrub, M., Walton, J.R., Systems of generalized Abel equations, SIAM J. Math. Anal. 10, pp. 794 - 807, 1979.

[7] Walton, J.R., Systems of generalized Abel integral equations with applications to simultaneous dual relations, SIAM J. Math. Anal. 10, pp. 808 - 822, 1979.

[8] Herman, G.T., Rowland, S.W., Three Methods for Reconstructing Objects from X Rays: A Comparative Study, Computer Graphics and Image Processing 2, pp.151 - 178, 1973.

[9] Bracewell, R.N., Strip integration in radio astronomy Aust. J. Phys 9, pp. 198 - 217, 1956.

[10] Das, Y., Boerner, W-M., On radar target shape estimation using algorithms for reconstructing from projections, IEEE Trans. Ant. and Prop. AP-26,pp. 274 - 279, 1978.

[11] Radon, J., Über die Bestimmung von Funktionen durch ihre Integralwerte längs gewisser Mannigfaltigkeiten, Ber. Verh. Sächs. Akad. Wiss. Leipzig, Math.Phys. Kl. 69, pp. 262 - 277, 1917.

[12] Wiechert, E., Geiger, L., Phys. Z. 11, pp. 294 - 311, 1910.

[13] Bullen, K.E., An Introduction to the Theory of Seismology, Cambridge University Press, New York, 1963.

[14] Kennett, B.L.N., Ray theoretical inverse methods in geophysics, in Applied Inverse Problems, ed. by P.C. Sabatier, Lecture Notes in Physics, Vol 85 (Springer, Berlin, Heidelberg, New York 1978).

[15] Bockasten, K., Transformation of observed radiances into radial distribution of the emission of a plasma, J. Opt. Soc. Am. 51, pp. 943 - 947, 1961.

[16] Cremers, C.J., Birkebak, R.C., Application of the Abel integral equation to spectrographic data, Appl. Opt. 5 pp. 1057 - 1064, 1966.

[17] Altschuler, M.D., Reconstruction of the global-scale three-dimensional solar corona, in Image Reconstruction from Projections, ed. by G.T. Herman, Topics in Applied Physics, Vol 32 (Springer, Berlin, Heidelberg, New York, 1979).

[18] Bohlin, J.D., Garrison, L.M., Numerical calculation of Thomson scattering from inhomogeneous models of the corona, and application to streamers of the 1970 and 1972 eclipses, Solar Physics 38,pp. 165 - 179, 1974.

[19] Hansen, R.T., Garcia, C.J., Hansen, S.F., Loomis, H.G., Brightness variations of the white light corona during the years 1964-67, Solar Physics 7, pp. 417 - 433, 1969.

[20] Phinney, R.A., Anderson, D.L., On the radio occultation method for studying planetary atmospheres, J. Geophys. Res 75 pp. 1819-1927, 1968.

[21] Kliore, A.J., Current methods of radio occultation data inversion, in Mathematics of Profile Inversion, ed. by L. Colin, NASA Tech. Mem. X-62, 150, 1972.

[22] Stewart, R.W., Hogan, J.S., Error analysis in the Mariner 6 and 7 occultation experiments, in Mathematics of Profile Inversion, ed. by L. Colin, NASA Tech. Mem. X-62, 150, 1972.

[23] Fjeldbo, G., Kliore, A.J., Eshleman, V.R., The neutral atmosphere of Venus as studied with the Mariner V radio occultation experiments, Astron. J. 76, pp. 123 - 140, 1971.

[24] Lusignan, B., Modrell, G., Morrison, A., Pomalaza, J, Ungar, S.G. Sensing the earth's atmosphere with occultation satellites, Proc. IEEE 57, pp. 458 - 467, 1969.

[25] Süß, H., Plasma-Fernerkundung durch Inversion von Messungen des Elektroneninhalts bei den Helios-Raumsonden (Okkultations-experiment), Dissertation, Bochum, 1981.

[26] Edenhofer, P., Esposito, P.B., Hansen, R.T., Hansen, S.F., Lüneburg, E., Martin, W.L., Zygielbaum, A.I., Time delay occultation data of the Helios Spacecrafts and preliminary analysis for probing the solar corona, J. Geophys. 42, pp. 673 - 698, 1977.

[27] Mac Doran, F., A first-principles derivation of the Differenced Range Versus Integrated Doppler (DRVID) charged-particle calibration method, JPL Space Progr. Sum. 37-62.II, pp. 28 - 34, 1972.

[28] Rosenbauer, H., Schwenn, R., Marsch, E., Meyer, B., Miggen-rieder, H., Montgomery, M.D., Mühlhäuser, K.H., Pilipp, W., Voges, W., Zink, S.M., A survey on initial results of the Helios plasma experiment, J. Geophys. 42, pp. 561-580, 1977.

[29] Esposito, P.B., Edenhofer, P., Lüneburg, E., Solar corona electron density distribution, J. Geophys. Res. 85, pp. 3414 - 3418, 1980.

I.11 (IM.1*)

SYMMETRY TECHNIQUES FOR DIRECT AND INVERSE SCATTERING

Paul F. Wacker

prepared in part at K. U. Leuven, BELGIUM
new address: 3037 Montrose Ave., Apt. 4
La Crescenta, CA 91214, USA

Abstract--Rigorous but practical procedures are presented for determining both overall and constituent symmetries of finite and infinite scatterers, guided mode junctions, and guided-spatial transducers (e.g., antennas). The systems may be lossy, anisotropic, and/or inhomogeneous. The procedures are based on the facts that a) all the familiar modal indices are symmetry indices, like even or odd, b) a linear operator may be expressed as a sum of operators, each designated by similar indices, and c) the symmetry of an operator (e.g., which EM multipoles are present in a scatterer) is indicated by its conversion of symmetry (modal) indices, just as a multiplier must be odd if it yields an even product from an odd multiplicand. Related mathematics reduces the number of measurements required to determine the pattern of a scatterer, radiator, or receiver of known symmetry. Many examples are given, with mathematical definitions, theorems, and references.

1. INTRODUCTION

(So that a general understanding can be obtained without wading through the complicated mathematics, the more complicated mathematical details and definitions are given in Section 5 and the notes of section 6. Further background is given in (26) and the references and will be given in a forthcoming paper; for more practical applications, see the latter two. Small differences between symbols should not be ignored; e.g., presence or absence of underlining, boldface or lightface, etc.) In this paper we provide practical techniques for determining both overall and constituent symmetries of both finite and infinite scatterers, of guided-mode junctions, and of guided-spatial transducers (e.g., antennas) from measurements

W.-M. Boerner et al. (eds.), Inverse Methods in Electromagnetic Imaging - Part 1, 203–222.

made at a distance. In addition to being of interest in their own right, knowledge of symmetries should speed inverse scattering analysis; further, related techniques reduce the number of measurements and computations required to determine scattering, radiation, or receiving patterns.

The procedures are based upon rigorous mathematics used in theoretical physics, particularly elementary particle theory. Specifically, the <u>modern</u> theory of group[1] representations[2] is used in combination with the invariances of natural law and the material properties of the <u>propagation</u> <u>medium</u> with respect to translation, rotation, reflection, and/or inversion of the coordinate system and possibly space-time transformations; we call this symmetry analysis (SA). Although little used in engineering and applied mathematics, this certainly constitutes one of the most powerful mathematical tools used in physical science (26). Although the <u>proofs</u> of many of the theorems require a sophisticated background indeed, the <u>results</u> of many do not and their power makes for global understanding, leading to simple direct proofs of practical problems, with rigor and sharpness not often seen in the engineering literature. Further, these techniques lead to rapidly converging modal sets, reducing the measurement and/or computational effort required for <u>comprehensive</u> information (see Section 3.3).

Because the general procedures are so basic, they apply equally well to many physical systems, including electromagnetic, acoustic, heat flow, and mixed, but, for simplicity, we confine <u>most</u> of our attention to the curl-curl and scalar Helmholtz equations. However, since a single differential equation may apply to several dozen physical systems (17),(16, pp. 265, 343-346), simple changes of variables can translate even the <u>detailed</u> discussions; see Note 8 for sonar and geophysical prospecting with both longitudinal and transverse waves. Both the propagation medium and the scatterer may be lossy and, subject to the <u>optional</u> assumptions, the scatterer may be anisotropic, inhomogeneous, and time varying. The presentation is <u>approximation</u> <u>free</u>, apart from assumed mathematical linearity and optional assumed invariances. Many examples are given, as well as mathematical definitions, theorems, and references.

The procedures are also based upon the facts that a) all the familiar modal indices are symmetry indices, like even or odd, b) a linear operator may be expressed as a sum of operators, each designated by similar indices, and c) the natural constraints upon the symmetry of a "product." For example, the propagation constants, TM,TE character, frequency, orders of Bessel functions, sin-cos distinction, both indices of associated Legendre functions, the azimuthal index m, and the multipole indices are all symmetry indices. Also for example, a multiplier must be odd if it yields an even product from an odd multiplicand. The multiplier may also be an operator, such as the scattering operator; thus, in the preceding example, the multiplier may be d^n/dx^n if n is odd, but not if n is even. The power of these techniques is illustrated by two examples: For acoustics with the scalar Helmholtz equation, a single ingoing mode may be used to determine the presence of each and every scalar multipole in the scatterer; the multipole is present if and only if the corresponding scattered mode is present. If the material properties of a capacitor (in a circuit or in space) are known to be symmetric with respect to a plane of symmetry and the single ingoing mode is even with respect to the plane, presence of any scattered mode odd with respect to the plane proves that the capacitor is differentially charged.

For near field determination of radiation, receiving, and scattering patterns, we use SA techniques, especially the orthogonalities, to determine the modal expansion coefficients and the far field patterns (25),(24),(27). In this paper, we emphasize techniques related to the presence or absence of modes; specifically we use a) a more detailed form of the mathematics which provides the selection rules of quantum mechanics and spectroscopy to reduce both the measurement and computational effort in determining the patterns of scatterers of known symmetries (25) and b) related techniques, used to determine the symmetries of molecules and the internal symmetries of crystals, to determine the symmetries of a macroscopic scatterer in inverse scattering (25).

2. MODES AND THEIR SYMMETRIES

2.1. Functions defined by symmetries

With an infinite number of symmetry operations, there are an infinite number of **symmetry types**[5] (each like even or odd) and an infinite number of constraints (each like $f(-x)=-f(+x)$ for an odd function) defining each symmetry type by specifying its transformation under each and every symmetry operation of the group.[1] (Technically, we say that a function is of a given symmetry type if it belongs to a given row[6] of a given irreducible[4,6] unitary[5] representation[2] of a given group.) The following groups and the functions they define are pertinent to our problems: The (continuum of) rotations about an axis, i.e., the two dimensional rotation (orthogonal) group C_∞, defines the functions $\exp(im\phi)$ and yields Fourier series analysis (9, pp. 322-325). The translations T_1 along the real line yield the functions $\exp(-i\omega t)$ and $\exp(ik_x x)$, the Fourier transform, and the double-sided Laplace transform (9, pp. 486-489). The addition of reflections in planes containing the axis ($C_{\infty v}$) or in planes perpendicular to the line (T_{1h}) yields the corresponding sines and cosines. The three dimensional rotation (orthogonal) group O_3^+ yields $P_n^m(\cos\theta)\exp(im\phi)$, the definition of the cross product (3), and the theories of spherical harmonics and angular momentum (21, pp. 130-168) (O_3 is not to be confused with the point groups O and O_h (12)).

The two dimensional Euclidean (translation-rotation) group E_2^+ yields $J_m(k_R R)\exp(im\phi)$, the analogous expressions with cylindrical Neumann and Hankel functions (21, pp. 189-214), and Hankel (Bessel function) transforms. Moreover, three separate theorems, each applying to a class of symmetry groups (2, pp. 166-176, 420-472), lead to the mathematical completeness, linear independence, and orthogonalities of the preceding sets of functions. (Lack of linear independence may cause oscillation or lack of convergence in iterative procedures.) The three dimensional Euclidean (translation-rotation) group E_3^+ yields $P_n^m(\cos\theta)\exp(im\phi)\, j_n(k_r r)$ and the analogous expressions with spherical Neumann and Hankel functions (21, pp. 215-233).

In general, there are three linearly independent three-vector functions for each of the preceding symmetry types. However, addition of gauge invariance (related to the absence of fixed charges and the divergence condition) requires that the field be solenoidal, reducing the number of independent polarizations in the EM cases to two. The addition of spatial inversion (and so reflections) to O_3^+ and E_3^+ yields the full 3D rotation and 3D Euclidean groups, $O_{\bar{3}}^+$ and $E_{\bar{3}}^+$, respectively, which separate the TM and TE modes into different symmetry types and so define them (19). E_3^+ and $E_{\bar{3}}^+$ define the scalar and EM multipoles, respectively, while the corresponding **O** groups define them apart from the radial dependence. The circular cylindrical ($E_2 X T_z$) and plane rectangular ($T_x X T_y$) modes and algebras are limits of the spherical on the pole and equator, respectively, as the radius goes to infinity (21); this explains the TM,TE separation for these cases. (The symbol X indicates the **direct product** of two groups; for our purposes, these are groups which occur simultaneously but involve independent sets of coordinates.) The fast Fourier transform arises from the group $\mathbf{C}_M$ of rotations by $2\pi q/M$ ($q=0$ to $M-1$) and separates the (discrete) terms of an exponential Fourier series into M types with an infinite number in each type; as such, the FFT is an approximation free symmetry decomposition; the approximations arise in replacing the continuum of terms in a transform by a Fourier series and assuming that there is only one significant term in each symmetry type. A fourth theorem leads to the orthogonalitity and (collective) completeness of the M symmetry types, as well as of even and odd functions. The preceding are merely pertinent examples; actually SA defines most of the named mathematical functions and gives their detailed properties, including some previously unknown (21),(15),(23). Further, it may be considered to be part of a generalization of Fourier series or transform analysis, with an expansion being a symmetry decomposition (2, pp. 420-472). We base our analysis upon the simple combination of the full Euclidean group $E_{\bar{3}}^+$ with the temporal translation-reflection group T_{th} and their subgroups (for most macroscopic problems, the space-time transformations of the Lorentz group of special relativity add only pointless complexity). For finite scatterers, we use subgroup $O_{\bar{3}}^+ X T_{th}$ and its subgroups, where $O_{\bar{3}}^+$ is the full 3D rotation group. (The groups of crystallography are discrete subgroups of E_3^+ subject to the law of rational indices, with the 32 point groups being such subgroups of O_3^+ (14, pp. 20-60); we apply neither constraint and so, for example, accept groups such as C_∞ and C_7.) For simplicity in treating plane rectangular and circular cylindrical modes in homogeneous isotropic media, we impose $k_x^2+k_y^2+k_z^2=k^2$ and $k_R^2+k_z^2=k^2$, respectively, as auxiliary constraints rather than basing these constraints upon the E_3^+ group.

2.2. Symmetry indices of functions

We partition the modal indices N=n;m;p into **n**, **m**, and **p** where those in **n** and **m** designate the **irreducible**[4,6] representation (major symmetry type) and its **row**[6] (minor, i.e., subsidiary, symmetry type), respectively, of the given group. A function is said to "belong to" a given row of a given irreducible representation (of a given group); together they define our "**symmetry type.**"[6] For example, for $m\neq0$, the $|m|$th representation of $C_{\infty v}$ has two rows and the **partner**[6] functions sin $m\phi$ and cos $m\phi$ belong to different rows. For any representation with more than one row, there are an infinite number of representations which are said to be "**equivalent**"[3]; thus, for the preceding example, the partner functions exp$(+im\phi)$ and exp$(-im\phi)$ belong to different rows of an "equivalent" representation (but neither the two sets of (partner) functions nor the two representations are equally suitable for expressing either linear polarization or circular polarization). Indices needed to specify the mode but not the symmetry relative to the given group are designated by **p**; our **p** indices are ordinarily symmetry indices of a larger group. The operations of the given group acting upon the N mode do not change the **n** or **p** indices (since we use linearly independent partner[6] functions) but yield linear combinations of modes with differing **m**.[6]

2.3. Eigenfunctions and eigenfunction equations

In addition to defining functions, SA yields, as eigenfunctions, the complete solutions of many eigenvalue equations, including solutions of the scalar Helmholtz and curl-curl equations; the latter include the familiar ones for unbounded media, waveguides, and cavities, as well as for unfamiliar waveguides and cavities. Further, it provides sets of such functions, each with its symmetry prescribed in detail, which are linearly independent but mathematically complete with respect to the eigenfunctions. We express both the incident and scattered field in terms of these functions to reduce the required measurement and computing effort; see Section 3.3. We consider primarily equations of the form

$$\underline{O}\ \mathbf{f}_N(\mathbf{x}) = \lambda_N\ \mathbf{f}_N(\mathbf{x}), \tag{1}$$

where $\underline{O}$ is a linear operator, boldface is used for symbols which are or may be compound (e.g., vectors), N represents the set of modal indices, and x represents the set of independent variables, e.g., the spherical coordinates and/or time. The fields may be scalar, vector, or tensor in n-space and the linear combinations may be sums, integrals, or sums of integrals over the eigenfunctions. For simplicity in presentation, we direct most of our discussion to the scalar Helmholtz equation

$$\nabla^2 f_N(x) = - k^2 f_N(x) \tag{2}$$

and the curl-curl equation

$$\nabla \times \nabla \times E_N(x) = k^2 E_N(x), \tag{3}$$

i.e., most of our discussion to homogeneous isotropic media of propagation and, in the electromagnetic case, to solenoidal fields (no fixed charges). We regard (3) as a special case of

$$\varepsilon^{-1}\nabla\times[(\mu^{-1}\nabla\times) E_N(x)] = \omega^2 E_N(x), \tag{4}$$

where ε and μ are complex tensor functions of position for the system as a whole; this automatically includes the boundary conditions at infinity, the walls of waveguides or cavities, etc.; step functions of course introduce generalized functions such as the Dirac delta "function." However, we use modes for the medium in the absence of the scatterer and probes, permitting treatment of the scatterer as a black box without even tentative assumptions concerning it, except for mathematical linearity. This limits the domain of validity of our modal expansions (say to outside a sphere which circumscribes the scatterer[9]) and causes our conclusions about overall and constituent symmetries to apply to a domain which includes the scatterer and usually some medium of propagation. As previously mentioned, the techniques and, with changes of variables, even the detailed discussions apply to many systems; see Section 2.4 for anisotropic systems, Sections 2.4 and 3.3 for lossy systems, and Note 10 for pulsed fields. Note that equation (4) also yields the solutions and eigenvalues of the Singularity Expansion Method.

2.4. Theorems relating symmetries and eigenfunctions

For modal symmetries, we assume that the operator $\underline{O}$ of Eq. (1) is linear, self-adjoint in the Hermitian sense, and **commutes** with every operation g of a group G (i.e., $\underline{O}g=g\underline{O}$) which is finite, Abelian,[1] or **compact**[1] (e.g., O_3) and consider a set of irreducible unitary[5] pairwise-inequivalent[3] representations. Then the set of eigenfunctions, each of which belongs to a row of a representation, is mathematically complete with respect to the eigenfunctions; furthermore, all linear combinations of a given set of partner eigenfunctions belong to the same real eigenvalue, and functions (not only eigenfunctions) belonging to different representations or to different rows of the same representation are linearly independent of each other and orthogonal in the Hermitian sense with respect to the sum and/or weighted integral over the group operations (14, p. 108ff),(2, p. 144),(9, p. 103).[6] Except for the eigenvalues being nonreal, the preceding theorems also apply if O̲ is a complex constant times a self-adjoint operator or **normal**, i.e., commutes with its Hermitian adjoint. Usually, for a large enough group, a) all the eigenfunctions of a given eigenvalue λ_N belong to the same representation, b) the number of linearly independent eigenfunctions belonging to that eigenvalue is equal to the **dimension** (number of rows[5]) of the representation, and c) no two eigenvalues belong to the same representation (14, p. 109 ff.).

We illustrate the preceding with the differential equations $i^{-n}\, d^n u/dz^n=\lambda u$. The operators $i^{-n}\, d^n/dz^n$ commute with all translations on the z axis and, if n is even, with reflection in every plane perpendicular to the z axis. Thus, we use T_z and, if n is even, T_{zh} and C_s (reflection in the z=0 plane). T_z yields the eigenfunctions $\exp(ik_z z)$ (for all these equations) and shows the set for k_z real to be complete with respect to the eigenfunctions, linearly independent, and orthogonal. For n even (but not odd), T_{zh} yields the partner eigenfunctions $\exp(+ik_z z)$ and $\exp(-ik_z z)$ (or alternatively $\sin k_z z$ and $\cos k_z z$), shows that every linear combination of them for a given k_z has the same eigenvalue (k_z^n), that the set for $k_z \geq 0$ is mathematically complete with respect to the eigenfunctions, and that all the functions of either set are linearly independent and orthogonal; T_z is not large enough to do this for even n, but it is large enough for odd n. For even n, C_s alone separates the eigenfunctions into even and odd (here $\cos k_z z$ and $\sin k_z z$) and shows that such a set is mathematically complete with respect to the eigenfunctions and that the even functions are linearly independent of and orthogonal to the odd ones; however, C_s is not large enough to a) separate the k_z values, b) show that the sine and cosine have the same eigenvalue, c) show that the complex exponentials are eigenfunctions, nor d) distinguish between even (or odd) functions which are and which are not eigenfunctions.

Our procedures do not assume that the operator representing the scatterer is self-adjoint; see Section 3.3 ff. Further, lossy media and evanescent modes usually cause no problem (at least if the medium is homogeneous and isotropic) if attention is confined to groups involving only two spatial coordinates and nonreal values are accepted only for the component of the propagation vector corresponding to the third coordinate, e.g., that perpendicular to a sphere, circular cylinder, or plane. The theory then ordinarily leads to surface eigenfunctions of a self-adjoint operator, from which one may easily determine the three dimensional eigenfunctions. The surface eigenfunctions are orthogonal in the Hermitian sense and both sets are mathematically complete with respect to their kind of eigenfunctions; however, symmetries involving only two coordinates do not provide orthogonalities between ingoing and outgoing modes.

In addition to giving the eigenfunctions and eigenvalues for unbounded media, SA provides complete exact solutions inside various waveguides and cavities. Since perfect conductors provide mirror images which "extend" the space of the medium outside, unbounded solutions consistent with the boundary conditions apply. In particular, SA provides the solutions for circular cylindrical guide and both spherical and right circular cylindrical cavities. Further, it provides the solutions for waveguides and cavities bounded by planes consistent with the law of rational indices (related to the Bravais lattices of crystallography), namely a) for waveguides with cross sections which are squares, rectangles, diamonds, parallelograms, equilateral triangles, and regular hexagons and b), for each of the 14 Bravais lattices of three space, the cavities bounded by planar walls which form a unit cell. The solutions provided by the mappings are mathematically complete inside all these structures and their extensions valid outside the structures, but complete outside only a sphere and an infinite cylinder.

3. THE SCATTERING AND PERTURBATION MATRICES AND THEIR SYMMETRY CONSTRAINTS

3.1. Eigenfunction expansions of fields

We express the fields in terms of linear combinations of modes which are eigenfunctions for the medium of propagation in the absence of both the scatterer and transmitting and receiving transducers, e.g., antennas. Thus,

$$f(x) = \sum\!\!\!\!\int_{N} f_N(x)\, C_N, \tag{5}$$

where the symbol $\sum\!\!\!\!\int$ indicates summation over discrete modal indices in N and/or integration over indices which assume continuous values, e.g., ω, k_x, k_y, k_z, k_R, and k_r, as appropriate.

3.2. Definitions of the scattering and perturbation "matrices"

We express scattering properties in terms of a "matrix" which relates two sets of modal coefficients, each from an expansion of the form of (5). We first define the scattering matrix, which applies to guided-mode junctions and guided-spatial transducers, e.g., antennas, as well as for scatterers. We use the generalized scattering "matrix" [S] (not the "power wave" or "source" scattering matrix); it relates the coefficients of modes which are outgoing with respect to a volume (possibly a half space) containing the scatterer to those of modes which are ingoing; thus,

$$b = \Gamma a + \underline{R}A \tag{6a}$$

$$B = \underline{T}a + \underline{S}A \tag{6b}$$

or

$$\begin{Bmatrix} b \\ - \\ B \end{Bmatrix} = \left[\begin{array}{c|c} \Gamma & \underline{R} \\ \hline \underline{T} & \underline{S} \end{array}\right] \begin{Bmatrix} a \\ - \\ A \end{Bmatrix}, \tag{6c}$$

where a and b are column submatrices of the detached coefficients of the ingoing and outgoing guided modes, respectively, and A and B are column submatrices of the detached coefficients of the ingoing and outgoing spatial modes, respectively. We commonly use individual modes with complex exponential dependence on ω, ϕ, and/or propagation constants, leading to coefficients and elements of [S] which are complex numbers. Then Γ represents the complex reflection coefficient(s) looking into the port(s), while $\underline{T}$, $\underline{R}$, and $\underline{S}$ are submatrices representing the transmitting, receiving, and scattering properties, respectively; it is assumed that the space is empty except for the propagation medium and the object which [S] represents; see (25). For a mathematically linear system, [S] expresses all the properties which can be observed at a distance and all the input information for an inverse scattering problem; see Note 10 for scattering of pulsed fields. $\underline{S}$ gives the properties of a scatterer, but application of the techniques of this paper to Γ, $\underline{R}$, and $\underline{T}$ yields similar results for guided-wave junctions, receiving transducers, and transmitting transducers, respectively. Ordinarily, translation, rotation, and/or reflection of the coordinate system will change [S]; in this paper we usually refer [S] and its submatrices to a coordinate system with its origin within the scatterer. In addition, the definition and even the form of the matrices will vary with the ordered sets of eigenfunctions used in Eq. (5).

Each row and each column is indicated by a set N of modal indices arranged in a dictionary order, i.e., each element is associated with a pair of eigenfunctions, one ingoing and the other outgoing. Depending upon the eigenfunctions involved, the row and column indices may be discrete, continuuous, or mixed; at least in a formal sense, the elements on the left hand side of equation (6) are commonly given by multiple summation and/or multidimensional integration. For a perfectly-matched perfect absorber $\underline{S}=\underline{0}$, and for a volume containing only the medium of propagation $\underline{S}=\underline{1'}$. If the rows and columns are discrete, $\underline{0}$ and $\underline{1'}$ are ordinary null and unit matrices, the latter provided that the same ordered set of modes is used on both sides of equation (6); however, for a continuous index, a row of $\underline{1'}$ may be proportional to a Dirac delta "function." Following Garbacz and Turpin (7), we also introduce a perturbation matrix $[P]=([S]-[\underline{1'}])/2$ with $\delta=(b-a')/2$ and $\Delta=(B-A')/2$, where the primed quantities differ from the unprimed ones only if different ordered sets are used for the ingoing and outgoing modes. This leads to the equation

$$\begin{Bmatrix} \delta \\ - \\ \Delta \end{Bmatrix} = [P] \begin{Bmatrix} a \\ - \\ A \end{Bmatrix} . \tag{7}$$

3.3. Symmetry and forms of the "matrices"

We now relate the detailed structures of the matrices [S], [Γ], [$\underline{R}$], [$\underline{T}$], [$\underline{S}$], and [P] to the symmetries of the scatterer, making use of the matrices and submatrices given or implied by Eqs. (6) and (7). The equation

$$u = \underline{P}v \tag{8}$$

may be interpreted in a number of ways. $\underline{P}$ may be taken as the perturbation operator, v the ingoing mode, and u the pertubation mode. Alternatively, $\underline{P}$ may be taken as the matrix [P] with u and v the column matrices of (7), or $\underline{P}$ may be taken as a submatrix of [P] with u and v the appropriate column submatrices. Since the formal algebra is the same, $\underline{P}$ may also be taken as [S] or one of its submatrices with u and v as the appropriate column matrices or submatrices. We discuss the perturbation and scattering matrices and their submatrices together, using the expression "the matrix" and, for the scattered or perturbation field, the "resultant" field.

We choose a group G which is convenient for expressing the properties of the scatterer as well as the medium (or media) of propagation, and base "the matrix" upon "symmetry adapted" functions, i.e., linear combinations of basis functions so that each combination belongs to a row (m) of an irreducible representation (n) of G, from a set of inequivalent[3] irreducible representations. However, we use, as basis functions, eigenfunctions (f_N's) of the differential equation for the medium (or media) of propagation <u>alone</u>, i.e., in the <u>absence</u> of the scatterer etc. Since G is ordinarily a subgroup of the group for the medium of propagation, judicious choice of the eigenfunctions ordinarily reduces the "symmetry adaptation" to mere grouping the eigenfunctions; see Koster's "compatibility tables" (12, pp. 18-20) for relating the representations of the spherical and point groups.

3.4. Scatterers known invariant under their symmetry groups

Prior to introducing the formal analysis, we introduce three simple but practical and enlightening special cases, one for scattering in this section and the other two for inverse scattering in the next section. (However, the treatments of these two sections may not yield <u>all</u> the symmetries available, e.g., completely determine the multipoles present in an EM (but not scalar) problem.) For scattering, we consider (in this section) scatterers, i.e., $\underline{P}$ operators, which are <u>invariant</u> with respect to (i.e., commute with) all the operations of the given group G; see (8),(9),(2). Then, just as multiplication by an even function yields a product which is even or odd as the multiplicand, so here P will change neither **n** nor **m**, i.e., the matrix will be diagonal in **n** and **m**, assuming no change in the coordinate system or ordered set of modes. Further, due to our use of partner[6] functions, the non-zero matrix elements will be independent of **m** (but not the **n**'s or **p**'s). See Section 2.1 for background of both this and the next section; for simplicity, the propagation media in the following examples d), e), and f) are assumed homogeneous and isotropic, but possibly lossy.

Invariant P's are adequate for the following examples, some of them familiar in terms of conventional viewpoints: a) If the material properties of the system (and all externally imposed fields and forces to which it is subjected) are independent of time, the system is invariant under the temporal translation-reflection group T_{th} and the matrices will be diagonal in the frequency ω and, for $\omega \neq 0$, the row index; further, the diagonal matrix elements will be independent of the row index. That is, if the mode v is proportional to any linear combination of sin ωt and cos ωt or of $\exp(+i\omega t)$ and $\exp(-i\omega t)$, then every resultant mode u (with nonzero amplitude) will have the that same dependence, regardless of the spatial dependence; further, the matrix elements (complex ratios of the final to the initial coefficients) may vary with ω, but not with the linear combination for a given ω. Unless otherwise stated, this symmetry and (suppressed) $\exp(-i\omega t)$ time dependence are assumed in the following. b) For two contiguous half spaces, each with properties which are invariant with respect to translation in x and y ($T_x X T_y$), neither k_x nor k_y will change, either on transmission or reflection. Further, if the material properties of the system are also even with respect to reflection in the plane of incidence, there is no conversion between modes which are even with respect to reflection in the plane (parallel E) and those which are odd (perpendicular E), explaining the separation in Fresnel's equations, which then follow from the boundary conditions. Note that either or both media may be lossy, anisotropic, and/or arbitrary functions of z, except as separately limited by the stated invariances. (Note that this applies to acoustics in a medium whose density varies with depth, reducing it to a one dimensional problem.) Hence, this provides generalizations of Snell's laws and Fresnel's equations. c) Application of these techniques to Γ provides a treatment of guided-mode junctions (11), but is not limited to ideal lossless waveguide leads; the leads may contain lossy inhomogeneous anisotropic media and the treatment of Sections 3.6 and 4.2 applies to much more general junctions. d) For a cylindrical scattering system invariant with respect to ϕ rotation and z translation ($C_\infty X T_z$) (but possibly lossy, anisotropic, and radially inhomogeneous), the matrix will be diagonal in the azimuthal index m and k_z. Further, if the system is also invariant with respect to reflection in all planes containing the cylindrical axis, the elements of the matrix will be independent of the sign of m, i.e., of the linear combination of sin $m\phi$ and cos $m\phi$

or of exp(imϕ) and exp(−imϕ). For a cylinder of finite conductivity, there will be 2x2 blocks on the diagonal due to possible TM,TE conversion. e) For a scatterer which is invariant with respect to both spatial inversion and the three Eulerian angles ($O_3^{\pm}$), there is no change in m or n of $P_n^m(\cos\theta)\exp(im\phi)$, the TM,TE character, or in the propagation constant k_r of the medium. Thus, the matrix has discrete rows and columns and is diagonal in all the modal indices, i.e., there is no modal conversion in the black box sense (except for phase and/or amplitude), even if the scatterer is lossy, radially inhomogeneous and locally anisotropic. Further, these elements, which represent the complex ratios of coefficients, are independent of m but may vary with n, the TM,TE character, or k_r. In example d), if the cylinder is perfectly conducting, the surrounding medium is "extended" into the cylinder by the mirror image, leading to $O_3^{\pm}$ symmetry (with an infinite spherical radius) and the matrix is diagonal in TM,TE character (see (20, pp. 524 ff.) for a conventional treatment). f) If the material properties of a transmitting antenna and feed are invariant with respect to rotation about the z axis (C_∞) and the mode in the feed has exp(imϕ) dependence from any combination of modes, the radiated field will have that dependence everywhere in space (with respect to the original axis). Further, if the system is invariant as well with respect to reflection in all planes containing the z axis ($C_{\infty v}$) and E_R of the feed has cos mϕ dependence from any combination of modes, then E_R, E_z, and H_ϕ of the radiated field will have cos mϕ dependence while E_ϕ, H_R, and H_z will have sin mϕ dependence, everywhere in space. This will be true regardless of the variation of the properties (or diameter) of the system with z or anisotropy consistent with the invariances, but TM and TE modes (here p indices) will mix. Thus, measurements of two components on a single radius vector determine an expansion which yields the phases and amplitudes of the six EM components everywhere in the forward half space bounded by an xy plane which just fails to intersect the antenna (see Note 9). Note that, in distinguishing between two airplanes or two missiles, it can usually be assumed that both of the former have a plane of symmetry and that both of the latter are essentially circularly symmetric.

3.5. Simple determinations of symmetry constituents of a scatterer

For simple detection of symmetry constituents of an arbitrary mathematically linear scatterer, we consider one or more ingoing modes which are invariant under all the operations of the given group G. Then each resultant mode will have the same symmetry (n, m) as the symmetry constituent of $\underline{P}$ which gives rise to it, and, for every symmetry type in the resultant modes, $\underline{P}$ will have a constituent of that symmetry. Note that a single resultant mode can show unambiguously the presence of a given symmetry constituent in $\underline{P}$. As previously noted, if the scatterer is a capacitor whose material properties are symmetric with respect to a plane and the incident mode is symmetric with respect to the same plane, a perturbed mode antisymmetric with respect to the plane will show that the capacitor is differentially charged. Note however, that if the group is not large enough for the symmetry type (n, m) to completely determine the ingoing mode, the lack of a resultant mode of a given type does not prove that that type does not occur in P.

Another simple procedure arises if G is Abelian, since the product of two symmetry types yields only a single type. For example, if G is C_∞ and the incident mode has $e^{im\phi}$ dependence, then the operator has $e^{ik\phi}$ dependence if there is any resultant mode with $e^{i(m+k)\phi}$ dependence. This is useful for studying the fins on a circularly symmetric rocket, since the diagonal blocks of the scattering or perturbation matrix corresponding to circularly symmetric part (k=0) can be disregarded and attention confined, e.g., to the blocks corresponding to k equal to the number of (equally spaced) fins and perhaps its multiples.

3.6. General treatment of mathematically linear scatterers

The procedures of Sections 3.4 and 3.5 are very useful, but in some cases they do not permit determination of all the symmetries. In particular, a) the product of two symmetry types, e.g., sin $m\phi$ sin $k\phi$, may yield a sum of types and b) for Section 3.5, there is no EM mode which is totally symmetric with respect to O_3, since E_3 here requires that $n \geq s=1$, i.e., the total angular momentum (n of the associated Legendre function) cannot be less than the spin momentum of the photon. We now discuss procedures for cases in which the preceding treatments are inadequate.

We express the scattering or perturbation operator as a linear combination of operators, each of a given symmetry type, i.e., each an irreducible tensor operator which belongs to a row of an irreducible unitary representation. This is possible for essentially arbitrary linear operators, self-adjoint or not (9, pp. 377-412),(2, pp. 242-276). The same set of indices which designate a mode or its symmetry designate a constituent operator or its symmetry.

The so-called Clebsch-Gordon coefficients express the permitted combinations of the symmetry types of the incident mode, the resultant mode, and the operator constituent; further, they give required relations between the coefficients of some of the permitted modes. These are tabulated for the three dimensional rotation group and the 32 point groups and may be easily determined for some other groups. With appropriate choice and ordering of the modes, the presence of a given symmetry constituent is indicated by the presence of one or more submatrices of the scattering or perturbation matrix. Further, the relative importance of symmetry constituents is roughly indicated by the relative magnitudes of elements of the submatrices. The details are given in Section 5.

4. PRACTICAL DETAILS

Conversion of the raw data from the usual plane wave representations
to the symmetrized modes used here involves at most a similarity
transformation, i.e., u, v, and $\underline{P}$ are replaced by Qu, Qv, and $Q\underline{P}Q^{-1}$,
respectively. In practice, these are matrix multiplications in
which the rows of Q are given by <u>natural</u> orthogonalities provided by
the theory; for spherical, circular cylindrical, and Cartesian
representations, these may be the orthogonalities with respect to
<u>summation</u> used in near field scanning, ordinarily the far field
forms of the ideal probe expressions (25),(24); more generally, the
symmetrizing combinations are given by Eq. (12) of Note 6. Further,
as explained Section 3.5, a single incident mode is commonly sufficient to determine the symmetry constituents of a scatterer. The
centers, axes, and planes of symmetry can often be located with
pulsed measurements, perhaps supplemented with cut and try or a
modified Newton–Raphson procedure.

5. MATHEMATICAL DETAILS: GENERAL SCATTERERS

For convenience in the use of the tables of Koster <u>et al</u>., we now
pattern our notation after theirs. We designate the resultant and
ingoing modes by u^γ_{pn} and v^β_{qj}, respectively, where the superscripts
indicate the representations, n and j indicate the rows of the
representations, and p and q indicate the partner sets. Similarly,
we use P^α_i to indicate the constituent of $\underline{P}$ corresponding to ith row
of the α representation (12, Eq. (3.18)). Thus, our n, m, and p
here correspond, respectively, to β, j, and q for the ingoing modes,
to γ, n, and p for the resultant modes, and to α, i, and − of $\underline{P}$. We
use lightface for the last eight indices to avoid confusion with the
first and third (our n and p). For a <u>finite</u> scatterer, the symmetry
operations of interest include rotation, reflection, and inversion,
but not translation, i.e., we are <u>commonly</u> concerned with the crys-
tallographic <u>point</u> groups. For the immediate purpose, we index on
the partner set before the row of the representation; thus, the row
and column locations of an individual submatrix are indicated by γ,p
and β,q respectively, and the numbers of rows and columns of the
submatrix are given by the numbers of rows in the γ and β representations, respectively. Should the operator $\underline{P}$ contain more than
one symmetry constituent, the following treats a constituent part of
the overall problem; the overall matrix is then the sum of the
constituent matrices.

We now express the constraints upon the scattering matrix elements
imposed by symmetry. From the theory of group representations (9),
(12),(8), the matrix elements are given by

$$\sum_r {}_r s^\alpha_{\gamma p, \beta q} \; {}_r U^{\alpha\beta, \gamma*}_{ij, n} . \qquad (9)$$

That is, each submatrix of the matrix is expressed as the sum of one or more products, each of a complex number s independent of i, j, and n (the rows of the representations) and a submatrix U^* which depends only upon the irreducible representations of the group, not upon the partner sets or other aspects of the physical problem. In various contexts with slight differences in definitions, the U's are called Clebsch-Gordon, coupling, Wigner 3j, vector addition, or Racah V coefficients. These are well known and tabulated for the three dimensional rotation group, but exist for other groups and are tabulated for the 32 point groups (12),(8). Only for two of the point groups (O and T_d) are there two products; for the other point groups there is only one product. In the tables (12), the U's for the first product are indicated by $\not{1}$ and, when they occur, the U's for the second product by $\not{0}$. A large fraction of the U's have no nonzero elements. Further, if the symmetry constituent of either v or $\underline{P}$ is totally symmetric, then U^* is zero unless the symmetry type of u is identical to that of $\underline{P}$ or v, respectively; Sections 3.4 and 3.5 follow from the fact. Some of the U's are well known. Thus, those for the one-dimensional translation group follow from the relation $\exp(ik_x x)\exp(ik'_x x) \propto \exp(i[k_x+k'_x]x)$. Similarly, the expressions for the products of sines and/or cosines in terms of sums or differences give all the U's for the one dimensional translation-reflection group.

In determining the matrix for a scatterer of known symmetry, the partner set may be indexed before the row of the representation; thus, the row and column locations of an individual submatrix are located by γ,p and β,q, respectively, and the numbers of rows and columns of the submatrix are given by the numbers of rows in the γ and β representations, respectively. The determination of "the matrix" thus usually reduces to determining a complex constant (the s) for each submatrix. Note that measurements on a *far field* sphere determine the EM field outside a sphere *just* circumscribes an antenna or scatterer; further, if the field is known to be outgoing or ingoing, only the *transverse* components of the electric field are needed since Hermitian orthogonalities are supplied by the O_3^+ group and the group operations do not mix the transverse with the radial components.

5. NOTES

[1]A **group** G is a set of elements {a,b,c,...} with a binary operation called **multiplication** which associates each ordered pair of elements a,b in G with a **product** ab also in G such that a) there exists in G a single element e, called the **identity**, such that ae=ea=a for *every* element a in G, b) for *each* a in G, there exists in G an **inverse** a^{-1} such that $aa^{-1}=a^{-1}a=e$, and c) the **associative law** (ab)c=a(bc) holds for all a, b, and c in G (14, p. 1). (The product of each element with itself is also in G.) The identity operation e and

reflection σ in the z=0 plane constitute the familiar reflection group $\mathbf{C}_s=\{e,\sigma\}$. A group is called **Abelian** if the effects of its operations are independent of the order of application. A group is said to be **compact** if the parameters designating the group operations vary over a finite range (9, p. 309). ⟨We enclose material not needed for general understanding in angular brackets and confine attention to locally compact, separable, unimodular topological groups (2, p. 134 ff.).⟩

[2]A **representation** $a\to T_a$ of a group is a set of operators such that a) each member of the group is associated with a single element of the representation, b) each operator of the representation is associated with one or more members of the group, and c) if members a, b, and c of the group are associated with operators A, B, and C of the representation, respectively, and ab=c, then AB=C (29, pp. 69, 72 ff). ⟨We confine our attention to representations whose elements are bounded linear operators in a separable complex Hilbert space (but our group elements and $\underline{O}$'s may be unbounded, e.g., differential, operators) (2, p. 134 ff.).⟩ A set of matrices, designated by D(a), is commonly used for a linear representation; they act upon vectors whose unit elements represent an orthonormal set of functions which provide a basis for the space in which the operators of T act.

[3]Two representations T and T′ of Hilbert spaces H and H′, respectively, are said to be **equivalent** if there exists a one to one linear mapping S of H onto H′ such that $T'_a=ST_aS^{-1}$ for all a of the group G, independent of a. (H′ may be identical to H.) For matrix representations, S is a nonsingular matrix.

[4]A subspace H_1 of H is said to be **invariant** (with respect to a representation T) if, for every element a of G and every function w of H_1, the function $T_a w$ is also in H_1. A representation T is said to be algebraically **irreducible** if it has no **proper** invariant subspace H_1, i.e., no invariant subspace H_1 other than the whole space H and the null space {0}. A matrix representation D is irreducible if no transformation S exists such that every element in every matrix of SDS^{-1} below a common form is zero; the form must consist of non-overlapping contiguous square blocks on the principal diagonal (9, p. 94 ff), (14, pp. 67-69). If the matrix representation is **unitary**,[5] reducible, and of finite **dimension** (matrix order), one may always obtain all elements zero above and below the blocks (2, p. 143); further, every representation of a finite ⟨or a compact linear Lie⟩ group is equivalent to a unitary representation. In either case, each set of submatrices corresponding to a given diagonal block of the form also constitutes a representation.

[5]A unitary representation consists of unitary operators. An operator $\underline{U}$ is unitary if $(\underline{U}f,\underline{U}g)=(f,g)$ for all function pairs f,g of the Hilbert space. A matrix is unitary if its complex conjugate transpose is its inverse.

[6]In each matrix representation, there is a matrix for each group operation. We indicate an individual element of such a matrix by $D^{(n)}_{rc}(R)$ where n indicates the representation, R indicates the group element (and so the matrix), and r and c indicate the row and column, respectively. We define R as an operator which acts upon a coordinate system (leaving objects or fields fixed in space) such that

$$x' = R\,x \tag{10}$$

where x' and x are the sets of coordinate values in the new and old coordinate systems, respectively; since R may represent a translation, it need not be representable as a matrix. Confusion exists in the literature concerning D's due to the failure to carefully define the associated R's (4),(32).

The symmetry of a function with respect to a given group is defined in terms of its transformations under each and every operation of the group. For a given irreducible representation n of a given group, a set of **partner functions** (29, p. 112) is defined as one which transforms according to the equation

$${}_pf^n_m(x) = P_R\;{}_pf^n_m(x') = P_R\;{}_pf^n_m(Rx) = \sum_{m'} D^{(n)}_{m'm}(R)\;{}_pf^n_{m'}(x) \tag{11}$$

for each and every operation R of the group. Note that the operator P_R acts upon the coordinate values, not the function proper, and reverses the change of coordinate values caused by R in (10). Given one partner function of a set, all the others are completely determined including complex normalization. The representation must be completely defined, not just up to a **similarity transformation** STS^{-1}.

We say that a function belongs to a given **symmetry type** if it is equal to a constant times a partner function belonging to a given row of a given (completely defined) irreducible representation. There may be more than one linearly independent partner function belonging to a single symmetry type (e.g., of a subgroup of a group which defines eigenfunctions). If so, there is more than one set of linearly independent partner functions and the sets do not mix under the operations of the given group. We use the index p to distinguish between partner sets; it consists of the modal indices required to specify the eigenfunction but not required to specify either the representation or row of the given group. For a finite group of e elements, a function f may be decomposed into its symmetry types by the equation

$$f^n_m(x) = (r/e)\sum_R D^{n*}_{mm}(R)\;P_R f(x), \tag{12}$$

where r is the number of rows in the representation (9, p. 113).

[7]Strictly speaking, some operators $\underline{O}$, e.g., differential operators, have no adjoint, i.e., there is no $\underline{A}$ such that $(\underline{O}u,v)=(u,\underline{A}v)$ for <u>all</u> functions u and v of the Hilbert space. However, there <u>may</u> be an $\underline{A}$ which fulfills the preceding equation for a set of functions which a) excludes some functions but b) is <u>dense</u> in the Hilbert space (2, pp. 641-645); in this sense the Laplacian and the curl-curl operators are "essentially" self-adjoint and the mathematical completeness of Section 2.4 is then with respect to the restricted set of functions.

[8]To illustrate how our detailed discussion carries over to other systems, we briefly treat geophysical prospecting and sonar. Assuming the displacement $\mathbf{s}$ is small, the longitudinal and transverse waves in an elastic medium split (16, p. 142 ff.) to yield

$$\nabla^2\psi = [\rho/(\lambda+2\mu)]\ \partial^2\psi/\partial t^2 \qquad (13)$$

and

$$-\nabla\times\nabla\times\mathbf{A} = (\rho/\mu)\ \partial^2\mathbf{A}/\partial t^2, \qquad (14)$$

respectively, where ρ is the density, μ is the shear modulus, $(\lambda+2\mu/3)$ is the compression modulus, $\mathbf{s}=\nabla\psi$ for the longitudinal waves, $\mathbf{s}=\nabla\times\mathbf{A}$ for the transverse waves, and the reciprocals of the numerical coefficients on the right hand sides are the squares of the respective velocities. For harmonic waves, the equations of course take the <u>forms</u> of Eqs. (2) and (3), respectively.

For sonar, assuming that the medium is incompressible and nonviscous, that the motion is irrotational and the displacements small, and that the surface and bottom discontinuities are unimportant, one obtains

$$\nabla^2 p = (\rho_0/K)\ \partial^2 p/\partial t^2, \qquad (15)$$

where $K=p\rho_0/(\rho-\rho_0)$ is the bulk modulus, p is the pressure, and ρ and ρ_0 are the density and its mean, respectively. Again it reduces to the scalar Helmholtz equation (2) for harmonic waves.

[9]For the curl-curl and scalar Helmholtz equations, the domain of validity of a modal expansion is similar to that of a function of a complex variable, e.g., of a Taylor or Laurent series (1),(22). The expansion is <u>not</u> valid for the whole analytic domain, but for a domain free of singularities (such as a scatterer or probe) between two concentric spheres, two coaxial cylinders, or two parallel planes, for spherical, circular cylindrical, and rectangular coordinates, respectively.

[10]With any <u>single</u> frequency, there is a continuum of exact solutions for any <u>general</u> inverse scattering problem. For example, one may

consider a scatterer to be an assemblage of multipoles at a point and the description is valid outside a sphere centered at the point and circumscribing the scatterer (assuming the medium to be homogeneous and isotropic); further, one may expand about any point whose sphere is within the domain in which no measurements are made. However, if one uses pulses of finite duration, localization may be obtained, giving (in principle) a unique exact solution (except for what is inside a perfectly conducting capsule) (31).

6. REFERENCES

(13) and (6) have introductory chapters on finite groups. (10) applies the point groups, C_∞, etc. to the derivation of selection rules. Theorems for compact Lie groups are similar to those for finite groups, but weighted integration replaces summation (9),(14). (23) and (2) are encyclopedic for application to special functions and elementary particle theory, respectively, but the latter assumes considerable mathematical background.

(1) Bers, L., Theory of Pseudo-analytic Functions. New York: N.Y.U. Inst. Math. Sci., 1953.

(2) Barut, A.O., and R. Raczka, Theory of Group Representations and Applications. Warsaw: Polish Scientific Publishers, 2nd ed., 1980.

(3) Boerner, H., Representations of Groups with Special Considerations for the Needs of Modern Physics. Amsterdam: North Holland, 1970.

(4) Bouten, M., "On the rotation operators in quantum mechanics": 1969, Physica 42, pp. 572-580.

(5) Emmerson, J.M., Symmetry Principles in Particle Physics. Oxford: Clarendon Press, 1972.

(6) Eyring, H., et al., Quantum Chemistry. New York: Wiley, 1944.

(7) Garbacz, R.J., and R. H. Turpin, "A generalized expansion for radiated and scattered fields,": 1971, IEEE Trans. Antennas Propagat. AP-19, pp. 348-358.

(8) Griffith, J.S., The Irreducible Tensor Method for Molecular Symmetry Groups. Englewood Cliffs, N.J.: Prentice-Hall, 1962.

(9) Hamermesh, M., Group Theory and its Application to Physical Problems. Reading, MA: Addison-Wesley, 1962.

(10) Herzberg, G., Infrared and Raman Spectra. New York: Van Nostrand, 1945.

(11) Kerns, D.M., "Analysis of symmetrical waveguide junctions,": 1951, J. Res. NBS 46, pp. 267-282.

(12) Koster, G.F., et al., Properties of the Thirty-Two Point Groups. Cambridge, MA: MIT Press, 1963.

(13) Margenau, H., and G. M. Murphy, The Mathematics of Physics and Chemistry, Princeton, NJ: Van Nostrand, 1959, Chap. 15.

(14) Miller, H., Jr., Symmetry Groups and their Applications, New York: Academic Press, 1972.

(15) ----, Symmetry and Separation of Variables. Reading, MA:

Addison-Wesley, 1977.
(16) Morse, P.M., and H. Feshbach, Methods of Theoretical Physics. New York: Academic Press, 1963.
(17) Olsen, H.F., Dynamical Analogies, New York: Van Nostrand, 1958.
(18) Opechowski, W., "Magnetoelectric Symmetry": 1974, Int. J. Magn 5, no. 4, pp. 317-325.
(19) Rose, M.E., Elementary Theory of Angular Momentum. New York: Wiley, 1967, pp. 127-139.
(20) Stratton, J.A., Electromagnetic Theory. New York: McGraw-Hill, 1941, pp. 486-488.
(21) Talman, J.D., Special Functions: A Group Theoretic Approach. New York: W. A. Benjamin, 1968.
(22) Vekua, I.N., Generalized Analytic Functions. Reading, MA: Addison-Wesley, 1962.
(23) Vilenkin, N.Ja., Special Functions and the Theory of Group Representations. Providence, RI: Am. Math. Soc., 1978.
(24) Wacker, P.F., "Non-planar near-field measurements: Spherical scanning": 1975, Natl. Bur. Stds., Washington, NBSIR 75-809.
(25) ----, "Unified theory of near-field analysis and measurement: Scattering and inverse scattering": 1981, IEEE Trans. Antennas Propagat. AP-29, pp. 342-351.
(26) ----, "Symmetry analysis and eigenfunction expansions in engineering and applied mathematics": 1983, Third Intl. Conf. Ant. Propagat., Norwich, UK, pp. 113-117.
(27) ----, "Unified theory of near-field analysis and measurement: Reply to Dr. Steven B. Berger": Nov. 1983, IEEE Trans. Antenna Propagat. AP-31.
(28) Weinberg, S., "Light as a fundamental particle": 1975, Physics Today 28, pp. 32-37.
(29) Wigner, E.P., Group Theory and Its Application to the Quantum Mechanics of Atomic Spectra. New York: Academic Press, 1959.
(30) ----, "Unitary representations of the inhomogeneous Lorentz group including reflections," in Group Theoretical Concepts an Methods in Elementary Particle Physics, Feza Gürsey, ed. New York: Gordon and Breach, 1964.
(31) Wohlers, M.R., "On the uniqueness of the sources of electromagnetic fields": 1970, IEEE Trans. Ant. Propagat. AP-1 pp. 825-826.
(32) Wolf, A.A., "Rotation operators,": 1969, Am. J. Phys. 37, pp. 531-536.

I.12 (OI.5)

RECONSTRUCTION OF REFRACTIVE INDEX PROFILE OF A STRATIFIED MEDIUM

E.Vogelzang, H.A.Ferwerda and D.Yevick*.

Department of Applied Physics, Nijenborgh 18, 9747 AG Groningen, The Netherlands, Institute of Optical Research, Stockholm, S-10044 Sweden*.

Abstract

In this paper we present a method for determining the permittivity profile of a stratified medium terminated by a perfect conductor from the (complex) reflectivity. The calculations are based on the Gelfand-Levitan and the Marchenko equations. The bound modes of the system are explicitly taken into account.

1. Introduction

In this contribution we shall describe how to reconstruct the refractive index profile of a stratified medium placed on top of a perfect conductor using the complex reflection coefficients for plane monochromatic waves, incident from different directions. Because of the monochromaticity of the incident waves we do not have to consider the dispersion of the refractive index.
It may be difficult to measure the phase of the reflection coefficient at optical frequencies. In that case a holographic recording of the reflected wave might be envisaged.

2. Outline of the method

A lossless dielectric layer of thickness d has the permittivity profile $\varepsilon_2(z)$, where z is measured along the normal on the layer. $\varepsilon_2(z)$ is assumed to be independent of the remaining coordinates x and y. Such a medium is called a stratified medium. The medium above the dielectric layer is supposed to be a dielectric with constant permittivity ε_1. We consider TE-polarized incoming plane waves, i.e. waves whose electric vector only has a non-vanishing component along the y-axis:

$$E_{y,inc} = E_o \exp[ik_1(x \sin\theta_1 + z \cos\theta_1)] , \quad (1)$$

W.-M. Boerner et al. (eds.), Inverse Methods in Electromagnetic Imaging - Part 1, 223–230.

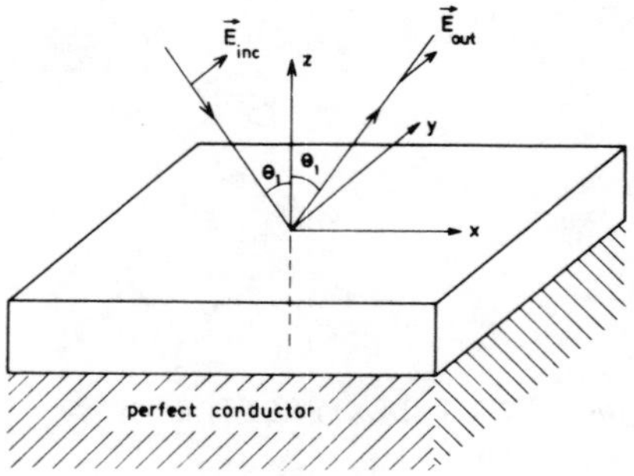

Figure 1

where k_1 is the wave number in the medium 1 and θ_1 is the angle of incidence (Fig.1). E_y has to satisfy the Helmholtz equation

$$[\partial^2/\partial x^2 + \partial^2/\partial z^2 + \omega^2 c^{-2}\varepsilon(z)]\, E_y(x,z)=o \tag{2}$$

where $\varepsilon(z) = \varepsilon_1$ for $z \geq d$
$= \varepsilon_2(z)$ for $o<z<d$
(the time-dependence $\exp(-i\omega t)$ of E_y has been suppressed). Substituting $E_y(x,z)=T(x)U(z)$ leads to

$$d^2T/dx^2+\beta^2\ T=o \quad \text{and} \quad d^2U/dz^2 + [\omega^2 c^{-2}\varepsilon(z)-\beta^2]\, U=o, \tag{3}$$

where β is a constant over all space . As $E_{y,inc}$ satisfies (2) in $z\geq d$ we immediately find that $\beta=k_1\sin\theta_1$ (4)
Eqn. (2) can be rewritten as

$$d^2U(z,m_1)/dz^2+ m_1^2U(z,m_1)=[k_1^2-k^2(z)]\ U(z,m_1), \tag{5}$$

where $m_1^2=k_1^2-\beta^2=k_1^2\cos^2\theta_1$
$k^2(z)=\omega^2c^{-2}\varepsilon(z)$.

The m_1-dependence of $U(z,m_1)$ is now explicitly denoted. (5) has the same form as the radial Schrödinger equation in Quantum Mechanics if we formally define a potential $V(z)$ by

$$V(z) = k_1^2-k^2(z) \tag{6}$$

It is clear that the potential is only non-vanishing on the interval $o\leq z\leq d$.
The basic equation (5) is henceforth written as

$$(d^2/dz^2+m_1^2)U(z,m_1)=V(z)U(z,m_1) \tag{7}$$

with the boundary condition $U(o,m_1)=o$, (8)
because the layer is placed on top of a perfect conductor.
The problem to be solved is the calculation of $V(z)$ from the complex reflection coefficiënt.

3. A brief reminder of the Gelfand-Levitan and Marchenko-formalisms. The problem discussed above shall be solved with the Gelfand-Levitan (GL) and Marchenko (M) formalisms [1]. We first introduce the regular solutions $\phi(z,m_1)$ of (7), which satisfy

$$\phi(o,m_1)=o \; ; [d\phi(z,m_1)/dz]_{z=o} = 1 . \tag{9}$$

Another, linearly independent solution of (5) is the singular solution (Jost solution) $f(z,m_1)$ specified by $\lim_{z\to+\infty} f(z,m_1)=\exp im_1 z$ (10)

It is straightforward to show that $f(z,m_1)$ and $f(z,-m_1)$ are linearly independent solutions. Consequently the regular solution $\phi(z,m_1)$ can be expressed in terms of $f(z,m_1)$ and $f(z,-m_1)$:

$$\phi(z,m_1)=i(2m_1)^{-1}[F(m_1)f(z,-m_1)-F(-m_1)f(z,m_1)] , \tag{11}$$

where the coefficient $F(m_1)$ is called the Jost-function and the symmetry property $\phi(z,m_1)=\phi(z,-m_1)$ has been used. The solution $U(z,m_1)$ of eqn. (7) is proportional to $\phi(z,m_1)$. On the other hand the asymptotic behavior of $U(z,m_1)$ can be written as

$$U(z,m\;) \underset{z\to+\infty}{\sim} \tfrac{1}{2}i[\exp(-im_1 z)-R(m_1)\exp(im_1 z)], \tag{12}$$

where $R(m_1)$ is the (complex) reflection coefficient at angle of incidence θ_1. Because of energy conservation $|R(m_1)|=1$ so that $R(m_1)$ can be written as

$$R(m_1)=\exp[i2\delta(m_1)]. \tag{13}$$

Combining (11) and (12) we find

$$R(m_1) = F(-m_1)[F(m_1)]^{-1} \tag{14}$$

$$U(z,m_1) = m_1[F(m_1)]^{-1}\phi(z,m_1) . \tag{15}$$

The bound modes correspond to poles of $R(m_1)$ and thus to zeroes of $F(m_1)$. It has been shown [2] that $F(m_1)$has simple zeroes on the positive imaginary axis of the complex m_1-plane corresponding to bound modes.

The set of regular solutions, including the bound modes form a complete set:

$$\int_{-\infty}^{\infty} d\rho(E)\phi(z,E)\phi(z',E)=\delta(z-z'), \tag{16}$$

where $E=m_1^2$ ("energy") and the spectral density $\rho(E)$ is given by

$$d\rho(E) = \begin{cases} \pi^{-1}E^{\frac{1}{2}}|F(E^{\frac{1}{2}})|^{-2}dE, & E\geq o \\ \sum_j C_j\delta(E-E_j)\,dE & , E<o \end{cases} , \tag{17}$$

$$C_j^{-1}= \int_o^{\infty}\phi^2(z,m_{1_j})dz.$$

m_{1j} is the value of m_1 for the bound mode with energy E_j. $|F(m_1)|$ can be calculated according to

$$|F(m_1)| = \prod_j |1-m_{1j}^2 \, m_1^{-2}| \qquad \exp[\pi^{-1} P \int_{-\infty}^{\infty} \delta(m_1')(m_1-m_1')^{-1} dm_1'] \tag{18}$$

The constants C_j cannot be inferred from experiment and still pose a problem in this approach, a problem which can be circumvented by using a matching medium such that the combined structure has no bound modes.
We will now introduce the generalized Gelfand-Levitan equation which allows to calculate the difference of two potentials $V_2(z)-V_1(z)$, of which one is assumed to be known, say $V_1(z)$. Let the regular solutions of $V_1(z)$ be denoted by $\phi_1(z,m_1)$. The <u>generalized Gelfand-Levitan equation</u> then reads:

$$K_{12}(z,t)+G_{12}(z,t) + \int_o^z K_{12}(z,s)G_{12}(s,t)ds=o, \tag{19}$$

where

$$G_{12}(z,t) = \int_{-\infty}^{\infty} \phi_1(z,m_1)\phi_1(t,m_1)\,[d\rho_2(E)-d\rho_1(E)]\,, \tag{20}$$

$\rho_j(E)$ being the mode density of $V_j(z)$, $j=1,2$. The potential is determined via

$$V_2(z) - V_1(z) = 2\ dK_{12}(z,z)/dz \tag{21}$$

The Gelfand-Levitan theory is based upon the regular solution of (7). We can also start from the Jost-solution $f(z,m_1)$, eqn.(10), which leads to the <u>Marchenko equation</u>:

$$A(z,t)=A_o(z+t)+\int_z^{\infty} A(z,s)A_o(s+t)ds \quad , \quad t>z, \tag{22}$$

where

$$A_o(t)=(2\pi)^{-1}\int_{-\infty}^{\infty} R(m_1)\exp(im_1 t)dm_1 - \sum_j s_j \exp(im_{1j}t)\ , \tag{23}$$

$$s_j^{-1} = \int_o^{\infty} f^2(z,m_{1j})dz\,. \tag{24}$$

$$V(z) \text{ follows from } V(z) = -2dA(z,z)/dz\,. \tag{25}$$

The Marchenko equation has the advantage over the Gelfand-Levitan equation of directly involving the measured quantity $R(m_1)$, while in the Gelfand-Levitan approach one first has to calculate the Jost function $F(m_1)$ which is quite involved! Eqn.(22) cannot directly be solved numerically because the s_j turn out to be large. This leads to numerical instabilities when the equation is discretisized. For that reason we treat the contributions of the continuous and discrete parts of the reflection spectrum separately (corresponding to both terms on the right hand side of (23)). In the next section we shall explain how this can be done.

4. Reconstruction of a potential with bound states

In order to separate the contribution of the continuous and the discrete part of the spectrum (corresponding to both terms of (23)) we construct from the experimentally determined reflectivity $R(m_1)$ another reflectivity $R_1(m_1)$ which has no poles and consequently corresponds to a potential without bound modes. Denoting the propagation constants of the bound modes by m_{1j} (the poles of $R(m_1)$) $R_1(m_1)$ is according to (14) given by

$$R_1(m_1) = \prod_j (m_1-m_{1j})^2(m_1+m_{1j})^{-2}R(m_1). \tag{26}$$

The corresponding potential is obtained by solving the Marchenko equation (22) with s_j=o. This is done by discretisizing the equation and solving the resulting matrix equation. $V_1(z)$ then follows from (25). $V_2(z)$, which only has bound modes, is solved using the generalized Gelfand-Levitan equation (19) where $d\rho_2(E)=\sum_j C_j\,\delta(E-E_j)dE$. (19) becomes in this case

$$K(z,t) + \sum_j C_j\phi_1(z,m_{1j})\phi_1(t,m_{1j}) +$$

$$+ \int_o^z K(z,s) \sum_j C_j\phi_1(s,m_{1j})\phi_1(t,m_{1j})ds=o. \tag{27}$$

We ultimately find

$$V_2(z) = -2\, d^2/dz^2\, [\log \det \mathbf{M}(z)], \tag{28}$$

$$\text{where}\quad \mathbf{M}(z) = \mathbf{1} + \int_o^z \mathbf{R}(s)ds. \tag{29}$$

The matrix $\mathbf{R}(s)$ has elements

$$R_{ij}(s) = C_j\phi_1(s,m_{1i})\phi_1(s,m_{1j}). \tag{30}$$

5. Numerical simulations

At this time of writing the main obstacle for the application of this approach to actual profiles as occurring in practical situations is the fact that the constants C_j cannot be determined experimentally. Therefore we test the method on some known profiles. For each profile the reflectivity, the bound modes and the normalization constants C_j are computed. Subsequently the profile is reconstructed from these data.

Example 1: a layer with profile $\varepsilon_2(z)=1.0025-0.001z$. Thickness of the layer d=1, wavelength λ=0.2 (arbitrary units). The layer has no bound modes. Fig.2 shows the result: the drawn line is the reconstructed profile, the original profile (input) is given by the dotted line. The wavelength is so small compared

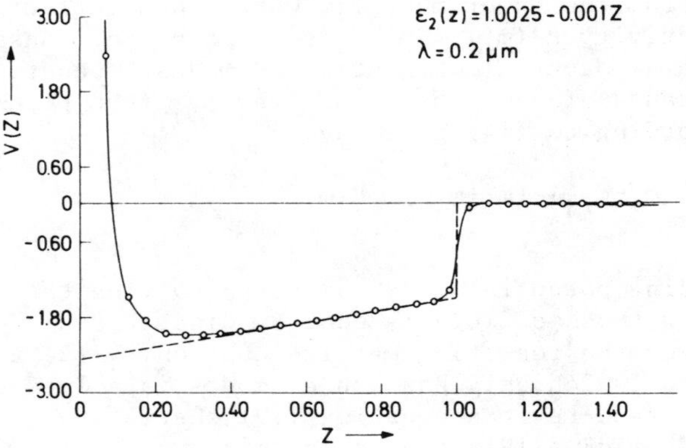

Figure 2

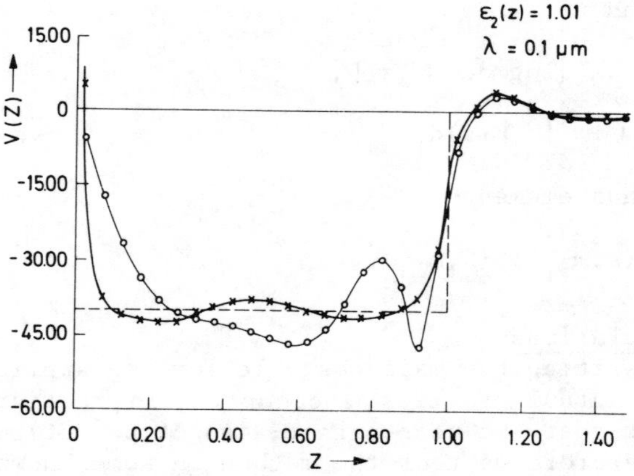

Figure 3

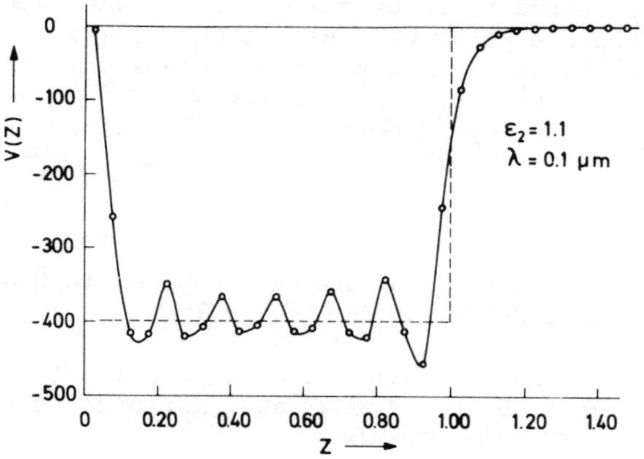

Figure 4

to the width d that the edge of the distribution is quite well reproduced. For larger values of λ the quality of the reconstruction deteriorates, as is to be expected.

Example 2: homogeneous layer $d=1, \varepsilon_2=1.01$, $\lambda=0.1$.
This layer supports 2 bound states. The thin line corresponds to the intermediate potential $V_1(z)$ while the thick line corresponds to the fully reconstructed potential $V(z)=V_1(z)+V_2(z)$.

Example 3: homogeneous layer $\varepsilon_2=1.1$, $\lambda=0.1$, $d=1$.
This system has 6 bound states. The reconstruction of the profile from the bound modes only is sketched in fig.4. The larger the number of bound modes the better the reconstruction.

6. Conclusions

The present method could lead to an alternative to the Newton-Kantorovich method for the determination of permittivity profiles from reflected intensities as developed by Roger [3]. In this approach it may happen that different permittivity profiles lead to the same reflected intensities, because the bound modes are not taken into consideration. In the numerical examples presented in the preceding section the importance of the bound modes has been established. At present the method defies application to practical problems because

1) the phase of the reflected wave has to be measured which, in principle, might be done holographically but which could be very difficult in practice
2) the normalisation constants C_j cannot be determined from experiment. Prior knowledge of the width of the layer may restrict the possible values of C_j

3) it is difficult to find a perfect conductor at optical frequencies. In some cases the finite conductivity can be accounted for by correcting the phase of the reflected wave with the phase shift due to the conductor of finite conductivity. This correction is only reliable if multiple reflections may be ignored, which is the case when the permittivity profile above the terminating conductor changes smoothly (so that the WKB-approximation applies).

References

[1] K. Chadan and P.C. Sabatier, Inverse Problems in Quantum Scattering Theory, Springer, Berlin 1977

[2] Ref. 1, p.22

[3] A. Roger, "Newton-Kantorovich Algorithm Applied to an Electromagnetic Inverse Problem", IEEE Trans. Antennas Propagat., vol. AP-29, pp. 232-238, March 1981

I.13 (IM.4)

INVERSE DIFFRACTION PROBLEM

A. G. RAMM

Math. Dept., Kansas State Univ., Manhattan, KS 66506

Abstract: Let $\mathcal{D}$ be a convex body in $\mathbb{R}^3$ with a smooth connected boundary Γ. A constructive procedure for recovering Γ from the scattering amplitude is given. The support function of $\mathcal{D}$ is recovered from the known values of the scattering amplitude at high frequencies. The support function defines explicitly the parametric equation of Γ. The stability of the recovered Γ with respect to small perturbations of the scattering data is established and the corresponding estimates are given. A construction of the surface Γ from the knowledge of its two principal curvatures is given.

Contents

I. INTRODUCTION

Let $\mathcal{D}$ be a convex bounded domain with a smooth boundary Γ, $\Omega = \mathbb{R}^3 \setminus \mathcal{D}$. Consider the problem

$$(\nabla^2 + k^2)u = 0 \quad \text{in} \quad \Omega,\ k > 0, \tag{1}$$

$$u\big|_{\Gamma} = 0, \tag{2}$$

W.-M. Boerner et al. (eds.), Inverse Methods in Electromagnetic Imaging - Part 1, 231–249.

$$u = u_0 + v,\ u_0 = \exp\{ik(\nu,x)\},\ r\left(\frac{\partial v}{\partial r} - ikv\right) \to 0, r = |x| \to \infty. \quad (3)$$

Here ν is a unit vector, u_0 is the incident field, v is the scattered field,

$$v \sim \frac{\exp(ikr)}{r} f(n,\nu,k),\ r \to \infty,\ \frac{x}{r} = n. \quad (4)$$

The function $f(n,\nu,k)$ is called the scattering amplitude, $f(n,-n,k)$ is the backscattering amplitude. The inverse diffraction (scattering) problem consists in finding Γ from the knowledge of $f(n,\nu,k)$ $n,\nu \in S^2$, $0 < k < \infty$, where S^2 is the unit sphere. It is an overdetermined problem: f depends on 5 variables while the shape of Γ depends on 2 variables, for example it is determined by the equation $r = r(\theta,\phi)$ of the surface Γ in the spherical coordinates. Therefore, not any function $f(n,\nu,k)$ is a scattering amplitude. A necessary and sufficient condition for a function $f(n,\nu,k)$ to be a scattering amplitude is not known. The uniqueness of the solution to this problem is established long ago.

Proposition 1. If $f(n,\nu,k)$, $n,\nu \in S^2$, $0 \le a \le k \le b$, $a < b$ is the scattering amplitude for Γ_1 and for Γ_2, where Γ_1 and Γ_2 are piecewise smooth, closed, connected (not necessarily convex) surfaces, then $\Gamma_1 = \Gamma_2$. Here $[a,b]$ is an arbitrary segment.

There are several ways to prove this statement. One way, due to M. Schiffer ([1]), is general but non-constructive. The second way requires an assumption about convexity and smoothness of Γ, Gaussian curvature is assumed to be strictly positive. This way is based on the high-frequency asymptotics of $f(n,\nu,k)$ and can be based on the results of Fock [2], although Fock did not discuss the inverse problems. In this way one finds the Gaussian curvature of the surface from the high-frequency asymptotics of $f(n,\nu,k)$. It is well known [3] that Minkowski's problem of recovering a convex compact surface Γ from the knowledge of its Gaussian curvature K as a function of the unit normal vector to Γ has not more than one solution. Let us formulate the corresponding result.

Proposition 2 [3, p. 286]. Let S_1 and S_2 be two compact oriented surfaces of class C^2 in $\mathbb{R}^3$. Suppose that $f: S_1 \to S_2$ is a diffeomorphism such that at each pair of corresponding points, S_1 and S_2 have the same unit normal vector and equal Gaussian curvatures K_1 and K_2, respectively. Then f is a translation in $\mathbb{R}^3$.

If one knows that the Gaussian curvature of a closed compact surface satisfies the inequality $K \ge a^{-2}$, $a > 0$, then the diameter d of the surface satisfies the inequality $d \le \pi a$, i.e. the body $\mathcal{D}$ is not very large [3, p. 305]. One can calculate the area S of the surface Γ by the formula

$S = \int_{S^2} \frac{dw}{K(w)}$, where dw is the element of the area of the unit sphere S^2, $w \in S^2$, and the area of the projection of $\mathcal{D}$ onto the plane perpendicular to the unit vector n, by the formula $S(n) = \frac{1}{2}\int_{S^2} \frac{|n\cdot w|}{K(w)}\,dw$. These formulas are well known in differential geometry and were used in inverse diffraction problem in [4]. In [5] a uniqueness theorem is proved. The theorem says that the shape of the obstacle and the potential in the exterior domain Ω can be uniquely determined from the scattering amplitude. In [6] the results of [5] are generalized, a constructive method for finding the shape of Γ from the scattering amplitude is given, and the stability of the solution to inverse diffraction problem is analyzed. Here we will follow the presentation in [6]. Some of the results of [4], [5] can be found in [7]. The role of high-frequency asymptotics in inverse scattering after work of Fock (1945-1950) [2] was discussed by many authors (see [8], [9], [10] and references in [9], [10]). The existence results in Minkowski's problem are given in [11]. Since the solution of the Minkowski's problem in [11] is non-constructive, it is of considerable practical interest to find a constructive procedure of recovering the surface Γ from the scattering amplitude. Such a procedure was given in [4] and its stability was analyzed in [7]. This procedure allows one to find the support function of $\mathcal{D}$ from the knowledge of $f(k,n,\nu)$ at one sufficiently large frequency k and the values of $n,\nu \in S^2$ such that the vector $l = \frac{n - \nu}{|n - \nu|}$ runs through all of S^2 (e.g. $\nu = -n$ and $n \in S^2$, i.e. backscattering).

In the literature [12] there were attempts to solve the inverse two-dimensional diffraction problem for a perfectly conducting cylinder assuming that $f(k,n,\nu)$ is known for $0 < k < k_0$, $n,\nu \in S^1$ (S^1 is the unit sphere in $\mathbb{R}^2$), where k_0 is a small number (low frequency data). From the practical point of view this is not very interesting because one needs to know all of the Taylor coefficients of $f(k,n,\nu)$ at $k = 0$, and it is difficult to measure even the second coefficient. The knowledge of the first non-zero coefficient in the case when the obstacle is much less than the wavelength ($ka \ll 1$, where a is the diameter of the obstacle; practically $ka < 0{,}2$ is often sufficient) allows one to recover 3 numbers only [13]. The obstacle under these assumptions scatters as an equivalent ellipsoid with the semiaxes given by the above 3 numbers [13]. Thus, the first coefficient (corresponding to dipole radiation) gives an idea about the shape of the obstacle but does not allow one to recover the shape in detail, and it is quite difficult to measure the other coefficients (corresponding to multipole radiation).

In [13, p. 109] an inverse radiation problem is solved. This problem consists in finding the initial field at the point x_0 from the knowledge of the field scattered by a small ($ka \ll 1$) obstacle placed in the initial field at the point x_0.

If the boundary condition is of the form

$$\frac{\partial u}{\partial N} + hu|_\Gamma = 0, \tag{5}$$

where h is constant (or a smooth function on Γ), $h > 0$ (or $\text{Re}h \geq 0$, $\text{Im}h \leq 0$, $|\text{Re}h| + |\text{Im}h| > 0$), then one can find the h and the shape of Γ from the knowledge of $f(k,n,\nu)$, $0 \leq a \leq k \leq b$, $b > a$, $n = n_0$, $\nu \in S^2$, where n_0 is a fixed unit vector (or $\nu = \nu_0$, $n \in S^2$, where ν_0 is a fixed unit vector).

In the physical literature ([2], [4], [9] etc.) the high frequency asymptotics of the scattering amplitude is obtained by using the Kirchhoff approximation (physical optics). This approximation and the high frequency asymptotics was rigorously justified in [14]-[16] for convex and for some non-convex obstacles.

This paper is organized as follows: in section 2 the support function and the shape of the surface are recovered from the scattering amplitude, in section 3 the stability of the solution to the inverse diffraction problem is proved, in section 4 a procedure for recovering the surface from the knowledge of its two principal curvatures K_1 and K_2 is given. This problem was formulated by the author and solved by F. Miller [17]. His results are presented in section 4, and the author thanks Professor Miller for permission to include these results in this paper. The interest in this problem comes from the fact that one can find the Gaussian curvature $K = K_1K_2$ ([2,4,8,9]) and the difference of the curvatures $K_1 - K_2$ ([18]) from the scattering data, and therefore find K_1 and K_2.

We conclude this introduction with a proof of the uniqueness of the solution to the inverse diffraction problem. In particular, from the argument below the validity of Proposition 1 follows. Suppose there are two obstacles $\mathcal{D}_1$ and $\mathcal{D}_2$ with the same scattering amplitudes $f(n,\nu_0,k)$, $n \in S^2$, $a < k < b$, $b > a$, and the boundary condition on $\Gamma_1 = \partial\mathcal{D}_1$ ($\Gamma_2 = \partial\mathcal{D}_2$) is (5) with $h = h_1$ ($h = h_2$). We want to prove that $\mathcal{D}_1 = \mathcal{D}_2$ and $h_1 = h_2$. Consider $v = v_1 - v_2$, where v_j, $j = 1,2$, are determined from (1), (3), (4), (5) with $\mathcal{D} = \mathcal{D}_j$. By the assumption $v = o(1/r)$ as $r \to \infty$ and v satisfies (1). This implies that $v \equiv 0$ in $B_R = \{x: |x| \geq R\}$, where R is large enough so that the ball $\mathbb{R}^3 \setminus B_R$ contains $\mathcal{D}_1$ and $\mathcal{D}_2$. This follows from the known result of Rellich ([7, p. 298]). Thus $u_1 = u_2$ in B_R. Suppose first that $\mathcal{D}_1 \cap \mathcal{D}_2 = \emptyset$. Then $u_1|_{\Gamma_2} = u_2|_{\Gamma_1} = 0$ for $a \leq k \leq b$, $b > a$, and $u_j \not\equiv 0$, $j = 1,2$, because $u_j = \exp\{ik(\nu_0,x)\} + O(r^{-1})$ as $r \to \infty$. Therefore any number $k \in [a,b]$ is an eigenvalue of the problem

$$(\nabla^2 + k^2)u = 0 \quad \text{in} \quad \mathcal{D}_2, \ \frac{\partial u}{\partial N} + h_2 u|_{\Gamma_2} = 0$$

and u_1 is the corresponding eigenfunction. But this is a

contradiction since the spectrum of this problem is discrete. If $\mathcal{D}_1 \cap \mathcal{D}_2 \neq \emptyset$ then consider a connected component of the set $\mathcal{D} = \mathcal{D}_2 \setminus (\mathcal{D}_1 \cap \mathcal{D}_2)$. The boundary Γ of this domain has, say, two parts: $\Gamma = \tilde{\Gamma}_1 \cup \tilde{\Gamma}_2$, where $\tilde{\Gamma}_j$ is a subset of Γ_j. Our argument is the same: u_1 is the eigenfunction of the problem

$$(\nabla^2 + k^2)u = 0 \quad \text{in} \quad \mathcal{D}, \quad \frac{\partial u}{\partial N} + h_2 u|_{\tilde{\Gamma}_2} = 0,$$

$$-\frac{\partial u}{\partial N} + h_1 u|_{\tilde{\Gamma}_1} = 0 \qquad (6)$$

for any $a \leq k \leq b$, which is a contradiction since the spectrum of the problem (6) is discrete. The sign minus in the last equation in (6) is taken because we assume that the domain $\mathcal{D}$ lies outside of $\mathcal{D}_1$ and therefore the exterior normal to the boundary $\tilde{\Gamma}_1$ of $\mathcal{D}$ differs by the sign from the exterior normal N to the part $\tilde{\Gamma}_1$ of the boundary of the domain $\mathcal{D}_1$. The contradiction proves that $\mathcal{D}_1 = \mathcal{D}_2$. Since $u_1 = u_2$ outside of $\mathcal{D}_1 = \mathcal{D}_2$ the quantities $h_j = -\frac{\partial u_j}{\partial N} \cdot \frac{1}{u_j}$, $j = 1,2$, must be equal. Note that u_j cannot vanish on the sets with positive measure because if $u_j = 0$ on $\tilde{\Gamma}_j$, meas $\tilde{\Gamma}_j > 0$, then from (5) it follows that $\frac{\partial u_j}{\partial N} = 0$ on $\tilde{\Gamma}_j$. If $u_j = \frac{\partial u_j}{\partial N} = 0$ on $\tilde{\Gamma}_j$, meas $\tilde{\Gamma}_j > 0$ then by the uniqueness of the solution to the Cauchy problem for elliptic equations one concludes that $u_j \equiv 0$ in Ω_j. This is impossible because of (3): $|u_j| \to 1$ as $r \to \infty$. This completes the proof.

II. RECOVERING OF THE SUPPORT FUNCTION FROM THE SCATTERING AMPLITUDE

1. The Green's formula gives the following expression for the scattering amplitude (for the problem (1)-(4))

$$f = -\frac{1}{4\pi} \int_\Gamma \exp\{-ik(n,s)\} h(s) ds, \quad h \equiv \frac{\partial u}{\partial N}, \qquad (7)$$

where

$$h + Ah = 2\frac{\partial u_0}{\partial N}, \quad Ah = \int_\Gamma \frac{\partial}{\partial N_s} \frac{\exp\{ik|s - t|\}}{2\pi|s - t|} h(t) dt,$$

N_s is the exterior unit normal to Γ at the point s,

$$u = u_0 - \int_\Gamma ghdt, \quad g = \frac{\exp(ik|x - y|)}{4\pi|x - y|} = g(x,y,k)$$

2. Let us fix the coordinate system with the origin outside of $\mathcal{D}$. Let us use the Kirchhoff approximation for h:

$$h = 0 \text{ on } \Gamma_- , \quad h = 2ik(N_s,\nu)\exp\{ik(\nu,s)\} \text{ on } \Gamma_+ , \tag{8}$$

where $\Gamma_+(\Gamma_-)$ is the illuminated (shadowed) part of Γ (i.e. $(N_s,\nu) < 0$ on Γ_+, $(N_s,\nu) > 0$ on Γ_-). Substitute (8) into (7) to obtain:

$$f = -\frac{2ik}{4\pi} \int_{\Gamma_+} \exp\{ik(\nu - n,s)\}(N_s,\nu)ds. \tag{9}$$

Define the support function of Γ to be

$$a(l) = -\max_{s\epsilon\Gamma_+} (l,s) \quad (\min_{s\epsilon\Gamma_+} |(l,s)|). \tag{10}$$

where l is a unit vector, and $(l,s) < 0$ on Γ_+.

3. Integral (9) can be calculated by the stationary phase method. The stationary point of the phase $(\nu - n,s) = -|n - \nu|(l,s)$, $l \equiv \frac{n - \nu}{|n - \nu|}$ is the point $s_0 \in \Gamma_+$ at which $l = N_{s_0}$. If Γ is convex then there is only one stationary point. The following result is well-known [19]:

Lemma 1. Let $f \in C_0^\infty(\Omega)$, $S(x) \in C^\infty(\Omega)$, $\Omega \subset R^n$ is a finite domain. Let $S(x)$ be a real valued function which has the only stationary (critical) point x_0 in Ω, and this point be nondegenerate (i.e. $dS(x_0) = 0$, $d^2S(x_0) \neq 0$). Then

$$\int_\Omega f(x)\exp\{ikS(x)\}dx = \left(\frac{2\pi}{k}\right)^{n/2} \exp\{i\lambda S(x_0) + $$

$$+ \frac{i\pi}{4} \operatorname{sgn} S''(x_0)\}|\det S''(x_0)|^{-1/2}[f(x_0) + O(\tfrac{1}{k})] \tag{11}$$

where

$$S''(x_0) = \left(\frac{\partial^2 S(x_0)}{\partial x_i \partial x_j}\right), \quad \operatorname{sgn} S''(x_0) = n_+ - n_-, \tag{12}$$

and n_+ (n_-) is the number of positive (negative) eigenvalues of the matrix $S''(x_0)$.

Let us calculate the term (l,s) around the stationary point.

$$(l,s) = (l,s_0) + (l,s - s_0) = (l,s_0) + \frac{1}{2}b_{ij}u^i u^j + o\{(u^1)^2 + (u^2)^2\},$$

where one should sum over the repeated indices, b_{ij} is the matrix of the second differential form of the surface Γ at the point s_0, u^i, $i = 1,2$ are the local coordinates in the tangential plane to Γ, $s_0 = (0,0)$ in these coordinates, and the third axis is directed along $l = N_{s_0}$, the unit normal vector to Γ at the point s_0.

Therefore $S''(x_0) = -|n - \nu| b_{ij}$, $n = 2$, and formula (11) applied to (9) yields

$$f = -\frac{ik}{2\pi}\frac{2\pi}{k}\exp\{-ik|n - \nu|(1,s_0) + \frac{i\pi}{4}\operatorname{sgn} b_{ij}(s_0)\}$$

$$|\det(-|n - \nu|b_{ij})^{-1/2}|[(1,\nu) + 0(\frac{1}{k})]. \tag{13}$$

It remains to calculate $\operatorname{sgn} b_{ij}(s_0)$ and $\det(-|n - \nu|b_{ij})$. One has $\det(-|n - \nu|b_{ij}) = |n - \nu|^2 \det b_{ij}$. If the surface is convex then $n_+ = 2$, $n_- = 0$, $\operatorname{sgn} b_{ij} = 2$. The eigenvalues of b_{ij} in the local coordinates described above are the principal curvatures K_1 and K_2 of the surface Γ at the point s_0, while $\det b_{ij} = K = K_1K_2 > 0$ is the Gaussian curvature of Γ at s_0. Thus the leading term of the asymptotics in (13) takes the form

$$f = \frac{\exp\{-ik|n - \nu|(1,s_0)\}(1,\nu)}{|n - \nu|\{K(s_0)\}^{1/2}} =$$

$$\frac{\exp\{-2ik(n,1)(1,s_0)\}}{2\{K(s_0)\}^{1/2}} \tag{14}$$

Here the formulas $|n - \nu| = 2(n,1),(1,\nu) = -(1,n)$ were used. Finally note that

$$-(1,s_0) = a(1) \tag{15}$$

since s_0 is the stationary point at which 1 is the normal to Γ. Thus

$$f = -\frac{\exp\{2ik(n,1)a(1)\}}{2\{K(s_0)\}^{1/2}} \tag{16}$$

is the leading term of the asymptotics of f as $k \to \infty$. Note that the sign of the phase in (16) differs from the sign of the phase in [4] because in [4] the scatterer was centrally symmetric and the origin 0 was placed at the center of symmetry, so that $|(1,s_0)| = (1,s_0) = a(1)$ (compare with (15)). If Γ is convex and $0 \in \mathcal{D}$ then $(1,s) > 0$ and $a(1)$ is defined as $a(1) = \max_{s\in\Gamma} (1,s)$. Let us formulate the result.

Theorem 1. Let Γ be a smooth convex surface. Then the leading term of the asymptotics of the scattering amplitude (7) as $k \to \infty$ is given by (16). In (16) $K(s_0)$ is the Gaussian curvature at the point $s_0 \in \Gamma$ at which $\frac{n - \nu}{|n - \nu|} = 1$ is the unit normal to Γ pointed outside of $\mathcal{D}$, and $a(1)$ is the support function of Γ defined in (10).

Remark 1. Vector s_0 defines in the fixed coordinate system the position of the scatterer. Translation of the scatterer $s' = s + d$ by a vector d will lead to the phase shift $2ik(n,l)\cdot(l,d)$ which reflects the new position of the scatterer.

For the back-scattering $l = n$, $\nu = -n$, $(l,s_0) = -a(l) < 0$ on Γ_+, $(l,s_0) = a(l) > 0$ on Γ_-.

4. Parametric equations of the surface can be written explicitly. Let the unit vector l have components $(\alpha_1,\alpha_2,\alpha_3)$, $\alpha_1^2 + \alpha_2^2 + \alpha_3^2 = 1$. Then

$$x_j = -\frac{\partial a(l)}{\partial \alpha_j}, \quad j = 1,2,3, \quad l = \sum_{j=1}^{3} \alpha_j e_j \tag{17}$$

is the parametric equation of Γ, the vectors e_1,e_2,e_3 are the orthonormal unit vectors defining the coordinate system with the origin 0.

To derive (17) let us write the equation of the tangent plane π to Γ orthogonal to the unit vector l:

$$\sum_{j=1}^{3} x_j\alpha_j = -a(l) \tag{18}$$

and consider $\alpha_1,\alpha_2,\alpha_3$ as parameters. The surface Γ is the envelope of the family of the planes (18) $\alpha_1,\alpha_2,\alpha_3$ being parameters. Thus, the equation of the surface can be obtained from the equations

$$\frac{\partial}{\partial \alpha_m}\left\{\sum_{j=1}^{3} x_j\alpha_j + a\left(\sqrt{\alpha_1^2 + \alpha_2^2 + \alpha_3^2}\right)\right\} = 0, \quad m = 1,2,3, \tag{19}$$

and (18). Equations (19) coincide with (17) and give the parametric equation of Γ.

Example 1. Consider a sphere $\sum_{j=1}^{3} (x_j - b_j)^2 = R^2$. The unit normal to the sphere $\frac{1}{R}\sum_{j=1}^{3} (x_j - b_j)e_j$. The tangent plane to the sphere orthogonal to the unit vector $l = \sum_{j=1}^{3} \alpha_j e_j$ is $\sum_{j=1}^{3} (x_j - b_j + R\alpha_j)\alpha_j = 0$. Therefore, assuming that $(b,l) < 0$ and the origin is outside of the sphere, i.e. $|b|^2 \equiv \sum_{j=1}^{3} b_j^2 > R^2$, one obtains $a(l) = -(b,l) + R$. Equations (17) are parametric equations of the sphere: $x_j = b_j - R\alpha_j$, $1 \le j \le 3$, $R = R(\alpha_1^2 + \alpha_2^2 + \alpha_3^2)^{1/2}$.

Remark 2. From the knowledge of the support function $a(l)$ only for $l \in \tilde{S}^2$, where $\tilde{S}^2 \subset S^2$ is a domain in S^2, one can reconstruct effectively the corresponding part $\tilde{\Gamma}$ of the surface Γ: equation (17) with $l = (\alpha_1, \alpha_2, \alpha_3) \in \tilde{S}^2$ is the parametric equation of this part of Γ.

5. Let us chose such set $\tilde{\Delta}$ of the unit vectors ν, n that $l = \frac{n - \nu}{|n - \nu|}$ runs through $\tilde{S}^2 \subset S^2$. For example, one can choose $n = -\nu$ and have n run over $\tilde{S}^2$. This gives the back scattering amplitude $f(n,-n,k)$. For back scattering $(n,l) = 1$ and (16) takes the form

$$f = -\frac{\exp\{2ik\, a(l)\}}{2\{K(s_0)\}^{1/2}} \tag{20}$$

Let us assume that $f(n,\nu,k)$ is known for $n,\nu \in \Delta$ and all sufficiently large k. Then formula (16) allows one to find $a(l)$ and the surface Γ by formulas (17). Let us formulate this as

Theorem 2. If Γ is a smooth convex surface and the scattering amplitude $f(n,\nu,k)$ is known for all n,ν, such that $\frac{n - \nu}{|n - \nu|} = l$ runs through $\tilde{S}^2 \subset S^2$, and all sufficiently large k, then the support function $a(l)$ is uniquely determined for $l \subset \tilde{S}^2$. Therefore the position and the shape of $\tilde{\Gamma} \subset \Gamma$ are uniquely determined. Moreover, explicit equation of the part $\tilde{\Gamma}$ of the surface Γ is given by formula (17).

6. Remark 3. The Kirchhoff approximation (7) has been rigorously justified in the literature (see e.g. [14]-[16]) as the leading term in the asymptotics of h as $k \to +\infty$. Therefore the argument in this paper is rigorous. Other boundary conditions can also be treated. For example, if $\frac{\partial u}{\partial N}\big|_\Gamma = 0$ then $f(n,\nu,k) = \frac{1}{4\pi} \int_\Gamma ds u \frac{\partial \exp\{-ik(n,s)\}}{\partial N_s}$. The Kirchhoff approximation gives $u = 0$ on Γ_-, $u = 2u_0$ on Γ_+. Thus $f(n,\nu,k) = -\frac{ik}{2\pi} \int_\Gamma ds \exp\{ik(\nu - n,s)\}(N_s,n)$ in this approximation (compare with (9)). Further study is similar to the given above.

III. STABILITY OF THE RECOVERED SURFACE

Suppose that the scattering amplitude is measured with an error $\delta > 0$, that is f_δ is known, $|f_\delta - f| \le \delta$. Let

$f_\delta = f(n,\nu,k_0) + \delta f$, where δf is noise, $|\delta f| \le \delta$, $\delta = \text{const} > 0$. Let us assume that $\sup_{s\in\Gamma} K(s) \le d^2$, $(n,l) \ge b > 0$, where b and d are some constants, and that k_0 is large, so that $0(\frac{1}{k_0})$ in (13) satisfies the inequality $0(\frac{1}{k_0}) << b$. Then, from (13) and (16) one obtains

$$f_\delta(n,\nu,k_0) = -\frac{\exp\{2ik_0(n,l)a(l)\}}{2\{K(s_0)\}^{1/2}}(1 + 0(\frac{1}{bk_0}) + 0(\delta d)). \quad (21)$$

Thus

$$a_\delta(l) \equiv \frac{1}{2ik_0(n,l)} \ln \frac{-f_\delta(n,\nu,k_0)}{|f_\delta(n,\nu,k_0)|} =$$

$$a(l) + 0(\frac{1 + k_0 b\delta d}{(k_0 b)^2}) \equiv a(l) + \eta,$$

$$\eta = 0((k_0 b)^{-2} + (k_0 b)^{-1}\delta d). \quad (22)$$

It follows from (22) that the noisy data f_δ determines explicitly the support function $a(l)$ with the error η of order given in (22). The stable reconstruction of the surface Γ is therefore reduced to the stable differentiation of $a(l)$ known with the error η. This problem was solved in [20]-[22].

The result is as follows. Assume that $|a''(\phi)| \le M$, $-\infty \le \phi \le \infty$ and a_η is given, $|a_\eta - a| \le \eta$. Let

$$\beta(\eta) \equiv (\frac{2\eta}{M})^{1/2}, \quad \hat{a}' \equiv \frac{a_\eta(\phi + \beta(\eta)) - a_\eta(\phi - \beta(\eta))}{2\beta(\eta)}. \quad (23)$$

Then

$$|\hat{a}' - a'| \le (2M\eta)^{1/2} \equiv \varepsilon(\eta). \quad (24)$$

Therefore $\hat{a}'$ is a stable approximation of $\hat{a}'$ with the error $\le \varepsilon(\eta)$. It can be proved that among all linear and non-linear estimates of a' the above estimate is the best possible (in a certain sense) on the class of all data given by the two numbers (M,η) (see e.g. [21]). Namely,

$$\inf_{\substack{T\varepsilon A,\ |a''| \le M, \\ |a_\eta - a| \le \eta}} \sup_\phi |Ta_\eta(\phi) - a'(\phi)| \ge (2M\eta)^{1/2}$$

where A is the set of all estimates and T is an estimate.

IV. RECOVERING OF THE SURFACE FROM THE KNOWLEDGE OF ITS PRINCIPAL CURVATURES

1. Let Γ be a closed compact oriented surface in $\mathbb{R}^3$ and $\psi:\Gamma \to S^2$ be such a map that the vector $\psi(s)$ when placed at the point s is the positive unit normal to Γ. Here $\psi(s)$ is the vector with the initial point at the origin (the center of the unit ball with the boundary S^2) and the end point at the $\psi(s)$. Let us assume that ψ is a diffeomorphism of Γ onto some part $\tilde{S}^2$ of S^2 and let $\Phi:\tilde{S}^2 \to \Gamma$ be the inverse mapping ψ^{-1}. Let $(e_1(s),e_2(s),e_3(s))$ be a moving orthonormal frame on Γ, $e_3(s) = \psi(s)$, $e_1(s)$ and $e_2(s)$ be the eigenvectors of $-de_3(s)$ with eigenvalues $K_1(s)$ and $K_2(s)$ respectively, where $K_j(s)$, $j = 1,2$ are the principal curvatures of Γ at the point s. Let us define differential forms ω_1,ω_2 by the equations $\omega_i(e_j) = \delta_{ij}$, $1 \le i, j \le 2$. Let $de_i = \omega_{ij}e_j$. Here and below one should sum over the repeated indices. It is known [3] that $d\omega_{ij} = \omega_{im} \wedge \omega_{mj}$, $d\omega_1 = -\omega_2 \wedge \omega_{12}$, $d\omega_2 = \omega_1 \wedge \omega_{12}$, where $\wedge$ denotes the wedge product of the differential forms, $\omega_{13} = K_1\omega_1$, $\omega_{23} = K_2\omega_2$, and the fundamental forms on Γ are $I = \omega_1^2 + \omega_2^2$, $II = K_1\omega_1^2 + K_2\omega_2^2$, $III = K_1^2\omega_1^2 + K_2\omega_2^2$. Let us pull back to S^2 the basic differential forms using the mapping $\Phi = \psi^{-1}$. The pulled back forms ω_i^*, ω_{ij}^* satisfy the equations

$$d\omega_{ij}^* = \omega_{im}^* \wedge \omega_{mj}^*, \quad d\omega_1^* = -\omega_2^* \wedge \omega_{12}^*, \quad d\omega_2^* = \omega_1^* \wedge \omega_{12}^*,$$

$$\omega_{ij}^* = -\omega_{ji}^*, \quad \omega_{13}^* = K_1\omega_1^*, \quad \omega_{23}^* = K_2\omega_2^* . \tag{25}$$

Here $K_i = K_i(\theta,\phi) = K_i(\Phi(\theta,\phi))$. These equations imply that

$$dK_1 \wedge \omega_1^* = \sigma_1(K_1 - K_2)\omega_1^* \wedge \omega_2^* \tag{26}$$

$$dK_2 \wedge \omega_2^* = \sigma_2(K_1 - K_2)\omega_1^* \wedge \omega_2^* \tag{26'}$$

where σ_1 and σ_2 are defined by the equation

$$\omega_{12}^* = \sigma_1\omega_1^* + \sigma_2\omega_2^* \tag{27}$$

Assume that $K_1(\theta,\phi)$ and $K_2(\theta,\phi)$ are given. If one finds the forms ω_i^*, then determines vectors e_i^* such that

$$\omega_i^*(e_j^*) = \delta_{ij}, \quad 1 \le i, j \le 2, \tag{28}$$

and finally uses the equation

$$d\Phi = -\frac{\omega_1^* e_1^*}{K_1} - \frac{\omega_2^* e_2^*}{K_2} \tag{29}$$

then, assuming that K_1 and K_2 are the principal curvatures of a surface, one concludes that $d\Phi$ is a closed differential. Integrating this differential one recovers the surface. In what follows the equations for the forms ω_i^*, $1 \leq i \leq 2$ are derived and analyzed and a constructive method for finding the surface is given under the assumption that K_1 and K_2 are the principal curvatures of a surface. Let

$$\omega_1^* = a_1 d\phi + a_2 d\theta, \quad \omega_2^* = b_1 d\phi + b_2 d\theta. \tag{30}$$

The e_j^* are the eigenvectors of the mapping $d\Phi$ with the eigenvalues $-1/K_j$. One can check, using the orthonormality of the vectors $\frac{e_j^*}{K_j}$ with respect to III* metric, or the fact that $III^* = \sin^2\theta (d\phi)^2 + (d\theta)^2$ (see Appendix), that

$$b_1 = \mp a_2 \frac{K_1}{K_2} \sin\theta, \quad b_2 = \pm a_1 \frac{K_1}{K_2} \frac{1}{\sin\theta}, \tag{31}$$

$$\frac{K_1^2 a_1^2}{\sin^2\theta} + K_1^2 a_2^2 = 1 \tag{32}$$

One has

$$\omega_{13}^* = K_1 \omega_1^*, \quad \omega_{23}^* = K_2 \omega_2^*, \quad \omega_{12}^* = \sigma_1 \omega_1^* + \sigma_2 \omega_2^* \equiv \alpha d\phi + \beta d\theta, \tag{33}$$

where α and β are to be found. Furthermore

$$d\omega_{12}^* = \omega_{13}^* \wedge \omega_{32}^* = -K\omega_1^* \wedge \omega_2^* = -K(a_1 b_2 - a_2 b_1) d\phi \wedge d\theta =$$

$$-K \frac{\sin\theta}{K} d\phi \wedge d\theta = -\sin\theta \, d\phi \wedge d\theta, \tag{34}$$

where $K = K_1 K_2$. If one uses the definition of ω_i^* in terms of a_i and b_i then one finds

$$a_1\sigma_1 - \frac{K_1}{K_2} \sin\theta \, a_2\sigma_2 = \alpha,$$

$$a_2\sigma_1 + \frac{K_1}{K_2 \sin\theta} a_1\sigma_2 = \beta.$$

Solving this system for σ_1 and $-\sigma_2$ one obtains:

$$\begin{pmatrix} \sigma_1 \\ -\sigma_2 \end{pmatrix} = T \begin{pmatrix} a_1 \\ a_2 \end{pmatrix}, \quad T \equiv \begin{pmatrix} \dfrac{K_1^2 \alpha}{\sin^2\theta} & K_1^2 \beta \\ -\dfrac{K\beta}{\sin\theta} & \dfrac{K\alpha}{\sin\theta} \end{pmatrix}. \tag{35}$$

Here the equation

$$\det \begin{vmatrix} a_1 - \frac{K_1}{K_2} \sin\theta \, a_2 \\ a_2 \quad \frac{K_1}{K_2 \sin\theta} a_1 \end{vmatrix} = \frac{\sin\theta}{K}$$

was used. Using the equations (26) one obtains

$$Q \begin{pmatrix} a_1 \\ a_2 \end{pmatrix} = \frac{K_2 - K_1}{K} \sin\theta \begin{pmatrix} \sigma_1 \\ -\sigma_2 \end{pmatrix} \tag{36}$$

where

$$Q = \begin{pmatrix} -\frac{\partial K_1}{\partial\theta} & \frac{\partial K_1}{\partial\phi} \\ \frac{K_1}{K_2 \sin\theta} \frac{\partial K_2}{\partial\phi} & \frac{K_1 \sin\theta}{K_2} \frac{\partial K_2}{\partial\theta} \end{pmatrix}. \tag{37}$$

Therefore

$$\left(Q - \frac{K_2 - K_1}{K} (\sin\theta)\, T\right) \begin{pmatrix} a_1 \\ a_2 \end{pmatrix} = 0. \tag{38}$$

This equation can be written as

$$\begin{pmatrix} A & B \\ C & D \end{pmatrix} \begin{pmatrix} a_1 \\ a_2 \end{pmatrix} = 0, \qquad A = -\frac{\partial K_1}{\partial\theta} - \frac{K_2 - K_1}{K} \frac{K_1^2 \alpha}{\sin\theta}, \quad B = \frac{\partial K_1}{\partial\phi} - \frac{K_2 - K_1}{K} \sin\theta \, K_1^2 \beta \tag{39}$$

where A and D are linear in α, B and C are linear in β. Since a_1 and a_2 are not zeros simultaneously, one has

$$AD - BC = 0. \tag{40}$$

If $K_2 - K_1 \neq 0$ this is a quadratic equation in α, β. Equations (39) can be written as

$$Aa_1 + Ba_2 = 0, \; Ca_1 + Da_2 = 0. \tag{41}$$

Therefore

$$A^2 a_1^2 = B^2 a_2^2, \; K_1^2 \sin^2\theta A^2 a_1^2 = B^2 \sin^2\theta \left(1 - \frac{K_1^2 a_1^2}{\sin^2\theta}\right). \tag{42}$$

Assume that $B \neq 0$. Then

$$a_1 = \pm \frac{\sin\theta}{K_1} \frac{1}{\left(1 + \frac{A^2}{B^2} \sin^2\theta\right)^{1/2}}, \; a_2 = -\frac{A}{B} a_1. \tag{43}$$

Therefore a_1, a_2 (and b_1, b_2) are determined by α, β.

From (34) it follows that

$$\omega^*_{12} = -\cos\theta\, d\phi + dg, \tag{44}$$

where $g(\theta,\phi)$ is a scalar function. Therefore

$$\alpha = \frac{\partial g}{\partial\phi} - \cos\theta,\ \beta = \frac{\partial g}{\partial\theta}, \tag{45}$$

so that α and β are determined by g. If one solves the quadratic equation (40) for α and write $\alpha = H(\theta,\phi,\beta)$, then (43) can be written as

$$a_j = F_j(\theta,\phi,\frac{\partial g}{\partial\theta}),\ j = 1,2. \tag{46}$$

The equation $d\omega^*_1 = -\omega^*_2 \wedge \omega^*_{12}$ can be written as

$$\frac{\partial a_2}{\partial\phi} - \frac{\partial a_1}{\partial\theta} = \frac{K_1}{K_2}\sin\theta\,(a_2\beta + a_1\alpha\,\frac{1}{\sin^2\theta}) = G(\theta,\phi,\frac{\partial g}{\partial\theta}) \tag{47}$$

From (46) and (47) one obtains

$$\frac{\partial F_2}{\partial\phi} + \frac{\partial F_2}{\partial\beta}\frac{\partial}{\partial\theta}(\frac{\partial g}{\partial\phi}) - \frac{\partial F_1}{\partial\theta} - \frac{\partial F_1}{\partial\beta}\frac{\partial^2 g}{\partial\theta^2} = G(\theta,\phi,\frac{\partial g}{\partial\theta}). \tag{48}$$

From (45) it follows that

$$\frac{\partial}{\partial\theta}\frac{\partial g}{\partial\phi} = -\sin\theta + \frac{\partial H}{\partial\theta}(\theta,\phi,\frac{\partial g}{\partial\theta}) + \frac{\partial H}{\partial\beta}\cdot\frac{\partial^2 g}{\partial\theta^2} \tag{49}$$

From (48) and (49) one obtains

$$(\frac{\partial F_2}{\partial\beta}\frac{\partial H}{\partial\beta} - \frac{\partial F_1}{\partial\beta})\frac{\partial^2 g}{\partial\theta^2} + \frac{\partial F_2}{\partial\phi} - \frac{\partial F_1}{\partial\theta} + (\frac{\partial H}{\partial\theta} - \sin\theta)\ .$$

$$\frac{\partial F_2}{\partial\beta} = G \tag{50}$$

This equation can be written as

$$\frac{\partial^2 g}{\partial\theta^2} = G_1(\theta,\phi,\frac{\partial g}{\partial\theta}). \tag{51}$$

Similarly, the equation $d\omega^*_2 = \omega^*_1 \wedge \omega^*_{12}$ leads to

$$\frac{\partial^2 g}{\partial\theta^2} = G_2(\theta,\phi,\frac{\partial g}{\partial\theta}), \tag{52}$$

where G_2 is explicitly determined. From (51) and (52) one obtains the equation for $\frac{\partial g}{\partial\theta}$

$$G_1(\theta,\phi,\frac{\partial g}{\partial\theta}) = G_2(\theta,\phi,\frac{\partial g}{\partial\theta}). \tag{53}$$

Assuming that this equation is solvable one finds $\frac{\partial g}{\partial \theta} = \beta$, then $\alpha = H(\theta,\phi,\beta)$, then A and B from (39), then a_1 and a_2 from (43), then b_1 and b_2 from (31), then ω_1^* and ω_2^* from (30), then e_1^* and e_2^* from (28), then $d\Phi$ from (29). Integrating (29) one finds the mapping Φ which determines the parametric equation of the surface Γ (parameters are θ,ϕ). The above construction is valid under the assumption that K_1, K_2 are the principal curvatures of a surface.

It was assumed above that $K_1 \neq K_2$. If $K_1 = K_2$ then it is easily seen that $K_1 = K_2 = \text{const}$. Indeed (36) shows, that in this case

$$Q \begin{pmatrix} a_1 \\ a_2 \end{pmatrix} = 0.$$

Since a_1 and a_2 are not both zero one concludes that

$$0 = \det Q = -\left(\frac{\partial K_1}{\partial \theta}\right)^2 \sin\theta - \left(\frac{\partial K_1}{\partial \phi}\right)^2 \frac{1}{\sin\theta}.$$ Thus

$\frac{\partial K_1}{\partial \theta} = \frac{\partial K_1}{\partial \phi} = 0$, $K_1 = \text{const}$. If $K_1 = K_2 = \text{const}$ then the surface is spherical, its radius is K_1^{-1}.

In conclusion of this section let us formulate some open problems:

1) Characterization problem: what are the necessary and sufficient conditions on $K_1(\theta,\phi)$, $K_2(\theta,\phi)$ in order that these functions be principal curvatures of a closed surface of class C^2?

The following problem is closely connected with characterization problem:

2) If K_1 and K_2 are the principal curvatures of a surface then $d\Phi$ (see (29)) is a closed differential. Is the converse true? What is the relationship between the condition that $d\Phi$ is closed and the solvability conditions for equations (40) and (53)?

APPENDIX I. Derivation of formulas (31-32).

The identity $III^* = \sin^2\theta(d\phi)^2 + (d\theta)^2 \equiv K_1^2(a_1 d\phi + a_2 d\theta)^2 + K_2^2(b_1 d\phi + b_2 d\theta)^2$ is the starting point. One obtains the system

$$K_1^2 a_1^2 + K_2^2 b_1^2 = \sin^2\theta$$

$$K_1^2 a_1 a_2 + K_2^2 b_1 b_2 = 0$$

$$K_1^2 a_2^2 + K_2^2 b_2^2 = 1$$

Let $\frac{K_1}{K_2} = \lambda$, $K_2 \neq 0$, $\frac{b_i}{\lambda} = c_i$. Then the system takes the form

$$a_1^2 + c_1^2 = \sin^2\theta, a_1 a_2 + c_1 c_2 = 0, a_2^2 + c_2^2 = 1.$$

One has $\frac{a_1}{c_2} = -\frac{c_1}{c_2}$, $c_1^2(1 + \frac{c_2^2}{a_2^2}) = \sin^2\theta$, $\frac{c_1^2}{a_2^2} = \sin^2\theta$,

$c_1 = \mp a_2 \sin\theta$, $c_2 = \pm \frac{a_1}{\sin\theta}$. Thus, $b_1 = \mp \frac{K_1 \sin\theta}{K_2} a_2$,

$b_2 = \pm \frac{K_1}{K_2 \sin\theta} a_1$, $\frac{K_1^2 a_1^2}{\sin^2\theta} + K_1^2 a_2^2 = 1$. The choice of the sign in (31) is $(- +)$. One could chose $(+ -)$ without any changes in the argument.

Remark. The unit sphere is the only compact closed surface such that $III^* = I^*$, that is the pull-backs of the third and first fundamental forms are equal on this surface. Indeed, $III^* = I^*$ iff $K_1^2 \omega_1^{*2} + K_2 \omega_2^{*2} = \omega_1^{*2} + \omega_2^{*2}$. The last equation implies that $K_1^2 = 1$, $K_2^2 = 1$. This means that $K_1 = 1$, $K_2 = 1$, i.e. the surface is the unit sphere. The case $K_1 K_2 = -1$ does not correspond to any compact closed smooth surface as is clear from the geometrical reasons.

APPENDIX 2.

A) Recovery of a convex surface from the sum of its two principal radii.

Let $h(\ell) = K_1^{-1} + K_2^{-1}$ be known on all of S^2. Then $\nabla_\alpha^2 a(\ell) = h$, $\ell = (\alpha_1, \alpha_2, \alpha_3)$. Let $Y_{nm}(\theta,\phi)$, $n \geq 0$, $-n \leq m \leq n$ be the normalized spherical harmonics, $h_{nm} = \int_{S^2} h\overline{Y}_{nm} \sin\theta \, d\theta \, d\phi$. Then $a(\ell) = \sum_{n \geq 0, n \neq -1} h_{nm}[(n-1)(n+2)]^{-1} Y_{nm}$. One proves that $h_{1m} = 0$, $m = 0, \pm 1$. If $a(\ell)$ is found then the surface is given by (17). If h_δ is given, $|h - h_\delta| \leq \delta$, then the coefficients $h_{nm}(\delta)$ are given, $|h_{nm}(\delta) - h_{nm}| \leq \delta(4\pi)^{1/2}$. One takes $h_{1m}(\delta) = 0$, $m = 0, \pm 1$. Assuming that $|h_{nm}| \leq \frac{A}{(n^2 + m^2)^a}$, A, a are constants, and knowing $h_{nm}(\delta)$ one estimates $a(\ell)$ with an error of order $\delta^{2-1/a}$ (see [20,p.355] for details) and finds Γ stably as in section III. This is presented in [22] and the relevant references are [23]-[27]. It is not clear if this method can be extended to the case when h is given on $\tilde{S}^2$ (and then extended in a certain way to S^2). On the other hand, the construction in section IV is entirely local.

B) A uniqueness theorem in scattering theory [28].

Theorem. (A. G. Ramm) The knowledge of $f(n, \nu_0, k_0)$, $n \in \tilde{S}^2$ defines a strictly convex reflecting obstacle uniquely.

Here f is defined in (4), ν_0 and k_0 are fixed, $\tilde{S}^2$ is any solid angle (open set in S^2), reflective obstacle means that (2) holds.

C) The author gave recently an analytical reconstruction of the three-dimensional refraction coefficient (velocity inversion in geophysics) [29].

REFERENCES

[1] P. Lax, R. Phillips, Scattering theory. New York: Acad. Press, 1967.

[2] V. Fock, Electromagnetic diffraction and propagation problems. New York: Pergamon Press, 1965.

[3] C. C. Hsiung, A first course in differential geometry. New York: Wiley, 1981.

[4] A. G. Ramm, "Determination of the shape of the reflecting body from the scattering amplitude", Radiofizika, Izvestija vuzov, vol. 13, N5, pp. 727-731, 1970.

[5] A. G. Ramm, "Reconstruction of the shape of an obstacle from the scattering amplitude", Radiotechnika i electronika, vol. 11, pp. 2068-2070, 1965.

[6] A. G. Ramm, "On inverse diffraction problem", J. Math. Anal. Appl., (to appear).

[7] A. G. Ramm, Theory and applications of some new classes of integral equations. New York: Springer-Verlag, 1980.

[8] J. B. Keller, "Diffraction by a convex cylinder", IRE Trans., vol. AP-4, N3, pp. 312-313, 1956.

[9] R. Lewis, "Physical optics inverse diffraction", IEEE Trans., vol. AP-17, pp. 308-314, 1969.

[10] W. Boerner, C. Ho, "Analysis of physical optics far field inverse scattering for the limited data case using radar theory and polarization information", Wave motion, vol. 3, pp. 311-333, 1981.

[11] L. Nirenberg, "The Weyl and Minkowski problems in differential geometry in the large", Comm. Pure Appl. Math., vol. 6, pp. 337-394, 1953.

[12] D. Colton, "The inverse electromagnetic scattering problem for a perfectly conducting cylinder", IEEE Trans., vol. AP-29, N2, pp. 364-367, 1981.

[13] A. G. Ramm, Iterative methods for calculating static fields and wave scattering by small bodies. New York: Springer-Verlag, 1982.

[14] M. E. Taylor, Pseudo-differential operators. Princeton: Princeton Univ. Press, 1981.

[15] H. D. Alber, "Justification of geometrical optics for non-convex obstacles", J. Math. Anal. Appl., vol. 30, pp. 372-386, 1981.

[16] V. Petkov, "High frequency assymptotics of the scattering amplitude for non-convex bodies", Comm. Part. Differ. Equations, vol. 5, pp. 293-329, 1980.

[17] F. Miller, The construction of a surface from its principal curvatures. Math. Dept. Kansas State Univ., Preprint, Feb. 1983.

[18] S. Chaudhuri, W. Boerner, B. Foo, A high frequency inverse scattering model to recover the specular point curvature from parametric scattering matrix data, Univ. of Illinois, Chicago, Preprint, 1983.

[19] M. Fedorjuk, Metod perevala, Nauka, Moscow: 1977. (Russian)
[20] A. G. Ramm, "Stable solution of some ill-posed problems," Math. Methods in the appl. sci., vol. 3, pp. 336-363, 1981.
[21] A. G. Ramm, "Estimates of the derivatives of random functions", J. Math. Anal. Appl. (to appear).
[22] A. G. Ramm, "On numerical differentiation", Mathem., Izvestija vuzov, 11, (1968), 131-135. 40 # 5130.
[23] A. G. Ramm, "Remarks about inverse diffraction problems".
[24] E. Christoffel, Uber die Bestimmung der Gestalt einer krummen Oberflache durch lokale Messungen auf Derselben, J. reine u. angew. Math., 64, 193-209, (1865).
[25] A. Hurwitz, Sur Quelques Applications Geometriques des Series de Fourier, Ann Ecol. Norm., 19, 357-408, (1902).
[26] W. Blaschke, Vorlesungen über Differential Geometrie, Springer-Verlag, Berlin, 1945, Vol. 1, 204.
[27] W. Firey, The Determination of Convex Bodies from their Mean Radius of Curvature Function, Mathematika, 14, 1-13, (1967).
[28] A. G. Ramm, A uniqueness theorem in scattering theory, (preprint Aug. 1983).
[29] A. G. Ramm, "Inverse scattering for geophysical problems".

I.14 (IM.5)

APPLICATION OF THE CHRISTOFFEL-HURWITZ INVERSION IDENTITY TO ELECTROMAGNETIC IMAGING

Brett Borden

Michelson Laboratory, Physics Division,
Naval Weapons Center, China Lake, California 93555

ABSTRACT. We review the theoretical framework of an algorithm which determines the shape of a conducting convex scatterer from the depolarization properties of the (monostatic) scattered signal and show it to be effective when the return signal is from a limited range of aspects. Numerical examples are included.

1. INTRODUCTION

The problem of interest in many applications of electromagnetic inverse scattering theory is that of determining the shape of a body from the scattered field it creates in interacting with sampling radiation, and the shape-recovery methods resulting from the many investigations in this field are well reviewed [1]. Of special importance for the present discussion are those techniques which use the vector nature of the problem as a means of extending the type of information available for such reconstruction algorithms [2].

Below we summarize and further develop a polarization/depolarization approach of the author to applications involving more realistic data sets than previously examined. These data sets are assumed to be made up of complete (i.e., full scattering matrix) information from perfectly conducting smooth convex surfaces over a limited range of aspects.

W.-M. Boerner et al. (eds.), Inverse Methods in Electromagnetic Imaging - Part 1, 251–259.

2. THE VECTOR INVERSE ALGORITHM

Our approach results from the observation that the principal radii of curvature local to the specular points of a perfectly conducting convex body are radar measurables [3]-[5]. Illuminating such a target with an impulse $\underline{H}_I = H_u\hat{u} + H_v\hat{v} = \underline{H}\delta(t - z/c)$ will effectively isolate the specular points, and it can be shown that the leading edge of the return in the far-field approximation obeys [3],[6]

$$\underline{H}_s = \frac{1}{2\pi r}\left[\frac{\partial^2 A_z}{\partial t^2}\underline{H} + \frac{1}{2}\frac{\partial A_z}{\partial t}(K_u - K_v)(H_u\hat{u} - H_v\hat{v})\right] ,$$

where $\underline{H}_s$ is the scattered magnetic field, r is the distance from the transmitter/receiver to the target, A_z is the target area projected onto the u-v plane (the so-called silhouette function), and K_u and K_v are the curvatures in the principal directions $\hat{u}$ and $\hat{v}$, respectively. This is the Kennaugh-Cosgriff formula [7],[8] with Bennett's first-order correction [6].

Denoting the Fourier transform of the m-n element of the scattering matrix by $s_{mn}(\omega)$, it is then straightforward to see that [3],[4]

$$K_u - K_v = -\frac{2\omega\tan(\phi_d/2)}{\left[1 + \left|\frac{s_{21}(\omega)}{s_{11}(\omega)}\right|^2 \csc^2(\phi_d/2)\right]^{\frac{1}{2}}} \equiv B , \tag{1}$$

where ϕ_d is the phase difference between $s_{11}(\omega)$ and $s_{22}(\omega)$. By virtue of the high-frequency relationship [9]

$$\sigma = \frac{\pi}{K_u K_v}$$

between the differential cross section σ and the principal curvatures, we can conclude

$$R_u + R_v = \frac{B\sigma}{\pi}\sqrt{1 + \frac{4\pi}{\sigma B^2}} . \tag{2}$$

($R_u = 1/K_u$ and $R_v = 1/K_v$ are the principal radii of curvature.)

The significance of this result lies in its relationship to a differential equation developed by Stoker to study the Christoffel-Hurwitz problem of classical differential geometry:

Can the surface be reconstructed from a knowledge of the sum of its principal radii of curvature as a prescribed function of direction of surface normal?

Let $\underline{r}(u,v)$ be a parametric representation of a convex surface $\mathcal{S}$. Denote the unit normal to $\mathcal{S}$ at $\underline{r}$ by $\hat{n} = (n_1,n_2,n_3)$. Then we can define the spherical image as the mapping induced by

$$h(n_1,n_2,n_3) = \underline{r} \cdot \hat{n} \quad .$$

Requiring for any vector $\underline{X} = (X_1,X_2,X_3)$ that

$$\mathcal{H} = \underline{r} \cdot \underline{X} = |\underline{X}|h$$

defines the Minkowski support function (X_1,X_2,X_3).

In [10], Stoker obtains a differential relationship between $\mathcal{H}$ and $\Phi = R_u + R_v$, a relationship which for the present problem reduces to [3]

$$\nabla^2\mathcal{H} = X^2\nabla^2 h + 2h = -\Phi \quad , \tag{3}$$

expressing h as a function of the spherical coordinates (θ,ϕ) of $\hat{n}$. Since $\underline{r} \cdot d\underline{X} = d\mathcal{H}$, the cartesian coordinates of $\mathcal{S}$ can then be written

$$x = h \sin\theta\cos\phi + \frac{\partial h}{\partial\theta} \cos\theta\cos\phi - \frac{\partial h}{\partial\phi}\frac{\sin\phi}{\sin\theta}$$

$$y = h \sin\theta\sin\phi + \frac{\partial h}{\partial\theta} \cos\theta\cos\phi - \frac{\partial h}{\partial\phi}\frac{\cos\phi}{\sin\theta}$$

$$z = h \cos\theta - \frac{\partial h}{\partial\theta} \sin\theta \quad ,$$

and so the target shape is recovered with the solution of (3).

In radar applications, Φ rarely will be determined as a prescribed function of aspect, and, because of this, a numerical solution to (3) is required. To this end, we investigate the functional

$$F = \int_V [(\nabla\mathcal{H})^2 - 2\mathcal{H}\Phi]dV - 2\int_{\mathcal{T}} \mathcal{H}\beta d\mathcal{T} \tag{4}$$

$$= \int_{\mathcal{D}} \left(\frac{\partial h}{\partial\theta}\right)^2 + \frac{1}{\sin^2\theta}\left(\frac{\partial h}{\partial\phi}\right)^2 - 2h(h + \Phi)\sin\theta d\theta d\phi - 2\int \mathcal{H}\beta d\mathcal{T} \ ,$$

where $\mathcal{D}$ is the intersection of V with a sphere of radius 1 centered on $\underline{r} = \underline{0}$. If $h(\theta,\phi)$ is a solution to (3) over $\mathcal{D}$ with boundary $\mathcal{T} = \partial\mathcal{D}$, then $h(\theta,\phi)$ also makes (4) stationary (β depends on the appropriate boundary conditions [11]). Given a discrete set of data $\{\Phi_i\}$ over a sparse set of aspects, we can divide $\mathcal{D}$ into N subdomains (one subdomain for each element of $\{\Phi_i\}$ and integrate a linearization $h_i(\theta,\phi)$ of $h(\theta,\phi)$ over each subdomain. Seeking the values $\{h(\theta_i,\phi_i)\}$ that satisfy $\delta F = 0$, we require

$$\frac{\partial F}{\partial h(\theta_i,\phi_i)} = 0 \qquad i = 1,\ldots, N \tag{5}$$

and solve the resulting set of linear algebraic equations for $\{h(\theta_i,\phi_i)\}$ by the usual methods. Of course, for aspect-limited data sets, the boundary integral would cause problems for most typical radar situations were it not for the following fortuitous circumstance.

We can deal with the boundary term in (4) by recognizing that two support functions which differ by only a linear function will have corresponding surfaces which differ by at most a translation [10]. Moreover, for the Christoffel-Hurwitz problem, such a linear function need only be specified on the boundary of $\mathcal{S}$, and the resulting translation will be a rigid body motion [12]-[15]. Because of this, we need only specify the boundary values of up to an <u>arbitrary</u> linear function.

Let $\mathcal{H}'$ be a support function differing from $\mathcal{H}$ by a linear function L,

$$\mathcal{H}' = \mathcal{H} + L \quad .$$

Suppose now that the boundary conditions corresponding to the Neumann problem are known for $\mathcal{H}$ [11],

$$\beta = \frac{\partial\mathcal{H}}{\partial\xi} = \hat{\xi} \cdot \nabla\mathcal{H}$$

($\hat{\xi}$ is the outward directed unit normal to $\mathcal{D}$.) Then the boundary conditions on $\mathcal{H}'$ become

$$\beta' = \frac{\partial\mathcal{H}'}{\partial\xi} = \beta + \hat{\xi} \cdot \nabla L \quad . \tag{6}$$

Clearly, L = 0 corresponds to our original problem.

Since L is arbitrary, then whenever

$$\int_{\mathcal{T}} \mathcal{H}' \hat{\xi} d\mathcal{T} \neq \underline{0} \quad ,$$

we can choose L so that

$$\nabla L \cdot \int_T \mathcal{H}'\hat{\xi} dT = - \int_T \mathcal{H}' \beta dT \quad . \tag{7}$$

Because the boundary condition (6) is natural, it will be satisfied automatically by any solution to (4) [11]. Consequently, we need never specify L any further than requiring it to satisfy (7)! Doing so, we may now consider the modified problem of determining a function $\mathcal{H}'$ that makes

$$\int_V [(\nabla\mathcal{H}')^2 - 2\mathcal{H}'\Phi] dV$$

stationary over aspect-limited data sets, and ignore the fact that such a solution will differ from $\mathcal{H}$ by some linear function.

3. NUMERICAL RESULTS

For convenience, all formulas will be developed in spherical coordinates. Dividing the domain of $h(\theta,\phi)$ into N subdomains $\mathcal{D}_i$ (with boundaries defined by constant θ and ϕ curves), we approximate over each $\mathcal{D}_i$

$$h = h_i(\theta,\phi) = h(\theta_i,\phi_i) + \frac{h(\theta_i,\phi_i+\Delta\phi_i) - h(\theta_i,\phi_i)}{\Delta\phi_i}(\phi-\phi_i)\sin\theta$$

$$+ \frac{h(\theta_i + \Delta\theta_i,\phi_i) - h(\theta_i,\phi_i)}{\Delta\theta_i}(\theta - \theta_i) \quad ,$$

where $\Delta\theta_i$ and $\Delta\phi_i$ are the differences between the θ curves and the ϕ curves bounding $\mathcal{D}_i$, respectively. Assuming $\Phi = \Phi_i$, a constant, on each $\mathcal{D}_i$, the functional of interest becomes

$$F = \sum_{i-1} \int_{\mathcal{D}_i} \left[\left(\frac{\partial h_i}{\partial\theta}\right)^2 + \frac{1}{\sin^2\theta}\left(\frac{\partial h_i}{\partial\phi}\right)^2 - 2h_i(h_i+\Phi_i)\right]\sin\theta d\phi d\phi \quad .$$

The set of equations (5) readily lends itself to numerical solution [16]. The Fortran code developed was tested on simulated data sets containing the appropriate aspect-limited curvature information for convex sample targets.

For an ellipsoid, this information takes the form [6]

$$R_u+R_v = -\{[a^2(b^2+c^2)\sin^2\theta\cos^2\phi+b^2(a^2+c^2)\sin^2\theta\sin^2\phi]$$

$$+[c^2(a^2+b^2)\cos^2\theta]\}/[a^2\sin^2\theta\cos^2\phi+b^2\sin^2\theta\sin^2\phi+c^2\cos^2\theta]^{3/2}$$

where a, b, and c are the semi-axes of the ellipsoid. Using this relation, data representative of what would be obtained by (2) were created, and the algorithm was tested for accuracy. Results when a = 1, b = 1, and c = 2 over the aspects $\theta \varepsilon (0,\pi/2)$ and $\phi \varepsilon (0,\pi)$ are displayed in figure 1.

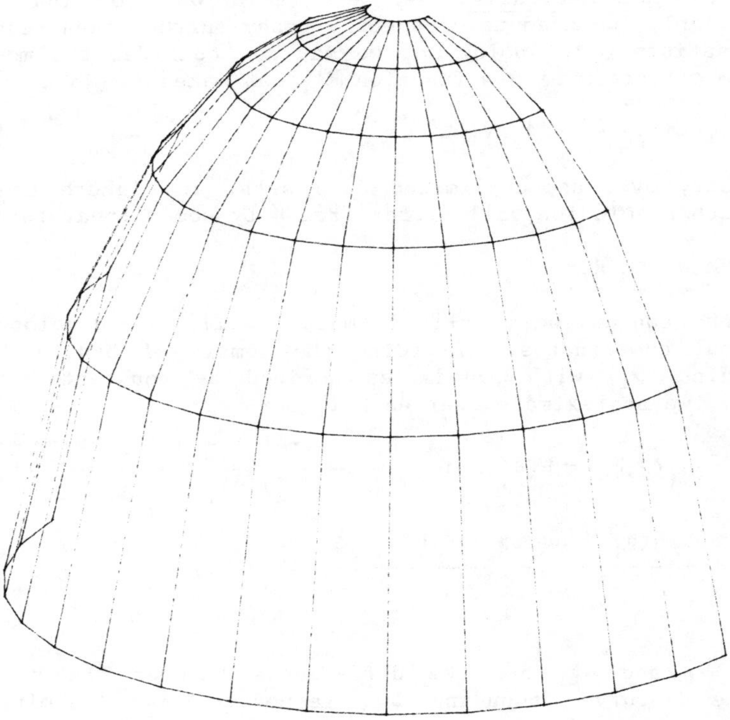

Figure 1. Reconstructed shape from synthetic data generated for an ellipsoid with semi-axes a = 1, b = 1, c = 2. The data were aspect-limited to the spherical directions $\theta \varepsilon (0,\pi/2)$ and $\phi \varepsilon (0,\pi)$.

In a similar way, a sphere-capped paraboloid was examined for data over $\theta \varepsilon (0,\pi)$ and $\phi \varepsilon (0,\pi/2)$ (see figure 2).

All calculations were performed in single precision on an HP-1000/F mini-computer.

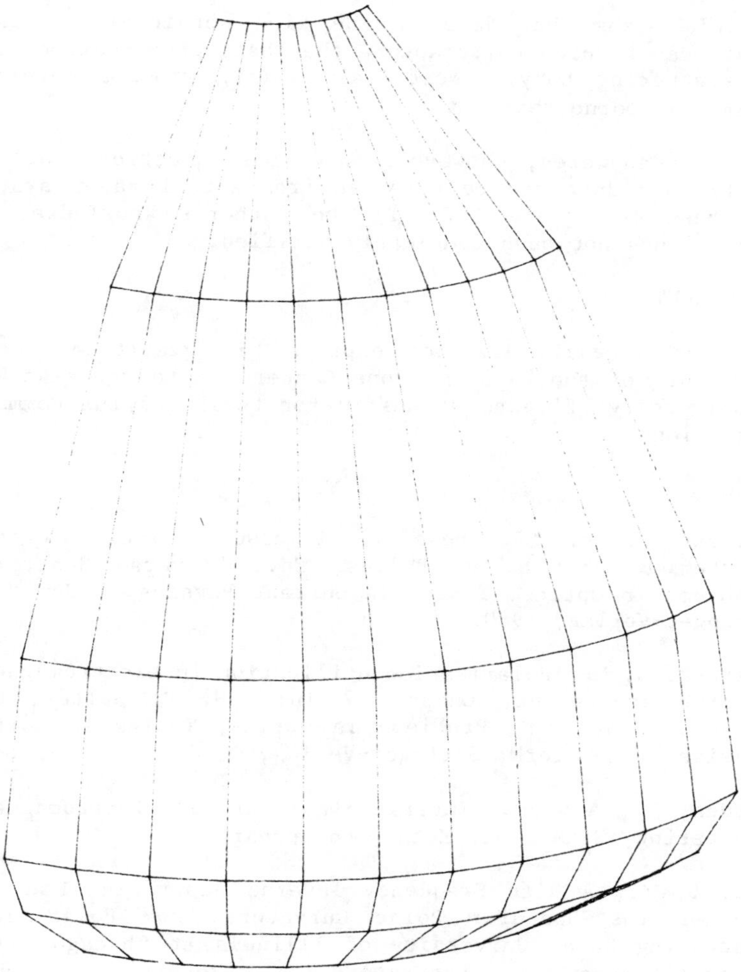

Figure 2. Reconstructed shape from synthetic data generated for a sphere-capped paraboloid. The aspect-limited data were created for the spherical directions $\theta\varepsilon(0,\pi)$ and $\phi\varepsilon(0,\pi/2)$.

CONCLUSION

The foregoing analysis and examples have demonstrated that the variational solution of the Stoker's differential equation approach to inverse scattering is not limited to full-aspect data. This means that data of the kind obtainable in radar encounters can be used to recapture the shape of convex portions of the scattering body. Numerical testing of the developed algorithms has borne this out.

Still unanswered, however, are the questions of how accurately this data can be obtained from actual radar systems and how best to obtain it. To the author's knowledge, the relation (1) has not been completely verified.

ACKNOWLEDGMENT

The author would like to express his gratitude to Dr. Charles Kenney of the Naval Weapons Center and to Dr. Alan Wolf of The University of Texas at Austin for their helpful comments and suggestions.

REFERENCES

[1] Baltes, H. P., Progress in inverse optical problems, Chapter 1 in: H. P. Baltes, Ed., "Inverse Scattering Problems in Optics, Topics in Current Physics." New York: Springer-Verlag, 1980.

[2] Boerner, W.-M., Polarization utilization in electromagnetic inverse scattering, Chapter 7 in: H. P. Baltes, Ed., "Inverse Scattering Problems in Optics, Topics in Current Physics." New York: Springer-Verlag, 1980.

[3] Borden, B., A vector inverse algorithm for electromagnetic scattering, SIAM Appl. Math., to appear.

[4] Foo, B. Y., A High Frequency Inverse Scattering Model to Recover the Specular Point Curvature from Polarimetric Scattering Data, University of Illinois at Chicago Circle Communications Laboratory Report Number EMID-CL-1982-05-21-01, 1982.

[5] Chadhuri, S. K., and Boerner, W.-M., A monostatic inverse scattering model based on polarization utilization, Appl. Phys. 11, p. 337, 1976.

[6] Bennett, C. L., Auckenthaler, A. M., Smith, R. S., and DeLorenzo, J. D., Space Time Integral Equation Approach to the Large Body Scattering Problem, Sperry Rand Research Center, Sudbury, MA, Final Report on Contract F30602-71-C-0162, RADC-CR-73-70, AD763794, May 1973.

[7] Kennaugh, E. M., and Cosgriff, R. L., The use of impulse response in electromagnetic scattering problems, IRE National Convention Record, Part I, pp. 72-77, 1958.

[8] Kennaugh, E. M., and Moffatt, D. L., Transient and impulse response approximation, IEEE Proc. 53, pp. 893-901, 1965.

[9] Silver, S., "Microwave Antenna Theory and Design." New York: McGraw-Hill, 1949.

[10] Stoker, J. J., On the uniqueness theorems for the embedding of convex surfaces in three-dimensional space, Commun. Pure Appl. Math. 3, pp. 231-257, 1950.

[11] Mikhlin, S. G., "Variational Methods in Mathematical Physics." New York: Macmillan, 1964.

[12] Hsiung, Ch.-Ch., A theorem on surfaces with a closed boundary, Math. Zeitschr. 64, pp. 41-46, 1956.

[13] Hsiung, Ch.-Ch., A uniqueness theorem for Minkowski's problem for convex surfaces with boundary, Illinois J. Math. 2, pp. 71-75, 1958.

[14] Busemann, H., Minkowski's and related problems for convex surfaces with boundaries, Michigan J. Math. 6, p. 259, 1959.

[15] Oliker, V. I., On certain elliptic differential equations on a hypersphere and their geometric applications, Indiana Univ. Math. J. 28, p. 35, 1979.

[16] Dongarra, J. J., Moler, C. B., Bunch, J. R., and Stewart, G. W., "LINPACK Users' Guide." Philadelphia, PA: SIAM, 1979.

II.1 (IS.1)

RECENT ADVANCES IN THE THEORY OF INVERSE SCATTERING WITH SPARSE DATA

V. H. Weston

Department of Mathematics
Purdue University

Abstract

Multifrequency inverse scattering by an object is considered for the case of sparse and non-redundent data, and where the measured object is not restricted to be a small perturbation of a known or comparison object. Previous quasi-linear techniques suffer from the severe restriction that the measured object is a small perturbation of a comparison object and each iterative step represents a perturbation of the same magnitude or else the data set must be complete. A non-linear iteration approach is discussed here which removes the restriction that the comparison body is close to the measured body, and the data set is sparse. This approach has the further advantage that the regularization process that must be used (for inverting almost singular matrices) is much simpler since there is only one constraint (involving the error in the matrix). In addition the method yields a condition that will tell whether the comparison body is close to the measured body and this can be used to terminate the iteration process. This is important because it has been shown that for sparse data, the solution of the exact system of equations is not an isolated set of points (values of the quantity to be determined) but a set of regions about these points (resolution cells). The details are given here for the case where the scattering object is characterized by the index of refraction $n(x)$ in the reduced wave equation $\Delta u + k^2n^2u = 0$. The method has application to other scattering problems, and applies to any spatial dimension.

W.-M. Boerner et al. (eds.), Inverse Methods in Electromagnetic Imaging – Part 1, 261–275.

1. Introduction

The multi-frequency inverse problem with sparse data for a three dimensional object in an uncontrolled environment is treated here. The uncontrolled environment implies that the scattering object or medium is not simply a small perturbation of a known object or medium, hence the usual linearizing techniques are invalid. In addition, in an uncontrolled environment, one would have little or no knowledge about the scatterer other than knowledge of its position and rough estimate of size. The data set itself is sparse and incomplete, as one would not be able to get all the measurements that would be desired. This could be due to physical interference or obstruction by other objects such as may occur in non-destructive testing, or else the object itself may be an uncooperative non-stationary target.

The main focus of this paper will be on the inverse problem connected with the reduced wave equation

$$\Delta u + k^2 n^2(x)u = 0, \quad x \in R^3 \tag{1}$$

(The method presented here can be abstracted and applied to other scattering problems such as scattering by objects characterized by Neumann or Dirichlet boundary conditions and vector problems. More details of the present paper are given in Weston (1984)).

In an uncontrolled environment associated with the reduced wave equation, all that can be specified a priori about the scattering object is that it is located in a prescribed bounded region D (with piecewise smooth (C^1) boundary), and is characterized by a real, sectionally continuous index of refraction $n(x)$ ($n(x)$ is continuous everywhere except for a finite set of piecewise smooth surfaces across which it has finite jump discontinuities). The host medium exterior to D is known, and for this paper will be taken to be homogeneous with $n(x) \equiv 1$.

The data consists of a finite set of measured values of the scattered field at different locations external to the scatterer, and at different frequencies. (It will be assumed that the real and imaginary parts of the scattered field are measured. The case where only the amplitude is measured can be treated in a similar fashion). Since the data set is sparse, the solution to the inverse problem is not unique. In particular it has been shown (Weston, 1984b) that the solution (of system (5)) is not an isolated set of points (values of $n(x)$) but a set of regions (resolution cells). Additional constraints have to be imposed to get uniqueness. Analytical results for this have been developed for case of fixed frequency scattering (Weston (1981, 1983). However the use of such constraints make the problem more complicated and less tractable for computation. Without the use of such constraints, the important thing to know is when one has an approximation that is close to the linearized region of a possible scatterer. Such conditions have been developed and are given by Eq (19).

It should be pointed out that other non-linear schemes based upon on-the-self techniques (conjugate-gradient, Newton's, etc.) have been employed for inverse scattering (Roger (1981), (1984) Coen et al (1981)). However these results are applied to the case where the data set is assumed complete, and no analysis has been done by these authors for the incomplete and sparse set of data.

2. Basic Mathematical Problem

To simplify analysis set

$$v(x) = n^2(x) - 1 \tag{2}$$

and note that $v(x) \equiv 0$ outside D. Let v_* and v_m denote the value of v corresponding to the comparison and measured body respectively.

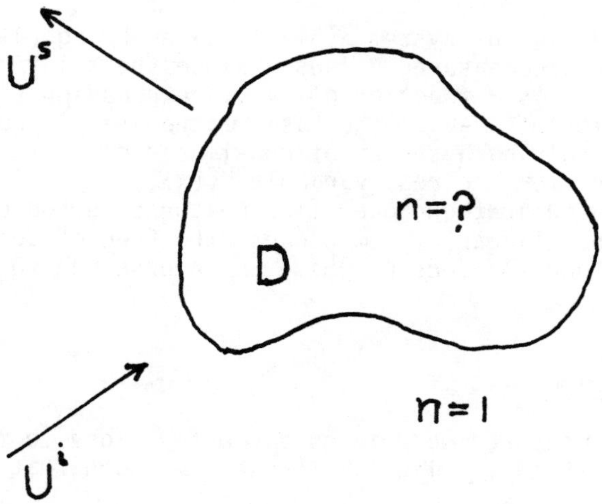

The data set will then consist of the set of N complex numbers $u_m^s(x_\ell,k_\ell)$ which is the measured value of the scattered field at a point x_ℓ outside D. The scattered field associated with each measurement is generated by a particular incident field $u^i(x,k_\ell)$ impinging upon the obstacle $(v=v_m)$ at a frequency ω_ℓ (corresponding to the wave number k_ℓ).

Let $G_*(x,y;k_\ell)$ denote the Green's function for the reduced wave equation associated with the comparison body $(n=n_*)$ and which satisfies the Radiation Condition. The total field $u(x,k_\ell)$ (where $u = u^i + u^s$) generated by the incident wave $u^i(x,k_\ell)$ impinging upon an arbitrary scatterer v (with support in D) will satisfy the integral equation

$$u(x,k_\ell) = u_*(x,k_\ell) + k_\ell^2 \int_D G_*(x,y;k_\ell)[v(y)-v_*(y)]u(y,k_\ell)dy \quad (3)$$

Here $u_*(x,k_\ell)$ is the total field produced by the same incident wave $u^i(x,k_\ell)$ impinging upon the comparison body $(v=v_*)$.

The scattered field at a point x_ℓ exterior to D is then given by

$$u^s(x_\ell,k_\ell) = u_*^s(x_\ell,k_\ell) + k_\ell^2 \int_D G_*(x_\ell,y;k_\ell)[v(y)-v_*(y)]u(y,k_\ell)dy \quad (4)$$

where the term in the integrand $u(y,k_\ell)$ is obtained by solving integral equation (3).

The inverse problem consists of finding the value of $v(x)$ such that

$$u^s(x_\ell,k_\ell) = u_m^s(x_\ell,k_\ell), \quad \ell = 1,2,\cdots N \quad (5)$$

where the left-hand side of system (5) is given by Eq.(4) and the right-hand side are measured values. Since the solution $u(y,k_\ell)$ of Eq. (3) is a function of $v-v_*$, expression (4) is a non-linear function of $v-v_*$. The basic mathematical problem is thus reduced to solving a system of non-linear complex functional equations for the real variable $v(x)$.

The usual non-linear techniques like Newton's method cannot be applied to the non-linear system because the Frechet derivative of the non-linear function does not have an inverse due to non-uniqueness.

3. Linearized Problem

If the actual measured body is sufficiently close to the comparison body so that v_m lies in the set v such that

$$\underset{\substack{k_\ell \\ \ell=1,2,\cdots N}}{\mathrm{Max}} \; k_\ell^4 \int\int_{D\times D} |G_*^2(x,y;k_\ell)|(v-v_*)^2 \, dxdy \ll 1 \quad (6)$$

then the solution of integral equation (3) could be approximated by

$$u(x,k_\ell) \sim u_*(x,k_\ell),$$

and system (5) in turn approximated by the linear system

$$b_\ell = k_\ell^2 \int_D G_*(x_\ell,y;k_\ell)u_*(y,k_\ell)w(y)dy \ , \ \ell = 1,2\cdots N \tag{7}$$

with

$$w = v - v_*, \tag{8}$$

where the complex number

$$b_\ell = u_m^s(x_\ell,k_\ell) - u_*^s(x_\ell,k_\ell) \tag{9}$$

is the difference between the measured value of the scattered field and the calculated value corresponding to v_*, at the point x_ℓ and wave number k_ℓ.

The system of N linear complex equations can be reduced to a system of $2N$ real equations. The complex quantities involved in the system are decomposed as follows

$$k_\ell^2 G_*(x_\ell,y;k_\ell)u_*(y,k_\ell) = H_\ell(y) + i\ H_{\ell+N}(y), \tag{10}$$

$$b_\ell = B_\ell + i\ B_{\ell+N}. \tag{11}$$

System (7) can then be reduced to the system of $2N$ real equations

$$(H_\ell,w) = B_\ell, \ \ell = 1,2,\cdots 2N \tag{12}$$

where the notation $(f,g) = \int_D f(x)g(x)dx$ is employed. With x_ℓ exterior to $\bar{D}$, both $G_*(x_\ell,y;k_\ell)$ and $u_*(y,k_\ell)$, and hence $H_\ell(y)$, will be continuous functions of y in $\bar{D}$, when v_* is a sectionally continuous function.

If the measured body v_m is in the linearized region of the comparison body v_* (i.e., condition (6) holds for $v=v_m$) then the system of $2N$ real linear equations (12) provides a useful model for solution. The system as it stands, is ill-posed, with a non-unique solution, and any resulting matrix that may be used to invert the system may be ill-conditioned. Moreover, due to errors in data, one would not want an exact solution of system (12), but a solution that satisfies the equation to within the errors of the data (errors in B_ℓ). The usual techniques of regularization may be employed. In particular the technique of Backus and Gilbert (1970) is most suitable.

4. Non-Linear Process

4.1. Conceptual Approach

If one has a catalogue of comparison bodies, then linear system (12) will be solved for each value of v_* belonging to a catalogue of prescribed comparison bodies $\{v_*\}$, obtaining a solution w which is a function of v_*. Then the comparison body that is closest to the measured body is selected. If this comparison body is in the linearized region of the measured body, then the usual techniques such as Backus and Gilbert or others may be employed to get the optimum solution. There are a number of criteria for closeness that can be used. The usual approach is to employ the following quantity

$$\phi(v_*) = \sum_{\ell=1}^{N} |u_m^s(x_\ell,k_\ell) - u_*^s(x_\ell,k_\ell)|^2$$

which is the sum of the squares of the differences in the measured and calculated scattered fields. This quantity by itself will not indicate when the linearization process is valid, but is useful to eliminate possible choices of comparison bodies.

The best choice from the theoretical viewpoint is to select the comparison value v_* such that

$$\underset{v_*}{\text{Min}} \quad \underset{\substack{k_\ell \\ \ell=1,2,\cdots N}}{\text{Max}} \quad k_\ell^4 \iint_{D\times D} |G_*^2(x,y;k_\ell)|\, w^2(v_*)\,dx\,dy \tag{13}$$

If the minimum value is much less than unity then the linearization process employing the value of v_* yielding the minimum, would be valid. A better choice from the practical computational standpoint would be to choose v_* so that

$$\underset{v_*}{\text{Min}} \int_D w^2(v_*)\,dx \tag{14}$$

This latter choice provides an obvious answer of non-uniqueness of linear system (12). Condition (14) immediately infers that one would want the least squares solution of system (12). The main difficulty in using condition (14) is that it will not signify directly whether the linearization process is valid. As it stands, it can be used only for comparison. However an additional condition to go along with (14) will be given.

If either one doesn't have a catalogue of comparison bodies to begin with, or else the comparison body belonging to a catalogue is not in the linear region of the measured body, a non-linear iterative scheme will be used based upon the above concepts. The idea is to obtain a sequence $\{v_*^n\}$ of comparison bodies which minimizes the function in expression (14). Note that when $v_* = v_m$ then the absolute minimum is obtained.

4.2 Iteration Procedure

An iterative procedure based upon a descent type process applied to the non-linear functional $f(v_*)$ where

$$f(v_*) = \int_D w^2(v_*)dx \tag{15}$$

will be used. A minimizing sequence v_*^n will be sought such that

$$f(v_*^{n+1}) < f(v_*^n).$$

This will require the Taylor expansion

$$f(v_*+\delta v_*) = f(v_*) + (F(v_*),\delta v_*) + \frac{1}{2}(\mathcal{F}'(v_*)\delta v_*,\delta v_*) + \cdots$$

where $F(v_*)$ is the gradient of $f(v_*)$ and $\mathcal{F}'$ the Frechet derivative of F

$$\delta F = \mathcal{F}'\delta v = (F'(v_*),\delta v_*). \tag{16}$$

Here $F'(v_*)$ denotes the kernel of the integral operator $\mathcal{F}'$. The descent approach (Vainberg (1973)) is given by the sequence

$$v_*^{n+1} - v_*^n = -\omega_n\alpha_n F(v_*^n) \tag{17}$$

where ω_n is the relaxation factor. The real positive coefficient α_n in the descent method is obtained by minimizing the function $g(t) = f(v_*^n - tF(v_*^n))$ of the positive real variable t, or in practical applications, the quadratic approximation to $g(t)$, (Ortega and Rheinboldt (1970), Rheinboldt (1974)). The latter yields the value

$$1/\alpha_n = ((F'(v_*^n),F(v_*^n)),F(v_*^n))/(F(v_*^n),F(v_*^n)) \tag{18}$$

However for the problem on hand, F' is not a positive kernel, and a modified version of (18) has to be used by splitting F' into two parts, one a positive kernel and the other, a non-positive kernel, and retaining the contribution from the positive kernels. Explicit expressions are given for $F(v_*;x)$ and $F'(v_*;x,y)$ in Weston (1983), where in particular it is shown that the gradient has the form

$$-\frac{1}{2}F(v_*;x) = w(v_*;x) + q(v_*;x),\ x \in R^3$$

When the comparison body v_* is close to the measured body v_m i.e., when $\|w(v_*;x)\|$ is small, then

$$-\frac{1}{2}F(v_*;x) = w(v_*;x) + 0(\|w\|^2)$$

and

$$\alpha\ (v_*) \sim \frac{1}{2}\ ,$$

hence the descent sequence approaches the usual linear approximation

$$v_*^{n+1} - v_*^n \sim w(v_*^n)$$

in the linear region. One consequence of this, is that a natural criterion exists to indicate when the comparison body is in the linear region of the measured body. It is given by

$$\|q\|/\|w\| << 1,\ ((F',F),F)/(F,F) \sim 2 \tag{19}$$

The descent process will terminate when a stationary point is reached. There are two types of stationary points, relative or absolute minimum, and saddle points, as expected. Of main interest is the absolute minimum point given by $\|w\| = 0$, yielding $B_\ell = 0$, $\ell = 1,2\cdots 2N$. There may be more than one such point due to the non-uniqueness of the problem. The descent approach can be modified to go through the saddle point. When the descent process has reached the linear region of the measured body (close to the absolute minimum of $f(v_*)$), it may be terminated and other techniques like Backus and Gilbert can then be employed to get the optimum fine structure.

4.3 Regularization

The critical juncture of the procedure is the solving of linear system (12). For the initial step (involving the initial choice of v_*) it is assumed that the data points $\{x_\ell\}$ and wave numbers $\{k_\ell\}$ are chosen so that $\{H_\ell(x)\}_{\ell=1}^{2N}$ is a linearly independent set of continuous functions over D; and any redundant set of measurements which violates this assumption will be culled out. Theoretically this only has to be assured of initially, since the descent process will tend to retain the linear independence of $\{H_\ell\}$ for each v_*^n. As a consequence the matrix $H(v_*)$ whose elements H_{ij} are given by

$$H_{ij} = \int_D H_i(x)H_j(x)dx \tag{20}$$

will be positive definite and invertible, since the quadratic form

$$\sum_{i,j=1}^{2N} c_i H_{ij} c_j = \int_D \Big(\sum_{i=1}^{2N} c_i H_i(x)\Big)^2 dx$$

will vanish only if $\sum_{i=1}^{2N} c_i H_i(x) \equiv 0$. The minimum $\mathcal{L}_2(D)$ norm solution of system (12) is given by setting

$$w(x) = \sum_{j=1}^{2N} \eta_j H_j(x) \tag{21}$$

yielding the algebraic system of equations

$$\sum_{j=1}^{2N} H_{ij}\eta_j = B_i,\ i = 1,2,\cdots,2N \tag{22}$$

In the numerical computation of the solution of system (22) constraints have to be imposed that involve some of the following:

(i) Errors in H
(ii) Errors in the data, or $\{B_j\}$
(iii) Bounds on the solution $\vec{\eta}$,

The errors in the matrix elements H_{ij} are the most serious. These errors arise from two sources, numerical errors (due to computational procedures, round-off, etc.) and model error due to the approximation of the space of continuous functions over D by a finite dimensional space. With H a positive definite matrix, the errors in H will distort the component of the solution belonging to the eigenspace spanned by the eigenfunctions corresponding to the small eigenvalues. If the error in the matrix H is denoted by $E = \{E_{ij}\}$, then the numerical solution of system (22) will be required to the following accuracy

$$\|H\vec{\eta} - \vec{B}\|_2 \le \|E\|_2\|\vec{\eta}\|_2 \tag{23}$$

where $\|\vec{\eta}\|_2 = \sum_{j=1}^{2N} \eta_j^2$.

No constraint need be placed upon the bound of the solution $\vec{\eta}$ for the iterative scheme developed here. This is in contrast to the schemes employed by Coen, Mei and Angelakos (1981), and Levenberg-Marquardt Algorithm (Moré (1978), Levenberg (1944), Marquardt (1963)), which require such a constraint due to a condition like (6) being imposed on their iteration process.

Errors in the data (provided they are not too large) can be ignored in the non-linear iterative scheme. They become important in the linear region where v_* is close to v_m. Generally, since no bounds are placed on $\|\vec{\eta}\|_2$, the error term $\|E\vec{\eta}\|_2$ may be larger than the error in the data.

With only the constraint (23) being considered, the system of equations can be regularized, Miller (1970) by replacing them by

$$(H + \varepsilon I)\vec{\eta} = \vec{B} \tag{24}$$

where $\varepsilon = \|E\|_2$.

Note the Levenberg-Marquardt iterative scheme is

characterized by a sequence of the form

$$v_*^{n+1} - v_*^n = w(\varepsilon_n, v_*^n)$$

with w found from Eq(21) and a regularized form of Eq(22) with the regularization parameter ε_n chosen at each step.

4.4 Constrained Iterative Process

The descent procedure can be modified to give a solution satisfying the constraint

$$n^2(x) \geq 1 \ , \ x \varepsilon D$$

This can be achieved by replacing the iterate

$$v_*^{n+1}(x) = v_*^n(x) - \alpha_n F(v_*^n, x) \tag{25}$$

by $v_*^{n+1}(x) = 0$, at all points x in D where the right hand side of Eq (25) is negative. This process is equivalent to adding a function $\psi(x)$ to the gradient F so that

$$v_*^{n+1} = v_*^n - \alpha_n(F(v_*^n, x) + \psi(x)) \geq 0$$

The sequence $\{v_*^n\}$ will still be a minimizing sequence provided that

$$-(F,\psi)/(F,F) << 1$$

When $-(F,\psi)/(F,F)$ is larger but still less than unity, α_n will have to be modified.

5. Computation

The non-linear process discussed here applies to any spatial dimension. At present, computations are being performed for the one-dimensional case where the domain D is given by $-1 \leq x \leq 1$. Preliminary results are given as follows.

Backscattered data corresponding to the "measured" body

$$n = 1 \ , \ -1 \leq x < 0 \ , \ n = 2, \ 0 < x \leq 1$$

are generated for the wave numbers $k = \pi/4$, $\pi/2$, $3\pi/4$ and π. Starting from the initial value $n \equiv 1.5$ for $-1 \leq x \leq 1$, the iteration procedure $(N = 4)$ employing the constraint $n \geq 1$, was used and the results for the 13th iterate are presented in Fig. 1. The values of $\int_D w^2(x)dx$ after each iteration are presented in Fig. 2.

The iteration procedure using an additional "measured" value of the backscattered field ($N = 5$) gave similar results but converged faster.

The effect of different initial choices of $n(x)$ was considered. The value $n \equiv 1.25$, $-1 \leq x \leq 1$, gave similar results (with slower convergence) as $n \equiv 1.5$. However the iteration procedure with the initial value $n \equiv 1.75$ for $-1 \leq x \leq 1$, converged to a different value of $n(x)$ for the case of four measurements ($N = 4$). The effect of additional "measurements" ($N = 5,6$) indicated that the value so obtained was a local minimum and not a true solution. However with the values $k = \pi/8, \pi/4, \pi/2, 3\pi/4$, the iteration procedure converged to the correct solution when the initial value $n \equiv 1.75$ was taken.

6. Acknowledgement

The research reported herein was supported in part through O.N.R. grant N14-83-K-0038.

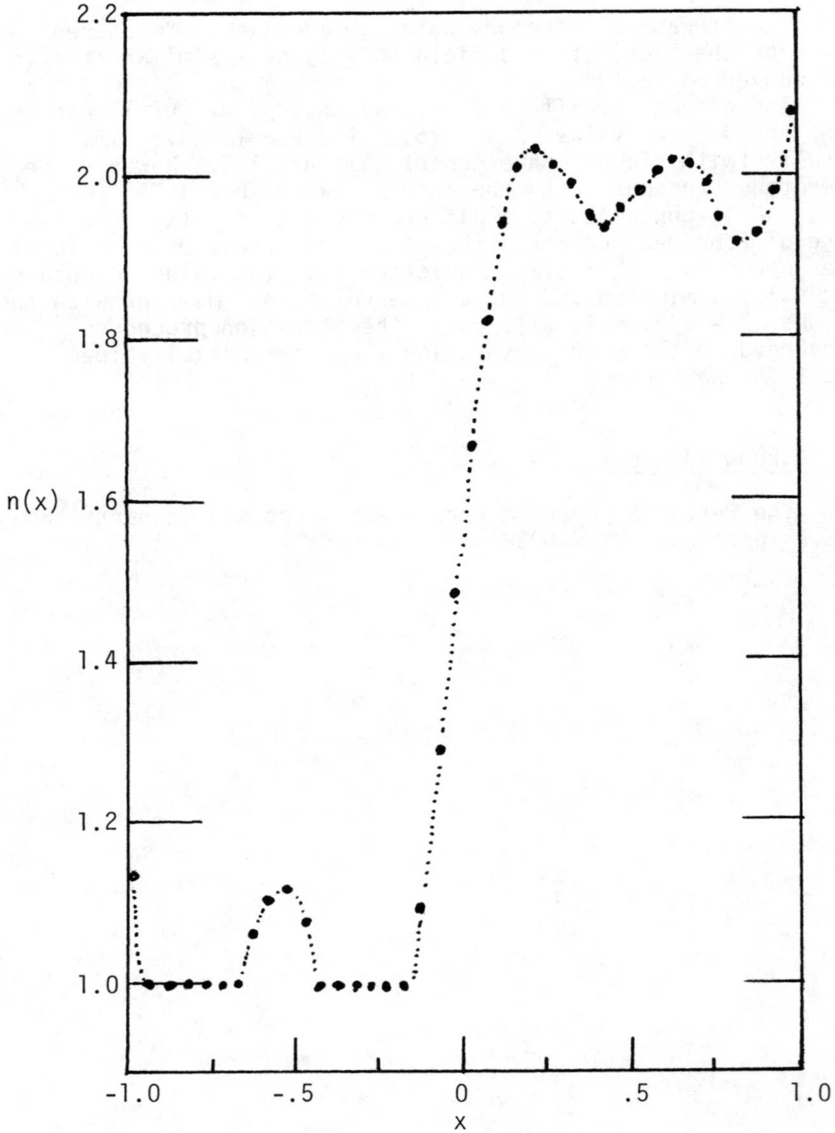

Fig. 1

Profile of the 13th iterate of $n(x)$, corresponding to the case where $N = 4$, the first iterate is $n(x) \equiv 1.5$, $-1 \leq x \leq 1$, and "measured" data corresponds to $n = 1$, $-1 \leq x < 0$, $n=2$, $0 < x \leq 1$.

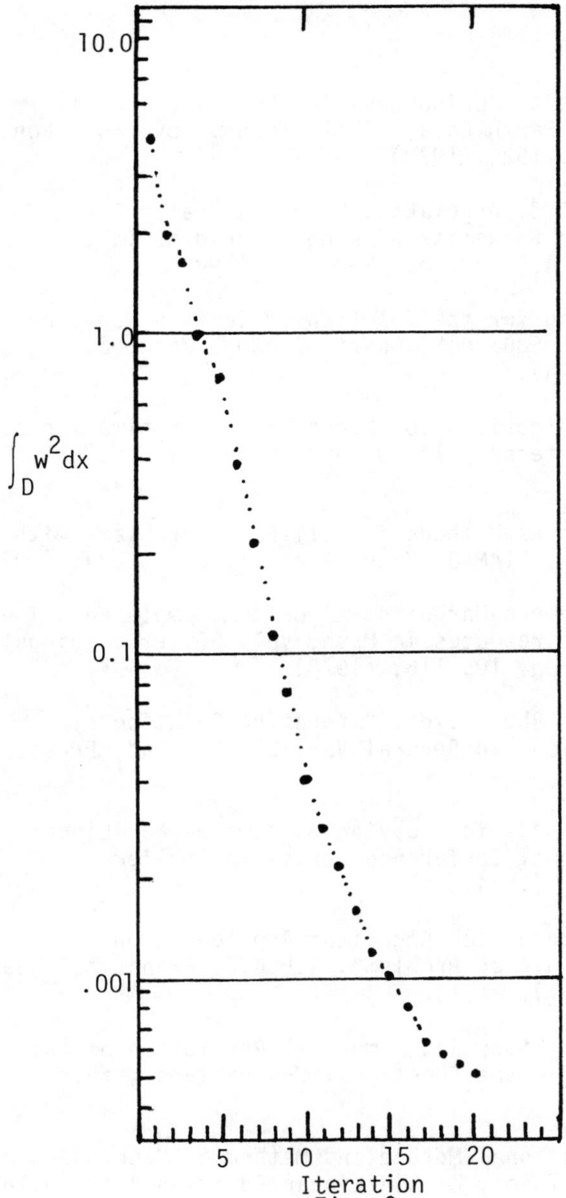

Fig. 2

Value of $\int_D w^2 dx$ after each iteration step for the case where $N = 4$, initial value of $n(x) = 1.5$, $-1 \leq x \leq 1$, and the "measured" data corresponds to $n = 1$, $-1 \leq x < 0$, $n = 2$, $0 < x \leq 1$.

References

G. Backus and F. Gilbert, "Uniqueness in the Inversion of Inaccurate Gross Earth Data", Phil. Trans. Roy. Soc. Lon. (a), 266, pp. 123-192, (1970).

S. Coen, K. Mei, and D.J. Angelakos, "Inverse Scattering Technique Applied to Remote Sensing Layered Media", I.E.E.E. Trans. A.P. 29, pp. 298-306, (1981).

K. Levenberg, "A Method for the Solution of Certain Non-Linear Problems in Least Squares", Quart. J. App. Math. 2, pp. 164-168, (1944).

D. W. Marquardt, "An Algorithm for Least Squares Estimation of Non-Linear Parameters", SIAM J. App. Math. 11, pp. 431-441, (1963).

K. Miller, "Least Squares Methods for Ill-Posed Problems with a Prescribed Bound", SIAM J. Math. Anal. 1, pp. 52-74, (1970).

J. J. More, "The Levenberg-Marquardt Algorithm, Implementation and Theory", Lecture Notes in Math. Vol. 30, Proceedings, Springer Verlag, pp. 105-116, (1978).

J. M. Ortega and W. C. Rheinboldt, "Iterative Solution of Non-Linear Equations in Several Variables", Acad. Press, New York (1970).

W. C. Rheinboldt, "Methods for Solving Systems of Non-Linear Equations", Regional Conference Series in Applied Mathematics, SIAM (1974).

A. Roger, "Newton-Kantorovitch Algorithm Applied to on Electromagnetic Inverse Problem", I.E.E.E. Trans. A.P. 29, pp. 232-238, (1981).

A. Roger, "Theoretical Study and Numerical Resolution of the Inverse Problem Via the Functional Derivatives", this volume.

M. M. Vainberg, "Variational Method and Method of Monotone Operators in the Theory of Non-Linear Equations", J. Wiley, New York (1973).

V. H. Weston, "Inverse Problem for the Reduced Wave Equation with Fixed Incident Field, Part II", J. Math. Phys. 22, 2523-2529, (1981).

V. H. Weston, "Inverse Problem for the Reduced Wave Equation with Fixed Incident Field, Part III", J. Math. Phys. 24, 828-833, (1983).

V. H. Weston, "Multi-Frequency Inverse Problem for the Reduced Wave Equation With Sparse Data", to be published (1984).

V. H. Weston, "The Resolution Cell for the Inverse Problem for the Reduced Wave Equation With Sparse Data", to be published (1984b).

II.2 (PT.2)

THE LIMITED ANGLE PROBLEM IN RECONSTRUCTION FROM PROJECTIONS

F. Alberto Grünbaum

Department of Mathematics
University of California, Berkeley

Abstract

The relation between "picture quality" and "amount of data" can be thoroughly analyzed both in X-ray and NMR imaging. The first case involves Euclidean geometry, the second one being three dimensional is best related to spherical geometry. In both of these cases one has the machinery of "prolate spheroidal wave functions". We consider the failure of this property for the case of hyperbolic geometry, except for Minkowsky space.

Table of Contents

[1]The research reported here was supported in part by National Science Foundation Grant MCS81-07086 and by the Director, Office of Basic Energy Sciences, Engineering, Mathematics, and Geosciences Division of the U.S. Department of Energy under contract DE-AC03-76SF00098.

W.-M. Boerner et al. (eds.), Inverse Methods in Electromagnetic Imaging – Part 1, 277–298.

1. Introduction

The motivation for this paper is the problem of reconstructing a real valued function f defined in R^n -- the unknown density -- from its projections $P_w f$, where the "directions" w are restricted to lie in a "limited range". In the case of the parallel beam geometry this can be put in the framework of "time and band limited functions". In some conditions this leads to a highly favorable and exceptional situation: the singular value decomposition of the corresponding "Finite Radon Transform" can be accomplished and the degree of illconditioning fully analyzed. The reason for this accident is poorly understood; we give a number of examples which include the cases of X-ray as well as NMR tomography.

After reviewing the "successful cases" we show how a most natural mathematical extension of this setup: from euclidean and spherical geometries to the hyperbolic case leads into a breakdown of "prolate spheroidal wave functions" bag of tricks. While we do not yet understand fully either the true reasons or the consequences of this mathematical fact, we present it with the hope that other people will find it useful. The last section of the paper gives indications of another piece of nineteen century mathematics lurking behind these matters.

2. Mathematical formulation

The problem of recovering, or estimating an unknown function $f(\underline{x})$, $\underline{x} \in R^n$, from

$$Pf(\underline{w},t) = \int_{(\underline{x},\underline{w})=t} f, \quad \underline{w} \in S^n$$

can be put in the familiar form

$$Af = b = \text{data}.$$

Here A is a linear map from $L^2(D)$ to $\oplus L^2(I)$, f denotes the unknown density, b stands for the collection of one dimensional projections.

The standard way to analyze a problem of this kind, although not necessarily the best way to go about solving it, is as follows: one considers the singular value decomposition of A. One brings in the vectors ψ_i, defined as the orthonormal eigenvectors of AA^*, with positive real eigenvalues μ_i^2, $\mu_i > 0$. Then setting $\phi_i = A^*\psi_i/\mu_i$ one gets for the smallest f which minimizes the error $Af - b$ the expression

$$\tilde{f} = \sum (b,\psi_i)/\mu_i \phi_i .$$

A small"singular value" μ_i spells troubles in obtaining the component of $\tilde{f}$ along ϕ_i if the data has any component along ψ_i.

We thus see that a complete analysis of the degree of ill-conditioning requires finding the triple $\{\phi_i,\psi_i,\mu_i\}$. For most problems of any interest this is the end of the road in that these quantities cannot really be found with any accuracy.

An example of the use of this "triple" is given in [1]. From some small scale numerical experiments reported there, one can see that the μ_i's satisfy the condition $0 < \mu_i < 1$ and split into two subsets: some of "close to one" and the rest are "close to zero." We saw in [1] that the effect of this split is that standard iteration methods for limited angle tomography -- without use of nonlinear constraints, like positivity -- give, instead of the desired function f, its projection into the linear span of the ϕ_i's with μ_i close to one. The analysis given in [1] was carried out in Fourier space and could not be used beyond small scale models because of numerical difficulties in obtaining the singular vectors and values.

3. X-ray and NMR imaging

To go beyond generalities one has to pick the situation very carefully or be very lucky. The simplest such case is 2 dimensional X-ray tomography where w becomes an angle and $P_w f \equiv P_\theta f$ becomes a collection of line integrals and finally

$$-\frac{\pi}{2} \le -a \le \theta \le a \le \frac{\pi}{2}$$

The next simplest case is given by 3 dimensional NMR imaging -- in one of its modalities -- where w becomes a pair of angles in R^3, $w = (\theta,\phi)$, and $P_w f$ becomes a collection of plane integrals.

B. Marr [2] pointed out that for X-ray tomography if one keeps the problem in its original form, one runs into an operator which is essentially identical to that studied by Slepian [3] in connection with "index and band limited sequences." For the case of X-ray tomography this analysis has been carried out by Davison [4,5], where the reader can find a detailed presentation.

The bottom line is that one can effectively compute (with high numerical accuracy) the required eigenvalues and eigenvectors.

The case of NMR tomography leads naturally to look for extensions of the work of Slepian, Landau and Pollak [6,7,8,9]. One considers the surface of the sphere instead of the circle and replaces the usual Fourier series by the expansion in spherical harmonics. The role of a segment $-a \leq \theta \leq a$ is now played by a spherical cap $0 \leq \theta \leq a$, ϕ arbitrary, and band limiting is taken to mean: keep only the components of f along $Y_{lm}(\theta,\phi)$ with $0 \leq l \leq L$. As it happens, this case is also exceptional in that the search for the appropriate singular vectors can be reduced to a numerically manageable problem. This, along with a host of related results, can be found in [10]. In all of these cases this reduction to a numerically feasible problem is accomplished by exhibiting a second order differential operator which has simple spectrum and commutes with the relevant operator AA^*. This latter one is typically a finite convolution integral operator, and A itself is the finite Radon transform alluded to in the abstract.

The phenomenon which was so well exploited in [6,7,8,9], and to be explained below, had already appeared in mathematical form in the work of Whittaker and Ince [11,12]. It also appears, in totally independent form in the work of Mehta and Gaudin on "random matrices" see [13].

In the notation used above take A to denote the finite Fourier Transform

$$(Af)(\lambda) = \int_{-T}^{T} e^{i\lambda t} f(t)\, dt, \quad \lambda \in [-\Omega,\Omega]$$

then the integral operator given by A^*A, namely

$$(kf)(t) = \int_{-T}^{T} \frac{\sin \Omega(t-s)}{t-s} f(s)\, ds \quad t \in [-T,T]$$

admits a second order differential operator commuting with it, namely

$$(Df)(t) = ((T^2 - t^2)f')' - \Omega^2 t^2 f.$$

Slepian proved that this same result holds if R^1 is replaced by R^n [9], and then in (14) -- mentioned earlier in connection with tomography -- extended this fact for functions defined on the circle or on the integers. We became interested in this problem and tried to extend its range. In [15] one finds an extension to the case of the nth roots of unity and then in [16] one finds a classification of all Toeplitz matrices which have a tridiagonal matrix, with single spectrum, in its commutator.

In [17] this property is seen to hold for the Hilbert matrix and in [24] one finds that the property (essentially) holds for the Discrete Fourier Transform matrix.

In trying to deal with the "limited angle" problem in R^3, we were led to consider the extension of Slepian's [14] results to the case of S^2, or more generally S^n. This was found to be true, as mentioned above.

Since the property holds for R^n and S^n it is only natural to expect that it should hold for hyperbolic space H^n. Indeed in [10] the result is shown to hold for H^3, see also [19] for a much more transparent proof.

And yet the result is not true for any H^n, $n \geq 2$, except when $n = 3$.

This is the topic of the next sections.

4. Harmonic analysis for hyperbolic space

Consider second order differential operators of the form

$$D \equiv \frac{1}{w(x)} \frac{d}{dx}\left(w(x) \frac{d}{dx}\right) \quad x \geq 0$$

and $w(x)$ given by

$$w(x) = (e^{x}-e^{-x})^p (e^{2x}-e^{-2x})^q$$

D is formally selfadjoint in $L^2(w(x)\,dx, R^+)$.

Operators of this form appear as the radial part of the Laplace Beltrami operator for noncompact symmetric spaces of rank one, with (p,q) a pair of nonnegative integers, see [20].

If we formally put $p = q = 0$, we obtain the Laplacian corresponding to the real line.

$$D = \frac{d^2}{dx^2} \; .$$

When $p \geq 0$, $q \geq 0$, $p + q > 0$ we are dealing with noncompact symmetric spaces arising from nonabelian groups.

The simplest examples come from $p = 0$, $q = 1$ and $p = 0$, $q = 2$ corresponding to $SO(2,1)/SO(2)$ and $SO(3,1)/SO(3)$

respectively. The first space is the upper half plane with the hyperbolic metric.

$$y^{-2}(dx^2 + dy^2)$$

-- or equivalently the unit disk with Poincaré metric. The second space is the Minkowski space of relativity theory obtained from factoring out the Lorentz group SO(3,1) by the group of spatial rotations SO(3).

We will consider only the cases $p \geq 0$, $q \geq 0$, $p + q > 0$. The compact and euclidean cases have been treated in [10].

Given D, let $\phi_\lambda(x)$ denote the unique solution to

$$D\phi_\lambda(x) = -(\lambda^2 + \rho^2)\phi \qquad x \geq 0 \tag{1}$$

satisfying $\phi_\lambda(0) = 1$ $\phi'_\lambda(0) = 0$. Here

$$\rho \equiv \frac{p + 2q}{2}$$

These functions take up the role of the radial (i.e. even) eigenfunctions of the Laplacian in R^1, namely cos x, and permit a harmonic (or Fourier) analysis of $L^2(w(x)\,dx)$ in terms of eigenfunctions of the operator D.

To obtain a useful analog of the Fourier analysis on R^1 one needs to find the "spectral measure" for the operator D. This gives an inversion formula and a Plancherel identity.

This is done below after some preparations.

Set

$$a \equiv a(\lambda) \equiv \frac{\rho + i\lambda}{2} \qquad b \equiv b(\lambda) \equiv \frac{\rho - i\lambda}{2} \qquad c \equiv \frac{p + q + 1}{2},$$

Then one has

$$\phi_\lambda(x) = F(a,b,c, -\sinh^2 x).$$

and setting

$$c(\lambda) = \frac{2^{2b}\Gamma(c)\Gamma(i\lambda)}{\Gamma(c-b)\Gamma(a)}$$

with F the usual hypergeometric function, one has

<u>Theorem</u>. For any $f \in L^2(w(x)\,dx,\ x \geq 0)$ and $\lambda > 0$ the integral

$$\tilde{f}(\lambda) = \frac{1}{\sqrt{2\pi}} \int_0^\infty f(x)\phi_\lambda(x)w(x)\,dx \tag{2}$$

is well defined in the $L^2(w(x)\,dx)$ sense. Moreover $\tilde{f}(\lambda) \in L^2(|c(\lambda)|^{-2}d\lambda,\ \lambda \geq 0)$ and the map $f \to \tilde{f}$ is a unitary operator between these two L^2 spaces. The inverse map is given by

$$(3) \qquad f(x) = \frac{1}{\sqrt{2\pi}} \int_0^\infty \tilde{f}(\lambda)\, \phi_\lambda(x)\, |c(\lambda)|^{-2}\, d\lambda$$

This is a particular case of the Weyl-Titchmarsh-Kodaira theory. Notice that if one considers the function

$$\tilde{\phi}_\lambda(x) = F\left(\frac{\rho + i\lambda}{2}, \frac{\rho - i\lambda}{2}, \frac{p + q + 1}{2}, -\sinh^2 x/2\right)$$

one has

$$D\tilde{\phi}_\lambda(x) = -\left(\frac{\lambda^2 + \rho^2}{4}\right)\tilde{\phi}_\lambda(x)$$

$$= -(\tilde{\lambda}^2 + \tilde{\rho}^2)\tilde{\phi}_\lambda(x)$$

with $\tilde{\lambda} = \frac{\lambda}{2}$ and $\tilde{\rho} = \frac{\rho}{2}$. In other words

$$\Psi_\lambda(x) \equiv F\left(\frac{p + 2q}{4} + i\lambda, \frac{p + 2q}{4} - i, \frac{p + q + 1}{2}, -\sinh^2 x/2\right)$$

satisfies

$$D\Phi_\lambda(x) = -\left(\lambda^2 + \frac{p + 2q}{4}\right)^2 \psi_\lambda(x).$$

For example if $p = 0$, $q = 1$ corresponding to $SL(2,R)/SO(2)$ ($= SO(2,1)/SO(2)$)

$$D\Psi_\lambda(x) = -(\lambda^2 + \frac{1}{4}\,\Psi_\lambda(x)$$

while if $p = 0$, $q = 2$, corresponding to $SO(3,1)/SO(3)$

$$D\Psi_\lambda(x) = -(\lambda^2 + 1)\Psi_\lambda(x).$$

The relation between $\Psi_\lambda(x)$ and $\phi_\lambda(x)$ above is given by

$$\Psi_\lambda(x) = \phi_{2\lambda}(x/2)$$

and using Ψ_λ or ϕ_λ is a matter of convenience. We will use ϕ_λ, but some people are more familiar with Ψ_λ.

5. A general formulation for "time and band limiting"

Given arbitrary values $T > 0$, $\Omega > 0$ consider an operator acting on functions $f \in L^2(w(x)\,dx,\ x \geq 0)$ as follows

1) Restrict f to $[0,T]$
2) Expand f in terms of the functions $\phi_\lambda(x)$ as in (3) above.
3) Restrict $\tilde{f}(\lambda)$ to $[0,\Omega]$
4) Invert the expansion
5) Restrict the resulting function to $[0,T]$.

In simpler terms the map in question can be defined, for $x \in [0,T]$ by means of

$$(Kf)(x) = \frac{1}{2\pi} \int_0^\Omega \left(\int_0^T f(y)\phi_\lambda(y)w(y)\,dy \right) \phi_\lambda(x)\,|c(\lambda)|^{-2}\,d\lambda$$

or exchanging the order of integration

$$(Kf)(x) = \frac{1}{2\pi} \int_0^T \left(\int_0^\Omega \phi_\lambda(x)\phi_\lambda(y)\,|c(\lambda)|^{-2}\,d\lambda \right) f(y)\,w(y)\,dy$$

This turns out to be a selfadjoint, compact operator in $L^2(w(x)\,dx,\ [0,T])$ whose eigenfunctions and eigenvalues play a crucial role in analyzing a number of practical extrapolation problems.

Even in thecase when $D = \frac{d^2}{dx^2}$, it is impossible to find these eigenfunctions "analytically", and one has to resort to approximate numerical methods. This sound practical but becomes impossible to do if one needs lots of eigenfunctions with any accuracy. The size of the corresponding matrix problem obtained from discretization becomes too large to handle since we are dealing with a "full" matrix.

The <u>ideal</u> solution to this problem would be to produce a second order differential operator D--with simple spectrum--which commutes with K and thus has the same eigenfunctions. While the problem still could not be analytically solved we would then have a feasible numerical problem which can be solved accurately and quickly using for instance (the last steps of) the Q-R algorithm.

In the case when

$$D = \frac{d^2}{dx^2}$$

one has

$$\tilde{D} = \frac{d}{dx}\left((T^2 - x^2)\frac{d}{dx}\right) - \Omega^2 x^2$$

This happens to be -- for totally mysterious reasons-- one of the three ordinary differential operators which appear when one separates the Helmholtz equation in "prolate spheroidal coordinates" in R^3. The proof of the relation

$$\tilde{D}K = K\tilde{D}$$

consists in writing out explicitly the kernel

(4) $$K_\Omega(x,y) = \int_0^\Omega \phi_\lambda(x)\, \phi_\lambda(y)\, |c(\lambda)|^{-2}\, d\lambda$$

which in this case becomes

$$\frac{\sin \Omega(x-y)}{x-y} + \frac{\sin \Omega(x+y)}{x+y}$$†

and checking that

(5) $$(\tilde{D}_x - \tilde{D}_y)\, K\, (x,y) =$$

Then with some integration by parts and using the fact that the leading coefficient of $\tilde{D}$ vanishes at $x = T$ one concludes that $\tilde{D}K = K\tilde{D}$.

The crucial property is (5), and for the rest of the paper we concentrate on it. Notice that the same property holds for the operator.

$$\tilde{D} = \frac{d}{dx} x^2 \frac{d}{dx} + \Omega^2 x^2$$

since this differs from $-\tilde{D}$ by the addition of a multiple of $\frac{d^2}{dx^2}$.

We now concentrate on the

---

†Strictly speaking this differs from the case in Slepian, Pollak, and Landau since we consider radial or even functions. This is an unimportant difference.

MAIN PROBLEM

Given D, as above, construct $K_\Omega(x,y)$ as in (4) and find functions

$$a(x,\Omega), \quad b(x,\Omega) \quad \text{with} \quad a(T,\Omega) = 0$$

such that for every Ω, the operators

$$(6) \qquad \tilde{D}_x \equiv \frac{1}{w(x)} \frac{d}{dx} \left(a(x,\Omega)\, w(x) \frac{d}{dx}\right) + b(x,\Omega)$$

$$(7) \qquad K\ (x,y) = \int_0^\Omega \phi_\lambda(x)\phi_\lambda(y)\,|c(\lambda)|^{-2}\, d\lambda$$

satisfy

$$(8) \qquad (\tilde{D}_x - \tilde{D}_y) K_\Omega(x,y) = 0$$

Observe that in the case of $D = \frac{d^2}{dx^2}$ we could choose $a(x,\Omega)$ independent of Ω, and $b(x,\Omega)$ with a simple factorized form.

This problem has a solution for higher dimensional Euclidean spaces, see [9], and for spheres and other compact examples in [10].

In [10] one finds a noncompact nonabelian example corresponding to the symmetric space $SO(3,1)/SO(3)$ -- then so called Minkowski space -- where $a(x,\Omega)$ could still be picked independent of Ω but $b(x,\Omega)$ no longer had a factorized form.

Now we move on to consider the general case of $p,q \geq 0$ $p + q > 0$.

6. The search for a commuting differential operator

1. Eliminating $b(x,\Omega)$

Since the functions $a(x,\Omega)$ and $b(x,\Omega)$ are only important up to an additive constant we can assume that

$$a(0,\Omega) = b(0,\Omega) \equiv 0$$

If we use (6) and (7) to write (8) in full we get

$$(8') \quad (a(x,\Omega) - a(y,\Omega)) \int_0^{\Omega} - (\lambda^2 + \rho^2)\phi_\lambda(x)\phi_\lambda(y)|c(\lambda)|^{-2}\, d\lambda$$

$$+ a'(x,\Omega) \int_0^{\Omega} \phi_\lambda'(x)\ \phi_\lambda(y)|c(\lambda)|^{-2}\, d\lambda$$

$$- a'(y,\Omega) \int_0^{\Omega} \phi_\lambda(x)\ \phi_\lambda'(y)|c(\lambda)|^{-2}\, d\lambda$$

$$= (b(y,\Omega) - b(x,\Omega)) \int_0^{\Omega} \phi_\lambda(x)\phi_\lambda(y)|c(\lambda)|^{-2}\, d\lambda$$

Here a prime means a derivative with respect to x or y. We now make the "ansatz" that a solution can be found with $a(0,\Omega) = a'(0,\Omega) = 0$.

Setting $y = 0$ in (8') we obtain for $b(x,\Omega)$ an expression in terms of $\phi_\lambda(x)$, $c(\lambda)$, ρ and the unknown function $a(x,\Omega)$. Plugging this expression back in (8') we conclude that satisfying (8) is equivalent to demanding that the function

$$\Phi(x,y,\Omega) = a(x,\Omega)\left\{\int_0^{\Omega} -(\lambda^2+\rho^2)\phi_\lambda(x)\phi_\lambda(y)|c(\lambda)|^{-2}\right.$$

$$\left. + \frac{\int_0^{\Omega} -(\lambda^2+\rho)\phi_\lambda(x)|c(\lambda)|^{-2} d\lambda}{\int_0^{\Omega} \phi_\lambda(c)|c(\lambda)|^{-2}\, d\lambda} \cdot \int_0^{\Omega} \phi_\lambda(x)\phi_\lambda(y)|c(\lambda)|^{2}\, d\lambda\right\}$$

$$+ a'(x,\Omega)\left\{\int_0^{\Omega} \phi_\lambda'(x)\phi_\lambda(y)|c(\lambda)|^{-2}\, d\lambda\right.$$

$$\left. - \frac{\int_0^{\Omega} \phi_\lambda'(x)|c(\lambda)|^{-2}\, d\lambda}{\int_0^{\Omega} \phi_\lambda(x)|c(\lambda)|^{-2}\, d\lambda} \int_0^{\Omega} \phi_\lambda(x)\phi_\lambda(y)|c(\lambda)|^{-2}\, d\lambda\right\}$$

should be symmetric in the pair (x,y).

It is convenient to introduce the following notation

$$\phi_\lambda(x) = \phi_0(x) + \frac{\lambda^2}{2}\phi_2(x) + \frac{\lambda^4}{4!}\phi_4(x) + \ldots$$

$$|c(\lambda)|^{-2} = p_2\lambda^2 + p_4\lambda^4 + \ldots$$

Of course functions $\phi_{2i}(x)$ and the coefficients c_{2i} depend on the choice of parameters (p,q) which determine every thing else.

A good part of the analysis below applies in total generality. Some finer points hinge on the dependence of the functions $\phi_0(x)$, $\phi_2(x)$, $\phi_4(x)$, introduced above, on the parametrs (p,q). Recall that $c = \frac{p+q+1}{2}$ and set $r \equiv \frac{\rho}{2} = \frac{p+2q}{4}$. We have, for later use,

$$\phi_0(x) = 1 - \frac{r^2x^2}{4c} \frac{(2r^2c - 3r^4 - 6r^3 - r^2)}{96(c^2 + c)} x^4 + \ldots$$

$$\phi_2(x) = \frac{x^2}{2c} = - \frac{x^2}{2c} - \frac{2c - 6r^2 - 6r - 1}{43(c^2 + c)} x^4 - \ldots$$

$$\phi_4(x) = \frac{3}{4} \frac{x^4}{c^2 + c} + \ldots$$

These explicit relations will not be used until much later. In this sense much of what follows can be used to analyze more general situations.

The expansions given above are not hard to do by hand, but they are better carried out using a symbol manipulator. We used VAXIMA.

We also used VAXIMA for some of the manipulations described below.

2. Looking for $a(x,\Omega)$

We try to find $a(x,\Omega)$ in the form

$$a(x,\Omega) = \sum_{n\geq 0} a_{2n}(x)\Omega^{2n}$$

such that the function $\Phi(x,y,\Omega)$ defined earlier turns out to symmetric in (x,y) for every Ω.

Setting

$$\Phi(x,y,\Omega) = \sum_{i\geq 0} \Phi_{2i+1}(x,y)\Omega^{2i+1}$$

one can see that $\Phi_1(x,y) \equiv \Phi_3(x,y) \equiv \Phi_5(x,y) \equiv 0$.

The first useful relation comes from $\Phi_7(x,y)$. We get

$$\Phi_7(x,y) = \left\{\frac{p_2}{175} a_0'(x)(\phi_2'(x) - \phi_2(x)\phi_0'(x)/\phi_0(x)) - 2a_0(x)\phi_0(x)\right\}\phi_2(y)$$

Notice that we get

$$\Phi_7(x,y) = \alpha^{(7)}(x)\beta^{(7)}(y)$$

which leads, from symmetry, to

$$\alpha^{(7)}(x) = A^{(7)}\beta^{(7)}(x).$$

Writing this out and dividing by $\phi_0(x)$ we get

$$\left(\frac{\phi_2(x)}{\phi_0(x)}\right)' a_0'(x) - 2a_0(x) = \frac{175}{p_2} A^{(7)} \frac{\phi_2(x)}{\phi_0(x0} \tag{9}$$

Going on to $\Phi_9(x,y)$ we obtain

$$\Phi_9(x,y) = \alpha_1^{(9)}(x)\beta_1^{(9)}(y) + \alpha_2^{(9)}(x)\beta_2^{(9)}(y)$$

and from the assumed symmetry we deduce

$$\alpha_1^{(9)}(x) = A_{11}^{(9)}\beta_1^{(9)}(x) + A_{12}^{(9)}\beta_2^{(9)}(x)$$

$$\alpha_2^{(9)}(x) = A_{21}^{(9)}\beta_1^{(9)}(x) + A_{22}^{(9)}\beta_2^{(9)}(x)$$

with a symmetric matrix $A_{ij}^{(9)}$.

This is an easy consequence of the symmetry of $\Phi_9(x,y)$ and the linear independence of $\beta_1^{(9)}(x)$, $\beta_2^{(9)}(x)$. In our case

$$\beta_1^{(9)}(x) = \phi_2(x),\ \beta_2^{(9)} = \phi_4(x)$$

The last coefficient which has to be looked at with care is $\Phi_{11}(x,y)$. One gets

$$\Phi_{11}(x,y) = \sum_{i=1}^{3} \alpha_i^{(11)}(x)\beta_i^{(11)}(y)$$

with $\beta_1^{(11)}(x) = \phi_2(x)$, $\beta_2^{(11)}(x) = \phi_4(x)$, $\beta_3^{(11)}(x) = \phi_6(x)$

and we conclude, as before, that

$$\alpha_i^{(11)}(x) = \sum_{j=1}^{3} A_{ij}^{(11)}\beta_j(x)$$

with a symmetric 3×3 matrix $A_{ij}^{(11)}$.

We display now two of the equations obtained in the manner indicated above.

From the equation involving $\alpha_1^{(9)}(x)$, after dividing by $\phi_0(x)$, we get

$$\frac{1}{1890} p_2 a_0'(x)\left(\frac{\phi_4(x)}{\phi_0(x)}\right)' - \frac{3}{1750} p_2 a_0'(x) \frac{\phi_2(x)}{\phi_0(x)}\left(\frac{\phi_2(x)}{\phi_0(x)}\right)'$$

$$(10) \qquad + \frac{23}{7875} p_4 a_0'(x)\left(\frac{\phi_2(x)}{\phi_0(x)}\right)' - \frac{23}{7875} p_2 a_0(x)\left(\frac{\phi_2(x)}{\phi_0(x)}\right)$$

$$- \frac{46}{7875} p_4 a_0(x) + \frac{p_2}{175} a_2'(x)\left(\frac{\phi_2(x)}{\phi_0(x)}\right)' - \frac{2}{175} p_2 a_2(x)$$

$$= A_{11}^{(9)} \frac{\phi_2(x)}{\phi_0(x)} + A_{12}^{(9)} \frac{\phi_4(x)}{\phi_0(x)}$$

From the equation involving $\alpha_2^{(11)}(x)$, after dividing by $\phi_0(x)$, we get

(11)

$$\frac{1}{19404} p_2 a_0'(x) \left(\frac{\phi_4(x)}{\phi_0(x)}\right)' - \frac{1}{6300} p_2 a_0'(x) \frac{\phi_2(x)}{\phi_0(x)} \left(\frac{\phi_2(x)}{\phi_0(x)}\right)'$$

$$+ \frac{73}{242550} p_4 a_0'(x) \left(\frac{\phi_2(x)}{\phi_0(x)}\right)' - \frac{73}{242550} p_2 a_0(x) \frac{\phi_2(x)}{\phi_0(x)}$$

$$- \frac{73}{121275} p_4 a_0(x) + \frac{p_2}{1890} a_2'(x) \left(\frac{\phi_2(x)}{\phi_0(x)}\right)' - \frac{p_2}{945} a_2(x)$$

$$= A_{21}^{(11)} \left(\frac{\phi_2(x)}{\phi_0(x)}\right) + A_{22}^{(11)} \frac{\phi_4(x)}{\phi_0(x)} .$$

Here we have used the fact that $A_{23}^{(11)} = 0$. This comes from the fact that the equation for $\alpha_3^{(11)}(x)$ combined with (9) give $A_{32}^{(11)} = A_{33}^{(11)} = 0$. The symmetry of $A^{(11)}$ does the job.

Eliminating $a_2(x)$ from (10) and (11) we get

$$- \frac{10}{2079} p_2 a_0'(x) \left(\frac{\phi_4(x)}{\phi_0(x)}\right)' + \frac{40}{693} p_2 a_0(x) \frac{\phi_2(x)}{\phi_0(x)}$$

$$+ 1890\, A_{22}^{(11)} \frac{\phi_4(x)}{\phi_0(x)} - 175\, A_{12}^{(9)} \frac{\phi_4(x)}{\phi_0(x)}$$

$$= \frac{40}{693} p_4 a_0'(x) \left(\frac{\phi_2(x)}{\phi_0(x)}\right)' - \frac{80}{693} p_4 a_0(x)$$

$$- 1890\, A_{21}^{(11)} \frac{\phi_2(x)}{\phi_0(x)} + 175\, A_{11}^{(9)} \frac{\phi_2(x)}{\phi_0(x)}$$

Now we claim that the right hand side of this equation vanishes identically. Using (9) one can replace the first two terms on the right hand side of (12), and then all the right hand side of (12), by a scalar multiple of ϕ_2/ϕ_0. We claim that this scalar is zero.

Indeed if $a_0(x)$ is regular at $x = 0$ it follows from (9) that $a_0(x) \approx \mathbb{C}x^2$ for small x. This combined with

$$\frac{\phi_2(x)}{\phi_0(x)} \sim - \frac{x^2}{2c} \quad \text{and} \quad \frac{\phi_4(x)}{\phi_0(x)} \sim \frac{3}{4}\, \frac{x^4}{c^2 + c}$$

gives the desired result by plugging in (12).

Using this and some scaling we have replaced (12) by

$$(12') \qquad a_0'(x) \left(\frac{\phi_4(x)}{\phi_0(x)}\right)' - 12 a_0(x) \frac{\phi_2(x)}{\phi_0(x)} = \tilde{k} \frac{\phi_4(x)}{\phi_0(x)}$$

From here one can proceed in many different ways.

If one is interested in the special case under discussion here one can solve for $a_0(x)$ in (9), and insert the resulting function in the left hand side of (12'). Expanding the ratio of this expression by ϕ_4/ϕ_0, setting equal to zero the coefficients of x^2 and x^4 one obtains a set of two algebraic equations in c, r with solutions

$$c = 0, \quad r = 0$$

$$c = -2, \quad r = -1 \pm i \frac{\sqrt{6}}{3}$$

$$c = -1, \quad r = -\frac{1}{2} \pm \frac{i}{2}$$

$$c = \frac{3}{2}, \quad r = 1$$

$$c = \frac{3}{2}, \quad r = \frac{1}{2}$$

$$c = 2, \ r = 1$$

$$c = \frac{1}{2}, \ r = 0$$

$$c = \frac{1}{2}, \ r = \frac{1}{0}.$$

Recalling that

$$c = \frac{p + q + 1}{2}; \quad r = \frac{p + 2q}{4}; \qquad p,q \geq 0; \quad p + q > 0$$

one can rule out the first three and the last two solutions.

The fourth and fifth correspond to $p = 0, \ q = 2$ and $p = 2, \ q = 0$ respectively.

The sixth solution gives $p = 2 \ q = 1$.

A look at the list in pages 30-32 of [8] shows that $p = 0$ $q = 2$ has no geometrical meaning, $p = 2 \ q = 0$ corresponds to

$SO(3,1)/SO(3)$

and finally, $p = 2 \ q = 1$ corresponds to

$SU(1,1)/SU(1)$

Unfortunately this case has to be ruled out since the coefficient of x^6 does not vanish, and we are left with $SO(3,1)/SO(3)$ as the only viable case.

A more general approach is to solve (9) and (12') for $a_0(x)$ and $a_0'(x)$ and obtain in this fashion an equation relating $\phi_0(x)$, $\phi_2(x)$ and $\phi_4(x)$. From the known behavior of these functions one can see again that only $(p,q) = (0,2)$ or $(2,0)$ are allowed. This method can be used in more general situations.

The conclusion is that at least for the class of second order differential operators considered here, only the case $SO(3,1)/SO(3)$ leads to a differential operator of order two which commutes with the integral operator K_Ω in (5).

We mention that for $p = 2, \ q = 0$ the eigenfunctions are given

$$\phi_\lambda(x) = \frac{\sin \lambda x}{\lambda \sinh x}$$

and we have

$$\frac{1}{\lambda^2} \frac{d}{d\lambda}(\lambda^2 \frac{d}{d\lambda} \phi(\lambda,r)) = -r^2 \phi(\lambda,r)$$

This "differential equation in the spectral parameter" does not hold for any other example of p,q: a remarkable coincidence. Such an equation is valid every time that the Slepian-Landau-Pollak phenomenon holds, see [10,21], and is the springboard for the next and last section.

7. Differential equations in the spectral parameter

Given a second order differential operator

$$L \equiv - \frac{d^2}{dx^2} + V(x)$$

denote by $\phi(\lambda,x)$ a family of eigenfunctions of L, $L\phi = \lambda^2\phi$. The previous consideration lead us to the following

Question. For what $V(x)$ does there exist a differential operator D_λ such that

$$D_\lambda \phi(\lambda,x) = \Theta(x)\ \phi(\lambda,x)$$

We have attacked this question in collaboration with H. Duistermatt [22].

We give below some few examples of this situation leaving a more complete discussion for [22].

(a) Take $V(x,t) = \frac{6x(x^3-2t)}{(x^3+t)}$, then if

$$\phi(\lambda,x) = \frac{(\lambda^2 x^3 + 3i\lambda x^2 - 3x + \lambda^2 t)}{\lambda^2 (x^3 + t)}\, e^{i\lambda x}$$

we've

$$-\left(\frac{d^2}{dx^2} + V(x,t)\right)\phi(\lambda,x) = \lambda^2\phi(\lambda,x)$$

and

$$\left(\left(\frac{d}{d\lambda} - \frac{2}{\lambda}\right)\left(\frac{d}{d\lambda} + \frac{2}{\lambda}\right)\right)^2 - 4it\frac{d\phi}{d\lambda} = (x^4 + 4tx)\phi$$

(b) Take $V(x,t,s) = \dfrac{12x^{10} + 324sx^5 + 450t^2x^4 + 300t^3x + 162s^2}{(x^6 + 5tx^3 + 9sx - 5t^2)^2}$

then for an appropriate family of eigenfunctions $\phi(\lambda,x)$ we've

$$\left[i\left(\frac{d}{d\lambda} - \frac{3}{\lambda}\right)\left(\frac{d}{d\lambda} - \frac{2}{\lambda}\right)\left(\frac{d}{d\lambda} - \frac{1}{\lambda}\right)\frac{d}{d\lambda}\left(\frac{d}{d\lambda} + \frac{1}{\lambda}\right)\left(\frac{d}{d\lambda} + \frac{2}{\lambda}\right)\left(\frac{d}{d\lambda} + \frac{3}{\lambda}\right)\right]\phi$$

$$+ \frac{63}{2}s\left(\frac{d}{d\lambda} - \frac{1}{\lambda}\right)\left(\frac{d}{d\lambda} + \frac{1}{\lambda}\right)\phi - 35it^2\frac{d}{d\lambda}\phi$$

$$+ \frac{35}{4}t\left[\left(\left(\frac{d}{d\lambda} - \frac{2}{\lambda}\right)\left(\frac{d}{d\lambda} + \frac{2}{\lambda}\right)\right)^2 - \frac{36}{\lambda^4}\right]\phi$$

$$= (x^7 + \frac{35}{4}tx^4 - \frac{63}{2}sx^2 - 35t^2x)\phi$$

We make the observation that the function $V(x,t)$ in example (a) is a solution of the Korteweg-deVries equation. Also $V(x,t,s)$ in example (b) can be written in the form

$$e^{tX_1+sX_2}\left(\frac{12}{x^2}\right)$$

where X_1, X_2 denote the Kdv flow and the fifth order differential operator given by $[(-L^{3/2})_+, L]$ and $[(-L^{5/2})_+, L]$ respectively, (up to scaling). See [23-28].

As one may suspect this works for the whole "hierarchy." More interesting still, not all examples are obtained this way. For much more detail about this whole development see [22,29].

References

[1] Grünbaum, F.A. A study of Fourier space methods for "limited angle" image reconstruction. Numerical Functional Analysis and Optimization 2 (1) (1980), 31-42.

[2] Marr, B. Private communication.

[3] Slepian, D. Prolate spheroidal wave functions, Fourier analysis and uncertainty. Bell System Tech. Journal 57, No. 5 (1978).

[4] Davison, M.E. The ill conditioned nature of the limited angle tomography problem. SIAM J. Applied Math., 43, 2, (1983) 428-448.

[5] Davison, M.E. A singular value decomposition for the Radon transform in n-dimensional euclidean space. Numer. Funct. Anal. and Optimiz. 3(3) (1981) 321-340.

[6] Slepian, D., Pollak, H.P. Prolate spheroidal wave functions, Fourier analysis and uncertainty I. Bell System Tech. Journal 40, No. 1 (1961) 43-64.

[7] Landau, H.J., Pollak, H.O. Prolate spheroidal wave functions, Fourier analysis and uncertainty II. Bell System Tech. Journal 40, No. 1 (1961) 65-84.

[8] Landau, H.J., Pollak, H.O. Prolate spheroidal wave functions, Fourier analysis and uncertainty III. Bell System Tech. Journal 41, No. 4 (1962) 1295-1336.

[9] Slepian, D. Prolate spheroidal wave functions, Fourier analysis and uncertainty IV. Bell System Tech. Journal 43, No. 6 (1964) 3009-3058.

[10] Grünbaum, F.A., Longhi, L., Perlstadt, M. Differential operators commuting with finite convolution integral operators: some nonabelian examples. SIAM J. Applied Math., 42, 5, (1982) 941-955.

[11] Whittaker, E.T. Proc. London Math. Soc. (2) 14,(1915) 260-268.

[12] Ince, E.L. 'On the connection between linear dif. systems and integral equations', Proc. of the Royal Society of Edinburgh (42),(1922) pp. 43-53.

[13] Mehta, M.L. Random matrices, Academic press, N.Y., 1967.

[14] Slepian, D. 'Prolate spheroidal wave functions, Fourier analysis and uncertainty', Bell System Tech. Journal 57, No. 5, (1978), 1371-1430.

[15] Grünbaum, F.A., 'Eigenvectors of a Toeplitz matrix: discrete version of the prolate spheroidal wave functions', SIAM J. Alg. Disc. Math. 2(2), (1981) 136-141.

[16] Grünbaum, F.A. 'Toeplitz matrices commuting with a tridiagonal matrix', Linear Algebra and its Applications 40, (1981) 25-36.

[17] Grünbaum, F.A., 'A remark on Hilbert's matrix', Linear Algebra and its Applications, 43, (1982), 119-124.

[18] Grünbaum, F.A. 'The eigenvectors of the discrete Fourier transform: a version of the Hermite functions', J. Math. Anal. Applic. 88, (1982), 355-363.

[19] Grünbaum, F.A., 'Doubly concentrated functions for three hyperbolic space', IHES preprint, Bures-sur Yvette, France (1982).

[20] G. Warner, Harmonic analysis on semisimple Lie groups, Vol. II, Springer Verlag (1972).

[21] Grünbaum, F.A. to appear, 'A new property of reproducting kernels for classical orthogonal polynomials', J. Math. Anal. Applic. 95, (1983).

[22] Duistermaat, J. and F.A. Grünbaum: in preparation, 'Differential equations in the eigenvalue parameter'.

[23] Lax, P. 'Integrals of nonlinear equations of evolutions and solitary waves', Communications on Pure and Applied Mathematics 21, (1968), 467-490.

[24] Gelfand, I. and Dikii, 'Fractional powers of operators and Hamiltonian systems', Funkts. Anal. Prilozhen, 10, 4, (1976), 13-39.

[25] H. Airault, H. McKean, and J. Moser, 'Rational and elliptic solutions of the Korteweg-deVries equation and a related many body problem', Communications in Pure and Applied Math.(30), (1977), 95-148.

[26] D. Chudnovski and G. Chudnovski, 'Pole expansions for nonlinear partial differential equations', Nuovo Cimento 40B, (1977), 339-353.

[27] M. Adler and J. Moser, 'On a class of polynomials connected with the Korteweg-deVries equation', Communications in Mathematical Physics (61), (1978), 1-30.

[28] M. Ablowitz and H. Airault, 'Perturbations finies et forme particuliere de certaines solutions de l'equation de Korteweg-deVries. C.R. Acad. Sci. Paris t. 292, (1981), 279-281.

[29] Grünbaum, F.A., Band-time limiting, recursion relations and nonlinear evolution equations, a Chapter in "Special functions: group theoretical aspects and applications, R.A. Askey et al. (eds.), D. Reidel, Holland (1984), p. 271.

II.3 (IM.1)

DIRECT AND INVERSE HALFSPACE SCALAR DIFFRACTION: SOME MODELS AND PROBLEMS

Giovanni Crosta

Istituto di Cibernetica dell'Università
Via Viotti 5 - I-20133 MILANO (Italy)

We consider herewith two main items: a) approximations of Helmholtz equation and b) aperture identification. We begin with Helmholtz equation , which is a model of the direct diffraction process. If we approximate Rayleigh-Sommerfeld's propagator in the spatial frequency domain we get the paraxial propagator , whereas if we approximate it in the spatial domain (far field) we get both the Fresnel propagator and the Fourier optics formula. We show the Fresnel propagator formally coincides with the paraxial one. We investigate to which extent approximate solutions are physically related to the exact one. Then we deal with some consequences on the so-called "superresolution" methods, which aim at identifying some subwavelength details in the aperture field.We show Fourier optics is no adequate model for superresolution. As of aperture identification, we deal with the system theoretical approach to the problem, which consists of minimising an adequate functional containing both an estimation error and a regularising term. We also show how duality theory helps in designing a minimisation algorithm based on the discrete gradient. We apply this method to identification of a Dirichlet boundary condition in two cases: where either complex field or intensity data are available.The latter is closely related to the phase reconstruction problem, for which we write the functional to be minimised . Several open problems are also presented.

W.-M. Boerner et al. (eds.), Inverse Methods in Electromagnetic Imaging - Part 1, 299–318.

TABLE OF CONTENTS

INTRODUCTION

The solution of any inverse problem consists of an interaction between a natural system (S) and a model (R), which on its turn is affected by the interaction process until model predictions are believed to be accurate with respect to observations performed on the natural system.

If we assume both systems , S and R , are described by maps, say σ and ϱ , which relate their respective inputs and outputs, then according to Conant and Ashby's definition (1970):

R is a model of S if there is an isomorphism between σ and ϱ. The same Authors also give an optimality criterion for system R to predict accurately the output of S when the input to both is known:

R must be a model of S.

These ideas affect our behaviour in describing natural processes related to electromagnetic phenomena and diffraction in particular, our aim being to solve what is named an inverse problem. This term means

either: the identification of the system parameters, in order to construct or improve a model,

or the identification of an input when both the model and the output are known.

The need for accurate models of the so-called direct problems, which consist of evaluating the electromagnetic field at a point when field sources, medium properties and boundary conditions are known, has paradoxically been emphasised by the failure of some inversion methods based on oversimplified models. As an example we shall deal with approximate solutions of Helmholtz equation in the halfspace.

1. MODELS OF DIRECT ELECTROMAGNETIC PROCESSES

The usual setting is in a classical framework, where field equations seem easy to solve , especially when the interaction between the fields and the detector is neglected. Actually, since experimental data are yielded by the detector, we expect some difficulty to arise precisely from this simplification when a data inversion procedure must be applied.

The ignorance about the photodetection process will therefore add "noise" to data , together with all other unknown processes for which we are too lazy to work out a model.

Let us assume Maxwell's equations for free fields make sense. At this point we choose

either to deal with the ($\underline{B}$,$\underline{E}$) tensor and be ready to solve the coupled equations it entails, by thus keeping track of polarisation effects (see e.g. Boerner,1980,1981 and the Radar Polarimetry section of this volume),

or to deal with a scalar equation , we believe to represent the process adequately. We thus come to an equation in a single function w, which may be e.g. a component of $\underline{E}$.

Then we get rid of the time dependence, which we assume to be $\exp(i\omega t)$ and arrive at Helmholtz equation. If only one component of $\underline{E}$, say E_z , were non-vanishing then Helmholtz equation alone would be sufficient. This corresponds however to two-dimensional (x,y) problems. Halfspace diffraction from an arbitrary aperture could not be dealt with. The assumption we make is then to use Helmholtz equation for the complex scalar field w instead of three such equations together with the supplementary condition on div E.

Actually w could also stand for a component, say A_z , of the vector potential A , but not the scalar potential Φ , as it can be easily shown by working e.g. in the Coulomb gauge, where div$\underline{A}$= = 0.

We assume our detectors are energy sensitive. Measured quantities are functions of $|w|^2$ both if $w = E_z$ or if $w = A_z$.

In the latter case $|E_z|^2 \approx \omega^2 |A_z|^2$, because Φ does not depend on time. All of this physically motivates energy norms in the Hilbert spaces we shall work with.

A further support towards adopting Helmholtz equation is the relative abundance of existence, uniqueness and regularity theorems for several geometries and Hilbert or Sobolev space problem formulations.

2. HELMHOLTZ EQUATION IN THE HALFSPACE

There is a well-known procedure which leads to a unique solution w of Helmholtz equation in the empty halfspace R^3_+:

$$(\text{divgrad} + k^2)\, w = 0 \qquad \text{in } R^3_+ ,$$

subject to Dirichlet boundary conditions (BCs) at the z = 0 plane (also named R^3_o, where position is denoted by (x_1,x_2)):

$$w|_{R^3_o} = v \in U ,$$

and to some asymptotic conditions (radiation conditions or RCs for short) which also affect the radial derivative w_r :

$$|w| = O(r^{-1}) \quad \text{as } r \to \infty, \quad \forall \text{ colatitude angle } \Theta ,\ \forall \varphi \text{ azimuth angle } \varphi ,$$

$$|w_r - ikw| = o(r^{-1}) \qquad 0 < \Theta \le \pi/2 - c, c > 0, \forall \varphi ,$$

$$|w_r| = O(r^{-1}) \qquad \pi/2 - c < \vartheta < \pi/2, \forall \varphi .$$

The Rayleigh-Sommerfeld propagator

$$g_{1n} := \partial_n g_1 = \partial_n \left[(1/r)e^{ikr} - (1/r^*)e^{ikr^*}\right]_{R^3_o}$$

is introduced, where ∂_n stands for the normal derivative operator and r* is the distance between the origin and the point x* symmetrical of x with respect to R^3_o. On its turn g_1 is the fundamental solution or Green's function of the problem, i.e. the right and left inverse (Hoermander,1963, pp 249-251) of the differential operator $(\text{divgrad} + k^2)$ plus BCs we are considering.

w is yielded by

$$w(\underline{x}_o) = (1/2\pi) g_{1n} * v , \tag{1}$$

where now * stands for convolution over R^3_o. Existence of w then is insured wherever the convolution integral makes sense. The set U of boundary conditions v can therefore be defined by giving e.g. a number of sufficient specifications. Let us assume the support of v is a bounded subdomain S of R_o^3, which corresponds to a finite aperture in a plane screen. Moreover we take the aperture field to behave regularly enough. One possible constraint is :

$$v \in W^{1,\infty}(S), \tag{2}$$

which implies the boundedness of v, its first partial derivatives and its energy integral, i.e. $v \in L^2(S)$. On the other hand, if the support of v is unbounded, some asymptotic conditions must hold for v, which are equivalent to radiation conditions in R_o^3. We shall however avoid this case, which to our present knowledge has nowhere been dealt with in a satisfactory way, particularly as far as the existence proof is concerned. Boundary conditions having an unbounded support can be alternatively dealt with by the limit amplitude or the limit absorption principles (Ramm, 1978). We stress that a solution w of Helmholtz equation yielded by either principle need <u>not</u> comply with the radiation conditions, because the asymptotic requirements on v may differ.

3. APPROXIMATIONS OF HELMHOLTZ EQUATION

3.1 General

It is well-known that the convolution integral (1) gives rise to a distinction between homogeneous and evanescent waves . The corresponding procedure is rather straightforward and we shall avoid the details. We need however some notations:

$\underline{y} := (y_1, y_2)$ is the two-dimensional spatial frequency vector, i.e. a position vector in the reciprocal domain,

D_k is the disk $\{ \underline{y} \mid |\underline{y}| \leq k \}$,

$Fw := \hat{w}$ is the Fourier transform (FT) of w with respect to x_1 and x_2.

The FT of all operands in (1) is defined, in particular because v is in U of (2) . By means of tables (Jahnke, Emde, Loesch, 1960) we get:

$$w = \begin{cases} F^{-1}\exp(iz\sqrt{k^2-y^2})Fv & ; |\underline{y}| \leq k \\ \\ F^{-1}\exp(-z\sqrt{y^2-k^2})Fv & ; |\underline{y}| > k . \end{cases} \tag{3}$$

Since S is bounded, the support of $\hat{v}$ is not, hence the aperture will contribute both homogeneous and evanescent waves to

w. We shall decompose $\hat{v}$ as the sum of two parts (Montgomery, 1968), according to the following diagram:

$$F : L^2(R^2) \supset W^{1,\infty}(S) \longrightarrow L^2(R^2) := B_H \oplus B_E$$
$$v \longmapsto \hat{v} := \hat{v}_H + \hat{v}_E$$

where supp $\hat{v}_H \subseteq D_k$, supp $\hat{v}_E \subseteq R^2 \setminus D_k$; therefore v_H and v_E mean the inverse FTs of $\hat{v}_H$ and $\hat{v}_E$ respectively.

The exponential operators appearing in (3) can be commuted with F , by extending the relationship between the multiplication operator ($Q^\wedge$) in the reciprocal domain and the differentiation operator ($P^\wedge$) in the spatial one (Jordan, 1969, p 64):

$$Q^\wedge F = FP^\wedge .$$

This leads to the following result:

$$w = w_H + w_E = \exp\left[ikz\sqrt{1+\mathrm{divgrad}_T/k^2}\right]. v_H + \exp\left[-kz\sqrt{-1-\mathrm{divgrad}_T/k^2}\right]. v_E ,$$

where $\mathrm{divgrad}_T$ is the laplacian with respect to x_1 and x_2.

It can now be easily understood that if the last term on the right hand is neglected, the corresponding field w_H need no longer satisfy the Helmholtz equation defined in Section 2. w_H belongs to the class of bandlimited fields , obtained by applying the Rayleigh-Sommerfeld propagator to a function v_B . The latter comes from restricting supp $\hat{v}$ of a BC v (which yields an acceptable solution w) to the bounded domain $D_B \subset R^2$. Bandlimited fields usually fail to satisfy the radiation conditions mentioned above. A sufficient argument is given by Baltes et al. (1982), who investigate the asymptotic behaviour of w in the z direction and on the R^3_o plane. A general proof relating to an arbitrary direction in R^3_o is not known to us.

Yet, it makes sense to introduce the following abstract evolution problem, of which w_H is the formal solution:

$$\begin{cases} -i\partial_z w_H = k\sqrt{1 + \mathrm{divgrad}_T/k^2} \; . \; w_H \\ w_H(.,z=0) = v_H \in B_H , \end{cases} \tag{4}$$

where we have chosen v_H in B_H to stress that BCs used here need not have a bounded support.

3.2 Approximations in the reciprocal domain

The above equation is the starting point for an approximation step in the reciprocal domain , leading to the paraxial equation. If

$$\text{supp } \hat{v}_H = D_b := \{\underline{y} \in R^2 \mid y^2/k^2 < b << 1\},$$

then we perturb the infinitesimal generator in equation (4) by expanding the square root and neglecting all but the first two terms. We then get another abstract evolution problem governed by the paraxial equation (Deschamps,1981) :

$$\begin{cases} -i\,\partial_z\, w_P(.,z) = k(1 + \text{divgrad}_T)\, w_P(.,z) & (5) \\ w_P(.,z=0)=v_P & (6) \\ \text{supp } v_P = D_b, & \end{cases}$$

the solution of which is :

$$w_P= e^{ikz(1+\text{divgrad}_T/2k\,)} .v_P := P_P . v_P. \quad (7)$$

The paraxial propagator P_P is both right and left inverse of the differential operator defined by (5) and (6) , as it could be easily shown.
The paraxial equation be also arrived at from Helmholtz's by a slightly different procedure (Bellman & Kalaba, 1959; Corones, 1975; Fock, 1960). We now examine the domain of

P_P: it is a non dense subset of $L^2(R^2)$, hence the operator P_P can

be extended to the whole of $L^2(R^2)$ in several ways. Let P_P^e denote the extended operator. Particularly relevant to inverse diffraction is the following alternative:

$$\text{e1)}\quad P_P^{(e1)} = P_P \text{ of } (7),\ \forall\, v \in L^2(R^2), \quad (8)$$

$$\text{e2)}\quad P_P^{(e2)} = P_P \quad \text{if supp } \hat{v} \subseteq D_b$$

$$= 0 \quad \text{elsewhere.}$$

Only extension e2) preserves the connection between the original Helmholtz equation and the approximated one.
As an example let us consider the following boundary condition:

$$v:= \cos(k_o x_1).\cos(k_o x_2).\ \ X_R(S),$$

where $k_o > k$, and $X_R(S)$ is a regularised characteristic function of S, such that v satisfies (2). If v is acted on by the Rayleigh

- Sommerfeld propagator, the energy of the corresponding $w(v)$ decays very rapidly as z increases, because most of the BC energy lies outside D_k. On the other hand if the extended propagator of (8), which is unitary, is used, the energy of the corresponding field $w_1(v)$ does not decrease. Then, at a given $z >> 0$ the relative estimation error :

$$\frac{\| w_1(z) - w(z) \|_{L^2(R^2)}}{\| w(z) \|_{L^2(R^2)}} \quad ; \qquad z >> 0$$

we make by taking $w_1(.)$ for $w(.)$, becomes very large.
Whenever we choose to model direct diffraction by the paraxial equation because, say, it leads to simpler computations, we must be aware of the hypotheses which led to it.
Let us now deal with another example of diffraction data inversion by the paraxial propagator. w_M is the measured field at a given plane $z=z_M$. After inverse propagation, the estimated $\hat{v}_M$ may well have a support extending outside D_b , as a consequence of the already mentioned processes we cannot model (photodetection ,apparatus behaviour etc.). The physical plausibility constraint we must enforce is then to reject the values $\{ v_M(\underline{y}) \mid |y| \notin D_b \}$. We have traded computational simplicity for accuracy in the reconstructed aperture field.

3.3 Approximations in the spatial domain

Another way of perturbing the operator relating v to w is to neglect some terms in g_{1n} of (1), which can be written as $\partial_z((1/s)e^{iks})$, where

$$s := \sqrt{(x_1-x_{1a})^2+(x_2-x_{2a})^2+z^2}; \quad (x_{1a},x_{2a}) \in R^3_o. \tag{9}$$

If we are interested in the far-field, we start with approximating $z/s \cong 1$ and leave the phase factor unchanged. We could easily show that we eventually get a propagator which still tells homogeneous waves apart from evanescent ones, but is undefined at $|\underline{y}| = k$. Moreover it solves the following Neumann problem:

$$\begin{cases} (\text{divgrad} + k^2)w=0 \\ w_n \big|_{R^3_o} = v \; ; \; \text{supp } v \subset\subset R^3_o \\ \text{radiation conditions.} \end{cases}$$

The consequence of a seemingly innocuous approximations yields a far-field solution which does not keep track of the type of BC we have to identify. If in addition we expand to first order the square root in (9) we arrive at the well-known Fresnel optics formula:

$$w_F = (-ik/2\pi z).\exp\left(ikz\left(1+ \frac{x_1^2 + x_2^2}{2z^2}\right)\right) * v(x_1,x_2).$$

By means of FT tables we can show that:

$$w_F = w_P$$

defined in Section 3.2, i.e. that the solution of the paraxial equation and the Fresnel approximate the original w in the same way. The same plausibility constraints defined in the preceeding Section hold here too. To greater extent this applies to the Fourier optics formula, which results from further simplifying (9). The consequences will be analysed in the following Section.

4. SUPERRESOLUTION, FOURIER OPTICS AND OTHER METHODS

In our case we define by superresolution any method aiming at estimating $\hat{v}_E$ starting from measured field data. $\hat{v}_E$ contains the so-called sub-wavelength details. In the realm of Fourier optics it has been often suggested that (Goodman,1968,pp 133-136) if the field w_M defined above were measured in an adequate way, then $\hat{v}_M$ could be estimated with arbitrarily high accuracy , even for $|\underline{y}| > k$. The procedure was believed to hold if a numerable set of values, the samples $\{w_M(\underline{x})\}$ were available . The link between the latter and the unknown function $\hat{v}_M$ with bounded support being due to Shannon's theorem. From the remarks of Section 3.3, nothing physically meaningful can however be achieved if supp $\hat{v}_M$ extends outside D_k. The constraint on supp $\hat{v}_M$ is a necessary condition for the Fourier optics solutions to approximate w. Then we see Shannon's theorem used in the wrong context. No wonder then if a superresolution algorithm based on Fourier optics yields some results in the values $\{\hat{v}_M(\underline{y}) \mid |\underline{y}| > k\}$ the dependence of which on slight perturbations of $\{w_M(.)\}$ cannot be physically motivated. A model is being used outside its validity range.

Shannon's theorem can be easily linked to other results available for analytic functions, i.e. Plancherel-Polya's theorem stating that the FT ,now denoted by u, of a function f with bounded support is analytic. Then we can apply in principle at least, some analytic continuation method to get u outside the region where its values have been given and eventually reconstruct f everywhere. This holds in particular if u satisfies Helmholtz equation. Necessary and sufficient conditions for the latter to have analytic solutions are also well-known: Sherman (1969) defines as source free a scalar field u which results from superposing homogeneous waves only , i.e. such that supp $\hat{u}$ $\equiv$ $\equiv$ supp f $\subset$ D_k . He then shows , we quote, u "has all the properties exhibited by an entire function", to which analytic continuation applies. We see that analiticity of the scalar field u is at variance with supp $\hat{u}$ extending outside D_k: this is another argument against superresolution based on analytic continuation.

Any approach liable to yield anything meaningful must therefore start with a different model , which does not neglect evanescent waves. As an example, let us mention the algorithms based on nonuniform plane wave expansion. The convolution (1) can be carried out with respect to the wavevector angles Θ and φ in the domain $0 \leq \varphi < 2\pi$, $0 \leq \mathrm{Re}\,\Theta < \pi/2$, $\mathrm{Im}\,\Theta < 0$, which are linked to the spatial frequencies by the class C^{∞} diffeomorphism:

$$\begin{cases} y_1 = (k/2\pi)\ \sin\Theta\ \cos\varphi \\ y_2 = (k/2\pi)\ \sin\Theta\ \sin\varphi . \end{cases}$$

Baltes,Schmidt-Weinmar and coworkers (see e.g. Schmidt - Weinmar,1978) have thoroughly investigated both theoretically and numerically, the role of this expansion in superresolution algorithms and have given some numerical noise sensitivity results, which should be interpreted by the general stability theory (Bertero,DeMol,Viano,1980).

5. SYSTEM THEORY AND INVERSE DIFFRACTION

5.1 Control and observation

Even in the crudest approach towards the model of a system,we need a distinction between input and output, i.e. between control and observation. Let us consider the direct problem for Helmholtz equation discussed in Section 2: since it consists of finding w everywhere in R^3_+ given the BC v, it makes sense to define v the input or the boundary control, and w the system state. Therefore U is the input set and W, to which w belongs, the state space. The direct problem is then a control problem, because the control map B :

$$B : U \longrightarrow W$$
$$v \longmapsto w(v)$$

is well-defined, thanks to radiation conditions which insure uniqueness of w(v). If we want the field at x_o , we perform the so-called point observation, hence we introduce the output or observation map C_o :

$$C_o : W \longrightarrow C$$
$$w \longmapsto w(\underline{x}_o).$$

The procedure which leads from v to $w(\underline{x}_o)$ can be interpreted in the system language and by introducing the adjoint system. Let us assume control and observation can be interchanged . Then we can define the following system:

$$\begin{cases} (\mathrm{divgrad} + k^2)\ p = \delta(\underline{x} - \underline{x}_o)\ ;\ \underline{x}_o \text{ in } R^3_+ \\ p|_{R^3_o} = 0 \\ \text{radiation conditions.} \end{cases}$$

A source term , i.e. an input, has appeared where we wanted to observe w(.) ; a homogeneous BC holds where w was controlled. We say p is the state vector adjoint to w. If we write the second Green's formula for w and p , we get precisely $w(\underline{x}_o)$ as a function of v. In this case p coincides with the already defined fundamental solution g_1 of Helmholtz equation. Green's formula is precisely the relationship which links w and p in a sesquilinear form (linear in w and antilinear in p). This extends to other control and observation maps and is a result of duality theory. Actually duality has a wider meaning either in cathegory theory (Goguen & Varela,1979), system theory (Delfour & Mitter,1972), or functional analysis and the calculus of variations in particular (Ekeland & Temam,1977;Lions,1971). We note in passing that if

the sesquilinear form is the inner product in $L^2(.)$, then the standard adjoint operators and vectors appear.

5.2 Variational approach

We have introduced the adjoint system because by solving it at the same time as the w-system we restate inverse diffraction problems as the variational approach suggests. The latter leads to the minimisation of some functional. For the general theory we refer to Lions's (1971) textbook.

As an example let us assume data, hereinafter denoted by z(.), have been measured in a domain D and that we have to find v. Then the problem becomes :

$$\text{minimise } J(v) := \int_D |Cw(v) - z|^2\, dD + a\, \|v\|^2_{Uad}$$

$$\forall\, v \in U_{ad}$$

where :

U_{ad} is the admissible input set, to be adequately defined,

w(v) is the field corresponding to an estimated v,
C is the observation map which turns w(v) into something comparable to z(.),
$a > 0$.

The input u in U_{ad} such that :

$$J(u) = \inf_{v \in U_{ad}} J(v)$$

if it exists, is named an optimal control and corresponds to a solution in the least-squares sense of the identification problem.

This setting is widely established and is known by the names of some Authors such as Lions and Chavent (see e.g. Chavent, 1977), Miller (1970), Tikhonov and Arsenin (1976). The standard optimal control theory is however based on real valued of real valued variables . In our case we have a real functional physically related to error and control "energies", which depends on complex valued fields. Since we must minimise J(.) by a gradient method, we first verify it is continuous and differentiable , then apply some differentiation rules which may easily be implemented by a numerical scheme. Let us assume the observation map is :

$$C_1 : W \longrightarrow Y_1$$

$$w \longmapsto C_1 w := w(\underset{\rightarrow}{x} \in D),$$

where D is a three-dimensional domain.
Let us the consider the first term in the functional $J_1(v)$ defined by:

$$J_1(v) := \int_D |w - z|^2 \, dD + a \|v\|^2_{Uad} := $$
$$= J_{11} + J_{12},$$

i.e. J_{11}. We write it as an inner product in $L^2(D)$, which makes sense if D is bounded. We recall that if D were unbounded, we would have to work in a weighted Hilbert space such as Vogelsang 's (1975), because a theorem (Ramm,1981) states Helmholtz equation in an unbounded domain , say T, has no nonvanishing solutions in $L^2(T)$. Physically available data refer however to bounded domains, hence our model is satisfactory. Then :

$$J_{11} = \int_D (w^* - z^*)(w - z) \, dD = \langle w - z \mid w - z \rangle_{L^2(D)},$$

where w^* is the complex conjugate of w. If we formally differentiate J_{11} with respect to w, we get:

$$\delta J_{11} = \int_D dD\ (w^*-z^*)\ \delta w + (w-z)\ \delta w^* =$$
$$= 2\ \mathrm{Re}\int_D dD\ (w^* - z^*)\ \delta w. \qquad (10)$$

We have no linear operator acting on δw, hence we cannot introduce the Gateaux derivative of J_{11} with respect to w. Nor it makes sense to introduce a direction independent derivative as it does for analytic functions. The way out of this difficulty is suggested by Lagrangian theory : we double the degrees of freedom and deal both with w^* and w as independent variables. Equivalently we may take the real (Re w) and imaginary (Im w) parts separately, according to a straightforward transformation, which can be made unitary:

$$\begin{bmatrix} w \\ w^* \end{bmatrix} = \begin{bmatrix} 1 & +i \\ 1 & -i \end{bmatrix} \cdot \begin{bmatrix} \mathrm{Re}\ w \\ \mathrm{Im}\ w \end{bmatrix}$$

Since Re w and Im w are real valued functions, we can properly define a <u>Gateaux derivative</u> of J_{11}, i.e. a bra vector :

$$J_{11} := \int_D \begin{bmatrix} \mathrm{grad}_w J_{11} & \mathrm{grad}_{w^*} J_{11} \end{bmatrix}^* \begin{bmatrix} w \\ w^* \end{bmatrix} dD.$$

From now on we shall denote the pair (w,w^*) by $\underline{w}$ and other pairs accordingly. Then

$$\delta J_{11} := < \underline{\mathrm{grad}_{\underline{w}} J_{11}} | \delta \underline{w} > \quad ,$$

which gives sense to the last term in (10). A similar procedure holds for another continuous and differentiable functional , J_{22} related to observation map C_2 :

$$C_2 : W \longrightarrow Y2$$
$$w \longmapsto C_2 w := |w(\underline{x} \in D)|^2,$$

whereby field <u>intensity</u> is given. The task is to find v, thus solving an inverse diffraction plus a phase reconstruction problem at the same time. J_2 reads:

$$J_2(v) := \int_D \left| |w|^2 - z \right|^2 dD + a \|v\|^2_{Uad} \cdot$$

The gradient of J_2 with respect to w or v can be readily evaluated by the same method. We have e.g.

$$\delta J_{21}(\underline{w}) = 4 \text{ Re} \int_D \left[|w|^2 - z\right] w^* \delta w$$

5.3 Functional minimisation

The functional gradient depends on w and p where p is the solution of an adjoint system. We shall deal with two examples, related both to J_1 and J_2.Given J_1 the adjoint field p_1 satisfies:

$$\begin{cases} (\text{divgrad} + k^2)p_1 = w-z \ ; \ \text{supp } (w-z)=D \\ p_1|_{R\bar{O}^3} = 0 \\ \text{radiation conditions.} \end{cases} \qquad (11)$$

For this to make sense we need (w-z) to be sufficiently regular, i.e. at least in $L^2(D)$. We see the state equation is error - driven. If the error were $\equiv 0$ then $p_1 \equiv 0$. Moreover the Dirichlet BC for p_1 is homogeneous.
The minimisation of a functional can be restated in an equivalent form , which is more easily translated into a practical algorithm. Reformulation in straightforward if the functional has a unique minimising element . this is the case for J_1, if D of (10) is a (bounded) three dimensional domain . It can be shown that u is unique : identification of u is related to solving in the least squares sense a set of Cauchy problems for Helmholtz equation. u i then characterised by the so-called <u>optimality system</u> made of :

the original Helmholtz equation or w-system,
the adjoint or p_1-system of (11),which needs w(v) and z(.) to be solved,
the variational inequality:

$$2 \text{ Re} \int_S dS \left[p^*_{1n}|_S \cdot (v-u) \right] + \langle \underline{\text{grad}}_{\underline{v}} J_{12}(v) \,|\, (\underline{v}-\underline{u}) \rangle_{Uad} \geq 0,$$

$$\forall v \in U_{ad} ,$$

where p_{1n} is the normal derivative of p_1 at S .
Minimisation of J_2 must be dealt with in a different way , because the minimising elements are not unique. Hence no optimality system in the strict sense can be introduced. Yet, if we solve:

the w-system as above,
the following p_2-system, defined by:

$$\begin{cases} (\text{divgrad} + k^2)p_2 = 2(w^*w-z)w^* \\ p_2|_{R\bar{O}^3} = 0 \\ \text{radiation conditions,} \end{cases}$$

and the variational inequality:

$$2\ \mathrm{Re}\left\{\int_S \rho^{*}_{2n\ S} \cdot (v-u)\ dS\right\} + \langle \underline{\mathrm{grad}}_{\underline{v}} J_{22}(v) \mid (\underline{v}-\underline{u}) \rangle_{Uad} \geq 0,$$

$$\forall\ v \in B_u\ , \tag{12}$$

where B_u is an adequate neighbourhood of an optimal control u, we find both minimising elements lying on the boundary of U_{ad} and all stationary points of J_2 which are strictly inside U_{ad}. Hence the above set is necessary but not sufficient to identify the u's. It gives however the functional gradient , which again will be used in an algorithm.

For the p_2-system to make sense we need the source term to be regular enough. A sufficient condition is the same required for J_2 to be well-defined, i.e. :

$$w \in L^4{}_{loc}(R^3_+),$$

otherwise J_2 would not be continuous. Hence the need to constrain v. In this case we may choose the input space

$$U_{ad} = C^o{}_S(R^3_o), \tag{13}$$

of continuous functions defined on R^3_o which have support in S.

Inverse diffraction in some other geometries are dealt with by (Crosta,1983), together with their optimality systems. Other applications of the discrete gradient method to electromagnetic imaging have been considered by Lesselier (1981) ,Block(1983), Roger(1983).

6. SIMULTANEOUS APERTURE IDENTIFICATION AND PHASE RECONSTRUCTION

6.1 Theory

We shall study the minimisation of J_2 with some more detail: it is strictly related to the phase problem (Fiddy,1983a), for some recent results about which we refer to Fiddy (1983b). Our aim is to give the available results and list some open problems about existence,uniqueness and continuous dependence (regularity) of the identified control v on the data set z(.) . The problem may be either given an infinite or a finite dimensional setting , the former being mainly of use for existence and regularity estimates, the latter leading to a practical algorithm. In principle the infinite dimensional approach could deal with data having any support, ranging from a countable set of points to a three dimensional domain ($D_3 \subset\subset R^3_+$). The former case is however best analysed by discretising the equations.

Let us briefly investigate the link between existence and uniqueness of v and the dimension of the data support D. If D is a two-dimensional regular surface D_2, hereinafter named Beckert's surface, which does not divide R^3_+ into two and does not intersect R^3_o, then it can be shown, by extending Beckert's (1960) theorem, that both the traces of w and w_n on D_2 approximate in the $L^2(D_2)$ sense any function in $L^2(D_2)$ by choosing an adequate v in U_{ad} of (13). If we are given Cauchy data on D_2, existence of the corresponding v is not insured in the strict sense but in the least-squares sense. If only intensity data are available, the least-squares solution is for sure not unique. Nor things improve if we take two nearby Beckert's surfaces , because they can be joined together and yield another Beckert's surface. This statement differs from the standard uniqueness result for the phase problem ,which requires data e.g. on two planes both parallel to R^3_o: they are not Beckert's surfaces because they are unbounded. We shall therefore choose a three dimensional bounded domain D_3 as data support.

In addition to having the properties discussed in Section 5.3, J_2 is coercive, because of the second term, i.e.:

$$\lim_{\|v\|_{Uad} \to \infty} J_2(v) = \infty$$

therefore it has at least one minimising element.

J_2 is not convex, i.e. does not satisfy :

$$J_2(\theta v_1 + (1-\theta)v_2) \leq \theta J_2(v_1) + (1-\theta) J_2(v_2),$$

$$0 \leq \theta \leq 1 \ , \ \forall \ v_1, v_2 \in U_{ad},$$

as it could be shown. Then we expect the minimum not to be unique, as announced in Section 5. From the above discussion, we try to reduce the multiplicity of minima by taking D_3.

The dependence of an optimal input u on z(.) is still an open problem : it could be easily investigated if an explicit relationship between u and z were available. We have however an implicit relationship, i.e. the variational inequality (12). If it were an equation, then by extending Dini's theorem to functionals, we could get something like $\partial u/\partial z$ and then carry out a stability estimate, but stationary points of J_2 satisfy a variational equation if they are strictly inside U_{ad}. Boundary minima satisfy the strict inequality. The separate study of inner and boundary optimal controls would then become unrewardingly complicated.

Another interesting open problem is how to combine the topological relationship a variational inequality gives about the minimising elements of J_2 with the non-local analysis of an inverse problem, recently suggested by Sabatier (1983) . This should reduce ambiguity and eventually lead to a better algorithm.

6.2 Discretisation

We denote by

$\{\underline{x}_n\}$ a finite countable set of points in $R^3_+, 1 \leq n \leq N$,

$\{\underline{x}_{an}\}$ a finite countable set of points in R^3_o,

$v_m := v(\underline{x}_{am})$ the estimated control values at R^3_o,

$w_n := w(\underline{x}_n)$ the estimated field values at $\underline{x}_n$, which depend on v

$z_n := z(\underline{x}_n)$ the measured intensity values at $\underline{x}_n$,

$J_3 := J_{31} + J_{32}$ the functional to be minimised ,which consists of (estimation error)+(quadratic form on v_m),and is the discretised counterpart of J_2.

In order to minimise J_3 we may:
either minimise J_{31} subject to the constraint that also J_{32} be a minimum,
or consider J_3 as a whole.

If we look for the set $\{w_n\}$ such that

$$J_{31} := \sum_{n=1}^{N} (w^*_n w_n - z_n)^2 = \text{minimum},$$

we have to solve the non-linear system of N equations

$$(w^*_n w_n - z_n) w^*_n = 0$$

in 2N unknowns. The study of the Hessian matrix does not prove useful because the ambiguity on w_n is not reduced. Then we may consider J_{32} as a constraint and follow a procedure similar to

Weston's (1980,1981, and in particular 1983,pp 830-32), who aims at identifying the refractive index starting from measured intensity data.

On the other hand if we consider the minimisation of J_3, we have better write v_m and v^*_m, which appear in the regularising

term,as a function of w_n (for simplicity we write this term as the discretised counterpart of the norm of v in $L^2(S)$; other cases could be dealt with similarly). This step involves a discretised inverse propagator , usually an ill-conditioned matrix.

Since:

$$w_n = \sum_m G_{nm} v_m,$$

where G_{nm} are the matrix elements of the discretised direct propagator (Rayleigh-Sommerfeld's ,the paraxial one or any other approximated propagator), we must choose the points $\underline{x}_n, \underline{x}_{am}$ in such a way as $G^{-1} := H$ exists. Then

$$v_m = \sum_n H_{mn} w_n$$

and J_{32} becomes:

$$\sum_m (\sum_j H^*_{mj} w_j)(\sum_n H_{mn} w_n) := \sum_{m,n} w^*_m E_{mn} w_m$$

where $E := (H^T)^* H$.

Then we get the non-linear system:

$$2(w^*_n w_n - z_n) w^*_n + \sum_m w^*_m E_{nm} = 0$$

of 2N equations in 2N unknowns, to which convergence theorems for Newton's method (see e.g. Demidovitch & Maron,1979,pp 454-475) apply. A stability estimate of the local solution may now be carried out according to Weston (1983,p 763) . The minimisation of J_3 can thus take advantage of the discrete gradient method (Chavent , 1977). The weak point of the procedure consists of the critical role the $\underline{x}_n$'s play in conditioning the matrix H. A tradeoff between the accuracy of data point location and of location of the v_m's, may then follow, according to (Bertero and DeMol,1981,p 370 ff).

ACKNOWLEDGEMENTS

The author thanks prof. Gianni Degli Antoni for encouragement, prof. Wolfgang M. Boerner for accepting this contribution and prof. Alexander D. Ramm for critically reading a preliminary version of this manuscript.

REFERENCES

Beckert H,1960,"Eine bemerkenswerte Eigenschaft der Loesungen des Dirichletschen Prob. bei lin. ellipt. Differentialgl.", Math. Ann.,139,255-64

Bellman R, Kalaba R,1959 "Functional equations,wave propagation & invariant imbedding", J. Math. Mech. 8, 683

Bertero M, De Mol C , 1981 "The stability problem in inverse diffraction", IEEE AP 29 368-372

Bertero M, De Mol C, Viano G A, 1980 "The stability of inverse problems",in Baltes H P "Inverse scattering problems in optics", pp 161-214, Berlin: Springer

Blok H,1983,"One dimensional time domain inverse scattering applicable to electromagnetic imaging", this volume

Boerner W M ,1980,"Polarisation utilisation in electromagnetic inverse scattering",in Baltes H P "Inverse scattering problems in optics",Berlin: Springer,

Boerner W M ,1981 "Polarisation dependence in electromagnetic inverse problems" IEEE AP 29 262-271

Chavent G,1977 "Identification problems for distributed systems",New Dehli IIT Winter School lecture notes, Rocquencourt: INRIA

Conant R C,Ashby W R,1970,"Every good regulator of a system must be a model of that system",Int.J.Syst. Scie.,1,pp 89-97,

Corones J,1975 "Bremmer series that correct parabolic approximations", J.Math.Analys.Appl.50 361-372

Crosta G,1982 "Inverse diffraction, duality and optimal control", J.Phys.A 15 645-660

Crosta G,1983 "Models for diffracting aperture identification: a comparison between ideal and convolutional observations", SPIE Proc.413, paper n° 24

Delfour M D & Mitter S K,1972,"Controllability and observability for infinite-dimensional systems",SIAM J.C.& O. 10,329-33

Demidovitch B & Maron I,1979,"Elements de calcul numerique", MIR:Moscow

Deschamps G A,1981 "Gaussian beams: paraxial theory", in J K Skwirzynski Ed."Theoretical methods for determining the interaction of electromagnetic waves with structures" , Rockville,MD:Sijthoff & Noordhoff

Ekeland I & Temam R,1974,"Analyse convexe et problèmes variationnels", Dunod,Paris

Fiddy M A,1983a,"The phase problem", SPIE Proc. 413,paper n° 26

Fiddy M A,1983b,"Object reconstruction from partial information", this volume

Fock V A,1960 "Electromagnetic diffraction & propag. problems", New York: McMillian

Goguen J A and Varela F J,1979,"Systems and distinctions: duality & complementarity",Int. Journ. Gen. Syst.,5,pp 31-43

Goodman J W,1968 "Introduction to Fourier optics",New York:McGraw

Hill
Hörmander L,1963 "Linear partial differential operators",Berlin: Springer, pp 249-51
Jahnke,Emde,Lösch,1960 "Tables of higher functions",New York: McGraw Hill
Jordan T F,1969,"Linear operators for quantum mechanics", Jordan: Duluth,MN
Lalor E,1968 "Conditions for the validity of the angular spectrum of plane waves",J.Opt.Soc.Amer. 58 1235-37
Lesselier D,1981 "Diagnostic des milieux inhomogénes unidimensionnels par échographie éléctromagnetique",Rev.du CETHEDEC 68 1-42
Lions J L,1971 "Optimal control of systems governed by partial differential equations",Berlin:Springer
Miller K,1970,"Least squares method for ill-posed problems with a prescribed bound",SIAM J. Math. Analys.,1,52-74
Montgomery W D,1968,"Algebraic formulation of diffraction applied to self imaging",J.O.S.A 58,pp 112-24
Ramm A D,1978,"Necessary and suff. conditions for the validity of the limit absorption principle", Math. Izv. Vizov. 5
Ramm A D,1981,"On some properties of solutions of the Helmholtz equation",J.Math. Phys. 22,pp 275-6
Ramsay W B,Baltes H P,Schmidt-Weinmar H G,1982 "Anomalies in bandlimited asymptotic fields",J.O.S.A. 72,1618-29
Roger A J,1985,"Theoretical study and numerical resolution of inverse problems via the functional derivatives", this volume
Sabatier P C,1985,"Critical analysis of the mathematical methods used in electromagnetic inverse theories", this volume
Schmidt-Weinmar H G,1978,"Spatial resolution of subwavelength sources from optical far field data",in H P Baltes,Ed. "Inverse source problems in optics",Berlin: Springer
Sherman G C,,1969,"Diffracted wave fields expressible by plane wave expansions containing only homogeneous waves", J.O.S.A,59,pp 697-711
Tikhonov A,Arsenin V,1976 "Mèthodes de résolution des problèmes mal-posés",Moscow:MIR
Vogelsang V,1975 "Das Ausstrahlungsproblem für elliptische Differentialgleichungen in Gebieten mit unbeschränktem Rand",Math. Zeit. 144 101-124
Weston V H,1980,"Inverse problem for the reduced wave equation with fixed incident field,part 1",J. Math. Phys. 21,pp 758-64
Weston V H,1981,ditto, part 2,J. Math. Phys. 22,pp 2523-30
Weston V H,1983,ditto,part 3,J. Math. Phys. 24,828-33.

II.4 (NM.2)

STABILITY AND RESOLUTION IN ELECTROMAGNETIC INVERSE SCATTERING - PART I

M.Bertero — Univ.Genova
C.De Mol — Univ.Bruxelles
E.R.Pike — RSRE,Malvern

ABSTRACT: A powerful technique for the solution of a number of experimental inverse problems, described by an underlying first-kind Fredholm equation, is presented. Such problems include, for example, diffraction-limited EM imaging and quasi-elastic laser light scattering. The technique requires the construction of the "singular system" of the problem which then provides exact orthonormal bases both for the description of sampled and truncated measured data and for the reconstructed continuous object "solution" of the inversion. The singular-system approach may be regarded as a theory of information which generalises in several directions the well known classical concepts of Shannon and Nyquist.

1. HISTORICAL INTRODUCTION

1.1. The Shannon Theory of Information

One of the most well studied inverse problems is that of the temporal frequency band-limited communication channel or, equivalently, the spatial frequency band-limited imaging system. The mathematical description of this problem is easily stated as follows, where we use the latter system in one dimension for illustration and notation. An "object", $O(x)$, $-1 \leqslant x \leqslant +1$, is resolved into its Fourier components $\tilde{O}(k)$ (by straightforward generalisations x and k may be multidimensional)

$$\tilde{O}(k) = \int_{-1}^{+1} O(x) \quad e^{-ikx} \, dx \tag{1.1}$$

W.-M. Boerner et al. (eds.), Inverse Methods in Electromagnetic Imaging - Part 1, 319–328.

and a "geometric" image, I(x'), is formed by re-summing only those spatial frequencies lying below a "band-limit" and which may be taken without loss of generality in $-1 \leq x' \leq +1$

$$I(x') = \frac{1}{2\pi} \int_{-\Omega}^{+\Omega} \tilde{O}(k)\, e^{ikx'}\, dk \quad . \tag{1.2}$$

Thus

$$I(x') = \frac{1}{2\pi} \int_{-1}^{+1} \left(\int_{-\Omega}^{+\Omega} e^{-ik(x-x')}\, dk \right) O(x)\, dx$$

$$= \int_{-1}^{+1} \frac{\sin[\Omega(x-x')]}{\pi\,(x-x')}\, O(x)\, dx \quad . \tag{1.3}$$

This has the form of a general Fredholm equation of the first kind

$$(Kf)(p) = g(p) \tag{1.4}$$

where

$$(Kf)(p) = \int_{-1}^{+1} K(p,t)\, f(t)\, dt \quad , \quad -1 \leq p \leq +1 \tag{1.5}$$

with the "kernel" K(p,t) of the integral operator, K, in this case a "sinc" function.

The remarkable solution of equation (1.3) in terms of its orthonormal eigenfunctions, namely the prolate spheroidal wave functions (Slepian and Pollack, 1961), helps to understand the classical theory of information and is the source of the accepted theory of resolution in diffraction-limited imaging problems. We recall that the eigenfunctions, φ_n, are the solutions of the homogeneous equation

$$\lambda_n \varphi_n(x') = \int_{-1}^{+1} \frac{\sin[\Omega(x-x')]}{\pi\,(x-x')}\, \varphi_n(x)\, dx \tag{1.6}$$

$$-1 \leq x' \leq +1$$

so that the object may be reconstructed as a series

$$O(x) = \sum_{n=0}^{\infty} \frac{c_n}{\lambda_n} \varphi_n(x) \qquad (1.7)$$

where

$$c_n = \int_{-1}^{+1} I(x') \varphi_n(x') \, dx' \qquad (1.8)$$

The mathematical theory will be discussed in more detail in part II of this paper but we note here that the use of this solution in the presence of noise is only possible by limiting the reconstructed object to components c_n which correspond to sufficiently high eigenvalues λ_n and therefore are not to be strongly affected by the noise, that is to say, those components which appear in the image above the ambient "noise level". For the prolate spheroidal functions the eigenvalue series λ_n drops sharply from approximately unity to nearly zero at a value of n known as the Shannon number $(2\Omega/\pi)$ which means that only this number of components may be recovered almost independently of the actual noise level in any particular situation. One may thus, with this problem, speak of a number of degrees of freedom, number of "bits" of information, Nyquist sampling limit and other concepts related only to the band limit and not to the noise. These concepts are used in what we may call the classical Shannon theory of information.

1.2. Laplace Transform Inversion

First-kind Fredholm equations appear in many more physical inverse problems, of course, than that with the sinc kernel discussed above. We shall consider in this section the case where the kernel is an exponential function, giving rise to the problem of the inversion of the Laplace transform

$$g(p) = \int_0^{\infty} e^{-pt} f(t) \, dt \quad , \quad 0 \leqslant p < +\infty \quad . \qquad (1.9)$$

This problem arises in diverse fields of science, our particular interest being its application to the determination of molecular size distributions from laser light scattering data (Ostrowsky et al., 1981). Many ad hoc methods of numerical inversion of eq. (1.9) have been proposed but it was not until five years ago that the problem was approached in the spirit of the information theory sketched above when McWhirter and Pike (1978) proposed the solution of (1.9) in terms of the generalised eigenfunctions, φ_ω (where ω

is now a continuous variable), given by

$$\varphi_\omega(t) = t^{-1/2-i\omega} \frac{\mathrm{Re}\,[\Gamma(\frac{1}{2}+i\omega)]^{1/2}}{[\Gamma(\frac{1}{2}+i\omega)]^{1/2}}, \quad \omega > 0 \tag{1.10}$$

$$\varphi_\omega(t) = t^{-1/2+i\omega} \frac{\mathrm{Im}\,[\Gamma(\frac{1}{2}+i\omega)]^{1/2}}{[\Gamma(\frac{1}{2}+i\omega)]^{1/2}}, \quad \omega < 0 \tag{1.11}$$

Furthermore the corresponding continuous spectrum of the operator was shown to have the simple form

$$\lambda_\omega = \frac{\omega}{|\omega|}\sqrt{\frac{\pi}{\cosh(\pi\omega)}} \tag{1.12}$$

This spectrum shows an unfortunate exponential fall-off causing a severe ill-posedness and a strong limitation in the presence of noise on the band of frequencies ω which may be recovered. It must be noted that in this case, contrary to the sinc-kernel case, the resolution limit depends on the noise level. The reason is that the spectrum (1.12) no longer has a plateau before a sharp fall as observed in the previous case.

1.3. The Discrete-Discrete Inversion

In order to use the McWhirter-Pike results for numerical inversion Ostrowsky et al.(1981) found that, provided a sufficiently low number of sample points of the solution were demanded, a discrete matrix inversion based on

$$g(p_i) = \sum_{n=1}^{S} a_n e^{-p_i t_n} \qquad i = 1,\dots,S \tag{1.13}$$

could be used. A geometric ratio was used for the spacing of the points p_i. The somewhat surprising fact emerged that the number S of sample points which could be recovered was somewhat larger than the number of degrees of freedom predicted by the McWhirter-Pike theory.

2. THE EFFECT OF FINITE OBJECT SUPPORT, SINGULAR VALUE ANALYSIS

Following the work of Ostrowsky et al.(1981), Bertero, Boccacci and Pike (1982) traced the increase of information capacity found by these authors to the implicit restriction of object support in the matrix inversion procedure used. The effect of a known finite support of the object in the Laplace transform was given a thorough investigation using an approach based on singular functions.

2.1. Continuous-Continuous Case

2.1.1. Laplace Transform. Bertero et al.(1982) considered the problem of the finite Laplace transform

$$(Kf)(p) = \int_a^b e^{-pt} f(t)\, dt \quad , \quad 0 \leq p < +\infty \qquad (2.1)$$

restricting the transformation to square integrable object functions on the interval (a,b). It is easy to show that the operator K is compact and injective and admits a singular system $\{\alpha_k; u_k, v_k\}$ (k = 0,1,...) given by the solutions of the coupled equations (see e.g. Miller, 1974)

$$Ku_k = \alpha_k v_k \quad , \quad K^* v_k = \alpha_k u_k \qquad (2.2)$$

where K^* is the adjoint operator given by

$$(K^* g)(t) = \int_0^\infty e^{-tp} g(p)\, dp \quad , \quad a \leq t \leq b \qquad (2.3)$$

Since both the object and ~~the~~ the image data are considered as functions of a continuous variable, we will call such a case a "Continuous-Continuous" inverse problem. In Bertero et al.(1982) it was shown that the singular value spectrum of operator (2.1) has a slower decrease rate than the corresponding spectrum for the problem where f(t) is assumed to have an infinite support. As a consequence, the knowledge that f(t) is a priori restricted to a finite known support allows to recover more information from the data and this in quantitatively the way observed by Ostrowsky et al.

2.1.2. Diffraction-Limited Imaging. The work of Bertero et al. described above gave rise to the suggestion that the increase in information transfer obtained in the Laplace inversion by using a known restricted support for the solution might have an analogy in the diffraction-limited imaging problem discussed in section 1 with coherent illumination, allowing an improvement in resolution over that defined by the Shannon number to be obtained. This possibility was investigated and proved to be correct in a paper by Bertero and Pike in 1982. In this paper it was shown that the restriction of the image to a support equal to that of the geometrical image, which is implicit in the eigenvalue analysis, causes a loss of information which can be regained by the use of the total diffraction image. It is then required to solve a first kind Fredholm equation as given by (1.4) where the integral operator K is now defined as follows

$$(Kf)(p) = \int_{-1}^{+1} \frac{\sin[\Omega(p-t)]}{\pi(p-t)} f(t)\, dt \ , -\infty < p < +\infty \quad (2.4)$$

The singular values of this operator are $\sqrt{\lambda_n}$ and clearly they fall off more slowly than the eigenvalues λ_n defined by (1.6), allowing more coefficients to be calculated in the reconstruction and hence a higher resolution to be obtained. The two-dimensional case with square and circular apertures was also considered in this work and in a later paper (Bertero et al., 1982) the analysis was carried out also for incoherent illumination. Furthermore it has been remarked in these problems that resolution beyond the classical diffraction limit (i.e. superresolution) is much easier to achieve when the space-bandwidth product is small (i.e. the constant Ω in equation (2.4)). A similar effect has also been demonstrated for the problem of inverse diffraction from plane to plane, when it is known a priori that the source is restricted to a small support with linear dimensions comparable to the wavelength of the EM field (Bertero et al., 1983).

2.2. Continuous-Discrete Case

<u>2.2.1. Laplace Transform</u>. In the last two sections we discussed the improvement in resolution in the Laplace transform and diffraction-limited imaging cases obtained when using a priori knowledge of the support of the object function. The singular systems used were constructed in bases of functions of continuous variables. In real experiments, however, clearly the data are measured over truncated regions of finite extent and at a finite number of sampled points. A possible route is then to discretise the unknown function. However, the singular value method offers a very interesting alternative since the types of functions, u describing the object and v describing the data, may be different. In particular, we may regard the data as a vector set but yet reconstruct the object in a basis of square integrable functions. In the previous discretisation mentioned above the object was constructed only on a set of sample points and, usually, an interpolation scheme was used between them. If we adopt the above suggestion the interpolation step is completely avoided.

The mathematics of this situation will be discussed in part II of this paper but here we may say that the method has been put into practice and will be the subject of a publication by Bertero et al. in the near future. It has been found that computations will produce accurately the required bases in the functional object space, which, of course, will be represented in the numerical calculations by a rather large number of discretisation points, and in the vector valued image space, which will normally be represented by rather smaller dimension.

The practical application of this method is under investigation at the present time in the context of the laser light scattering problem mentioned above.

2.2.2. Diffraction-Limited Imaging. In the diffraction-limited imaging problem, just as in the Laplace problem discussed in the last subsection, the same idea of using a singular vector basis for the data and a singular function basis for the object can be applied. Successful numerical computation of such bases have been performed in Genoa and at Malvern and are currently in the process of being applied to practical problems in microscopic and other imaging. The theory looks particularly promising for application, for example, in EM imaging problems in the microwave region.

An attractive feature of the method is the speed at which the reconstruction can be effectuated, all that is required is to form a scalar product of the data set with each of the basis data vectors in turn and these scalar products are used as the coefficients, divided by the corresponding singular value, of the singular function expansion of the object. Both the basis vector set and the basis function set are, of course, stored numerically once and for all for a given problem.

3. EXTENSION TO OTHER PROBLEMS

So far in this paper we have only mentioned two cases of the kernel K(p,t) of the integral equation, namely, the sinc kernel and the Laplace exponential kernel. There are, however, many other problems of a similar type and in this last section we will mention briefly some problems which are under study using the singular system technique.

3.1. Fraunhofer Diffraction

In this problem the kernel has the form $K(p,t) = J_1^2(pt)/(pt)^2$. This represents the form of the intensity scattered in the far field by a spherical particle illuminated by a coherent laser beam and the solution of the problem will be the distribution of sizes of the scattering particles. The problem is of great importance in many industrial and scientific applications and the technique of particle sizing by Fraunhofer diffraction is quite widely used.

As a first attempt to quantify the information content in a typical data set in such an experiment we have used the results of the paper of McWhirter and Pike in 1978 and found the following spectrum for the present problem (Bertero and Pike, 1983)

$$\lambda_\omega^\pm = \pm \left| \frac{2^{-\frac{5}{2}} \Gamma\left(\frac{5}{2} - i\omega\right) \Gamma\left(\frac{1}{4} + \frac{i\omega}{2}\right)}{\Gamma^2\left(\frac{7}{4} - \frac{i\omega}{2}\right) \Gamma\left(\frac{11}{4} - \frac{i\omega}{2}\right)} \right| \tag{3.1}$$

This spectrum can be used to calculate the number of resolution cells in particle size, which can be found in a certain level of experimental noise. For example we can ask how many size fractions can be resolved in the range 2 to 100 microns when the signal-to-noise ratio is 10^2. The result is that 7 or 8 fractions on a geometrical scale may be resolved in this range. Further work is required on this problem to extend the calculations to the use of singular functions and vectors, which will provide a very practical approach to the inversion in this type of experiments.

3.2. Laser Anemometry

Yet another example of this sort of inversion problem is provided by laser anemometry. In this type of experiment a laser beam scattered off a particle in uniform motion at velocity v, gives rise to a measured data set with the form of an exponentially damped cosine wave. In turbulent flows or non-uniform flows the data represent the motion of particles with a probability distribution of velocities which one needs to recover as the solution. The kernel in this case has the form

$$K(p,t) = e^{-(pt)^2/2\sigma^2} \quad (1 + a\cos(bpt)) \tag{3.2}$$

where a,b and σ are constants. A certain amount of work with this kernel has been done using the concepts of information theory discussed above and one finds, in this case, that the fall-off of the eigenvalue or singular value spectrum is a function of the value of σ, characterising the damping of the cosine signal. In the limit of small damping a large amount of information is available from the data, while if the damping is more severe there is much less velocity information to be found. Again one can introduce a concept of resolution associated with a number of components of the solution which may be transmitted above the noise level. The pressure to push the analysis of this particular kernel to the same degree as has been done for the Laplace problem above is not so strong since in many cases the damping term may be arranged to be small. However in critical circumstances, as for example are met in very high speed supersonic flows, or where very high spatial resolution is required it will be important to extract the maximum information from data of this type and further work along the lines described above will allow this to be done.

3.3. Gaussian-Beam Illumination

In this last section we come back to the diffraction-limited imaging problem but consider the case where the object is illuminated by a coherent laser beam of infinite spatial extent but with a Gaussian amplitude profile. This is not the place to report detailed results of this calculation but it will suffice to say that singular value spectra may be calculated numerically (Bertero et al., in preparation) and these allow a discussion of such an imaging problem in terms of the other concepts of number of degrees of freedom as a function of noise and the implications of this in terms of the limits of resolution which may be achieved. Our preliminary results indicate that the singular value spectrum for this problem falls off a little more slowly than for the associated diffraction-limited imaging problem in which the illumination is assumed to have a "top hat" form (sinc kernel) of the same total power and the same width. However, the zeroes of the singular functions do not increase in density at the same rate as for the "top hat" case and therefore the overall resolution limits might be worse.

ACKNOWLEDGEMENT

C. De Mol is Research Associate with the National Fund for Scientific Research (Belgium).

REFERENCES

Bertero, M., Boccacci, P. and Pike, E.R., 1982, Proc. Roy. Soc. Lond. A383 pp. 15-29.

Bertero, M., Boccacci, P. and Pike, E.R., 1982, Optica Acta 29 pp. 1599-1611.

Bertero, M., De Mol, C., Gori, F. and Ronchi, L., 1983, Optica Acta 30 pp. 1051-1065.

Bertero, M., De Mol, C., Pike, E.R. and Walker, J., in preparation.

Bertero, M. and Pike, E.R., 1982, Optica Acta 29 pp. 727-746.

Bertero, M. and Pike, E.R., 1983, Optica Acta 30 pp. 1043-1049.

Mc Whirter, J.G. and Pike, E.R., 1978, J. Phys. A 11 pp. 1729-1745.

Miller, G.F., 1974, Numerical Solution of Integral Equations (eds L.M. Delves and J. Walsh) , p. 175, Oxford: Clarendon Press.

Ostrowsky, N., Sornette, D., Parker, P. and Pike, E.R., 1981, Optica Acta 23 pp. 1059-1070.

Slepian, D. and Pollack, H.O., 1961, Bell Syst. Tech. J. 40 pp. 43-64.

II.5 (NM.3*)

STABILITY AND RESOLUTION IN ELECTROMAGNETIC INVERSE SCATTERING - PART II

M. Bertero C. De Mol E.R. Pike

Univ. Genova Univ. Bruxelles RSRE, Malvern

ABSTRACT: Singular function expansions, whose applications to EM inverse problems have been reviewed in Part I, provide a very simple method for obtaining stable solutions and for defining resolution limits in a number of inverse problems. Their extension to cover the case of problems with discrete data is discussed and the connection between the case of continuous data and the case of discrete data is clarified. Then the general method is applied to the particular case of the finite Laplace transform inversion, a problem which is important both from the pedagogical and from the practical point of view. By means of this example we show how an appropriate choice of a small number of data points can provide almost the resolution achievable with continuous data.

1. THE MATHEMATICAL THEORY OF SINGULAR SYSTEMS

As was shown in the Part I of this paper, many inverse problems in EM imaging can be reduced to the solution of a Fredholm integral equation of the first kind

$$\int_a^b K(p,t)f(t)dt = g(p) \ , \ c \leqslant p \leqslant d. \tag{1.1}$$

When the kernel satisfies suitable conditions (for instance it is an integrable function) then the solution of eq.(1.1) is a classical example of an ill-posed problem.

W.-M. Boerner et al. (eds.), Inverse Methods in Electromagnetic Imaging - Part 1, 329–340.

The mathematical analysis is greatly simplified if we assume that both the "object" f and the "image" g are elements of L^2-spaces, respectively denoted by X and Y.

1.1. Eigenfunction expansions

The solution of problem (1.1) can be easily investigated when one can introduce eigenvalues and eigenfunctions of the integral operator

$$(Kf)(p) = \int_a^b K(p,t)f(t)dt, \quad c \leqslant p \leqslant d \; . \tag{1.2}$$

If the supports of the "object" f and of the "image" g coincide (namely, a = c and b = d) and if the kernel K(p,t) is square integrable and satisfies the condition: $K(p,t) = K^*(t,p)$ (here the star denotes complex conjugation), then K is a compact, self-adjoint operator in $L^2(a,b)$ and the set $\{\phi_k\}_{k=0}^{+\infty}$ of the eigenfunctions associated with the eigenvalues λ_k is an orthogonal basis of the orthogonal complement of the null space of K and therefore an orthogonal basis of $L^2(a,b)$ when K is injective (we will consider only this case). By expanding both f and g as series of the ϕ_k we obtain the solution of eq.(1.1) in the absence of noise

$$f(t) = \sum_{k=0}^{+\infty} \frac{g_k}{\lambda_k} \phi_k(t) \tag{1.3}$$

where

$$g_k = \int_a^b g(p)\, \phi_k^*(p)dp \quad . \tag{1.4}$$

Since $\lambda_k \rightarrow 0$ when $k \rightarrow +\infty$ (this is a general property of the eigenvalues of compact operators) the series (1.3) generally is not convergent when the "image" g is affected by noise or experimental errors. Then the most simple technique for obtaining stable approximate solutions is to truncate, at a suitable value of k, the series (1.3). This technique, often called <u>numerical filtering</u> (Twomey, 1965), can be justified both from the point of view of regularization methods and from the point of view of Wiener filter methods (Bertero et al., 1980). If we assume, as usual, that the eigenvalues are ordered in a decreasing sequence, $\lambda_0 \geqslant \lambda_1 \geqslant \lambda_2 \geqslant \ldots$, then we take in the expansion (1.3) only the eigenvalues which are greater than a prescribed constant C and if $K_0 = \max\{k \,|\, \lambda_k \geqslant C\}$, then we get the filtered (regularized) solution

$$\tilde{f}(t) = \sum_{k=0}^{K_o} \frac{g_k}{\lambda_k} \phi_k(t) \quad . \tag{1.5}$$

The quantity 1/C can be often interpreted as a "signal-to-noise ratio" and the number K_o as the "number of degrees of freedom" of the "image" g.

An important property of the eigenfunctions of many integral operators encountered in experimental inverse problems is that the eigenfunction ϕ_k has exactly k zeros inside the interval $[a,b]$. Then, if the distribution of the zeros is approximately uniform one can define a resolution distance, associated with the solution (1.5), as follows

$$d = (b-a)/K_o \tag{1.6}$$

or, if the zeros of ϕ_k form approximately a geometric progression ,a resolution ratio given by (Bertero et al., 1982)

$$\delta^{K_o} = b/a \quad . \tag{1.7}$$

In those cases where the eigenvalues drop to zero very rapidly, the number of degrees of freedom K_o is practically independent of the signal-to-noise ratio and the same property holds for the resolution distance d or the resolution ratio δ .

1.2. Singular function expansions

The previous analysis does not apply to the case where the supports of the "object" f and of the "image" g do not coincide. But if the operator K satisfies the condition

$$\int_c^d dp \int_a^b dt \left| K(p,t) \right|^2 < +\infty \tag{1.8}$$

(more generally, if K is compact), a powerful generalization of the eigenfunction method is provided by the singular function method (Picard, 1910). The singular system $\{\alpha_k ; u_k , v_k\}_{k=o}^{+\infty}$ of the operator K is the set of the solutions of the coupled equations

$$Ku_k = \alpha_k v_k \quad , \quad K^* v_k = \alpha_k u_k \tag{1.9}$$

where K* is the adjoint of K

$$(K^*g)(t) = \int_c^d K^*(p,t)g(p)dp \quad , \quad a \leqslant t \leqslant b. \tag{1.10}$$

The numbers α_k are called the singular values of K (usually they are ordered in a decreasing sequence) and the functions u_k, v_k are called the singular functions of K associated with α_k. The singular functions u_k are also eigenfunctions of the operator K*K associated with the eigenvalues α_k^2 and they form an orthogonal basis of the orthogonal complement of the null space of K (of $L^2(a,b)$ if K is injective). Analogously the singular functions v_k are eigenfunctions of KK*, also associated with the eigenvalues α_k^2, and they form an orthogonal basis of the orthogonal complement of the null space of K* (of $L^2(c,d)$ if K* is injective).

In the absence of noise the solution of problem (1.1) is now given by

$$f(t) = \sum_{k=0}^{+\infty} \frac{g_k}{\alpha_k} u_k(t) \tag{1.11}$$

where

$$g_k = \int_c^d g(p)v_k^*(p)dp \ . \tag{1.12}$$

This solution has a structure very similar to that of the solution (1.3), which applies to the case of a self-adjoint operator, except for the fact that now two orthogonal systems, instead of one, are used. Again one can introduce a filtered solution as follows

$$\tilde{f}(t) = \sum_{k=0}^{K_0} \frac{g_k}{\alpha_k} u_k(t) \tag{1.13}$$

and therefore it is clear that the concepts of "number of degrees of freedom" and of resolution distance (or resolution ratio) can be easily extended to the present case.

1.3. The case of discrete data

A very important property of the singular function methods is their flexibility which allows their extension to problems with discrete data.

Let us assume that the "image" $g(p)$ is measured only on a finite set of points, $p_1, p_2, \ldots, p_N$, so that eq(1.1) is replaced by

$$g(p_n) = \int_a^b K(p_n,t)f(t)dt \; ; \; n = 1,\ldots,N. \qquad (1.14)$$

Now the data space Y is an N-dimensional vector space equipped with the scalar product

$$(g,h)_Y = \sum_{n=1}^{N} w_n \, g(p_n)h^*(p_n) \qquad (1.15)$$

where the weights w_n must be introduced in order to take into account a non-uniform distribution of the data points p_n.

Let K_N be the operator, defined by eq.(1.14), which transforms an "object" of $L^2(a,b)$ into a vector of Y. The adjoint K^*_N transforms a vector of Y into a function of $L^2(a,b)$ and it is given by

$$(K^*g)(t) = \sum_{n=1}^{N} w_n \, g(p_n)K^*(p_n,t) \; . \qquad (1.16)$$

Then $K^*_N K_N$ is a self-adjoint integral operator of finite rank in $L^2(a,b)$, whose kernel is given by

$$T_N(t,s) = \sum_{n=1}^{N} w_n \, K^*(p_n,t)K(p_n,s) \; , \qquad (1.17)$$

while the operator $K_N K^*_N$ is a matrix N x N whose matrix elements are given by

$$t^{(N)}_{nm} = w_m \int_a^b K(p_n,t)K^*(p_m,t)dt. \qquad (1.18)$$

We denote by $\{\alpha_{N,k}; u_{N,k}, v_{N,k}\}_{k=0}^{N-1}$ the singular system of K_N, the $u_{N,k}$ being the eigenfunctions of $K^*_N K_N$ and the $v_{N,k}$ being the eigenvectors of the matrix $K_N K^*_N$.

The singular value $\alpha_{N,k}$ provides a good approximation of the corresponding singular value α_k of the problem with continuous data, if the integral operator $K^*_N K_N$ is a good approximation of the integral operator K^*K. Indeed, let $T(t,s)$ be the kernel of K^*K:

$$T(t,s) = \int_c^d K^*(p,t)K(p,s)dp \qquad (1.19)$$

and let us write

$$\epsilon_N(t,s) = T(t,s) - T_N(t,s) \tag{1.20}$$

$T_N(t,s)$ being defined in eq.(1.17). Clearly $K_N^* K_N$ is a good approximation of K^*K if $\epsilon_N(t,s)$ is small, and by the Weyl-Courant lemma (Riesz & Nagy, 1955) we have

$$\left|\alpha_{N,k}^2 - \alpha_k^2\right| \leqslant \left(\int_a^b dt \int_a^b ds \left|\epsilon_N(t,s)\right|^2 \right)^{1/2} . \tag{1.21}$$

Therefore if K_o is the number of degrees of freedom in the problem with continuous data, we can look for a distribution of the data points which allows a good approximation of the singular values α_k up to $k = K_o$. If K_o is small, it is reasonable to argue that, using a small number of data points, conveniently placed, it is possible to obtain such a good approximation.

We conclude by remarking that the filtered solution of the problem with discrete data is

$$\tilde{f}_N(t) = \sum_{k=0}^{K_o} \frac{g_{N,k}}{\alpha_{N,k}} u_{N,k}(t) \tag{1.22}$$

where

$$g_{N,k} = \sum_{n=1}^{N} w_n\, g(p_n) v_{N,k}^*(p_n) . \tag{1.23}$$

In the last equation $v_{N,k}(p_n)$ denotes the n-th component of the singular vector $v_{N,k}$.

2. THE FINITE LAPLACE TRANSFORM INVERSION

An example of the applications of the method presented in Sect. 1 is the inversion of the finite Laplace transform which, as shown in the Part I of this paper, is important, for example, in the problem of polydispersity analysis in quasi-elastic light scattering.

Given a function $f \in L^2(a,b)$ $(0 < a < b < +\infty)$, we define the <u>finite Laplace transform</u> of f as

$$g(p) = \int_a^b e^{-pt} f(t)dt \quad , \quad 0 \leqslant p < +\infty \ . \tag{2.1}$$

The function g(p) is square integrable and therefore eq.(2.1) defines an operator K from $L^2(a,b)$ into $L^2(0,+\infty)$. It is easy to verify that the kernel K(p,t) = exp(-pt) satisfies the condition (1.8) and therefore K is compact.

2.1. Laplace inversion with continuous data

As a consequence of the scaling properties of the Laplace transform the singular values of the operator K depend on the upper bound, b, and on the lower bound, a, of the support of f through the ratio

$$\gamma = b/a \ . \tag{2.2}$$

Therefore we find convenient to take as a support of f the interval $[1,\gamma]$ and to define the operator K as follows

$$(Kf)(p) = \int_1^\gamma e^{-pt} f(t)dt \ . \tag{2.3}$$

The singular values α_k and the singular functions $u_k(t)$ are given by the solutions of the eigenvalue problem

$$\int_1^\gamma T(t+s)u_k(s)ds = \alpha_k^2 u_k(t) \tag{2.4}$$

where

$$T(t) = \frac{1}{t} \ . \tag{2.5}$$

Then the singular functions $v_k(p)$ are given by the finite Laplace transforms of the singular functions $u_k(t)$:

$$v_k(p) = \frac{1}{\alpha_k} \int_1^\gamma e^{-pt} u_k(t)dt. \tag{2.6}$$

All the eigenvalues α_k^2 have multiplicity one (Bertero et al., 1982) and therefore they can be ordered in a strictly decreasing sequence. They tend to zero exponentially fast as follows from general results on the asymptotic behaviour of the eigenvalues of integral operators (Hille & Tamarkin, 1931). As a consequence, for moderate values of γ, very few singular values are significant. For instance, in the case $\gamma = 5$, we have (Bertero et al., 1982):

$$\alpha_o = 8.751 \times 10^{-1},\ \alpha_1 = 1.935 \times 10^{-1},\ \alpha_2 = 3.827 \times 10^{-2} \tag{2.7}$$

$$\alpha_3 = 7.343 \times 10^{-3},\ \alpha_4 = 1.435 \times 10^{-3},\ \alpha_5 = 2.765 \times 10^{-4},\ldots$$

Accurate computations of the singular values α_k and of the singular functions u_k can be performed along these lines (Bertero et al., 1983): i) approximation of the kernel T(t+s) by means of a tensor product of splines (Hämmerlin & Schumaker, 1980); the approximating kernel $\tilde{T}(t,s)$ is of finite rank and its eigenvalues coincide with the eigenvalues of a non-symmetric matrix Q - ii) the largest eigenvalues of Q are computed by a combination of the power method and of the deflation method (Ralston, 1965); this procedure works very efficiently when the eigenvalues are well separated as in our case - iii) once the eigenvalues have been computed, the corresponding eigenvectors of Q are determined by means of the inverse power method (Wilkinson, 1965) - iv) finally the eigenfunctions $u_k(t)$ are computed as linear combinations of splines, the coefficients being provided by the components of the eigenvectors of Q.

It was confirmed by these computations, that the singular functions $u_k(t)$ have the property mentioned at the end of Sect.1.1, namely $u_k(t)$ has exactly k zeros in the interval $[1,\gamma]$. Furthermore the distribution of the zeros is approximately geometric and therefore one can define a resolution ratio δ as in eq.(1.7).

2.2. Laplace inversion with discrete data

The analysis of Sect.1.3 can be immediately applied to the problem of the finite Laplace transform inversion and, due to the simplicity of the kernel, some computations can be done analytically. For instance the matrix $t^{(N)}_{nm}$ of eq.(1.18) takes the very simple form

$$t^{(N)}_{nm} = \frac{w_m}{p_n+p_m}\left\{\exp\left[-(p_n+p_m)\right] - \exp\left[-\gamma(p_n+p_m)\right]\right\}. \tag{2.8}$$

In the case of equidistant data points, $p_n=(n-1)\sigma$, (n=1,..,N), we can take $w_n = \sigma$ so that the matrix (2.8) becomes

$$t_{nm}^{(N)} = \frac{1}{n+m-2}\left\{\exp\left[-\sigma(n+m-2)\right] - \exp\left[-\sigma\gamma(n+m-2)\right]\right\} \tag{2.9}$$

(remark that $t_{11}^{(N)} = \sigma(\gamma-1)$). The largest eigenvalues and the corresponding eigenvectors of this matrix can be easily computed by means of a combination of the power method and of the Hotelling deflation method (Ralston, 1965), since the matrix is symmetric. When we have the eigenvalues $\alpha_{N,k}^2$ and the corresponding singular vectors $v_{N,k}$, the singular functions $u_{N,k}(t)$ can be obtained by means of the relation

$$u_{N,k}(t) = \frac{\sigma}{\alpha_{N,k}} \sum_{n=1}^{N} v_{N,k}(p_n)e^{-p_n t} , \tag{2.10}$$

where $v_{N,k}(p_n)$ denotes the n-th component of $v_{N,k}$.

The singular values $\alpha_{N,k}$ can provide a good approximation of the singular values α_k of the problem with continuous data. The kernel of eq.(1.20) is given by $\epsilon_N(t,s) = \epsilon_N(t+s)$ where

$$\epsilon_N(t) = \frac{1}{t} - \sigma\frac{1 - \exp(-N\sigma t)}{1 - \exp(-\sigma t)} \tag{2.11}$$

and this function tends to zero uniformly over the interval $2 \leq t \leq 2\gamma$ when $\sigma \rightarrow 0$ and $N \rightarrow +\infty$. Therefore from eq.(1.21) it is expected that $\alpha_{N,k} \sim \alpha_k$ if one takes a sufficiently large number of data points at a sufficiently small distance σ. The following singular values

$$\alpha_{N,0}=9.075 \times 10^{-1}, \alpha_{N,1}=1.983 \times 10^{-1}, \alpha_{N,2}=3,555 \times 10^{-2} \tag{2.12}$$

have been obtained in the case $\gamma = 5$ using $N = 64$ and $\sigma = 0.03$ (Bertero et. al., 1983). If one compares with the singular values of the continuous case, eq.(2.7), one finds an average error of about 4%. This is the best approximation which can be obtained using 64 data points.

3. THE EFFECT OF GEOMETRIC SAMPLING OF DATA

In the case of Laplace inversion the use of equidistant points is not the best one. Such a choice is only convenient for deconvolution problems, i.e. for the solution of problems which have simple transformation properties with respect to the translation group. In the case of Laplace transform inversion the underlying group is the dilation group and for this reason a geometric distribution of data points seems to be more natural, at least from the theoretical point of view.

Therefore we consider now the following data-point distribution

$$p_n = p_1 \Delta^{n-1} \quad ; \; n = 1,\ldots,N \tag{3.1}$$

where p_1 and Δ can be considered as free parameters. As concerns the choice of the weights w_n, the following result can be proved (Bertero et al., 1983): if the p_n are as in eq.(3.1) and if

$$w_n = (\ln \Delta)\, p_n \tag{3.2}$$

then, given ε, it is possible to choose Δ and N so that, for any $t \in [2,2\gamma]$

$$|\epsilon_N(t)| = \left| \frac{1}{t} - \sum_{n=1}^{N} w_n e^{-p_n t} \right| \leq \varepsilon \, . \tag{3.3}$$

This result, combined with eqs.(1.17),(1.20), suggests that we introduce the weights (3.2) into the scalar product (1.15).

As concerns the computation of the eigenvalues and eigenvectors of the matrix (2.8) we remark that it is not symmetric in the usual sense ($t^{(N)}_{nm} \neq t^{(N)}_{mn}$) since it is self-adjoint with respect to the weighted scalar product (1.15). Then, for computational convenience, we transform the matrix (2.8) into a symmetric one having the same eigenvalues

$$t^{(N)}_{nm} = \frac{\sqrt{w_n w_m}}{p_n + p_m} \left\{ \exp\left[-(p_n+p_m)\right] - \exp\left[-\gamma(p_n+p_m)\right] \right\}. \tag{3.4}$$

We also remark that if $\bar{v}_{N,k}$ is an eigenvector of this matrix, the corresponding eigenvector of the matrix (2.8) is given by

$$v_{N,k}(p_n) = w_n^{-1/2}\, \bar{v}_{N,k}(p_n) \, . \tag{3.5}$$

Therefore singular values and singular vectors can be computed again by a combination of the power method and of the Hotelling deflation method.

The fundamental property of geometric sampling of data is that, using a very small number of data points, it is possible to obtain extraordinarily good approximations of the singular values of the problem with continuous data (and also of the singular functions). For instance, again for the case $\gamma = 5$, using 5 data points, $\Delta = 5.5$ and $p_1 = 1.366 \times 10^{-3}$ we have obtained the following singular values (Bertero et al., 1983)

$$\alpha_{N,0}=8.742 \times 10^{-1},\ \alpha_{N,1}=1.913 \times 10^{-1},\ \alpha_{N,2}=3.827 \times 10^{-2}. \qquad (3.6)$$

If we compare with the singular values of the problem with continuous data, we find an average error of about 1/2% . Also the singular functions $u_{N,k}(t)$ (k=0,1,2) provide an excellent approximation of the singular functions $u_k(t)$ (k=0,1,2) corresponding to the problem with continuous data.

Now, if we notice that, when the "signal-to-noise ratio" is equal to 100, the "number of degrees of freedom" in the case $\gamma=5$ is just 3, we can conclude that using 5 data points optimally placed, the filtered solution (1.22) practically coincides with the filtered solution (1.13), the main error deriving from the different noise contribution to the "image" components g_k and $g_{N,k}$. Therefore 5 data points allow almost the same resolution as continuous data.

We believe that the previous result can be extended to any severely ill-posed problem (namely a problem where the "number of degrees of freedom" is small): a number of data points, optimally placed, approximately equal to the "number of degrees of freedom" is practically equivalent to continuous data.

R E F E R E N C E S

Bertero, M., De Mol, C. and Viano G.A., 1980, Inverse Scattering Problems in Optics, ed. H.P. Baltes, Topics in Current Physics, Vol.20 (Berlin: Springer), pp.161-214.

Bertero, M., Boccacci, P. and Pike, E.R., 1982, Proc. R. Soc. Lond. A383, pp. 15-29.

Bertero, M., Brianzi, P. and Pike, E.R., 1983 (in preparation).

Hämmerlin, G. and Schumaker, L.L., 1980, Numer. Math. 34, pp.125-141.

Hille, E. and Tamarkin, J.D., 1931, Acta Math. 57, pp.1-76.

Picard, E., 1910, R.C. Mat. Palermo 29, pp. 615-619.

Ralston, A., 1965, A First Course in Numerical Analysis (New York: McGraw-Hill)

Riesz, F. and Sz. Nagy, B., 1955, Functional Analysis, (New York: Ungar).

Twomey, S., 1965, Jour. Franklin Inst. 279, pp.95-109.

Wilkinson, J.H., 1965, The Algebraic Eigenvalue Problem (Oxford: Claredon Press).

II.6 (OI.3)

OBJECT RECONSTRUCTION FROM PARTIAL INFORMATION

M.A. Fiddy

Physics Department, Queen Elizabeth College,
University of London, Campden Hill Road,
Kensington, London W8 7AH, UK.

ABSTRACT. The first Born and Rytov approximations are briefly described and compared. If they provide a suitable model, limited data on the Fourier transform of the object can be calculated. Optimum use of these data through the incorporation of prior knowledge is described. The problem when only intensity data are available is also discussed.

1. INTRODUCTION

There are two approaches to determining the internal structure of objects from scattered-field data. The first relies on a geometrical-optics description which requires the assumption that the wavelength used is considerably smaller than the structural detail of interest. This assumption is adequate at X-ray wavelengths, for example, when used in computerised tomography. When the wavelength is not relatively small, scattering and diffraction effects become important and object information, e.g. a scattering potential, is approximately reconstructed by some method of inverse scattering. These methods usually rely on the assumption that the scattering is weak, and have become known as backpropagation algorithms for diffraction tomography [e.g. 1]. Inversion formulae based upon the first Born or Rytov approximations [2,3,4] are essentially Fourier-reconstruction techniques and the questions of uniqueness and stability of the estimate for the object obtained in this way have received much attention [e.g. 5]. We briefly describe here the nature of these approximations (§ 2), the inversion problem when limited complex Fourier data are available (§ 3) and the phase retrieval problem when only limited intensity data are available.

W.-M. Boerner et al. (eds.), Inverse Methods in Electromagnetic Imaging - Part 1, 341–349.

2. THE BORN AND RYTOV APPROXIMATIONS

A solution to the wave equation is difficult to calculate if a small perturbation approximation is not taken. Consider the scaler wave equation

$$[\nabla^2 + k^2(1 + \delta n)^2]\, U = 0 \tag{1}$$

where k is the average wave-number and δn the fluctuation of, say, the refractive index about the mean. It is this function, δn, the function describing the object, that we wish to reconstruct. It is usual [e.g. 6] to choose a series solution for U in terms of an ordering parameter ε having either of the forms

$$U = \sum_{j=0} \varepsilon^j U_j = \exp\left(\sum_{j=0} \varepsilon^j \Psi_j\right) \tag{2}$$

Converting equation (1) into an integral equation and using the Liouville-Neumann method of successive substitutions gives [7]

$$\begin{aligned} U &= U_o + U_s \\ &= U_o + U_{s1} + U_{s2} + \ldots \\ &= U_o + k^2 \int_V (2\,\delta n + \delta n^2) G\, U_o\, dr + k^2 \int_V (2\,\delta n + \delta n^2) G\, U_{s1}\, dr + \ldots \end{aligned} \tag{3}$$

where U_o is the incident field, U_s the scattered field and V the scatterer volume, and the Green function G is defined by

$$[\nabla^2 + k^2]\, G = -4\pi\,\delta(r) \tag{4}$$

If we write δn as $\varepsilon\delta n$ and U_{sj} as $\varepsilon^j U_j$ we find that

$$\begin{aligned} U_1 &= k^2 \int_V 2\,\delta n\, G\, U_o\, dr \\ U_2 &= k^2 \int_V \delta n^2\, G\, U_o\, dr + k^2 \int_V 2\,\delta n \int_V 2\,\delta n\, G\, U_o\, dr\, G\, dr \quad \text{etc.} \end{aligned} \tag{5}$$

upon equating equal powers of ε; U_1 rather than U_{s1} is frequently taken to be the first Born approximation to U_s. From equation (2) it is straightforward to show that [8]

$$\begin{aligned} U_o &= e^{\Psi_o} & \Psi_o &= \ln U_o \\ U_1 &= U_o \Psi_1 & \Psi_1 &= U_1/U_o \\ U_2 &= U_o(\Psi_1^{\,2}/2 + \Psi_2) & \Psi_2 &= U_2/U_o - \tfrac{1}{2}(U_1/U_o)^2 \end{aligned} \tag{6}$$

and, obviously, a sufficient condition for the convergence of the series in equation 2 is that successive terms get smaller sufficiently quickly.

If one represents the scattered field U, with unwrapped phase, by $U = \exp(\Psi)$, where Ψ is a complex phase, then equation (1) gives

$$\nabla^2 \Psi + |\nabla \Psi|^2 + k^2(1 + \delta n)^2 = 0 \tag{7}$$

which has the solution [e.g. 6,9]

$$\begin{aligned}\Psi &= \Psi_o + \Psi_s \\ &= \Psi_o + \frac{k^2}{U_o} \int_V (2\,\delta n + \delta n^2) G\, U_o\, dr + \frac{1}{U_o} \int_V |\nabla \Psi_s|^2\, G\, U_o\, dr\end{aligned} \tag{8}$$

The assumption that $k^2(2\,\delta n + \delta n^2) \gg |\nabla \Psi_s|^2$ is the Rytov approximation which can be seen to be equivalent to taking, as an approximation to Ψ, the expression $\Psi_o + \Psi_1$ (from equation (6) if the term in δn^2 is neglected). The complex phase Ψ within the Rytov approximation is simply related, therefore, to the first Born approximation to U_s. Formally, the neglect of the δn^2 term gives a Rytov solution which is equivalent to the first order perturbation solution, as was pointed out by Fried [9]. While this mathematical equivalence has been emphasised (in particular by Devaney [1]) the validity of the two approximations for U_s depends upon the terms neglected. If one considers an exponentiated Born series as in equation (2), one might deduce that the Born and Rytov approximations have the same domain of validity by comparing the second order terms [10]. However, one could obtain an iterative solution by substituting Ψ_1 for Ψ_s in equation (8) as suggested by Ishimaru [6]. If we again write δn as $\varepsilon\,\delta n$ and make this substitution in equation (8), we obtain a sequence of Ψ_j's ordered by powers of ε which bear no simple relation to the Ψ_j's in equations (2) and (6), the two expressions for Ψ_2 being of similar magnitude only if

$$\int_V |\nabla \Psi_1|^2\, G\, U_o\, dr \simeq \int_V 2\,\delta n\, U_o\, \Psi_1\, G\, dr - \frac{\Psi_1^2}{2\, U_o} \tag{9}$$

One could expect that Rytov will be preferable when fluctuations in δn are large compared to the wavelength and for larger distances of propagation [9].

In either the first Born or Rytov approximations, it follows [11, 2] that the Fourier transform of the scattered field U_1 (Born) or $U_o\,\Psi_1$ (Rytov) over any plane normal to the incident field direction allows one to determine $\widehat{\delta n}$, the Fourier transform of δn, on the surface of an Ewald sphere of radius k and whose centre is determined by the incident field direction. By varying the angle of incidence one can calculate $\widehat{\delta n}$ on sets of semi-circular arcs and, when data are sufficiently dense, inverse transform. Alternatively, rather than interpolating in Fourier space, for each angle of incidence one can propagate either U_1 or $U_o \Psi_1$ back into object space using the "inverse" Green function (assuming no evanescent waves) and superpose [1].

Three problems can arise with these procedures. The first concerns the consequences of having an object δn which does not conform to the requirements for the validity of the Born or Rytov approximations. There is much to be done here by comparison with exact solutions or data from known objects. The second concerns the consequences of having only limited incident angles and limited samples of the scattered field. Thirdly, the required phase unwrapping of U in order to make use of the Rytov approximation is non-trivial.

3. LIMITED COMPLEX DATA AVAILABLE

The restoration or reconstruction of an object when only limited (incomplete) discrete data are available is a common problem with any radiation. Such problems are known to be ill-posed but regularisation theory defines a set of admissible or approximate estimates for the object by imposing constraints on these estimates [5]. Even in the absence of noise, incomplete sampled data means that there are infinitely many object estimates consistent with the Fourier data and the object support, if known. This ambiguity is overcome by using prior knowledge to reduce the class of allowed solutions and to design an optimality criterion. For finite data it is always necessary to use models to find an optimum estimate $\widetilde{\delta n}$ for δn, and a simple Fourier series expression may not be the best. A commonly used and quite successful technique for interpolating or extrapolating Fourier data is the iterative procedure of Gerchberg-Papoulis [12]. By repeatedly Fourier transforming between object and data spaces, imposing the anticipated object support and measured data values at each iteration, convergence is to a minimum norm solution estimate. This solution is the orthogonal projection onto the subspace of appropriately space limited and data consistent object functions. The numerical performance of such an algorithm will depend upon the noise but can be regularised and the corresponding error assessed [13]. We let the function p encode the support of the object and perhaps some profile information also. One can choose an optimum estimator for δn which incorporates this. Suppose

$$\widehat{\delta n}(K_j) = \int_V \delta n(r) e^{-iK_j . r} d^3 r$$

$$= (\delta n, g_j) \qquad (10)$$

If we consider H, the weighted L_2 space with inner product

$$(f,g)_H = \int_V p^{-1} f g^* d^3r \tag{11}$$

where p is real and positive and * denotes complex conjugate then

$$\widehat{\delta n}(K_j) = (\delta n, pg_j)_H \tag{12}$$

The object estimate incorporating p and given by

$$\widetilde{\delta n} = p \sum_j a_j g_j \tag{13}$$

minimises $\| \delta n - \widetilde{\delta n} \|_H$ and is a unique estimate of minimum norm when the a_j found by solving the linear equations [14]

$$\widehat{\delta n}(K_j) = \sum_m a_m (pg_m, pg_j)_H = \sum_m a_m \hat{p}(K_j - K_m) \tag{14}$$

The $\widetilde{\delta n}$ thus obtained will have features driven by p and will be data consistent wherever the data $\widehat{\delta n}(K_j)$ are located. If p is a binary function encoding only the object support then equation (13) gives the closed form solution of the Gerchberg-Papoulis iterative method. Clearly, if p is incorrect or inappropriate a data consistent $\widetilde{\delta n}$ will still be found. Also, any broadband noise on the data can cause spurious oscillations in $\widetilde{\delta n}$ because of the requirement that it be zero outside the object support. However, a more realistic function p would take on a small value outside the expected object support and this ensures stability. This modification to p has been shown to be equivalent to a form of Miller regularisation [15].

Examples of the application of this technique to limited angle limited sampled Fourier data are shown in [15]. Extensions of this approach to provide a non-linear estimator for which Burg's maximum entropy method is a special case can be found in [16].

4. LIMITED INTENSITY DATA AVAILABLE

If only the intensity of the field U or the scattering amplitude F, can be measured then its phase must be found; we assume in this section that some kind of holographic phase recording is not possible. The phase retrieval problem has been the subject of several recent reviews [e.g. 17,18] and can be conveniently divided into one dimensional and more than one dimensional problems. It is well known that in one dimension many equally acceptable solutions for the phase could occur even with complete and noise free data. The reason for this is that if U or F is a bandlimited function (entire function of exponential type) it can be specified by its zero locations like a polynomial (the Hadamard product). The intensity data has the zeros of U and, in conjugate positions those of

U* leading to a zero symmetry about the real axis. If U had M complex zeros then 2^M possible phase functions could be constructed consistent with the intensity data. In higher dimensions a product formation still exists for U but now it is a product of irreducible factors (the Osgood product) having from one to an infinity of factors [18].

Clearly, with finite data and arbitrary phases it is always possible to find an infinity of δn having any support and consistent with the modulus samples of the scattering amplitude. A technique such as that described in § 3 would do this. By enforcing, with the weighting function p (equation (13)), that δn should be identically zero outside some support, an extrapolation of the available data is realised regardless of the phases associated with that data. If the support information encoded by p is correct, and in general it cannot be deduced from the intensity data [19], still no constraints can be brought to bear on the choice of phases to associate with the data.

The problem becomes somewhat more tangible if we do not impose the constraint that δn is zero or very small outside its support. If a discrete model is adopted for δn of the form

$$F(x,y,z) = \sum_{j,k,l}^{M-1} \delta n_{jkl} e^{ijx} e^{iky} e^{ilz} \tag{15}$$

we can write

$$F(u,v,w) = \sum_{j,k,l}^{M-1} \delta n_{nkl} u^j v^k w^l \tag{16}$$

which is a polynomial of finite degree. We wish to find M^3 object samples δn_{jkl} from at least this number of samples of F or $|F|^2$, with some $\delta n_{jkl} = 0$ defining the object support.

In this form, it has been shown [20] that for the set of multivariable polynomials of finite degree the set of reducible polynominals is of measure zero. This implies that F is very likely to be a single irreducible polynomial and hence that a unique phase can be associated with $|F|$. To be certain of phase uniqueness for this model there is the following necessary condition. If the nonzero coefficients δn_{jkl} of equation (16) are replaced by unity and the corresponding polynomial [F] known as the form of F [21] is irreducible then F is irreducible. In terms of objects and their scattered fields this means that if the scattering amplitude from an aperture or uniform volume is irreducible then it remains so whatever δn is introduced into that support. Actually demonstrating irreducibility is non-trivial but a suffient condition was recently suggested based on Eisenstein's criterion [22]. The constraints on the object for this criterion to hold are that two points A and B in a plane are non zero, the object support lying anywhere within the box shown in the figure. Point A may be anywhere to the right of the location shown and when sufficiently far would result in $|F|^2$ being an off-axis hologram from which ϕn would be found.

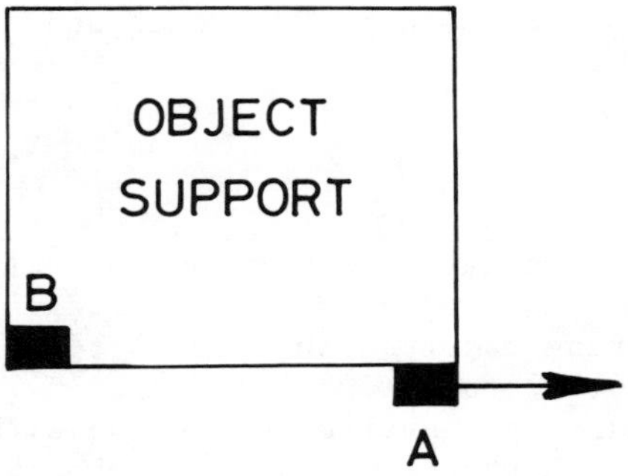

Ensuring that there is a unique phase in this way does not help us find it. An algorithm such as Fienup's [23] which repeatedly Fourier transforms between spaces, imposing the object support and intensity data each iteration, is moderately successful. This is particularly the case for object supports lacking symmetry which may relate to the comments made above. However the method only succeeds when Eisenstein's criterion is satisfied when the reference point A is large; this corresponds to an almost holographic situation, Fienup's algorithm converging because a simple initial phase guess is close to the actual phase.

Thus apart from making additional and different measurements of some kind [e.g. 18], knowing that the phase is unique does not offer any obvious way for determining it.

5. CONCLUSIONS

This paper was concerned with the problem of object reconstruction from partial information and three main aspects of this problem were discussed.

(i) Inverse scattering modelled by the first Born or Rytov approximations is computationally feasible provided the phase unwrapping can be reliably carried out for Rytov. However, conditions for the validity of these approximations and their accuracy remain vague.

(ii) The first Born or Rytov approximations can be regarded as requiring a Fourier inversion for object reconstruction. The difficulties associated with having sampled limited (angle) data can be reduced by constructing an estimator for the object, which can incorporate whatever prior knowledge about it is available.

(iii) If only single intensity data sets are available, prior knowledge about the object such as its support can be incorporated but appears unhelpful unless a discrete model for the object is adopted. A unique phase can then be assured but finding that phase appears to be an open problem.

6. ACKNOWLEDGEMENTS

The author would like to acknowledge the support of NATO (in particular NATO grant 0556/82) and the U.K. S.E.R.C. He would also like to acknowledge his collaboration on the work presented here with C.L. Byrne, A.M. Darling, J.C. Dainty, S. Leeman and L. Zapalowski.

7. REFERENCES

1. Devaney, A.J., 'A filtered backpropagation algorithm for diffraction tomography', Ultrasonic Imaging, 4, 336-350 (1982).
2. Devaney, A.J., 'Inverse scattering theory within the Rytov approximation', Opt. Letters, 6, 374-376 (1981).
3. Devaney, A.J., 'Inversion formula for inverse scattering within the Born approximation', Opt. Letters, 7, 111-112 (1982).
4. Kaveh, M., Soumekh, M. Lu, Z.Q., Mueller, R.K. and Greenleaf, J.F., 'Further results on diffraction tomography using Rytov's approximation', Acoustical Imaging, 12, Eds. E.A. Ash and C.R. Hill, 599-608 (1982).
5. Bertero, M., De Mol, C. and Viano, G.A., 'The stability of inverse problems', ch.5 of Inverse Scattering Problems in Optics, Ed. H.P. Baltes, 20, Topics in Current Physics, Springer-Verlag, Berlin, 161-214 (1980).
6. Ishimaru, A., 'Wave propagation and scattering in random media', Vol. II, ch. 17, Academic Press, New York (1978).
7. Morse, P. and Feshbach, H., 'Methods of Theoretical Physics', Vol. II, McGraw-Hill, New York (1953).
8. Sancer, M.I. and Varvatsis, A.D. 'A comparison of the Born and Rytov methods', Proc. IEEE, 58, 140-141 (1970).
9. Fried, D.L., 'Diffusion analysis for the propagation of mutual coherence', J. Opt. Soc. Amer., 58, 961-969 (1968).
10. Yura, H.T., Sung, C.C., Clifford, S.F. and Hill R.J., 'Second-order Rytov approximation', J. Opt. Soc. Amer., 73, 500-502 (1983).
11. Wolf, E., 'Three-dimensional structure of semi-transparent objects from holographic data', Opt. Communications, 1, 153-156, (1969).
12. Papoulis, A., 'A new algorithm in spectral analysis and band-limited extrapolation', IEEE CAS-22, 735-742 (1975).
13. Abbiss, J.B., Defrise, M., De Mol, C. and Dhadwal, H.S., 'Regularised iterative and non-iterative procedures for object restoration in the presence of noise: an error analysis', J. Opt. Soc. Amer., 73, November issue (1983).
14. Hall, T.J., Darling, A.M. and Fiddy, M.A., 'Image compression and restoration incorporating prior knowledge', Opt. Letters, 7, 467-468 (1982).
15. Darling, A.M., Hall, T.J. and Fiddy, M.A., 'Stable, non-iterative, object reconstruction from incomplete data using

prior knowledge', J. Opt. Soc. Amer., 73, November issue (1983).
16. Byrne, C.L., Fitzgerald, R.M. Fiddy, M.A., Hall, T.J. and Darling, A.M., 'Image restoration and resolution enhancement', J. Opt. Soc. Amer., 73, November issue (1983).
17. Taylor, L.S., 'The phase retrieval problem', IEEE AP-29, 406-416 (1981).
18. Fiddy, M.A., 'The phase retrieval problem', SPIE 413, Inverse Optics, Ed. A.J. Devaney, paper 26 (1983).
19. Fienup, J.R., Crimmins, T.R. and Holsztynski, W., 'Reconstruction of the support of an object from the support of its autocorrelation', J. Opt. Soc. Amer., 72, 610-624 (1982).
20. Hayes, M.H. and McCllean, J.H., 'Reducible polynomials in more than one variable', Proc. IEEE, 70, 197-198 (1982).
21. Dainty, J.C. and Fiddy, M.A., 'The essential role of prior knowledge in phase retrieval', submitted for publication.
22. Fiddy, M.A., Brames, B.J. and Dainty, J.C., 'Enforcing irreducibility for phase retrieval in two dimensions', Opt. Letters 8, 96-98 (1983).
23. Fienup, J.R., 'Phase retrieval algorithms: a comparison', App. Optics, 21, 2758-2769 (1982).

II.7 (NM.5)

APPROXIMATION OF IMPULSE RESPONSE FROM TIME LIMITED INPUT AND OUTPUT: THEORY AND EXPERIMENT

Tapan K. Sarkar[1]
Soheil A. Dianat[1]
Bruce Z. Hollmann[2]

Abstract. Since it is practically difficult to generate and propagate an impulse, often a system is excited by a narrow time domain pulse. The output is recorded and then a numerical deconvolution is often done to extract the impulse response of the object. Classically, the fast Fourier transform technique has been applied with much success to the above deconvolution problem. However, when the signal to noise ratio becomes small, sometimes one encounters instability with the FFT approach. In this paper, the method of conjugate gradient is applied to the deconvolution problem. Unlike the FFT, this approach solves the deconvolution problem entirely in the time domain with any desired degree of accuracy. The conjugate gradient method has the advantage of both a direct method (i.e., the solution is obtained in a finite number of steps) and that of an iterative method (the round-off and truncation are generally limited to the last stage of iteration). Moreover, the method converges for any initial guess. Also for the application of this method the time samples need not be uniform as in the FFT. In addition, the dc component can be recovered with this method, but not with FFT. Computed impulse responses utilizing this technique are presented for measured incident and scattered fields from a sphere and a cylinder.

---

[1] Tapan K. Sarkar
[1] Soheil A. Dianat
Department of Electrical Engineering
Rochester Institute of Technology
Rochester, New York 14623 U.S.A

[2] Bruce Z. Hollmann
Naval Surface Weapons Center
Dahlgren, Virginia 22448 U.S.A.

This work has been supported in part by the U.S. Office of Naval Research at Rochester Institute of Technology, and by the U.S. Naval Air Systems Command at the Naval Surface Weapons Center.

W.-M. Boerner et al. (eds.), Inverse Methods in Electromagnetic Imaging - Part 1, 351–364.

1. Introduction

Consider an electromagnetic system which is excited by a causal time domain waveform x(t), and the recorded output is y(t). Then we know they are related by the convolution integral

$$y(t) = \int_0^{\infty} x(t - \tau)h(\tau)d\tau = \int_0^{t} x(t - \tau)h(\tau)d\tau, \tag{1}$$

where $h(\tau)$ is the impulse response of the system.

We could write (1) as an operator equation

$$\hat{A}\, h(t) = y(t) \tag{2}$$

where $\hat{A}$ is the convolution operator $\int_0^{\infty} x(t - \tau)(\cdot)d\tau$. Even though (1) is a very simple integral equation, the numerical solution of (1) is an extremely difficult problem. The difficulty is that the solution h(t) does not depend continuously on the data.

To illustrate this point, consider two solutions to (1)

$$h_1(t) = h(t) \;\Rightarrow\; \text{exact solution} \tag{3}$$

and

$$h_2(t) = h(t) + B\cos\omega t \tag{4}$$

where B and ω are two arbitrary constants. Now observe that

$$|y_2(t) - y_1(t)| = |B\int_0^{t} x(t - \tau)\cos\omega\tau d\tau| \tag{5}$$

As an example, consider the causal input to be

$$x(t) = e^{-t} \qquad \text{for } t > 0 \tag{6}$$

By utilizing (6) in (5) we obtain

$$|y_2(t) - y_1(t)| = \left| B \cdot \frac{\omega \sin \omega t + \cos \omega t - e^{-t}}{\omega^2 + 1} \right| \leqslant |B| \cdot \left[\frac{2 + \omega}{\omega^2 + 1} \right] \tag{7}$$

Equation (7) tells us that for a fixed B, we can choose ω large enough so that the absolute difference between $y_2(t)$ and $y_1(t)$ would be very small. Since we do computation with a finite number of digits, the difference in $y_2(t) - y_1(t)$ can be interpreted as the digital representation of the exact output. Therefore, we could have two inpulse responses $h_1(t)$ and $h_2(t)$ which are radically different and yet produce the same output y(t) when convolved with x(t). It is important to point out that the problem is not in the digital representation of the analog signal nor in the numerical technique used to obtain the solution. The problem is inherent in equation (2). Operator equation (2) belongs to the class of problems defined as ill-posed problems [1-3]. What the problem actually is, is that in (2) the operator does not have an inverse. We now give a mathematical proof for the cause of instability of equation (1) for any input x(t).

Since $\hat{A}h = y$, symbollically we can write the solution

$$h(t) = \hat{A}^{-1} y(t) \tag{8}$$

However, if $\hat{A}^{-1}$ does not exist, then of course (8) has no meaning. We now explain mathematically that $\hat{A}^{-1}$ does not exist. Since x(t) is the input to the system and also as the input is of finite energy, x(t) is square integrable, i.e.

$$\int_0^T x^2(t)dt < \infty \tag{9}$$

In Stakgold's notation [4], x(t) is the kernel of the integral operator $\hat{A}$. Since the kernel is square integrable, the operator is Hilbert Schmidt [p. 300, 4]. It is shown [p. 336, 4] that for a Hilbert Schmidt operator, the smallest eigenvalue approaches zero. In other words, zero is the accumulation point of the smaller eigenvalues of $\hat{A}$ and hence $\hat{A}^{-1}$ does not exist. This is not of any solace to a person who is interested in obtaining a solution to (1).

In summary, even though theoretically $\hat{A}^{-1}$ does not exist, numerically we can still find a solution and now we have to be very careful in what we mean by a solution. We explain in the forthcoming section how we obtain a solution and what is our definition of the solution. It is important to stress that for the deconvolution problem, the concept of a solution is subjective.

2. Historical Perspective

Interestingly enough, the convolution type integral equation arises in almost all areas of engineering and researchers in electrical engineering have been solving this problem for quite a few years. Yet there is no clearcut solution to this problem for reasons mentioned above.

A conventional way of solving (1) is to transform it to the frequency domain by the application of the Fast Fourier Transform. So we obtain:

$$H(j\omega) = \frac{Y(j\omega)}{X(j\omega)} \tag{10}$$

and then obtain h(t) by the inverse Fourier Transform

$$h(t) = \frac{1}{2\pi} \int_{-\infty}^{\infty} H(j\omega) e^{j\omega t} d\omega \tag{11}$$

Even when there is no noise in the data, the above approach for deconvolution of waveforms arising in electromagnetic systems runs into problems for three specific reasons. The reasons, as observed by other researchers [5-7], are as follows.

(a) It is quite possible that at a particular frequency ω, $X(j\omega)$ becomes zero in (10) and $H(j\omega)$ blows up.

(b) Often in electromagnetic systems, h(t) has an impulse at $t = 0$. The conventional z-transform techniques cannot handle this jump discontinuity at $t = 0$. Special care has to be taken to extract this impulse at the origin. Even though methods exist to do this type of extraction, they are problem dependent and sensitive to signal to noise ratio.

(c) In most cases, the transient responses are time limited and, hence, by definition they are not band limited. Therefore, if one obtains the transform $X(j\omega)$ and $Y(j\omega)$ without paying proper attention to the sampling frequency, aliasing could be a severe problem.

It is for these reasons that researchers, for deconvolving transient waveforms, have applied techniques which are entirely in the time domain, i.e. the deconvolution is done without going to the frequeency domain [7-10]. In this paper we apply the conjugate gradient method to solve the deconvolution problem entirely in the time domain. Since we are solving the problem entirely in the time domain, we do not have to worry about the problem of aliasing. However, the problem about which we have to be careful is that the samples are close enough so we can approximate the time domain waveform by linear segments. The conjugate gradient method has been applied previously for the ill-posed image enhancement problems [11]. In [12] it has been shown that if the number of iterations in the conjugate gradient method is determined properly, then the undue oscillations in the solution do not appear.

Another advantage of the conjugate gradient method is that we always recover the impulse of the impulse response at the origin, since we do not deal with discrete transforms.

3. The Method of Conjugate Gradient

In this method, the basic philosphy is that instead of solving for $h = \hat{A}^{-1}y$ directly, we try to minimize the functional F(h) given by

$$F(h) = \langle R, R \rangle = \langle \hat{A}h - y, \hat{A}h - y \rangle \tag{12}$$

where the inner product is defined as

$$\langle C,D \rangle = \int_z C(z)D(z)dz \tag{13}$$

and the norm of C is given by

$$\| C \|^2 = \langle C, C \rangle = \int_z |C^2(z)| dz \tag{14}$$

The conjugate gradient method starts with an initial guess h_0 and generates

$$P_0 = - b_0 \hat{A}^* R_0 = - b_0 \hat{A}(^* \hat{A} h_0 - y) \tag{15}$$

where $\hat{A}^*$ represents the adjoint operator. With reference to equation (1), the adjoint operator $\hat{A}^*$ is the advance convolution operator. We now define the adjoint operator as [4]

$$\hat{A}^* z = \int_0^\infty x(t - \tau)z(t)dt = \int_0^\infty x(t)z(t + \tau)dt \tag{16}$$

In defining (16) it has been assumed that x(t) is causal. The conjugate gradient method then develops

$$h_{k+1} = h_k + \alpha_k P_k \tag{17}$$

$$R_{k+1} = R_k + \alpha_k \hat{A} P_k \tag{18}$$

$$\alpha_k = \frac{1}{\| \hat{A} P_k \|^2} \tag{19}$$

$$P_{k+1} = P_k - b_{k+1} \hat{A}^* R_{k+1} \tag{20}$$

and

$$b_k = \frac{1}{\| \hat{A}^* R_k \|^2} \tag{21}$$

Equations (15)-(21) are somewhat different from the conventional Hestenes and Steifel algorithms [13-14]. The above method brings about a certain saving in memory. This is because in [14] one needs to store the four vectors h_k, P_k, R_k and $\hat{A}P_k$ whereas in the modified algorithm it is not necessary to store $\hat{A}P_k$.

The method of conjugate gradient always converges for any initial guess and for any functional equations with a bounded operator as long as (15)-(21) are implemented. The actual rate of convergence is given in [15]. It is important to point out that even when y is not in the range of the operator $\hat{A}$ (this may happen when y is contaminated with noise), the method of conjugate gradient yields the minimum norm solution. In addition, it is shown in [15] that the method of conjugate gradient solves the normal equation

$$\hat{A}^* \hat{A} h = \hat{A}^* y \tag{22}$$

exactly.

The last question we address in this section is how to terminate the iteration. This is because, for ill-posed problems, too many iterations will minimize the error $\hat{A}h - y$ no doubt, but would introduce spurious oscillations in the solution. The reason for this is explained by the example of section 1. On the contrary, too few iterations may give a solution that is not good enough. The criterion that we utilize to stop the iterative process in the conjugate gradient method is that the magnitude of the largest elemtent in the residual must be below a certain value, i.e.

$$| R_n(t) | \leqslant M \text{ (a prefixed constant)} \tag{23}$$

Now we discuss the problem of choosing M.

For most problems of interest, the elements of the input and output are not sharply defined. This is because there is always some discretization or quantization error associated with each sample of x(t) and y(t). It is now assumed that all that is known about the input x(t) and output y(t) is that they are within the following intervals. Here the subscript e denotes exact quantities.

$$x_e(t) - \Delta x(t) \leqslant x(t) \leqslant x_e(t) + \Delta x(t) \tag{24}$$

$$y_e(t) - \Delta y(t) \leqslant y(t) \leqslant y_e(t) + \Delta y(t) \tag{25}$$

It is further assumed that $\Delta x(t_n)$ and $\Delta y(t_n)$ at the sampled values of the input and output are independent distinct quantities due to round-off error and, moreover, one does not depend on the other. Thus,

$$\int_0^T x_e(t - \tau)h_e(\tau)d\tau = y_e(t) \tag{26}$$

and

$$R(t) = \int_0^T x(t - \tau)h(\tau)d\tau - y(t) \tag{27}$$

Observe in (26) and (27) we have replaced the upper limit of ∞ in (1) by a finite value T. This is because in actual practice we can record the data only up to a finite time T. We further assume that

$$|\Delta x(t)| \leqslant C \text{ (constant)} \tag{28a}$$

and

$$|\Delta y(t)| \leqslant C \text{ (same constant as in (28a))} \tag{28b}$$

We then check at each interation whether

$$C\left[\int_0^T |h(\tau)|d\tau + 1\right] \geqslant |R(t_n)| \tag{29}$$

where t_n represents the sampling instances. As was shown by Oettli and Prager [16] the above inequality in (29) is both a necessary and sufficient condition for h(t) to be a solution of (26) under (24) and (25). So for the conjugate gradient method, where the residuals are computed routinely, if for a certain residual R(t) the inequality in (29) is satisfied, we have obtained an excellent solution under the conditions of (24) and (25). Also, note that the introduction of the conditions (24) and (25) does not make the solution of (26) unique. In fact, there are many solutions of (26) under the constraints of (24) and (25). However, only those solutions are acceptable which satisfy inequality (29).

The next question is how to determine the constant C in (28). The value of C is obtained from the number of bits used in the quantization process. In our experiments, eight bit quantization has been utilized in recording the digital form of the signal. However, the number of effective bits in the signal would be much less than eight, because of additional noise in the data. This could be stray pick-up, base line shift, time sampling jitter and so on. In all our theoretical computations we have utilized 4.5 effective bits, rather than eight. The figure 4.5 has been chosen after careful analysis of the equipment and experimental set up. For our case, the constant C is given by

$$C = \frac{1}{2^{4.5}} = .04419417 \qquad (30)$$

This value of C has been utilized in all the computations reported in section 6.

A brute force universal way of making this algorithm perform well over a wide range of problems is to stop the iteration interactively when the solution starts diverging. This is because iterative methods for ill-posed problems provide asymptotic convergence, i.e. the solution improves with the first few iterations and then becomes worse as more iterations are done.

4. Description of the Experimental Setup

The experiments were conducted on an indoor aluminum time domain range, 9.14 meters (30 ft) in diameter. The conical transmitter antenna, in the center of the ground plane, has a half angle of 5.27 degrees, and is 4.19 meters (13 ft. 9 in.) tall. A sub-nanosecond Gassian pulse is transmitted and strikes an object on the ground plane. The return from the object is received by a curved-mouth TEM horn, whence it goes to a sampling oscilloscope and a microprocessor for processing and storage. A detailed description of the development of this antenna has been given elsewhere [17]. The system is bistatic with the transmitter (T) and the receiver (R) separated by 1.52 meters (60 in.). The distance from the object position (0) to the transmitter (T) and receiver (R) is 2.03 meters (80 in.). That is, TOR is an isosceles triangle.

A diagram of the apparatus beneath the ground plane is shown in Figure 1. The impulse generator is a spark tube device which generates a sub-nanosecond Gaussian pulse about 1 kilovolt in amplitude. A capacitive pickoff probe feeds a trigger pulse to the sampling oscilloscope. The signal pulse goes through a 15.24 meter (50 ft) delay line of RG-331 semirigid cable to the transmitter. The return from the object is fed to the sampling oscilloscope. The microprocessor controls the horizontal position of the sampling scope trace. The position of the trace is held in one position on the screen while the microprocessor receives as many samples as desired, computes the average, stores the result in memory, then moves the trace to the next point on the screen where the averaging procedure is repeated. The microprocessor does this at 256 equally spaced points across the sampling scope screen. As many samples as desired (from 1 to 255) may be averaged at each position. After a set of 256 averaged values is collected, the results may be plotted and/or recorded on digital tape by command from the terminal. In all recorded data runs, 50 samples were averaged at each of 256 positions on the waveforms.

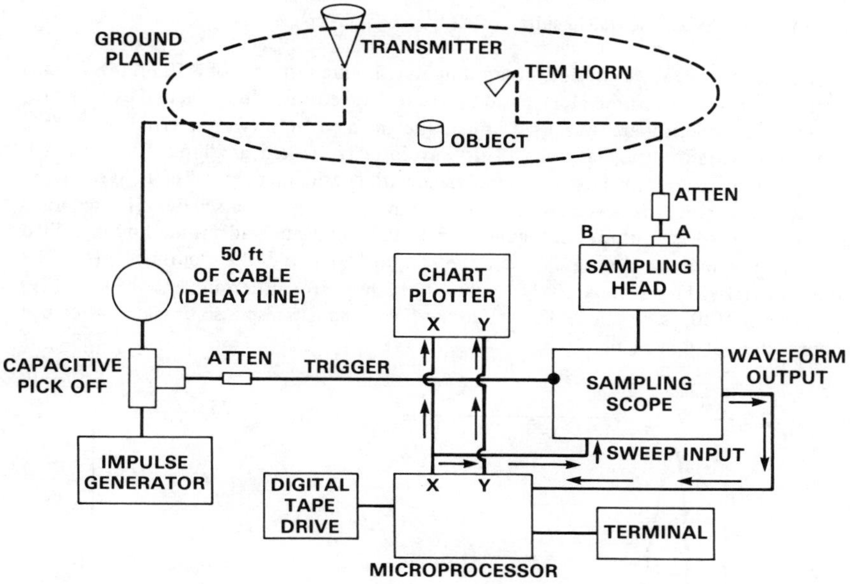

FIGURE 1. EXPERIMENTAL APPARATUS

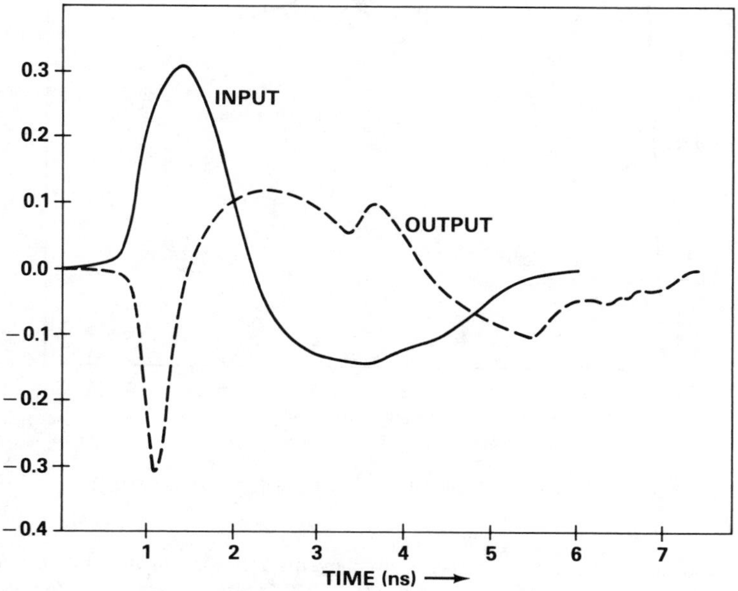

FIGURE 2. INPUT AND OUTPUT WAVEFORMS FOR A 19″ DIAMETER CONDUCTING SPHERE

5. Numerical and Experimental Results

As a first example, we consider computing the impulse response of a 19 in. (48.26 cm) diameter sphere from measured incident and scattered waveforms. In the actual experiment, a hollow 19 in. hemisphere was used. First, the radiated pulse was received by a TEM horn facing the transmitting antenna. This waveform is marked as "input" in Figure 2. Next the hemisphere was taped to the ground plane with conducting tape. The field scattered by the hemisphere was received by the TEM horn facing the hemisphere. The recorded waveform is marked as "output" in Figure 2. Next the conjugate gradient method is applied to compute the impulse response, given the "input" x(t) and the "output" y(t). The equations (15) - (21) are successively applied until the error criterion described by (29) with C given by (30) is satisfied. The estimate of the impulse response obtained after five iterations is presented in Figure 3.

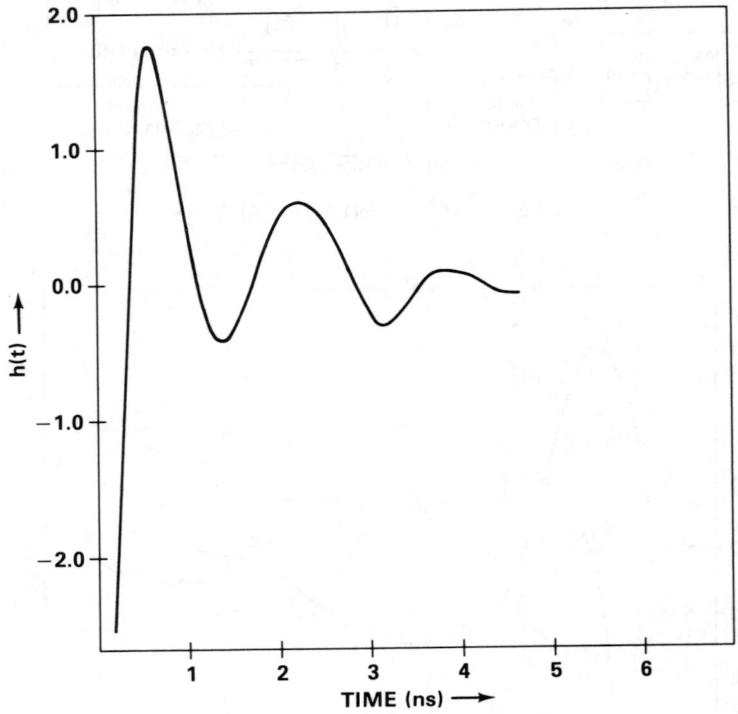

FIGURE 3. IMPULSE RESPONSE OF A 19″ DIAMETER SPHERE

The reasonances of a solid sphere are given by the following modes and their cut-off frequencies. The modes are [p.271, 18], $TM_{m,1,1}$ (543 MHz); $TM_{m,2,1}$ (766 MHz), $TE_{m,1,1}$ (889 MHz) and so on. The frequencies in brackets represent the cut-off frequencies of the modes. Observe the lowest order modes are, therefore, the three $TM_{m,k,1}$

modes. Except for a rotation in space, these three modes have the same mode pattern. It is expected that the scattered fields may have these resonant frequencies. The computed impulse response is shown in Figure 3. From the impulse response, it is seen that indeed some of these frequencies are present. It is important to point out that the impulse response indeed contains an impulse at the origin. We have been able to recover this information without incorporating any additional constraints on the solution procedure. It is important to point out that the recovery of the impulse is often not possible by the classical z-transform and FFT procedures.

As a second example, we consider computing the impulse response of a finite length conducting cylinder of diameter 15 in. (38.10 cm) and length 60 in. (152.4 cm). The half cylinder was taped to the ground plane and was irradiated by an incident field from the broadside direction with the electric field perpendicular to the axis of the cylinder. The incident and the scattered waveforms for this case are shown in Figure 4. If the cylinder were of infinite length, this type of excitation then would typically represent a TE excitation. However, since the cylinder is of finite length, it is expected that in addition to TE modes, there will be TM modes. The dominant resonant frequencies for an infinite length cylinder are TE_{11} (461 MHz), TM_{01} (603 MHz), TE_{21} (765 MHz), TE_{01} (960 MHz) and TM_{11} (960 MHz) - a degeneracy. The computed impulse response is presented in Figure 5. Again, the impulse at the origin has been reproduced. Also, the resonant frequencies mentioned earlier are present in the impulse response.

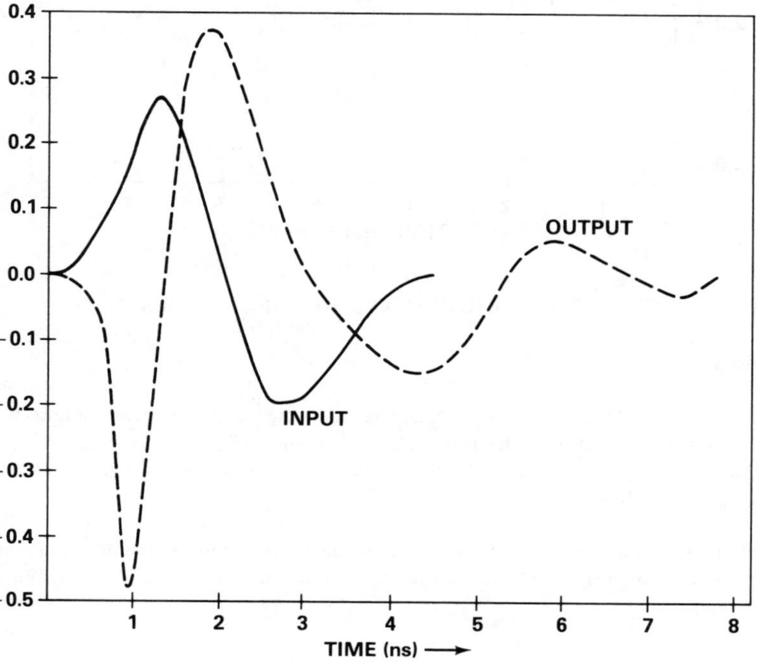

FIGURE 4. INPUT AND OUTPUT WAVEFORMS FOR A 15" DIAMETER, 60" LONG CONDUCTING CYLINDER

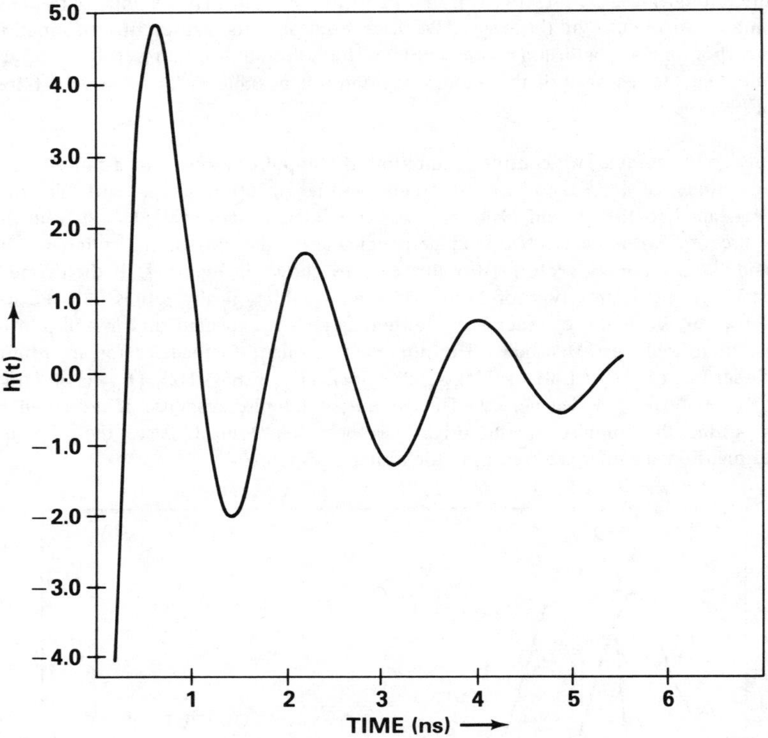

FIGURE 5. IMPULSE RESPONSE OF THE CYLINDER

6. Conclusion

The conjugate gradient method is presented to compute the impulse response from measured input and output data. Even though the deconvolution problem is ill-posed, the conjugate gradient method would yield stable results as long as the number of effective bits in the signal is correctly chosen.

Computed impulse responses from measured data have been computed to illustrate the ability of the conjugate gradient method to perform automatic stable deconvolution of noisy data.

7. References

[1] A. N. Tikhonov and V. Y. Arsenin, "Solutions of I11-Posed Problems," New York, Wiley, 1977.

[2] M. Z. Nashed, "Some Aspects of Regularization and Approximation Solutions of I11-Posed Operator Equations," in Proc. 1972 Army Analysis Conference, pp. 163-181.

[3] T. K. Sarkar, D. D. Weiner and V. K. Jain, "Some Mathematical Considerations in Dealing with Inverse Problems," IEEE Trans. of Antennas and Propagation, March 1981, pp. 373-379.

[4] I. Stakgold, "Green's Functions and Boundary Value Problems," John Wiley, New York, 1979.

[5] E. M. Kennaugh and D. L. Moffatt, "Transient and Impulse Response Approximations," Proc. IEEE, Vol. 53, Aug. 1965, pp. 893-901.

[6] C. L. Bennett, R. Hieronymus and H. Micras, "Impulse Response Target Study," Sperry Research Center, Subdury, MA, Final Report, F30602-76-C-0209.

[7] N. Nahman and M. E. Guillaume, "Deconvolution of Time Domain Waveforms in the Presence of Noise," N.B.S. Technical Note 1047, National Bureau of Standards, Boulder, Colorado, October 1981.

[8] F. Ba Hli, "A Time Domain Approach: Aspects of Network and System Theory," R. E. Kalman and N. Declaris, Eds., New York: Holt, Rinehart and Wisston, 1971.

[9] M. P. Ekstrom, "A Spectral Characterization for the I11-Conditioning in Numerical Convolution," IEEE Trans. Acoustics, Speech, Signal Processing, Aug. 1973, pp. 344-347.

[10] T. K. Sarkar, D. D. Weiner, S. A. Dianat and V. K. Jain, "Impulse Response Determination in the Time-Domain Theory," IEEE Trans. on Antennas and Propagation," July 1982, pp. 657-663.

[11] C. Johnson, "Numerical Solution of I11-Posed Problems Using the Conjugate Gradient Method," Int. Symposium on I11-Posed Problems: Theory and Practiace, Oct. 1979.

[12] O. M. Alifanov and S. V. Rumyanstev, "Regularizing Gradient Algorithms for Inverse Thermal Conduction Problems," Inzhenerno-Fisicheskii Zhurnal," Vol. 39, No. 2, Aug. 1980, pp. 253-258.

[13] M. Hestenes and E. Stiefel, "Method of Conjugate Gradient for Solving Linear Systems," J. Res. Nat. Bur. Standards, Vol. 49, pp. 409-436. 1952.

[14] T. K. Sarkar, K. Siarkiewicz, and R. Stratton, "Survey of Numerical Methods for Solution of Large Systems of Linear Equations for Electromagnetic Field Problems," IEEE Trans. on Antennas and Propagation, Vol. 29, Nov. 1981, pp. 847-856.

[15] U. J. Kammerer and M. Z. Nashed, "On the Convergence of the Conjugate Gradient Method for Singular Linear Operator Equations," SIAM Journal Numer. Anal., Vol. 4, 1972, pp. 165-181.

[16] W. Oettli, "On the Solution Set of a Linear System with Inaccurate Coefficients," J. SIAM Numerical Analysis, Ser. B., Vol. 2, No. 1, 1965, pp. 115-119.

[17] B. Z. Hollmann, "A New Curved-Mouth TEM Horn for Time Domain Data Acquistiion," National URSI Meeting, 5-7 Jan 1983, p. 154.

[18] R. F. Harrington, "Time Harmonic Electromagnetic Fields," McGraw Hill, New York, 1961.

II.8 (NM.4)

APPROXIMATION SCHEMA FOR GENERALIZED INVERSES: SPECTRAL RESULTS AND REGULARIZATION OF ILL-POSED PROBLEMS

Virginia V. Jory

Microwave Systems Division
Engineering Experiment Station
Georgia Institute of Technology
Atlanta, Georgia 30332, USA

ABSTRACT

Let H_1 and H_2 denote Hilbert spaces and let B denote a closed linear operator mapping a dense subset of H_1 into H_2. Spectral methods are used to construct approximation schema for the Moore-Penrose generalized inverse B^+. The approximators are shown to provide regularization operators for the equation $Bx = y$. Some of the results are extended to dissipative operators on reflexive and general Banach spaces.

1. Introduction

We let H_1 and H_2 be Hilbert spaces and let B denote a closed linear operator mapping a dense subset of H_1 into H_2. The operator equation

$$Bx = y \qquad (1.1)$$

is said to be well-posed in the sense of Hadamard provided it has a unique solution which depends continuously on the data "y"; otherwise it is ill-posed or improperly-posed. Examples of ill-posed problems in electromagnetic theory are: (1) the antenna synthesis problem in which a desired far-field radiation pattern is specified and the aperture distribution required to yield the far-field pattern is to be determined, and (2) the meteorological reconstruction problem in which spatial distributions of temperature and water vapor are to be determined from radiometric measurements, (3) medical tomography, (4) nuclear magnetic resonance imaging, and (5) more generally, almost any problem represented by the integral equation

$$B\,x(t) = \int_a^b K(t,\xi)\,x(\xi)\,d\xi,\ t \in [a,b],\ K \in L^2([a,b] \times [a,b]).$$

W.-M. Boerner et al. (eds.), Inverse Methods in Electromagnetic Imaging – Part 1, 365–374.

Methods of regularizing such ill-posed problems (that is, redefining an associated well-posed problem) include changing the definition of a solution, changing the space in which the problem is formulated, introducing regularization operators, and utilizing probabilistic techniques. Here emphasis is placed on changing the definition of a solution to include the notion of a generalized inverse and on introducing regularization operators.

Suppose, for example, B is an nxn matrix and y is an nx1 column vector over the real or complex numbers; then exactly three possibilities exist:

(i) there is a unique solution to Equation (1.1),
(ii) there is an infinite set of solutions, or
(iii) there is no solution.

In case the first possibility holds, (that is, B is one-to-one), then it remains to calculate the inverse B^{-1}. When the second possibility occurs, (that is y is in the range of B but the null space of B is nontrivial), the problem is one of choosing the "best" solution, say the solution u of minimal norm. In case the third possibility holds, (that is, y is not in the range of B), it still might be possible and even desirable to assign a "best solution." One might, for example, assign a vector u which minimizes $|Bx-y|$, a so-called *least squares solution*. (Again, such a vector exists since the range of B is a closed subspace.) Notice that since the null space of B also is nontrivial, the problem again reduces to choice. This final issue may be settled as before by selecting the vector u_0 of minimal norm which minimizes $|Bx - y|$. that is, the *least squares solution of minimal norm*. The operator $B^\dagger$ which makes the assignment is referred to as the Moore-Penrose generalized inverse of the matrix B. Of course, the definition of the Moore-Penrose generalized inverse has been extended to much more general operators B.

2. Preliminary Results, Definitions

Again, let H_1 and H_2 denote Hilbert spaces and let B denote a closed linear operator mapping a dense subset of H_1 into H_2. Let D(B), N(B) and $R(B)^{CL}$ denote the domain, null space, and the closure of the range of B, respectively. Then, in the Hilbert space setting, the Moore-Penrose or orthogonal generalized inverse $B^\dagger$ exists and is the unique solution to the following four equations

$$BB^\dagger B = B, \quad B^\dagger B = 1 - P, \text{ on } D(B),$$
$$B^\dagger BB^\dagger = B^\dagger, \quad BB^\dagger = Q, \text{ on } D(B^\dagger),$$

where P and Q are the orthogonal projections onto N(B) and $R(B)^{CL}$, respectively **[6]**. Moreover, if y is in the domain of $B^\dagger$, then $B^\dagger y$ is the least squares solution of minimal norm of Equation (1.1).

The notion of a least squares solution leads to a less restrictive definition of well-posedness than well-posedness in the sense of Hadamard.

Definition 2.1 The Equation (1.1) is said to be well-posed in (H_1, H_2) if for each y in H_2, (1.1) has a unique least squares solution of minimal norm; otherwise the equation is ill-posed.

It can be shown that the following statements are equivalent (see [5], [7]):

(i) the operator equation (1.1) is well-posed in (H_1, H_2);
(ii) B has closed range in H_2; and
(iii) the generalized inverse $B^\dagger$ is a bounded operator on H_2 into H_1.

Thus, in general, ill-posed problems are characterized by noise in the data, y, causing large errors in the solution, x, of Equation (1.1).

Recall from Hilbert space theory that if B is closed and densely defined from H_1 into H_2, then $N(B) = N(B^*B)$, $N(B^*) = N(BB^*)$, $R(B)^{CL} = R(BB^*)^{CL}$, $R(B^*)^{CL} = R(B^*B)^{CL}$, $H_1 = R(B^*)^{CL} \oplus N(B)$, $H_2 = R(B)^{CL} \oplus N(B^*)$, and $D(B^\dagger) = R(B) + N(B^*)$ (see, for example, [5], [6], [9]). In the Hilbert space setting, spectral methods from the operational calculus [9] provide an excellent tool for construction of approximators to the generalized inverse $B^\dagger$. Groetsch [1] utilized the operational calculus to obtain the following result for bounded operators.

Theorem 2.2 (Groetsch [1]). Suppose T is bounded linear, $T : H_1 \to H_2$. Let S be an unbounded subset of $(0, \infty)$. Let $\{U_\beta(t) : \beta \in S\}$ be a family of real valued functions such that each U_β is continuous on $[0, \|T\|^2]$ and such that there exists M >0 with

$$|tU_\beta(t)| \le M \quad \text{for all } t \text{ and } \beta, \text{ and}$$

$$U_\beta(t) \to t^{-1} \text{ as } \beta \to \infty \quad \text{for each } t \ne 0. \text{ Then,}$$

$$\lim_{\beta\to\infty} U_\beta(T^*T)T^*b = T^\dagger b \quad \text{for all } b \in D(T^\dagger).$$

Examples of families $\{U_\beta : \beta \in S\}$ satisfying the assumptions of Theorem 1 are shown by Groetsch [1] to include

$$(i) \quad U_\beta(t) = \int_0^\beta e^{-\beta t}dt, \; for\, t \in S = (0,\infty),$$

$$(ii) \quad U_\beta(t) = (1+\beta t)^{-1}, \; for\, t \in S = (0,\infty),$$

$$(iii) \quad U_\beta(t) = \sum_{k=0}^{\beta} (1+at)^k, \; for\, a > 0 \; and\, t \in Z^+,$$

$$(iv)\quad U_\beta(t) = \sum_{k=0}^{\beta} \frac{1}{k+1} \prod_{j=1}^{k} (1 - \frac{1}{1+j} t),\ for\ t \in Z^+$$

Lardy [4] established the following result by use of spectral methods for unbounded self-adjoint operators.

Theorem 2.3 (Lardy [4]). Let A be a densely defined, closed linear operator mapping H_1 into H_2. If y is in $D(A^\dagger)$, then

$$A^\dagger y = \sum_{k=1}^{\infty} A^*(1+AA^*)^{-k} y.$$

Moreover, if R(A) is closed then

$$A^\dagger = \sum_{k=1}^{\infty} A^*(1+AA^*)^{-k}.$$

Definition 2.4 Suppose E is any Banach space and $A: D(A) \subseteq E \to E$ is a linear densely defined operator. Then A is said to be dissipative only in case for each $c>0$, $|(1-cA)| \geq |x|$ for each x in D(A) and $R(1-cA) = E$.

3. Approximation Schema for $B^\dagger$ as Regularization Operators for Bx = y

Let E_{B^*B} denote the spectral family for B^*B (see, for example, [9]), where B^* is the Hilbert space adjoint of B, and let S be an unbounded subset of $(0, \infty)$. We establish a representation theorem for the generalized inverse $B^\dagger$ and show that the resulting approximation schema provide regularization operators for Equation (1.1).

Theorem 3.1 Suppose for each β in S, f_β is a continuous function on $[0, \infty]$ such that for $\lambda>0$, $\lim_{\beta\to\infty} f_\beta(\lambda) = 1/\lambda$. Suppose also that there exists $M>0$ such that $|\lambda f_\beta(\lambda)| \leq M$ almost everywhere with respect to E_{BB^*}. Let h_β be defined by:

$$h_\beta(\lambda) = \phi(\lambda) - \lambda f_\beta(\lambda),\ where$$

$$\phi(\lambda) = 0,\quad if\ \lambda = 0$$

$$= 1,\quad if\ \lambda > 0.$$

Suppose there exist a real-valued function h and a number γ such that

$$a)\quad \int_0^\infty |h(\lambda)|^2 d\|E_{B^*B}(\lambda)x\|\ exists\ for\ all\ x\ in\ D(B),\ and$$

b) $if\beta \geq \gamma, then \|h_\beta\| \leq \|h\|.$.

Suppose, finally, that

$$c)\quad B^* f_\beta(BB^*) = f_\beta(B^*B)B^* \text{ on } D(B^*) \text{ and},$$

$$d)\quad Bf_\beta(B^*B) = f_\beta(BB^*)B \text{ on } D(B)$$

Then $\quad s\text{-}\lim_{\beta\to\infty} B^* f_\beta(BB^*) = B^\dagger.$

(This result was presented at a conference on approximation theory honoring Professor George G. Lorentz on the occasion of his seventieth birthday [3]. The proof is given here for completeness and to represent the flavor of the mathematics involved.)

__Proof.__ Let y be in $D(B^\dagger)$. Since $H_2 = R(B)^{CL} \oplus N(B^*)$, there exists n in $N(B^*)$ such that $y = BB^\dagger y + n$. Thus for each β:

$$\begin{aligned} |B^\dagger y - B^* f_\beta(BB^*)y|^2 &= |B^\dagger y - B^* f_\beta(BB^*)y|^2 \\ &= |B^\dagger y - B^* B f_\beta(B^*B)B^\dagger y|^2 \\ &= |[\phi(B^*B) - B^*Bf_\beta(B^*B)]B^\dagger y|^2 \\ &= \int_0^\infty |\phi(\lambda) - \lambda f_\beta(\lambda)|^2 d|E_{B^*B}(\lambda)B^\dagger y|^2 \end{aligned} \tag{3.1}$$

The last equality in (3.1) obtains as a result of the operational calculus for (unbounded) self-adjoint operators (see, for example, [9, Chapter XI]) Furthermore, by the Lebesgue dominated convergence theorem, the integral in (3.1) has limit zero as β approaches infinity.

__Definition 3.2__ Let Λ be an indexing set. A __regularization family__ (or __regularization algorithm__ as in [7]) for (1.1) is a family of bounded linear operators $\{F_\alpha : \alpha \in \Lambda\}$ from H_2 into H_1 with the property that for every y_0 in $D(B^\dagger)$, there is an element x in the set of all least squares solutions of (1.1) such that for every $\varepsilon > 0$, there exists $\delta > 0$ and α in Λ satisfying $|F_\alpha y - x| \leq \varepsilon$, whenever $|y - y_0| \leq \delta$.

The effect of a regularization family is illustrated in Figure 3.1.

__Theorem 3.3__ Under the hypotheses of Theorem 3.1, the family of operators

$$\{B^* f_\beta(BB^*) : \beta \in S\}$$

is a regularization algorithm for Equation (1.1).

<u>**Proof**</u>. Let $\varepsilon > 0$ and y in $D(B^\dagger) = R(B) + N(B^*)$ be given. Let β be such that

$$|B(B^\dagger y - {}^*f_\beta(BB^*)\,y_0| \le \varepsilon/2.$$

Since $\lambda f_\beta(\lambda)$ is bounded almost everywhere with respect to E_{BB^*}, $B^*\,f_\beta\,(BB^*)$ is bounded. Thus there exists δ such that if $|y - y_0| \le \delta$, then

$$|B^* f_\beta(BB^*)\,y_0 - B^*\,f_\beta\,(BB^*)\,y\,| \le \varepsilon/2.$$

Realization that $B^\dagger y$ is a least squares solution of (1.1) yields the result.

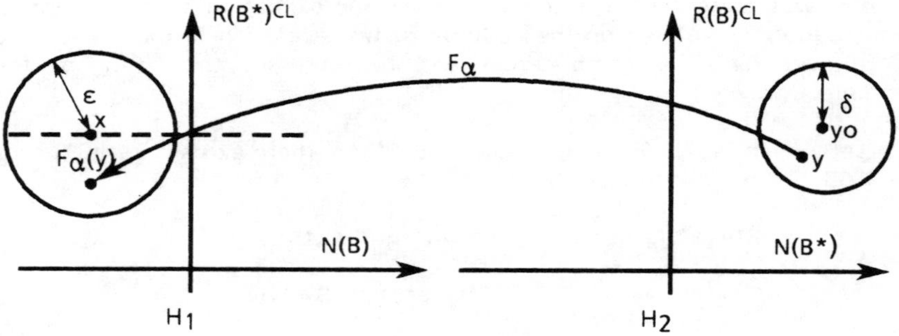

FIGURE 3.1 Mapping By a Regularization Operator.

4. Applications

Examples of families of real-valued functions which satisfy the hypotheses of Theorem 3.1 are given next.

<u>**Theorem 4.1**</u> The sets of functions $\{f_\beta : \beta \in S\}$ defined by the following equations provide regularization families for Equation (1.1)

$$(i)\quad f_\beta(\lambda) = \frac{1}{\beta}\sum_{j=1}^{\beta}\sum_{i=1}^{j}(1+\lambda)^{-i},\ \beta \in S = Z^+,$$

$$(ii)\quad f_\beta(\lambda) = \int_0^\beta e^{-\xi\lambda}d\xi,\ \beta \in S = R^+$$

(iii) $f_\beta(\lambda) = \beta(1+\beta\lambda)^{-1},\ \beta \in S + R^+$

(iv) $f_\beta(\lambda) = \sum_{p=1}^{\beta} \lambda_p \sum_{j=1}^{p} (1+\lambda_j\lambda)^{-1},\ \beta \in S = Z^+$, *where*

$\{\lambda_p\}$ is a positive number sequence such that $\Sigma\lambda_p = +\infty$.

For bounded operators B, the family in (ii) results in the Showalter integral formula for $B^\dagger$ and the family in (iii) yields the Tikhonov regularization algorithm of order zero. The family in (iv) includes the result of Lardy with $\lambda_p = 1$ for all p.

Remark 4.2 The convergence in Theorem 3.1 is uniform when B has closed range; moreover, the convergence is of order $1/\beta$ for the approximators to $B^\dagger$ constructed from the families in Theorem 4.1 (i) - (iii). In (iv), if $\Sigma\lambda_p^2 = +\infty$, the convergence is of order $(\Sigma_{p=1}^{\beta} \lambda^2_p)^{-1}$.

Remark 4.3 The following lemma is utilized in the proof of Theorem 4.1, (see [8]) for (i), (ii), [2] for (iii), (iv)).

Lemma 4.4 Let $\beta,\lambda > 0$, z be in D(B*), x be in D(B), and T_{-B^*B}, T_{-BB^*} be the C_0 semigroups generated by -B* B, -BB* respectively. Then

(i) $(1+\lambda B^*B)^{-1}B^*z = B^*(1+\lambda BB^*)^{-1}z$

(ii) $(1+\lambda BB^*)^{-1}Bx = B(1+\lambda B^*B)^{-1}x$

(iii) $T_{-B^*B}(t)B^*z = B^*T_{-BB^*}(t)z$, for all $t > 0$, and

(iv) $$\frac{1}{\beta}\int_0^\beta \xi T_{-B^*B}(\beta-\xi)B^*z\,d\xi = B^*\frac{1}{\beta}\int_0^\beta \xi T_{-BB^*}(\beta-\xi)z\,d\xi.$$

Example 4.5 For an example of an operator whose range is not necessarily closed and to which the computational procedures of this section apply, we shall consider an integral operator of the first kind B: $H_1 \to H_1$ defined for x in $L^2[a,b]$ by

$$Bx(.) = \int_a^b K(.,\xi)\,x(\xi)\,d\xi$$

where the kernel K is such that

$$\int_a^b\int_a^b |K(\xi,\sigma)|^2 d\xi d\sigma < 0.$$

Since B is compact, R(B) is closed if and only if R(B) is finite dimensional. Of course $B^\dagger$ is bounded only in case R(B) is closed. To compute $B^\dagger$, let $U_\beta = -\beta(1+\beta B^*B)^{-1}$, according to the family f_β in Theorem 4.1 (iii). Then B^* is given by

$$B^*x(.) = \int_a^b K(\xi, .)\, x(\xi)\, d\xi$$

for all x in H_1. Thus, if f_β is in H_1 then for $\beta > 0$, $-U_\beta B^* f_\beta$ is the solution ω_β of the integral equation of the second kind

$$\omega_\beta(.) + \beta \int_a^b \left[\int_a^b K(\xi,.)\, K(\xi,\sigma)\, d\xi \right] \omega_\beta(\sigma) d\sigma = \beta \int_a^b K(\xi,.) f_\beta(\xi) d\xi. \tag{4.1}$$

Since $-B^*B$ is dissipative in H_1, in the sense of Definition 2.4, then a unique solution to Equation (4.1) exists for each $\beta > 0$. Furthermore by Theorem 3.1 if f_β is in $D(B^\dagger)$ then $B^\dagger f_\beta = \lim_{\beta\to\infty} \omega_\beta$. Thus we see that the generalized inverse of an integral operator of the first kind may be obtained as the pointwise or strong limit of solutions of integral equations of the second kind.

5. Regularization Operators in Banach Space

With less structure in a general Banach space than in a Hilbert space, we lose the powerful operational calculus which allows us to construct regularization *operators* from *functions* of real numbers in the Hilbert space setting. Moreover, since the notion of a least squares solution may not persist in the general Banach space, it is necessary to modify Definition 3.2 slightly.

Throughout this section we assume that E_1 and E_2 are Banach spaces with A a closed linear operator mapping a dense subset of E_1 into E_2. We assume also that there are subspaces S and M so that $E_1 = N(A) \oplus S$ and $E_2 = R(A)^{CL} \oplus M$. Let P be the projector onto N(A) along S, Q be the projector onto $R(A)^{CL}$ along M, and let $A^\dagger = A^\dagger_{P,Q}$.

Definition 5.1. Let J be an indexing set. A regularization family relative to (P, Q) for

$$Ax = y \tag{5.1}$$

is a collection of bounded linear operators $\{F_\alpha : \alpha \in J\}$ from E_2 into E_1 with the property that for any y_0 in $R(A) + M$, there is an element n in N(A) such that, for every $\varepsilon > 0$, there exists $\delta > 0$ and a in J, so that $|F_\alpha y - (A^\dagger + n)| \le \varepsilon$, if $|y - y_0| \le \delta$.

Definition 5.1 agrees with Definition 3.2 in case E_1 and E_2 are Hilbert spaces and P and Q are the orthogonal projectors onto N(A) and $R(A)^{CL}$, respectively; the element $A^\dagger y + n$ is then a least squares solution of (5.1).

The main result of this section follows.

Theorem 5.2 Let S be a subset of the real numbers such that $\sup_{\beta \in S} \beta = \infty$ Suppose $E = E_1 = E_2 = R(A)^{CL} \oplus N(A)$ and $\{U_\beta: \beta \in S\}$ is a family of bounded linear operators defined on E satisfying Properties 1-4.

Property 1. The members of the family $\{1\text{-}A\, U_\beta: \beta \in S\}$ map E into D(A).

Property 2. For each β in S the operators A and U_β are commutative on D(A).

Property 3. The family of operators $\{1\text{-}AU_\beta: \beta \in S\}$ is uniformly bounded on E.

Property 4. On $R(A)^{CL}$, $\text{s-lim}_{\beta\to\infty}(1\text{-}AU_\beta)$ exists and is the zero element of E.

Then $\{U_\beta Q: \beta \in S\}$ is a regularization family relative to (P, Q) for $Ax = y$, where P is the projector onto N(A) and $Q = 1\text{-}P$.

Proof. Let y be in R(A) + N(A). Then $|A^\dagger y - U_\beta Q y| = |A^\dagger y - U_\beta A A^\dagger y| = |(1\text{-}AU_\beta) A^\dagger y|$. Thus $\text{s-lim}_{\beta\to\infty} U_\beta Q = A^\dagger$, and, since for each β, $U_\beta Q$ is bounded, the result follows.

If E is a reflexive Banach space and A is dissipative in E, then regularization families satisfying the hypotheses of Theorem 5.2 can be constructed for Equation (5.1).

Theorem 5.3. Suppose E is a reflexive Banach space and A is dissipative on E. Then each of the following families provides regularization operators relative to (P, Q) for $Ax = y$, where P and Q are as in Theorem 5.2.

$$(i)\quad \{-\beta(1-\beta A)^{-1}Q: \beta \in R^+\},$$

$$(ii)\quad \{-\frac{1}{\beta}\sum_{j=1}^{b}\sum_{i=1}^{j}(I-A)^{-i}Q: \beta \in Z^+\},\ \text{and}$$

$$(iii)\quad \{-\frac{1}{\beta}\int_0^\beta \xi T(\beta\xi)Q\,d\xi: \beta \in R^+\},\ \text{where}$$

T is the C_0 semigroup generated by A.

6 Additional Work

Work is underway to identify linear penalty functions q_α associated with the regularization operators F_α described herein and with others under investigation so that in the presence of noisy data in the non-closed range case the problem becomes one of minimizing

$$|F_\alpha y - A^\dagger y_o| \leq |F_\alpha y_o - A^\dagger y_o| + |F_\alpha(y_o - y)| \leq |F_\alpha y_o - A^\dagger y_o| + |q_\alpha(y_o - y)|.$$

where y and y_o are as in Figure 3.1.

ACKNOWLEDGMENTS

The author wishes to express appreciation to those who have influenced and encouraged her work in inverse problems, especially Dr. James V. Herod, Dr. M. Zuhair Nashed, and Dr. Wolfgang M. Boerner.

REFERENCES

1. Groetsch, C. W. , Generalized Inverses of Linear Operators, Representation and Approximation, Marcel Dekker, New York, 1977.

2. Jory, V. V., Approximators to generalized inverses as regularizers of ill-posed problems, Numer. Funct. Anal. and Optimiz., 3(4), 1981, pp. 477-492.

3. Jory, V. V., Spectral representation and approximation of the generalized inverse of a closed linear operator, Approximation Theory III, Editor: E. W. Cheney, Academic Press, New York, 1981, pp. 543-548.

4. Lardy, L. J., A series representation for the generalized inverse of a closed linear operator, Atti Acad. Naz. Lincei Rend. Cl. Sci. Fis. Mat. Natur., Ser.VIII,58, 1975, pp. 152-156.

5. Nashed, M. Z., Singular operator equations, Nonlinear Functional Analysis and Applications, Editor: L.B. Rall, Academic Press, New York, 1971, pp. 311-359.

6. Nashed, M. Z. and Votruba, G. F., A unified operator theory of generalized inverses, Generalized Inverses and Applications, Editor: M. Z. Nashed, Academic Press, New York, 1976, pp.1-109.

7. Nashed, M. Z. and Wahba, G., Generalized inverses in reproducing kernel spaces: An Approach to regularization of linear operator equations, MRC Technical Summary Report #1200, University of Wisconsin Mathematics Research Center, Madison, Wisconsin, 1972. See also SIAM J. Math. Anal. 5, 1974, pp. 974-987.

8. Riesz, F. and Sz.-Nagy, B., Functional Analysis, Frederick Ungar, New York, 1955.

9. Yosida, K., Functional Analysis, Springer-Verlag, New York, 1978.

II.9 (SP.1)

MAXIMUM ENTROPY METHODS IN ELECTROMAGNETIC/GEOPHYSICAL/ ULTRASONIC IMAGING

R. M. Bevensee

Lawrence Livermore National Laboratory
Livermore, California, USA

Maximum Entropy (ME) methods of resolving underdetermined electromagnetic images are reviewed. A new non-Burg method of resolving complex amplitudes of coherent radiators in wave-number space (direction) is described. Several ME methods in geophysics are surveyed, based on Burg's method for spatial array data processing and a Boltzmann method for parameter distributions in the earth. A Boltzmann-ME method is described for resolving anomalies in materials ultrasonically, where the data is governed by a Fredholm integral equation of the second kind. Some issues of uniqueness, confidence, and noise are briefly assessed.

1. ME METHODS IN ELECTROMAGNETICS

Maximum Entropy methods of resolving underdetermined objects from image data with or without noise have proven effective in both real space and wavenumber space. We first review several Boltzmann-type ME methods for resolving incoherent sources in real space and wavenumber space, respectively. Then we outline a special Boltzmann-ME method, unrelated to the Burg or Capon methods of spatial array data processing (see Sec. 2.2), for resolving an angular distribution of coherent sinusoidal radiators in amplitude and phase from receiver data or, alternatively, a spatial distribution of coherent antennas from far-field data. We describe numerical results from a special computer code for two-dimensional distributions of infinitely long, parallel radiators. We deduce an approximate "super-resolution" limit of this method for resolving distant coherent radiating

W.-M. Boerner et al. (eds.), Inverse Methods in Electromagnetic Imaging - Part 1, 375–395.

sources. Fundamental limitations on the resolution of various radiating sources are discussed.

1.1 Reconstruction From Photon Images

A Boltzmann-ME method has been applied successfully to enhance a photon image in a square aperture system [1]. If $\hat{p}_i(r_i)$ is the estimated probability of photon emission from the i'^{th} object pixel at position $\bar{r}_i$ and $t_{ik}(\bar{r}_i, r'_k) = \sin^2(x-x')\sin^2(y-y')/[(x-x')(y-y')]^2$ (normalized coordinates) is the transmission factor to $\bar{r}' = (x',y')$ of the image, and $B(\bar{r}')$ is the measured brightness, the quantity

$$\sum_{i,\text{object}} [\hat{p}_i(\bar{r}_i)t_{ik}(r_i,\bar{r}'_k) - B_k(\bar{r}'_k)]^2 \tag{1}$$

is the squared deviation between predicted and measured brightness due to noise in the k' pixel. The quantity

$$E\{\hat{p}_i\} = \sum_{k,\text{image}} \sum_{i,\text{object}} [\hat{p}_i t_{ik} - B_k]^2 \tag{2}$$

is the noise "energy" in the image. The ME method of [1] seeks the most probable distribution of the $\hat{p}_i$ subject to the constraint of E being less than the known noise level.

Entropy H measures the logarithm of the total number of distinct combinations of photon intensities on the object,

$$H = -\sum_i \hat{p}_i(\ell n\ \hat{p}_i - 1) \quad . \tag{3}$$

In maximizing H relative to the $\hat{p}_i$ subject to the constraints it is usually convenient to employ Lagrange's method of undetermined multipliers and maximize an "objective function." In this case it is

$$F = H + \beta E\{\hat{p}_i\} + \gamma(\sum_i \hat{p}_i - 1) \quad . \tag{4}$$

Here β is an a priori weight multiplier for the noise and γ is the multiplier for the normalization $\sum \hat{p}_i = 1$. The set of equations obtained by setting $\partial F/\partial \hat{p}_i = 0$ for all i and the constraint equation must be solved to obtain all the $\hat{p}_i$.

For large β this technique enables resolution of incoherent sources separated by ≃ λ/2L along x or y, which is half the Zernike-Van Cittert limit. The angular resolution limit for broadside antennas is also $\lesssim \lambda/L$.

Figure 1 shows the rather impressive resolution possible for a flat object on a 20 x 20 grid of binary pixels from its diffraction-limited image. The ME resolution is extremely sensitive to ß for large ß; the resolution shown is for $ß = 10^7$.

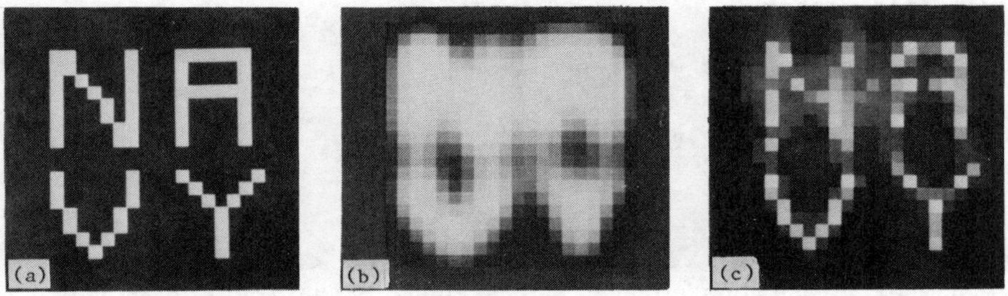

Fig. 1. ME resolution of a flat object from its diffraction limited image. (a), object defined on a 20x20 grid of binary pixels, (b), image, (c) resolution with $ß = 10^7$ [1, p. 51].

1.2 Reconstruction in Wavenumber Space

Two-dimensional digital image reconstruction is important in many physical sciences, notably radiography, radio astronomy, optics, and microscopy. An example is the problem of resolving radio stars in close angular proximity against the Fourier-transform background noise.

Let the relative intensity or brightness of a distant point source be $f(\bar{r})$; the wavenumber spectrum S is the transform

$$S(\bar{k}) = \int_{-\infty}^{\infty} dx \int_{-\infty}^{\infty} dy \; f(\bar{r}) e^{-j\bar{k}\cdot\bar{r}} \quad . \tag{5}$$

The measured brightness $B(\bar{k})$ differs from S because of noise.

Discretization of $\bar{r}$- and $\bar{k}$-space and the assumption that the $S(\bar{k}_m)$ are independent, zero-mean random variables with known variances σ_m enables us to write an "energy" constraint as [3]

$$E\{\hat{f}_i\} = \sum_{m=1}^{M} \frac{1}{\sigma_m^2} \left| B(k_m) - \Delta A \sum_{i=1}^{I} \hat{f}_i(x_i,y_i) e^{-j2\pi(u_m x_i + v_m y_i)} \right|^2 \leq M, \tag{6}$$

ΔA being the object pixel area, $\hat{f}_i$ an ME estimate of the brightness, and $2\pi(u_m,v_m) = (k_{mx},k_{my})$, the transverse wavenumber vector.

The entropy H is written

$$H = +\Delta A \sum_i \ln \hat{f}_i \tag{7}$$

because for photons $1 <$ (degrees of freedom) $\ll \bar{n}_i$, $\bar{n}_i$ = average number of photons emitted per second. Equation (7) is the correct entropy for the classical wave (field) limit for Bose-Einstein statistics [2].

The objective function analogous to (4) is

$$F = H + \beta E\{\hat{f}_i\} \tag{8}$$

which has a unique local maximum. Wernecke and D'Addario [3] describe an algorithm for maximizing F relative to variations of the $\hat{f}_i$ subject to a stated initial estimate of ß. An impressive example of this Fourier synthesis is the reconstruction of the radio double star Cygnus A (3C405). The direct Fourier transform resolves the star whose parts are separated about 120 arc secs somewhat ambiguously, with considerable background noise. The ME method, using only 50 data points on the nine concentric measurement ellipses in the u-v wavenumber plane, resolves them more sharply and with essentially no background noise!

1.3 Resolution of Two-Dimensional Coherent Radiators in Amplitude and Phase

The preceding sections treated the resolution of distant radiating incoherent sources by direct spatial transformation from object to image and by Fourier transformation. We now describe an ME method we have developed for resolving coherent radiators, all at sinusoidal frequency ω. Unlike the Burg method, ours does not rely on measured and extrapolated correlation coefficients. We have applied the method in two dimensions to resolve angular distributions of distant radiators from receiver data and--with a minor change in the transfer functions--to resolve spatial distributions of antennas from far-field angular measurements. All radiators are infinitely long and parallel. We shall infer some general properties of our ME method from synthetic data examples studied.

We further assume that the coherent nature of the sources is known and that receiver noise has been correlated out by mixing with a low-noise local oscillator, for example.

Figure 2 illustrates the "distant source" problem of resolving K possible radiators in amplitude and phase, the kth being $A_k e^{i\psi_k}$, from measurements at I receivers, $I < K$, the ith being $B_i e^{i\phi_i}$. In amplitude all $R_{ik} \simeq R_o$, which is absorbed into the B_k-amplitude. The equation for the ith receiver is, with neglect of the common phase factor $\exp(-i\omega R_o/c)$, c = light velocity, is

$$\sum_{k=1}^{K} A_k e^{i\psi_k} e^{i\omega\rho_i \cos(\Theta_i - \Theta_k)/c} = B_i e^{i\phi_i} \text{ measured} \qquad (9)$$

$$1 \leq i \leq I < K,$$

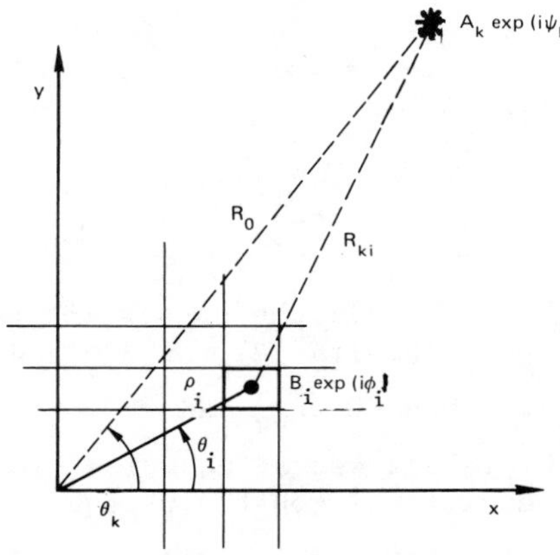

Fig. 2. Description of the "distant source" resolution problem in two dimensions. Coherent source $A_k e^{i\psi_k}$ at frequency ω radiates to the receivers in its far field; receiver i records a total field $B_i e^{i\phi_i}$.

and this is separated into real and imaginary parts, $A_k e^{i\psi_k} = A_{iR} + iA_{iI}$, etc., as

$$\sum_k (A_{kR}C_{ki} - A_{kI}S_{ki}) = B_{iR} \; ; \quad \sum_k (A_{kR}S_{ki} + A_{kI}C_{ki}) = B_{iI} \tag{10}$$

with

$$C_{ki} = \cos[\omega\rho_i \cos(\Theta_i - \Theta_k)/c], \quad S_{ki} = \sin[\omega\rho_i \cos(\Theta_i - \Theta_k)/c] \tag{11}$$

This formulation of the "distant source" problem is identical to the reciprocal problem of resolving K coherent antennas in the receiver region of Fig. 2 from I far-field measurements, with the proviso that receiver position vector ρ_i in Fig. 2 must be changed to source vector ρ_k. The solution to this reciprocal problem by our method has been published [4] and we summarize it for both problems below.

We regard each A_{kR}, A_{kI} as constructed of an enormous number of building blocks of size Δ (real), so $A_{kR} = n_{kR}\Delta$, $A_{kI} = n_{kI}\Delta$. We choose to define the entropy of the sources without regard for phase and with respect to amplitude squared. The Boltzmann H, with no prior probabilities on the $n_k = n_{kR} + in_{kI}$ is therefore

$$H = \ell n \frac{N^2!}{\prod_k |n_k|^2!} \quad , \quad k \leq K, \quad \sum |n_k|^2 = N^2 \tag{12}$$

Stirling's approximation yields

$$H = -N^2 \sum_k p_k \ell n p_k \quad , \quad p_k = |n_k|^2/N^2 \tag{13}$$

The objective function is F = H + all the constraint equations (10), each with its Lagrange multiplier β_{iR} or β_{iI} and including the constraint $\Sigma p_k = 1$. Then $\partial F/\partial n_{kR}$ and $\partial F/\partial n_{kI}$ set to zero for each $k \leq K$ yield a set of equations for all the n_{kR}, n_{kI}, and the I+1 additional constraint equations furnish enough information for determining all the multipliers as well.

One can show that $\partial^2 H/\partial n_{kR}^2 < 0$, $\partial^2 H/\partial n_{kI}^2 < 0$ for all k and fixed N^2 at the solution point; i.e., $H = H_{max}$ and there is only one H_{max}.

The essential equations for the A_k and $ß_i$ are

$$A_{kR} = (1/2L_k) \sum_{j=1}^{I} (ß_{jR}C_{kj} + ß_{jI}S_{kj}), \; k \leq K \;, \tag{14a}$$

$$A_{kI} = (1/2L_k) \sum_{j} (-ß_{jR}S_{kj} + ß_{jI}C_{kj}) \tag{14b}$$

along with

$$L_k = \ell n[A_{kR}^2 + A_{kI}^2)/A_T] \;, \quad A_T = \sum_k (A_{kR}^2 + A_{kI}^2) \;, \tag{15}$$

the constraint relations (10), and an implied constraint from (10) and (14) which is useful in scaling the A_k^2:

$$-H = \sum_k (A_{kR}^2 + A_{kI}^2)L_k = \frac{1}{2} \sum_j (ß_{jR}B_{jR} + ß_{jI}B_{jI}) \tag{16}$$

Matrix equations for the $ß_{jR}, ß_{jI}$ are obtained by multiplying (14a) by $C_{kj'}$, (14b) by $S_{kj'}$, subtracting and taking $\sum_k$. Use of (10a) yields

$$\sum_{j=1}^{I} [F_{j'j}ß_{jR} + G_{j'j}ß_{jI}] = b_{j'R} \;, \quad j' \leq I \tag{17a}$$

Similarly we obtain

$$\sum_i [-G_{j'j}ß_{jR} + F_{j'j}ß_{jI}] = b_{j'I} \;, \tag{17b}$$

in which

$$F_{j'j} = \frac{1}{2} \sum_k (C_{kj}C_{kj'} + S_{kj}S_{kj'})/L_k \;,$$

$$G_{j'j} = \frac{1}{2} \sum_k (S_{kj}C_{kj'} - C_{kj}S_{kj'})/L_k .$$

The algorithm iterates through a loop in which the $ß_j$ determine the A_k in (14), which are then solved to satisfy (16), after which the $ß_j$ are solved by (17), the B_k are solved by (10), and if they are not close enough to the measured B_k the loop is repeated. The initial L_k of (15) are all set equal and the initial β_j solved by (17) accordingly. Typical problems of interest require ten or so iterations to achieve a few percent accuracy in the ME distribution.

As the examples will show, this ME algorithm tends to yield the relative coherent source amplitudes. Spatial correlation coefficient data have not been incorporated as constraints; they would tend to fix the absolute source amplitudes.

The ME equations (14), (15) are actually solved relative to a total parameter of A_T' instead of A_T, with A_T' greater. This procedure corresponds to a convenient entropy shift which allows us to avoid the double-valued nature of the equation solved for each A_{kR}, after which A_{kI} is found. That equation is obtained by dividing (14b) by (14a) to give $\alpha_k \triangleq A_{kI}/A_{kR}$, whereupon (14a) is solved as

$$A_{kR} \ln[A_{kR}^2(1+\alpha_k^2)/A_T'] = T_{kR}(\beta) \quad . \tag{18}$$

The left side is a double-valued function of $|A_{kR}|$, but if A_T' is large enough A_{kR} may be found in the range $0 \leq |A_{kR}| < \sqrt{A_T'/(1+\alpha_k^2)}$ and the relative A_{kR} determined correctly. After converging on the ME solution A_T is obtained by (15).

The solution for the β_k in (17) is by Gauss elimination and back substitution, in which the pivot elements often are $< 10^{-5}$. However, this does not, by itself, introduce non-negligible error in the iterative solution or slow convergence.

We comment at this point about an apparent complication in the analogous situation with entropy formulated in terms of amplitude and not amplitude squared. Then the entropy is $-N \sum_k p_k \ln p_k$ of (13) but with $p_k = |n_k|/N = \sqrt{(n_{kR}^2+n_{kI}^2)}/N$. The equations replacing (14) are

$$A_{kR} = A_k \cos\psi_k = (A_k/L_k') \sum_j (\beta_{jR} C_{kj} + \beta_{jI} S_{kj}) \tag{19a}$$

$$A_{kI} = A_k \sin\psi_k = (A_k/L_k') \sum_j (-\beta_{jR} S_{kj} + \beta_{jI} C_{kj}) \tag{19b}$$

with

$$L_k' = \ln(A_k/A_T'), \quad A_T' = \sum A_k \quad , \tag{20}$$

and these two equations for angle ψ_k (A_k cancels) may be inconsistent, there being more A_k than $ß_j$, unless the algorithm is modified nontrivially.

According to the analogy between this spatial resolution of sources in wavenumber space and temporal resolution of sources in frequency space by Fourier transform, our ME method serves as an alternative to the Burg method of waveform spectral analysis.

We studied the angular resolution capability of a linear array of 10 receivers spaced $\lambda/2$ apart in Fig. 2 for resolving two distant coherent sources $< \lambda/L$ apart in angle. Figure 3 shows the resolution for four cases of two sources separated by 6.3° = .52 λ/L : $1/\underline{0}°$ and $2/\underline{90}°$ nearly broadside and 45° to broadside and two $1/\underline{0}°$ sources nearly broadside and 45° to broadside. In each case, the scan region was ± 12° on either side of one of the sources, in steps of 1.2°; i.e., we assumed 21 possible sources.

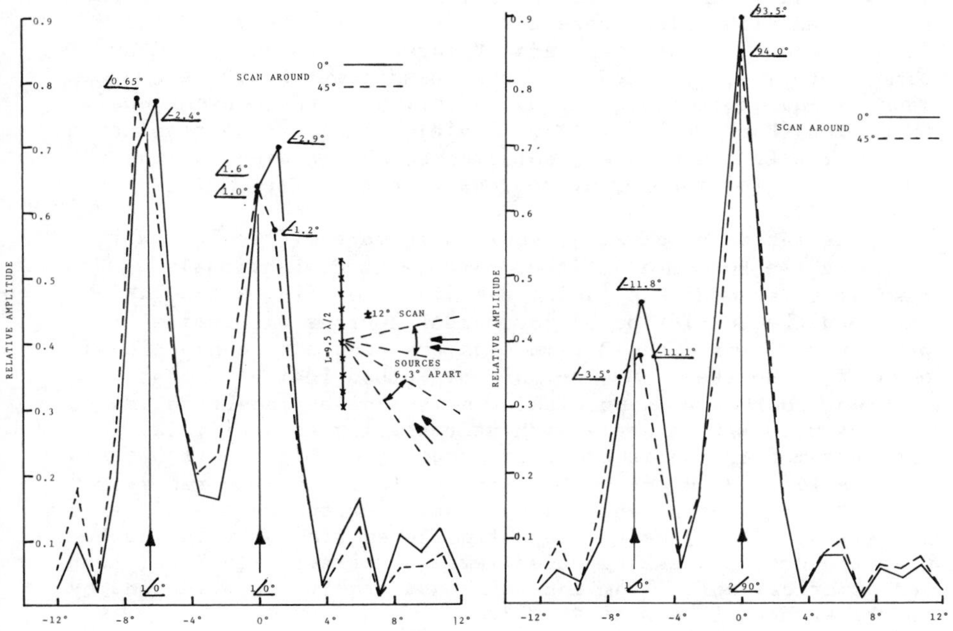

Fig. 3 "Superresolution" of two coherent sources 0.52 λ/L apart in angle, by a 10-receiver array of length $L=9.5\lambda/2$. Left: both sources $1/\underline{0}°$ in a sector broadside and at 45°. Right: sources $1/\underline{0}°$, $2/\underline{90}°$ in a sector broadside and at 45°.

In all four cases depicted in Fig. 3 the relative amplitudes were well resolved. The phases of the $1/\underline{0}°$ and $2/\underline{90}°$ sources were resolved within 12°; those of the two $1/\underline{0}°$ sources, within about 3°. In our opinion we have obtained superresolution of coherent sources, and the nearly 3° half-maximum of each curve indicates the resolution could be better with this linear array. The resolution of two $1/\underline{0}°$ sources separated by .25 λ/L at broadside, scanned within a $\pm$ 6° sector at 21 angles, showed two peaks at the actual source positions. Although the phase resolution was excellent (-0.7°, -0.1°) the relative peaks differed by a ratio of 1.5.

1.4 Fundamental Limitations on the Resolution of Radiating Sources

There are four situations to consider: incoherent sources, with and without additive noise and coherent sources, with and without noise. Gabriel [5] has explained that incoherent sources without noise can be resolved arbitrarily closely in angle by conventional techniques (viz, adaptive arrays) and gives an empirical universal curve for resolution versus SNR with noise present. However, Steinberg and Luthra [6,6A] show that this "universal" curve can be surpassed by time averaging, provided that the noise power N_o is removed from the autocorrelation $R_n(0)$, after which the conventional ME method (Burg method [7]) will yield angular superresolution to $\lesssim 0.2\ \lambda/L$. Such superresolution can be attained by the Burg method for incoherent sources with no noise.

Resolution of coherent sources is more difficult. With no noise the Burg method is comparable to conventional resolving methods; with noise the time-averaging method which improved the resolution of incoherent sources with noise present won't make the Burg method preferable to conventional ones [6]. However, our non-Burg ME method indicates that coherent sources with or without noise can be superresolved to perhaps 0.25 λ/L in angle without resorting to artificial aperture modulation in order to create a Doppler shift between sources [5]. With this method one could reduce the receiver noise to an acceptable level by modulating the received signals by a local oscillator of noise sufficiently low as to make the data processing noise level satisfactorily lower than the numerical noise from the ME method itself, as evidenced by the background levels in Fig. 3.

Luthra and Steinberg believe that there is <u>no</u> fundamental resolution limit for a receiver of finite size aperture, either for coherent and/or incoherent sources, despite the S/N ratio [8], provided that the noise is thermal or its statistical equivalent.

2. ME METHODS IN GEOPHYSICS

We review the work of several researchers in resolving the density distribution of the earth from its mass, moment of inertia and gravity coefficients. Spatial array data processing by the Burg and Capon methods is briefly surveyed. We conclude by summarizing our work with a particular Boltzmann-ME technique for resolving two-dimensional underground parameter distributions from cross-borehole data. An eigenvalue method for improving the numerical stability of our iterative technique is explained.

2.1 Density Distribution of the Earth

Rietsch [9] has obtained an ME estimate of the spherically uniform component of earth density, $\rho(r)$ from radius, mass, and moment of inertia data. He postulates a "Gibbsian" probability distribution $p(\rho)$ and defines the entropy as $H = - \int p(\rho)\ell n[p(\rho)/w(\rho)]dv$, $w(\rho)$ being an "invariant measure function," which in this problem is irrelevant because no "a priori" probability distribution is assigned to the discretized ρ_n assigned to the equivolume shells into which the earth is subdivided. He solves for the expectation values $\bar{\rho}_n$ over the $p(\rho)$ and is able to incorporate assumptions about lower and upper limits ρ_ℓ, ρ_u in the ME expression for $\bar{\rho}_n$.

We have also obtained a Boltzmann-ME solution to the problem with no assumptions about ρ_ℓ, ρ_u. Comparison of the two plots of Rietsch's $\bar{\rho}_n(r_n)$ for $\rho_\ell = 0$ or $1\ g/cm^3$ and $\rho_u = N\bar{\rho}$ (N = no. shells, $\bar{\rho}$ = average earth density) with each other and with our curve--all for $N = 100$--show surprisingly little difference down to a depth of about 4000 km below ground. The ME formulation and assumptions have the most effect in the inner core.

A lateral, non-spherical component of earth density has been inferred from spherical harmonic gravity field coefficients by Rubincam [10], assumed to be a small correction to the spherically symmetric $\rho(r)$ assumed known. He too used a Gibbsian-ME method which effectively obtains the population numbers of indistinguishable particles following Bose-Einstein statistics. An advantage to this method is the fact that the (grand) partition function can be factored, yielding a concise Bose-Einstein function for the ME lateral density distribution. The solution implies density anomalies concentrated near the earth's surface and decreasing with depth, in contrast to other non-entropy studies. We believe it would be more difficult to extend such a Gibbsian-ME treatment of density to include seismic constraints than to extend the corresponding Boltzmann-ME method.

2.2 Spatial Array Data Processing

ME spatial array analysis conventionally means estimation of the wavenumber spectrum (i.e., the directions of distant sources) from a sampled, finite spatial autocorrelation function by the Burg method [7]. This method is analogous to the Burg method of spectral analysis of a finite temporal sample, with frequency analogous to wavenumber and time to the space dimension. In spatial array processing one usually can average over time, with the consequence that the wavenumber spectrum is not the Fourier transform of the spatial autocorrelation function but rather its power spectrum (its Fourier transform over time). Aside from this distinction, both array processing and the corresponding power spectral analysis by the Burg method are equivalent to all-pole autoregressive modeling. We regard the Burg method as a solution to a slightly underdetermined ME problem, in the sense that the space or time record is extended only one more point in order to derive the ME wavenumber or frequency spectrum, respectively.

Lacoss [11] has investigated the MLM (Maximum Likelihood Method) of Capon [12] and the MEM for power spectra, or equivalently, wavenumber spectra. He reports that both methods are particularly valuable for resolving narrow frequency peaks, and neither is unduly sensitive to small noise in the underlying correlation functions. This is because both adapt to the actual noise by predicting power at a given frequency with least disturbance from other frequencies.

2.3 Resolution of Two-Dimensional Underground Parameter Distributions

The problem is to invert a Fredholm integral equation of the first kind, subject to constraints, in order to infer an underground distribution of electrical conductivity, dielectric constant or reciprocal seismic wave velocity from cross-borehole measurements of sinusoidal-frequency attenuation, phase shift, or transit time, respectively, of the rays traversing the region. The description of our Boltzmann-ME method of inferring an ME parameter distribution in discrete cells (rather than a Gibbsian probability distribution over possible earth models) has already been published [13], along with synthetic data examples. We merely summarize results here.

The discretized form of the Fredholm equation is written

$$T_i = \sum_{k=1}^{k} D_{ik}\sigma_k \qquad 1 \leq i \leq I < K \quad , \tag{21}$$

where T_i is the ith ray observable (eg., ray attenuation), D_{ik} is the ith path distance in the kth cell, and σ_k is the unknown parameter (eg., conductivity) in the kth cell. There are n_k blocks of $\Delta\sigma$ in the kth cell, $\sigma_k = n_k\Delta\sigma$, the total number of blocks is $N = \sum n_k$ and the entropy is

$$H = \ell n \frac{N!}{\prod n_k!} \simeq -N \sum_L (n_k/N)\ \ell n(n_k/N) \tag{22}$$

and the objective function to be made an extremum is

$$F = H + \gamma \sum n_k + \sum_{i=1}^{I} \beta_i [\sum_{k=1}^{K} D_{ik} n_k \Delta\sigma - T_i] \quad . \tag{23}$$

The ME solution is

$$\hat{\sigma}_k = \hat{\sigma}_T \exp[\sum_i \beta_i D_{ik}]/Z \quad , \qquad \hat{\sigma}_T = \sum \hat{\sigma}_k \tag{24}$$

partition function Z being

$$Z = \sum_k \exp[\sum_i \beta_i D_{ik}] \quad , \tag{25}$$

and the β_i must be found to satisfy the I constraints of (21).

We were able to resolve anomalies rather well in various synthetic data examples with an algorithm which neglected noise explicitly in the T_i, assuming $\hat{\sigma}_T$ was known to within 10% or so, and assuming straight rays (known D_{ik}). The algorithm computed consecutively (A) estimates $\hat{\sigma}_k$ from a $\bar{\beta}$ vector by (24)--starting from a simple β_o, (B) estimates $\hat{T}_j$ from the $\bar{\beta}$ by (21), with (24) in place of σ_k, and (25), (C) a correction to $\hat{\sigma}_T$ by a factor obtained from the implied constraint

$$\sum_i \beta_i \hat{T}_i = \sum_i \beta_i T_i \qquad (\sum \hat{T}_i = \sum T_i \text{ if } \bar{\beta}_o = 0) \quad , \tag{26}$$

in which $\hat{T}_i$ is proportional to the $\hat{\sigma}_k$ by (21), (D) $\hat{T}_i$ and $\hat{\sigma}_k$ scaled by the factor just obtained, (E) values of the derivatives $\partial\hat{T}_i/\partial\beta_j = \partial\hat{T}_j/\partial\beta_i$ obtained analytically from (21) to (25), (F) the changes $\Delta\beta_i$ necessary to improve the computed ray data T_i according to

$$T_i - \hat{T}_i = \sum_{j=1}^{I} \frac{\partial\hat{T}_i}{\partial\beta_j} \Delta\beta_j \quad , \qquad 1 < i \leq I \tag{27}$$

which in matrix form is

$$\Delta T_{I\times 1} = M_{I\times I}(\Delta\beta)_{I\times 1} \tag{27'}$$

by LU decomposition and back substitution. The algorithm then returned to step (A) and iterated until convergence was obtained to three significant digits or better. $\hat{\sigma}_T$ also converged to σ_T with this accuracy if it was specified initially with 10% or so error. In fact, once the solution determines the $\hat{\sigma}_k/\hat{\sigma}_T$ (21) will yield the correct $\hat{\sigma}_T$.

If we wanted to include the effect of ray bending according to ray optics we could do so between steps (D) and (E) above and correct the D_{ik}. Regardless of the constraints, there is only one extremum of H and it is a maximum, since $\partial^2 H/\partial n_k^2 < 0$ for all k subject to whatever constraints are imposed.

One can expect to resolve regions of K cells effectively with $I \simeq K/2$ rays, provided each cell is probed by at least one ray. In one example, a 48-cell region with one cell of $\sigma' = 50$ and the rest of $\sigma = 5$ was resolved by 25 rays to have $\hat{\sigma}' = 39.5$, one neighbor of 15.8 and all other cells with $\hat{\sigma} < 8.15$.

In such problems certain rays can be removed if they "overprobe" the region <u>without changing the ME solution for the $\hat{\sigma}_k$</u>. However, the solution to (27) becomes more ill-conditioned because of the nature of the Fredholm integral equation itself.

A typical solution to (27) tends to show very small ($< 10^{-5}$, for example) pivotal elements in the triangularization of the M-matrix. This can be alleviated by employing eigenvalue techniques. Generally, the best procedure is to resolve M into eigenvector-eigenvalue form and remove the vectors of very small eigenvalues:

$$M \to M^c = W_{I\times L}\Lambda_{L\times L}W^t_{L\times I} \quad , \qquad L < I \quad , \tag{28}$$

and solve for ($\Delta\beta$) using M^c in (27'). The contraction may vary from iteration to iteration.

However, it may be necessary to use the full M-matrix in (27') for the final iterations when the norm of ($\Delta\beta$) is very small in order to converge to the correct β!

3. ME ULTRASONIC IMAGING

Ultrasonic imaging is a nondestructive evaluation technique of inferring the presence of an anomaly such as a void or impurity within an object by irradiating it with ultrasonic waves and measuring the scattered wave field. Usually the object can be irradiated from only one side, and shear/compressional wave mixing occurs on discontinuity surfaces. We report below the results of inverting monochromatic scatter data in two-dimensional, simplified, synthetic-data problems by our Boltzmann-ME method [14].

3.1 The Method of Inverting the Fredholm Integral Equation

Gubernatis [15] has presented the exact equations for three-dimensional scattering of an incident shear or compressional wave by a volume anomaly. We assume a simplified, two-dimensional situation in which a compressional (or shear) wave characterized by one cartesian component is scattered from a homogeneous anomaly in an infinite medium, with neither coupling between components nor wave mixing. The Fredholm-II equation for the monochromatic total scalar displacement $u(\bar{r})$ is, in three dimensions,

$$u(\bar{r}) = u^\circ(\bar{r}) + \int dV' \, g(\bar{r},\bar{r}') \, v(r')u(\bar{r}') \quad , \tag{29}$$

with u° the incident (plane) wave and green's function g being

$$g(\bar{r},\bar{r}') = \frac{k_o^2}{4\Pi\rho\omega^2} \frac{e^{-ik_oR}}{R(\bar{r},\bar{r}')} \quad , \qquad k_o = \frac{\omega}{v_p} = \frac{2\Pi}{\lambda} \tag{30}$$

ρ = density, ω = circular frequency ($\exp(i\omega t)$ time dependence), v_p = phase velocity, λ = wavelength. $v(\bar{r}')$ describes the anomaly as

$$v(\bar{r}') = \delta\rho\omega^2 \, \Theta(\bar{r}') \quad , \tag{31}$$

where $\delta\rho$ = change in density from the background value and $\Theta(\bar{r}')$ = 0 outside the anomaly and 1 inside.

The problem is to infer the area and shape of the anomaly via $\Theta(\bar{r}')$ from the scattered far-field displacement in direction $\bar{k}$, given by u^s.

$$u^s(\bar{k}) = S e^{-ik_o r}/r \quad , \tag{32}$$

$$S = \frac{k_o^2}{4\Pi\rho\omega^2} \int dV \exp(i\bar{k}\cdot\bar{r})\, v(\bar{r}) u(\bar{r}) \quad . \tag{33}$$

This synthetic problem includes the essential ingredients of the full coupled-wave problem formulated in tensor notation.

For the ME analysis we discretized the 2D region into square cells Δ on a side and interpreted (29) per unit length in the third direction so as to read

$$u(\bar{r}_i) = u^\circ(\bar{r}_i) + \sum_k \frac{\Delta e^{-ik_o R}}{R(\bar{r}_i,\bar{r}_k)} u(\bar{r}_k)\, \tilde{\Theta}(\bar{r}_k) \quad , \tag{34}$$

with $\tilde{\Theta} = k_o^2 \Delta\delta\rho\Theta/(4\Pi\rho)$ and special evaluation of the summand for the $k = i$ cell. We wrote (33), $\bar{r}$ of cell $k = \bar{r}_o - \bar{r}_k$ and $\bar{k}_i = \bar{k}$, as

$$S = \Delta \sum_k e^{i\bar{k}_i\cdot\bar{r}_k} u(\bar{r}_k)\, \tilde{\Theta}(\bar{r}_k) \quad , \text{ the observable.} \tag{35}$$

We interpreted S as the scattered field from a "radiating source" distribution, with source $A_k \exp(i\psi_k) = u(\bar{r}_k)\, \tilde{\Theta}(\bar{r}_k)$ at the center of the kth cell in the anomaly, where $u(\bar{r}_k)$ is determined by the Fredholm equation (34). We then solved for the $A_k \exp(i\psi_k)$ by the same Boltzmann-ME method employed in Sec. 1.3 to solve the problem reciprocal to the "distant source" resolution problem [4]. The A_k-amplitude distribution indicates the subjective probability of $\Theta(\bar{r}_k)$ being zero (no anomaly) or one (anomaly) within the region examined. Note that A_k measures $|u(\bar{r}_k)\, \tilde{\Theta}(\bar{r}_k)|$, so several scattering experiments would resolve $\Theta(\bar{r}_k)$ more reliably.

In the synthetic data problems explored with this ME method the Fredholm equation was only used to obtain the exact S corresponding to a measurement, not in the ME solution. We observed in Sec. 1.3 there is only one maximum entropy solution to this problem.

3.2 Synthetic Data Examples

The very first example selected arbitrarily to test this method is the problem of resolving the 2D anomaly shown in Fig. 4. In the 100 cells subdividing the 2λ x 2λ region the anomaly of $\tilde{\Theta} = -1$ occupied 10 cells. For an incident plane wave u° propagating in the -x direction ($\phi_s = 0$) we fed into the ME algorithm scatter data u^s in 20 directions, at equal angles over $-81° \leq \phi_s \leq 90°$. In solving the equivalent of (17) for the β_i each iteration by LU decomposition and back substituting some of the pivot elements were miniscule ($\approx 10^{-5}$) but eigenvector decomposition was unnecessary on the CDC 7600 and the algorithm converged in about 10 iterations. For this u° the 21 cells of largest magnitude $.0315 \leq |u_k \Theta_k| \leq .074$ are indicated by the lines slanting upward and to the right in Fig. 4. The magnitudes in the other cells are all < .0295.

We repeated the analysis but for $\tilde{\Theta} = -10$ and got the same resolution with the 21 cells of largest $|u_k \tilde{\Theta}_k|$.

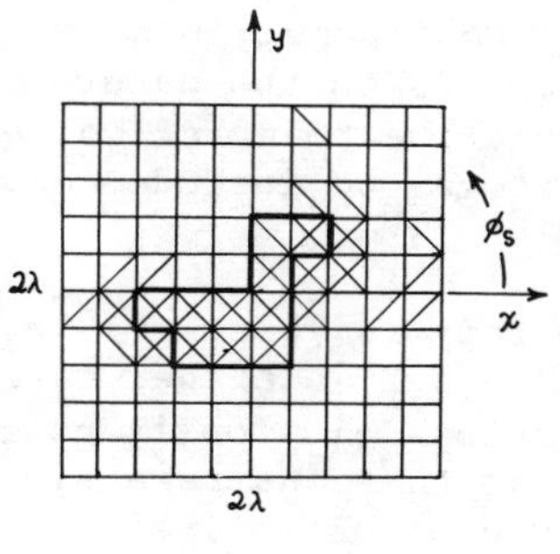

Fig. 4. Two-dimensional anomaly and its resolution by the "radiating source" ME method [14].

We then processed the data for $\tilde{\Theta} = -1$ and u° incident from the $\phi_s = 30°$ direction, and the 21 cells of largest magnitude $.025 \leq |\hat{u}_k \tilde{\Theta}_k| \leq .102$ are indicated by the lines slanting upward and to the left.

Note that all but one of the anomalous-region cells are included in both ME groups of 21 cells each. Six "normal" cells are indicated as possibly anomalous.

Such synthetic data examples as these suggest guidelines for applying this ME method to more complicated problems. If the anomaly within N_A cells is known to lie within a region of K cells, the number of scatter data (complex) J_s should be increased until reasonable convergence to a (fuzzy) shape is attained. This will occur when

$$N_A < J_s \lesssim K/2 \quad .$$

The overall size of the region picked is not crucial but the cell size should be $\Delta/\lambda \lesssim .25$. Scatter data should be taken over as wide an angle as possible, which may be < 180°; data from several incident waves should be processed. The value of Θ, which measures both the density anomaly $\delta\rho/\rho$ and the cell size Δ/λ, seems not to be crucial for resolution. Finally, the method does not inherently favor single-volume anomalies; disjoint ones of different Θ might very well be resolved.

4. THEORETICAL AND PRACTICAL ISSUES IN ME INVERSION

In this section we address the issues of uniqueness, confidence, and noise in ME inversion. We have already mentioned the uniqueness of our Boltzmann-ME solution to problems in various areas. Confidence in an ME solution being close to the correct one depends on the nature of the system. There seems to be no rigorous treatment of noise within any ME framework at present, but it may be included empirically quite satisfactorily.

The measurement of entropy as $\sum \ln p_i$ or $-\sum p_i \ln p_i$ depends on the nature of the system [2], but in either case, the uniqueness of a maximum entropy solution is guaranteed by the convexity of H <u>independent</u> of the constraints. However, there is usually more than one type of ME analysis for a problem; we have contrasted to some extent the Gibbsian versus our Boltzmann-ME treatment of some problems. The Burg method is of the Gibbsian type. We have also observed that, in principle, there are at least two Boltzmann-ME philosophies to follow in solving coherent radiation problems of interest to us--one with entropy formulated in terms of amplitude squared (power) and the other based on amplitude. Objective criteria for preferring one ME analysis are simply not available.

The confidence one can place on an ME solution yielding predictions close, in some objective sense, to the correct ones depends critically on whether the system analyzed is "fluid" or not. The second law of thermodynamics applies to systems not only with an astronomical number of configurations (microstates) but also with the freedom to assume any of these states, or at least a subensemble of states with H close to H_{max}, during an observational period. And so one places high confidence in H being within ΔH of H_{max}, as given by the Concentration Theorem [15], loosely written as

$$N\Delta H = \chi^2(s) \quad , \tag{36}$$

where the right side designates the Chi-squared distribution at significance level s. N is the number of trials of an experiment or elements in a configuration (number of molecules, symbols in a possible message, elements of luminescence, building blocks of a parameter distribution, etc.)

However, suppose the system is not "fluid," such as a parameter distribution in a solid volume. There is no guarantee that the system will be near its maximum entropy state, (36) is merely a description of concentration about H_{max}, and an ME solution merely furnishes a most probable, very rational guess about the system configuration. Since we cannot relate such a physical system to the average of an ensemble of mathematical systems we must say there is no defined confidence that the system will be within ΔH of the ME system.

Despite this unsettling situation it is of some comfort to know that, as N increases, the ME system becomes more probable according to (36). We can say more about the distribution of individual elements about their H_{max} values as $N \to \infty$ from a Darwin-Fowler (D-F) analysis, appropriate to a Boltzmann-entropy description. Suppose the system has K cells, with n_k elements in the kth cell, $\Sigma\, n_k = N$. The D-F analysis for the first two moments of n_k yields

$$\bar{n}_k = n_k^{ME} \tag{37a}$$

$$\overline{(n_k - \bar{n}_k)^2}/\bar{n}_k^2 \cong 1/\bar{n}_k \quad , \qquad N \gg \bar{n}_k \ . \tag{37b}$$

In the problems we have discussed all n_k were enormous, which made our Boltzmann-ME solutions overwhelmingly most probable--but not more possible.

Jaynes [16] has discussed the modification of an ME solution so as to include noise in a full Bayesian solution. In essence, noise constraints are added in the manner of Eq.

(6), and any ME solution which satisfies them as well as the other constraints would be a member of an ensenble of most probable solutions with noise included. The noise blurs the distribution of each parameter n_k about its noiseless n_k^{ME} value.

5. REFERENCES

1. Soffer, B. H. and Kikuchi, R., "Maximum Entropy Image Estimation Analysis of Image Formation with Thinned Random Arrays," Hughes Research Laboratories, Malibu, CA, 90265, 1981. Sponsored by the Air Force Office of Scientific Research.

2. Kikuchi, R. and Soffer, B. H., Maximum Entropy Image Restoration I The Entropy Expression, Jour. Opt. Soc. Am., 67, pp. 1656-1665, 1977.

3. Wernecke, S. J. and D'Addario, L. R., Maximum Entropy Image Reconstruction, IEEE Trans. Comput., C-26, pp. 351-364, 1977.

4. Bevensee, R. M., "Maximum Entropy Resolution of Two-Dimensional Antenna Distributions," IEE Conf. Pub. 195, Antennas and Propagation, Part 1: Antennas, pp. 383-387, 1981.

5. Gabriel, W. F., Spectral Analysis and Adaptive Array Superresolution Techniques, Proc. IEEE, 68, pp. 654-666, 1980.

6. Steinberg, B. D. and Luthra, A. J., "Single and Multiple Snapshot Maximum Entropy Imaging of Two Targets," Valley Forge Research Center, Quarterly Progress Report, No. 36, University of Pennsylvania, Valley Forge, PA, pp. 22-29, 1981.

6A. Luthra, A. J. and Steinberg, B. D., "The Maximum Entropy Image of Two Targets," Proc. IEEE, 70, pp. 98-99, 1982.

7. Burg, J. P., Maximum Entropy Spectral Analysis in "Modern Spectral Analysis," ed. by D. G. Childers, IEEE Press, pp. 34-41, 1978.

8. Steinberg, B. D., Private Communication.

9. Rietsch, E., The Maximum Entropy Approach to Inverse Problems, J. Geophys., 42, pp. 489-506, 1977.

10. Rubincam, D. P., "Information Theory Lateral Density Distribution for Earth Inferred From Global Gravity Field," Journal Geophys. Res., 87, pp. 5541-5552, 1982.

11. Lacoss, R. T., "Data Adaptive Spectral Analysis Methods," Geophys., 36, pp. 661-675, 1971.

12. Capon, J., "High-resolution Frequency-wavenumber Spectrum Analysis," Proc. IEEE, 57, pp. 1408-1418, 1969.

13. Bevensee, R. M., "Solution of Underdetermined Electromagnetic and Seismic Problems by the Maximum Entropy Method," IEEE Trans. Ant. Prop., AP-29, pp. 271-274, 1981.

14. Bevensee, R. M., "The Maximum Entropy Formulation of Inverse Problems of NDE," Review of Progress in Quantitative Non Destructive Evaluation, 2, D. O. Thompson and D. E. Chimenti, eds., pp. 937-949, Plenum Press, NY, 1983.

15. Gubernatis, J. E., "Elastic Wave Scattering Calculations and the Matrix Variational Pade Approximation Method," Proceedings of the DARPA/AFWAL Review of Progress in Quantitative NDE, AFWL-TR-81-4080, Rockwell International Science Center, pp. 300-310, 1981.

16. Jaynes, E. T., "On the Rational of Maximum-Entropy Methods," Proc. IEEE, 70, pp. 939-952, 1982.

Work performed under the auspices of the U.S. Department of Energy by the Lawrence Livermore National Laboratory under contract number W-7405-ENG-48.

II.10 (NM.1)

DETAILED NEAR/FAR FIELD MODELING OF COMPLEX ELECTRICALLY LARGE THREE DIMENSIONAL STRUCTURES

Allen Taflove
Korada Umashankar

IIT Research Institute
10 West 35th Street
Chicago, Illinois 60616

ABSTRACT

The finite-difference time-domain (FD-TD) method is shown as a means of accurately computing electromagnetic scattering by arbitrary-shaped, extremely complex metal or dielectric objects excited by an external plane wave. In the present method, one first uses the FD-TD method to compute the near total fields within a rectangular volume which fully encloses the object. Then, an electromagnetic field equivalence principle is invoked at a virtual surface of this rectangular volume to transform the tangential near scattered fields to the far field. To verify the applicability of this method, the surface currents, near scattered fields, far scattered fields, and radar cross section of canonical two-dimensional and three dimensional objects are presented. For these cases, it is shown that the FD-TD method provides magnitude of current and field predictions which are within ±2.5% and further phase values within ±3° of values predicted by the method of moments (MOM) at virtually every point including in shadow regions. The FD-TD method is presently being extended to three-dimensional structures as large as twenty wavelengths.

This work was supported partly by Air Force Contract No. F30602-80-C-0302, Rome Air Development Center, GAFB, NY and partly by Internal Research of the Electronics Division, IIT Research Institute, Chicago, IL

W.-M. Boerner et al. (eds.), Inverse Methods in Electromagnetic Imaging – Part 1, 397–430.

1. INTRODUCTION

General electromagnetic scattering problems have been difficult to treat with either analytical [1] or numerical methods because of the complicating effects of curvatures, corners, apertures, and dielectric loading of structures. In an attempt to gain insight into scattering mechanisms using analytical and numerical approaches it has been necessary to use canonical structures, rather than realistic models. A potential alternate approach is the finite-difference time-domain (FD-TD) method [2-17] which allows the computation of internal and external near fields by direct modeling of realistic structures.

In order to treat realistic scattering problems effectively, a method has been developed [14,15,17] which involves combining the FD-TD method with a near-field to far-field transformation using field equivalences [18]. In this method, the scattering problem is analyzed in two steps by treating the relatively complex near-field region and the relatively simple far-field region separately. The method involves first the determination of equivalent electric and/or magnetic currents tangential to a virtual surface surrounding the scatterer of interest by using the FD-TD method for a given external illumination. The computed near-field equivalent currents are transformed then to derive the far-field scattering pattern and the radar cross section. Since the FD-TD method can deal with dielectric, permeable, and inhomogeneous materials in a natural manner, it is possible to incorporate most of the physics of wave interaction with any complex scatterer of interest.

To validate applicability of this method, the FD-TD computed surface electric current distribution and near scattered electric and magnetic fields are presented for the case of a two dimensional rectangular metal cylinder and a three dimensional metal cube subject to plane-wave illumination at normal incidence. These currents and scattered fields are compared to method of moments (MOM) computed results. The far scattered field pattern and radar cross section are derived from the FD-TD data using a near to far-field transformation and are also compared to the results obtained using MOM [19]. It is shown that a very high degree of correspondence is obtained using this present proposed method.

The FD-TD method is presently being extended to obtain detailed modeling of near/far fields of complex three-dimensional scatterers as large as twenty wavelengths. Experimental verifications are also being obtained for radar cross section of complex material scatterers including composites.

2. FORMULATION

A brief enumeration of the steps to analyze scattering by complex objects is presented in this section, based on the application of the FD-TD method to obtain the near fields, and then the transformation of the near fields to far fields to obtain the scattering cross section. In Figure 1 is shown a geometry of an arbitrary-shaped scattering object. This object can be either conducting, dielectric, or permeable as well as inhomogeneous. It can have apertures/cavities containing internal dielectric loading [9-10]. The incident field($\vec{E}^i$, $\vec{H}^i$) excites the scatterer to produce simultaneously some interior field penetration and also an exterior scattered field ($\vec{E}^s$, $\vec{H}^s$).

2.1 Near-field analysis

As shown in Figure 1, the scatterer is enclosed in a rectangular volume with a boundary surface, S_a, for analyzing the near total fields based on FD-TD [9,10,14,15,17]. The finite difference formulation of the FD-TD method allows a straightforward modeling of the surfaces and interiors of arbitrary complex structures. The structure of interest is mapped into the space lattice, Figure 1, by first choosing the space increment [10] and assigning values of permittivity and conductivity to each component of total electric field $\vec{E}$. No special handling of electromagnetic boundary conditions at media interfaces is required because the curl equations generate these conditions in a natural way [10]. The various details of the structure are modeled with a maximum resolution of one unit cell, with thin surfaces being modeled as stepped-edge sheets. The explicit numerical formulation of the FD-TD method is particularly suited for programming with minimum storage and execution time using recently developed array processing computers. The required computer storage and running time increases only linearly with N, where N is the total number of unknown field components [9]. Since all FD-TD operations are explicit and can be performed in parallel, rapid array-processing techniques can be readily applied. As demonstrated [9,10], these can be used to solve for more than 10^6 field components in a single FD-TD problem (current array processing capability).

As discussed in reference [9,10] the FD-TD method is a direct numerical solution to the Maxwell's time-dependent curl equations useful for studying propagation of an electromagnetic wave into a volume of space containing an arbitrary shaped dielectric or conducting body. By time-stepping or repeatedly implementing a finite-difference analog of the curl equations at each cell (Figure 1) of the corresponding space lattice, the incident wave is tracked as it first propagates to the

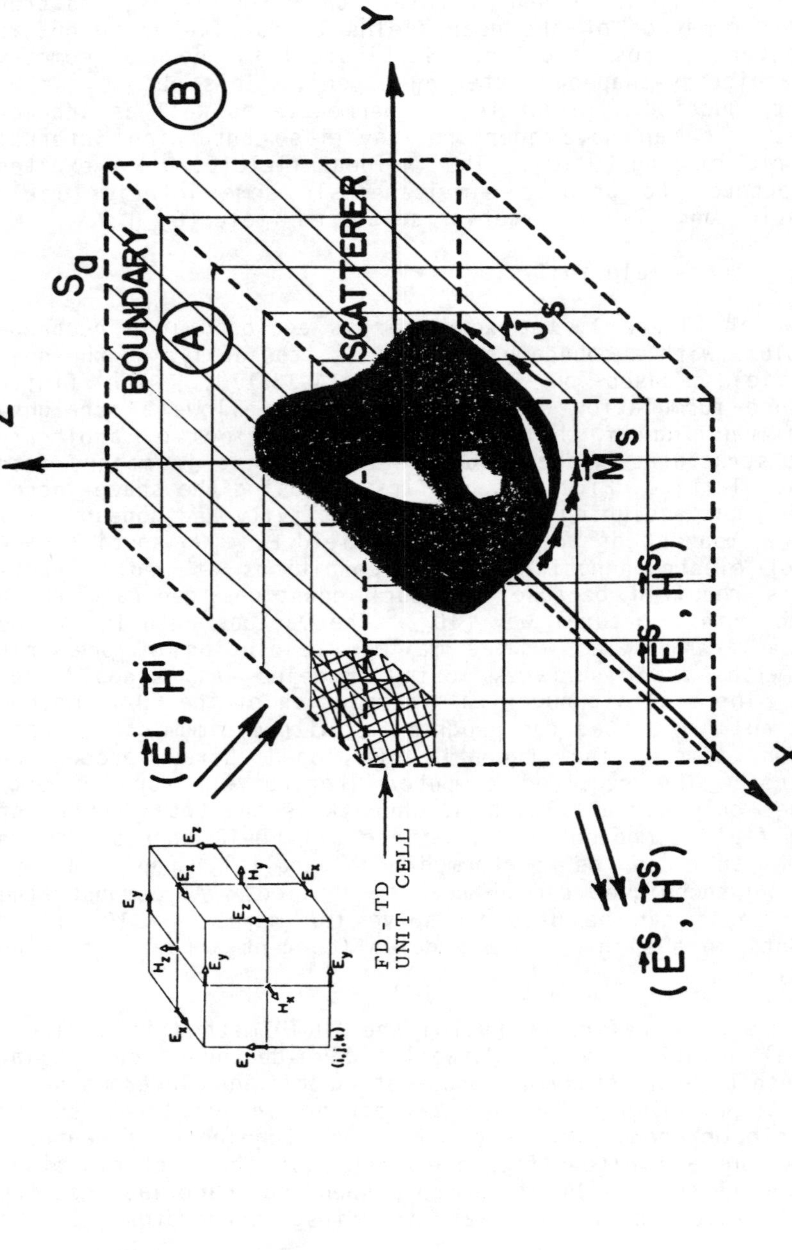

FIGURE 1. GEOMETRY OF A GENERAL SCATTERER IN FREE SPACE MEDIUM AND FD-TD LATTICE ARRANGEMENT.

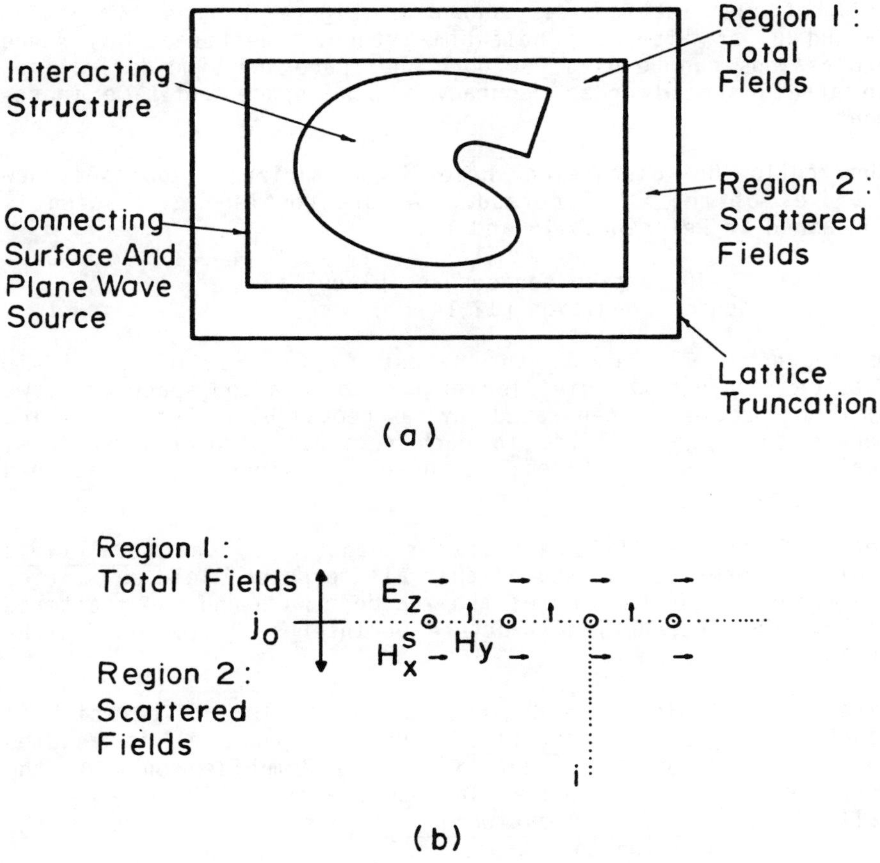

FIGURE 2. DIVISION OF FD-TD LATTICE INTO TOTAL-FIELD AND SCATTERED-FIELD REGIONS: (a) LATTICE DIVISION; (b) TYPICAL CONNECTING CONDITIONS BETWEEN REGIONS.

structure and then interacts with it via penetration and diffraction. Wave tracking is completed when the desired late-time or sinusoidal steady-state behavior is observed at each lattice cell. The time stepping for the FD-TD method is accomplished by positioning the components of $\vec{E}$ and $\vec{H}$ about a unit cell of the lattice, as shown in Figure 1, and evaluating $\vec{E}$ and $\vec{H}$ at alternate half-time steps. Centered difference expressions can be used for both the space and time derivatives to attain second order accuracy in the space and time increments.

The following discussion briefly summarizes important new features of the FD-TD method. A more complete discussion is contained in References 15 and 17.

2.2 FD-TD lattice regions and plane wave source condition [12,14-17]

As shown in Figure 2a, the latest formulation of the FD-TD lattice involves the division of the computation space into two distinct regions, separated by a rectangular surface which serves to connect fields in each region. In two dimensions, the surface has four faces; in three dimensions the surface has six faces.

Region 1 of the FD-TD lattice is denoted as the total-field region. Here, it is assumed that all computed field quantities are comprised of the sum of the incident wave and the scattered field. The interacting structure of interest is embedded within this region.

External to Region 1 is Region 2 of the FD-TD lattice which is denoted as the scattered-field region. Here, it is assumed that all computed field quantities are comprised only of the scattered field. The outer lattice planes bounding Region 2, called the lattice truncation planes, serve to implement the free-space radiation condition.

The rectangle faces comprising the boundary between Regions 1 and 2 contain $\vec{E}$ and $\vec{H}$ field components which require the formulation of various field-component differences across the boundary planes for proceeding one time step. These computations serve to connect the total-field and scattered-field regions and simultaneously generate the desired plane wave of arbitrary polarization and angle of incidence [14-17]. An alternative approach for incident-wave generation based on the Huygen's source formulation is discussed in Reference 12.

There are a number of key advantages to this methodology: (a) The total-field formalism is retained for the entirety of the interacting structure, permitting accurate computations of low-level fields penetrating into cavities through apertures, and in the shadow regions of scatterers. (b) The scattered-field formalism is retained for the lattice truncation region, permitting a very accurate simulation of the radiation condition. (c) The incident wave contribution need be computed or stored only for the field components at the rectangular surface connecting Regions 1 and 2. This results in much less computation or storage than if the incident field were to be computed at all points within the interacting structure to implement a pure scattered-field formalism. (d) The scattered near field in Region 2 can be easily integrated to derive the far-field scattering and radar cross section, as discussed later in this paper.

2.3 Lattice truncation conditions [3-5,8,15-17,20,21].

A basic consideration with the FD-TD approach to solve electromagnetic field problems is that most such problems are usually considered to be "open" problems where the domain of the computed field is ideally unbounded. Clearly, no computer can store an unlimited amount of data, and the field computation zone must be limited. The computation zone must be large enough to enclose the structure of interest, and a suitable boundary condition on the outer perimeter of the computation zone must be used to simulate the extension of the computation zone to infinity. Outer boundary conditions of this type have been called either radiation conditions, absorbing boundary conditions, or lattice truncation conditions.

In three dimensions, an outgoing scattered-wave field component, F^s (either an electric or magnetic field) has a (r, θ, ϕ) variation of the type [22],

$$F^s = F_o e^{j(\omega t - k_o r)} \left[\frac{A(\theta,\phi)}{r} + \frac{B(\theta,\phi)}{r^2} + \ldots \right] \qquad (1)$$

Here, the bracketed infinite series represents in effect a multipole expansion of the scattered field, where A, B, . . . are initially unknown functions of angular position.

First-order FD-TD simulations of the outer lattice boundary condition approximate the $A(\theta,\phi)/r$ dependence only. Second and higher-order approximations simulate the $B(\theta,\phi)/r^2$ and higher-order r-dependent behavior of F^s in addition to the $A(\theta,\phi)/r$ term [15]. A typical FD-TD computation realizing a second order correct radiation condition and its accuracy is given in references 16 and 20.

2.4 Sinusoidal steady-state information

Such data can be obtained either by (a) directly programming a single-frequency incident plane wave or (b) performing a separate Fourier transformation step on the pulse waveform response. Both methods require time-stepping to a maximum time equal to several wave periods at the desired frequency. The second method has two additional requirements. First, a short-rise-time pulse suffers from accumulating waveform error due to overshoot and ringing as it propagates through the space lattice. This leads to a numerical noise component which must be filtered before Fourier transformation. Second, Fourier transformation of many lattice-cell field-versus-time waveforms would significantly add to the total requirements for computer storage and execution time [10].

Recent work has shown that very accurate magnitude and phase information for sinusoidal steady state FD-TD problems can be obtained by method (a) above and observing the peak positive and negative-going excursions of the fields over a <u>complete cycle</u> of the incident wave (after having time-stepped <u>through 2-5</u> cycles of the transient period following the beginning of time stepping). For certain two and three-dimensional scattering problems, a dc offset of particular computed field components can be possible. This leads to the following requirements to obtain correct magnitude and phase data :

1. The peak-to-peak value of the sinusoidal response at any point must be observed to eliminate the effects of any dc offset upon the computation of the phasor magnitude.

2. The zero-crossing of the field waveform may not be useful in determining relative phase. Rather, it may be necessary to locate the zero-derivative points of the waveform for this purpose, possibly incorporating an interpolation algorithm to enhance resolution of the relative phase computation without requiring a smaller time step.

As will be shown, this methodology has been found to be successful in achieving ±2.5% (or better) and ±3° correspondence of FD-TD results with moment-method computations for two-dimensional canonical problems; and similar correspondence is obtained for three-dimensional canonical problems as well.

2.5 Derivation of far-field scattering data :

Far-Field scattering data can be obtained with the FD-TD method by applying a powerful and flexible near-field to far-field transformation [14,17,18]. Referring to Figure 3, a rectangular virtual surface, S_a, which fully enclose the scatterer, is located in the scattered field region (Region 2 of Figure 2). The tangential components of the scattered fields $\vec{E}^s$ and $\vec{H}^s$, are first obtained at S_a using FD-TD. Then, as indicated in Figure 3b, an equivalent problem is set up which is completely valid for Region B, external to S_a. The new excitation data are $\vec{J}_{s_{eq}}$ and $\vec{M}_{s_{eq}}$, the equivalent surface electric and magnetic currents respectively, on S_a which are obtained according to [22,23]:

$$\vec{J}_{s_{eq}}(\vec{r}) = \hat{n} \times \vec{H}^s(\vec{r}) \tag{2a}$$

$$\vec{M}_{s_{eq}}(\vec{r}) = -\hat{n} \times \vec{E}^s(\vec{r}) \tag{2b}$$

where $\hat{n}$ is the outward unit normal vector at the surface S_a.

The scattered far fields are given by the transform of the equivalent currents of equations (2a) and (2b) over the free space Green's function [23]. If (μ_o, ε_o) are the region B medium characteristics with $k_o = 2\pi/\lambda$ and $\eta_o = 120\pi$, for θ and ϕ polarizations the following scattered far-field expressions are obtained,

$$E_\theta = (-jk_o\mu_o)\left[A_\theta + \frac{F_\phi}{\mu_o}\right] \tag{3a}$$

$$E_\theta = (jk_o\mu_o)\left[A_\phi - \frac{F_\theta}{\mu_o}\right] \tag{3b}$$

where

$$A_\theta = A_x \cos\theta \cos\phi + A_y \cos\theta \sin\phi - A_z \sin\theta \tag{4a}$$

$$F_\theta = F_x \cos\theta \cos\phi + F_y \cos\theta \sin\phi - F_z \sin\theta \tag{4b}$$

$$A_\phi = - A_x \sin\phi + A_y \cos\phi \tag{4c}$$

$$F_\phi = - F_x \sin\phi + F_y \cos\phi \tag{4d}$$

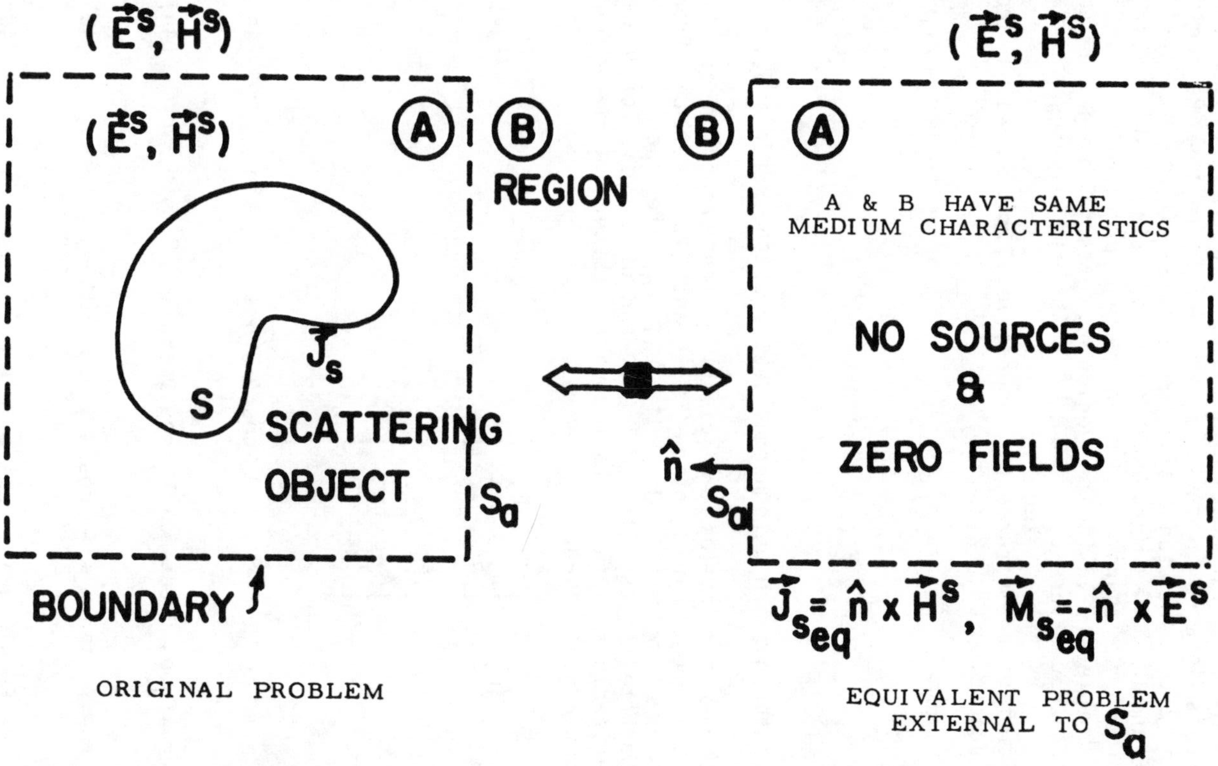

FIGURE 3. ELECTROMAGNETIC EQUIVALENCE TO TRANSFORM NEAR-FIELDS TO FAR-FIELDS.

and the potentials in the far-field region are given by

$$\begin{bmatrix} \vec{A}, \\ \vec{F} \end{bmatrix} = \left(\frac{e^{-jk_o r}}{4\pi r}\right) \iint_{S_a} \begin{bmatrix} \vec{J}_{s_{eq}}, \\ \vec{M}_{s_{eq}} \end{bmatrix} e^{jk_o r' \cos \xi} ds'_a \qquad (5a)$$

$$r' \cos \xi = (x' \cos \phi + y' \sin\phi) \sin \theta + z' \cos \theta \qquad (5b)$$

The radar cross section is calculated as the ratio of

$$RCS = 4\pi r^2 \left[\frac{E_\theta^2 + E_\phi^2}{{E_\theta^i}^2 + {E_\phi^i}^2} \right], \quad r \to \infty \qquad (6)$$

E_θ^i and E_ϕ^i are the corresponding components of the incident plane wave excitation, for θ and ϕ polarizations.

This approach to computing the far scattered fields is very promising since 1) the near-field data for arbitrary scatterers can be obtained in a straight-forward manner using the FD-TD method; and 2) the transformation of the near-field data to the far field is computationally simple and independent of the nature of the scatterer which resides within the integration surface, S_a. The complete bi-static radar cross section is a natural result of this procedures, for a given incident angle and polarization of the illuminating wave. For problems involving variable incidence and/or polarization of the illuminating wave, new FD-TD data is required for each selected incident-wave condition.

3. BASIS OF THE MOM MODEL USED FOR FD-TD VERIFICATION

General arbitrary-shaped scattering bodies can be analyzed based on a frequency domain integral-equation/boundary-value problem approach. A popular technique for the numerical solution which implements this approach is the method of moments [19]. This method is particularly suited for low-frequency scattering problems, but can be applied to bodies spanning approximately 1 wavelength in three dimensions [26-33].

The following formulations of the method of moments have been found to be generally suited for certain scattering bodies based upon their geometry and material characteristics :

1. Conducting Scatterers (homogeneous, isotropic)

   a. Electric field integral equation formulation (EFIE) for closed and open bodies

   b. Magnetic field integral equation formulation (MFIE) for closed bodies

2. Dielectric Scatterers (homogeneous, isotropic)

   Combined field integral equation formulation (CFIE)

Recent work has concentrated on Case 1 [31, 32], especially in the development of the triangular patch model for arbitrary scatterers. Further extension to dielectric scatterers is presently being completed.

To treat arbitrary-shaped conducting bodies [31], the EFIE formulation is generally used. Referring to Figure 1, if S denotes the surface of an open or closed body, the scattered electric field is given by

$$\vec{E}^s = j\omega\vec{A} - \nabla\phi \tag{7}$$

where the magnetic vector potential and the scalar potential are defined as

$$\vec{A}(\vec{r}) = \frac{\mu_o}{4\pi} \iint_S \vec{J}(\vec{r}\,') \, G(\vec{r},\vec{r}\,') \, ds' \tag{8}$$

$$\phi(\vec{r}) = \frac{1}{4\pi\varepsilon_o} \iint_S \rho^e(\vec{r}\,') \, G(\vec{r},\vec{r}\,') \, ds' \tag{9}$$

$$G(\vec{r},\vec{r}\,') = \frac{e^{-jk_o|\vec{r} - \vec{r}\,'|}}{|\vec{r} - \vec{r}\,'|} \tag{10}$$

where (μ_o, ε_o) are scalar constants representing the permeability and permittivity of the surrounding medium; $\vec{J}$ and ρ^e are the unknown electric current and charge distributions induced on the conducting scatterer; and $R = |\vec{r} - \vec{r}\,'|$ is the distance between an arbitrarily-located observation point $\vec{r}$ and a source point $\vec{r}\,'$ on S. Both $\vec{r}$ and $\vec{r}\,'$

are defined with respect to a global coordinate origin. An integro-differential equation for $\vec{J}$ is derived [19,32] by enforcing the boundary condition $n \times (\vec{E}^i + \vec{E}^s) = \vec{0}$ on S

$$-\vec{E}^i_{tan} = (-j\omega\vec{A} - \nabla\Phi)_{tan}, \quad \vec{r} \text{ on } S \tag{11}$$

In arbitrary surface modeling the above EFIE has the advantage that it applies to both open and closed bodies, whereas the MFIE applies only to closed surfaces [31]. On the other hand, for arbitrarily-shaped objects the EFIE is considerably more difficult to apply than the MFIE.

For modeling arbitrarily-shaped surfaces, planar triangular patch models have been found to be particularly appropriate. Since triangular patches are capable of accurately conforming to any geometrical surface and boundary, the patch scheme is easily described to the computer, and a varying patch density can be used according to the resolution required in the surface geometry or current. This patching method permits the straightforward construction of MOM basis functions defined on the triangular patches which are free of line charges.

In References [31] and [32], planar triangular patch modeling and the MOM [19] were applied to develop numerical procedures using the EFIE formulation to treat scattering by arbitrarily-shaped conducting bodies. An efficient computer code based on this formulation was described which is capable of handling either open or closed and arbitrarily-curved structures of finite extent. Discounting computer limitations, the code can, in principle, treat any conducting object whose surface is orientable, connected (or multiply-connected), and free of intersecting surfaces. An updated version of this computer program [34] was used to obtain benchmark data for the surface currents and scattered fields for the conducting cube scatterer discussed next.

4. NUMERICAL RESULTS - TWO DIMENSIONAL CASE

In order to validate the feasibility of this hybrid method to analyze electromagnetic scattering, a canoniocal two-dimensional conducting structure is studied [25]. The numerical results of the FD-TD-computed surface electric current distribution, and near electric and magnetic fields are presented for the case of a two-dimensional rectangular metal cylinder subject to a TM-polarized illumination at normal (broadside) incidence. These electric currents and near fields are compared to the MOM-computed results [19, 24]. The scattered-field pattern and the corresponding radar cross section (RCS) are derived from the near-to-far field transformation of the FD-TD data. These are then compared to the results obtained by using the MOM. It is shown that a very high degree of correspondence is obtained using this method.

4.1 Square cylinder

The scattering by a two-dimensional conducting cylinder of arbitrary cross section is considered first to validate feasibility of the present method. This cannonical problem, shown in Figure 4, is well documented [19,22,24]. We first consider the example of the scattering of a plane wave by a square conducting cylinder. The cylinder has the electrical size $k_o A_s = 1$, where A_s is the half-width of the side of the cylinder. The plane-wave excitation is TM polarized, with components E_z^i and H_x^i, and propagates in the +y direction so that it is at normal incidence to one side of the cylinder ($\phi^i = 90^0$). An 84-point MOM solution of Equation (11) is used as the benchmark for comparison with all FD-TD results, with pulse current expansion and point matching [19,24].

For the FD-TD analysis, the square cylinder is embedded in a two-dimensional lattice as shown in Figure 2. Each side of the cylinder spans 20 lattice-cell divisions. The connecting virtual surface between the FD-TD total-field and scattered-field regions is located at a uniform distance of 5 cells from the cylinder surface. Figures 4a and 4b graph the comparative results for the FD-TD and MOM analyses of the magnitude and phase of the cylinder surface electric current distribution for this case. Here, the FD-TD-computed surface current is taken as $n \times \vec{H}_{tan}$, where n is the unit normal vector at the cylinder surface, and $\vec{H}_{tan}$ is the FD-TD value of the magnetic field parallel to the cylinder surface, but at a distance of 0.5 space cell from the surface. (The displacement of the $\vec{H}$ component from the cylinder surface is a consequence of the spatially-interleaved nature of the $\vec{E}$ and $\vec{H}$ components of the FD-TD lattice, indicated in Figure 1. In Figure 4a, the magnitude of the FD-TD-computed surface current agrees with the 84-point MOM solution to better than ±1% (±0.09 dB) at all comparison points more than 2 cells from the cylinder edges (current singularities). In Figure 4b, the phase of the FD-TD solution agrees with the MOM solution to within ±3° at all comparison points, including the shadow region. The uncertainty bars shown in this figure indicate the present level of imprecision in using the FD-TD method to locate the constant-phase points of the computed time-domain $\vec{H}_{tan}$ waveform (equivalent to ±1 time step). This imprecision can be reduced in future FD-TD programs by incorporating a simple interpolation algorithm to achieve fractional time-step resolution of points of constant phase.

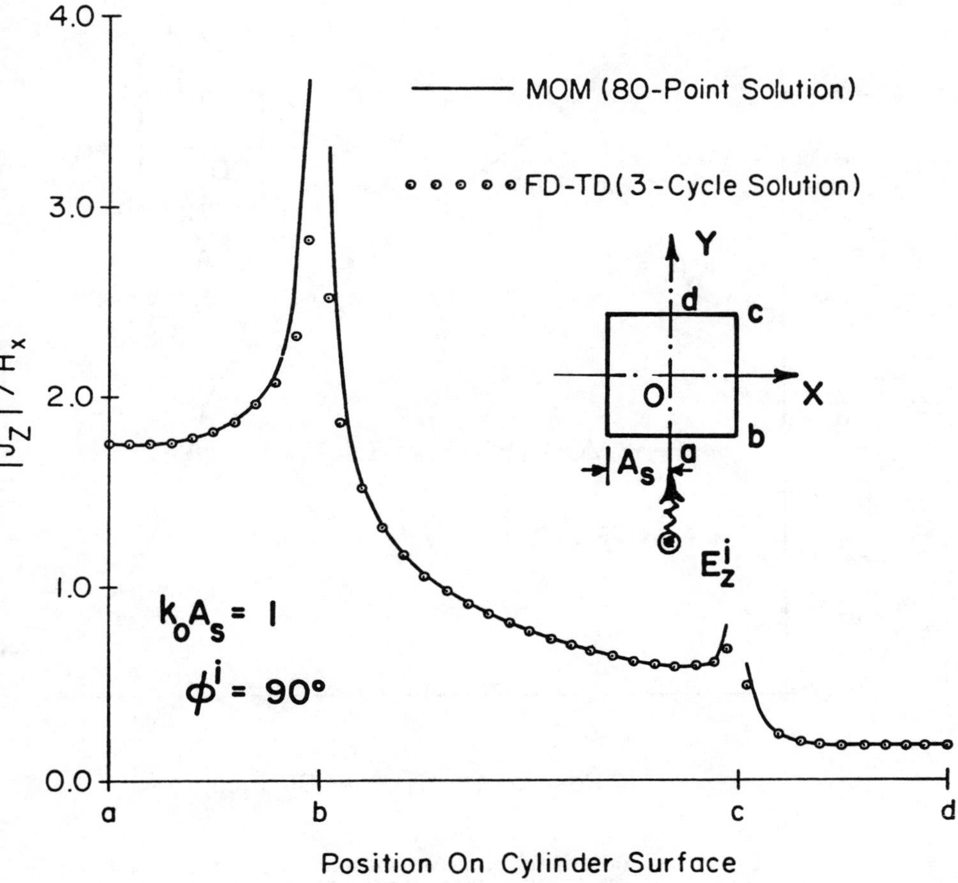

FIGURE 4a. COMPARISON OF MOM AND FD-TD RESULTS FOR MAGNITUDE OF ELECTRIC CURRENTS ON SURFACE OF SQUARE CYLINDER.

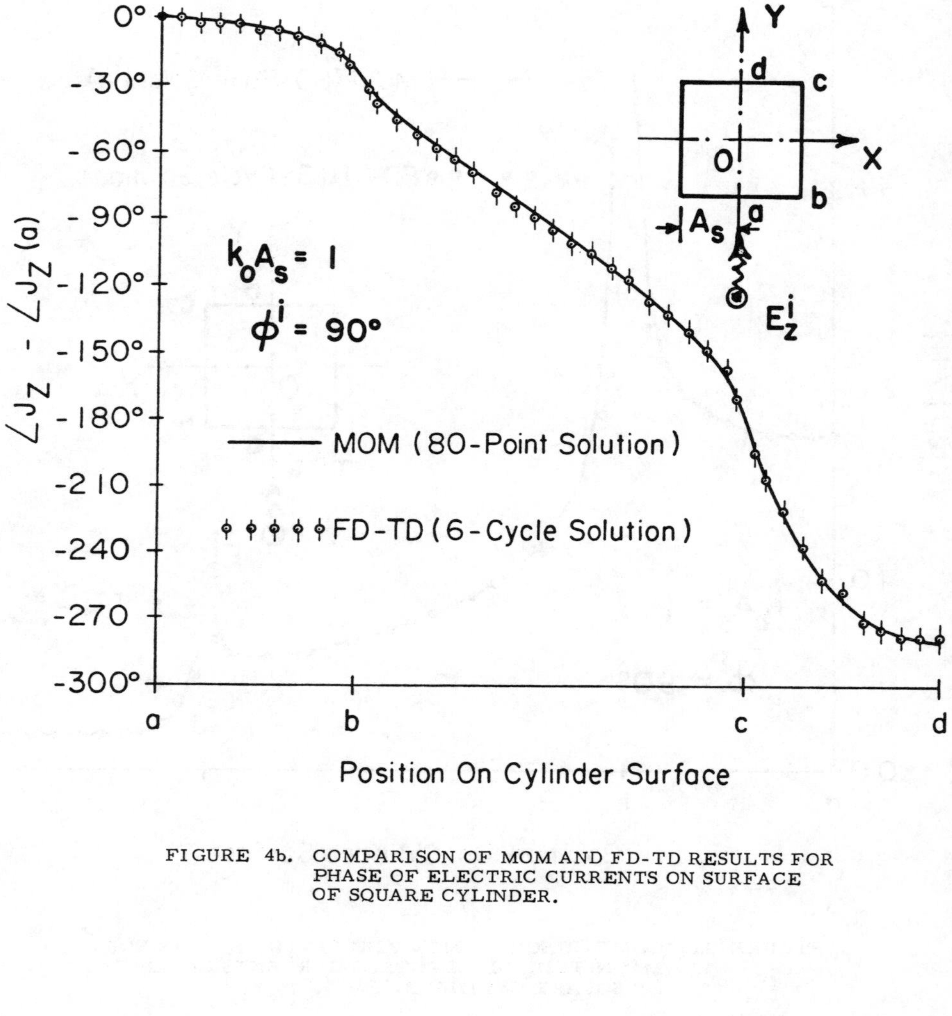

FIGURE 4b. COMPARISON OF MOM AND FD-TD RESULTS FOR PHASE OF ELECTRIC CURRENTS ON SURFACE OF SQUARE CYLINDER.

Figure 5a and 5d show the comparison of the magnitude and phase of the near-scattered electric and magnetic fields computed by the FD-TD method and MOM. The electric field is tangential to a virtual surface, S_a, located at a uniform distance of $h_a = 7$ space cells from the cylinder surface; and the magnetic field is tangential to a virtual surface, S_a, located at 6.5 cells from the cylinder surface. Both virtual surfaces are embedded in the scattered-field region of the FD-TD lattice. The level of correspondence between the FD-TD and MOM results is ±2.5% (±0.2 dB) and ±3°.

In order to obtain the far field pattern and the scattering cross section, the near-field to far-field transformation discussed in 2.4 is followed for two-dimensional structures. The near scattered fields shown in Figures 5a to 5d are converted into the corresponding equivalent electric and magnetic current distributions along S_a according to Equations 2a and 2b.

Figure 6 shows the scattering cross section of the square conducting cylinder obtained by the near field (FD-TD data) to far field transformation. The results agree very well with the scattering cross section derived directly from the electric current on the square cylinder.

5. THREE DIMENSIONAL CASE NUMERICAL RESULTS

This section presents comparative numerical results obtained using the most recent formulations of the FD-TD [14, 15] and MOM triangular surface patch techniques [31, 32] for the surface currents and far scattered fields of a canonical three-dimensional scatterer, a conducting cube, illuminated by a plane wave. It is shown that a very high degree of correspondence exists between the results obtained by these two disparate approaches.

5.1 Scatterer geometry

Figure 7 illustrates the basic geometry of the cube scattering problem. The cube was assumed to have the electrical size $k_o s = 2$, where s = side width of cube. The plane-wave excitation was assumed to have the field components E_z^i and H_x^i and propagate in the +y direction (normal incidence to one side of the cube). For the FD-TD analysis, each side of the cube was spanned by 400 square cells (20 x 20 lattice-cell divisions). For the MOM analysis, each face of the cube was spanned by either 18 triangular patches or 32 triangular patches (to test the convergence of the patch model). The 18-patch/face MOM model required the filling and inversion of a 162 x 162 system matrix. The 32-patch/face MOM model required the filling and inversion of a 288 x 288 system matrix. Comparative results

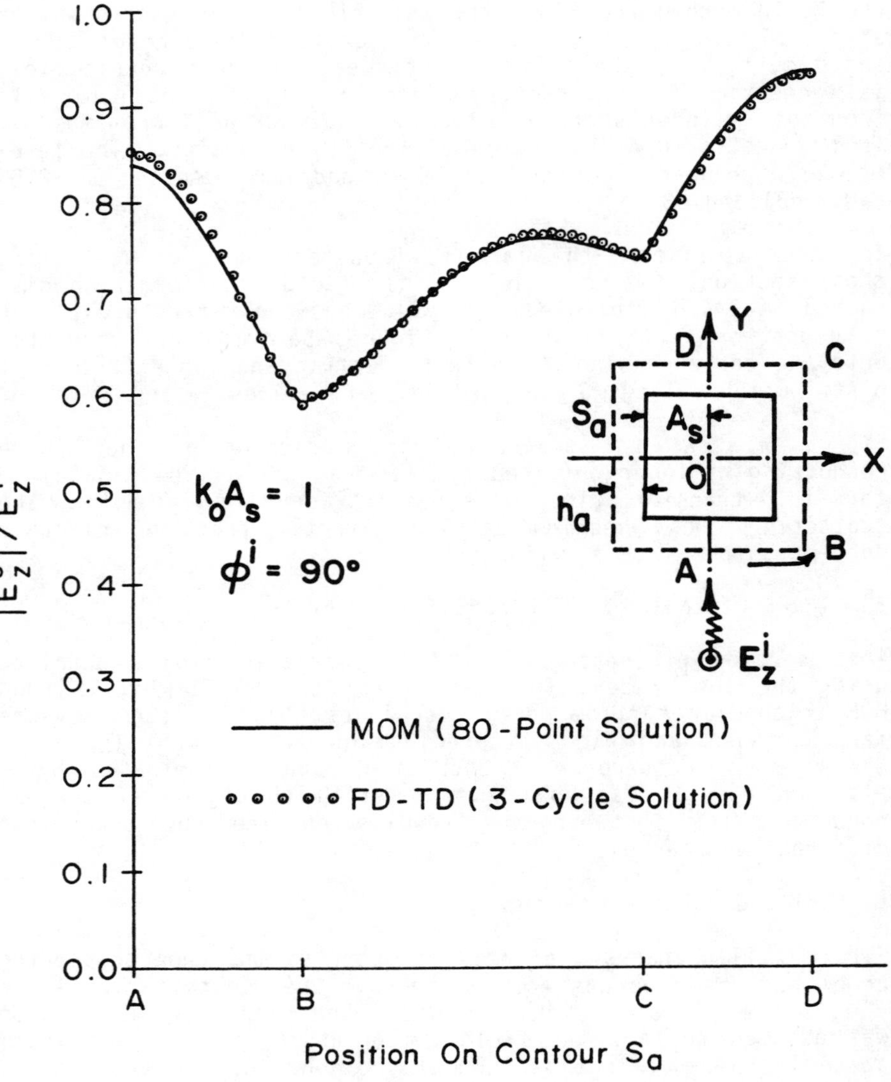

FIGURE 5a. COMPARISON OF MOM AND FD-TD RESULTS FOR MAGNITUDE OF NEAR ELECTRIC FIELD TANGENTIAL TO CONTOUR S_a.

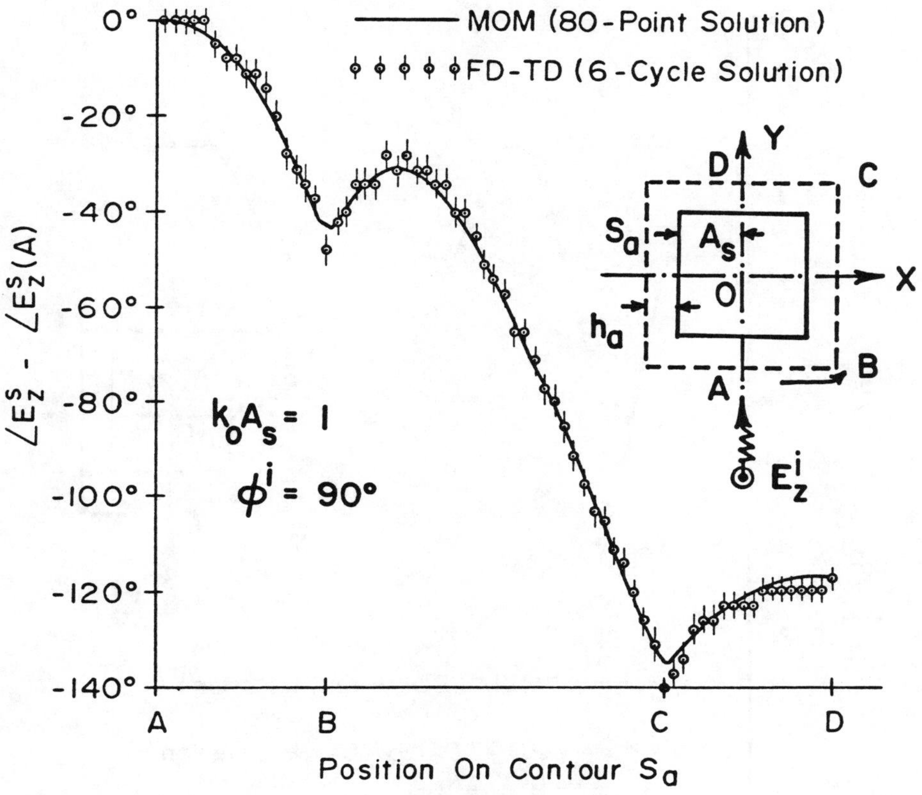

FIGURE 5b. COMPARISON OF MOM AND FD-TD RESULTS FOR PHASE OF NEAR ELECTRIC FIELD TANGENTIAL TO CONTOUR S_a.

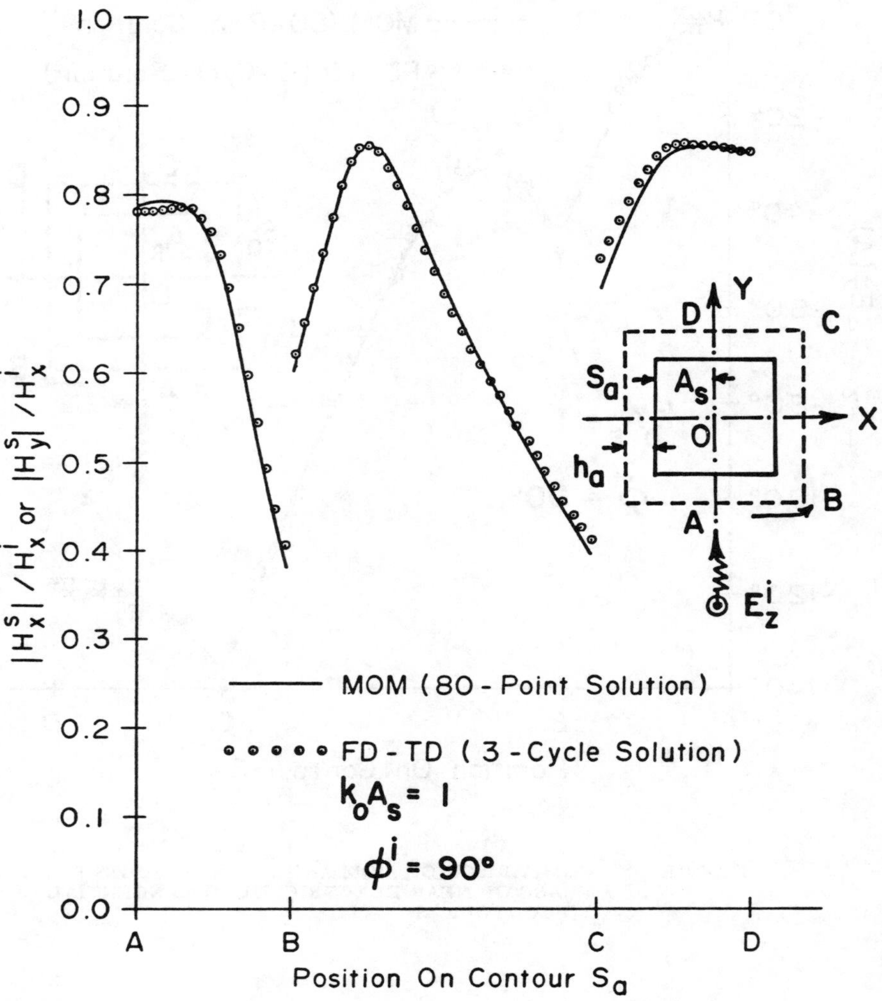

FIGURE 5c. COMPARISON OF MOM AND FD-TD RESULTS FOR MAGNITUDE OF NEAR MAGNETIC FIELD TANGENTIAL TO CONTOUR S_a.

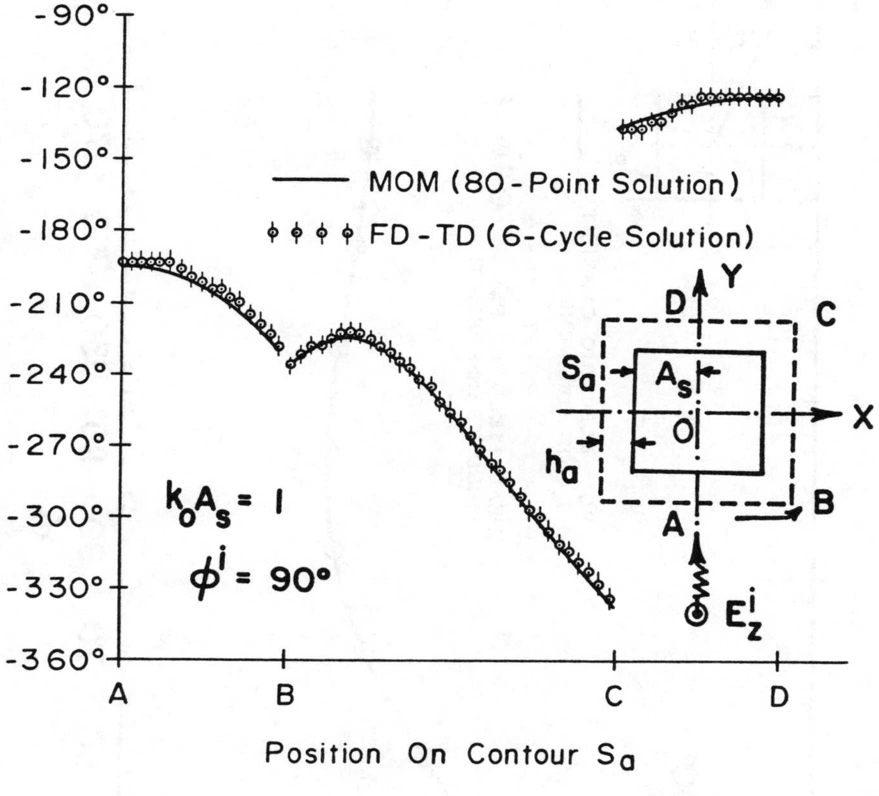

FIGURE 5d. COMPARISON OF MOM AND FD-TD RESULTS FOR PHASE OF NEAR MAGNETIC FIELD TANGENTIAL TO CONTOUR S_a.

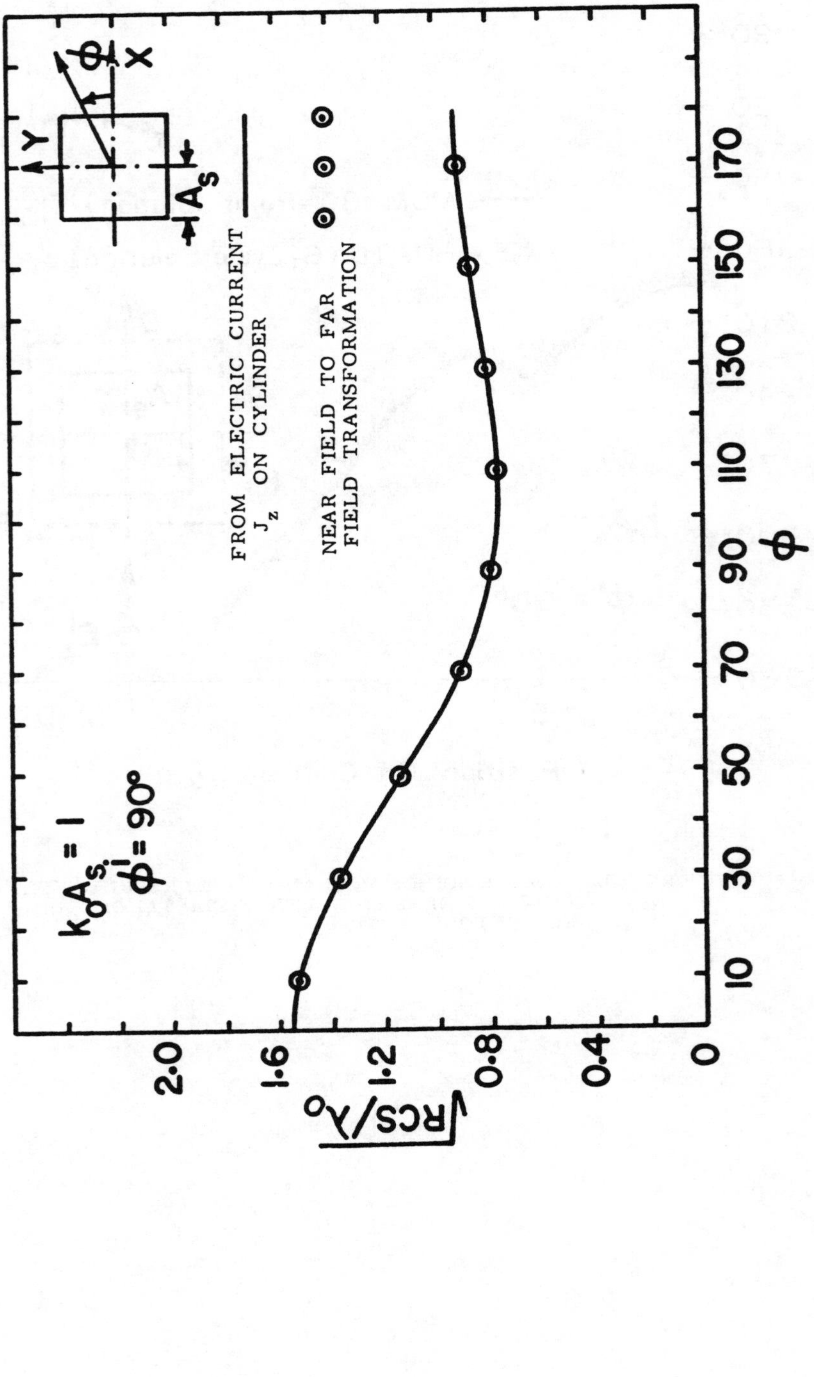

FIGURE 6. SCATTERING CROSS SECTION OF CONDUCTING SQUARE CYLINDER.

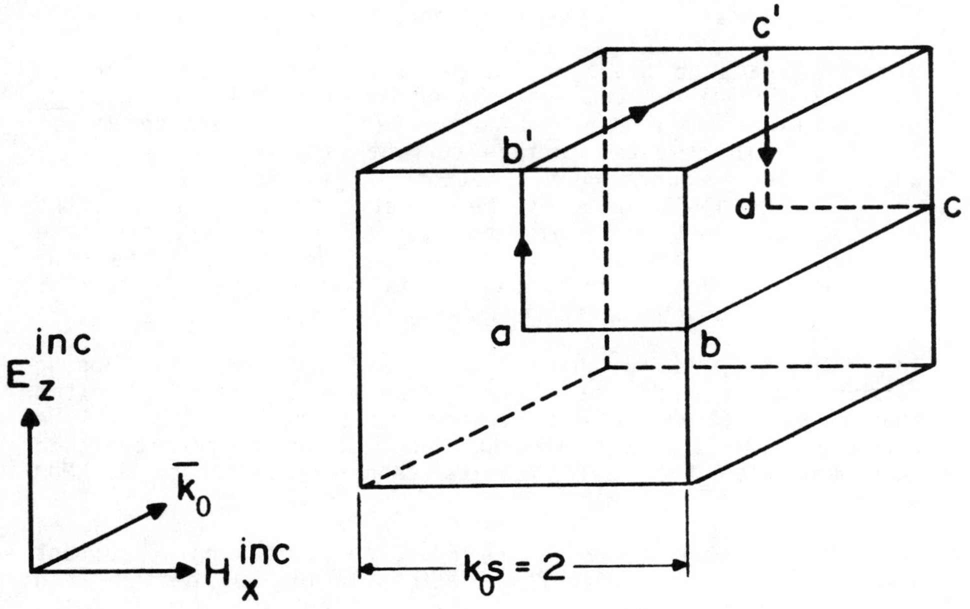

FIGURE 7. GEOMETRY OF METAL CUBE SCATTERER.

are graphed along two straight-line loci along the cube surface, as shown in Figure 7. Locus abcd is parallel to the incident magnetic field. Locus ab'c'd is parallel to the incident electric field.

5.2 Comparative surface currents

Figures 8a and 8b graph the comparative results for the FD-TD and MOM/triangular patch analyses of the magnitude and phase of the looping surface electric current (in the direction ab'c'd). The FD-TD computed surface current is taken as $\hat{n} \times \vec{H}_{tan}$, where $\hat{n}$ is the unit normal vector at the cube surface, and $\vec{H}_{tan}$ is the FD-TD value of the magnetic field parallel to the cube surface, but at a distance of 0.5 space cell from the surface. (The displacement of the H-component from the cube surface is a consequence of the spatially-interleaved nature of the E- and H-components of the FD-TD lattice).

In Figure 8a, the magnitude of the FD-TD computed looping current agrees with the 32-patch/face MOM solution to better than ±2.5% (0.2 dB) at all comparison points. It should be noted that the 32-patch/face MOM model yields better magnitude agreement with the FD-TD results than the 18-patch/face MOM model.

In Figure 8b, the phase of the FD-TD computed looping current agrees with the 32-patch/face MOM solution to better than ±1°. It should be noted that the lower-resolution, 18-patch/face MOM model yields a phase anomaly of almost -100° in the shadow region. The MOM phase is corrected to coincide with the FD-TD results upon going to the higher-resolution, 32-patch/face MOM model. Evidently, the MOM triangular patching approach has a slow convergence in the shadow region, even for quite an electrically small three-dimensional object.

Figures 9a and 9b graph the comparative results for the FD-TD and MOM/patch analyses of the magnitude and phase of the z-directed current along the cut abcd. In Figure 9a, the FD-TD computed magnitudes agree with the edge-corrected [34] 32-patch/face MOM model to better than 2.5% (0.2 dB) at all comparison points. Anomalous behavior is noted for the 32-patch/face MOM model without edge correction, especially near the cube edges b and c. Evidently, the MOM patching approach requires edge correction to permit proper resolution to the current singularity behavior for physically realizable numbers of patches.

In Figure 9b, the phase of the FD-TD computed z-directed current agrees with the 32-patch/face MOM model to better than ±10°. The 18-patch/face MOM model is again seen to have severe

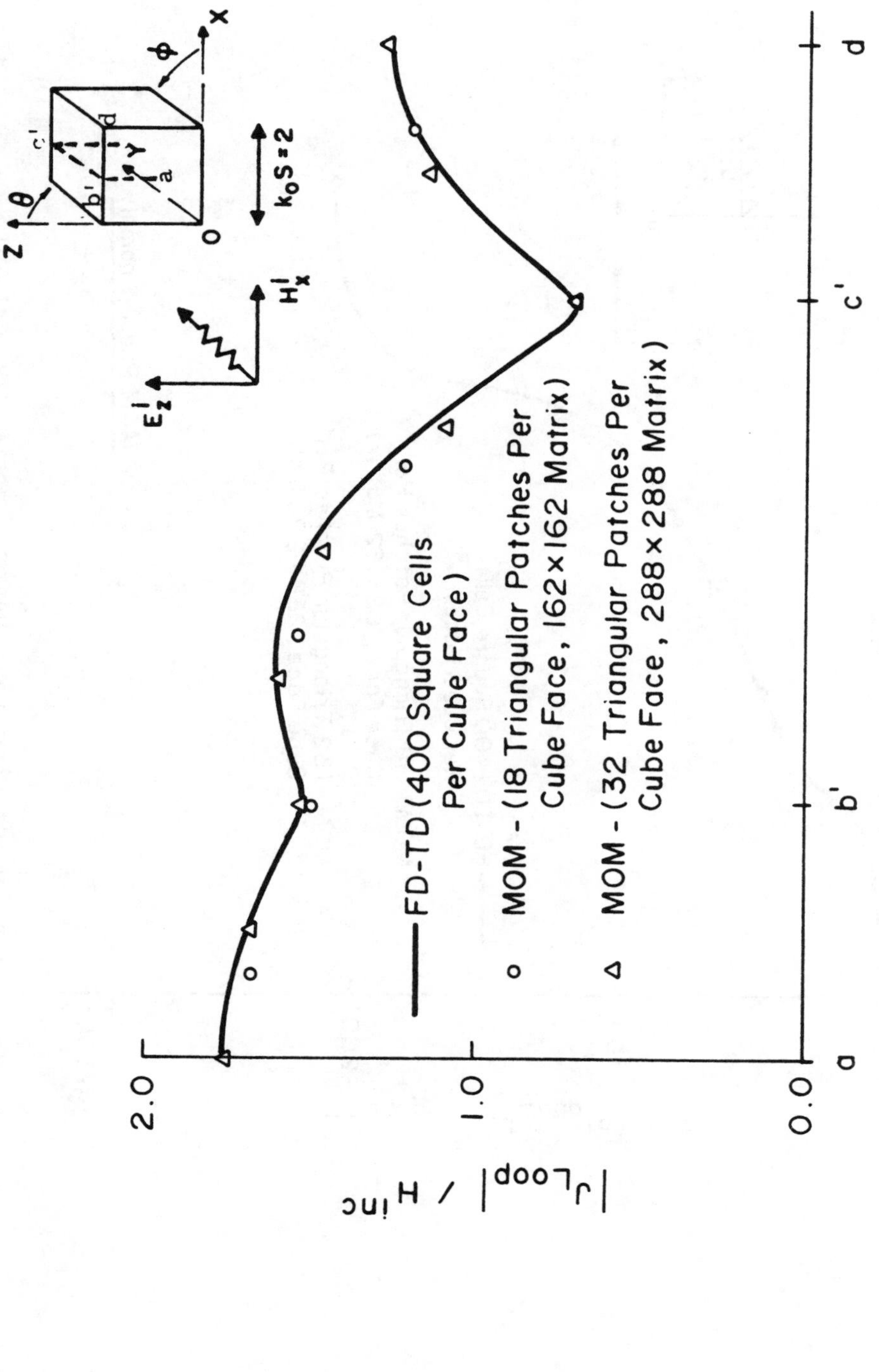

FIGURE 8a. MAGNITUDE OF LOOPING CURRENT, CENTER VERTICAL CUT.

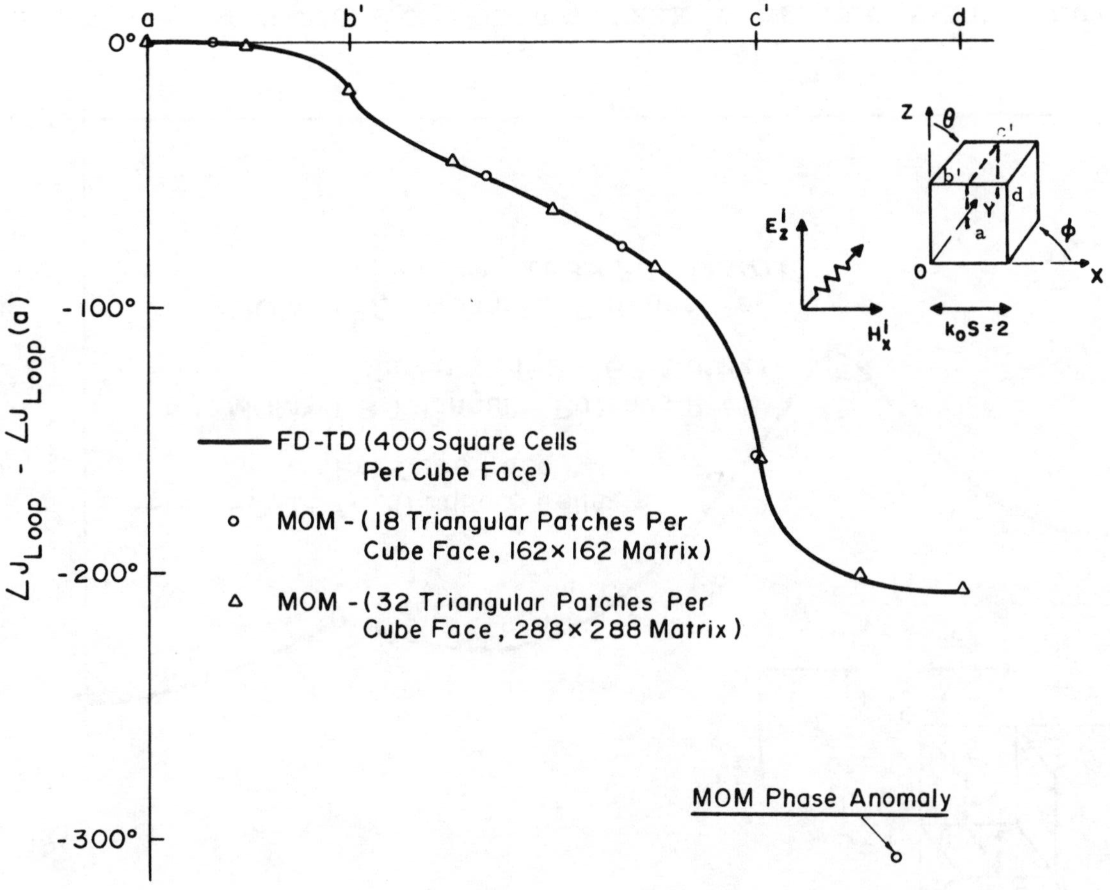

FIGURE 8b. PHASE OF LOOPING CURRENT, CENTER VERTICAL CUT.

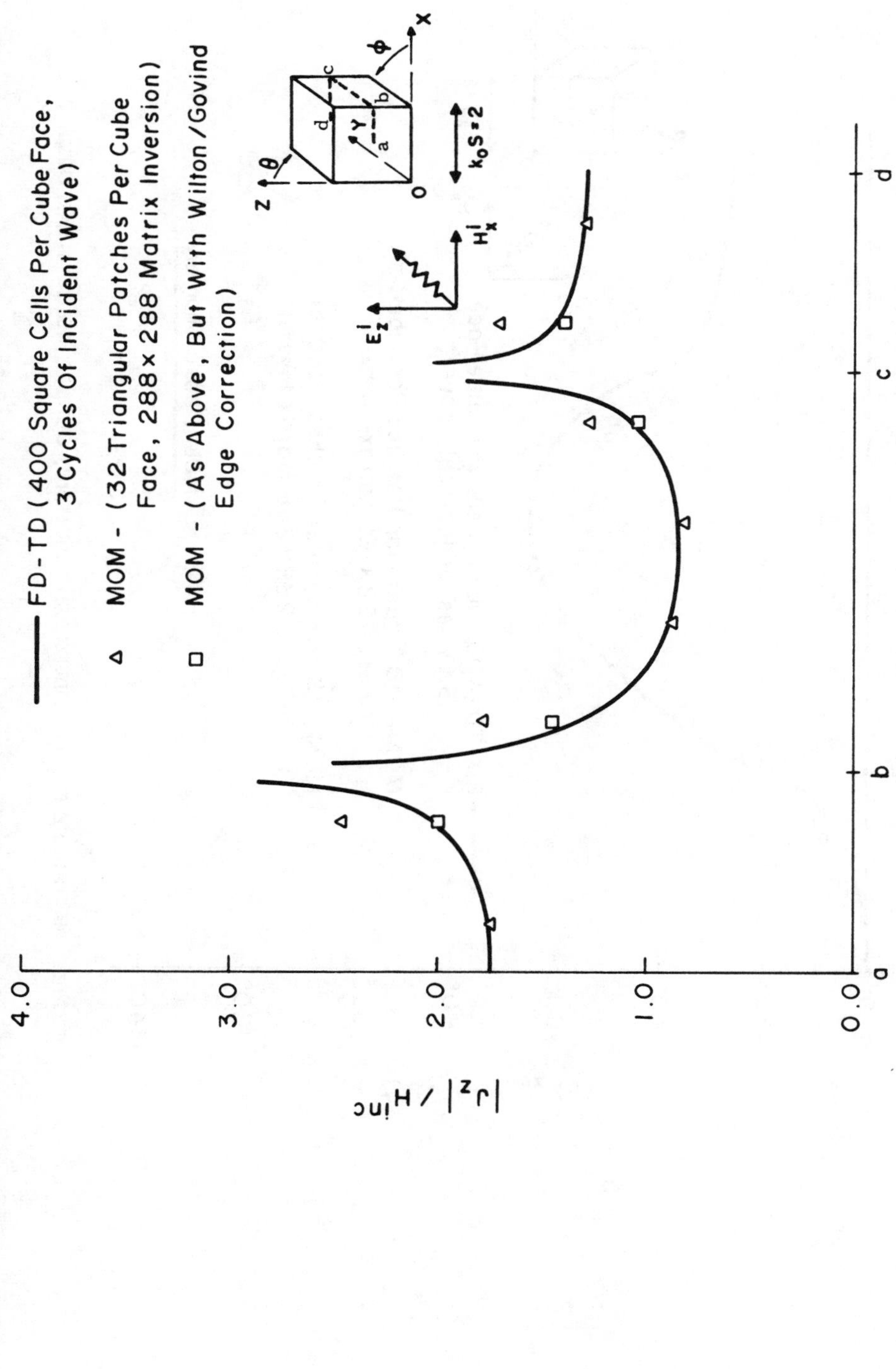

FIGURE 9a. MAGNITUDE OF +Z -DIRECTED CURRENT, CENTER HORIZONTAL CUT.

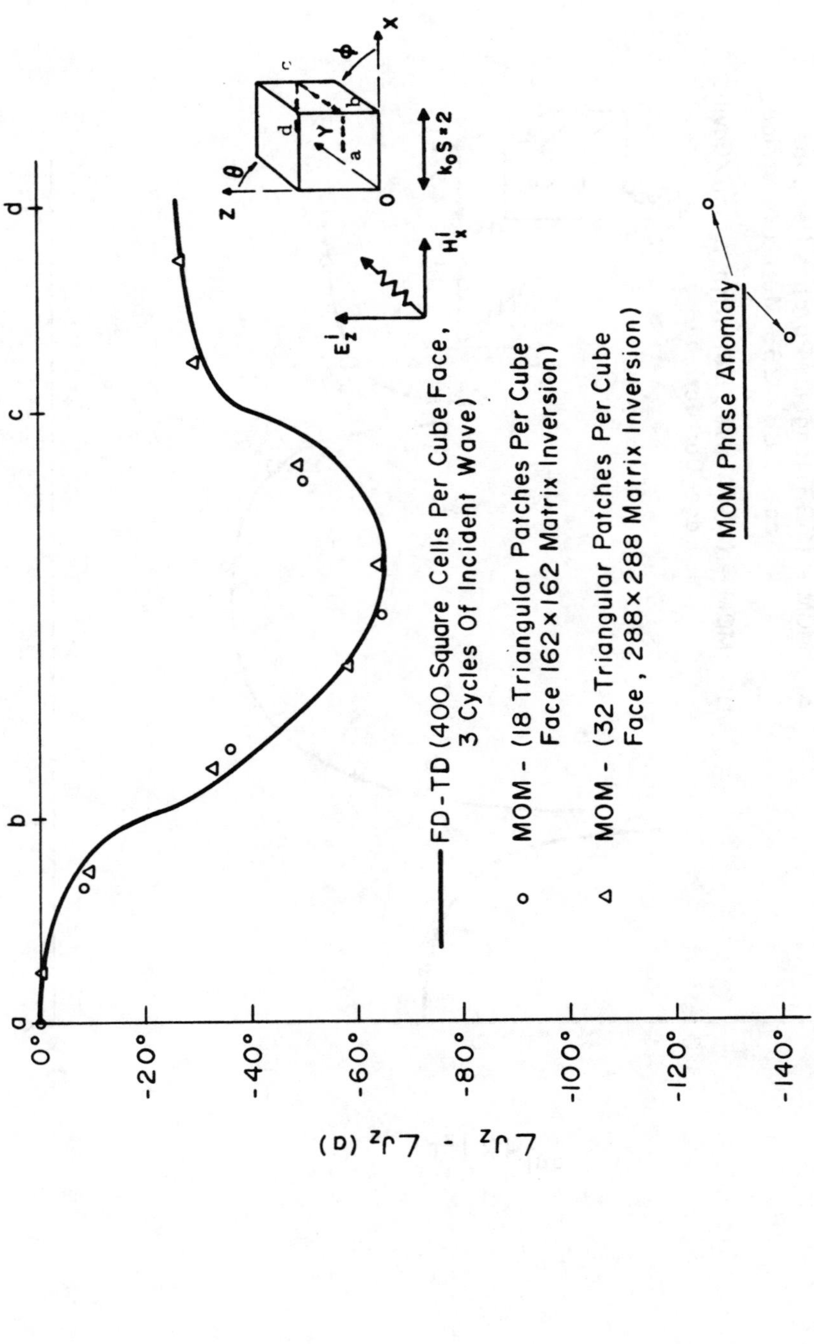

FIGURE 9b. PHASE OF +Z -DIRECTED CURRENT, CENTER HORIZONTAL CUT.

phase problems in the shadow region. In fact, the switch from 18 patches/face to 32 patches/face is seen to result in better phase agreement with the FD-TD model at non-shadow points as well.

5.3 Comparative scattered far-fields

Figure 10 shows radar cross section (RCS) for the conducting cube scatterer with normal plane wave excitation on side of the cube. The RCS result is based on the FD-TD near field data and transformed to far field based on equations (3) to (6). The results are in well agreement with the RCS computation based on the induced surface electric currents on the cube.

5.4 Comparative computer resources

In analyzing the surface current results, it is clear that the MOM triangular patching model approaches the results of the FD-TD model as the number of triangular patches per cube side is increased. It should be noted that the 32-patch/face MOM model required a large expenditure of computer resources to fill and invert the 288 x 288 matrix (over \$200 on the Control Data Cybernet). Application of the MOM patch model to three-dimensional objects larger than the $k_o s = 2$ of the cube would result in increasingly severe computer resource requirements. On the other hand, the FD-TD model of the cube required very comparable computer resources (\$200), and yet can be directly applied to three-dimensional scatterers as large as $k_o s \simeq 25$ with little loss of accuracy for the radar cross section calculations.

The MOM approach thus seems to be best suited for three-dimensional structures no larger than about 1 wavelength, so that matrix filling and solving is tractable. FD-TD, however, gives promise of modeling structures as large as 20 or more wavelengths as the capabilities of vector-array processing computers advance. This promise is currently being explored.

6. CONCLUSIONS

General electromagnetic scattering problems are difficult to treat. This paper presents a method to analyze scattering by complex objects by combining the FD-TD method to obtain near scattered fields with a near-field to far-field transformation based on electromagnetic equivalences. To validate the applicability of this method, canonical two-dimensional and three-dimensional structures are analyzed and the results are verified with respect to the method of moments. This potentially alternative approach has wide applications to analyze

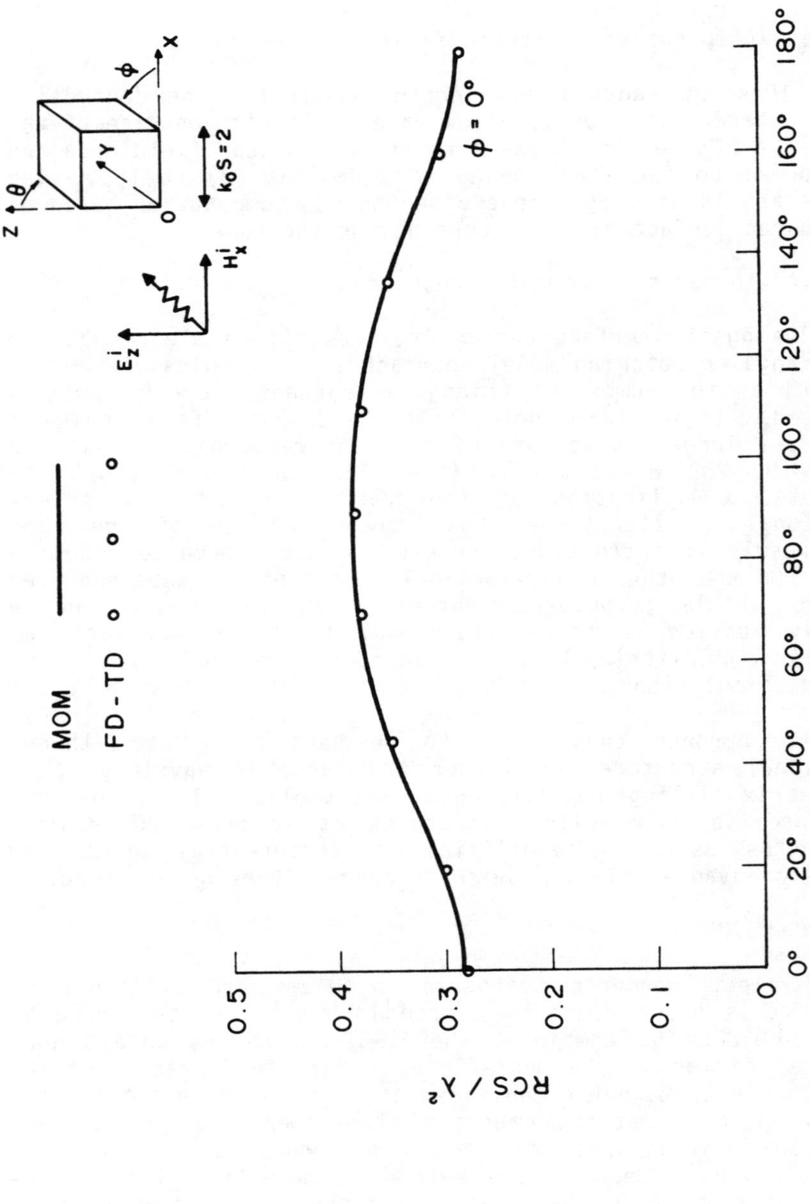

FIGURE 10. SCATTERING CROSS SECTION OF CONDUCTING CUBE SCATTERER.

scattering by structures with complex apertures and dielectric or permeable loadings.

Recently, the FD-TD method has been extended to three-dimensional structures as large as 20 wavelengths. Measurement verification of the far scattered fields are currently being obtained for such scatterers. Initial results indicate accuracy of the FD-TD computed radar cross section to within 0.5 dB.

REFERENCES

1. J.J. Bowman, T. B. A. Senior and P. L. E. UsLenghi, eds., Electromagnetic and Acoustic Scattering by Simple Shapes. John Wiley and Sons, New York, 1969.

2. K.S. Yee, "Numerical Solution of Initial Boundary Value Problems Involving Maxwell's Equation in Isotropic Media", IEEE Trans. Antennas Prop., Vol. AP-14, May 1966, pp. 302-307.

3. C. D. Taylor, D. H. Lam, and T. H. Shumpert, "Electromagnetic Pulse Scattering in Time-Varying Inhomogeneous Media", IEEE Trans. Antennas Prop., Vol. AP-17, September 1969, pp. 585-589.

4. D.E. Merewether, "Transient Currents Induced on a Metallic Body of Revolution by an Electromagnetic Pulse", IEEE Trans. Electromagnetic Compatibility, Vol. EMC-13, May 1971, pp. 41-44.

5. A. Taflove and M. E. Brodwin, "Numerical Solution of Steady-State Electromagnetic Scattering Problems Using the Time-Dependent Maxwell's Equations", IEEE Trans. Microwave Theory Tech., Vol. MTT-23, August 1975, pp. 623-630.

6. A. Taflove and M.E.Brodwin, "Computation of the Electromagentic Fields and Induced Temperatures Within a Model of the Microwave-Irradiated Human Eye," IEEE Trans. Microwave Theory Tech., Vol. MTT-23, November 1975, pp. 888-896.

7. R. Holland, "Threde : A Free-Field EMP Coupling and Scattering Code" IEEE Trans. Nuclear Science, Vol. NS-24, December 1977, pp. 2416-2421.

8. K. S. Kunz and K. M. Lee, "A Three-Dimensional Finite-Difference Solution of the External Response of an Aircraft to a Complex Transient EM Environment : Part I - The

Method of its Implementation", IEEE Trans. Electromagnetic Compatibility, Vol. EMC-20, May 1978, pp. 328-333.

9. A. Taflove and K.R. Umashankar, "A Hybrid Moment Method/Finite Difference Time Domain Approach to Electromagnetic Coupling and Aperture Penetration into Complex Geometries," Chap. 14 in Applications of the method of Moments to Electromagnetic Fields, B.J. Strait, ed., SCEEE Press, Orlando, Florida, February 1980.

10. A. Taflove, "Application of the Finite-Difference Time-Domain Method to Sinusoidal Steady-State Electromagnetic-Penetration Problems," IEEE Trans. Electromag. Compat., Vol. EMC-22, August 1980, pp. 191-202.

11. R. Holland, L. Simpson, and K.S. Kunz, "Finite-Difference Analysis of EMP Coupling to Lossy Dielectric Structures," IEEE Trans. Electromag. Compat., Vol. EMC-22, August 1980, pp. 203-209.

12. D. E. Merewether, R. Fisher and F. W. Smith, "On Implementing a Numeric Huygen's Source Scheme in a Finite-Difference Program to Illuminate Scattering Bodies", IEEE Trans. Nucl. Sci., Vol. NS-27, December 1980, pp. 1819-1833.

13. R. Holland and L. Simpson, "Finite-Difference Analysis of EMP Coupling to Thin Struts and Wires," IEEE Trans. Electromag. Compat., Vol. EMC-23, May 1981, pp. 88-97.

14. A. Taflove and K.R. Umashankar, "A Hybrid FD-TD Approach to Electromagnetic Wave Backscattering", 1981 URSI/APS - International Symposium Proceedings, Los Angeles, California, June 1981, p. 82.

15. A. Taflove and K.R. Umashankar, "Solution of Complex Electromagnetic Penetration and Scattering Problems in Unbounded Regions," in Computational Methods for Infinite Domain Media - Structure Interaction, AMD-Vol. 46, American Society of Mechanical Engineers, November 1981, pp. 83-113.

16. G. Mur, "Absorbing Boundary Conditions for the Finite-Difference Approximation of the Time-Domain Electromagnetic-Field Equations", IEEE Trans. Electromag. Compat., Vol. EMC-23, November 1981, pp. 377-382.

17. A. Taflove and K. R. Umashankar, "User's Code for FD-TD", Final Report RADC-TR-82-16 by IIT Research Institute,

Chicago, IL to Rome Air Development Center, Griffiss AFB, NY, on Contract F30602-80-C-0302, February, 1982.

18. S.A. Schelkunoff, "Field Equivalence Theorems," Comm. Pure Appl. Math., Vol. 4, June 1951, pp. 43-59.

19. R. F. Harrington, Field Computation by Moment Methods., MacMillan, New York 1968.

20. B. Engquist and A. Majda, "Absorbing Boundary Conditions for the Numerical Simulation of Waves", Math Comp., Vol. 31, July 1977, pp. 629-651.

21. G. A. Kriegsmann and C. S. Morawetz, "Solving the Helmholtz Equation for Exterior Problems with Variable Index of Refraction: I", SIAM J. Sci. Stat. Comput., Vol. 1, September 1980, pp. 371-385.

22. J. Van Bladel, Electromagnetic Fields., McGraw Hill, New York 1964.

23. R. F. Harrington, Time Harmonic Electromagnetic Fields. McGraw Hill, New York 1961, Chapter 3.

24. K.K. Mei and J. Van Bladel, "Scattering by Perfectly Conducting Rectangular Cylinders," IEEE Trans. Antennas Prop., Vol. AP-11, March 1963, pp. 185-192.

25. K. R. Umashankar and A. Taflove, "A Novel Method to Analyze Electromagnetic Scattering of Complex Objects", IEEE Trans. Electromagnetic Compatibility, Vol. EMC-24, November 1982.

26. N. C. Albertsen, J. E. Hansen, and N. E. Jenson, "Computation of Radiation from Wire Antennas on Conducting Bodies", IEEE Trans. Antennas Prop., Vol. AP-22, March 1974, pp. 200-206.

27. L. L. Tsai, D. G. Dudley, and D. R. Wilton, "Electromagnetic Scattering by a Three-Dimensional Conducting Rectangular Box", J. Applied Physics, Vol. 45, October 1974, pp. 4393-4400.

28. A. Sankar and T. C. Tong, "Current Computation on Complex Structures by Finite Element Method", Electronics Letters, Vol. 11, October 1975, pp. 481-482.

29. E. H. Newman and D. M. Pozar, "Electromagnetic Modeling of Composite Wire and Surface Geometries", IEEE Trans. Antennas Prop., Vol. AP-26, November 1978, pp. 784-789.

30. J. Singh and A. T. Adams, "A Non-Rectangular Patch Model for Scattering from Surfaces", IEEE Trans. Antennas Prop., Vol. AP-27, July 1979, pp. 531-535.

31. D. R. Wilton, S. M. Rao, and A. W. Glisson, "Electromagnetic Scattering by Surfaces of Arbitrary Shape", in Applications of the Method of Moments to Electromagnetic Fields, B. J. Strait, ed., SCEEE Press, Orlando, Florida, 1980, pp. 139-170.

32. S. M. Rao, D. R. Wilton, and A.W.Glisson, "Electromagnetic Scattering by Surfaces of Arbitrary Shape", IEEE Trans. Antennas Prop., Vol. AP-30, May 1982, pp. 409-418.

33. E. K. Miller and A. J. Poggio, "Moment-Method Techniques in Electromagnetic from an Application Viewpoint", Chapter 9 in Electromagnetic Scattering, P.L.E. Uslenghi, ed., Academic Press, New York, 1978.

34. D.R. Wilton and S. Govind, "Incorporation of Edge Conditions in Moment Method Solutions", IEEE Trans. Antennas Prop., AP-25, November 1977, pp. 845-850.

II.11 (TS.5)

APPLICATIONS OF TIME-DOMAIN INVERSE SCATTERING TO ELECTROMAGNETIC WAVES

W. Tabbara

Laboratoire des Signaux et Systèmes
Groupe d'Electromagnétisme
CNRS - ESE, Plateau du Moulon
91190 GIF-sur-YVETTE, FRANCE

ABSTRACT : Over the past few years, interest has developed in the use of time-domain methods or data in order to solve inverse scattering problems. This approach has many attractive features among them the large amount of information contained in a single pulse response and simple physical interpretations. The improvement in time-domain technology increases the interest in such an approach. The purpose of this paper is to present a short review of some of the techniques that have been used in the analysis of time-domain inverse scattering.

I. INTRODUCTION

One may say that the time-domain approach is the most natural one to solve inverse scattering problems. The measurement of the travel time of a pulse will help locate a target or a fault in a material; the impulse response of a target is a characteristic feature of its shape and constitution, and provides us with wide-band information on the target. Furthermore, time-domain analysis allows simple physical interpretation of scattering phenomena. In fact, in many inverse scattering problems a rapid inspection of the scattered pulses will provide usefull information on the number of discontinuities (or scattering centers), their position, or in the case of dielectric bodies, their reflecting power. The advances in digital computer capability made computations directly in the time-domain possible, and triggered the development of powerful inverse methods (iterative ones), where a fast and accurate solution of the direct problem (computation of the scattered field) is needed at each step.

In this paper we shall consider three large groups of inverse

W.-M. Boerner et al. (eds.), Inverse Methods in Electromagnetic Imaging - Part 1, 431–439.

problems : location, identification and reconstruction of a target. The target will be a perfect conductor or a dielectric (with losses or not). It is not our purpose to provide an exhaustive inspection of time-domain inverse scattering methods, but to present few ones illustrating the main approaches used in the past few years.

II. TARGET LOCATION

Detection and location of a target are basic problems in inverse scattering. Pulses have been used for many years to detect and locate targets in extended media. Once a pulse is emitted, the reception of a pulse indicates the existence of a target (the term is considered in a very broad sense, as a local modification of the properties of the propagating medium); the inspection of the shape of the returned pulse, will provide some information on the nature of the target. Measuring the time delay between emission and reception, and considering the medium outside the target to be known, one may obtain the distance from the emitter to the target.One common application is the detection by means of radars and sonars. Other applications concern the detection of voids or buried objects [1], the location of discontinuities in dielectric media [2] or the location of faults along a propagating device. In order to detect two or more closely located targets, short pulses are required. The propagation in a dissipative or dispersive medium, introduces overlaps between returns from different scatterers and deconvolution methods are needed in order to separate the various contributions to the total pulse.

III. TARGET IDENTIFICATION

It is not necessary to have an image of a target to be able to recognize it. One may think that the determination of some characteristic parameters of the target will allow a fairly accurate identification.

1. Waveform identification (Pattern recognition)

A natural manner to identify a target belonging to a given class of objects, is to compare the waveform from the unknown target to the one of each object in the given class and identify the target with the object which provides the waveform that looks like the one that has been measured.

First, one builds a data set which contains as much information as possible on specific objects : waveforms as function of aspect angle, polarization...

Next, one defines the logic used in comparing the measured waveform with the ones taken from the data set. The simplest one is the nearest neighbor logic where a euclidian metric is used to compute the distance between the two waveforms. More efficient methods can be developped based on various stastical approaches. These methods, while requiring more elaborate algorithms, are more accurate and fast. With the identification logic in place, one may then start the classification process. An interesting application of this approach is given in [3], where the power of this method clearly appears and suggests further developments and investigations.

2. Poles identification

Instead of keeping the whole waveform related to a given object, one may look at extracting from it some parameters which are characteristics of the object. Convenient sets of parameters are built with poles, or with the natural resonant frequencies of the object. They depend on the geometrical and electromagnetical parameters of the object.

The basic idea is that the time-domain response f(t) of a target can be written in the form :

$$f(t) = \sum_{\ell=1}^{N} A_\ell \, e^{\gamma_\ell t}$$

Where N can be finite or not, γ_ℓ is the pole of the target and A_ℓ the residue at the pole γ_ℓ. The resonant frequencies are obtained from a frequency-domain analysis using moment method and the dominant ones coïncide with the dominant poles [4], [25].

Pole extraction is a problem in itself since, as it has been shown in [5], different approaches based on Prony's method [6], Kalman filter and maximum likelihood gives different results. Furthermore, higher order poles (large values of ℓ) are very sensitive to noise. But it seems that Prony's method is a convenient one. A catalog of poles can be built for objects of interest and stored in a computer.

Two different identification processes are possible. In the first one, when a waveform is received from an unknown target (illuminated by the same pulse that has been used to extract the poles in the catalog), the corresponding poles are extracted and compared by means of an appropriate parameter (correlation coefficient, mean square error...) to those in the catalog. The nearest ones found identifie the target. In some applications [7] it is possible to compute the poles analytically and then compute from their expression the geometrical and electromagnetical parameters of the target. This is a powerful approach but of limited scope.

In the second approach, the poles of a given object in the catalog are used to synthesize its response to the pulse used to illuminate the unknown target. The synthesized response f_S is compared to the received waveform and the target is identified with the object giving the nearest synthesized response. Comparison can be made by means of correlation techniques[8]. This approach has been extensively studied and provides very good results . The effects of polarization, the choice of the discretization step Δt and the number of poles used to compute f_s have been examinated [8, 4 , 9].

3. Natural mode identification

In the preceding section we mentionned the fact that the natural frequencies of a scatterer can be used to characterize it. To each natural frequency we associate a waveform called the natural mode. The purpose of the method developed by Chen et al. and described in [10, 11], consists in synthesizing a waveform for the incident signal which excites the target in such a way that the return signal consists of only a single-natural mode of that target in the late-time period. The incident signal is synthesized from a sum of damped sinusoïds using the values of the natural frequencies of the target. The duration T_e of the signal is taken to be equal to the period of the first natural mode. The late time-period begins for $t > t_e + T_t$ where T_t is the transit time of the scatterer. The method requires the knowledge of a large number of natural frequencies. Its usefulness seems limited to targets of simple shape.

IV. TARGET RECONSTRUCTION (Imaging)

An unambiguous method in target recognition is the formation of an image of that target form the measurement of its scattered field. Various approaches have been developped, but they usually apply to simple geometries.

1. Perfectly conducting targets

1.1. Exact method

From the investigation of time-domain scattering by means of integral equations, Bennett [12] developped, for the case of rotationally symmetric target illuminated by a wave from the axial direction, an iterative method that allows the reconstruction of the generating curve defined by the function $\rho(z,t)$. The basic equation is :

$$\rho(z,t) = \left[2\, r_o\, \vec{a}_H . H_R^s(\vec{r},t) - \frac{1}{2\pi}\, \vec{a}_H . \int_S (\vec{J}_R - 2\vec{a}_{n\Lambda} \vec{H}_R^i)(\vec{r}\,',\tau) \Lambda \vec{a}_r \, dS \right]^{1/2}$$

$$\tau = t - R/C, \quad \vec{a}_H = \vec{H}^i / |\vec{H}_i| , \quad R = |\vec{r}_o - \vec{r}\,'| , \quad \vec{a}_r = \frac{\vec{r}_o - \vec{r}\,'}{R} \qquad (1)$$

where $\vec{H}_r^i$ is the incident ramp field, $r_o\vec{H}_R^s$ the ramp response, $\vec{J}_R$ the induced current, r_o the distance of the far-field observer, $\vec{a}_n$: unit normal. An initial value of ρ , ρ_o, is given, then the solution of the direct problem gives $\vec{J}_R$, which by means of (1) provide a new value ρ_1 of ρ. The process in then continued until it converges. Example in [12] for a sphere capped cylinder shows rapid convergence and good reconstruction in the shadow region. A basic requirements for a fast reconstruction is the existence of a fast algorithm for the solution of the direct problem at each step of the iterative process.

1.2. Approximate methods

An early method based on the physical optics approximation linked the backscattered ramp response $f_R(t)$ of the target to its cross-sectional area by a plane perpendicular to the direction of the incident wave [13] :

$$f_R(t) = \frac{A(r)}{\pi C^2} \qquad o < r = \frac{Ct}{2} < r_o \tag{2}$$

$r = o$ identifies the tip of the target where $A(o) = o$ and r_o the shadow boundary. This identity requires low-frequency data, apparently violating the physical optics approximation. In order to obtain an image of the target, an intuitive approach using three look angles has been developped [14] and gave rather accurate results for complex bodies.

Later, generalizing a physical-optic frequency-domain method established by Bojarski, Bleinstein[15] developed a time-domain approach allowing the reconstruction of the characteristic function $\gamma(\vec{x})$ ($\gamma(\vec{x}) = 1$ inside the target, zero outside) from the measurement of the backscattered waveform $f(\vec{x}_o,t)$ for all directions of incidence :

$$\gamma(\vec{x}) = \frac{1}{4\pi^2} \int_\Omega f(\vec{x}_o, -\frac{2\,\vec{x}_o \cdot \vec{x}}{C})\, d\vec{x}_o \tag{3}$$

Where Ω denotes the unit sphere with variable $\vec{x}_o$. This method has been mostly investigated in the frequency domain, but the main results remain valid in the time-domain. It must be noted that (2) can be derived from (3). The main drawback of this method is the necessity of measuring the backscattered field for all direction in space. Both approaches described above, deserve further investigations and specially the shadow region, or those of non symmetric target are strongly dependent on polarization. Corrective terms must be introduced as mentionned in [16].

The third method is not really a time-domain one. It is a

high frequency approach based on the fact that the specular radar cross-section is simply related to the Gaussian curvature K through $\sigma_{sp.} = \pi/K$. In order to isolate the specular return, short pulses are used [17] and in the case of body of revolution, $\sigma_{sp.}$ measured at various angles of incidence leads to a simple reconstruction of the target shape. Here too, polarization information will increase the accuracy of the method.

2. Dielectric targets (lossy or not)

2.1. Exact methods

A widely used method is the Time-domain Reflectometry (TDR). A pulse illuminates the unknown dielectric sample placed in a transmission line, the reflected pulse is used to determine the permittivity and tangent loss of the sample as a function of frequency. The reflection coefficient of the sample is first determined, then simple formulas allow the computation of the unknown parameters [18]. This broad-band tehnique can also be usedto analyse networks or characterize discontinuities in transmission lines.

The determination of the permittivity ε and conductivity σ of stratified medium has triggered the development of various methods. Among them, one based on an integral representation of the field directly in the time domain [19, 20, 21]. In [21] a step by step approach of the integral equation gives ε or σ, it is a very fast approach but very sensitive to noise and thus does not allow very deep investigation. On the other hand in [19, 20] optimization thechniques are used. Initial values are given for ε and σ, the reflected field is computed and compared by means of a convenient distance to the measured field. The iterative process is stopped whenever computed and measured values are sufficiently close. These techniques are powerful, accurate and not very sensitive to noisy data. They allow fairly deep investigation of the medium and the reconstruction of ε or σ. The determination of both parameters is not yet possible. Another advantage of the methods described in [19, 20] is the existence of a very fast algorithm for the solution of the direct problem, this is a fundamental point for real-time analysis.

2.2. Approximate methods

A large class of methods is based on the Born approximation, where the field in the scatterer is assumed to be equal to the incident field. This method has been developed in many domains : diagnostic of strafied media, analysis of tapered transmission lines... . The purpose is the reconstruction of a permittivity profile (lossless medium) or an impedance profile. In the case of stratified media, it has been shown [22] that the step response of the medium, conveniently corrected, allows the determination of the interfaces and the values of ε in each layer. The method behaved accurately for values of ε up to 10 ε_o. Above this limit the shape of the profile was correctly determined but the values of ε where false. The approach was not very sensitive to noise. The analysis of thin layers requires short pulses.

Another class of methods uses the time of flight measurement to retrieve the refractive index profile of the target. These methods are based on a ray propagation hypothesis which leads to the solution of an Abel integral equation when spherical symmetry is assumed [23]. It has been used in the inversion of radio occulation data, and in plasma diagnosis, among other applications. More recent applications of time of flight measurement are related to ultrasonic and NMR tomography [24] where very successful results have been obtained.

V. CONCLUSION

From this brief review one may note the wide spectrum of application of time-domain measurements in inverse problems. They provide flexible approaches with simple and fast physical interpretation. There is a need to improve actual methods by taking into accounts polarization effects, extension to three dimensional structures lacking rotational symmetries, and to develop new approaches in the case of imperfectly conducting targets.

REFERENCES

[1] Moffatt D.L., Puskar R.J., Peters L. Jr. : 1973, Electromagnetic pulse sounding for geological surveying with application in rock mechanics and the rapid excavation program. Final Report 3408-2, Electroscience Laboratory, The Ohio State University, Columbus, Ohio 43212, U.S.A..

[2] Robinson L.A., Weir W.B., Young L. : 1974, Location and recognition of discontinuities in dielectric media using synthetic RF pulses, Proc. IEEE, Vol. 62, n° 1, January, pp. 36-44.

[3] Bennett C.L., Auchenthaler A.M., Smith R.S., De Lorenzo J.D. : 1973, Space-time integral equation approach to the large body scattering problem. Final Report RADC-TR-73-70, Sperry Research Center,Sudbury, Massachussets, U.S.A..

[4] Chuang C.W., Moffatt D.L. : 1976, Natural resonances of radar targets via Prony's method and target discrimination, IEEE Trans. Aerosp. Electron. Syst., Vol AES-12, n° 5, Sept., pp. 583-589.

[5] Gavel D.T., Candy J.V., Lager D.L. : Parameter estimation from noisy transient electromagnetic measurement, Lawrence Livermore Laboratory, Livermore, CA 94550, U.S.A..

[6] Hildebrand F. : 1956, Introduction to numerical analysis, Mc Graw-Hill, New York.

[7] Miller E.K., Lager D.L. : 1979, Inversion of one-dimensional scattering data using Prony's method, Report n° UCRL-52667, Lawrence Livermore Laboratory, Livermore, CA 94550 U.S.A..

[8] Moffatt D.L., Mains R.K. : 1975, Detection and discrimination of radar targets, IEEE Trans. Ant. Prop., Vol. AP-23, N° 3, May, pp. 356-367.

[9] Moffatt D.L., Rhoads C.M. : 1982, Radar identification of naval vessels, IEEE Trans. Aerosp. Electron. Syst., Vol. AES - 18, N° 2, March, pp. 182-187.

[10] Chen K.M., Westmoreland D. : 1982, Radar waveform synthesis for exciting single-mode backscatters from a sphere and application for target discrimination, Radio-Science, Vol. 17, N° 3, May-June, pp. 574-588.

[11] Chen K.M., Nyquist D.P., Westmoreland D., Chuang C.I., Drachman B. : 1982, Radar waveform synthesis for single-mode scattering by a thin cylinder and application for target discrimination, IEEE Trans. Ant. Prop., Vol AP-30, N° 5, Sept., pp. 867-880.

[12] Bennett C.L. : 1981, Inverse scattering in Theoretical methods for determining the interaction of electromagnetic waves with structures, J.K. Skwirzynski editor, Sijthoff and Noordhoff, The Netherlands.

[13] Kennaugh E.M., Moffatt D.L. : 1965, Transient and impulse response approximations, Proc. IEEE, vol. 53, n° 8, August, pp. 893-901

[14] Shubert K.A., Young J.D., Moffatt D.L. : 1977, Synthetic radar imagery, IEEE Trans. Ant. Prop., Vol. AP-25, n° 4, July pp. 477-483.

[15] Bleistein N. : 1976, Physical optics farfield inverse scattering in the time domain, Jour. Acoust. Soc. Amer., Vol. 60, n° 6, Dec., pp 1249-1255.

[16] Boerner W.M. : 1979, Development of physical optics inverse scattering using Radon projection theory, Conference on Mathematical methods and applications of scattering theory, Washington, D.C., May.

[17] Hong S., Borison S.L. : 1968, Short-pulse scattering by a cone-direct and inverse, IEEE Trans. Ant. Prop., vol. AP-16, N° 1, Jan., pp. 98-102.

[18] Weir W.B. : 1974, Automatic measurement of complex dielectric constant and permeability at microwave frequencies, Proc. IEEE, Vol. 62, N° 1, Jan., pp. 33-36.

[19] Lesselier D. : 1981, Optimization theory and time-domain inverse scattering, Radio Science, Vol. 16, Nov. Dec 1981, pp 1059-1063.

[20] Tijhuis A.G. : 1981, Iterative determination of permittivity and conductivity profiles of a dielectric slab in the time domain, IEEE Trans. Ant. Prop., Vol. AP-29, March, pp. 239-245.

[21] Lesselier D. : 1978, Determination of index profiles by time-domain reflectometry, Jour. Optics, Vol. 9, n° 6, pp. 349-358.

[22] Tabbara W. : 1979, Reconstruction of permittivity profiles from a spectral analysis of the reflection coefficient, IEEE Trans. Ant. Prop., Vol. AP-27, n° 2, March, pp. 241-244.

[23] Weston V. : 1978, Electromagnetic inverse problem, in Electromagnetic Scattering, PLE Uslenghi editor, Academic Press, New York.

[24] Cho Z.K., Kim H.S., Song H.B., Cumming J. : 1982, Fourier transform nuclear magnetic resonance tomographic imaging, Proc. IEEE, Vol. 70, n° 10, Oct., pp. 1152-1173.

[25] Sarkar T.K., Nebat J., Weinder D.D., Jain V.K. : 1980, Suboptimal approximation/identification of transient waveforms from electromagnetic systems by pencil-of-function method, IEEE Trans. Ant. Prop., Vol. AP-28, n° 6, November, pp. 928-933.

II.12 (TS.4)

TIME-DEPENDENT RADAR TARGET SIGNATURES, SYNTHESIS AND DETECTION OF ELECTROMAGNETIC AUTHENTICITY FEATURES

David L. Moffatt and Ted C. Lee

The Ohio State University
Department of Electrical Engineering
2105 Neil Avenue
Columbus, Ohio 43210

A rational function approximant for extracting the complex natural resonances of a scatterer from complex measured or calculated multiple frequency scattering data is described. The extraction procedure incorporates preprocessing filtering and windowing techniques as well as a sum operator. Utility of the extraction process is illustrated for simulated data on a spherical scatterer, to demonstrate noise sensitivity, and for other objects using measured or calculated data. A method for using the complex natural resonances to improve target imaging techniques is suggested.

W.-M. Boerner et al. (eds.), Inverse Methods in Electromagnetic Imaging - Part 1, 441–460.

INTRODUCTION

It is well known that the time-dependent electromagnetic backscatter responses elicited by ramp-type interrogating signals can be used to produce three-dimensional isometric images of a target. The approach is an approximate one and handicapped because the physical optics approximation, with attendent lack of shadow region information, is exploited. This paper basically explores how complex natural resonances, related to circumferential paths on the target, can be extracted and utilized.

Currently available experimental co-polarized and cross-polarized scattering data on both classical objects such as spheres, spheroids and discs and more complicated scatterers are in the form of stepped-swept frequency complex data spanning at least 10:1 bandwidths. A rational function model for fitting these scattering data and extracting complex natural resonances is described. Filtering and windowing are utilized to minimize the problem of noisy data. A method for incorporating certain of the complex natural resonances into improved image models of the target is suggested.

Canonical Responses and an Area Function

There is no loss of generality if a perfectly conducting scatterer is assumed to be located in the forward hemisphere ($z \geq 0$) of a rectangular coordinate frame and an incident plane wave field is assumed to be linearly polarized in the $\hat{x}$ direction* and propagating in the +z direction. It has been repeatedly** demonstrated that the physical optics approximation predicts a relationship [1,2]

$$A_z = -\pi c^2 F_R(2z/c)(U(z) - u(z-z_1)) \quad , \tag{1}$$

between the target's cross sectional area function, A_z, and the backscattered ramp response waveform F_R. The cross sectional area function A_z is the projection of the scatterer surface between z=0 and a cutting plane at z onto the xy plane. The distance z_1 is the distance from the origin to the shadow boundary, assumed for simplicity to be orthogonal to the propagation direction. With normalizations as in [2], the ramp response is given using Fourier synthesis as [3]

* the carat denotes a unit vector.

** see [1] for perhaps the earliest derivation and [2] for a latter more readily accessible publication.

$$F_R(t'/T_0) = -T_0^2 /(2\pi^2) \sum_{n=1}^{N} A(2\pi n)/(n^2)\cos(2\pi nt/T_0+\phi(2\pi n)), \quad (2)$$

where $A<\phi$ is the scattered field in phasor notation and T_0 is the fundamental period.

The previous paper [3] demonstrated that extending the physical optics approximation to encompass the entire scatterer generally yielded improved ramp response estimates and correspondingly improved imaging results were inferred. It is desirable however to develop more formal methods for extending the utility of ramp response waveforms for imaging as well as other applications. When an aperiodic signal interrogates a scatterer the time-dependent response waveform consists of first a forced response as the signal front moves across the object and then a free or natural response as the signal front moves beyond the scatterer. Thus the ramp response consists of an early time forced response as a ramp-type excitation moves across the scatterer and then a later time natural response. It follows that an expansion of the total ramp response waveform in terms of complex natural resonances would not be correct.* For early times the complete geometry of the scatterer has not yet been defined and the global properties are therefore inappropriate. It has been demonstrated [4] that the first electric and magnetic mode complex natural resonances of a conducting sphere can be extracted from the ramp response waveform using samples for times corresponding to returns coming from the shadow boundary and beyond. It was found however that attempts to extract resonances corresponding to higher order modes were unsuccessful, i.e., the poles were highly inaccurate. There is a good reason for this failure. As is clearly demonstrated by the remarks of Kennaugh and Moffatt in [5], the first 19 electric mode pole-pairs of the sphere serve to completely define the creeping wave of the sphere as seen in the backscattered impulse response waveform. Equating the known ratio of required bandwidths to define the ramp and impulse response waveforms to the ratio of poles which could reasonably be expected to be extracted from the creeping wave portion of the response yields approximately three for the ramp response. The inverse frequency squared weighting appropriate for the ramp response essentially negates the effects of the higher poles. This has been seen previously in another context [2], where it is stated that there is little discernable difference between the ramp response synthesized from data spanning a 10:1 bandwidth and data spanning a 60:1 bandwidth.

---

*It is assumed that finite summations without an entire function, as are inevitably used in practice, are utilized.

Rational Function Model

Numerous methods and procedures for extracting complex natural resonances of an object from measured or calculcated radiation or scattering data have been suggested. A rational function model was first suggested in [2]; some initial work using such a model was reported in [6] and later models were also proposed [7]. The rediscovery of Prony's method and attendent eigenanalysis [8,9] directed most effort toward time domain approaches. A rational function model however has been found to be sometimes advantageous because nonharmonically related data samples can be used and this in turn permits extensions of the model to much higher frequencies than previously tested. Also, the restriction to a free response region is avoided. The rational function model is written as

$$G(jx_i) = \frac{\sum_{n=0}^{N} a_n (jx_i)^{n+2}}{1 + \sum_{n=1}^{M} b_n (jx_i)^n} \qquad i = 1,2, \ldots, I \ . \tag{3}$$

where

$$x_i = k_i L \quad . \tag{4}$$

is an electrical length, k the wavenumber and L some convenient linear reference length for the object. It is assumed that the

$$G(jx_i) = \alpha_i + j\beta_i \quad , \tag{5}$$

are known, generally from measurements, for I values of x_i. The order of the numerator and denominator polynomials can be selected to make the model Rayleigh-like or not as appropriate. In this paper, a numerator polynomial of order N+2 is called model I and of order N model II. From Equation (3), I complex equations in the unknown polynomial coefficients can be generated. The equations can be solved in a least squared error sense or in an exact sense. Some of our initial comparisons of extracted poles using both approaches indicate that on some occasions exact solutions can be advantageous. Similarily, the normalization shown in Equation (3) should be arbitrary, i.e., mathematically any coefficient could be normalized. It is found, however, that numerically the normalization shown achieves superior pole results. Also, double precision computer programs are inevitably needed.

We have found that rational function models are most effective when sample values $(G(jx_i))$ are taken from near relative maxima and minima of the amplitude scattering data. This means that in general nonharmonically related samples are used with other samples somewhat arbitrarily taken between the maxima and minima. When extending the rational function models to relatively high frequencies, a preponderance of the nonextremum samples should be at lower frequencies. It has also been noticed that the fits to data obtainable with rational function models fail abruptly beyond the highest frequency sample. The reason is that a rational function model is not an appropriate approximation at higher frequencies.

Consider the first few poles and residues (backscatter) of a conducting spherical scatterer. The sphere is admittedly an overworked example. We choose, however, to work with real poles and residues even when essentially using simulated data (finite number of poles) to avoid bias. Much too often, simulated data are generated to test a proposed pole extraction scheme and then the assumption is made that all of the residues are of equal magnitude. This type of assumption is unrealistic as will be seen in Table I. An alternative approach is to let the residues be quasi-random [10]. Our simulated data (finite number of poles) are generated as, ($k_i a = 0.2(0.02)4.0$)

$$E^S(jk_i a) = \sum_{i=1}^{2M'} \frac{R_i}{jk_i a - p_i} \quad , \tag{6}$$

where R_i are the residues, M' the number of pole-pairs, p_i the poles and a the radius of the sphere. A comparison of the exact and extracted poles and residues when 4 pole-pairs are used and noiseless data are assumed is shown in Table I. To 5 s there is zero error in the extracted poles and residues. Figure 1 shows the amplitude and phase fits that were achieved using the rational function model. It is not apparent but there are actually two curves in Figure 1. The fit however is so good that only one curve is visable. The crosses shown on the amplitude curve show those frequencies used for input data (samples) for the fit. The results in Table I are by no means surprising and are indeed equivalent to a number of Prony-type extraction methods [10] when noiseless data are used. As has been shown however [10], even eigenanalysis becomes ineffective as a pole extraction method when the data are contaminated by more than a moderate amount of noise. While the results discussed in this paper are much improved ones over previous efforts [10] we still maintain that the pole extraction from real world full scale measurements will usually not be feasible. However, as is illustrated in this paper, pole extraction is feasible from controlled laboratory measurements. Since exploitation of the complex natural resonances in identification and imaging need not from the various windows are listed. Table IV demonstrates two points. First, if the system poles lie close to the ends of the window then large errors due to truncation can be anticipated. Second, these errors tend to increase at higher frequencies. The reason for the truncation errors are obvious. The gradual failure at higher frequencies is due to the inadequacy of the model. Rational approximants are at best reasonable in the Rayleigh and low resonance ranges. At higher frequencies, while they can be forced to model, they fail abruptly beyond the highest frequency sample. It is also apparent from these and other examples that when windowing is used, the assumed system order should be roughly twice the suspected system order. As a general rule we discard as curve fitting poles which are not obtained by two or more windows, right half-plane poles, poles with very small residues, real poles and poles with very large negative real parts. Only when such poles consistently reappear (several windows) are they reconsidered. We also discard poles which are approximately cancelled by zeros of the rational approximant. Much of the need for double precision programs involves pole-zero cancellations.

To test the effects of noise on the pole extraction process, we add two uncorrelated pseudo-random Gaussian processes with zero mean and standard deviation σ to the real and imaginary parts of the simulated sphere scattering data. The signal to noise ratio in dB is defined as

$$S/N = 20 \log \left| \frac{1}{N\sigma} \sum_{k=1}^{N} A(k) \right| , \qquad (7)$$

where A(k) is the sphere scattering plus noise at the frequency $j\omega_k$ and N is the total number of samples used. An example of the noisy scattering data is shown in Figure 5 for a noise standard deviation of 0.1. Figure 5 shows the signal plus noise (solid curve) and the rational approximant fit (dashed curve). The poles extracted by an exact fit and windowing for noise standard deviations of 0.01, 0.02 and 0.1 are shown in Table V. The results are not good and obviously pole extraction fails for signal-to-noise ratios much less than roughly 20 dB.

To preprocess the data we implement a tenth order zero phase shift Butterworth lowpass filter by using the product of a fifth order filter and its complex conjugate. Details of the filter and its implementation are given in [13]. The utility of the filter is shown in Figures 6, 7 and 8. Figure 6 shows the noisy amplitude scattering data of the sphere with a signal-to-noise ratio of 14.81 dB, Figures 7 and 8 compare the amplitude and phase data respectively after filtering with the original noiseless amplitude and phase curves. Clearly much of the deleterious effects of the noise have been removed without seriously distorting the true amplitude and phase data.

Preprocessing with a Sum Operator

Least squared error solutions (eigenanalysis) have the advantage that all of the available data samples can be used but are sensitive to noise and may have a convergence problem if iteration is used. Another way to use more of the available data is with row echelon algorithms [13,14] as are used in Gauss elimination to reduce N rows of a matrix to one row. In practice, the data are first low pass filtered and then the rational function approximant is applied using a sum operator. an example of the result is shown in Figure 9 where the processed result of the noisy data in Figure 6 ($S/_N$ = 14.81 dB) is compared to the noiseless amplitude data. Improvements over the result on Figure 7 (filtering alone) may appear slight but more of the data with some target information have been used. The pole-pairs extracted using filtering alone and filtering plus a sum operator are given in Tables VI and VII respectively. Difficulty with the highest frequency pole, as shown, could be anticipated because of truncation which forces both the filter and sum operator to be one-sided at the high frequency end.

The Thin Circular Disc

For scatterers of sufficient geometrical simplicity the rational function approximation (low frequencies) can be combined with an analytical method developed by Kennaugh [15] (high frequencies) to obtain estimates of the complex natural resonances of the object. Table VIII illustrates the overlap, for broadside backscatter which can be obtained to correct the K-pulse approach (which uses Geometrical Theory of Diffraction estimates) at the low frequency end. There are two dominant strings of complex natural resonances for the disc. One set of resonances is from diffraction at the edges of the disc, the second comes from a creeping wave around the circumference of the disc. For backscatter the diffraction mode dominants at broadside incidence and the creeping wave mode at off-broadside incidence. Table IX gives poles obtained using rational function approximants of Hodge's calculated data [16]. Table X is an extended list of the K-pulse estimated [13] creeping wave poles obtained by using Knott et al. creeping wave analysis [17]. In Table IX, the angle given is the angle off broadside, PHI polarization is in the plane of the disc and THETA polarization is perpendicular to the plane of the disc.

necessarily involve direct pole extraction [10,11,12], the poles remain a valuable tool.

Processing

When the rational approximant in Equation (3) is applied to real scattering data, the proper order of the numerator and denominator polynomials (N,M) as well as the optimum sample points (jx_i) are unknowns. It follows that even with noiseless data, some error in the pole extraction procedure will occur. It was noted above that best results are achieved when the frequency samples are at a near relative maxima and minima of the amplitude scattering data and all of the results shown in this paper utilize this fact. In Tables II and III, the effect of numerator and denominator orders, Rayleigh (model I) or non-Rayleigh (model II) rational approximants and minor adjustment of sample positions on the error of the extracted poles and residues are shown. The simulated input data are the same as for Table I. There is one far-reaching conclusion which can be drawn from the results in Tables I, II and III. When the assumed system order (M) is not correct, curve fitting poles are obtained. In these simulated examples the proper system order and the correct poles are known a priori and can be identified. Thus while the poles are extracted with satisfactory accuracy even with improper assumed system order, a method is needed for identifying the curve fitting poles for the general problem. This procedure, windowing, and a filtering process to reduce the effects of noise and clutter are described next. We note that if six sphere pole-pairs and residues are used to calculate the simulated data over the same spectral range (k_ia = 0.2(0.02)4.0) then two of the poles have oscillatory parts greater than the highest input data. These last two poles are not extracted satisfactorily by any of the models. One can conclude therefore that the assumed system order should always be equal to or greater than the actual suspected (from measured amplitude spectrum) system order.

Windowing, i.e., applying the model repeatedly over overlapping spectral spans to cover the spectral range covered by the available data is a well-known technique [7]. In principle, only those poles which repeatedly appear from various windows are considered true poles of the system. The degree of effort devoted to windowing is essentially dictated by the available processing time. Since we have a priori concluded that noise dictates pole extraction in nonreal time, windowing can be a very effective procedure. We show one example using the simulated data generated by six sphere poles and residues over the k_ia = 0.2(0.02)6.0 spectral range. The windows are shown in Figures 2, 3 and 4, where the rational approximant fits to the amplitude data are shown. Within each window the fits, amplitude and phase (not shown) are excellent. In Table IV, the poles extracted

TABLE I
COMPARISON OF TRUE POLES AND EXTRACTED POLES, TRUE RESIDUES AND EXTRACTED RESIDUES (M=8,N=7); FOUR PAIRS OF POLES AND RESIDUES ARE USED IN THE SIMULATED DATA

Poles (M=8,N=7)

true	extracted	% of error real	% of error imag.
-0.50000+/-j0.866025	-0.50000+/-j0.866025	0.E0	0.E0
-0.701964+/-j1.80740	-0.701962+/-j1.80740	1.04E-4	2.07E-4
-0.842862+/-j2.75786	-0.842859+/-j2.757866	1.03E-4	4.97E-5
-0.954230+/-j3.71478	-0.954238+/-j3.714782	2.08e-4	4.97E-5

Residues

True	extracted	real	imag.
-0.0946447+/-j0.516674	-0.094644-/+j0.516674	1.33E-4	0.E0
-0.633323-/+j0.0853256	0.633320-/+j0.08533	4.76E-4	3.75E-4
0.0802221+/-j0.733736	0.08024+/-j0.73373	2.42E-3	8.07E-4
-0.822075+/-j0.07674781	-0.822088+/-j0.07676	1.57E-3	1.48E-3

TABLE II
COMPARISON OF TRUE POLES AND EXTRACTED POLES, TRUE RESIDUES AND EXTRACTED RESIDUES (M=20,N=19); FOUR PAIRS OF POLES AND RESIDUES ARE USED IN THE CALCULATION OF THE SIMULATED DATA

True Poles	Extracted poles	% of error real	% of error imag.
-0.50000+/-j0.866025	-0.50000+/-j0.86602	0.E0	4.95E-4
-0.701964+/-j1.80740	-0.70196+/-j1.80741	2.06E-4	5.16E-4
-0.842862+/-j2.75786	-0.84283+/-j2.75791	1.11E-3	1.74E-3
-0.954230+/-j3.71478	-0.95421+/-j3.71492	5.22E-4	3.65E-3

Curve fitting poles	Curve fitting zeros
0.19396+/-j0.78060	-0.19396+/-j0.78060
0.23181+/-j1.51326	-0.23181+/-j1.51327
-0.39652+/-j1.86546	-0.39649+/-j1.86548
0.12284+/-j2.97965	0.12284+/-j2.97967
0.18009+/-j3.84572	0.18008+/-j3.84573
0.43016+/-j2.16078	0.43016+/-j2.16079

Extracted zeros

-0.38086+/-j3.01887
-0.16461+/-j1.68440
-1.16102+/j0.
1.29822+/-j0.
-1.42429+/-j0.

True residues	Extracted residues	% of error	
-0.0946447-/+j0.516674	-0.09463-/+j0.51668	2.79e-3	1.15E3
0.633323-/+j0.0853256	0.63332-/+0.0854	4.76E-4	1.16E-2
0.0802221+/-j0.733736	0.08046+/-j0.7335	3.22E-2	3.197E-2
-0.822075+/-j0.07674781	0.8219+/-j0.07706	2.119E-2	3.781E-2

*all residues corresponding to the curve fitting poles are in the order of !.E-5or less for both real and imaginary parts.

TABLE III
THE EXTRACTED POLES AND THE ERRORS BETWEEN THE RATIONAL FUNCTION FIT AND THE ORIGINAL DATA (SIMULATED DATA I)

TRUE POLES	MODEL I M=14,N=13	Percent error real part	imag. part
-0.500000+/-J0.866025	-0.50000+/-J0.866022	0.E0	2.98e-04
-0.70196+/-J1.80739	-0.7019523+/-J1.80743	6.06E-4	1.55E-3
-0.842849+/-J2.75786	-0.84282+/-J2.75741	1.457E-3	1.56E-2
-0.954299+/-J3.714787	-0.9648+/-J3.7253	0.27	0.27

MAX. AMP. ERROR 0.19E-5
AVE. AMP. ERROR 0.2E-6
MAX. PHA. ERROR 0.1068E-3
AVE. PHA. ERROR 0.155E-4

MODEL II M=14,N=13	Percent error real part	imag. part
-0.499998+/-J0.86606	2.E-4	3.505E-3
-0.70196+/-J1.80738	2.06E-4	1.03E-3
-0.84295+/-J2.75778	3.05E-3	2.77E-3
-0.95515+/-J3.71550	2.4E-02	1.877E-2

MAX. AMP. ERROR 0.5E-6
AVE. AMP. ERROR 0.1E-6
MAX. PHA. ERROR 0.916E-4
AVE. PHA. ERROR 0.148E-4

Model I M=16,N=15	percent error real	imag.
-0.499997+/-j0.866023	3.01E-4	1.97E-4
-0.701981+/-j1.80739	8.76E-4	5.16E-4
-0.84286+/-j2.75810	7.03E-5	8.33E-3
-0.95048+/-j3.71478	9.78E-2	0.E0

MAX. AMP. ERROR 0.19E-5
AVE. AMP. ERROR 0.2E-6
MAX. PHA. ERROR 0.176E-3
AVE. PHA. ERROR 0.187E-4

TABLE IV
A COMPARISON OF TRUE POLES AND POLES EXTRACTED VIA WINDOW TECHNIQUE

Exact poles of sphere	Poles extracted via RFA & windows(three) technique		PERCENT ERROR REAL PART	IMA. PART
-0.500000+/-j0.866025	-0.50038+/-j0.86610	(W I)	3.80E-2	7.50E-3
-0.701964+/-j1.80740	-0.7086+/-j1.8133	(W I)	0.342	0.304
-0.842862+/-j2.75786	-0.8766+/-j2.7483	(W I)	1.17	0.332
-0.954230+/-j3.71478	-0.9277+/-j3.7388	(W II)	0.692	0.626
-1.04764+/-j4.67641	-0.9996+/-j4.6191	(W III)	1.002	1.196
-1.12891+/-j5.64163	-1.1338+/-j5.5328	(W III)	8.50E-2	1.891

TABLE V
A COMPARISON OF EXACT AND EXTRACTED POLES OF SPHERE VIA RATIONAL FUNCTION APPROXIMATION; A GAUSSIAN UNCORRELATED WHITE NOISE WITH STANDARD DEVIATION IS ADDED TO THE BACKSCATTERING DATA OF SPHERE

	poles extracted via RFA	
exact poles	σ =1.E-2	σ =2.E-2
-0.500000+/-j0.866025	-0.5362+/-j0.9616	-0.4386+/-j0.80129
-0.701964+/-j1.80740	-0.4745+/-j1.9577	-0.6482+/-j1.8509
-0.842862+/-j2.75786	-0.38518+/-j2.643	-0.9795+/-j2.7384
-0.954230+/-j3.71478	-0.4745+/-j3.538	-0.6285+/-j3.5081

σ =1.E-1
-0.439+/-j0.5497
-0.295+/-j1.6033
-0.2574+/-j2.784
not located

TABLE VI
POLES EXTRACTED VIA RATIONAL FUNCTION APPROXIMATION AND AVERAGING TECHNIQUES (S/N=14.81 dB)

TRUE POLES
-0.50000+/-J0.866025
-0.701964+/-J1.80740
-0.842862+/-J2.75786
-0.954230+/-J3.71378

EXTRACTED POLES (σ =2E-1)	Percent Error TRUE	IMAG.
RUN 1		
-0.2989+/-J0.8643	20.11	0.17
-0.2300+/-J1.6802	24.34	6.56
-0.1827+/-J2.2712	22.89	16.88
-0.2455+/-J3.6937	18.48	0.55
RUN 2		
-0.4571+/-J0.9676	4.29	10.16
-0.4727+/-J2.088	11.82	14.47
-0.2651+/-J3.296	20.03	18.66
-0.1739+/-J3.674	20.34	1.063
RUN 3		
-0.4695+/-J0.9854	3.05	11.94
-0.2530+/-J2.160	23.15	18.19
-0.5323+/-J2.5157	10.77	8.397
-0.4474+/-J3.6547	13.21	1.57
AVERAGE POLES		
-0.405+/-J0.9391	9.5	7.307
-0.318+/-J1.976	19.80	8.70
-0.3267+/-J2.6943	17.90	2.20
-0.2889+/-J3.674	17.35	1.06

TABLE VII
POLES EXTRACTED VIA RATIONAL FUNCTION APPROXIMATION USING DIGITAL FILTER AND SUM OPERATOR (S/N=14.81 dB)

poles extracted via digital filter and sum operator M=16,N=15	percent error real	percent error imag.
run 1		
-0.568+/-j0.915	6.8	4.89
-0.664+/-j1.790	1.96	0.89
-0.565+/-j3.001	9.63	8.43
run 2		
-0.392+/-j0.927	10.8	6.10
-0.455+/-j1.927	12.73	6.17
-0.453+/-j3.08	13.52	11.17
run 3		
-0.568+/-j0.8235	6.80	4.25
-0.789+/-j1.538	4.49	13.89
-0.455+/-j2.902	13.45	4.99
run 4		
M=14,N=13		
-0.531+/-j0.926	3.10	6.037
-0.446+/-j1.730	13.21	3.992
-0.321+/-j2.784	18.08	0.893
-0.246+/-j3.587	18.46	3.33
Average		
-0.515+/-j0.897	1.499	3.097
-0.589+/-j1.759	5.826	2.455
-0.429+/-j2.923	14.35	5.72

TABLE VIII
COMPARISONS OF POLES OF DISK AT BROADSIDE BACKSCATTERING USING RATIONAL FUNCTION APPROXIMANT AND GTD METHOD

	RFA String I	RFA String II	GTD method
1.	-2.22+/-j1.656	-0.49+/-j1.23	
2.	-1.04+/-j4.87		-1.04+/-j5.0542
3.	-1.16+/-j8.10		-1.1616+/-j8.2116
4.	-1.24+/-j11.28		-1.2418+/-j11.3611
5.	-1.27+/-j14.4		-1.3024+/-j14.5075

TABLE IX
POLES EXTRACTED BY RATIONAL FUNCTION APPROXIMANTS
THE RATIONAL FUNCTION APPROXIMANTS ARE FITTED TO ka=6.5

poles

PHI POL	P1	P2	P3	P4
0.	-0.49+/-J1.23	-1.04+/-J4.87	-1.16+/-J8.10	-1.24+/-J11.28
15.	-0.49+/-J1.23	-0.64+/-J2.33	-1.38+/-J3.3	-0.98+/-J4.82
30.	-0.49+/-J1.23	-0.61+/-J2.32	-1.39+/-J3.71	-0.95+/-J4.52
45.	-0.49+/-J1.23	-0.62+/-J2.32	-0.67+/-J3.40	-0.75+/-J4.3
60.	-0.49+/-J1.23	-0.63+/-J2.3	-0.97+/-J3.8	
75.	-0.49+/-J1.25	-0.61+/-2.33	-0.725+/-J3.35	-0.67+/-J4.5
90.	-0.49+/-J1.23	-0.60+/-J2.31	-0.74+/-J3.42	-0.65+/-J4.28

THETA POL.				
15.	-0.49+/-J1.22	-0.64+/-J2.27	-1.08+/-J3.30	-0.82+/-J5.02
30.	-0.49+/-J1.23	-0.62+/-J2.33	-1.23+/-J3.52	-0.94+/-J4.57
45.	-0.49+/-J1.23	-0.64+/-J2.34	-0.87+/-J2.9	-0.86+/-J4.8
60.	-0.49+/-J1.23	-0.62+/-J2.32	-0.65+/-J3.21	-1.52+/-J5.4
75.	-0.49+/-J1.23	-0.63+/-J2.32	-0.64+/-J3.24	-1.26+/-J5.8

TABLE X
POLES OF DISK USING SENIOR'S CREEPING WAVE MODE

1.	(-0.5057350,-1.208781)
2.	(-0.6030402,-2.282039)
3.	(-0.6746498,-3.331739)
4.	(-0.7328901,-4.370643)
5.	(-0.7826892,-5.403155)
6.	(-0.8265909,-6.431364)
7.	(-0.8660754,-7.456467)
8.	(-0.9021299,-8.479179)
9.	(-0.9354153,-9.499999)
10.	(-0.9664097,-10.51928)
11.	(-0.9954736,-11.53727)
12.	(-1.022884,-12.55416)
13.	(-1.048859,-13.57012)
14.	(-1.073574,-14.58525)
15.	(-1.097172,-15.59966)
16.	(-1.119781,-16.61343)
17.	(-1.141471,-17.62662)
18.	(-1.162356,-18.63929)
19.	(-1.182501,-19.65149)
20.	(-1.201968,-20.66326)
21.	(-1.220811,-21.67463)
22.	(-1.239077,-22.68564)
23.	(-1.256811,-23.69632)
24.	(-1.274049,-24.70668)
25.	(-1.290820,-25.71675)
26.	(-1.307164,-26.72656)
27.	(-1.323088,-27.73611)
28.	(-1.338675,-28.74543)
29.	(-1.353844,-29.75452)
30.	(-1.368698,-30.76340)

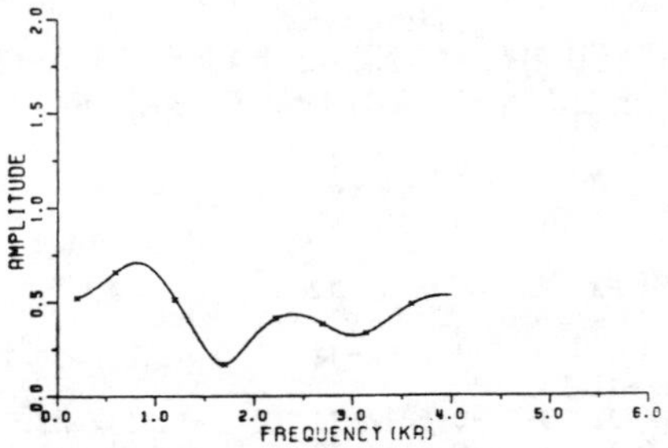

Figure 1a. Comparison of amplitude of the simulated data (____) and the rational function approximant (...) obtained with ka=0.2 to ka=0.4, M=8,N=7. The x's are the __ selected data points.

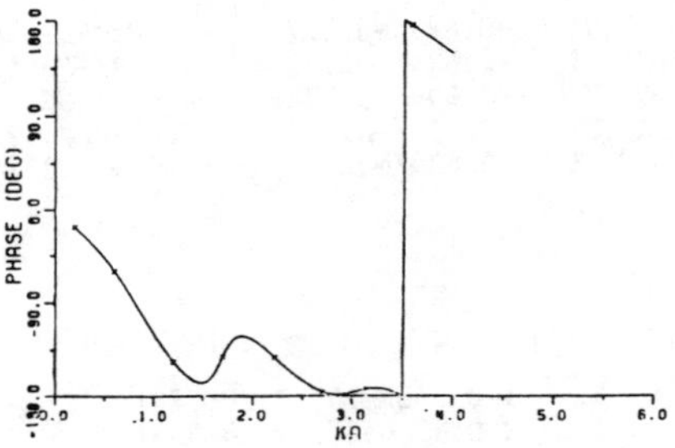

Figure 1b. Comparison of phase of the simulated data (____) and the rational function approximant (...) obtained with ka=0.2 to ka=0.4, M=8,N=7. The x's are the __ selected data points.

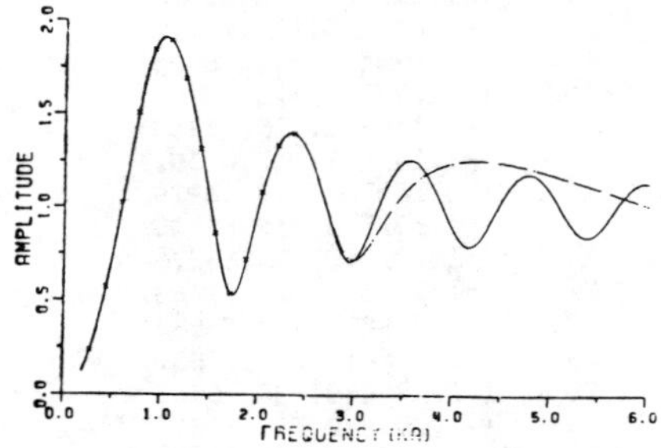

Figure 2. Amplitude plot of a rational function approximant and the window technique at M=14,N=13. X's are the data points used in the exact solution.

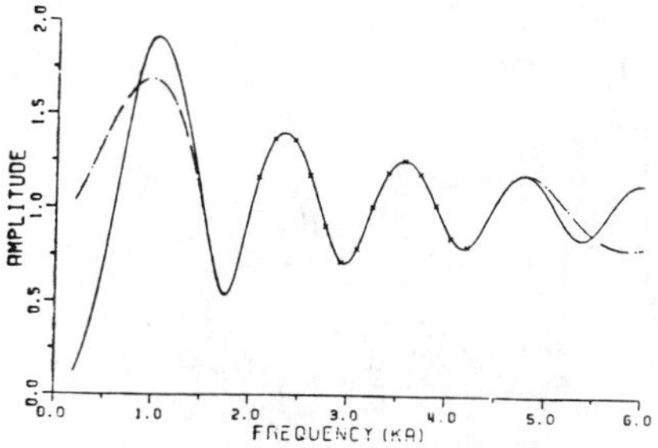

Figure 3. Amplitude plot of a rational function approximant and the window technique at M=14,N=13. X's are the data points used in the exact solution.

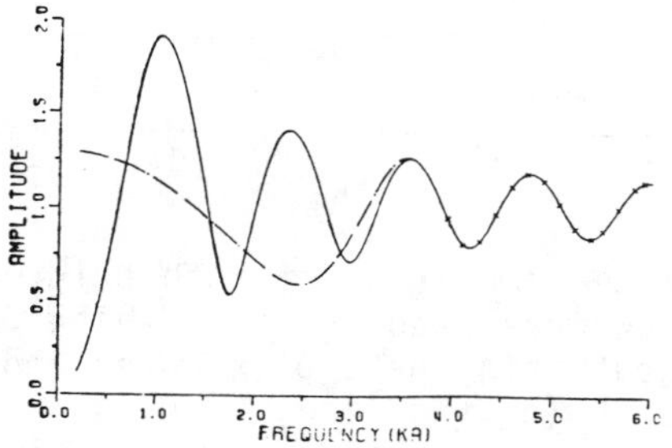

Figure 4. Amplitude plot of a rational function approximant and the window technique at M=14,N=13. X's are the data points used in the exact solution.

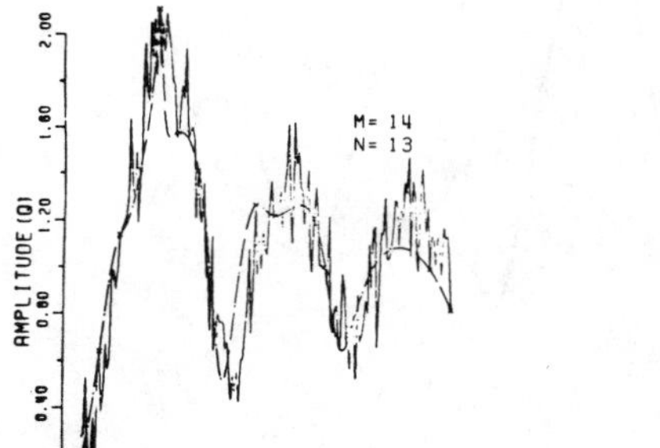

Figure 5. A plot of the amplitude of both original data (solid line, an uncorrelated Gaussian white noise is added to the backscattering data of sphere) and the rational function fit (dotted line). σ=0.1.

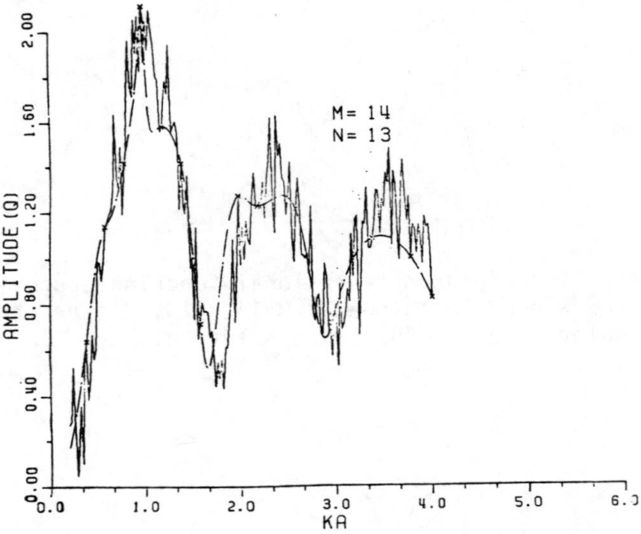

Figure 6. A plot of the amplitude of both original data (solid line, an uncorrelated Gaussian white noise is added to the backscattering data of sphere) and the rational function fit (dotted line). σ=0.1.

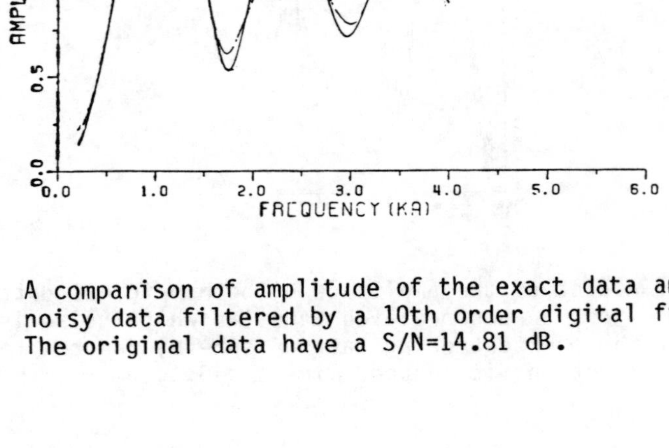

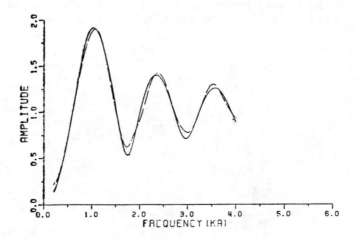

Figure 7. A comparison of amplitude of the exact data and the noisy data filtered by a 10th order digital filter. The original data have a S/N=14.81 dB.

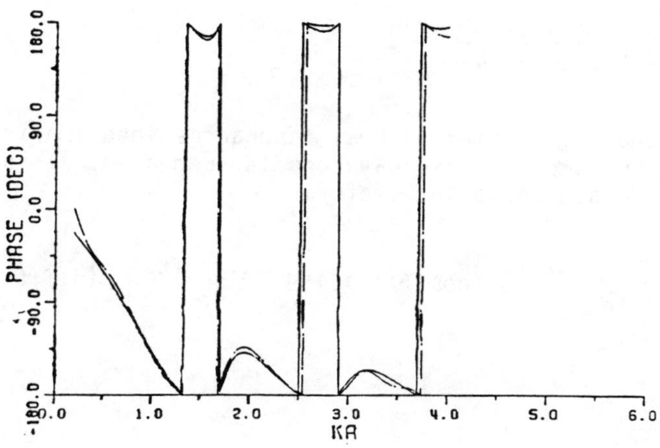

Figure 8. A comparison of phase of the exact data and the noisy data filtered by a 10th order digital filter. The original data have a S/N=14.81 dB.

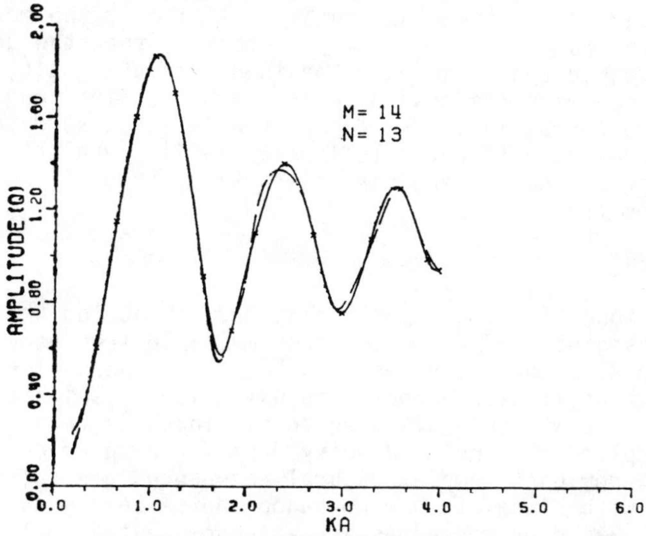

Figure 9. The application of rational function approximant to the prefiltering noisy data of sphere where 10th digital filter and sum operator are used. (S/N=14.81 dB) x's' are the selected data points for the exact fit.

Poles and Ramp Response

An approximate model

$$G(s) = \frac{a + bs}{1 + ds + es^2} e^{-\gamma s} \quad , \tag{8}$$

for incorporating complex natural resonances into imaging methods utilizing the ramp response waveform is suggested. This model has a time domain response waveform

$$g(t) = \left[\frac{b}{e} e^{-\alpha(t-\tau)} \cos(\beta(t-\tau) + (\frac{a-b\alpha}{e\beta} e^{-\alpha(t-\tau)} \sin(\beta(t-\tau) \right] u(t-\tau) \quad . \tag{9}$$

where $\alpha = \frac{d}{2\alpha}$

and $\beta = \frac{1}{2} \sqrt{\frac{4}{e} - (\frac{d}{e})^2}$.

If the delay, τ, is taken as roughly the time to the maximum value of the ramp response (always negative) then the unknown constants a,b,d and e can be determined by fitting g(t) to four values of the ramp response between τ and the first negative to positive zero crossing. The dominant pole $-\alpha+j\beta$ so determined, when compared to a target pole library, will reveal at least to first order the dominating circumferential shape of the target shadow geometry.

Conclusions

A rational function approximant for extracting the complex natural resonances of an object from multiple frequency complex scattering data for the object has been developed. The approximant utilizes preprocessing using low pass filtering and a sum operator. Windowing is used to eliminate pattern fitting poles. Applied to simulated noisy data for a spherical scatterer, dominant complex natural resonances are satisfactorily extracted with signal to pseudo random noise levels as low as 14.82 dB. Based on these results we conclude that pole extraction from measured scattering data is only feasible under controlled conditions, i.e., using laboratory measurements. This in no way detracts from use of the poles in targe identification and possibly in target imaging. It has also been demonstrated that poles obtained using rational function approximants can be used to correct, at low frequencies, poles obtained from a K-pulse approach. This was demonstrated for a thin circular disc.

An emperical technique for using the dominant complex natural resonances of a scatterer to deduce the basic geometrical shape of the scatterer in the shadow region has been suggested.

REFERENCES

[1] J.W. Eberle, "Extension of the Physical Optics Approximation to Small Bodies", Report 827-6, The Ohio State University ElectroScience Laboratory, Department of Electrical Engineering; prepared under Contract No. AF 19(604)-3501 for Bistatic Reflection Characteristics, Air Force Cambridge Research Center, November 1, 1959.

[2] E.M. Kennaugh and D.L. Moffatt, "Transient and Impulse Response Approximations", Proc. IEEE, Vol. 53, No. 8, pp. 893-901, 1965.

[3] D.L. Moffatt, "Ramp Response Radar Imagery Spectral Content", IEEE Trans. Antenna Propagat., Vol. AP-29, No. 2, pp. 399-401, 1981.

[4] C.W. Chuang and D.L. Moffatt, "Natural Resonances of Radar Targets via Prony's Method and Target Discrimination", IEEE Trans. Aerospace Electron Sys., Vol. AES-12, No. 5, pp. 583-589, 1976.

[5] E.M. Kennaugh and D.L. Moffatt, "Comments on Impulse Response of Conducting Sphere Based on Singularity Expansion Method", Proc. IEEE, Vol. 70, No. 3, pp. 294-295, 1982.

[6] D.L. Moffatt, "Time Domain Electromagnetic Scattering from Highly Conducting Objects", Final Report 2971-2, The Ohio State University ElectroScience Laboratory, Department of Electrical Engineering; prepared under Contract No. F19628-70-C-0125 for the Air Force Systems Command, Laurence G. Hanscom Field, Massachusetts, May 1971.

[7] J.N. Brittingham, E.K. Miller and J.L. Willows, "Pole Extraction from Real Frequency Information", Proc. IEEE, Vol. 68, No. 2, pp. 263-273, 1980.

[8] M.L. Van Blaricum, "Techniques for Extracting the Complex Natural Resonances of a System Directly from its Transient Response", Ph.D. dissertation, Department of Electrical Engineering, University of Illinois at Urbana-Champaign, 1976.

[9] ElectroScience Laboratory, "Time Domain Interpretations of Scattering Phenomena", Final Technical Report 2144-2, The Ohio State University ElectroScience Laboratory, Department of Electrical Engineering; prepared under Grant No. GK-708 for the National Science Foundation, Washington, D.C., July 11, 1968.

[10] D.L. Moffatt, J.D. Young, A.A. Ksienski, H.C. Lin and C.M. Rhoads, "Transient Response Characteristics in Identification and Imaging", IEEE Trans. Antennas Propagation, Vol. AP-29, No. 2, pp. 192-205, 1981.

[11] L.C. Chan, D.L. Moffatt, and L. Peters, Jr., "A Characterization of Subsurface Radar Targets", Proc. IEEE, Vol. 67, No. 7, pp. 991-1000, 1979.

[12] L.C. Chan, L. Peters, Jr. and D.L. Moffatt, "Improved Performance of a Subsurface Radar Target Identification System through Antenna Design", IEEE Trans. Antenna Propagation, Vol. AP-29, No. 2, pp. 307-311, 1981.

[13] Ted C. Lee, "Approximate Methods for Obtaining the Complex Natural Electromagnetic Oscillations of an Object", Ph.D. dissertation, The Ohio State University, Department of Electrical Engineering, Columbus, Ohio, 1983.

[14] J.R. Auton and M.L. Van Blaricum, "Investigation of Procedures for Automatic Resonance Extraction from Noisy Transient Electromagnetics Data", Vols. I, II and III, Final Report on Contract N00014-80-C-0299, Effects Technology Inc., 1981.

[15] E.M. Kennaugh, "The K-Pulse Concept", IEEE Trans. Antenna Propagation, Vol. AP-29, No. 2, pp. 327-331, 1981.

[16] D.B. Hodge, "Spectral and Transient Response of a Circular Disc to Plane Electromagnetic Waves", IEEE Trans. Antennas Propagation, Vol. AP-19, pp. 558-562, 1971.

[17] E.F. Knott, T.B.A. Senior, and P.L.E. Uslenghi, "High-Frequency Backscattering from a Metallic Disc", Proc. IEEE, Vol. 118, No. 12, 1971.

II.13 (SR.3)

SINGULARITIES IN QUASI-GEOMETRICAL IMAGING*

G. Dangelmayr & F.J. Wright

Institute for Information Sciences
University of Tübingen
Koestlinstr. 6, 7400 Tübingen, FRG

Various geometrical and quasigeometrical techniques for determining the shape of an unknown scatterer by analyzing backscattered acoustic or electromagnetic signals are reviewed. A geometrical imaging method is proposed for localizing edges by means of mid field data taken along a curve in the observation surface. The focusing phenomena associated with edge diffracted rays are discussed in terms of catastrophe theory. Closed surfaces are reconstructed from far field travel times by utilizing Minkowski's support fuction and the far field focusing directions are identified with singularities of the spherical mapping. A quasigeometrical imaging method is presented which is based on electromagnetic cross-polarized amplitude measurements.

1. INTRODUCTION

This paper reviews various geometrical and quasigeometrical techniques developed in [7] and [8] for the reconstruction of scattering surfaces from scalar (acoustic) or electromagnetic backscattered (zero-offset) signals. For high frequencies of the incident field the principal radii of the surface are large compared with the wavelength. Then the backscattered field at a given observation point depends not on the field over the entire scattering surface but only over restricted portions of that surface centered around the specular points. The contribution to the field coming from a neighborhood of a specular point is embodied in an asymptotic expansion in inverse powers of the frequency which consists of a phase function (the eikonal) and a hierarchy of amplitudes corresponding to the sequence of increasing inverse powers. In homoge-

* Work supported in part by the Stiftung Volkswagenwerk, FRG

W.-M. Boerner et al. (eds.), Inverse Methods in Electromagnetic Imaging – Part 1, 461–472.

neous media the phase function is essentially the specular distance, i.e., the spatial distance from the observation point to the specular point on the surface which determines the geometric travel time. Any inversion technique which uses only this phase information is referred to as "geometrical imaging". The term "quasigeometrical imaging" stands for techniques which incorporate information about geometric optics or higher order amplitudes into the inversion process.

The paper is organized as follows. In Sec. 2 the basic scattering formulae are established. In the scalar case the physical optics or Kirchhoff-approximation is presented for the backscattered wave in the mid field. In the electromagnetic case we confine ourselves to far field backscattering and perform a two term approximation of the backscattered field where the physical optics approximation is supplemented by a non-Kirchhoff correction. In Section 3 we consider a surface with an edge and present a geometrical imaging method which permits one to reconstruct the edge alone from mid field data taken along a curve in the observation surface. A geometrical imaging method for the smooth part of the surface is reviewed in [11]. Special consideration is given to the caustics associated with specular points on the edge which are identified with topological singularities of the distance function and classified by means of catastrophe theory. In Section 4 a geometrical imaging method based on far field travel times is presented which relies on Minkowski's support function. The far field focusing directions are interpreted as singularities of the spherical mapping and their local geometry is elucidated. In Section 5 we present a quasigeometrical imaging method based on electromagnetic first order far field amplitudes. The complete quadratic approximation of the surface at a specular point is derived from cross-polarized amplitude measurements which allows for a pointwise reconstruction of the scattering surface.

2. BASIC DIFFRACTION FORMULAE

2.1 Scalar Diffraction

A time periodic point source with frequency ω is localized at $\underline{x}_o$ and emits a spherical wave

$$u_I(\underline{x},\underline{x}_o,\omega)=(1/4\pi R)\exp(i\omega R/c), \quad R=|\underline{x}-\underline{x}_o| \tag{1}$$

which propagates through a homogeneous medium with velocity c and is incident upon an acoustically hard scatterer bounded by a surface S, where a scattered field $u_S(\underline{x},\underline{x}_o,\omega)$ is induced. We confine ourselves to zero offset and denote the backscattered field by $U(\underline{x}_o,\omega)=u_S(\underline{x}_o,\underline{x}_o,\omega)$. In the high frequency limit the field U can be approximated by the physical optics or Kirchhoff approximation [2]

$$U(\underline{x}_o,\omega) \sim -(i\omega/8\pi^2 c)\int_{L(\underline{x}_o)} dS(\partial R^{-1}/\partial n)\exp(2i\omega R/c) \qquad (2)$$

where $L(\underline{x}_o)$ is the geometrical lit side of S relative to the source point $\underline{x}_o$ and $\partial/\partial n$ denotes the outward normal derivative. The relation (2) must be understood in a true asymptotic sense, i.e., the leading terms of the asymptotic series of U and the r.h.s. of (2) in inverse powers of the frequency coincide, irrespectively of the types of the underlying specular points. If one attempts to improve (2) for higher order asymptotics, non-Kirchhoff correction terms must be taken into account [7] resulting from an iterative solution of the integral equations of diffraction theory [14]. For our purposes the leading order asymptotic relation (2) is sufficient.

2.2 Electromagnetic Far Field Diffraction

Consider an electromagnetic plane wave with magnetic field vector

$$\underline{h}_I = h_o\underline{a}\exp(i\omega\underline{x}\cdot\hat{\underline{x}}/c), \quad |\hat{\underline{x}}| = |\underline{a}| = 1, \quad \hat{\underline{x}}\cdot\underline{a} = 0 \qquad (3)$$

which propagates in the direction of $-\hat{\underline{x}}$ (the time harmonic factor $\exp(i\omega t)$ has been suppressed in (3)) and is incident upon a perfect conducting target bounded by a closed surface S. The origin of the coordinate system is assumed to be inside the target. For large r the magnetic vector of the backscattered field $\underline{h}_S(\underline{x}_o,\omega)$ received at $\underline{x}_o = r\hat{\underline{x}}$ can be written as [5]

$$\underline{h}_S(\underline{x}_o,\omega) = (1/4\pi r)\exp(i\omega r/c)\underline{m}(\hat{\underline{x}},\omega) + O(r^{-2}) \qquad (4)$$

where $\underline{m}$ is given by [13]

$$\underline{m}(\hat{\underline{x}},\omega) = (i\omega/c)\hat{\underline{x}}\times\int_S dS\underline{J}(\underline{x},\hat{\underline{x}},\omega)\exp(-i\omega\underline{x}\cdot\hat{\underline{x}}/c) \qquad (5)$$

with $\underline{J}$ being the surface current induced on S by the incident field. Defining

$$V_K(\hat{\underline{x}},\omega) = (-2i\omega/c)\int_{L(\hat{\underline{x}})} dS(\hat{\underline{x}}\cdot\underline{n})\exp(-2i\omega\underline{x}\cdot\hat{\underline{x}}/c), \qquad (6)$$

$$\underline{m}_1(\hat{\underline{x}},\omega) = h_o\hat{\underline{x}}\times\int_{L(\hat{\underline{x}})} dS(\kappa_2-\kappa_1)\{(\underline{a}\cdot\underline{t}_2)\underline{t}_1 + (\underline{a}\cdot\underline{t}_1)\underline{t}_2\}\exp(-2i\omega\underline{x}\cdot\hat{\underline{x}}/c) \qquad (7)$$

where $L(\hat{\underline{x}})$ is the lit side relative to the direction $\hat{\underline{x}}$, the reduced far field magnetic vector $\underline{m}$ is approximated by

$$\underline{m}(\hat{\underline{x}},\omega) \sim h_o\underline{a}V_K(\hat{\underline{x}},\omega) + \underline{m}_1(\hat{\underline{x}},\omega). \qquad (8)$$

In Eqs. (6) and (7) $\underline{n}$ is the outward unit normal, $\underline{t}_1$ and $\underline{t}_2$ are unit vectors pointing in the principal directions of S at $\underline{x}$ such that $\underline{t}_1 \times \underline{t}_2 = \underline{n}$ and κ_1, κ_2 are the principal curvatures corresponding to the directions $\underline{t}_1$, $\underline{t}_2$, respectively. The relation (8) is a high frequency approximation in the sense that the first two terms of the asymptotic expansions of $\underline{m}$ and the r.h.s. of (8) in inverse powers of ω coincide, i.e., (8) is a first order asymptotic identity. The first term $h_o \underline{a} V_K$ in (8) is the Kirchhoff approximation which results from the standard physical optics approximation of the current $\underline{J}$ and determines the leading order asymptotics. The second term $\underline{m}_1$ is obtained from a first order correction [4,6,7] to the physical optics current derived from the integral equations of diffraction theory [14]. For further details we refer to [7]. Contrary to the scalar leading order asymptotic relation (2) we have utilized a first order asymptotic approximation in the electromagnetic case because $\underline{m}_1$ contains useful information about the surface's curvatures at the specular points (Section 5).

3. MAPPING OF EDGES

In this Section it is assumed that the scatterer has the form of a wedge as shown in Fig.1 where the observation point $\underline{x}_o$ varies on the surface Σ above the wedge. The scattering surface S is identified with the upper boundary of the scatterer because the lower one does not contribute to the asymptotic backscattered field. We assume further that S is only slightly curved so that for any $\underline{x}_o$ on Σ the lit side $L(\underline{x}_o)$ is the whole surface S. The edge E of the wedge is parametrized via $\underline{x} = \underline{x}_e(l)$ where l is the arclength along E. The unit tangent is denoted by $\dot{\underline{x}}_e(l) = d\underline{x}_e(l)/dl$ and the surface's upward unit normal at $\underline{x}_e(l)$ by $\underline{n}_e(l)$ which, in general, differs from the edge's principal normal.

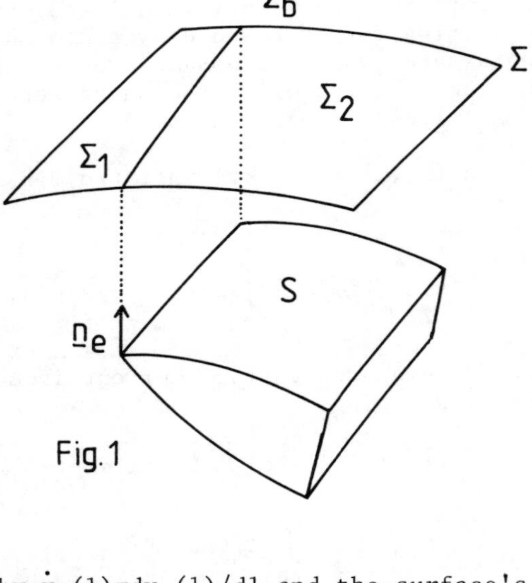

Fig.1

The task is to reconstruct the edge from measurements of backscattered scalar fields on Σ where the frequency response $U(\underline{x}_o, \omega)$ is given by (2) with $L(\underline{x}_o)$ replaced by S. The leading order asymptotics of U are governed by the specular points on S for which the distance $R = |\underline{x}_o - \underline{x}|, \underline{x} \varepsilon S$, has a stationary value. Due to the surface's edge the integration domain in the integral (2) has a boundary. Consequently, two types of specular points contribute to U, viz., interior ones and specular points on the boundary. Properties of the former and their use for imaging the smooth part of S have been discussed in [8,11]. Here we confine ourselves to specular points on the boundary and present a method which permits one to reconstruct the edge.

3.1 Geometric Imaging of the Edge

The stationarity condition for the distance $R_e=|\underline{x}_o-\underline{x}_e|$ from $\underline{x}_o$ to the edge $\underline{x}_e=\underline{x}_e(l)$ requires that the distance vector $\underline{R}_e=\underline{x}_o-\underline{x}_e$ lies in the normal plane of the edge point $\underline{x}_e$. For any point $\underline{x}_o$ on Σ there may be several specular points $\underline{x}_{es}=\underline{x}_{es}(\underline{x}_o)$ on E and we label them and further geometrical quantities evaluated at $\underline{x}_{es}$ by the subscript s, e.g., $\underline{R}_{es}=\underline{x}_o-\underline{x}_{es}$, $R_{es}=|\underline{R}_{es}|$. We assume now that the specular edge-distances $R_{es}=R_{es}(\underline{x}_o)$ have been inferred from backscattered signals. Then the geometric reconstruction of the edge is based on the identity

$$R_{es}(\underline{x}_o)\nabla_{\underline{x}_o}R_{es}(\underline{x}_o)=\underline{x}_o-\underline{x}_{es}(\underline{x}_o). \tag{9}$$

For a generalization of (9) to rays in an arbitrary inhomogeneous medium see [15]. Since $|\nabla R_{es}|=1$, only two components of ∇R_{es} are needed. Moreover, all points $\underline{x}_o$ on the curve C_e defined by the intersection of the normal plane of $\underline{x}_e$ and Σ yield the same edge point. It suffices, therefore, to vary $\underline{x}_o$ on a curve C_o on Σ which does not coincide with one of the curves C_e. Let C_o be parametrized by the arclength σ, $\underline{x}_o=\underline{x}_o(\sigma)$, and let $\dot{\underline{x}}_o(\sigma)$ be its unit tangent. Let $\underline{t}(\sigma)$ be a unit vector in the tangent plane of $\underline{x}_o(\sigma)$ perpendicular to $\dot{\underline{x}}_o(\sigma)$. Then, defining

$$\alpha=dR_{es}/d\sigma, \quad \beta=\frac{\partial}{\partial\rho}R_{es}(\underline{x}_o+\rho\underline{t})\Big|_{\rho=0} \tag{10}$$

where all quantities in (10) are evaluated along C_o, the edge is reconstructed via

$$\underline{x}_e=\underline{x}_o-R_{es}\{\alpha\dot{\underline{x}}_o+\beta\underline{t}+(1-\alpha^2-\beta^2)^{1/2}\underline{n}_o\} \tag{11}$$

$\underline{n}_o$ being the downward unit normal of Σ at $\underline{x}_o(\sigma)$. Hence, measurements of R_{es} in a small neighborhood of a curve C_o on Σ allows a reconstruction of the edge E.

3.2 Edge-Focusing

In general, focusing occurs if two or more specular points (resp. their rays) coalesce in a degenerate specular point $\underline{x}_{sc}$ which may be in the interior of S or on the edge. The codimension of a focusing point $\underline{x}_o=\underline{x}_{oc}$ is defined as the number of coalescing specular points minus one and is identical with the catastrophe theoretic codimension of the topological singularity the distance $R=|\underline{x}_{oc}-\underline{x}|$, $\underline{x}\varepsilon S$, exhibits at $\underline{x}=\underline{x}_{sc}$ [12,16,18]. The highest possible codimension encountered generically in multi-parameter families of functions - here the distance function $R=|\underline{x}_o-\underline{x}|$, $\underline{x}\varepsilon S$ - is the number of parameters involved which in our case is just the number of degrees of freedom the observation point $\underline{x}_o$ possesses. Because $\underline{x}_o$ varies on a two-dimensional observation surface Σ, only singularities with codimension less than or equal to two can generically occur [12,16,18]. We consider here only singularities on the edge. For degenerate specular points in the interior of S see [8,11]. Rather than listing the catastrophe normal forms we confine ourselves to a geometric description of the various singularities. For a more detailed discussion including codimension-three singularities see [19].

(i) Shadow boundary. The straight lines in the normal directions $\underline{n}_e$ sweep out the geometrical shadow boundary which generically intersects the observation plane Σ in a curve Σ_b. We denote by Σ_1 and Σ_2 those parts of Σ lying to the right and to the left of Σ_b, respectively, (cf. Fig. 1). Hence, in Σ_1 there are interior and edge speculars while in Σ_2 interior speculars fail to exist. On Σ_b an edge- and an interior specular point coalesce, i.e., Σ_b is a line of codimension one singularities. The specular distances R_s and R_{es} in the vicinity of Σ_e are sketched in Fig. 2 where $R_s=|\underline{x}_o-\underline{x}_s|$ is the distance associated with an interior specular $\underline{x}_s$. The R_s-branch terminates tangentially in the R_{es}-branch at the shadow boundary Σ_b.

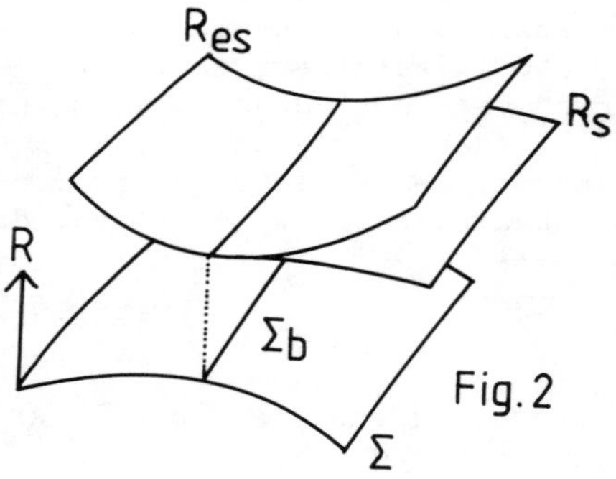

Fig. 2

(ii) Edge-caustic. We define EV to be the envelope of the normal planes of the edge, i.e., the surface in three-dimensional space where neighboring normal planes intersect. EV is a ruled surface with rulings made up by straight lines in the normal planes which are parallel to the binormal directions and pass through the curvature centres of the underlying edge points. Generically EV intersects the observation surface Σ in a curve Σ_e. On Σ_e two specular edge points coalesce, i.e., two branches of specular edge distances join along a cusped line above Σ_e, similar to what happens when two interior speculars coalesce [8,11]. Hence, EV plays the role of the edge's caustic and Σ_e is a line of codimension-one singularities.

(iii) Caustics on the shadow boundary. A codimension two singularity takes place if EV or the interior caustic (the surface's evolute) meets the shadow boundary giving rise to the coalescence of three rays, two from the edge and one from the interior in the former and conversely in the latter case. One possible arrangement of the various branches of R_s and R_{es} is sketched in Fig. 3 for an edge caustic and in Fig. 4 for an interior caustic meeting the shadow boundary. In Fig. 4, Σ_i denotes the line where the surface's evolute intersects Σ.

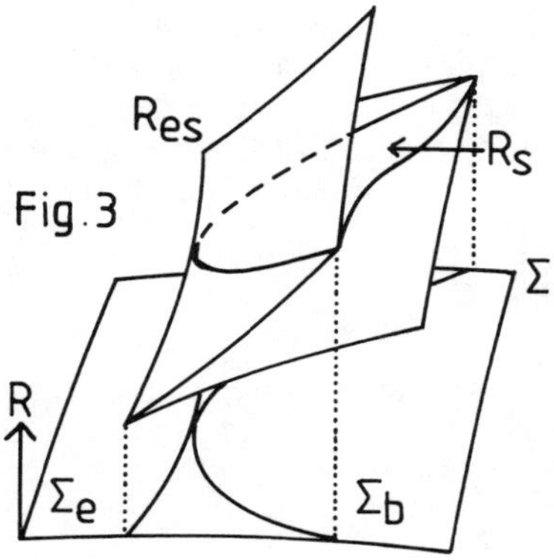

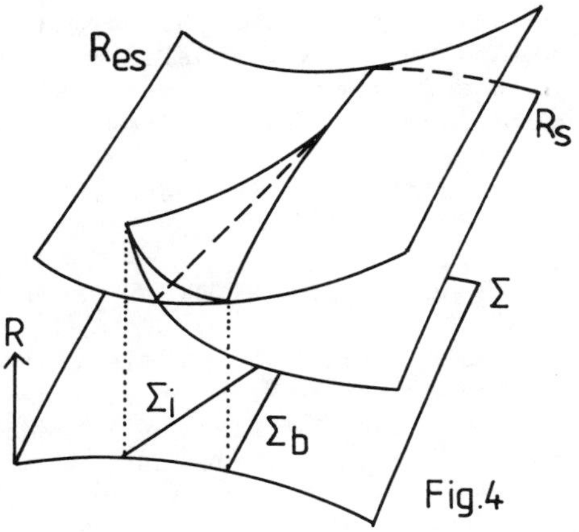

(iv) Cusps of the edge caustic. The edge of regression ER of the envelope of the normal planes of E is defined as the curve on EV which is everywhere tangent to the rulings. Alternatively, ER may be defined as the curve formed by the centres of the osculating spheres of E [10]. Along ER two smooth sheets of EV are glued together, i.e., ER plays the role of a rib line [8,15] of the edge caustic EV. Generically ER intersects the observation plane in isolated points where Σ_e exhibits a cusp. Near this cusp point the specular edge distances behave qualitatively like the swallowtail bifurcation set, similar to cusp points associated with interior caustics [8, 11].

The phenomena described so far exhaust all generic singularities up to codimension two which can occur on the edge. Special examples of surfaces where these and higher singularities take place are presented in [19].

3.3 Contributions to the Backscattered Field from Specular Edge Points

Let $\underline{x}_{es}=\underline{x}_{es}(\underline{x}_o)$ be a non-degenerate specular point on the edge. The contribution U_{es} to the backscattered field U coming from a neighborhood of $\underline{x}_{es}$ is given by [7]

$$U_{es} \sim \gamma A_{es}|2c/\omega|^{1/2}\exp\{2i\omega R_{es}/c+\tfrac{1}{4}\pi\mu\,\mathrm{sgn}\,\omega\}(1+O(\omega^{-1})) \qquad (12)$$

with amplitude

$$A_{es}=(8\pi R_{es})^{-3/2}(R_{\shortparallel}/R_{\perp})\,|1-\underline{R}_{es}\cdot\ddot{\underline{x}}_{es}|^{-1}. \tag{13}$$

Here, $\gamma=1(-1)$ if $\underline{x}_o$ is on Σ_2 (Σ_1) and $\mu=\mathrm{sgn}(1-\underline{R}_{es}\cdot\ddot{\underline{x}}_{es})$. We have used the notation $R_{\shortparallel}$ and $R_{\perp}$ for the components of the distance vector $\underline{R}_{es}$ parallel and perpendicular to $\underline{n}_{es}$, respectively, and $\ddot{\underline{x}}_e=d^2\underline{x}_e/dl^2$ denotes the curvature vector pointing in the direction of the edge's principal normal. The delta-response $G'(\underline{x}_o,t)$, obtained by Fourier inversion of $U(\underline{x}_o,\omega)$, diverges like $(t-t_{es})^{-1/2}$ at the geometric two way travel-time $t_{es}=2R_{es}/c$. This permits one to detect the travel-times t_{es}, e.g., from an aperture limited Fourier inversion of the backscattered field U [7] or from the spacing of peaks in the high frequency interference spectrum [1].

4. GEOMETRIC INVERSION IN THE FAR FIELD

We assume here the standard far field configuration, i.e., the scatterer occupies a bounded region which contains the origin of the coordinate system and is enclosed by a smooth surface S. In the scalar case we define

$$V(\hat{\underline{x}},\omega)=\lim_{r\to\infty}\{(4\pi r)^2\exp(-2i\omega r/c)U(r\hat{\underline{x}},\omega)\}, \tag{14}$$

so that the far field approximation of U for large r is $U\sim\exp(2i\omega r/c)V/(4\pi r)^2$. Since asymptotically, up to leading order in $1/\omega$ [7]

$$V(\hat{\underline{x}},\omega)\sim V_K(\hat{\underline{x}},\omega) \tag{15}$$

where the Kirchhoff approximation V_K is defined by (6), the leading order asymptotics of both, the electromagnetic and the scalar backscattered far fields are governed by the same function V_K. In order to deduce the actual leading order asymptotic term of V_K we have to perform a stationary phase integration of the integral in (6). The stationary condition selects those points $\underline{x}_s=\underline{x}_s(\hat{\underline{x}})$ on the lit side $L(\hat{\underline{x}})$ for which the outward normal points in the direction of $\hat{\underline{x}}$. We use the same notation as in Sec. 3, i.e., the subscript s refers to a geometric quantity evaluated at $\underline{x}_s$. Assuming then that the Gaussian curvature of S at $\underline{x}_s$ does not vanish, the contribution V_{Ks} to V_K coming from a neighborhood of the specular point $\underline{x}_s$ takes the form

$$V_{Ks}\sim\nu A_s\exp(-2i\omega\rho_s/c)(1+O(\omega^{-1})) \tag{16}$$

where

$$\rho_s=\underline{x}_s\cdot\hat{\underline{x}},\quad A_s=2\pi|\kappa_{1s}\kappa_{2s}|^{-1/2} \tag{17}$$

and ν =1(-1) or $-i\,\mathrm{sgn}\,\omega$ if the principal curvatures κ_{1s} and κ_{2s} are both negative (positive) or of opposite signs corresponding to a convex (concave) or saddle-type region of S around $\underline{x}_s$, respectively. From (14) and (16) we discover that the geometric two-way travel-time for a wave travelling from the source $r\hat{\underline{x}}$ to S and then back to the source again is given by $t_s=2(r-\rho_s)/c$. Using methods described in [1,8,11] (e.g., an aperture limited Fourier inversion of the frequency response) permits one to infer t_s and consequently ρ_s from backscattered signals. The task is now to reconstruct S from this phase information.

4.1 Utilizing Minkowski's Support Function

A genuine surface S can be divided into several regions S_j where S_j is either convex, concave or of saddle-type. In general, the regions S_j are separated by certain smooth lines where the Gaussian curvature vanishes. Let Ω_j be the spherical image of S_j on the unit sphere (the outward normal directions). Because the spherical image of S_j is single valued (the Gaussian curvature does not vanish), S_j can uniquely be parametrized in the form $\underline{x}=\underline{X}_j(\underline{n})$, $\underline{n}\varepsilon\Omega_j$, where $\underline{n}$ is the outward unit normal of S_j at $\underline{x}$, i.e., the polar angles of $\underline{n}$ are treated as independent variables. Let C_j be the cone in three dimensional space subtended by the directions in Ω_j. Then Minkowski's support function [17]

$$M_j(\underline{\xi})=\underline{\xi}\cdot\underline{X}_j(\underline{\xi}/|\underline{\xi}|),\quad \underline{\xi}\varepsilon C_j \tag{18}$$

associated with S_j is a single valued homogeneous function of degree one defined in C_j. From M_j the region S_j can be reconstructed via [17]

$$\underline{x}=\nabla_{\underline{\xi}}M_j(\underline{\xi})=\underline{X}_j(\underline{\xi}/|\underline{\xi}|),\quad \underline{\xi}\varepsilon C_j \tag{19}$$

We note that if the gradient in (19) is written out in polar coordinates, then the angular and radial derivatives give the components of $\underline{X}_j$ tangent and normal to S, respectively.

Recalling that the far field specularity condition requires that $\hat{\underline{x}}$ coincides with the outward unit normal at the specular points we may label these by $\underline{x}_{sj}(\hat{\underline{x}})=\underline{X}_j(\hat{\underline{x}})$ where $\hat{\underline{x}}\varepsilon\Omega_j$ and $\underline{x}_{sj}\varepsilon S_j$. From (17) and (18) we find that $M_j(\xi\hat{\underline{x}})=\xi\rho_{sj}(\hat{\underline{x}})$, $\xi>0$, where $\rho_{sj}=\underline{x}_{sj}\cdot\hat{\underline{x}}$ is the phase function corresponding to the specular point $\underline{x}_{sj}$. The regions S_j are then readily reconstructed by virtue of (19).

4.2 Topological Far Field Singularities

Let the surface be parametrized by two coordinates, $\underline{x}=\underline{X}(\xi,\eta)$, e.g., the polar angles of the rays coming from the origin and intersecting S at $\underline{x}$. Then the phase function $\rho=\underline{x}\cdot\hat{\underline{x}}$ in the integral (6) becomes a function of two "state variables" (ξ,η) and two "control parameters" parametrizing the unit sphere, $\rho=\rho(\xi,\eta,\hat{\underline{x}})$. The solutions of the equations $\partial\rho/\partial\xi=\partial\rho/\partial\eta=0$ are clearly the parameter values (ξ_s,η_s) of the specular points $\underline{x}_s=\underline{X}(\xi_s,\eta_s)$. In the terminology of catastrophe theory, ρ exhibits a singularity at (ξ_s,η_s). Furthermore, catastrophe theory tells that the only singularities which can generically occur are Morse singularities and degenerate singularities of the fold- and cusp-type [8,11,16,18]. The underlying topology on the surface is the following. There are smooth closed lines on S, called L-lines, where one of the two principal curvatures vanishes, i.e., the spherical map is locally not invertible. These L-lines are the boundaries between adjecent regions S_j and their normals constitute the focusing directions. For $\underline{x}$ on L let $P(\underline{x})$ be the principal curvature line corresponding to that principal curvature which vanishes at $\underline{x}$. For "almost all" points $\underline{x}$ on L the curvature line $P(\underline{x})$ intersects L transversely and ρ exhibits a fold-singularity. For isolated points $\underline{x}_c$ on L, however, $P(\underline{x}_c)$ has a tangential contact with L (the zero principal curvature direction is tangent to L) and ρ encounters a cusp-singularity. In Fig. 5a we have sketched an L-line (solid) with a cusp-point $\underline{x}_c$, two principal curvature lines $P_j=P(\underline{x}_j)$ and $P_c=P(\underline{x}_c)$ (broken). The images L' and P_j', P_c' of resp. L and P_j, P_c on the unit sphere under the spherical mapping are shown in Fig. 5b. The various branches of the "specular distances" ρ_s near $\underline{n}(\underline{x}_c)$ behave qualitatively like the bifurcation set of the swallowtail catastrophe. For further details about cusps of the spherical mapping we refer to [3]. The geometry of the additional singularities which generically occur in a one parameter family of surfaces are discussed in [9].

Fig.5

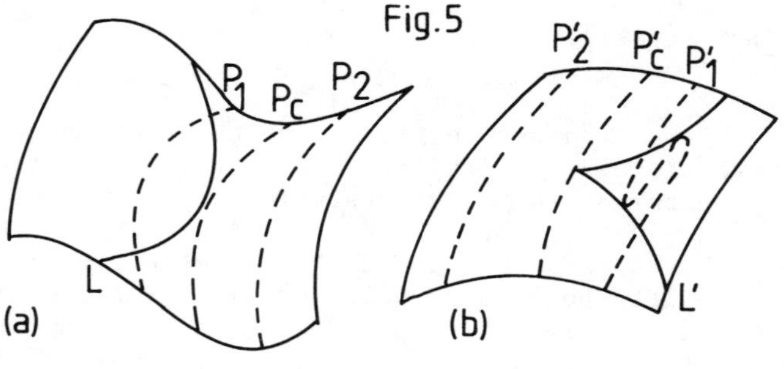

5. INVERSION BY MEANS OF FIRST ORDER ASYMPTOTIC ELECTROMAGNETIC FAR FIELD AMPLITUDES

Recalling that (8) is a first order asymptotic relation for the reduced far field magnetic vector $\underline{m}(\hat{\underline{x}},\omega)$, Eq. (5), the first two terms of the asymptotic contribution $\underline{m}_s$ to $\underline{m}$ coming from a neighborhood of the specular point $\underline{x}_s$ are found by a stationary phase integration of (6) and (7). The result is [7]

$$\underline{m}_s \sim h_o \nu A_s \exp(-2i\omega\rho_s/c)\{\underline{a}+(c/2i\omega)(A_1\underline{a}+B\underline{b})+O(\omega^{-2})\} \tag{20}$$

where A_s, ρ_s and the sign factor ν are defined in Sec. 4, Eq. (17), and $\underline{b}=\hat{\underline{x}}\times\underline{a}$, i.e., $\underline{a}$ and $\underline{b}$ are orthogonal unit vectors in the tangent plane. The amplitude A_1 results from the first order asymptotic expansion of V_{Ks} [7] and needs not concern us here further. The amplitude B is given by

$$B=(\kappa_{1s}-\kappa_{2s})\sin(2\alpha) \tag{21}$$

where α is the angle enclosed by $\underline{a}$ and the principal direction $\underline{t}_1$. By measuring the polarizations in the directions of $\underline{a}$ and $\underline{b}$ one infers ν, the absolute value of the Gaussian curvature and the first order amplitude B. If the same experiment is performed with an incident polarization direction $\tilde{\underline{a}}$ obtained by a 45°-rotation of $\underline{a}$, the corresponding first order amplitude $\tilde{B}$ is given by

$$\tilde{B}=(\kappa_{1s}-\kappa_{2s})\cos(2\alpha) \ . \tag{22}$$

Eqs. (21) and (22) permit one to determine the curvature difference $\kappa_{1s}-\kappa_{2s}$ and the angle α, i.e., the principal directions, from measurable quantities. Using this and the geometric optics amplitude A_s one infers the principal curvatures κ_{1s}, κ_{2s} [7].

The method described above allows for a deduction of the complete quadratic approximation of the surface at the specular points from cross-polarized amplitude measurements. The task is now to reconstruct the shape of the scattering surface which is achieved as follows. Assume first that the curvature information is given for a specular point in a convex region where the principal curvatures are negative. Fix a (x_1,x_2,x_3)-coordinate system and consider an ellipsoid $E:(x_1/c_1)^2+(x_2/c_2)^2+(x_3/c_3)^2=1$ with yet unknown semi-axes c_j. There is exactly one point $\tilde{\underline{x}}$ on E which has the same outward normal direction as the prescribed one $\hat{\underline{x}}$. Equating the quadratic approximation of E at $\tilde{\underline{x}}$ with the measured one yields three equations which determine the three semi-axes c_j. From this one finds the coordinates of the point $\tilde{\underline{x}}$ on E which is identified with the underlying specular point $\underline{x}_s$. The technique resembles the methods of equivalent ellipsoids introduced by Chaudhuri and Boerner [6] for relating the time domain response of convex scatterers to the projected area function. If the principal curvatures are positive (the specular point is in a concave region) one perfoms the same computation with $(\hat{\underline{x}},\kappa_1,\kappa_2)$ replaced by $(-\hat{\underline{x}},-\kappa_1,-\kappa_2)$ and then reflects the resulting point on E at the origin. In saddle-type regions the same technique may be utilized as in convex regions, but with a hyperboloid rather than an ellipsoid.

REFERENCES

[1] Achenbach, J.D., and Norris, A., "Crack characterization by the combined use of time-domain and frequency-domain scattering data", 1981 AF/DARPA-Review of Quantitative NDE, in press.
[2] Baker, B.B. and Copson, E.T., "The mathematical theory of Huggens' principle", 2nd ed., Oxford University Press 1950.
[3] Banchoff, T., Guffney, T. and McCrory, C., "Cusps of Gauss mappings", Pitman, London 1982.
[4] Bennet, C.L., Auckenthaler, A.M., Smith, R.S. and De Lorenzo, J.D., "Space time integral equation approach to the large body scattering problem", Sperry Rand Research Centre, Sudbury, Mass., Final Report on Contract F 30602-71-C-0162, RADC-CR-73-70, AD76394 (May 1973).
[5] Boerner, W.-M., "Polarization utilization in electromagnetic inverse scattering", in H.P. Baltes (ed.), "Inverse scattering problems in optics", Springer 1980.
[6] Chaudhury, S.K. and Boerner, W.-M., Appl. Phys. 11, pp. 337-350 (1976).
[7] Dangelmayr, G., "Asymptotic inverse scattering", submitted for publication (1983).
[8] Dangelmayr, G. and Güttinger, W., Geophys.J.R.Astr.Soc. 71, pp. 79-126 (1982).
[9] Dangelmayr, G. and Armbruster, D., "Singularities in phonon focusing", submitted for publication (1983).
[10] Goetz, A., "Introduction to differential geometry", Addison Wesley 1970.
[11] Güttinger, W., "Topological approach to inverse scattering in remote sensing", this Volume.
[12] Güttinger, W. and Eikemeier, H. (eds.), "Structural stability in physics", Springer 1979.
[13] Jackson, J.D., "Classical Electrodynamics", 2nd ed., Wiley 1975.
[14] Maue, A.W., Zeitschr. Phys. 126, pp. 601-618 (1949).
[15] Norris, A.N. and Achenbach, J.D., "Mapping of a crack by ultrasonic methods", to be published in J.Acoust.Soc.Am.
[16] Poston, T. and Stewart, I., "Catastrophe theory and its applications", Pitman, London 1978.
[17] Stoker, J.J., Comm. Pure Appl. Math. 3, pp. 231-257 (1950).
[18] Thom, R., "Structural stability and morphogenesis", Benjamin 1975.
[19] Wright, F.J., and Dangelmayr, G., in preparation.

II.14 (RS.5)

DECONVOLUTION OF MICROWAVE RADIOMETRY DATA

R. H. Dittel

Deutsche Forschungs- und Versuchsanstalt
für Luft- und Raumfahrt
Institut für Hochfrequenztechnik
8031 Oberpfaffenhofen, FRG

ABSTRACT

The methods of deconvolution by multiple convolution, maximum-likelihood and matrix inversion have been investigated for the purpose of data reconstruction for multispectral microwave radiometry data. The principle objective was the use of a-priori-information in order to reduce the variety of the solution manifold, to force the process to converge and to decrease computer time. By application of some algorithms on real data, a contrast enhancement by the factor 5 could be reached for objects 1/10 of the 3-dB-halfwidth of the antenna wide.

1. INTRODUCTION

The microwave signal of a body, measured for the purpose of earth-scientic investigations is altered in a characteristic manner on the way from the radiation source to the output of the detection equipment. The different steps are as follows:

a) addition of reflected radiation energy from the environment to the emitted radiation energy of the object

b) alteration of the thus integral generated radiation energy by obsorption and attenuation on passing through the atmosphere.

c) convolution of the radiation patterns of the resolution elements of the scene detected by the imaging characteristics of the antenna system.

W.-M. Boerner et al. (eds.), Inverse Methods in Electromagnetic Imaging - Part 1, 473–492.

d) addition of systematical and statistical intensity fluctuations to the detected radiation energy by the electronic components of the detection equipment.

In order to determine the nature of a detected object from its radiation signature, it is necessary to eliminate or at least to reduce the influence of the distortions pointed out under items a) to d) and thus obtain the genuine radiation signature of the object. The processing in general is performed in a way reverse to the signal sequence, starting with the elimination of the noise and systematic errors and ending with the atmospheric correction of the signal.

The purpose of this paper is to demonstrate the results of an investigation concerning the practical possibilities of signal reconstruction by deconvolution and thus correcting the influence indicated under item c).

The necessity for such a reconstruction is mandatory for the following reasons:

a) the convolution effect leads to unsharp imaging of the scene and thus to difficulties in object boundary identification.

b) the convolution effect leads to contrast reduction and thus to the impossibility of discriminating geometrically small objects and object induced intensity fluctuations. The term "small" is assumed for object sizes representing a filling factor < 0.1. The filling factor is defined as the ratio of the object projection area relative to the projection area of the 3-dB-halfwidth of the antenna pattern.

c) the convolution effect leads to signatures at the boundaries of areas with different radiation characteristics. The generation of radiation intensities not really existing at the different locations leads to wrong identification of object signatures and properties and in consequence to misclassifications of objects and their location in the scene.

2. BASIC METHODS OF DECONVOLUTION SOLUTIONS

There are basically two different approaches for the reconstruction of the scene T from the image B, generated by convolution

$$A * T = B$$

with the antenna characteristics A applying deconvolution operations.

a) in the time domain by multiplying the above term with the inverse of the antenna characteristics:

$$T = A^{-1} B$$

b) in the frequency domain, by multiplying the transformed terms

$$A(\omega) \cdot T(\omega) = B(\omega)$$

with the term $1/A(\omega)$

$$T(\omega) = \frac{B(\omega)}{A(\omega)} .$$

The transformation of $T(\omega)$ into the time domain leads to the temperature map T of the scene.

The application of the deconvolution algorithms for either one method involves the following problems:

a) the reconstruction in the time domain requires even for modest scene sizes and geometrical resolutions data matrices of several hundred elements in line and column direction.

The matrix operations are characterized by a large number of multiplications. These lead to rounding errors and in consequence to numerical inaccuracies of the inverse matrix. The effects of the rounding error in a digital computer are enhanced by the special property of the antenna characteristics. Its function is approximately normal distributed and contains therefore small weighting values in the region off the symmetry axis. It is this set of numerically small values that lead to relatively large rounding errors during the process of matrix inversion. The result is an enhancement of just those elements, whose value in the inverse matrix should have the value 0.

The consequence for the reconstruction is an exceeding enhancement of the small deviations from the ideal undisturbed result of a convolution operation. An increase in the numerical accuracy (by increasing the number of bits) only results in higher memory requirements and computer time, but does not necessarily lead to higher stability of the deconvolution process, as statistical numerical fluctuations are not eliminated by an increase of the number of digits. A comparable problem also arises at the reconstruction of images in the frequency domain.

b) A general point of controversy, concerning the deconvolution process, is the assumption that an exact reconstruction of a scene cannot be obtained due to the effects of missing knowledge about the data elements outside the image boundaries.

An analysis of the antenna - /scene-relation, however, shows, that the real deconvolution of a scene does not require the full information content of the infinite large projection area of the antenna characteristics, but only needs a minor part thereof around the symmetry axis. Outside this central region, the weighting factors rapidly decrease to the value 0 and can therefore be fully ignored for the deconvolution process. Of importance is the requirement, that a large number of independent data elements of the scene are available.

3. DATA ERRORS AND SELECTION OF ALGORITHM

For the derivation of the convolution and deconvolution terms, the presence of undistorted data has been assumed. The real data, however, are disturbed by a variety of systematical and statistical errors, whose origin is not necessarily induced by and related to the radiation properties of the object detected. The influence of the errors can be defined in the form

$$B = A(T + n_1) + n_2 + n_3$$

which leads to the deconvolution term

$$A^{-1} B = A^{-1} A(T + n_1) + A^{-1} n_2 + A^{-1} n_3 .$$

The different contribution habe the following properties:

a) the contribution n_1 which represents the incertainty ΔT of the radiation temperature prior to the deconvolution process by the receiving antenna. It is influence by the convolution process. Typical examples for n_1 are system external disturbances of technical radiation sources and atmospheric phenomena

b) the contribution of the system induced systematical error n_2 and the statistical error n_3. Examples for n_2 is the switching cycle to the reference radiation source in the radiometer system, n_3 is the thermal noise of the receiver section. They are not convolved by the system.

In order to facilitate the deconvolution process, as many as possible error contributions should be eliminated prior to the data reconstruction process. The problem is to properly identify and eliminate the different contributions in such a way, that the object information is distorted only to a minor degree.
In most cases, n_1 is characterized by weak shorttime effects, which are eliminated through the digitation procedure. Only influences of exterme radiation intensity and long duration, such as signals from radar stations, cannot be eliminated by a simple filtering process and therefore requires special treatement. In general, $n_1 \ll T$ and therefore $T + n_1 \sim T$ can be assumed.

No difficulties occur for the elimination of the contribution n_2, as the properties of systematic errors in general are very well known and can be replaced from the data by application of filter procedures. Care has to be taken on the necessary of the filter characteristics.

Under this assumption the deconvolution term is reduced to

$$T = A^{-1} B + A^{-1} n_3.$$

For practical cases, the reproducibility of the deconvolution process depends on the statistical properties of the term $A^{-1} n_3$, which generally leads to diverging behaviour of the deconvolution process. If n_3 is of minor influence to the signal value T, then the application of recursive methods many lead to exceeding radiation temperature values.

For most of the algorithms described in textbooks and applied for practical cases absolute ignorance of the object and data properties and the image distortions is assumed. In general there is, however, no complete ignorance of the scene properties, but there exists in most cases basic knowledge on the approximate average radiation temperature, kind and extension of statistical data errors and the approximate temperature profiles along the flight path. This knowledge is being used as a first approximation to the most probable solution of the algorithm applied. The a-priori-knowledge not only reduces the amount of computer time, but also the high variety of possible solutions, which otherwise would be generated due to the statistical errors of the individual data elements. Thus the application of a-priori-knowledge also sets the preliminaries for divergence of the reconstruction process. The question for the optimum method of deconvolution for a given data set can be answered in such a way, as the best algorithm is the one, which best uses the a-priori-knowledge offered. An for all possible cases best algorithm for the solution of all kinds of problems does not exist.
Depending on the properties of the data, a-priori-knowledge can be used to quantify deterministical and statistical processes. The parameter may be single values, or value sets for the characterization of single or global data properties. Typical examples of a-priori-parameters may be:

- the arithmetic mean or the median as the most probable value of a physical quantity
- the standard deviation as a value for the most probable degree of data, fluctuation (e.g. thermal noise)

- upper and lower limit of a data range to define the dynamics of deterministical and statistical properties (e.g. intensity dynamics of a scene)

- the degree of derivation between single data elements of a time or spatial data sequence (e.g. for the characterization of common temperature fluctuations caused by statistical processes).

4. INVESTIGATED DECONVOLUTION ALGORITHMS

The following algorithms have been investigated for the purpose of deconvolving radiometry data:

a) deconvolution by multiple convolution,

b) deconvolution by maximum likelihood,

c) deconvolution by matrix inversion.

4.1 Deconvolution by multiple convolution

The deconvolution process is maintained on the basis of the convolution term

$$B = A * T$$

This form does not indicate the presence of noise, which statistically influence the original image

$$B = A * T + N$$

and thus the result of the deconvolution process. For practical applications a distinction between noise and undisturbed signal and their seperation is not possible. Therefore the algorithm has to be designed in such a way, as to converge when applied on noise disturbed data.

The best approximation of the first iteration step n = 0 is the original noise distorted data set T itself.

$$T^{(n=0)}(x) = B^{(o)}(x).$$

The following iterations steps are maintained as demonstrated in the following loop [1]:

$$
\begin{aligned}
B^{(n)}(x) &= T^{(n)}(x) * A(x) \\
T^{(n+1)}(x) &= T^{(n)}(x) + K(x)\,[B^{(0)}(x) - B^{(n)}(x)] \\
|T^{(n+1)}(x) &- T^{(n)}(x)| < \varepsilon \rightarrow \text{End} \\
n &= n + 1
\end{aligned}
$$

The contribution of the different iteration steps leads to a solution in the form

$$T^{(n)} = (-1)^{n-1} \binom{n}{1} B^{(n-1)}.$$

This term indicates, that the reconstruction is a non-linear operation. Convergence is only obtained if the elements of the deconvolution term,

are $\sum_i |a_i| \leq 1$

for all weighting elements a_i of the antenna characteristics, which exactly is the case. This performance does, however, give no support for the depression of noise in the data. Of importance for this intention is the term

$$K(x) = C\left[1 - 2\left(\frac{T^{(n)}(x) - \frac{H(x) + L(x)}{2}}{H(x) - L(x)}\right)\right],$$

where H(x) and L(x) are the upper and lower boundary value of the intensity dynamics [1]. The purpose of the term is also to decrease ascillatory fluctuations at the location of abrupt intensity changes and the enhancement of statistical data fluctuation such as those generated by noise. C is considered as on amplification factor. The disadvantage of the corrective term K(x) is its deterministic restriction to the boundary values. This leads to a depression of naturally excessive intensities and problably to diverging resolution if values $T > H(x)$ or $T < L(x)$ are present. In order to prevent this phenomenon, an alternative corrective coefficient

$$K(x) = C\left[1 - e^{-1/2}\left(\frac{T^{(n)}(x) - \frac{H(x) + L(x)}{2}}{V(x)\,\sigma(x)}\right)^2\right]$$

was defined. This term has a probabilistic characteristic as it continuously decreases the step width with increasing deviation from the most probable intensity value without deterministic restriction by the now most probable treshold values H(x), L(x) of the intensity dynamics. V(x) is a parameter to control the influence of stochastic events.

The investigation of this and the other algorithms were performed on two different types of intensity profiles:

a) a temperature profile, consisting of only one abrupt intensity step varying between 100 K and 400 K

b) a highly periodical fluctuating to test the feasibility of the algorithms to reconstruct objects which require high geometrical resolution capabilities for recognition. (The figures present the convolved and the deconvolved data for comparison. The synthetic and the real data are used to demonstrate the capabilities of the algorithms on one-dimensional intensity profiles.)

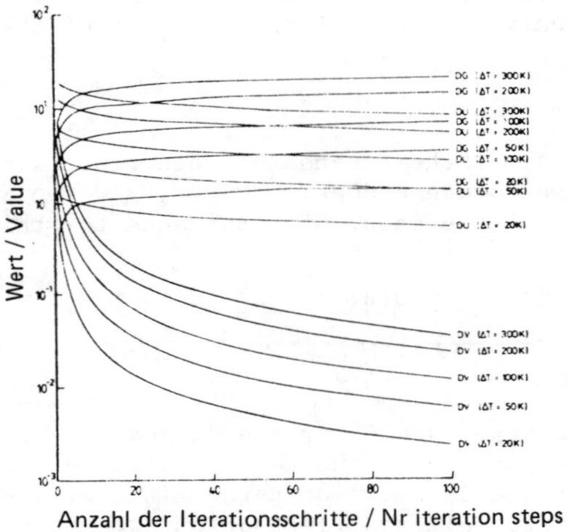

Legend:
DV: Sum of absolute values for solution enhancements for succeeding iteration steps
DU: Sum of absolute values of deviation of the solution from unconvolved data set
DG: Sum of absolute values of deviation from convolved data set
(All values refer to 80 data elements)

Fig. 1

Fig. 1 shows the convergence properties. They demonstrate, that the first 25 reconstruction steps bring significant improvement, so that further trials to improve the result can be abandoned. The good convergence behaviour for noise free data (Fig. 2) is

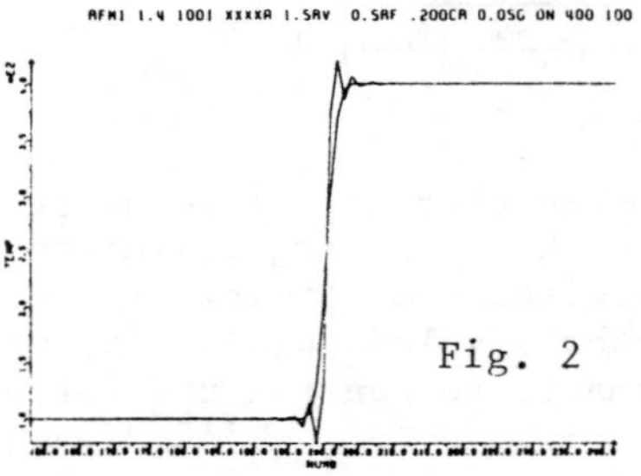

Fig. 2

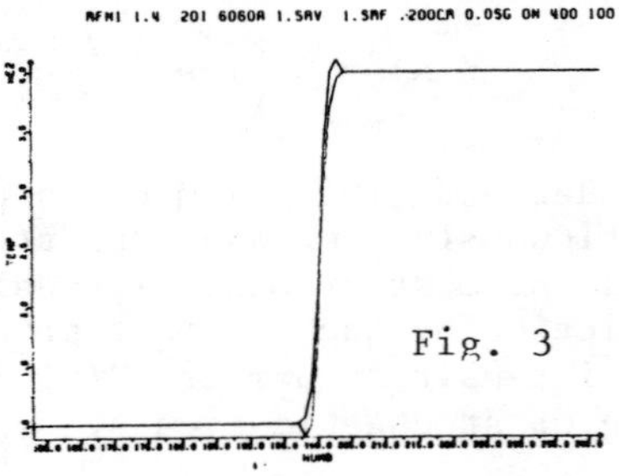

Fig. 3

due to the fact, that even for missing a-priori-knowledge the image values are the best estimation for the reconstruction process. Of significance is the oscillatory effect at the location of abrupt intensity changes. This effect can be suppressed by using a-priori-knowledge for the upper and lower boundary of the most probable dynamic range of the radiation intensity, symmetric or asymmetric to the real intensity value (Fig. 3, Fig. 4).

Fig. 5 shows deconvolved noise distorted data. The error reducing effect is demonstrated by variation of V in the Fig. 6a - 6c. The influence of varying parameter C to the transition behaviour is presented in Fig. 7a - 7d.

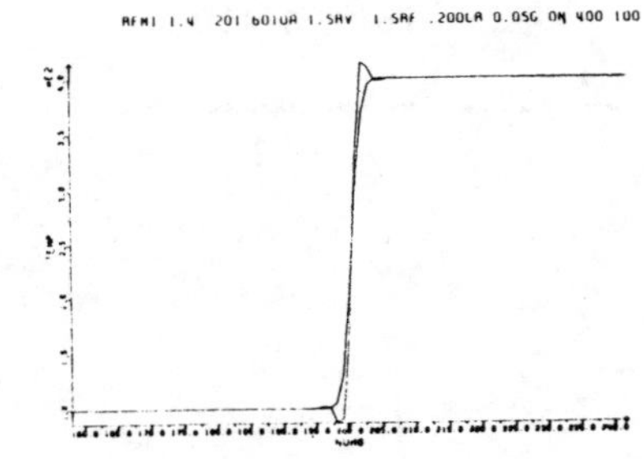

Fig. 4

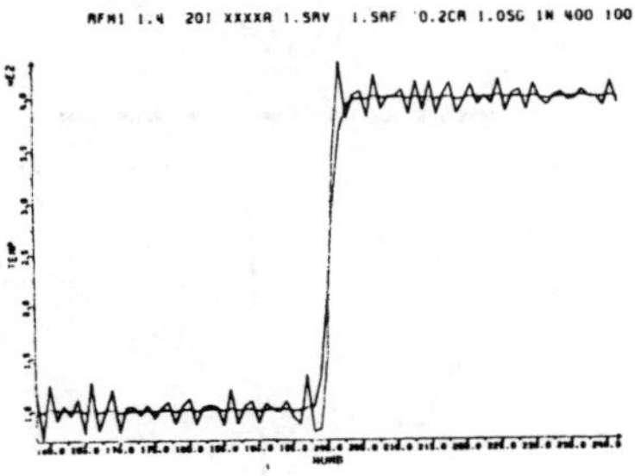

Fig. 5

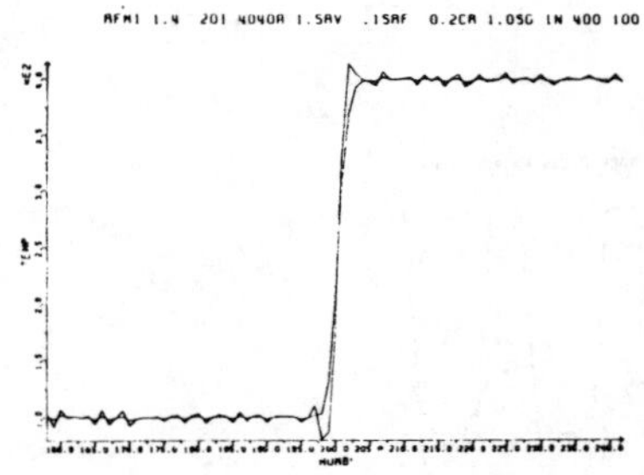

Fig. 6a V=0.1

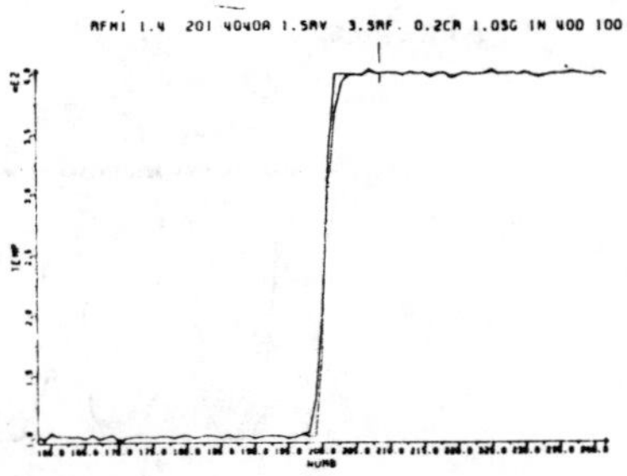

Fig. 6b V=3.0

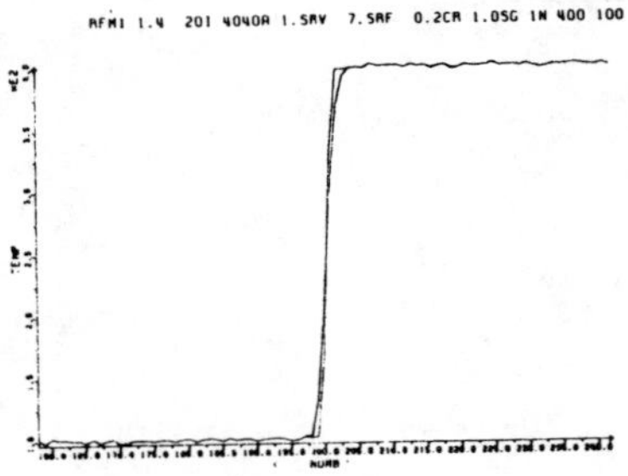

Fig. 6c V=7.0

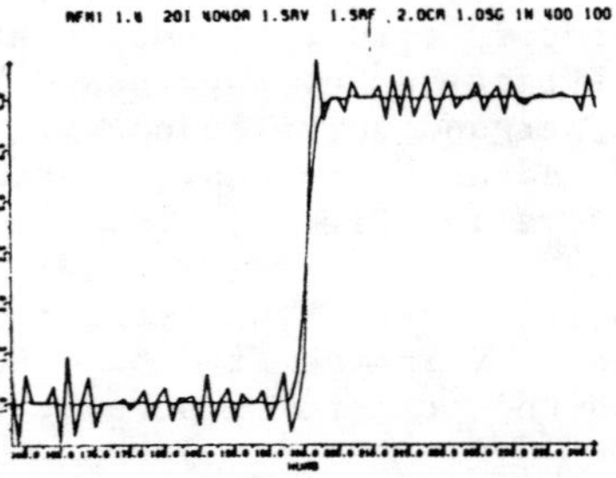

Fig. 7a C=2.0

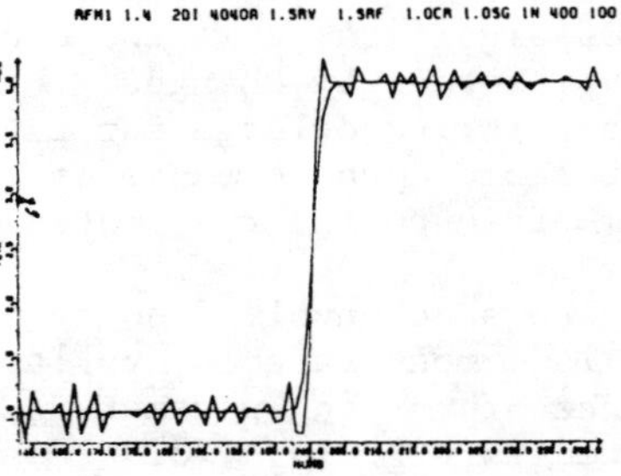

Fig. 7b C=1.0

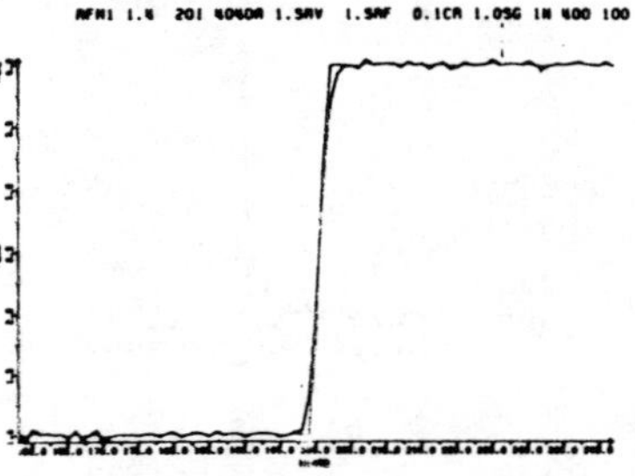

Fig. 7c C=0.1

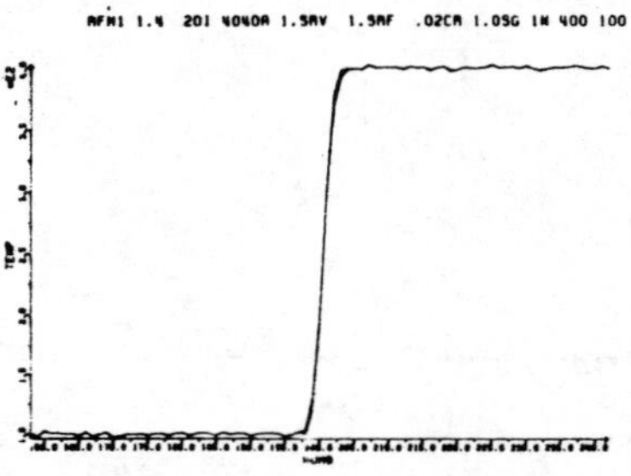

Fig. 7d C=0.02

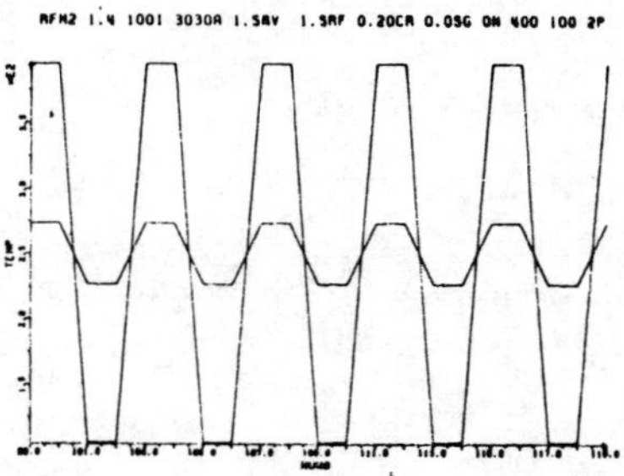

Fig. 8

Good reconstruction could be obtained with a set of data, consisting of intensity fluctuations only two elements wide, although the width of the antenna characteristics over the region from -10 to +10 degrees consisted of 81 resolution elements, each representing an angular resolution of 0.25 degrees (Fig. 8).

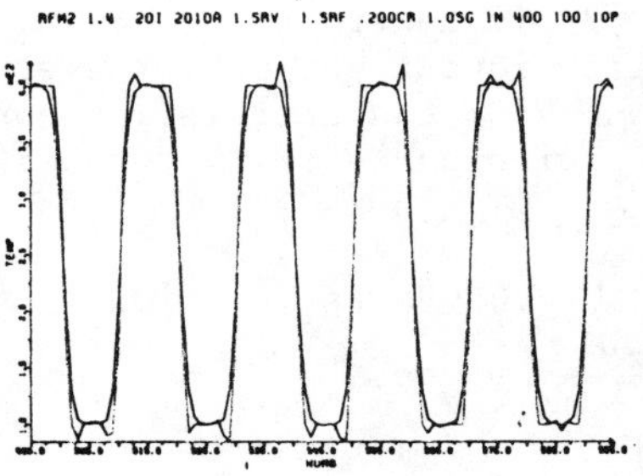

Fig. 9a C=0.2

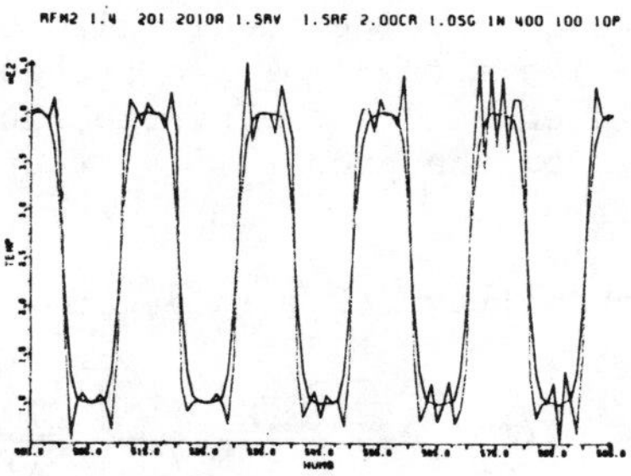

Fig. 9b C=2.0

Difficulties only occured for noise disturbed data. In these cases, it was not always possible to successfully suppress the noise influence to the data elements. A proper selection of the control parameters leads to an increase of the quality of the result (Fig. 9a, 9b).

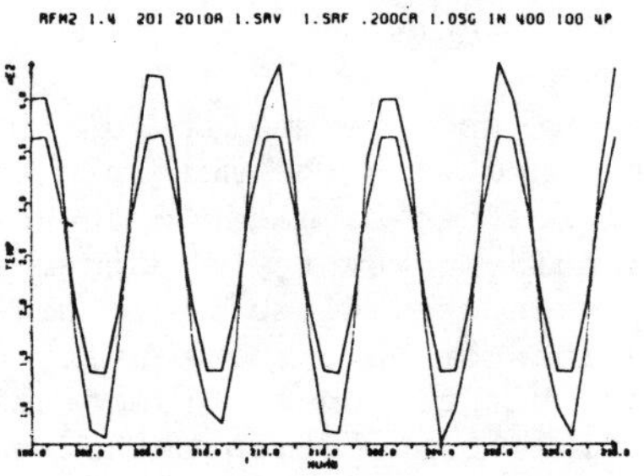

Fig. 10 C=0.2

The fact, that high periodicity fluctuations do not permit expressed transient behaviour of the reconstruction curve between the individual elements, causes sometimes a signigicant deviation from the ideal result (Fig. 10).

4.2 DECONVOLUTION BY MAXIMUM LIKELIHOOD

Checking the convolved data it is very difficult to distinguish between object- and noise-generated intensity peaks. Only the form and the texture of the neighbour data element permit to identify the reason for an expressed peak.

The method of principle likelihood [2] in contrary to the multiple convolution method makes a probability investigation for every data element and its vicinity. The antenna characteristics can be compared to the probability function with an overall probability of 1.

The probability $P(T_i)$ is determined by the Bayes-relation

$$P(T_i) = \sum_K \frac{P(B_K/T_i)\ B(T_i)\ P(B_K)}{\sum P(B_K/T_j)\ P(T_j)} ,$$

where i, j, k are the indices of the vectors B, T. This term indicates, that the profile T can interatively be determined to a probability $P(T_j)$ by application of a Bayes-Operator.

The iteration process derives the temperatur T_i for the iteration step (r + 1) from the value T_i at the preceding step r:

$$T_{i,\ r+1} = T_{i,\ r} \sum \frac{A_{i,k}\ B_k}{\sum_j A_{j,\ k} \cdot T_{j,\ r}} .$$

The convergence behaviour and the quality of the results are identical to those obtained with the multiple convolution method. For this reason, reference can be made to the different figures published there. It should, however, be pointed out, that the algorithm is not strong enough, to sufficiently suppress intensity excesses at locations of abrupt intensity changes or due to noise distortions. For this reason a correcting factor K(x), as described for the multiple convolution method has also been used here.

For some examples, the algorithm showed for the first few iteration cycles a slight divergend trend, but after not more than the third step switched to converging behaviour. The main advantage of this algorithm compared to the multiple convolution method is evidently its greater stability and trend in convergence, even for the case of very uncertain a-priori-knowledge.

4.3 Deconvolution by Matrixinversion

The term

$$T = A^{-1} AT = A^{-1} B + A^{-1} N$$

is definitely only converging in the cases N = 0.

The reason for solution instability are stochastic data behaviour and numerical problems due to quasi-singular properties of the convolution matrix. The effect are highly oscillating solutions even for data with on N/S-ratio of 10^{-4} to 10^{-5}.

The solution by matrix inversion was obtained with the term [3]

$$T = (A^t A + \gamma G)^{-1} (A^t B + \gamma W).$$

The individual parameters are:

G matrix to reduce the solution variety and to prevent oscillatory solution,

B matrix of convolved data,

W matrix of a-priori-knowledge,

γ stabilization term of the solution,

A, A^t antenna matrix and its transposed, (smoothed by second and higher difference terms of neighbour matrix elements).

For the general case, the convolution matrix A is point symmetric normal distributed and all elements are $|a_i| < 1$. The best solutions were obtained for G = I (unitary matrix).

The algorithm is no iterative method, so that the deconvolution result is obtained in one step. (This could be verified for different examples.) Fig. 11 shows the results of the deconvolution, using undisturbed data. They are characterized by a very good reconstruction of the unconvolved data set. The investigation showed, however, that the result is very much depending on the quality of the a-priori-information. Unfortunately the solution tends in any case to the assumed, estimated temperature profile, and if the assumption is not correct, to a wrong deconvolution result.

Of general interest is the effect of the value of the parameter γ for the solution stability. A decrease of the value γ shows an increase in the solution stability even for undistorted data. The effect of instability is expressed in form of a high-amplitude oscillation (Fig. 12a - 12n).

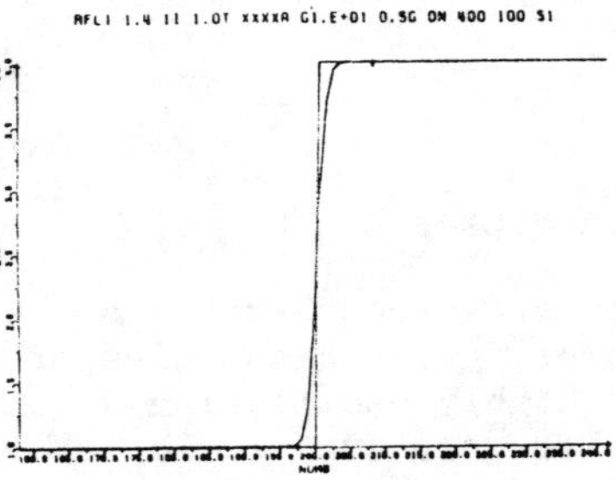

Fig. 11

1 Iterationstep

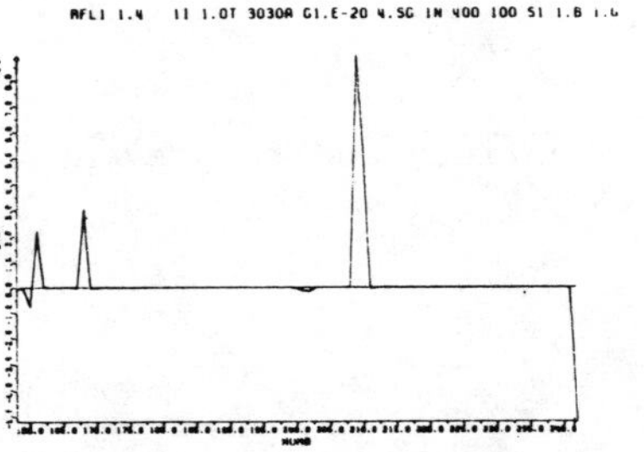

Fig. 12a $\gamma=10^{-20}$

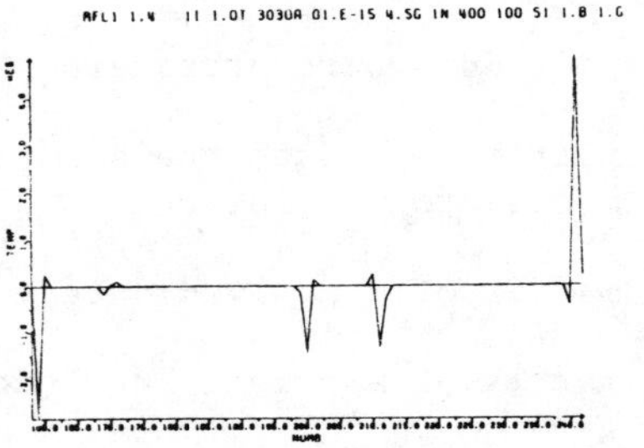

Fig. 12b $\gamma=10^{-15}$

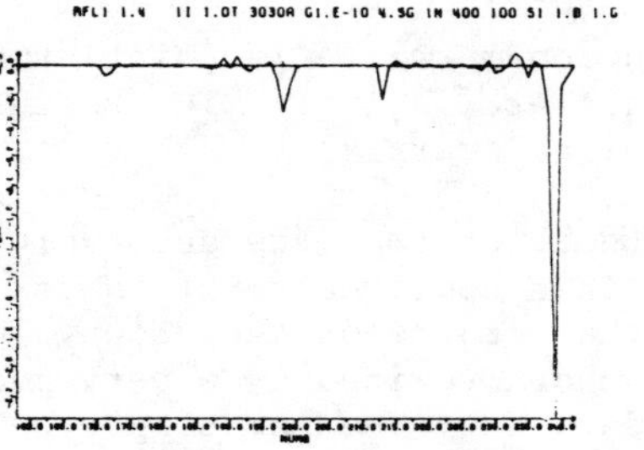

Fig. 12c $\gamma=10^{-10}$

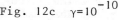

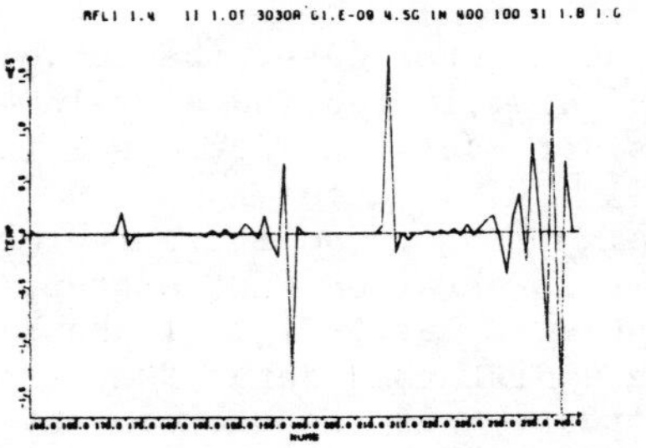

Fig. 12d $\gamma=10^{-9}$

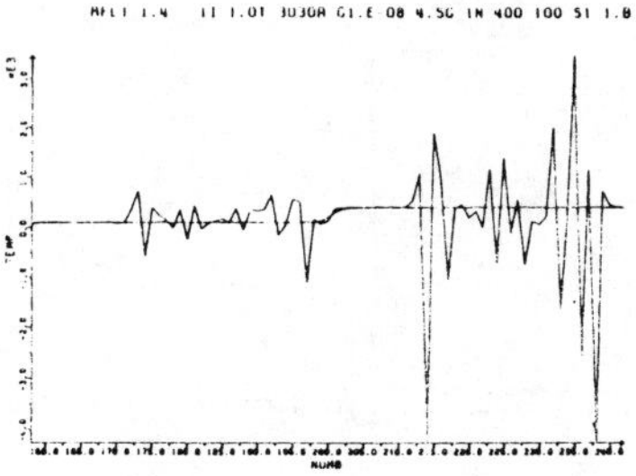

Fig. 12e $\gamma=10^{-8}$

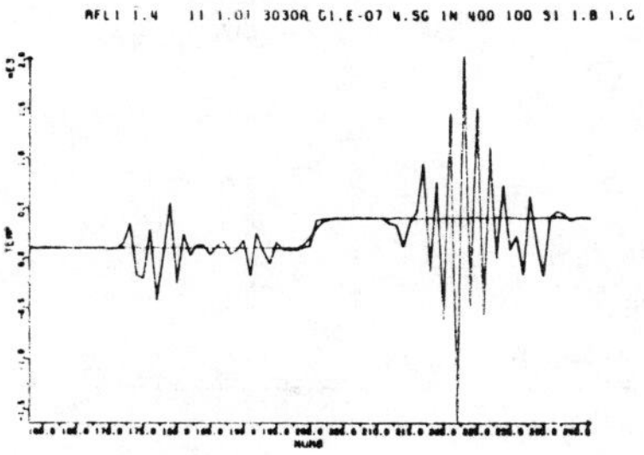

Fig. 12f $\gamma=10^{-7}$

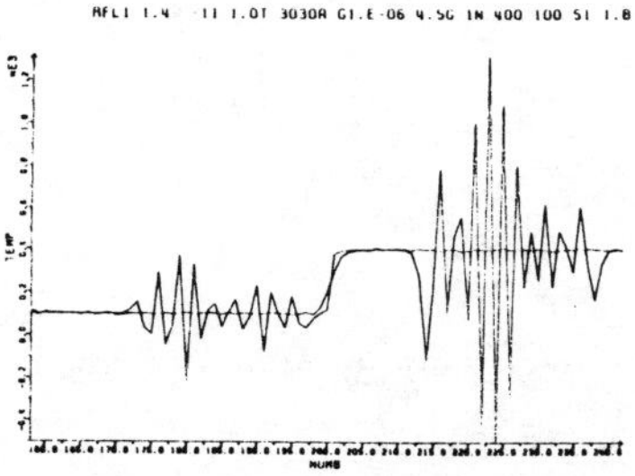

Fig. 12g $\gamma=10^{-6}$

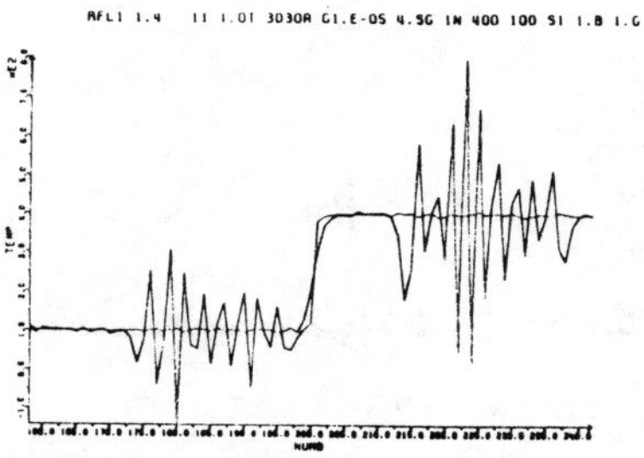

Fig. 12h $\gamma=10^{-5}$

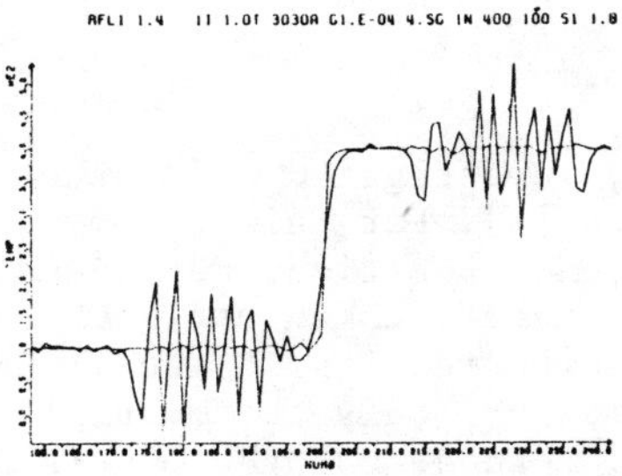

Fig. 12i $\gamma=10^{-4}$

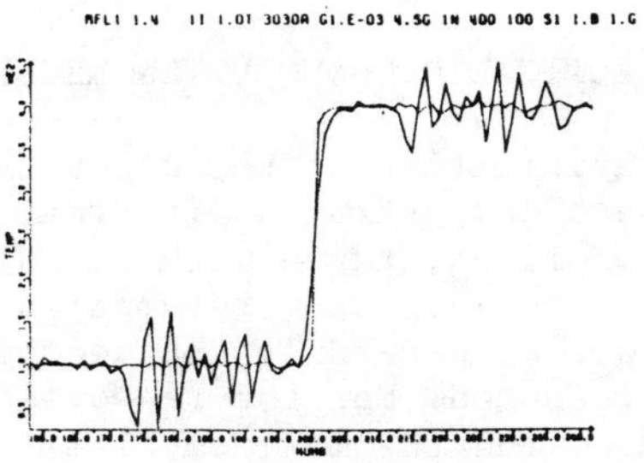

Fig. 12j $\gamma=10^{-3}$

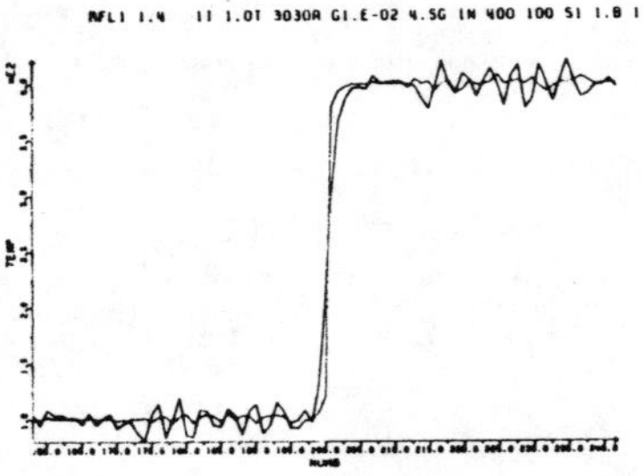

Fig. 12k $\gamma=10^{-2}$

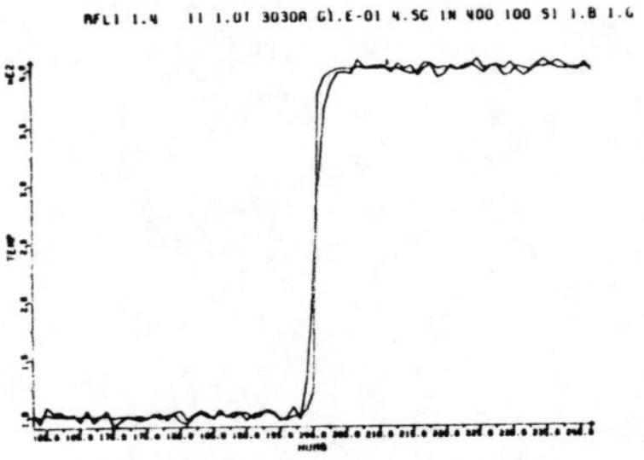

Fig. 12l $\gamma=10^{-1}$

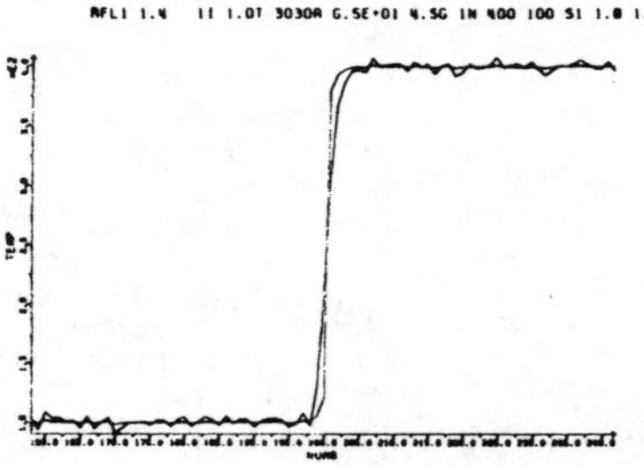

Fig. 12m $\gamma=5$

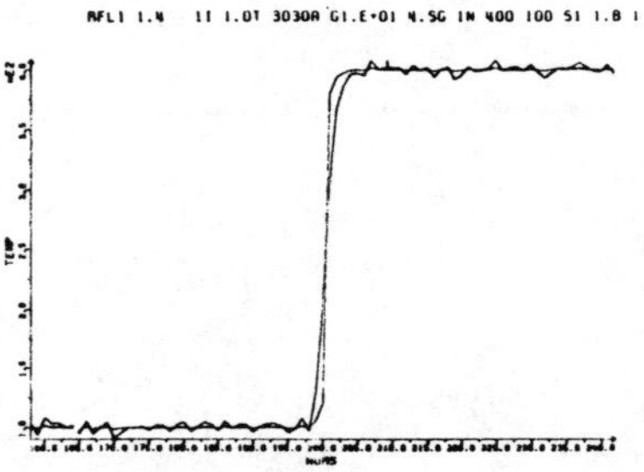

Fig. 12n $\gamma=10$

5. PRACTICAL DECONVOLUTION RESULTS

The capibilities of the algorithms, investigated on synthetic data, has been demonstrated with real data [4]. For this purpose, intensity profiles, taken with fixed antenna position, were used. Although the methods were applied to scenes, taken at 11 GHz, 32 GHz and 90 GHz, here only the deconvolution results of the 11 GHz band are presented. For the reasons indicated, only the multiple convolution and the maximum likelihood method, not the matrix inversion method, were verified. For the purpose of data reconstruction, only the segment of the antenna characteristic just over the profile was used.

Fig. 13 shows a boundary region of a water/land area. The deconvolved profile unveils features, that in the convolved signal cannot be recognized. The objects of interest are a group of trees at the rim of the waterbody and on area of minor snow coverage. In the bank region an area of relatively to the water surface elevated intensity can be identified, probably consisting of a thin ice cover over the water. As a significant drop appears the supposed water/ice boundary. The waterbody shows not the expected constance in radiation behaviour. This is definitely not due to statistical errors, but to stochastic variations of the radiation temperature.

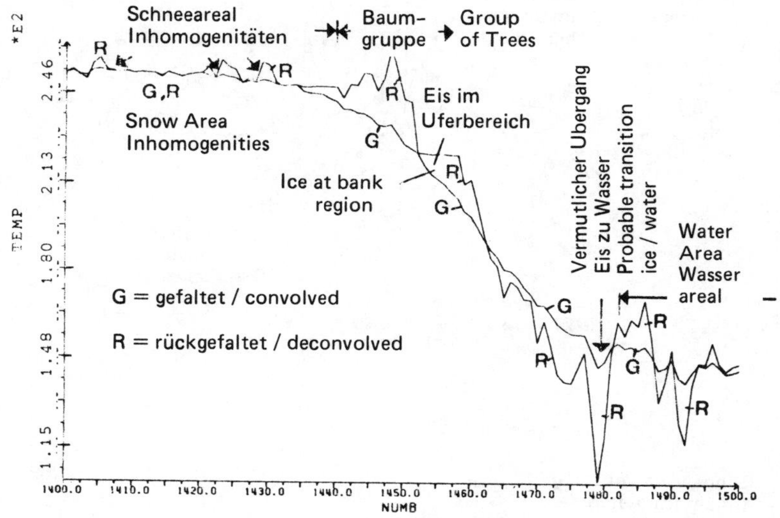

Fig. 13

Fig. 14 shows the radiation profile of an partial snow covered area. The flight path went over a street and a parking lot. On both sides of the street was no snow, so that this region appears at a higher radiation intensity than the snow covered area. The low intensity of the street is due to a ice cover. Of particular interest is a snow covered area at the left side of the profile. Part of this region was illuminated by the sun and is evidently covered with a thin film of melting water. It therefore appears, due to the higher degree of reflexion, at a lower radiation temperature than the region at the extreme left which was in the shadow region of a forest and had therefore a greater surface roughness. On the extreme right the radiation structure of a parking lot can be seen.

Fig. 15 shows the bank area of a lake. Of particular interest is here a small path which could not be seen in the convolved region. The path was only one picture elements wide and can in the convolved data not be recognized as an object different to snow. The deconvolved region also shows a small area of ice covered water area immediately right to the bank.

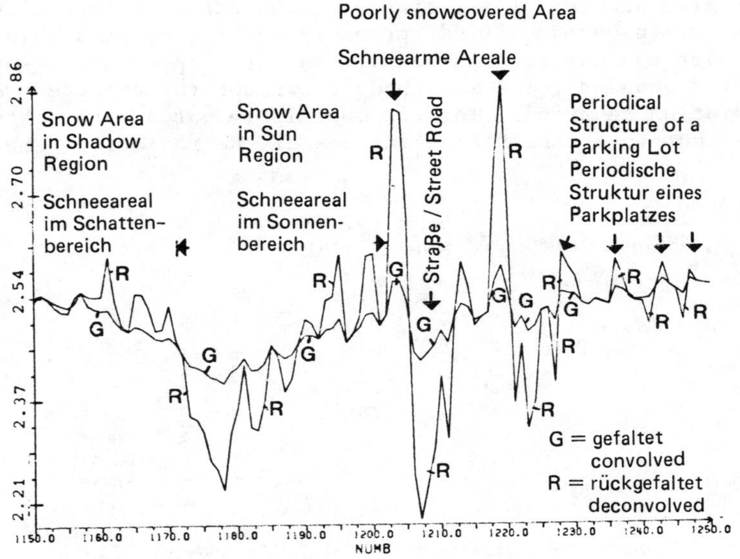

Poorly snowcovered Area
Schneearme Areale
Snow Area in Shadow Region
Schneeareal im Schatten-bereich
Snow Area in Sun Region
Schneeareal im Sonnen-bereich
Straße / Street Road
Periodical Structure of a Parking Lot
Periodische Struktur eines Parkplatzes
G = gefaltet convolved
R = rückgefaltet deconvolved
TEMP
*E2
NUMB

Fig. 14

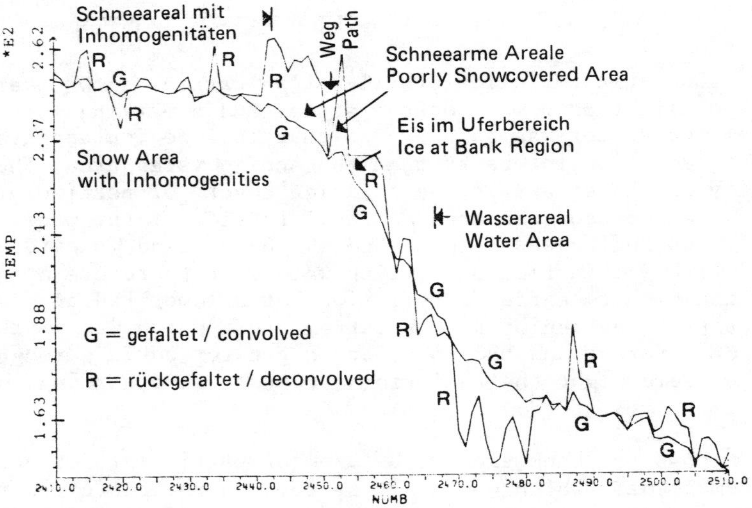

Schneeareal mit Inhomogenitäten
Weg Path
Schneearme Areale
Poorly Snowcovered Area
Eis im Uferbereich
Ice at Bank Region
Snow Area with Inhomogenities
Wasserareal
Water Area
G = gefaltet / convolved
R = rückgefaltet / deconvolved
TEMP
*E2
NUMB

Fig. 15

Fig. 16 shows a scene, where the flight path, crossing a water body, is approximating a land area. Although the central line of the flight profile is still over the water, the side lobes are already detecting the land area. Although the radiation pattern of an housing area is not visible in the convolved data and although the flight path did not go directly over this region, the patterns of the housing area appears as zig-zag line in the deconvolved profile. This example demonstrate the influence of objects outside a scene to the pattern at the rim region in a deconvolved image. Special care has to be taken at the interpretation of such phenomena.

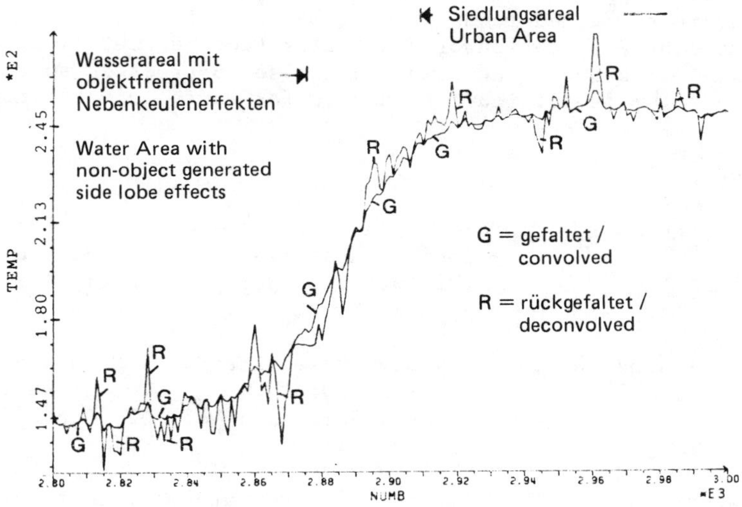

Fig. 16

6. SUMMARY

The investigation of three algorithms (multiple convolution, maximum likelihood, matrix inversion) for the deconvolution of microwave radiometry data showed that converging solution could (expect for the third algorithm) always be obtained by use of a-priori-knowledge on data properties. Here, global information, such as the dynamic range or the approximate treshold values, were already sufficient. At the same time the computer time could be reduced up to the factor 10 for the first algorithms. For the matrix inversion method, a comparison of the computer time is irrelevant, as the method consists only of the step of matrix multiplications and matrix inversions. This algorithm reacts very sensitive on any kind of a-priori-knowledge in such a way that it has a large tendency also towards wrong information. It is therefore not in any case applicable for earth scientific purposes.

The methods of deconvolution by multiple convolution and maximum likelihood obtain good results even than, if the geometrical resolution of a single event was in the size of 0.25 degrees for an antenna halfwidth of 2.8 degrees. A contrast enhancement to obtain the correct intensity profile up to the factor 5 could be observed. The needed computer time was thus, that real time deconvolution can be considered as possible.

A significant characteristics of the first and second algorithm was their capability to reconstruct high periodicity events which require high resolution gain, even for random distorted data. In these cases a-priori-information are, however, mandatory. The matrix inversion method tends to oscillatory effects, which could be reduced ba application of a stabilization parameter. At the same time all other fluctuations are also suppressed, so that this algorithm is difficult to handle for earth scientific purposes.

LITERATURE

1 Frieden, B.R. Image Enhancement and Restoration in: Topics in Applied Physics, Vol. 6 (Picture Processing and Digital Filtering), pp. 21-203, 227-229.

2 Richardson, W.H. Bayesian-Based Iterative Method of Image Restoration I. Opt. Soc. Am., Vol. 62, Jan. 1972, pp. 55-59.

3 Twomey, S. Introduction to mathematics of Inversion in Remote Sensing and Indirect Measurements (Development in Geomathematics, Vol. 3) Amsterdam: Elsevier, 1977, pp. 122-144.

4 Rudolf H. Dittel Basic Investigation on the feasibility of multispectral radiation information from microwave radiometry measurements for earth scientific objectives of remote sensing. Köln: DFVLR Forschungsbericht 83-27, 1983.

Data Source:
Remote Sensing Radiometry Measurements of DFVLR, 1974-1977.

III.1 (RP.1*)

POLARIZATION DESCRIPTORS. UTILIZATION OF SYMMETRIES

Georges A. Deschamps

Department of Electrical Engineering, University of Illinois, Urbana, Illinois 61801, USA.

ABSTRACT. This paper reviews the tools used for handling polarization problems in electromagnetics and shows how they can be used to take into account symmetries of a scatterer. Various representations of polarization states are described and related. They are applied to plane waves and to radiation fields due to an antenna or to scattering of a plane wave. Symmetries of a scatterer help in measuring its pattern. Conversely symmetries of the pattern can be a means of identification. A few general remarks are made about the bearing of polarization on inverse scattering problems.

0. INTRODUCTION

The first part of this paper consists of a brief description of several known representations of polarization states and their relations. It is mostly expository but it brings out a few subtle points that are frequently ignored. This presentation is of necessity brief and the reader should consult the references for more detailed discussions. The second part treats polarization transformations and their representation. The third part discusses various "pattern" descriptions of the far field of an antenna or of a scatterer. Part four treats the effect of symmetries of a scatterer on its scattering properties and the utilization of these symmetries in measurements and target identification problems. Part five offers a few general comments about the importance of polarization for inverse problems.

The reader should consult Appendix A for a selected bibliography and Appendix B for a list of some notations, most of them conventional, but a few peculiar to this paper.

W.-M. Boerner et al. (eds.), Inverse Methods in Electromagnetic Imaging – Part 1, 493–520.

1. POLARIZATION DESCRIPTORS AND THEIR RELATIONS

1.1 Concept of Polarization

Any field represented, at point r and time t, by a vector $\mathcal{V} \in \mathbb{R}^n$ can be specified by its magnitude $|\mathcal{V}|$ and its direction. The latter, taken alone, is sometimes called the polarization of vector $\mathcal{V}$. This is done in particular if the direction of the vector (vertical for instance) is independent of t and/or of r within some region of space: the field would be called "vertically polarized" over that region.

For a real vector field however it is hardly justified to introduce a special name synonymous of direction. In contrast, for a time harmonic field, $\mathcal{V}(t)$ at some point r describes in general an ellipse as a function of t, and therefore has a variable direction. Expressing $\mathcal{V}(t)$ as usual by

$$\mathcal{V}(t) = \mathrm{Re}\left[V \exp(-i\omega t)\right] \tag{1.1.1}$$

we consider the direction of the complex vector

$$V = V' + iV'' \in \mathbb{C}^n \tag{1.1.2}$$

as defining the polarization of $\mathcal{V}$.

Since V is complex an appropriate definition of direction (or parallelism) is required. We say that V_1 and V_2 have the same direction[1] if they differ only by a complex factor $z \in \mathbb{C}$

$$V_2 = zV_1 \tag{1.1.3}$$

which may also be expressed by

$$V_2 \in \mathbb{C}V_1 \tag{1.1.4}$$

where $\mathbb{C}V_1$ is the set of products of V_1 by all numbers $z \in \mathbb{C}$. The vectors $\mathcal{V}_1(t), \mathcal{V}_2(t)$ (and by 'abus de langage' V_1, V_2) are said to be in the same polarization state if V_1 and V_2 have the same direction.

[1]In other contexts parallelisms defined by $V_2 = \mathbb{R}V_1$ or $V_2 = \mathbb{R}_+V_1$ might be appropriate (WANG and DESCHAMPS, 1974). Those are not relevant to the present considerations.

In this article we shall consider only time-harmonic fields and very briefly quasi-harmonic and random fields. The time convention will be that of equation 1. Although most considerations apply to vectors in an n-dimensional linear space $\mathbb{C}^n$, whose elements may be collection of tensors,[2] we shall mainly discuss the case of an electric field E in $\mathbb{C}^2$ or $\mathbb{C}^3$.

To isolate a description of the polarization state of E one can at first divide E by its magnitude $|E|$. The magnitude $|E|$ is the positive square root of $E^*\!\cdot E$ and represents an energy density. The ratio

$$\hat{E} = E / |E| \tag{1.1.5}$$

is a unitary vector ($|\hat{E}| = 1$) which has the same polarization as E.

It has been called the "unit polarization vector". One should note that any vector $E\, e^{i\alpha}$ ($\alpha \in \mathbb{R}$) is also a unit polarization vector and one should take the set of these vectors as the proper description of polarization alone. Such sets have been called rays (not be confused with rays in geometrical optics) in the context of quantum mechanics where a ray defines the state of a quantum system [Bargman, 1964].

At this point a general remark is in order, namely that the techniques for handling polarization states are but a special case of techniques for handling quantum states [e.g. Fano, 1957]. The analogies are particularly clear with representations of the spinning electron by a pair of complex-valued wave functions.

1.2 Polarization Ellipse

Instead of considering the complex vector $V = V' + iV''$ one may examine the locus of

$$\mathcal{V}(t) = \mathrm{Re}\left[V \exp(-i\omega t)\right] = V' \cos\omega t + V'' \sin\omega t \tag{1.2.1}$$

It is easily seen that this point describes an ellipse $\mathcal{E}$. The characteristics of this ellipse that are common to all field vector having the same polarization are (1) the polarization plane Π defined by vectors (V', V''); (2) the direction of the major axis, which makes an angle β with some reference direction L (being an angle between two straight lines, the tilt angle β is defined

[2]For instance in a warm magneto plasma V may represent electric and magnetic field, the velocity field and the scalar pressure. Then $V \in \mathbb{C}^{10}$.

modulo π or can be chosen between 0 and π); (3) the axial ratio minor over major axis, which can be expressed as tan α with $0 < |\alpha| < \frac{\pi}{4}$; (4) the sense of travel, which is called + or - by comparison with a reference sense of rotation which makes Π into an oriented plane Π. This orientation of Π can also be specified by a vector V normal to Π and is called right-hand or left-hand according to the well known rule. If the vector E is multiplied by a complex number $m\,e^{i\alpha}$ the ellipse $\mathcal{E}$ is expanded in the ratio m and the vector $V(0)$ is moved on the ellipse so as to sweep an area $\alpha/2\pi$ times the total area of the ellipse. These transformations leave the characteristics (1) to (4) invariant. For the limiting case of linear polarization: $\alpha = 0$, Π and sense are undetermined, and for circular polarization: $\alpha = \pm \frac{\pi}{4}$, and tilt angle β is undetermined.

1.3 Poincaré Sphere

As shown for instance in [Deschamps, 1951] the Poincaré sphere $\mathcal{P}$ provides a one-to-one representation of states of polarization. Using geographic notations, and (β, α) defined in 1.2, the longitude of P, measured from a reference point L on the equator is 2β, the latitude of P is 2 where α has been given the sign $\pm$ according to the sense of polarization. The pairs

$$(\alpha, \beta) \in [-\tfrac{\pi}{4}, +\tfrac{\pi}{4}] \times [0, \pi] \qquad (1.3.1)$$

are mapped uniquely on $\mathcal{P}$. The limiting cases fall in place without disrupting the continuity.

The Poincaré sphere is at the crossroad of various polarization descriptions. A point $P \in \mathcal{P}$ is related to the slope (or polarization ratio) $W = Y/X$ by sterographic projection (1.4). It is related to 2x2 Hermitian matrices $H_o(2)$ as shown in (1.5). Then through a map $H_o(2) \to \mathbb{R}^4$ to a vector space, in 4 dimension endowed either with the Minkowski metric or, for some computations with the Euclidean metric (see Stokes parameters in 1.5).

The main properties of Poincaré sphere are: 1) It provides a one-to-one representation of the states of polarization defined by the polarization ellipse. 2) It relates through formulas of spherical trigonometry various parameters of this ellipse: tilt, axial ratio, polarization ratio, etc. 3) It displays a "natural" measure of distance between polarization states P_1 and P_2, namely the arc $P_1 P_2$. 4) Linear transformation of a vector field (with values on $\mathbb{C}^2$) correspond to inversions of the sphere. 5) When these transformations are lossless their matrices belong to SU(2) which correspond to rotation O(3) of the sphere $\mathcal{P}$.

An inner product between polarization states (or quantum system states) can be defined in general when ψ_1 and ψ_2 belongs to $\mathbb{C}^n$ or to some Hilbert space where the inner product $(\psi_1, \psi_2) = \psi_1^* \cdot \psi_2$ [Bargman, 1964]. This is done by setting

$$P_1 \cdot P_2 = |(\psi_1, \psi_2)| \tag{1.3.2}$$

where ψ_1, ψ_2 are unitary vectors [their phases do not matter]. For $n = 2$ it can be shown [Deschamps, 1951] that this product, if P_1 and P_2 are the representative points on the Poincaré sphere, is

$$P_1 \cdot P_2 = \cos \delta \tag{1.3.3}$$

if 2δ is the arc length of P_1P_2. This gives the magnitude of the orthogonal projection of state ψ_2 along state ψ_1. The square of this quantity

$$\cos^2 \delta = \frac{1}{2}(1 + \cos 2\delta) \tag{1.3.4}$$

can be recovered by using properties of the Hermitian matrix representation (1.5)

$$(P_1 \cdot P_2)^2 = \psi_1^+\psi_2 \psi_2^+\psi_1 = \mathrm{tr}(\Psi_2\Psi_1) = \frac{1}{2} \vec{\Psi}_1 \cdot \vec{\Psi}_2 \tag{1.3.5}$$

The last term is half the Euclidean scalar product of the Stokes' vectors $\overline{\Psi}_1\ \overline{\Psi}_2$

$$\vec{\Psi}_1 \cdot \vec{\Psi}_2 = I_1I_2 + M_1M_2 + C_1C_2 + S_1S_2 = 1 + \vec{OP}_1 \cdot \vec{OP}_2 \tag{1.3.6}$$

which recovers (1.3.4). When $\delta = \frac{\pi}{2}$ the vectors ψ_1, ψ_2 are orthogonal or cross polarized. The points P_1 and P_2 are antipodal in $\mathcal{P}$.

1.4 Polarization Ratio

Let the complex vector E in a plane Π be expressed as

$$E = X\bar{x} + Y\bar{y} \tag{1.4.1}$$

in terms of its coordinates

$$\psi = \begin{bmatrix} X \\ Y \end{bmatrix} \in \mathbb{C}^2 \tag{1.4.2}$$

and a reference frame $\bar{f}$ consisting of two (non parallel) vectors in

$$\bar{f} = (\bar{x}, \bar{y}) \tag{1.4.3}$$

Equation (1) may be written as a matrix product of (1.4.3) by (1.4.2)

$$E = \tilde{f}\psi \tag{1.4.4}$$

A natural way of expressing the direction, hence the polarization, of E is to form the polarization ratio

$$W = Y/X \tag{1.4.5}$$

This a faithfull (one-to-one) representation of the polarization state of E (provided $W = \infty$ is added to the set **C** of complex numbers). Although $\bar{x}$ and $\bar{y}$ are arbitrary (non parallel) and may be complex it is convenient to take them as orthonormal and even real. A change of reference frame

$$\tilde{f}' = \tilde{f}M \tag{1.4.6}$$

where $M = GL(2,C)$ corresponds to the change of coordinates

$$\psi' = M^{-1}\psi \tag{1.4.7}$$

so that

$$E = \tilde{f}'\psi' = \tilde{f}\psi \tag{1.4.8}$$

The effect of M on the polarization ratio is of the form

$$W \to W' = \frac{aW + b}{cW + d} \tag{1.4.9}$$

if

$$M^{-1} = \begin{bmatrix} d & c \\ b & a \end{bmatrix} \tag{1.4.10}$$

Given the polarization ratio W in an orthonormal frame, a unitary vector having the same polarization as E may be taken as $\tilde{f}\psi_W$ with

$$\psi_W = \frac{1}{1 + |W|^2}\begin{bmatrix} 1 \\ W \end{bmatrix} \tag{1.4.11}$$

Of course the phase of E is lost in the construction, that of ψ_W has no particular meaning.

The transition from polarization ratio W to a point P on Poincare sphere $\mathcal{P}$ is through a stereographic projection from the pole P_x of $\mathcal{P}$ that represents the linear polarization $\bar{x}$ ($W = \infty$) onto the equatorial plane. The projection is one-to-one, conformal, transforms circles into circles. The cross-ratio of any four points equals that of their images.

1.5 Representation by Hermitian Matrices

Let $\psi = \begin{bmatrix} X \\ Y \end{bmatrix} \in \mathbb{C}^2$ be the representation of the complex electric field $\vec{E}$ in an orthonormal reference frame $\bar{f}$. We associate to ψ the 2x2 matrix

$$\Psi = \psi\psi^+ = \begin{bmatrix} XX^+ & XY^+ \\ YX^+ & YY^+ \end{bmatrix} \tag{1.5.1}$$

This matrix is Hermitian

$$\Psi = \Psi^+ \in H(2) \tag{1.5.2}$$

and has zero determinant

$$\Psi \in H_0(2) \tag{1.5.3}$$

Any Hermitian 2x2 matrix $Q \in H(2)$ can be expressed in terms of 4 real numbers I,M,C,S called Stokes' parameters, by letting

$$Q = \frac{1}{2}\begin{bmatrix} I + M & C - iS \\ C + iS & I - M \end{bmatrix} \tag{1.5.4}$$

This defines a map

$$\mu : \quad H(2) \rightarrow \mathbb{R}^4 : \quad Q \mapsto \vec{Q} \tag{1.5.5}$$

We have

$$\det Q = \frac{1}{4}(I^2 - M^2 - C^2 - S^2) = \frac{1}{4}\check{Q}^2 \tag{1.5.6}$$

where $\check{Q}$ is the "length" of vector $\vec{Q}$ according to Minkowski metric. A vector $\vec{Q}$ corresponding to a matrix $Q \in H_0(2)$ is on the positive light cone Γ_+.

$$I^2 = M^2 + C^2 + S^2 \tag{1.5.7}$$

The trace of Q, $I = \psi^+\psi = X^2 + Y^2$ is positive.

If ψ is a unitary vector, I = 1 and (M,C,S) are precisely the coordinates of the point P on Poincare sphere $\mathcal{P}$ that represent the polarization of ψ. We note that Ψ, hence Ψ = (I,M,C,S), are insensitive to a change of phase.

The representation by matrices in H(2) is particularly convenient for quasi-harmonic field vectors (i.e. those that occupy a narrow frequency band about ω), or statistical mixtures of purely polarized waves. The average matrix $\langle\Psi\rangle$ becomes a sum of pure state matrices $\langle\psi\rangle$ with positive weight ρ_i whose sum is 1 (ρ_i = probability of, or relative time spent in, state ψ_i). The vectors $\langle\overline{\Psi}_i\rangle$ are combined in the same manner hence $\langle\overline{\Psi}\rangle$ lies inside the cone Γ_+. The point P of coordinate $\frac{1}{I}$ (M,C,S) lies inside the Poincaré sphere. See Deschamps and Mast [1973]. Similar considerations apply for $\psi \in \mathbb{C}^n$ or for wave functions in quantum mechanics [Fano, 1957].

A useful property of the Hermitian matrices Q and associated vectors $\overline{Q} \in \mathbb{R}^4$ is that they form a Hilbert space through the two equivalent definition of a scalar product. In H(2)

$$Q_1 \cdot Q_2 = \mathrm{tr}\,(Q^{1\,T} Q_2) = \mathrm{tr}\,(Q_1{}^* Q_2) = \Sigma\, Q_1{}^{ij} Q_2{}^{ij} \quad (1.5.8)$$

This equals half the Euclidian scalar product of the corresponding vectors

$$Q_1 \cdot Q_2 = \frac{1}{2}\,\overline{Q}_1 \bullet \overline{Q}_2 \quad (1.5.9)$$

This may be applied to Hermitian forms of a vector $\mathbb{C}^2$

$$\psi^+ Q \psi = \mathrm{tr}(\psi^+ Q \psi) = \mathrm{tr}(Q\psi\psi^+) = \mathrm{tr}(Q\Psi) \quad (1.5.10)$$

2. EFFECT OF A LINEAR TRANSFORMATION OF A FIELD VECTOR ON ITS POLARIZATION

We shall consider only the case of fields in a plane since they are most important from a practical point of view. This analysis applies to the situation where a plane wave incident on a scatterer produces at large distances, in some direction, a scattered field which is locally a plane wave.

Another example of linear transformation of a vector in C^2 amenable to the same analysis, is provided by the junction of two wave guides each supporting two propagating modes. If (X,Y) are the (complex) amplitudes of these modes incident on one side of the junction while (X',Y') are the amplitudes of the transmitted modes emerging from the other side we have a relation

$$\begin{bmatrix} X' \\ Y' \end{bmatrix} = \begin{bmatrix} a & b \\ c & d \end{bmatrix} \begin{bmatrix} x \\ y \end{bmatrix}, \quad \psi' = T\,\psi \quad (2.1.1)$$

where T is part of the junction scattering matrix (2.1.2a).

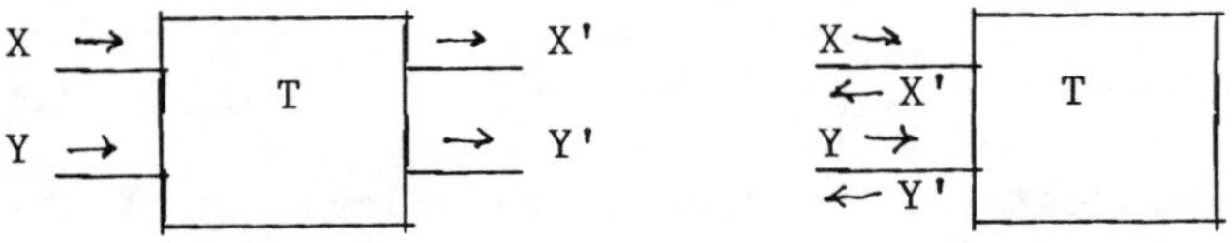

(2.1.2)

a. Junction b. Termination

A similar relation describes the reflection at the end of a two modes waveguide (2.1.2b).

These networks are an exact model of a linear field transformation. They are most convenient when the modes are chosen to represent linear polarizations.

The description of the field E by its components $\begin{vmatrix} X \\ Y \end{vmatrix}$ which form a vector $\psi \in \mathbb{C}^2$ depends of course on the choice of a reference frame $\bar{\bar{f}} = (\bar{x}, \bar{y})$

$$E = \bar{f}\psi = X\bar{x} + Y\bar{y} \qquad (2.1.3)$$

and similarly for the transformed field E' leading to the representation (2.1.1)

$$\psi \to \psi' = T\psi \qquad (2.1.4)$$

A change of reference frame where $E \to E' = E$ leads to a relation of the same type between the coordinates (see 1.4.7).

We shall consider the effect of T on the field polarization in its various representations - namely the slope W = Y/X, the point P on the Poincaré sphere $\mathcal{P}$, the matrix $\Psi = \Psi^+$, and the Stokes parameters (I,M,C,S) components of a vector $\bar{\Psi}$ in $\mathbb{R}^4$.

To the matrix T in (2.1.1) corresponds the fractional linear (or Möbius) transformation for the slope W.

$$\tilde{T}: \mathbb{C} \to \mathbb{C}: W \mapsto W' = \frac{dW + c}{bw + a} \qquad (2.1.5)$$

(Here $\mathbb{C}$ must be understood as the complex axis completed by a point at infinity.) The transformation T is the projective aspect of

$$T: \mathbb{C}^2 \to \mathbb{C}^2: \psi \mapsto \psi' \qquad (2.1.6)$$

It is a conformal map that transforms circles into circles. Passing to the Poincaré sphere

$$C \to \mathcal{P}: \quad W \mapsto P \tag{2.1.7}$$

by stereographic projection, leads to a transformation of the Poincaré sphere onto itself

$$T: \mathcal{P} \to \mathcal{P} : P \mapsto P' \tag{2.1.8}$$

which is also circular, conformal, and preserves the cross-ratio of four points. All the T form the group $\mathcal{I}$ of inversions of the sphere $\mathcal{P}$.

The group $\mathcal{I}$ operating on $\mathcal{P}$ can be extended to a group $\bar{\mathcal{I}}$ operating on the Poincaré space $\bar{\mathcal{P}}$ i.e., the space $\mathbb{R}^3$ into which $\mathcal{P}$ is imbedded. This is done by associating with each circle on $\mathcal{P}$ the plane that contains it. A point in $\mathcal{P}$ is then associated with the family of circles whose planes contains it. Transformations in $\bar{\mathcal{P}}$ are projective: they carry straight line into straight lines, planes into planes, preserve the cross ratio. Furthermore they transform the sphere $\mathcal{P}$ into itself.

If we identify the space $\bar{\mathcal{P}} = \mathbb{R}^3$ with the hyperplane $I = 1$ in the space $\mathbb{R}^4$ of Stokes vectors (I,M,C,S) every point of $\mathcal{P}$ defines a line through the origin and a unit vector (with Minkowski metric) along that line. In this manner it is seen that the Lorentz group, which leaves $I^2 - M^2 - C^2 - S^2$ invariant, is isomorphic to the group $\mathcal{I}$ and its extension $\bar{\mathcal{I}}$.

Since there is a one-to-one correspondence between Hermitian matrices and points of the space $\mathbb{R}^4$ (see 1.5) this tells us how the transformation T acts upon the Hermitian matrices $\Psi = \psi\psi^+$, that describe states of polarization.

A direct representation of transformation T acting of $\psi \in \mathbb{C}^2$ by a transformation T acting on $\Psi = \psi\psi^+ \; H_0(2)$ is

$$T: \Psi \to \Psi' = \psi'\psi'^+ = T\psi\psi^+T^+ = T\Psi T^+ \tag{2.1.9}$$

This extends naturally from $H_0(2)$ to $H(2)$ being linear with respect to Ψ. If T is unitary $T^+ = T^{-1}$

$$\Psi' = T\Psi T^{-1} \tag{2.1.10}$$

In general Ψ being expressed in term of IMCS by

$$\Psi = \frac{1}{2} (I\sigma_0 + C\sigma_1 + S\sigma_2 + M\sigma_3) \tag{2.1.11}$$

where the σ_i are Pauli matrices, the coefficients of the matrix that acts on the Stokes vector result from the $T\Psi T^+$ expanded in the basis (σ_i) [Mueller 1943, Soleillet, 1929].

3. FAR FIELD PATTERNS OF AN ANTENNA OR A SCATTERER

3.1 Plane Electromagnetic Waves

In an homogeneous isotropic medium a plane electromagnetic wave is defined by its wave vector k and its electric field E(0) at some reference point taken as the origin. This field is a unit plane wave if E(0) is a unitary vector u. Using subscripts 1, 2, .., i, .. to distinguish various unit waves we shall say that (u_1,k_1) defines the plane wave $|1\rangle$ whose components at point r are

$$\langle r | 1\rangle = \begin{bmatrix} E_1 (r) \\ H_1 (r) \end{bmatrix} = \begin{bmatrix} u_1 \\ Y_0 \; k_1 \times u_1 \end{bmatrix} \exp (k_1 \cdot r) \tag{3.1.1}$$

where $Y_o = (\varepsilon/\mu)^{1/2}$ is the medium admittance and $k_1 = k_1/k_0$ is the unit vector in direction of k_1. All fields $\mathbf{C}|1\rangle$ are said to be in the same state of direction and polarization, u_1 may be any point of the ray $\mathbf{C}\, u_1$. If u_1 is chosen equal to $E(0)/|E(0)|$ it will give besides the polarization, some phase information by telling that the direction of the real field at time 0 is that of Re u_1. Note that the phase "information" carried by a choice of u_1 is not expressible by a number. In other words it is not possible to give a natural meaning to the phase difference between two field vectors having different polarizations. The phase information becomes important when interferences are considered. It cannot be separated from the polarization information. Any planewave in state $|1\rangle$ being $a|1\rangle$ with $a = Ae^{i\alpha}$ the plase α can be absorbed in whole a part by u_1

$$Ae^{i(\alpha-\phi)} | u_1 e^{i\phi}, k_1\rangle \tag{3.1.2}$$

3.2 Far Field Patterns of an Electric Current Distribution

Let us assume that the antenna or the scatterer are of finite extent and that their far fields (for a given excitation) is that of some equivalent current distribution J(r) in free space. The radiated field at point kr (distance r is direction of some wave vector k) is given by

$$E(\hat{k}r) = i k_0 Z_0 G(r) p(k) \tag{3.2.1}$$

where $k_o = 2\pi/\lambda$, Z_0 is the free space impedance, $G(r) = (\exp ik_0 r)/4\pi r$ and p is the current moment, or strength, of a equivalent electric dipole. This dipole depends on k as follows: it is Fourier transform $\tilde{J}(k)$ of J(r), restricted to the sphere S: $|k| = k_0$ and projected on the plane $T_k S$ tangent to the sphere S at point k.

The far field pattern is completely described by the complex vector field p(k) over the sphere S. One can also write (3.2.1) in the form

$$E(\hat{k}r) = \sqrt{Z_0}(e^{ikr}/r)A(k) \tag{3.2.2}$$

where

$$A(k) = \sqrt{Z_0}\,\frac{ik_o}{4\pi}\,p(k) \tag{3.2.3}$$

is now such that $|A|^2$ is the radiation intensity in direction of vector k.

Expressions similar to (3.2.1) (3.2.2) also hold when the source of the field contains magnetic currents K(r). Then the contribution of these currents to p(k) is obtained by forming the Fourier transform K(k) evaluated on the sphere S, projecting it on the tangent plane $T_k S$, multiplying it by the admittance Y_0 and rotating by $-\frac{\pi}{2}$ about k. Thus

$$p(k) = -Y_0\,\hat{k} \times K(k) \tag{3.2.4}$$

3.3 Radiation Intensity and Phase Patterns

Sometimes a complete description of the far field by a complex vector such as p(k) or A(k) is not needed. This is the case if we desire only to know the radiation intensity $|A|^2$ (power radiated per unit angle) as a function of direction. This lead to concepts of gain and directivity.

If we attempt to describe the phase by a function of direction we encounter the difficulties already mentioned (3.1). There is no naturally defined number that can be called the phase of the polarized field or that can be called phase difference between two field vectors of different polarizations. Polarization and phase information are linked in the unit vector $\hat{E} = E/|E|$.

One may however consider phases of a pair of components of the field having specified polarization (linear for convenience) and plot two phase patterns. The problem of specifying reference polarizations for different direction is considered in 3.4 and 3.5.

3.4 Polarization Patterns - The Poincaré Bundle

It is not easy to measure a complete scattering or radiation pattern in magnitude, polarization, and phase. The phase is particularly elusive while the intensity measured through a polarizor allows to extract easily the polarization. This is essentially a measurement of Stokes parameters or of a set of Hermitian forms which are represented in Poincaré space (The space $\mathbb{R}^3$ that contains the Poincaré sphere $\mathcal{P}$) by distances to a set of planes. See Deschamps [1952, 1953], Deschamps and Mast [1973]. Of course we encounter here a difficulty already discussed: How to compare the polarization of non coplanar fields?

Using the Poincaré sphere representation we should plot the point P on $\mathcal{P}$ as a function of the vector $k \in S$. However for two different directions k_1 and k_2 the field vectors p_1 and p_2 lie in different planes $\Pi_1 = T_{k_1}S$ and $\Pi_2 = T_{k_2}S$. Thus we need a convention to transcribe polarizations in Π_1 and in Π_2 to a single sphere $\mathcal{P}$.

This convention, as described in (1.2) (1.3) consists of attaching to every point $k \in S$ an oriented plane $\overline{\Pi}$ (This will be the plane T_kS oriented by the normal vector k), and a reference direction L (a particular linear polarization lying on Π). On the sphere we choose a point L representing polarization L and we use the hemisphere $\mathcal{P}_+$ for ellipses travelled in the sense associated with k by the right hand rule. In brief we can state that the convention requires a field of straight lines tangent to the sphere S. The set of all such straight lines form a manifold PS which is known as a bundle over the base space S.

$$\begin{array}{ccc} L & \in & PS \\ \pi\downarrow & & \downarrow\pi \\ k & \in & S \end{array} \qquad (3.4.1)$$

To each $L \in PS$ corresponds a projection πL on S which is the point k where L touches S. All lines tangent to S at point k form the fiber over k. The field of reference lines or polarizations is in fact a section of the bundle i.e., a map

$$L_0 : S \to PS : k \mapsto L_0(k) \qquad (3.4.2)$$

such that $\pi \circ L_0$ is the identity over S.

We have here a simple example of the bundle concept which has come to play an important role in physics. It should be noted that PS is not a cartesian product of the base space S by a fiber (projective space P(1)). The field L_0 is not a map $S \to P(1)$. No such continuous map exists because "one cannot comb a sphere!" The bundle is not trival. To handle this situation one can cover the sphere by two overlapping spherical caps S_a and S_b. The bundle is trivial over each of these caps and some transition function relates the reference directions in the overlap region.

A variant of this construction would use Poincaré spheres as fibers leading to the Poincaré bundle $\mathcal{P}S$, a structure simply related to PS. A polarization pattern is a section of $\mathcal{P}S$.

3.5 Concept of Copolarization

Two vectors that have the same polarization are said to be copolarized (or copolar). This concept creates no problem when applied to field vectors that lie in the same (oriented) plane $\tilde{\pi}$. However this term is more currently used with reference to a polarization pattern (3.4) where statements are made about the "sameness" of the field polarization in different directions, i.e., for fields that lie in different planes Π. We have discussed in (3.4) the necessity of specifying a field of straight lines L (section of the bundle LS) which serves as reference linear polarizations and determines how polarizations, for each particular direction, are plotted on a single Poincare sphere. Then two fields are copolar if they are plotted at the same point. This insure that they have the same axial ratio and the same sense of travel (relative to the oriented plane Π). But their tilt angles have to be measured from some reference line $L_0(k)$ which depend on the direction of propagation, described by the wave vector k. The function

$$L_0 : S \to LS : k \mapsto L_0(k) \qquad (3.5.1)$$

is a section of the bundle LS. It is largely arbitrary, except for a smoothness condition, and there is no "natural" way to choose it. There would be no problem if some parallelism could be defined on the sphere S but this unfortunately is impossible. One must resort to some arbitrary convention such as taking direction tangent to θ (or ϕ) circles in spherical coordinates. Even this cannot be done globally on the sphere as there exist no continuous field of vectors on S. One might think of using parallel transport of the vector u_1 at k_1 to the vector u_2 at k_2. The result however would depend on the path from k_1 to k_2. It seems natural to chose the shortest path i.e., an arc of great circle joining k_1 to k_2 and to call u_2 copolar to u_1.

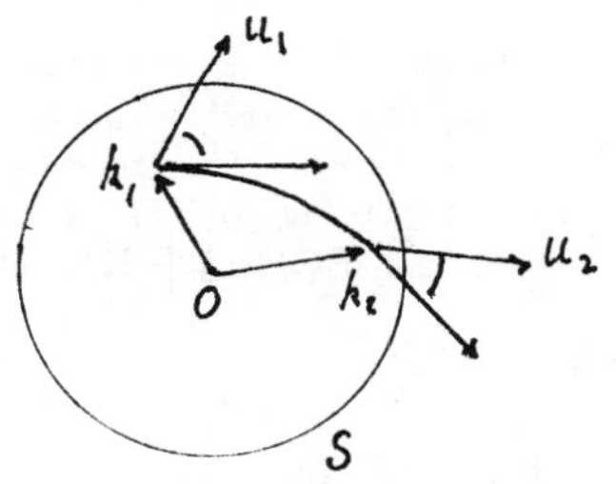

Then u_1 is also copolar to u_2. The relation is reflective. To extend it to directions in the main beam of an antenna one can choose k_1 along the axis of the beam, direction of maximum radiation or direction of interest for other particular reason. We take u_1 arbitrary as defining $L_0(k_1)$. For any direction k_2 within the beam, we construct $L_0(k_2)$ as support of u_2 (k_2). This construction can also be described by considering the stereographic projection of the vicinity of k_1 from the point k_2 on the sphere then the direction of u_2 projects on a parallel to the direction of u_1. It can be shown, with a bit of trigonometry, that this is essentially the solution proposed by Ludwig [1973]. Ludwig emphasizes the concept of cross polarization which in fact is subordinate to that of copolarization: $|u_2,k_2\rangle$ is cross polarized to $|u_1,k_1\rangle$ if it is cross polarized to $|u_3,k_2\rangle$ and u_3 is copolarized to u_1. Ludwig's solution is reasonable that it is good in the vicinity of k_1 where the sphere does not differ much from its tangent plane. It is made definite by the particular choice of k_1 and has the basic flaw that if $|u_2,k_2\rangle$ and $|u_3,k_3\rangle$ are copolar to $|u_1,k_1\rangle$, $|u_3,k_3\rangle$ is not copolar to $|u_2,k_2\rangle$ (unless k_1 k_2 k_3 are coplanar).

3.6 Scattering Operator

The scattering operator, or scattering matrix, of a target defines the scattered far field due to any incident plane wave and by integration, the field due to any incident field expanded into plane waves.

Consider first the incident plane waves $|E_1,k_1\rangle$ with a fixed wave vector k_1 and the scattered far field in direction k_2. This field can be expressed as discussed on 3.2 by (3.2.1) or (3.2.2). At point k_1 and in its vicinity this field is a plane wave $|E_2,k_2\rangle$, where E_2 depends lineary on E_1. Thus

$$E_2 = \bar{\bar{S}}(k_2,k_1)\, E_1, \tag{3.6.1}$$

Choosing reference frames in planes $\Pi_1 \perp k_1$ and $\Pi_2 \perp k_2$, $\bar{\bar{S}}$ is represented by a 2x2 matrix. The effect of (3.6.1) on the polarization were discussed in (2).

An alternative description of the scattering is to take a unit incident field $|1\rangle = |u_1,k_1\rangle$ and a unit reference scattered field $|2\rangle = |u_2,k_2\rangle$. The scattered field due to $|1\rangle$ being $E_S = \bar{\bar{S}}(k_2k_1)u_1$, its component along state 2 will be $u_2^* \cdot E_S$. The scattered field in state $|2\rangle$ is $S(2,1)\,|2\rangle$ with

$$S(2,1) = u_2 \cdot \bar{\bar{S}}(k_2,k_1)u_1 \qquad (3.6.2)$$

Varying $|1\rangle$ and $|2\rangle$ overthe manifold of plane wave states, S is the scalar-valued function

$$S: \mathcal{M} \times \mathcal{M} \to \mathbf{C}: (i,j) \mapsto S(i,j) \qquad (3.6.3)$$

4. UTILIZATION OF SYMMETRIES

This part of the paper analyzes the effect of symmetries of the scatterer on its scattering properties. The general principle that symmetries of the causes imply symmetries of the effects (P. Curie) is applied to the scattering problem direct and inverse. Knowing some symmetries of the scatterer leads to some simplification in the measurement of its scattering operator. Conversely observing the symmetries of the scattering operator leads to a presumption of corresponding symmetries of the scatterer. Unfortunately, one cannot draw this conclusion with certainty. This is shown by simple counter-examples, e.g. nonidentical cross dipoles may scatter as if they were identical. Symmetries of the scattering operator however remain a means of identifying or classifying targets.

The idea of making use of symmetries in fairly obvious, and is used extensively in Physics and Engineering. In relation to the scattering problem see a recent paper by Davidovitz and Boerner [1983].

The present discussions could be considered as an extension of this paper to arbitrary groups of isometries.

4.1 Symmetries of the Scatterer

A symmetry is defined by a transformation g of space onto itself which carries a point r of coordinates (x,y,z) to the point r' = gr of coordinates (x',y',z'). Thus:

$$g: \mathbb{R}^3 \to \mathbb{R}^3: r \mapsto r' = gr: (x,y,z) \mapsto (x'y'z') = g(x,y,z) \qquad (4.1.1)$$

We use the same letter g for the map $r \mapsto r'$ and for the associated transformation of coordinates. To be strictly correct one should introduce the chart

$$\phi : r \mapsto (x,y,z) \tag{4.1.2}$$

from points to their coordinates. The coordinates would then transform according to

$$\bar{g} = \phi \circ g \circ \phi^{-1} \tag{4.1.3}$$

which is represented by the commutative diagram

$$\begin{array}{lccc} \text{Points} & r & \xrightarrow{\;g\;} & r' \\ & \downarrow & & \downarrow \\ \text{Coordinates} & (x,y,z) & \xrightarrow{\;\bar{g}\;} & (x'y'z') \end{array} \tag{4.1.4}$$

We shall omit the overbar, relying on the context to tell us which is meant, g or $\bar{g}$.

The transformation g is extended naturally to a set of points A by taking gA as the set of the transforms (or images) gr of all points $r \in A$.

If a scatterer is a perfect conductor occupying the volume V, its image occupies the volume gV. The image of its boundary $\partial V = S$ is the boundary of its image $\partial(gV)$.

$$\begin{array}{ccc} V & \xrightarrow{\;g\;} & V' \\ \partial \downarrow & & \downarrow \partial \\ S & \xrightarrow{\;g\;} & S' \end{array} \tag{4.1.5}$$

The scatterer admits the symmetry g, or g in a symmetry of the scatterer, if V is invariant under g:

$$gV = V \tag{4.1.6}$$

Then in view of (4.1.5)

$$gS = S \tag{4.1.7}$$

A scatterer may be a region of finite extent surrounded by free space and characterized by scalar functions ε, μ, σ that reduce to ε_o, μ_o, 0 outside of the scatterer. We like to extend g to these functions. Let f be any scalar-valued function of point r in space, the action of g on the function f.

$$g: f \mapsto f' = (gf) \qquad (4.1.8)$$

is defined by

$$f'(r) = f(g^{-1}r) \qquad (4.1.9)$$

This is the <u>pullback</u> of functions f under g^{-1} and could be denoted by

$$f' = (g^{-1})^* f \qquad (4.1.10)$$

or when g is a diffeomorphism by the <u>push forward</u>

$$f' = g_* f \qquad (4.1.11)$$

See Deschamps [1981].

The scatterer defined by ε, μ, σ <u>admits the symmetry</u> g when these three scalar functions are invariant under g, e.g. $(g\varepsilon) = \varepsilon$ (or with the push forward notation $g_*\varepsilon = \varepsilon$). Note that in these formulas ε, $(g\varepsilon)$, $g_*\varepsilon$ are functions, not numbers.

4.2 Isometries

The symmetries that are of interest in this paper are those that leave Maxwell's equations invariant. It can be shown that only isometries, i.e. transformations that preserve distances, have this property. For any two points r_1, r_2 and the isometry g

$$|gr_1 - gr_2| = |r_1 - r_2| \qquad (4.2.1)$$

Isometries also preserve angles. They map straight lines and planes respectively into straight lines and planes. Isometries comprise translations, rotations and reflections through a point or a plane. They form a group $\mathcal{J}$. Those that leave a point 0 invariant are represented by orthogonal 3x3 matrices (in Euclidean coordinates). They form a subgroup of $\mathcal{J}$ designated by O(3). By definition an orthogonal matrix equals the inverse of its transpose. Consequently its determinant equals ± 1. The matrices in O(3) having determinant +1 are called positive (or proper). They form the subgroup SO(3). The other isometries, for instance reflections through a plane or a point have determinant -1 and are called negative or improper.

It is obvious that the symmetries of a scatterer, or of any object, form a group. If these symmetries are isometries, and if the object is of finite extent, it can be shown that this group G leaves some point 0 invariant. Thus G is a subgroup of the group O(3). These groups have been extensively studied and classified, in crystallography. An easy and entertaining introduction of this topic is a set of lectures by H. Weyl [1952].

4.3 Isometric Transformations of a Vector Field

A vector field V is a function

$$V: \mathbb{R}^3 \rightarrow T\mathbb{R}^3: r \mapsto V(r) \tag{4.3.1}$$

where $T\mathbb{R}^3$ the tangent bundle to $\mathbb{R}^3$ is essentially $\mathbb{R}^3 \times \mathbb{R}^3$.

An arbitrary smooth transformation g of $\mathbb{R}^3$ in $\mathbb{R}^3$ can be extended to vectors ($T\mathbb{R}^3 \rightarrow T\mathbb{R}^3$) by considering them as directional derivatives or infinitesimal vectors. If $g: r \mapsto r'$ and V is a vector at r, V' at r' = gr depends linearly on V via the Jacobian matrix of g

$$g_*: V \mapsto V' = MV \tag{4.3.2}$$

When g is an isometry - which is the case of interest here - the matrix M is the same as that which describes g. Thus the vector field may be considered as the geometrical object, a collection of vectors each attached to a point r of space $\mathbb{R}^3$. The action of g on V is to produce a field V' = (gV) whose value at point r is

$$V'(r) = (gV)(r) = g \circ V \circ g^{-1} r \tag{4.3.3}$$

Thus

$$V' = (gV) = g \circ V \circ g^{-1} \tag{4.3.4}$$

The parenthesis about (gV) is to avoid confusion with the composition $g \circ V$.

4.4 Action of Isometries on an Electromagnetic Field

Consider the field F defined by the pair of an electric field E and a magnetic field H. In a source free region these two vectors satisfy Maxwell's equation

$$\begin{aligned} \text{curl } H + i\omega\varepsilon \, E &= 0 \\ \text{curl } E - i\omega\mu \, H &= 0 \end{aligned} \tag{4.4.1}$$

It is readily verified that if one applies an isometry g to the curl of a vector field V one obtains $\pm$ the curl of (gV)

$$(g \text{ curl } V) = \pm \text{ curl } (gV) \tag{4.4.2}$$

according to the sign of the isometry.

If ε and μ are invariant and if g is a positive isometry the system 4.3.1 remains unchanged when E,H are replaced by E' = (gE) and H' = (gH). When g is a negative isometry however the system becomes

$$\begin{aligned} \text{curl } H' - i\omega\varepsilon E' &= 0 \\ \text{curl } E' + i\omega\mu H' &= 0 \end{aligned} \qquad (4.4.3)$$

The field (E',H') is not Maxwellian. To remedy this situation we define the action of g on F by

$$F \to F': \begin{bmatrix} E \\ H \end{bmatrix} \mapsto \begin{bmatrix} E' = (gE) \\ H' = -(gH) \end{bmatrix} \qquad (4.4.4)$$

If the field F satisfies the boundary conditions $E_{tan} = 0$ at the surfaces of a conducting volume V, and if V is invariant under the isometry g (positive or negative), it follows from invariance of angles that the field F' = (gF) also satisfies the same boundary condition.

4.5 Alternative Approach Through Exterior Differential Forms

The quantities associated with an electromagnetic field are conveniently represented by differential forms [Deschamps, 1981]. Their relations are then expressed by means of the exterior differential operator d and through the Hodge's star operator $*$, the latter being combined with the material constants ε, μ, and σ. This makes it particularly simple to express the effect of space transformations, or changes of variables, on the electromagnetic quantities and their relations. The final results agree with those of preceding sections so that this new approach is not essential. However it still presents some interest being formally simpler and more general since it is not restricted to isometries. For the sake of brevity we shall rely heavily on references to Deschamps [1981].

A point-to-point transformation of space

$$g: \quad r \mapsto r' = gr \qquad (4.5.1)$$

is expressed in terms of coordinates by

$$g: \quad (x,y,z) \mapsto (x'y'z') \qquad (4.5.2)$$

(We omit the overbar, as explained in 4.1) If a field quantity is represented by a differential p-form β in the (x,y,z) system of coordinates it will become β' in the system of coordinates $(x',y',z') = g^{-1}(x,y,z)$. Thus β' is a pull-back of β under g^{-1}

$$\beta' = (g^{-1})^*\beta \qquad (4.5.3)$$

When g is non singular (diffeomorphism) this is also expressed by the push forward:

$$\beta' = g_* \beta \tag{4.5.4}$$

Such an expression has already been stated (4.1.10, 4.1.11) in the case where β is a scalar function f. It does carry over to p-forms with the extremely useful properly that g_* commutes with d. That is, if $\beta = d\alpha$ and $\beta' = g_*\beta$, $\alpha' = g_*\alpha$, then $\beta' = d\alpha'$. This is also expressed by the commutative diagram

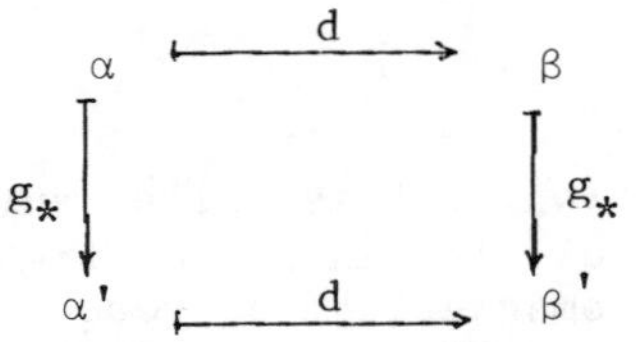

For example the relation B = dA between the magnetic field B (2-form) and its potential 1-form A becomes B' = dA' under an arbitrary transformation. In contrast to this the vector relative B = curl A transforms in a simple manner B' = $\pm$ curl A' but only when g is an isometry.

For the differential forms approach the special role played by isometries reappears when considering the star operator. A relation such as D = *E between electric field one-form E and displacement two-form D becomes D' = *'E' where $D' = g_*D$, $E' = g_*E$ and *' can be deduced from the Jacobian matrix of g. For an isometry, in Euclidean coordinates, the Jacobian being $\pm$ 1 the star operator * is transformed into $\pm$ *. The positive sign occurs for positive (proper) isometries such that the determinant of the Jacobian is + 1. The positive isometries that leave a point 0 fixed form the group SO(3), "Special orlhogonal in 3 dimensions," a subgroup of O(3). Negative (improper) isometries such as reflection in a point or in a plane transform * unto -*.

The conclusion is that a positive isometry that leaves invariant a given medium (e.g. a scatterer) will transform a field in that medium, due to sources or incident fields, into another field due to the transform of the sources or of the incident fields.

For a negative isometry the transformed field satisfies Maxwell's equation in a medium where ε, μ are replaced by $-\varepsilon$, $-\mu$. As one wishes to keep the medium invariant one may define the transform of some of the electromagnetic quantities as $+g_*$ and that of others by $-g_*$.

For instance consider the equation

$$dE + i\omega\mu H = 0 \qquad (4.5.5)$$

where $\mu = \mu_o *$ contains the star operator. Applying g_* to E, H, gives the equation

$$dE' - i\omega\mu H' = 0 \qquad (4.5.6)$$

between $E' = g_* E$, $H' = g_* E$. To recover the original equation (4.3.5) we take the transform of H as

$$H' = -g_* H \qquad (4.5.7)$$

calling H an odd quantity (while E is called an even quantity). Referring to Deschamps [1981] it can be verified that the entire set of equations of electromagnetics is preserved provided quantites in table I are even and those in table II are odd.

As an illustration consider the effect of a mirror reflection $g = g^{-1}$

$$g: \ (x,y,z) \mapsto (x,y,-z) \qquad (4.5.8)$$

on an electric field

$$E = X(x,y,z)dx + Z(x,y,z)\,dz \qquad (4.5.9)$$

The pull-back

$$g^* E = E' = X(x,y,-z)dx - Z(x,y,-z)dz \qquad (4.5.10)$$

This corresponds to the vector transformation

$$(g\bar{E}) = g \circ \bar{E} \circ g^{-1} = g \circ \bar{E} \circ g\ . \qquad (4.5.11)$$

The effect on the two-form

$$D = U(x,y,z)dydz + V(x,y,z)dzdx + W(x,y,z)dxdy \qquad (4.5.12)$$

$$g^* D = -U(x,y,-z)dxdz - V(x,y,-z)dzdx + W(x,y,z)dxdy \qquad (4.5.13)$$

We are led to define $D' = -g^* D$ to make the star operator invariant and this corresponds to taking

$$(g\bar{H}) = -g \circ \bar{H} \circ g \qquad (4.5.14)$$

as shown in (4.3) and (4.4).

4.6 Symmetries of Scattered Fields

This section discusses the conclusions that can be drawn about the scattering operator S(i,j) when the scatterer has known symmetries. The general result is simply expressed: if the scatterer is invariant under an isometry g and if the incident field F_i produces a scattered held F_s, then the incident field (gF_i) will produce the scattered field (gF_s). This applies in particular when F_i is a plane wave $|1\rangle = |u_1,k_1\rangle$ and F_s is the far field $S(2,1)|2\rangle$ radiated in state $|2\rangle = |u_2,k_2\rangle$. We note that the application of g to a plane wave $|1\rangle$ gives a plane wave $|1'\rangle$ such that

$$u'_1 = g\, u_1 \qquad k'_1 = g\, k_1 \tag{4.6.1}$$

and similarly for the plane wave $|2\rangle$. The symmetry of the scattering operator is expressed

$$S(2',1') = S(2,1) \tag{4.6.2}$$

To obtain a similar simple statement about the 2x2 matrix $S(k_2,k_1)$ which relates the field $\Pi_1 \perp k_1$ to the field in $\Pi_2 \perp k_2$ we can take arbitrary reference frame $\bar{f}_1$ in Π_1, $\bar{f}_2$ in Π_2. Then if we use their images under the isometry g

$$f'_1 = gf_1, \quad f'_2 = gf_2 \tag{4.6.3}$$

the matrices $S(k_2,k_1)$ and $S(k'_2, k'_1)$ gk_1 will be <u>identical</u>

$$\bar{\bar{S}}(gk_2,gk_1) = \bar{\bar{S}}(k_2,k_1) \tag{4.6.4}$$

A particular case of the general statement 4.6.2 and 4.6.3 results when $|1'\rangle = |1\rangle$ i.e. $gu_1 = u_1$, $gk_1 = k_1$. Then the scattered field is also unvariant under g. As an illustration of that remark consider the back scattering of a target having the following symmetries: (1) invariance under reflection P through a plane P that contains an axis A through O, (2) invariance under a rotation $A^{\pi/2}$ through angle $\pi/2$ about A. A consequence of the group property is that

$$Q = A^{\pi/2} \circ P \circ A^{-\pi/2} \tag{4.6.5}$$

reflection though the plane $Q = A^{\pi/2}(P)$, is also a symmetry of the target. Take the incident field $|1\rangle$ defined by k_1 along A, u_1 in P. It is linearly polarized. The back scattered field $|2\rangle$ admits the reflection P, hence it is also linearly polarized in plane P. The incident field $|3\rangle = A^{\pi/2}|1\rangle$ has similar properties. Using (u_1, u_3) as reference frame the matrix $S(-k_1, k_2) = 1$ as for an infinite reflecting plane. Its backscattering cross section for circular polarization is zero.

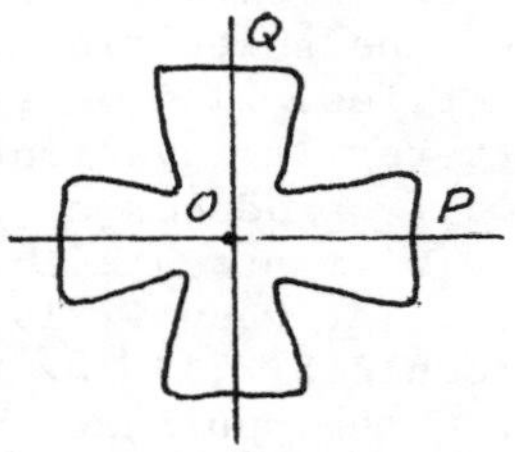

4.7 Application to Measurements of the Scattering Operator

A consequence of the results in 4.6 is that if the target admits known isometries the measurement of its scattering properties can be simplified accordingly. The symmetries of an object obviously form a group G that acts on the manifold $\mathcal{M}$ of plane wave states. This group defines a fundamental domain $\mathcal{M}_0$ such that $G\mathcal{M}_0$ (union of $g\,\mathcal{M}_0$ for $g \in G$) covers exactly $\mathcal{M}$. Note that because of linearity $\mathcal{M}$ and $\mathcal{M}_0$ may be limited to states of linear polarization and the scattering operator defined by

$$S: \quad \mathcal{M} \times \mathcal{M} \to \mathbb{C}: \ (i,j) \mapsto S(i,j) \tag{4.7.1}$$

The measurement may be limited to that of S over $\mathcal{M}_0 \times \mathcal{M}$. Combined with reciprocity this may produce substantial reduction of labor.

5. POLARIZATION AND INVERSE PROBLEMS

One particular but still typical inverse problem is the determination of the sources (electric and/or magnetic currents) that produce an observed radiation pattern. The far field being given otherwise this becomes an antenna synthesis problem [Deschamps and Cabayan, 1972]. Referring to 3.2 the problem may be to find J(r) knowing the function $J_S(k)$ restricted (pulled back) to the sphere S.

A more general problem is to deduce shape, size or other parameter of a scatterer from the observation of its scattering properties, expressed for instances by the scattering operation S(i,j). An essential and more difficult step is to formulate the relation between the incident field $|j\rangle$, the scatterer's parameter and the induced currents: J(r,j). At this point some approximations are usually necessary even to solve the direct problem and to make its inversion (finding the parameters) possible.

The difficulties of the inverse problems are many and will be discussed in depth by other members of the workshop. Prior knowledge and approximations must be used. One of these is to treat the fields as if they were scalar-valued. It is clear that something is lost in that process. One should make every effort to look at the vector problems, hence to take polarization into account. This of course does not mean that only polarization should be observed. Neglecting intensity (and phase!) is throwing away information. The only excuse for doing this may be the difficulty of the measurements, of phase in particular and the fact that sometimes polarization alone (easier to obtain) may be revealing of some characteristics of the target. In that case identification may be possible, at least if aided by prior knowledge.

There are difficulties intrinsic to holographic solutions connected with the concept of phase front reconstruction. What may be done is to "look" at the field through different polarizors and to observe the images thus obtained to gain some familiarity with their meaning. Whatever is done quantitatively it is hoped that its techniques reviewed in this paper may be of some use.

APPENDIX A. SELECTED BIBLIOGRAPHY

1. Bargman, V. (1964). Note on Wigner's Theorem on Symmetry Operations. J. Math Phys., vol. 5, pp. 862-868.

2. Boerner, W. M. (1980). Polarization Utilization in Electromagnetic Inverse Scattering, pp. 237-297 in Inverse Scattering Problems in Optics, (R. H. Baltes ed., Springer Verlag).

3. Davidovitz, M. and W.-M. Boerner (1983). Reduction of bistatic scattering matrix measurements for inversely symmetric radar targets. IEEE Transactions, vol. AP-31, pp. 237-242.

4. Deschamps, G. A. (1951). Geometrical representation of the polarization of a plane electromagnetic wave. Proc. IRE, vol. 39, pp. 540-544.

5. Deschamps, G. A. (1952). Geometric viewpoints in the representation of waveguides and waveguide functions. Proc. of Symposium on Modern Network Synthesis PIB, New York.

6. Deschamps, G. A. (1953). "A Hyperbolic Protractor for Microwave Impedance Measurements and Other Purposes." International Telephone and Telegraph Co., New York.

7. Deschamps, G. A. and H. Cabayan (1972). Antenna synthesis and solution of inverse problems by regularization methods. IEEE Trans. Antennas Propagat., vol. AP-20, pp. 268-274.

8. Deschamps, G. A. and P. E. Mast (1973). Poincare sphere representation of partially polarized fields. IEEE Proceedings, vol. AP-21, pp. 474-478.

9. Deschamps, G. A. (1981). Electromagnetics and differential forms, Proc. IEEE, vol. 69, pp. 676-698.

10. Fano, U. (1957). Description of states in quantum mechanics by density matrix and operator techniques. Rev. of Modern Physics, vol. 39-1, pp. 74-93.

11. Feynman, R. P., F. L. Vernon and R. W. Hellwarth (1957). Geometrical representation of the Schrodinger equation for solving Maser problem. J. Apply. Phys., vol. 28-1, pp. 49-52.

12. Jones, R. C. (1942). A new calculus for the treatment of optical systems. JOSA 31, pp. 443-503.

13. Ludwig, A. C. (1973). The definition of cross polarization. IEEE Trans. Antennas Propagat., vol. AP-21, pp. 116-119.

14. Mueller, H. (1948). Foundations of optics. J. Opt. Soc. Am., vol. 38, pp. 661(A).

15. Neumann, J. von. (1932). Mathematische Grunlagen der Quantenmechanik. (Berlin).

16. Perrin, F. (1926). Jour. de Physique, vol. 7, p. 390.

17. Perrin, F. (1942). Polarization of light scattered by isotropic opalescent media. J. Chem. Phys., vol. 10, pp. 415-427.

18. Soleillet, P. (1929). Ann. de Physique, vol. 12, p. 23.

19. Wang, W. D. and G. A. Deschamps (1974). Application of complex ray tracing to scattering problems. Proc. IEEE, vol. 62, pp. 1541-1551.

20. Weyl, H. (1952). Symmetry. Princeton University Press.

21. Wiener, V. (1929). Harmonic analysis and the quantum theory. J. Franklin Institute, vol. 207, pp. 525-539.

APPENDIX B. NOTATIONS

Vectors. We have abstained from using the vector sign over quantities that have been defined as vectors such as V, E, the wavevector k, the position vector r, etc. Length of k and r are written k and r. The vector sign (or the unit vector sign ($\hat{\ }$)) are used occasionally when the same letter is used to designate both a vector (or a unit vector) and some other associated quantity e.g. matrix Q, or Ψ and Stokes vector Q or ψ, vector E and unit vector $\hat{E}$.

Vector Spaces. $\mathbb{R}^n$ and $\mathbb{C}^n$, sets of n-tuples of real and complex numbers, respectively

unit vector $u \in \mathbb{C}^n$, $u \cdot u = 1$
unitary vector $\in \mathbb{C}^n$ $u^* \cdot u = 1$
(called unit vector if no confusion results)

For a matrix M the transpose is M^T, complex conjugate M^*, Hermitian transpose M^+, H(2) Hermitian matrices over C^2, $H_0(2)$ the same with zero trace.

Groups of Transformation and nxn matrices. $GL(n,\mathbb{R})$, $GL(n,\mathbb{C})$, non singular, nxn matrices with elements in $\mathbb{R}$, $\mathbb{C}$.

U(n) unitary matrices $M^+ = M^{-1}$
O(n) orthogonal matrices $M^T = M^{-1}$

Suffix S in SL, SU, SO means subgroup of matrices with determinant 1.

A statement "T: A $\to$ B: a $\mapsto$ b" means that T maps the set A in B, carrying $a \in A$ to $b \in B$.

Composition. If g and g' are two transformations such that the range of g' is in the domain of g, $g \circ g'$ designates their compositions

$$g \circ g' : r \to g(g'(r))$$

Special Symbols.

$\mathcal{P}$ Poincaré sphere (1.3)
$\bar{\mathcal{P}}$ Poincaré space (1.5)
S sphere of radius k in k-space
T_kS plane tangent to S at point k
TS tangent bundle to S
LS bundle of lines tangent to S (3.5)
$\mathcal{P}$S Poincaré bundle (3.4)

III.2 (RP.5)

POLARIZATION INFORMATION UTILIZATION IN PRIMARY RADAR: An Introduction and Up-date to Activities at SHAPE Technical Centre

A.J. Poelman, M.Sc., Mem. NERG
and
J.R.F. Guy, Ph.D., Mem. IEEE

SHAPE Technical Centre, The Hague, The Netherlands

Relatively little has been published on how polarization information in the backscattered radiation from targets and clutter could be exploited to enhance the detection capability of primary radar systems. The historical development of a long-term STC research project based upon this topic is described.

The main theoretical analyses and the vector-signal processing schemes developed for practical real-time applications are summarized and explained. The principal features of the project's experimental polarimetric X-band radar system are described, and typical performance figures given for the methods developed for processing in the polarization domain; the adverse environment was represented by experimentally-derived rain data. Finally there is a brief outline of future study areas.

1. INTRODUCTION

1.1 Background

In the early seventies STC began a long-term research project which had, as its broad terms of reference, the investigation of the applicability of wave polarization as a source of information for improving the performance of radar systems of the future. This task was undertaken because of the continuing need to improve the detection capability of radar systems which operate in adverse environments whilst emitting a minimum of energy. Such an improvement would undoubtedly result if more of the information contained in the backscattered and/or interference fields were utilized.

W.-M. Boerner et al. (eds.), Inverse Methods in Electromagnetic Imaging - Part 1, 521–572.

Therefore, it was felt that in addition to studies dealing with the use of the information associated with the frequency and spatial domains, attention had to be devoted to the applicability of the use of the information inherently available in the polarization state of an electromagnetic wave.

The utilization of polarization information is a topic that is currently receiving attention not only as a potential means of improving target detection, but also as an aid in target imaging applications [1]. The operational radar systems of the early seventies made no attempt to exploit the information available in the polarization state of electromagnetic waves, with the exception of the use of a circularly-polarized antenna to enhance target detection in precipitation-clutter environments [2, 3]. All these systems in essence apply scalar-signal processing, since at the antenna the incident vector-field is firstly converted to a scalar-field prior to further processing, and as a consequence the polarization vector is lost. To take advantage of the vectorial nature of an electromagnetic field, there has to be a means of enabling vector-signal processing, which implies a system that has mutually-coherent orthogonally-polarized receive antennas and channels. Such a configuration is referred to here as a dual-channel receiver system.

In general, an optimum detection receiver is one which uses all of the information carried by an electromagnetic wave, and this information is contained in the basic wave characteristics: amplitude (or power), phase (or direction), frequency, and polarization. To date the STC project has dealt only with polarization-vector signal processing, in order to obtain a deep insight of its full potential. Furthermore, the project has tended to concentrate on those radar problem areas where this type of processing is complementary to the well-established processing techniques used in the other domains. Examples of such areas are homogeneous clutter (weather, sea, dipole-clouds) and main-beam interference. Because of the applications-oriented nature of the research, it was the intention from the beginning to support the theoretical studies with experimental work. However, from the outset the project was faced with two main problems:

(a) a lack of vector-signal processing algorithms, and

(b) a lack of suitable measured polarization data.

Conceivably, the lack of promising vector-signal processing algorithms leading to practically realizable systems may well be the reason that over the years little attention has been paid to signal processing in the polarization domain relative to processing in the frequency and spatial domains.

1.2 Basics

The usefulness of polarization information in processing schemes which attempt to improve target enhancement is based upon the fact that the response of an antenna system is polarization sensitive, i.e., the antenna acts as a filter. This phenomenon has been mathematically formulated in [4]. In [5], it is shown that the response of an antenna system in terms of polarization to a monochromatic or completely-polarized incident field (i.e., a field whose polarization state does not vary with time) can simply be represented by the spherical distance between two points on the surface of a sphere which represents polarization space and is known as the Poincaré sphere.

> The polarization state of a field or a transmit antenna is defined at a point in space when looking into the propagation direction: the receive antenna polarization is the polarization which would obtain if the antenna were transmitting.

Figure 1 shows the response of an antenna in terms of its polarization at receive (M_r) to an incident field with polarization M_i.

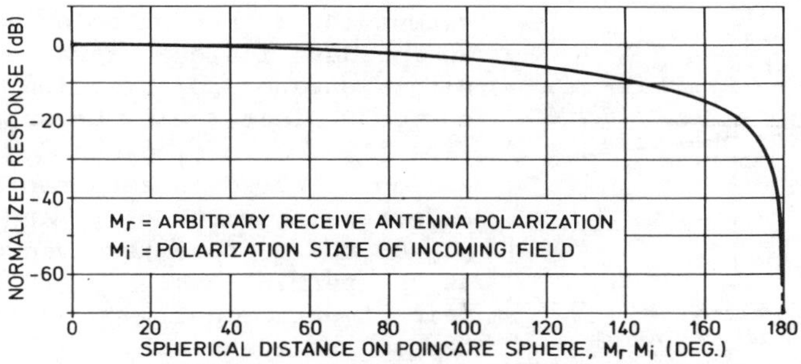

Fig.1 Incident-field polarization and receive antenna response [5]

It follows from Fig. 1 that:

(a) the maximum response is obtained when the polarization states of the incident field and the receive antenna are parallel;

(b) the maximization process is a very insensitive process;

(c) the minimum (i.e., zero) response is obtained when the polarization states of the incident field and the receive antenna are orthogonal;

(d) the minimization process is a very sensitive process; and

(e) loci of constant response are circles on the surface of the Poincaré sphere, concentric around the receive polarization.

It can be concluded from these observations that in adverse environments polarization filtering techniques are far more effective when they concentrate on the suppression of unwanted signals rather than upon maximizing the response of target returns. Further, unwanted signal suppression by means of a single antenna polarization, or a single polarization filter (the conventional approach used, for example, for rain-clutter suppression), is a single-notch nulling process and as such is inherently sensitive. Hence, when the backscatter/interference waves have fluctuating polarizations (i.e., the polarization state varies with time) the average amount of achievable suppression is limited. Such waves are said to be partially-polarized, and a decomposition theorem [6] states that any partially-polarized wave may be uniquely regarded as the sum of a completely-polarized wave and a completely-unpolarized (or randomly-polarized) wave. The ratio of the average power contained in the completely-polarized portion to the average power contained in the wave is called the "degree of polarization", p, of the wave ($0 \leqslant p \leqslant 1$). Figure 2 shows the extreme average responses versus the degree of polarization of an incident field achievable with a single antenna polarization. The curves indicate that the suppression capability of a single-notch polarization filter decreases rapidly with a decreasing value of p, and when p = 0, the average received powers at the terminals of any arbitrary pair of orthogonally-polarized antennas are identical.

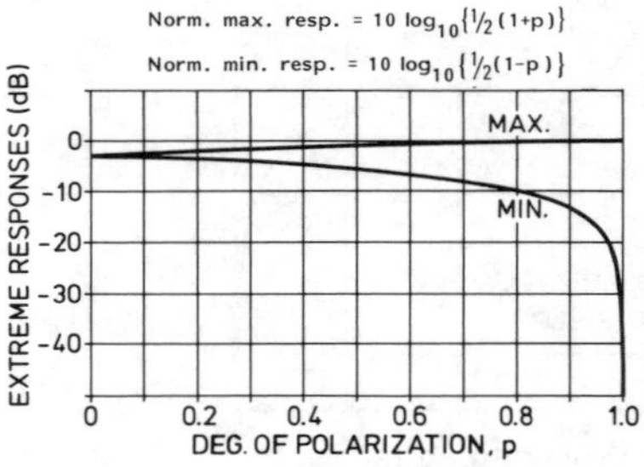

Fig. 2 Extreme average responses at receive against degree of polarization (p) of incident field [6]

By analogy with the use of frequency (i.e. the Doppler effect) as a discriminant for target enhancement, the use of the backscatter-field polarization as a discriminant to improve target detection in (homogeneous) clutter environments implies that basically the quality of the detection of a target depends on the extent to which the polarization characteristics of the target and clutter returns deviate. Independent of the actual locations of the target and clutter returns in polarization space, vector-signal processing schemes based upon polarization filtering can, in principle, be developed to take advantage of such deviations provided that a dual-channel receiver is available. The polarization state of the return from an object is determined by its backscatter operator, the scattering matrix [7,8] (a 2 x 2 matrix defined in terms of polarization) and the polarization state of the illuminating wave (i.e., the radar's transmit polarization). Consequently, it is obvious that the use of polarization as a discriminant for target enhancement is particularly attractive in those environments where the transmit polarization can either be chosen arbitrarily or selected on the basis of "a priori" knowledge, because no on-line learning-process is then required. The conditions under which the use of polarization information is extremely beneficial are illustrated on the "modified polarization charts" of Fig. 3.

The polarization chart is an orthogonal projection of the Poincaré sphere on a plane, having polar coordinates $\rho = \cos(2\tau)$ and $\Phi = 2\varphi$, where τ is the ellipticity angle and φ the orientation angle of the polarization ellipse. The modified polarization chart has polar coordinates $\rho = 1 - |r|$ and $\Phi = 2\varphi$, where $r (= \tan\tau)$ is the ellipticity ratio of the polarization ellipse $(-1 \leqslant r \leqslant 1)$; here $0 < r \leqslant 1$ represents right-handed elliptical polarization (RHEP), $r=0$ linear polarization (LP), and $-1 \leqslant r < 0$ left-handed elliptical polarization (LHEP).

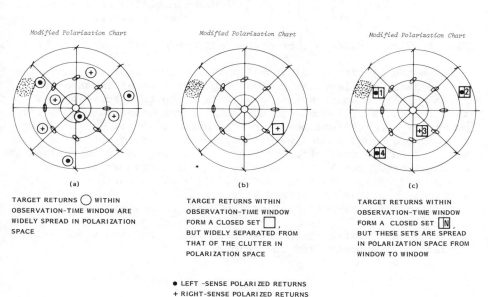

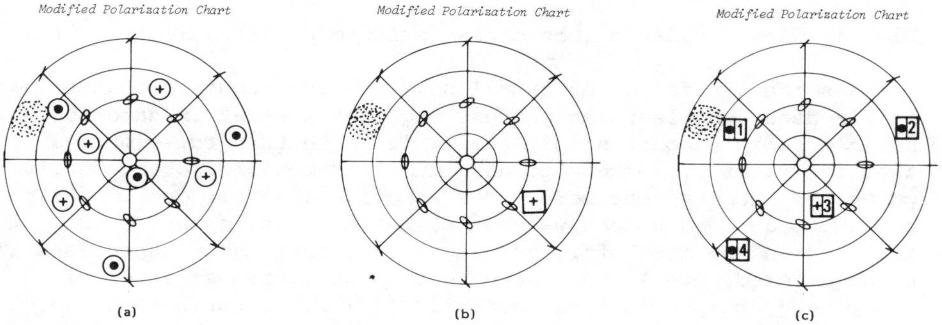

(a) TARGET RETURNS ◯ WITHIN OBSERVATION-TIME WINDOW ARE WIDELY SPREAD IN POLARIZATION SPACE

(b) TARGET RETURNS WITHIN OBSERVATION-TIME WINDOW FORM A CLOSED SET ☐, BUT WIDELY SEPARATED FROM THAT OF THE CLUTTER IN POLARIZATION SPACE

(c) TARGET RETURNS WITHIN OBSERVATION-TIME WINDOW FORM A CLOSED SET [N], BUT THESE SETS ARE SPREAD IN POLARIZATION SPACE FROM WINDOW TO WINDOW

● LEFT -SENSE POLARIZED RETURNS
+ RIGHT-SENSE POLARIZED RETURNS

Fig. 3 Conditions under which the use of the polarization state of the returns as a discriminant for target enhancement is extremely beneficial; clutter returns within observation-time window form a closed set and the transmit polarization used is arbitrary, or is based on a-priori knowledge, or is derived from an on-line learning-process

However, in the case of an 'unknown' clutter environment, a suitable transmit polarization would be one obtained from a "learning-process"-derived time/window-averaged characteristic null polarization of the clutter (e.g., in the case of homogeneous clutter, averaging over a range window would be the obvious method, since any cell in the window might contain a target). This approach is based upon the physical phenomenon that an object's characteristic null polarizations [7,9] (see Fig. 4) are the two polarizations which most closely characterize the object as far as its actual (or apparent) geometry, or structure, is concerned. These characteristics are reflected in its scattering matrix. Consequently, the selected average characteristic null polarization, which can be termed the "minimum" polarization, would, in the absence of "a priori" knowledge of the target polarization signatures, be the "best" with regard to discriminating between target and clutter returns.

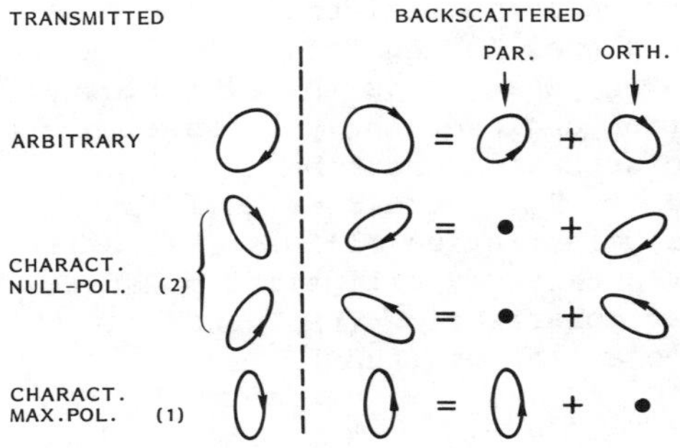

Fig. 4 Single radar object characteristic polarizations [7,9]

As the derivation of a null polarization requires scattering matrix data, the learning process requires the matrix data to be in principle continuously available. In turn this implies that the radar system has to have mutually-coherent orthogonally-polarized receiver channels and a specific transmit mode. With regard to uncoded pulsed waveforms, the scattering matrix is measured in the most efficient way (i.e., fast and with no loss in backscattered power) when there is sequential transmission of two orthogonally-polarized waveforms [10] e.g., horizontal- and vertical-polarization (HP and VP), right-hand circular- and left-hand circular-polarization (RHCP and LHCP), etc.

The basic differences between scalar-signal and vector-signal processing are summarized in Table 1.

1.3 Approach

The aim of the early studies initiated at STC was to demonstrate the flexibility of a system with orthogonally-polarized receiver channels, on the basis of a simple signal processing concept not feasible with a conventional single-channel receiver system. A theoretical study [11,12,13], analogous to the approach used in frequency-diversity radar systems, examined the effectiveness of the linear addition of the video signals in the

Table 1 Comparative summary between vector- and scalar-signal processing

	IN VECTOR APPROACH	IN SCALAR APPROACH
POLARIZATION INFORMATION	is used	is not used
OBSERVED	scattering matrix (+ Doppler shift)	cross-section (+ Doppler shift)
ANTENNA SYSTEM	dual-polarized	single/dual polarized (single in principle)
RECEIVER	dual-channel	single-channel
TRANSMIT MODE	(a) learning: alternate orthogonally-polarized uncoded pulse waveforms (b) operating: waveform polarization adapted to unwanted environment	fixed polarization (for example: HP, VP, or CP) or switching between a limited number of polarizations for example: HP⟷CP

two channels prior to the detection stage. The antenna polarization was considered to be fixed, but adapted to a clutter environment for minimum response. The point of interest was the effectiveness of this processing concept in range cells outside the clutter, that is, in clear as well as quiet environments. However, there was no exploitation of polarization-vector information, as the only parameter processed was the total target backscattered power.

A second area of study concentrated on a theoretical assessment of the effectiveness of "adaptive antenna polarization" for target enhancement in a moving (i.e. time-varying), homogeneous-type of distributed clutter. In the early seventies, because there was no model suitable for describing the scattering properties of homogeneous clutter in terms of polarization (with the exception of an incomplete model for a rain-cloud), a simple model of a random dipole-cloud was developed at STC [14]. The first study using this model was devoted to the effectiveness of adapting identical transmit and receive antenna polarizations for a minimum dipole-cloud response, using as a variable the probability space of individual dipole-axis direction [14]. A second study concerned the antenna polarization at transmit and receive which gives rise to maximum correlated and uncorrelated orthogonally-polarized backscatter components in the receiver channels [15]. The observations from this latter study may have significance in processing schemes which utilize the correlation between orthogonally-polarized backscatter components from targets and clutter as a discriminant for obtaining target enhancement. In addition to the above-mentioned dipole-cloud model, a simple model of a moving multiple-component body was investigated.

As the studies progressed into the mid-seventies it became apparent that in practice "actual" adaptation of the antenna polarizations for target enhancement had severe limitations in the practical situation where the polarization properties of the "dynamic" adverse environment are varying in both the spatial and temporal domains [16,17,18].

These limitations prompted the next avenue of study, which concentrated on the concept of "virtual polarization adaptation" (VPA) for target enhancement. This technique, which is equivalent to appropriately adapting the polarization states of the transmit and receive antennas to the unwanted environment for minimum response, allows optimal adaptation of the transmit and receive polarizations to each individual radar cell, since the adaptation takes place within the processor and <u>not</u> at the antenna. Of practical importance is the fact that the VPA concept allows the sensor head to be relatively simple in the sense that the system need only be capable of emitting <u>two</u> specific orthogonally-polarized uncoded pulsed waveforms, instead of every type of polarization.

As investigation progressed the power and flexibility offered by the general VPA concept became apparent, and a natural consequence was the development of algorithms for vector signal processing within the VPA concept. These algorithms exploit the polarization information as a discriminant for target enhancement to an extent not realizable by adaptation of the antenna's actual polarization. One of these algorithms is principally concerned with increasing the ability to suppress unwanted signals that have fluctuating polarization, and so enhance target detection. As the conventional single-notch nulling process is inherently sensitive (see Fig. 1), the average amount of achievable suppression is limited when the unwanted incident signals have fluctuating polarizations (see Fig. 2). This was considered such a serious drawback that a means of desensitizing the nulling process was sought, and resulted in the concept of a vector-processor called "the multinotch logic-product polarization suppression filter", or MLP filter. With this filter the location, the form, the size of the region of deep suppression in polarization space, and the average level of deep suppression are all under design control [18,20].

The location of the MLP filter in polarization space (as determined by the average incident-field polarization of closed sets of unwanted signals) will vary in both the spatial and temporal domains in a typical adverse environment. Consequently, its adaptation to the dynamic environment would place a heavy load on the processor in terms of processing time, and therefore an algorithm which is economical in terms of processing time has been devised. This algorithm considers the MLP-filter at an arbitrary, fixed location in polarization space and involves the mathematical transformation of the polarization states of the incident-field samples to be suppressed, so that they are in effect brought to the location of the filter for processing. The process is called "polarization-vector translation" (PVT), and because it is so economical it is eminently suitable for real-time implementation [19].

Studies into the design of an effective MLP filter have revealed that to cover only a moderate area of polarization space (typically 2% to 3%) at least 61 notches are required to make up a filter giving, on average, a 40 dB suppression level in this area. The real-time realization of such a filter would, using current technology, result in quite a bulky piece of hardware. This practical limitation stimulated the need to devise a processing algorithm that would reduce the area of spread (in polarization space) of the sets of incident-field samples that were to be suppressed, such that the MLP filter could be configured with a reduced number of notches but the desired average level of suppression maintained. The resulting vector-process is also a PVT type. However, since it is essentially non-linear it is referred to as the non-linear PVT-process [21] to distinguish it from the linearly operating PVT-process.

It was evident from an early stage that it would be necessary to do experimental work in support of this project, in order to:

(a) assess the polarization signatures of clutter and targets (there was, and still is, a lack of measured data in terms of scattering matrices);

(b) study, off-line, the effectiveness of advanced vector-signal processing schemes with live data; and

(c) demonstrate the usefulness of performing signal processing in the polarization domain in real-time.

For this reason work began in 1971 on the realization of a

multi-purpose experimental X-band radar facility for polarization studies. Since then this facility has gradually evolved and at present has the capability to obtain and record mass live polarization data from targets and clutter, and perform off-line evaluations of vector-signal processing schemes with the aid of this data.

From this brief overview the emphasis of the STC studies so far should be clear. The ultimate objective of this project has been, and currently still is, the investigation of new vector-signal processing schemes suitable for real-time implementation..

1.4 Layout of Paper

The format of this paper reflects the historical development of the principal topics, and accordingly the evolution of the ideas involved.

Chapter 2 contains a summary of the evaluation of a simple system which exploits polarization information by using a dual-channel receiver system. In such a system the transmit antenna polarization is considered to be matched to suppress the clutter, and linear video recombination is applied in clutter-free (and quiet) areas to maintain the system detection performance.

Prompted by the lack of a suitable model describing the signatures of general scatterers in terms of polarization, the study described in Chapter 3 was initiated, and derived a simple random dipole-cloud model which subsequently enabled studies into the applicability of discriminants useful for improving target detection.

A major development which has had a significant influence upon the direction of this project is the concept of virtual polarization adaptation and the related vector-signal processing algorithms. The freedom of not having to adapt the actual polarization of the antenna has enormous advantages when attempting to improve target detection by means of unwanted signal suppression. Chapter 4 highlights these advantages and an outline of the associated adaptive vector-signal processing algorithms is given in Chapter 5.

In order to evaluate the vector-signal processing methods which had been devised, experimentally-derived data had to be obtained. Therefore, an X-band experimental radar facility was realized, and its main features are discussed in Chapter 6. The subsequent evaluation and examples of the performances which can be realistically expected from the processing methods (using data derived from rain) are contained in Chapter 7.

To conclude, Chapter 8 discusses future studies on the utilization of polarization information for target enhancement in primary radar systems.

2. USE OF LINEAR VIDEO RECOMBINATION IN DUAL-CHANNEL RECEIVERS

2.1 General

Radar systems that use linear polarization at transmit and receive generally have difficulty in detecting airborne targets in precipitation clutter. However, a well-known technique [2] for improving detection capability in rain clutter is to use circular polarization (CP) or an almost-circular elliptical polarization (EP). The physical explanation for this improvement is that the drops in a rain cloud have a more or less spherical geometry, and thus the specular reflections by far dominate the non-specular reflections, while for aircraft targets these reflections are on average of the same order.

A property of any conventional radar system is that at receive it cancels incident waves which have a polarization orthogonal to the polarization used on transmission. This means that although the use of CP or EP gives an improvement in rain clutter visibility of typically 15 to 30 dB [2] (because most of the scattered power is in the orthogonal polarization), there is also always a loss in target echo strength due to the cancellation by the system of part of the target reflected power (the part in the orthogonal polarization). This means that for a radar operating in a CP mode, the target-detection performance in precipitation-free sectors is approximately 3 dB [3] less than that obtained in an HP (horizontal polarization) mode.

The conflicting requirements to obtain the best performance of target detectability in precipitation <u>and</u> precipitation-free sectors indicate that some form of polarization mode switching is desirable. However, instantaneous switching between HP and CP at transmit is difficult to achieve in practice, mainly because of the high power levels involved. Moreover, mode switching at transmit has the disadvantage that there is a loss in target-detectability in precipitation free-sectors on the same bearing as the cloud. Therefore, faced with these problems switching at receive is a better approach.

2.2 Circular Polarization Transmission and Processing in Precipitation-free Sectors

As a means of reducing the loss in target detection in precipitation-free sectors, the study reported in [11] proposed a system which uses fixed CP on transmit and a dual-channel receiver system. A selector determines whether detection-performance is better with the video signal derived from the received parallel-polarized backscatter component, or with the recombined video signal obtained from the linear addition of the video signals derived from the received parallel-polarized and orthogonally-polarized backscatter components. Since selective switching between the two types of video is now performed at low power levels it can be done electronically, and hence instantaneously. The application of linear video addition here is optimum (from the detectability point of view) when the received orthogonally-polarized components of the target returns (RHCP and LHCP) are on average of equal strength.

To evaluate the proposed system operating in the linear video recombination mode it was compared with a conventional system that had a single-channel receiver and operated in the CP transmit mode. Schematic diagrams of these two systems are shown in Figs. 5(a) and 5(b).

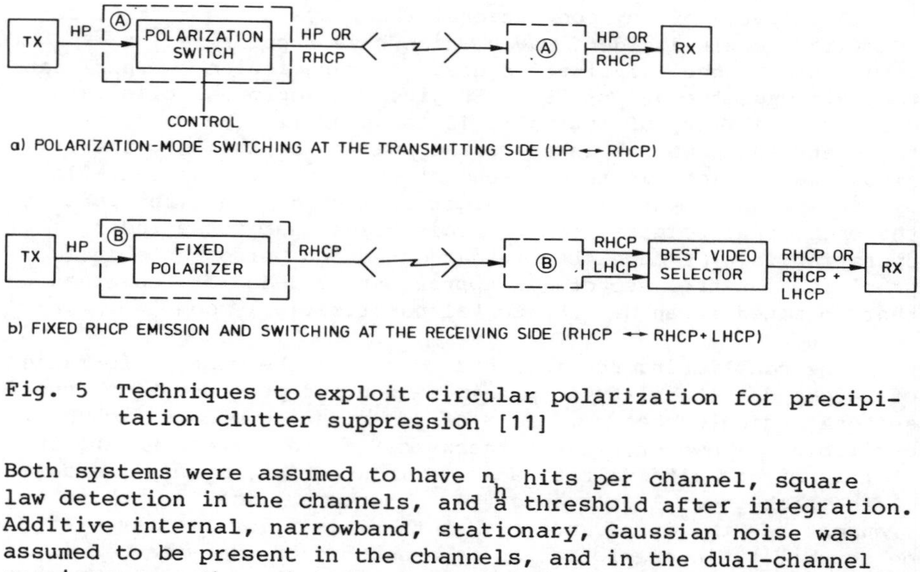

Fig. 5 Techniques to exploit circular polarization for precipitation clutter suppression [11]

Both systems were assumed to have n_h hits per channel, square law detection in the channels, and a threshold after integration. Additive internal, narrowband, stationary, Gaussian noise was assumed to be present in the channels, and in the dual-channel receiver case the noise components were independent and had equal average power. Although [11] treats Swerling Case-1 and Case-2 target scintillation models, the results presented here are only for Swerling Case-1 targets (slow scintillation) for the situation where the orthogonally-polarized components of the target returns are statistically independent.

The threshold levels, Y_b, are related to the probability of false alarm, P_f, by

$$P_f = 1 - \left[I \frac{Y_b}{\sqrt{mn_h}}, mn_h - 1 \right] \quad , m = 1, 2 \tag{1}$$

where I is the incomplete gamma function as defined by Pearson and m is an index equal to 1 for the single-channel receiver and equal to 2 for the dual-channel receiver. The probablity of detection, P_d, is given by

$$P_d = \int_{Y_b}^{\infty} f_x(x)\, dx = 1 - \int_0^{Y_b} f_x(x)\, dx \tag{2}$$

where x is the signal after integration and, where relevant, recombinination, and $f_x(x)$ is the probability density function (pdf) of x.

An efficiency parameter, E, was used to directly compare the two systems. This is defined as the gain in the required signal-to-noise power ratio per hit over the signal-to-noise power ratio per hit in a conventional HP radar system; the transmitted power being the same for both systems.

The efficiency E_1 for the single-channel receiver is given by

$$E_1(\mathrm{dB}) = 10 \log_{10} \left[\frac{\overline{x}_r}{\overline{x}_o} \right] = 10 \log_{10} \left[\frac{K_t^4 K_s^2 + 1}{(K_t^2 + 1)^2} \right] \tag{3}$$

where $\overline{x}_r$ is the average signal-to-noise power ratio per hit when operating in the CP mode and $\overline{x}_o$ is the average signal-to-noise power ratio per hit wheh operating in the HP mode. Parameter K_t is introduced to take account of the practical observation that an antenna polarization different from CP would result in maximum suppression of rain clutter, and

$$K_t = \frac{\text{vertically-polarized power transmitted}}{\text{horizontally-polarized power transmitted}}$$

Also, a target parameter, K_s, relating to practical observation is introduced:

$$K_s = \frac{\text{average target cross-section to vertically-polarized waves}}{\text{average target cross-section to horizontally-polarized waves}}$$

Since the efficiency E_2 for the dual-channel system is dependent on n_h, P_d and P_f, a simple expression similar to (3) cannot be obtained. However, E_2 is related to:

$$10 \log_{10}\left[\frac{\overline{x}_a}{\overline{x}_o}\right] \text{(dB)} = 10 \log_{10}\left[\frac{K_t^2 K_s^2 + 1}{2(K_t^2 + 1)}\right] \quad (4)$$

where $\overline{x}_a = \overline{x}/2$ and $\overline{x}$ is the average recombined signal-to-noise power ratio per hit. A detailed derivation of the above relationships and of the efficiency E_2 over a wide range of n_h, P_d, and P_f is given in [11].

For an objective comparison it is sufficient to consider average values $\overline{E}_1$ and $\overline{E}_2$ of the efficiencies, with parameters K_t and K_s uniformly distributed over a practical range of 0.5 to 2. Table 2 shows average efficiency values for both systems under the same conditions of target type, P_d, P_f, etc. These figures do not include system losses, which typically are approximately 2 dB (polarizer loss) for the single-channel system and 3 dB (2 dB polarizer loss and 1 dB video recombination loss) for the dual-channel receiver. The performance figures obtained showed that, relative to the conventional radar system switched to the CP mode at transmit, the dual-channel receiver system should give an overall improvement of roughly 3 dB for $P_d = 0.8$ and $n_h = 10$ in precipitation-free sectors.

Table 2 Efficiency ($\overline{E}_m$) of the two systems of Fig. 5 in precipitation-free sectors relative to horizontally-polarized radar systems (system losses not included) [11]

$P_f = 10^{-6}$, $n_h = 10$, $0.5 \leqslant K_t^2 \leqslant 2$, $0.5 \leqslant K_s^2 \leqslant 2$		
Swerling Case-1 scintillating target	$P_d = 0.8$	$P_d = 0.5$
Single-channel receiver system in the RHCP mode (m = 1)	$\overline{E}_1 = -2.6$ dB	$\overline{E}_1 = -2.6$ dB
Dual-channel receiver system in the linear video addition mode (m = 2)	$\overline{E}_2 = 1.4$ dB	$\overline{E}_2 = 0.0$ dB
Gain of the dual-channel system over the single-channel system	4.0 dB	2.6 dB

2.3 Arbitrary Polarization Transmission and Detection in Gaussian Noise

The dual-channel receiver of Section 2.2 can be reconsidered [12,13] in a more general manner, this time with an arbitrary transmit polarization which is matched to the unwanted part of the environment for minimum average response. Figure 6 shows the system in more detail. It is non-coherent, processes both orthogonally-polarized received backscatter components, and uses linear video addition (i.e., non-weighted addition). The assumptions made about the receiver in the previous section are assumed to apply again, but to maintain generality the mathematical model uses the average "orthogonally-polarized" (or "cross-polarized") cross-section of the target normalized to its average "parallel-polarized" (or "co-polarized") cross-section as a parameter. This is denoted by F. Thus, for equal average noise powers in each channel,

$$F = \overline{x}_{\perp} / \overline{x}_{\parallel} \qquad (5)$$

where $\overline{x}$ is the average target signal power to average noise power and '$\parallel$' and '$\perp$' denote parallel and orthogonal respectively.

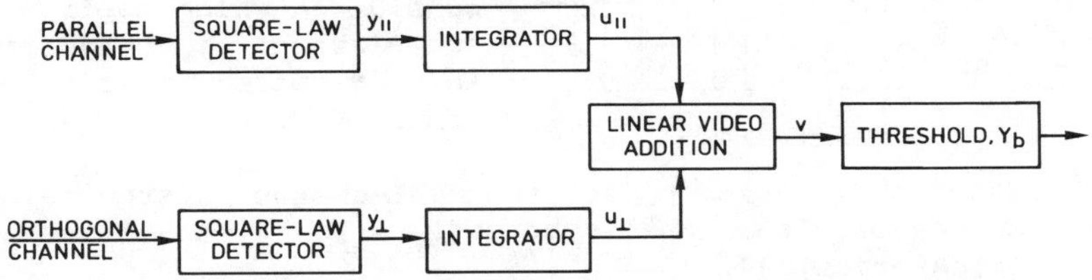

Fig. 6 Block diagram of the dual-channel detection receiver; target detection in unpolarized Gaussian noise [12,13]

The range of F considered was 0 to 1, because of the wide variation of target returns in terms of polarization relative to the transmit polarization. In [12] and [13] Swerling Case-1, -2, -3 and -4 target-scintillation models are treated, but again the results presented here are only for Swerling Case-1 for the case where the orthogonally-polarized components of the target returns are statistically independent.

The probability of false alarm and detection are again determined by equations (1) and (2) respectively. A detectability factor D_s was derived to demonstrate the performance of the system. For a single-channel receiver the detectability factor is denoted by D_{s1} and defined as the required ratio of average signal power to average noise power at the input to the threshold-device, for specified values of P_f, P_d and n_h. For the dual-channel receiver, the detectability factor is denoted by D_{s2}, but the power ratio is considered in the parallel channel at the input to the video addition unit for specific values of not only P_f, P_d, n_h, but also F.

To compare the performance of the dual-channel system to the single-channel receiver, a "figure-of-merit", ΔD_s, defined as the gain in D_{s2} relative to D_{s1}, is used. Figure 7 shows, for Swerling Case-1 targets, ΔD_s plotted against n_h for a P_f of 10^{-6}, two values of P_d (0.5 and 0.8), and values of F ranging between 0 and 1. The results indicate that when the background noise is Gaussian the use of the orthogonally-polarized component in systems operating with a small number of hits per target ($n_h \leqslant 3$) and $P_d \geqslant 0.8$ is advantageous (for Swerling Case-1 targets) when $F \geqslant 0.2$. It appears that if F = 0, the relative loss in target detectability is a maximum of 1 dB under the conditions investigated and for $n_h \leqslant 20$. Thus, it may be concluded that the noncoherent processing scheme considered which applies 'unweighted' video addition is near optimal over the full range of F.

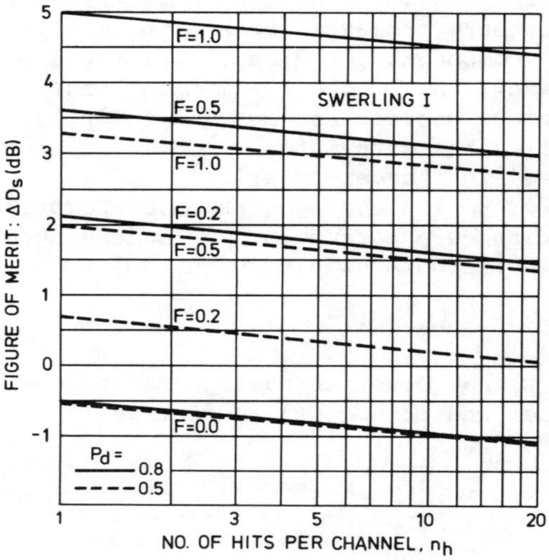

Fig. 7 Gain in D_s versus n_h for the dual-channel system relative to the single-channel system ($P_f = 10^{-6}$; non-coherent integration) [12,13]

3. POLARIZATION SIGNATURES OF A RANDOM DIPOLE CLOUD

3.1 Dipole Cloud Model

A theoretical assessment of the effectiveness of 'adaptive antenna polarization' with respect to target enhancement in a dipole-cloud clutter environment is of importance to the design of radar systems with improved sub-clutter visibility. The first step in such an assessment is the determination of which polarization signatures of a dipole cloud are suitable for exploitation. A simple statistical model was developed at STC [14] because no model of the scattering properties of a random dipole cloud in terms of polarization could be found.

The main assumptions under which the model was derived are as follows:

(a) The space angle describing the probability space of dipole axis direction is identical for all dipoles and is determined by $0 \leqslant \chi \leqslant \pi$ and $\eta_o \leqslant \eta \leqslant \pi-\eta_o$ (see Fig. 8)

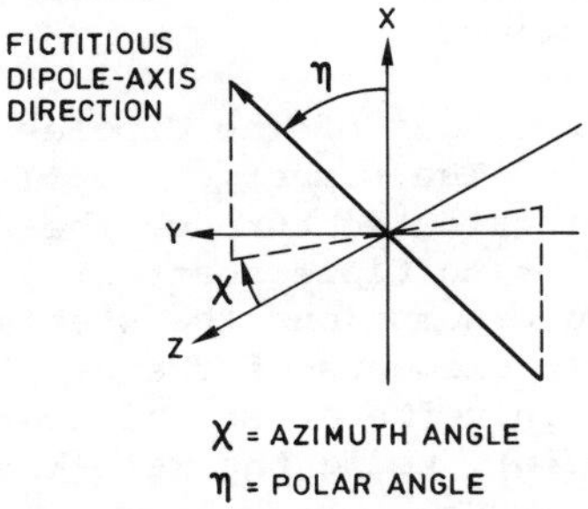

Fig. 8 The dipole in its coordinate frame [14,15]

(b) Each dipole is excited only by that component of the illuminating electric field which is linearly polarized in the plane formed by the dipole axis direction and direction of propagation. In other words, its normalized scattering matrix for $\chi = \eta = \pi/2$ on a basis defined by a pair of horizontally - and vertically - polarized unit vectors is given by,

$$T(\chi = \pi/2,\ \eta = \pi/2) = e^{j\beta}\begin{bmatrix} 1 & 0 \\ 0 & 0 \end{bmatrix}$$

(c) The far-field backscattered pattern of each dipole is proportional to $\sin^2(\gamma)$, where γ is the angle between the dipole axis direction and the direction of propagation of the illuminating field. The normalized general scattering matrix $T(\chi,\eta)$ of the individual dipole can be derived by means of simple matrix algebra, and due to this assumption it takes the form

$$T(\chi,\eta) = \begin{bmatrix} t_{yy} & t_{yx} \\ t_{yx} & t_{xx} \end{bmatrix} = e^{j\beta}\begin{bmatrix} \sin^2(\chi)\sin^2(\eta) & \sin(\chi)\sin(\eta)\cos(\eta) \\ \sin(\chi)\sin(\eta)\cos(\eta) & \cos^2(\eta) \end{bmatrix} \quad (6)$$

(d) The axis direction of the individual dipole is uniformly distributed over the specified space angle. Under this assumption the joint probability density function of the fictitious dipole axis direction, see Fig. 8, is

$$f_{\chi,\eta}(\chi,\eta) = \sin(\eta)/\{2\pi\cos(\eta_o)\} \quad ,(0\leqslant\chi\leqslant\pi, \text{ and } \eta_o\leqslant\eta\leqslant\pi-\eta_o)$$

Further, it is assumed that the absolute phase-angle β is uniformly distributed between $-\pi$ and π, and is independent of χ and η. Under this assumption the first-order statistics (mean values) of the elements of $T(\chi,\eta)$ are zero, and the statistical properties of the scattering matrix elements can be determined in terms of second-order statistics.

(e) The axis directions of the dipoles are independent of one another. The number of dipoles with positions uniformly distributed through the cloud is large, and the effects of multiple scattering can be neglected. Due to these assumptions the statistical properties of the matrix elements of the dipole cloud can be readily derived in terms of second moments (first moments all equal zero), using the well-known central-limit theorem.

The dipole cloud model is characterized by

$$\overline{|t_{yy}|^2} = \sigma^2_{yy} = \frac{1}{8}\left\{15 - 10\cos^2(\eta_o) + 3\cos^4(\eta_o)\right\} \tag{7a}$$

$$\overline{|t_{xx}|^2} = \sigma^2_{xx} = \cos^4(\eta_o) \tag{7b}$$

$$\overline{|t_{yx}|^2} = \overline{t_{yy}t^*_{xx}} = \sigma^2_{yx} = \frac{1}{6}\left\{5\cos^2(\eta_o) - 3\cos^4(\eta_o)\right\} \tag{7c}$$

$$\overline{t_{yy}t^*_{yx}} = \overline{t_{xx}t^*_{yx}} = 0 \tag{7d}$$

where σ^2_{yy} and σ^2_{xx} are the normalized average cross-section for horizontally- and vertically-polarized illumination respectively,

σ^2_{yx} is the normalized average cross-polarized cross-section for horizontally- and vertically-polarized illuminations,

and η_o, where $0 \leqslant \eta_o \leqslant \pi/2$, is the model parameter.

The dipole cloud model has an average axis of symmetry which is oriented along the x-axis (see Fig. 8). Although in a practical case this axis may have an arbitrary orientation, this specific orientation has been chosen to simplify the analysis without affecting the basic signatures.

3.2 Polarization Dependence of Received Backscattered Power

One potentially-exploitable polarization signature of any radar object is that antenna polarization which is identical at transmit and receive and gives a minimum response; the two possibilities are referred to as the average null-polarizations. Obviously such polarizations can be exploited to improve target detection by suppressing the unwanted part of the environment. The effectiveness of adapting identical transmit and receive antenna polarizations for minimum dipole-cloud response, with parameter η_o, has been assessed as follows [14].

The instantaneous normalized received backscattered power from the random dipole cloud is determined by [7,10]:

$$P = |v|^2 = |\underline{T}\underline{h}_T \cdot \underline{h}_R|^2 = |\underline{T}\underline{h} \cdot \underline{h}|^2 = |\underline{E}_s \cdot \underline{h}|^2 \tag{8}$$

where subscripts 'T' and 'R' refer to transmit and receive respectively, the generalized scattering matrix T is given by (6), and the normalized identical transmit and receive antenna polarizations ($\underline{h}$), on a basis defined by a pair of horizonally-polarized (H) and vertically polarized (V) unit vectors ($\underline{i}_y = \underline{i}_H$ and $\underline{i}_x = \underline{i}_V$), is given by:

$$\underline{h} = \begin{bmatrix} \cos(\theta) & ; \underline{i}_H \\ \sin(\theta)\exp(j\delta) & ; \underline{i}_V \end{bmatrix} \tag{9}$$

with $0 \leqslant \theta \leqslant \pi/2$ and $-\pi < \delta \leqslant \pi$ and with right-handed elliptical polarization (RHEP) defined by $0 < \delta < \pi$. The average received power $\overline{P}$ follows from substitution of (6) and (9) into (8) and the use of (7), namely:

$$\overline{P} = \sigma_{yy}^2 \cos^4(\theta) + \sigma_{xx}^2 \sin^4(\theta) + 2\sigma_{yx}^2 [2 + \cos(2\delta)] \cos^2(\theta)\sin^2(\theta) \tag{10}$$

The antenna polarizations (θ, δ) which give minimum and maximum values for $\overline{P}$ in (10) are determined by the following conditions:

$$\frac{\partial \overline{P}}{\partial \theta} = 0 \text{ and } \frac{\partial \overline{P}}{\partial \delta} = 0 \ .$$

The results obtained are presented in Table 3.

Table 3 Dipole-cloud model: the parameters of the 'minimum' antenna polarization and maximum average backscattered power suppression relative to horizontal polarization; dipole orientation parameters $0 \leqslant \chi \leqslant \pi$ and $\eta_o \leqslant \eta \leqslant \pi - \eta_o$ [14]

Dipole-cloud parameter η_o (degrees)	Complex polarization factor representing the minimum antenna polarization $R_{min} = [\tan(\theta)]_{min} \exp(j\delta_{min})$		Maximum average suppression for "minimum" antenna polarization (dB)
	$[\tan(\theta)]_{min}$	δ_{min}	
0	1	$\pm\pi/2$	1.8
15	1.14	$\pm\pi/2$	2.1
30	1.92	$\pm\pi/2$	3.4
41.5	∞	-	6.2
45	∞	-	7.3
60	∞	-	14.0
75	∞	-	26.0
90	∞	-	∞

From this table it can be concluded that 'adaptive antenna polarization' is very promising for cases where the dipole-cloud parameter η_o is larger than 60°.

3.3 Cross-correlation of Orthogonally-polarized Backscatter Components

The cross-correlation of received orthogonally-polarized backscatter components depends both on the transmit and receive antenna polarizations. Extreme values for the cross-correlation obtainable with controllable antenna polarizations can be regarded as radar object polarization signatures, and therefore find application in radar systems which utilize cross-correlation to improve target detection.

The derivation of the antenna polarizations at transmit and receive that give rise to maximum correlated and uncorrelated orthogonally-polarized backscatter components in the receiving channels was the subject of the study presented in [15]. Two mathematical models of complex radar objects were considered: one representing the random dipole cloud and the other a moving, multiple-component body.

In this area of study coherency theory is applied, and so a brief review is given here together with some results obtained for the dipole cloud model.

According to (8) and (9) the electric-field vector of the backscatter wave $\underline{E}_s$ is given by,

$$\underline{E}_s = \begin{bmatrix} E_y ; \underline{i}_H \\ E_x ; \underline{i}_V \end{bmatrix}_s = T\, \underline{h}_T \tag{11}$$

and when the object is time-varying (moving), a time-varying elliptically-polarized backscatter wave is observed with components $E_y(t)$ and $E_x(t)$. In this study the time variation of the backscatter field are attributed to random processes in which the absolute phases of the components $E_y(t)$ and $E_x(t)$ are uniformly distributed and independent of the moduli. Hence the time-averaged values $\langle E_y \rangle$ and $\langle E_x \rangle$ are zero, and in accordance with [6] it is usual to define a time-varying partially-polarized plane wave in terms of the coherency matrix of the complex field components E_y and E_x, i.e.,

$$J = \begin{bmatrix} \langle E_y E_y^* \rangle & \langle E_y E_x^* \rangle \\ \langle E_x E_y^* \rangle & \langle E_x E_x^* \rangle \end{bmatrix} = \begin{bmatrix} J_{yy} & J_{yx} \\ J_{xy} & J_{xx} \end{bmatrix} . \tag{12}$$

The average backscattered power $\langle P_s \rangle$ is then,

$$\langle P_s \rangle = \langle \underline{E}_s \cdot \underline{E}_s^* \rangle = \langle (T\underline{h}_T)\cdot(T\underline{h}_T)^* \rangle = J_{yy} + J_{xx}. \qquad (13)$$

As far as the polarization state of the backscattered wave is concerned, two extreme cases [6] are of special interest:

(a) the wave is completely-unpolarized or randomly-polarized if $J_{yx} = J_{xy} = 0$ and $J_{yy} = J_{xx}$; and

(b) is completely-polarized if $|J| = J_{yy}J_{xx} - J_{yx}J_{xy} = 0$.

In general the backscattered wave will be partially-polarized, and can be uniquely expressed as the sum of two independent waves: one completely-polarized and the other completely-unpolarized. The ratio of the average power of the completely-polarized portion to the total average backscattered power is called the degree of polarization 'p' of the backscattered wave. In terms of coherency-matrix components this ratio [6] is

$$p = 1 - \left\{[4|J|/(J_{yy} + J_{xx})^2]\right\}^{1/2}, \text{ with } 0 \leqslant p \leqslant 1. \qquad (14)$$

Note that the value of 'p' given by (14) depends on $\underline{h}_T$, but is independent of the orthogonal reference frame against which the backscatter wave is measured.

The complex correlation factor (or complex cross-correlation coefficient) of the received voltages (v and $v_\perp$) at the terminals of the orthogonally-polarized receive antennas is determined by

$$\mu = \langle vv_\perp^* \rangle / \left\{\langle P\rangle\langle P_\perp\rangle\right\}^{1/2} = |\mu|\exp(j\nu), \quad 0 \leqslant |\mu| \leqslant 1 \text{ and } -\pi < \nu \leqslant \pi, \qquad (15)$$

where

$$v = T\underline{h}_T \cdot \underline{h}_R \text{ and } v_\perp = T\underline{h}_T \cdot \underline{h}_{R_\perp}, \qquad (\underline{h}_R \cdot \underline{h}^*_{R_\perp} = 0) \qquad (16a)$$

and

$$\langle P\rangle = \langle vv^* \rangle \text{ and } \langle P_\perp \rangle = \langle v_\perp v_\perp^* \rangle \qquad (16b)$$

Furthermore, it is shown in [6] that,

$$|\mu| \leqslant p,$$
$$|\mu| = p \neq 1, \text{ if } \langle P\rangle = \langle P_\perp\rangle, \qquad (17)$$
$$|\mu| = 1, \text{ if } p = 1; \text{ generally indepedent of } \underline{h}_R.$$

Expressions (11) to (17) provide the mathematical basis for the coherency analysis, which is mainly concerned with the determination of the conditions of antenna polarization that result in extreme values of $|\mu|$ for the cases $\underline{h}_R = \underline{h}_T$ and $\underline{h}_R \neq \underline{h}_T$.

By applying (6), (7), (9), (11), (12) and (14), an expression for the degree of polarization of the backscattered wave from the random dipole cloud is obtained as a function of θ_T and δ_T (the parameters which define the transmit antenna polarization) and dipole-cloud parameter η_o.

Subsequently the conditions of θ_T and δ_T for $p = p_{max}$ and $p = p_{min}$ are determined from

$$\partial[|J|/(J_{yy} + J_{xx})^2]/\partial\theta_T = 0 \text{ and } \partial[|J|/(J_{yy}+J_{xx})^2]/\partial\delta_T = 0 . \quad (18)$$

The extreme degree-of-polarization values $p_{max}(\eta_o)$ and $p_{min}(\eta_o)$ are determined by substituting the solutions of (18) into the equation derived for p. Note that p determines the upperbound for $|\mu|$, see (17). In Fig. 9 curves of p_{max}, p_{min}, and p for horizontally polarized (p_{HP}) and vertically polarized (p_{VP}) emission versus η_o are presented. In the case where $\underline{h}_R = \underline{h}_T = \underline{h}$, conditions of $\underline{h}$ that give rise to $\mu = 0$ always exist if $p \neq 1$, and these can be determined analytically by setting $\langle vv_{\perp}^* \rangle$ in (15) equal to zero. Alternatively, $|\mu|_{max}$ and the corresponding antenna polarizations can be determined, by means of a numerical optimization process. For comparison the graph of $|\mu|_{max}$ versus η_o is also shown in Fig. 9.

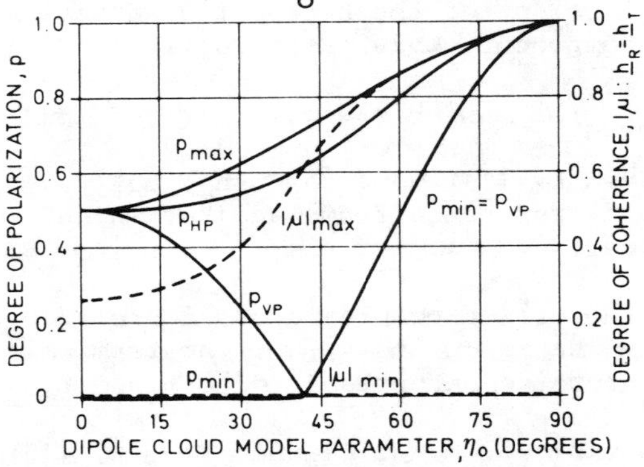

Fig. 9 Dipole-cloud model: particular values of the degree of polarization of the backscattered wave, and for the case $\underline{h}_R = \underline{h}_T$, extreme values of the complex cross-correlation coefficient of received orthogonally-polarized backscatter components versus dipole-cloud parameter η_o; dipole orientation parameters $0 \leqslant \chi \leqslant \pi$ and $\eta_o \leqslant \eta \leqslant \pi-\eta_o$ [15].

In [15] the same analysis that was applied to the dipole cloud model was performed for a model of a moving, multiple-component body representing a simple target. It was concluded that these two models were totally different as far as the cross-correlation of received orthogonally-polarized backscatter components and their dependence on transmit antenna and receive antenna polarization are concerned. This observation might stimulate further studies related to the applicability of this 'cross-correlation' as a controllable parameter for the purpose of improving target detection.

4. VIRTUAL POLARIZATION ADAPTATION FOR TARGET ENHANCEMENT

4.1 Comparison of Actual and Virtual Adaptation

In practice a radar system can be confronted with:

(a) spatially-distributed clutter, either fixed or moving, for which the polarization properties differ from area (volume) to area (volume) or even from resolution cell to resolution cell, and

(b) a number of spatially-distributed interference sources, which moreover have time-varying polarization characteristics.

Therefore in such an adverse environment actual adaptation of the antenna polarizations has severe limitations because:

(a) the severe time-constraints imposed on the adaptation process would be extremely difficult to meet; a particularly stringent condition would be separate adaptation of the transmit and receive-antenna polarizations;

(b) it allows only optimum adaptation to one resolution cell per beam direction, or at best to one group of resolution cells per beam direction when the polarization properties of the cells are strongly correlated (e.g., homogeneous clutter);

(c) when there is no a-priori knowledge available on the optimum antenna polarization to be used, adaptation cannot take place during the periods that the system is in its "learning" mode; and

(d) the antenna sub-system must be capable of generating, under control, every type of wave polarization.

These observations strongly indicate that in a "complex" adverse environment actual adaptation of the antenna polarization is impractical, and so an appropriate concept to overcome the limitations was developed.

In the case of a "complex" clutter environment, the solution is to give the transmit polarization a fixed pattern (in the case of uncoded pulsed waveforms, e.g., alternate horizontal and vertical polarization) and to use orthogonally-polarized receiving channels. Then measured scattering-matrix data are available, which allows the exploitation of the property that when the scattering matrix is known the response for any combination of transmit and receive polarizations is determined. Consequently it allows the adaptation to each individual resolution cell for minimum clutter response to take place within the processor and not at the antenna [16,17,18], and therefore this concept is called "virtual polarization" adaptation (VPA). Consequently, the VPA concept is "learning-process" based, where the learning process runs concurrently with the adaptation process. The essential differences between actual and virtual polarization adaptation are presented in the block diagrams of Fig. 10. It should be remembered here that the improvement in target-to-clutter ratio obtainable through suppression of the clutter returns using polarization filtering depends on the degree of polarization of the set of clutter returns and on the extent to which target polarization signatures differ from those of the clutter (see Section 1.2).

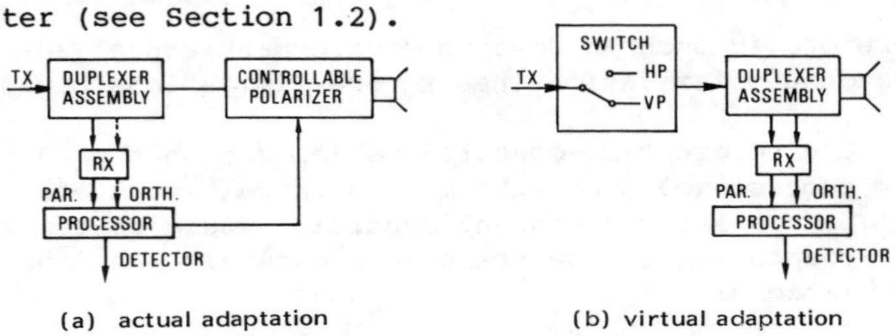

Fig. 10 Adaptive polarization methods for obtaining target enhancement [18]

4.2 Nature of Vector-signal Processing

The simplest form of VPA in a clutter environment involves a type of vector-signal processing which is realized as the measurement and storage of scattering-matrix data (T_i), the derivation of a "minimum polarization ($\underline{h}_{min}$)" from this data (see Section 1.2), and the combination of the stored (or updated) scattering matrix data with the derived minimum polarization to give minimum unwanted signal response (v_i). By way of illustration, Fig. 11 presents a simplified blockdiagram outlining the principles of a VPA processor which adapts identical transmit and receive polarizations to suppress homogeneous clutter. The stage at which virtual adaptation takes place can be described mathematically by $v_i = T_i\underline{h}^T_{min}.\underline{h}^R_{min}$ (see Section 3.2) (where superscripts 'T' and 'R' signify transmit and receive respectively); the term $T_i\underline{h}^T_{min}$ represents the backscatter vector signals ($\underline{E}_i$) resulting from the transmit polarization $\underline{h}^T_{min}$ and is the input to the polarization suppression filter represented by the receive polarization $\underline{h}^R_{min}$; and the mathematical condition for minimum average response is $\langle\underline{E}_i\rangle.\underline{h}_{min} = 0$.

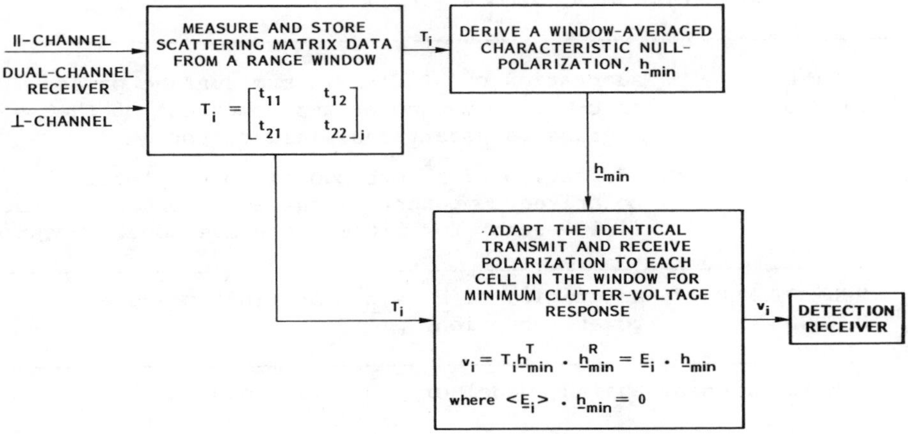

Fig. 11 Simplified block diagram illustrating the original 'virtual polarization adaptation' (VPA) concept; virtual adaptation of identical transmit and receive polarizations to suppress homogeneous clutter [19]

A more detailed mathematical description of the processing steps involved in VPA processing is presented in [16] and [17]. VPA as an operational technique is still in the development phase, since fast algorithms for the calculation of optimum transmit and receive polarizations suitable for real-time implementation have yet to be developed.

From the discussion in Section 1.2 it will be clear that optimum suppression of unwanted returns is obtained when the transmit and receive polarizations are adapted separately. The transmit polarization has to be set such that the degree of polarization of the set of returns is a maximum, so that the average power contained in the completely-unpolarized component of the returns is a minimum (as it is physically impossible to suppress this component); the receive polarization is then set orthogonal to the polarization of the "optimized" completely-polarized component of the returns. The optimum and sub-optimum methods in adaptive clutter suppression are summarized in Table 4.

Note: the optimum adaptive-polarization suppression technique does not necessarily lead to a maximum value of the ratio of average wanted-signal power to average clutter-signal power.

Table 4 Adaptive-polarization clutter suppression; the optimum and sub-optimum methods [16,17,18]

OPTIMUM METHOD	* adaptation of $\underline{h}^T$ for minimum average power in completely-unpolarized component of the average backscattered field-vector * adaptation of $\underline{h}^R$ orthogonal to completely-polarized component of the average backscattered field-vector for minimum average power reception
SUB-OPTIMUM METHOD	* adaptation of $\underline{h}^T = \underline{h}^R$ for minimum average power reception.

4.3 Effectiveness against Model of a Dipole Cloud

An example of the effectiveness of the optimum clutter-suppression method is shown in Fig. 12 for the model of a random dipole cloud reviewed in Section 3.1 [16]. For comparison, the results for the sub-optimum suppression method from Section 3.2 are also included in this figure. From the graphs it can be concluded that considerable suppression of dipole-cloud clutter can be achieved when the dipole-cloud parameter (η_o) is sufficiently large, and that the difference in effectiveness between the optimum and sub-optimum suppression becomes less with increasing values of η_o. As an illustration assume that the dipole cloud is characterized by $\eta_o = 60^o$, and further assume that the adaptation process results in an average loss of 3 dB in target echo strength relative to the use of horizontal polarization. Then from Fig. 12 it can be seen that optimum adaptation results in a value of 14 dB for the improvement factor.

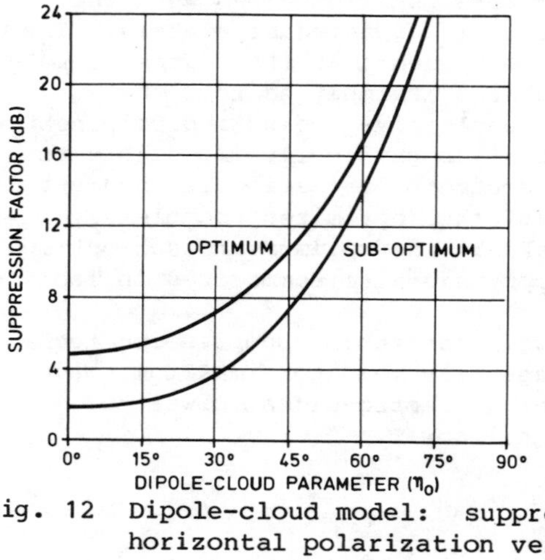

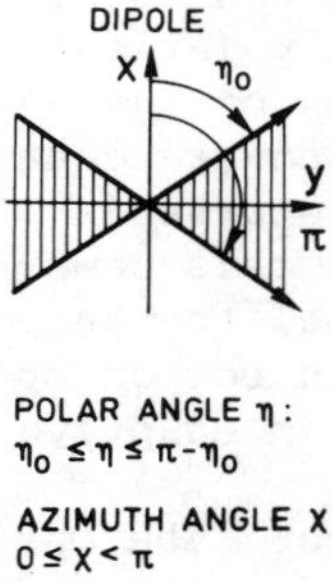

Fig. 12 Dipole-cloud model: suppression factors relative to horizontal polarization versus dipole-cloud parameter η_o [16,17,18]

5. VECTOR SIGNAL PROCESSING ALGORITHMS WITHIN THE VPA CONCEPT

5.1 General

The vector-signal processing inherent in the VPA concept calls for algorithms which are realistically realizable as functional hardware. As well as having to manipulate the states of the signals in polarization space, the algorithms must be fast, and able to cope with system errors. Also, where no "a priori" information about the polarization characteristics of the unwanted background is available, the nature of the algorithm must be based upon a learning-process.

The three processes described in this chapter were derived with the aim of improving the detectability of targets by suppressing backscatter returns from clutter and/or interference signals in the polarization domain. Although the processes are quite general in nature, precipitation is the clutter medium used here to demonstrate their effectiveness.

5.2 The Multinotch Logic-product (MLP) Polarization Suppression Filter

The basis for the design of a filter to suppress unwanted signals in the polarization domain is the fact that the antenna's polarization response to a monochromatic, or completely-polarized wave, can be made equal to zero (Fig. 1). However, when the backscatter/interference signals have fluctuating polarization and the radar system imperfections affect the polarization purity, the average amount of suppression achievable is limited, because the sharp roll-off of this single-notch response makes it a very sensitive nulling process, as is obvious from Figs. 1 and 2.

The solution to this drawback is to desensitize the nulling process by using a filter which covers a much larger area in polarization space. This filter's response should allow a specified minimum average level of suppression over a specified coverage area to be obtained under design control. This has been accomplished [18] by a processor which uses multiple polarization filters and a logic-circuit arrangement. The composite configuration is called a multinotch logic-product polarization suppression filter (or MLP filter) and a schematic diagram of it is shown in Fig. 13. The arrangement consists of a bank of single-notch filters, each matched to a slightly different receive polarization. The backscatter/interference vector-signal samples fed to the bank are processed and presented to a logic-product device to select the smallest response for further consideration.

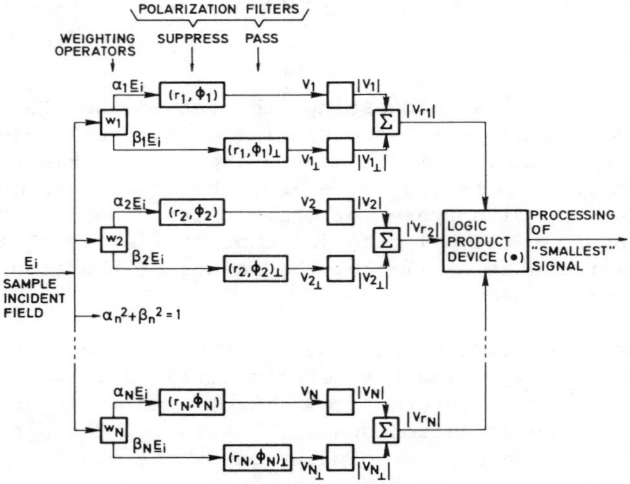

Fig. 13 Schematic representation of a general MLP filter [18]

In general, the location of the MLP filter in polarization space is flexible, in that it can be made to coincide with any 'closed' set of incident field samples to be suppressed. However, in the study reported in [20] precipitation was the principal clutter medium to be suppressed, so the most effective polarization was then CP (as discussed in Chapter 2); therefore in this case the most logical choice for the centre (i.e., location in polarization space) of the MLP filter was either LHCP or RHCP. A further advantage is that this choice simplifies the physical interpretation of the mathematics involved in the design, and so in [20] LHCP was chosen as the filter's centre; RHCP then became the centre of the area of receive polarizations.

Ultimately, the design aim is to achieve an optimum improvement factor (i.e., improvement in the ratio of average wanted-signal power to average unwanted-signal power). This is obviously determined by the spread of the unwanted returns in polarization space about the centroid of the set of samples and the variation of the location of the target returns in polarization space. However, the study described in [20] only addressed the design of an MLP filter in relation to its size and shape in polarization space in order to achieve a specific average suppression level, and the resulting effort required for practical implementation.

Although the shape of the filter in polarization space is essentially free, investigations with returns experimentally derived from rain illuminated with RHCP waves have shown the samples to be approximately circularly distributed about the centroid of the set. This shape may well be regarded as a natural phenomenon for all sorts of homogeneously distributed clutter. Hence, it seems a logical initial choice to consider a filter in which the notches are equi-spaced and circularly distributed in equi-spaced rings about LHCP.

The average level of suppression within the circular area centred on LHCP is determined by the required improvement factor. The size of the MLP filter, the number of notches per ring, and the number of rings needed to cover a specific area, are inter-related under the constraint 'desired level of suppression'. The design aim is to maintain an area in polarization space where the response is not greater than a desired level below the maximum MLP filter response, where the maximum response results from a RHCP incident field.

A cross-section of the normalized logic-product power response on the half great-circle from RHCP through HP to LHCP in polarization space for a typical multi-ring (p-ring) filter is given in Fig. 14. A suppression level of P_s can be achieved over the circular area around LHCP and bounded by the circle of ellipticity ratio r_f.

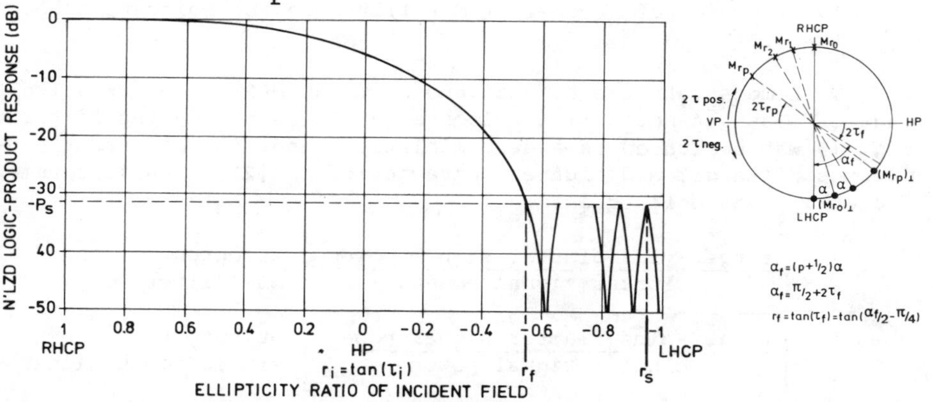

Fig. 14 Logic-product power response for receive polarizations $M_{r_q}(\tau_{r_q} = \pi/4 - q\alpha/2, \varphi_{r_q} = \pi/2;\ q = 0, 1, 2, ..., p)$ and M_i located on half-great circle through HP [20]

One possible layout of the notches making up a MLP filter is shown in Fig. 15. This consists of 5 rings with 6q notches in each ring (where q = 1, ... p ; p = 5). It is obvious from Fig. 15 that, since P_s is to be maintained over the whole area of deep suppression, the critical design parameter is the longest of the inter-cardinal distances between notches in adjacent rings. Coupled with this fact and because the roll-off of the MLP filter is governed by the fundamental response depicted in Fig. 1, the area covered in polarization space and the level of suppression achievable tend to be fixed for each ring. Thus, to cover a larger area, but maintain the value of P_s, one or more rings have to be added.

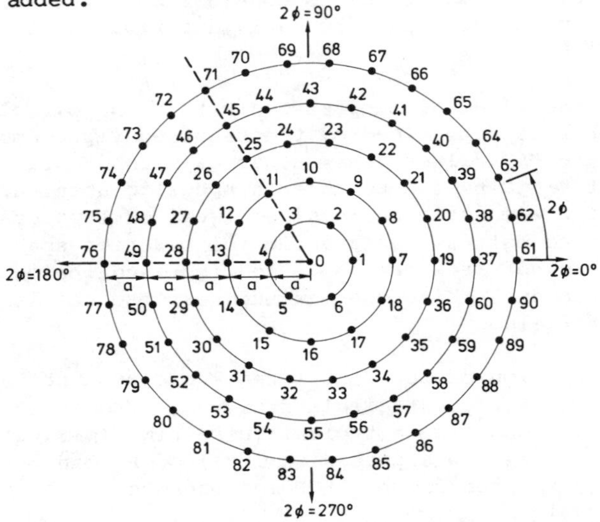

Fig. 15 Location of notches of MLP filter on the Poincaré sphere [20]

To demonstrate the effectiveness of the MLP filter relative to that of a single-notch filter located at LHCP, the 5-ring filter of Fig. 15 was evaluated in a deterministic manner for different conditions of target and clutter characteristics [21]. An improvement factor I_{MF} was defined as,

$$I_{MF} = \frac{\left[\dfrac{\text{target-plus-clutter signal power}}{\text{clutter signal power}}\right] \text{at output of the MLP filter}}{\left[\dfrac{\text{target-plus-clutter signal power}}{\text{clutter signal power}}\right] \text{at output of the single-notch filter}} \qquad (19)$$

and used to quantify the comparison. For simplicity the ellipticity ratio of the fixed clutter return (r_c) was chosen at the edge of the MLP filter (i.e., $r_c = r_f$) and that of the target return (r_t) was chosen outside of the area of deep suppression (i.e., $r_f < r_t \leqslant 1$).

Figure 16 shows I_{MF} versus r_t for a typical set of parameters. Values for the effective "insertion loss" of the MLP filter are also given. This loss is expressed by the ratio of total target-plus-clutter signal power at the output of the system to that at the input, and is an indicator of the usefulness of the processor with regard to signal detectability after processing. The results show that under the stated conditions a 5-ring MLP filter gives a performance in clutter suppression typically 20 dB better than a single-notch filter, and is relatively insensitive to changes in the input signal-to-clutter power ratio (S/C). However, it must be remembered that the 5-ring MLP filter consists of 91 notches, and therefore with currently-available devices is only realizable by a tremendous amount of hardware. When such a filter is to be implemented, this hardware factor will have to be considered even though difficult to quantify [20].

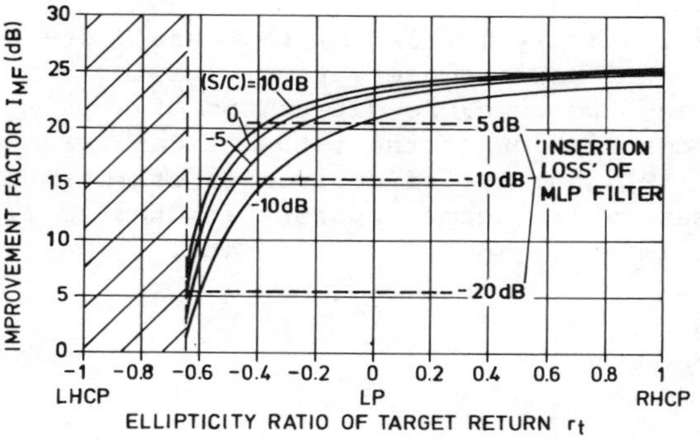

Fig. 16 Theoretical improvement of MLP filter over single-notch filter for different values of the ratio of target signal power and clutter signal power (S/C) at the input to the systems [21]

(i) MLP filter:- 5-ring filter centred about LHCP (see Fig. 15)
- edge of filter $r_f = -0.648$
- average response in deep suppression area is -39.13 dB

(ii) single-notch filter at LHCP

(iii) ellipticity ratio of the clutter return $r_c = -0.649$

5.3 Linear Polarization-vector Translation (Linear PVT)

The previous section stated that the location of a MLP filter can be varied in an adaptive manner to coincide in polarization space with closed sets of incident field samples to be suppressed. Conversely, however, the filter's location can remain fixed and the data sets translated to the area in polarization space that it covers (to achieve the same end). A suitable process to perform this manipulation has been developed called linear polarization-vector translation (PVT) [19]. This is a vector-signal processing concept which translates the polarization of a signal from one state to another whilst conserving total power.

A diagram illustrating the action of linear PVT in translating polarization states on the Poincaré sphere is shown in Fig. 17. Three closed sets of incident field samples to be suppressed by the MLP filter are shown. The transformation operator for each set is determined by:

i) either $\underline{h}_{min}$ or the average incident-field polarization of the set (since $<\underline{E}_i>.\underline{h}_{min}=0$, see Fig. 11); and

ii) the polarization state corresponding to the fixed centre of the MLP filter, denoted by $(\underline{h}_{MLP})_{\perp}$.

This operator is a 2 x 2 unitary matrix (M) that has a determinant $|M| = \pm 1$ (condition of power conservation), giving two solutions: one for each of the two possible directions of rotation. A more detailed description of the mathematical representation is given in [19], but the effectiveness of this simplistic yet elegant process will become apparent in Chapter 7, where it is applied to experimentally-derived rain data.

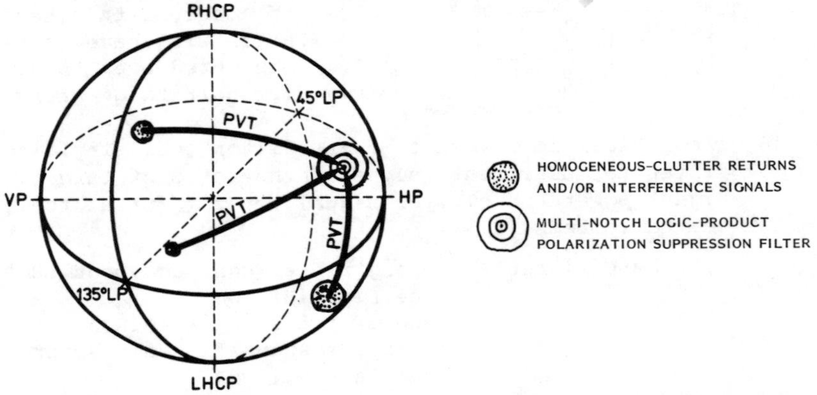

Fig. 17 Poincaré sphere representation of the linear PVT of closed sets of incident-field samples (to be suppressed by the MLP filter at a fixed location in polarization space) [19]

5.4 Non-linear Polarization-vector Translation (Non-linear PVT)

It was pointed out in Section 5.2, that to cover only a moderate circular area on the Poincaré sphere a large number of notches, and hence a large amount of hardware, is required to make up a MLP filter. If this could be reduced, but the degree of suppression maintained, target enhancement would be achieved for a more optimum, acceptable balance between the amount of hardware and the operational benefits. On this premise the requirement is for a vector-process that

i) reduces the area of spread in polarization space of any data set to be suppressed;

ii) separates polarizations such that any difference between the polarization states of wanted and unwanted signals can be further accentuated and exploited (at least when target and clutter are spatially separated); and

iii) is not restricted by the fixed roll-off of the basic response (Fig. 1).

A process that fulfils these requirements is discussed in [21] and is briefly described here.

The desired transfer function of the envisaged vector-process was such that the overall signal processing system would have an equivalent normalized power response of the form shown in Fig. 18. This gives control over the gradient of roll-off, and therefore the process is non-linear in nature. Because the process is still a PVT type, it has been termed the non-linear PVT process.

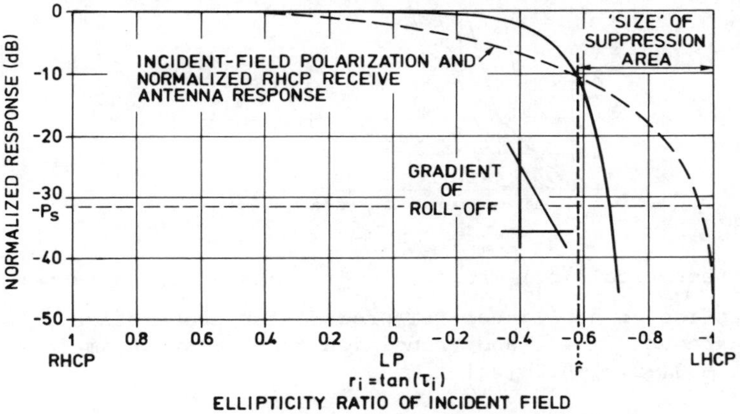

Fig. 18 Desired normalized power response of polarization suppression filter [21] with control of:

(a) 'size' of suppression area (transition $\hat{r}$)
(b) average level of suppression
(c) gradient of roll-off

It is assumed that any data sets to be processed by the non-linear PVT have already been manipulated by the linear PVT such that they are circularly distributed about LHCP (the centre of the MLP filter). This circular distribution means that the design of any process which compresses the set of unwanted samples in polarization space around LHCP need only be governed by the ellipticity ratio of the samples (independent of orientation angle). The non-linear PVT process therefore was designed to translate the ellipticity ratios of signals at the input of the system to:

values closer to LHCP when $-1<(r_i)_{T_L}<\hat{r}$, and

values closer to RHCP when $\hat{r}<(r_i)_{T_L}<1$.

It is obvious that $\hat{r}$ is a process control parameter which determines the transition

$$(r_i)_{T_L,\, T_{NL}} = (r_i)_{T_L},$$

where subscript T_L denotes the linear-PVT processed sample and subscript T_{NL} the non-linear-PVT processed sample. The translation gradient of the process is controlled by a parameter m (see [21]). The characteristics of this process for m = 2 are shown in Fig. 19.

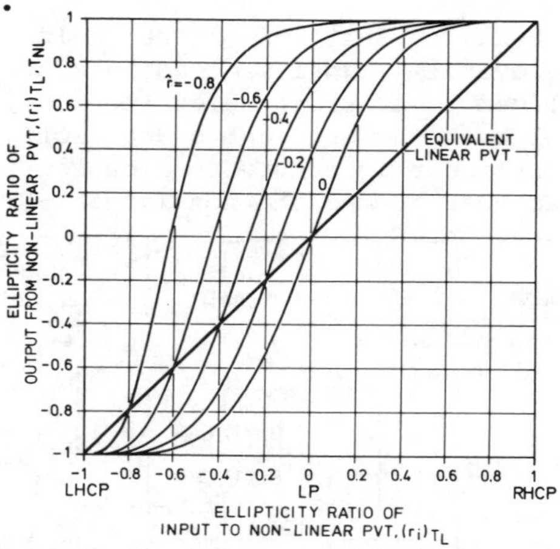

Fig. 19 Theoretical $(r_i)_{T_L}$ to $(r_i)_{T_L,T_{NL}}$ translation characteristics of non-linear PVT process for translation gradient parameter m = 2 and transition constraint parameter $\hat{r}$ = 0(-0.2)-0.8 [21]

A theoretical evaluation of the non-linear PVT combined with a single-notch filter system, relative to the use of a single-notch filter alone, was again performed in a deterministic manner similar to that described in Section 5.2. Improvement factors typical of those that can be expected are shown for two representative sets of parameters in Fig. 20. The non-linear nature of the process means its performance is heavily dependent upon (S/C). These graphs show that an improvement can be expected when using the process compared to a single-notch filter alone, given that $(r_c)_{T_L} < (r_t)_{T_L} \leqslant 1$ and (S/C) > -10 dB. When (S/C) > -5 dB the improvement is roughly 20 dB, i.e. comparable to that obtained with the 5-ring MLP filter of Section 5.2

Figures 20(a) and 20(b) suggest that a larger and larger improvement factor (especially for (S/C) ⩾ 0 dB) can be obtained if $\hat{r}$ is further and further increased, tending in the limit to 1. However, this improvement is obtained at the expense of a reduction in the level of target-plus-clutter signal power at the output of the system. This is shown in Fig. 20 by the "insertion loss" of the system, from which it is clear that for fixed values of (S/C) the area in polarization space for which the insertion loss is less than the indicated levels (-5 dB, -10 dB and -20 dB) in each case becomes smaller as $\hat{r}$ is increased towards 1. This means that although a significant improvement factor has been achieved, the output from the system might become so low that in a practical system target detection would become unreliable.

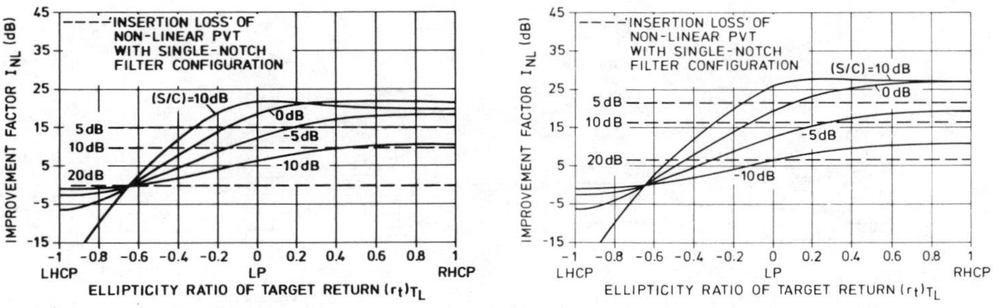

(a) $\hat{r} = -0.2$ (b) $\hat{r} = 0$

Fig. 20 Theoretical improvement factor of non-linear PVT processing combined with the single-notch filter over the single-notch filter alone, for different values of the ratio of target signal power to clutter signal power (S/C) at the input to the systems [21]

(i) non-linear PVT:- translation gradient parameter $m = 2$
- transition constraint parameter $\hat{r} = -0.2$ and 0

(ii) single-notch filter at LHCP

(iii) ellipticity ratio of clutter return $(r_c)_{T_L} = -0.649$

5.5 A Typical VPA Processor

The culmination of the three processes described in Sections 5.2 to 5.4 is the proposed VPA processor shown as a schematic diagram in Fig. 21. It would use the linear and non-linear PVTs in conjunction with a MLP filter. The vector-signals derived from a dual-channel receiver would be fed in parallel to algorithms which derive the control parameters that best configure and match the system to the data to be processed.

Section 5.4 suggested that under specific conditions the use of such a processor in conjunction with a single-notch filter should give an improvement of about 20 dB in the wanted-to-unwanted signal power ratio. Thus, a further improvement should be possible when a MLP filter is incorporated. The application of a 2-ring MLP filter would require typically 19 notches, which with current technology probably represents an optimum compromise between the detection performance in adverse environments and the amount of hardware needed for realization.

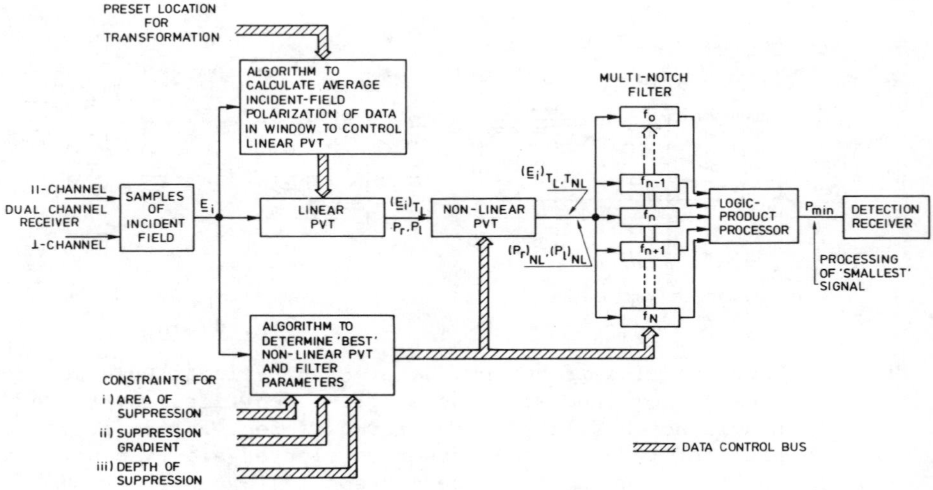

Fig. 21 Schematic diagram of a VPA processor which uses a MLP filter in conjunction with the linear PVT and the non-linear PVT

6. STC's EXPERIMENTAL X-BAND RADAR FACILITY

6.1 Major Features

The general requirements for a radar system with the ability to measure the polarization characteristics of backscatter-returns/ interference have been discussed in the earlier sections of this paper. A block diagram of the system which has evolved at STC to meet these needs is shown in Fig. 22. It is a multi-purpose X-band radar system which has two 6-ft diameter dish antennas, one with a dual-polarized horn feed, the other a monopulse feed. The former transmits and receives signals from which the experimental data is derived; the latter operates only in a receive mode and is used to track targets.

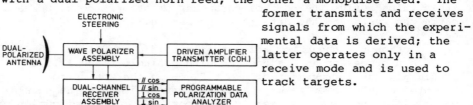

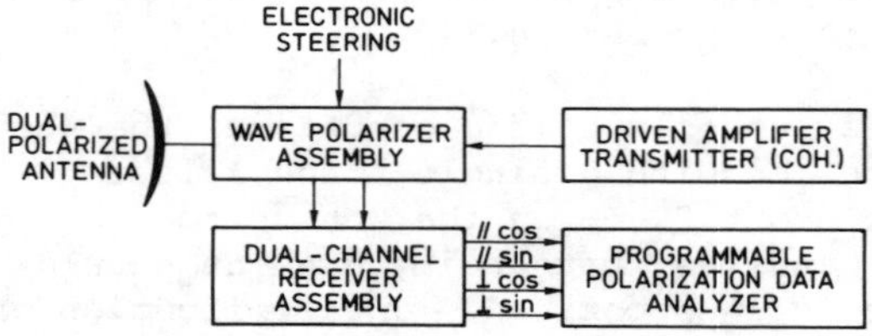

Fig. 22 Simplified block diagram of experimental X-band radar facility for polarization studies [18]

The antennas are mounted side-by-side on a 3-D pedestal which permits full steering in azimuth and elevation. A photograph of the antenna system is shown in Fig. 23.

Fig. 23 The X-band antenna system; dual-polarized antenna and monopulse antenna mounted side-by-side on a 3-D pedestal

The heart of the system is the wave polarizer which allows the radar to operate with many different states of polarization. Figure 24 is a simplified block diagram of the "RF head", with the wave polarizer assembly and control. This microwave portion of the system is housed in a box mounted directly behind the dual-polarized antenna. At transmit the signal passes through a duplexer and is split equally in phase by a hybrid tee. The two

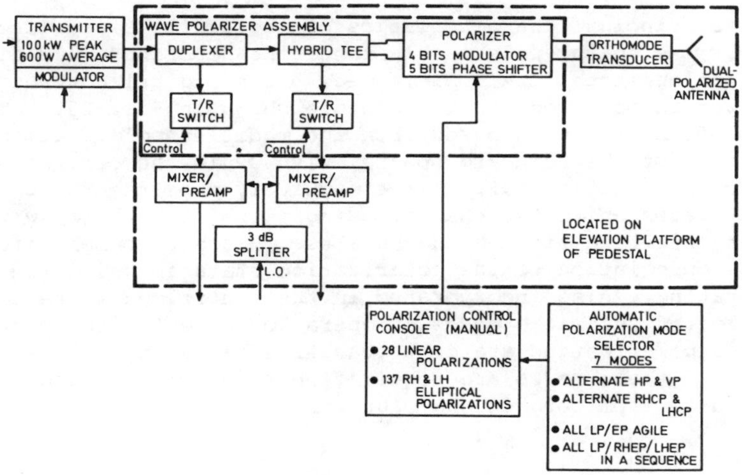

Fig. 24 Block diagram of RF head with wave polarizer assembly and control

outputs are fed to the polarizer, in each arm of which there is a 4-bit ferrite phase shifter covering the range 0° to 90° in steps of $6\frac{3}{7}^{\circ}$. The phase-shifted signals are then cross-coupled by a hybrid sidewall coupler, and each of its two outputs connected to a 5-bit ferrite phase shifter. This has a range of 0° to 180°, again covered in steps of $6\frac{3}{7}^{\circ}$.

The final two outputs (horizontally-and vertically-polarized) feed an orthomode transducer which launches power into the dual-polarized horn feed. With this arrangement combinations of phase shift produced by the phase shifters result in weighted (in amplitude and relative phase) horizontally- and vertically-polarized components that recombine in the orthomode transducer and antenna. The result is a transmitted wave whose polarization state can be controlled.

The situation is similar at receive, except that the phase shifters are non-reciprocal devices. This can, however, be compensated at the control stage such that at the output of the hybrid tee two channels are available: one parallel-polarized with respect to the setting of the polarizer at transmit, the other orthogonally-polarized. These two channels, having passed through the duplexer and transmit/receive switches to achieve transmit/receive isolation, are amplified and down-converted to a 120-MHz IF. Subsequently, these signals are down converted to a second IF of 30 MHz and the video signals detected and digitized such that their phase and amplitude are obtained. This results in four 8-bit I and Q channels.

The principal characteristics of the facility are listed in Table 5. With the polarizer arrangement described, 302 different polarization states are implemented. The digital control of the polarizer is achieved via the console shown in Fig. 25. The design of this unit is based upon the modified polarization chart projection of the Poincaré sphere, with CP in the centre as described in Section 1.2. It effectively maps the 302 polarization states on to two charts, using a "shift-key" to switch between left- and right-handed states. Such a console gives a direct appreciation of the polarization state in which the system is operating. Also incorporated in the control unit are several pre-selected automatic modes of operation which switch between specific polarization states at variable rates up to 10 kHz. Several of these modes are required when deriving scattering matrix data from targets and clutter.

Table 5(a) The STC multipurpose experimental X-band radar facility: features of the antenna system and the transmitter

ANTENNA SYSTEM			
* Dual-polarized antenna	: pencil-beam 3 dB beamwidth		~1.2°
	: max. sidelobe level		-25 dB
	: gain		42 dB
	: cross-polarization isolation		> 25 dB
	: controllable polarization		
	(a) manual	: 28 linear (LP)	
		: 137 RH/LH elliptical (RHEP/LHEP)	
	(b) automatic	: alternate HP/VP	
		: alternate RHCP/LHCP	
		: all LP/RHEP/LHEP in sequence	
		: all LP/EP agile	
* Tracking antenna	: monopulse feed		
* 3-D pedestal	: antennas mounted side-by-side, allowing target tracking and controllable 2-D/3-D scanning modes		
TRANSMITTER			
* Peak/average power	: 100 kW/600 W		
* Frequency	: 9532 (4) 9568 MHz		
* Pulsewidth	: 0.4 to 2.4 μs, variable		
* Pulse repetition time	: 0.1 (0.001) 10 ms		
* Frequency modes	: fixed (diversity and agility to be implemented in near-future)		

Table 5(b) The STC multipurpose experimental X-band radar facility: features of the receiver and the polarization data analyser/processor

RECEIVER		
* Type	:	dual-channel (with orthogonally-polarized channels)
* Video outputs	:	I and Q linear analogue and digitized 8-bit
	:	analogue logarithmic
POLARIZATION DATA ANALYSER/PROCESSOR		
* Coherent and non-coherent processing		
* Measurement window controllable in range and azimuth/elevation		
Sampling rate	:	currently 650 kHz max. (5 MHz max. to be implemented in near-future)
Window measurement repetition time	:	0, 5, 10 or 20 s
Number of sweeps in window	:	1(1)99
Samples in range	:	8, 16 or 32
* PDP 11/34 minicomputer		
Memory capacity	:	256 kbytes
Disk storage capacity	:	60 Mbytes
Plotter and hardcopy outputs		
Colour graphics facility for dynamic display of output		

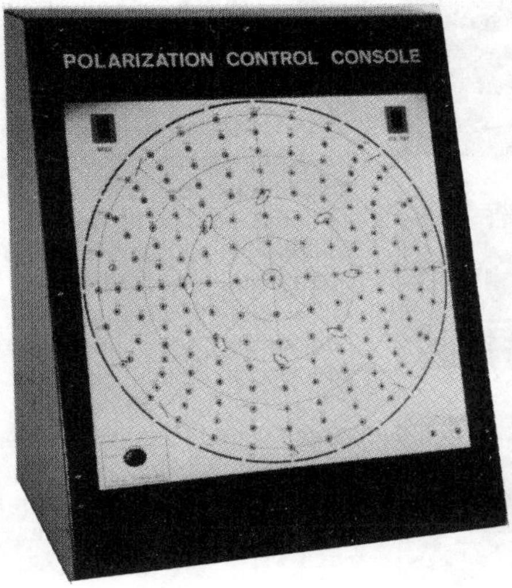

Fig. 25 STC developed 'polarization control console' (2nd generation)

6.2 Some Early Experimental Results

By early 1977 the STC X-band radar facility had evolved to the stage where the 30-MHz IF channel responses could be displayed on an A-scope, and thus were available for preliminary investigations regarding the polarization dependence of received backscattered signals from radar objects. Of principal interest for the project at that stage was the "practical" confirmation of the theory [7,9] stating that any single radar object (i.e., an object of which the backscattering properties can be represented by a scattering matrix) possesses so-called "characteristic polarizations". These are defined as those polarizations of identical transmit and receive antennas which produce maximum and minimum (i.e. zero) responses at the receiver terminals; in general for any object there exists two characteristic minimum polarizations and one characteristic maximum polarization. At that time experimental data of this nature were very scarce in the open literature (and currently still are).

Some experimental results have been published in [22]. As a typical example, the responses for specific polarizations obtained from an isolated boresight tower are presented in Fig. 26. The characteristic polarizations were obtained by trial and error using the polarization control console of Fig. 25, and the responses appeared to be stable over a prolonged period (many hours, depending on weather conditions). It should be noted that, in contrast to the maximum response, the minimum response is very sensitive to a change of antenna polarization. Furthermore, since the boresight tower represented a swaying multiple-component body, the back-scattered waves were partially-polarized and thus a zero response could not be obtained. Moreover it should be noted that since only discrete polarizations were available, deep nulling of the response in the parallel channel was limited, even when the incident field was completely polarized.

These results indicated that for man-made objects (such as bore-sight towers, buildings, etc.) the maximum to minimum response ratio in the parallel channel was between 15 and 30 dB.

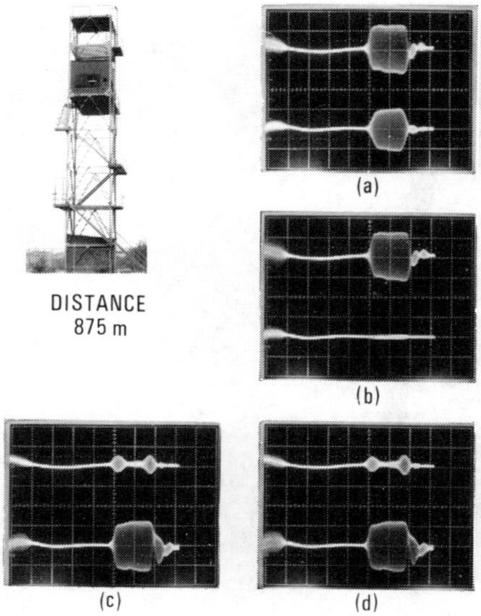

Fig. 26 Boresight tower and responses at 30 MHz IF level for specific polarizations; the upper-trace presents the parallel-channel response and the lower-trace the orthogonal-channel response [22,17]

(a) Horizontal polarization (reference)
(b) Maximum polarization
(c) Null-polarization 1
(d) Null-polarization 2

It was concluded that the results were therefore encouraging enough to justify further investigations of the applicability of adaptive polarization for clutter cancellation and consequent target enhancement.

With the implementation in late 1979 of video channels (analogue and digital), a digital core store, and a limitedly-programmable desk calculator (as an intermediate solution to the inclusion of a PDP-11 minicomputer in 1982), the facility was capable of small-scale measurements and subsequent analysis of the data in order to derive object characteristic polarizations. Characteristic polarizations calculated from single-cell scattering-matrix measurements made at a fixed antenna bearing are presented on modified polarization charts in Figs. 27 and 28 for an isolated building and a foliage area respectively; they give an impression of how object polarization signatures change with increasing observation time.

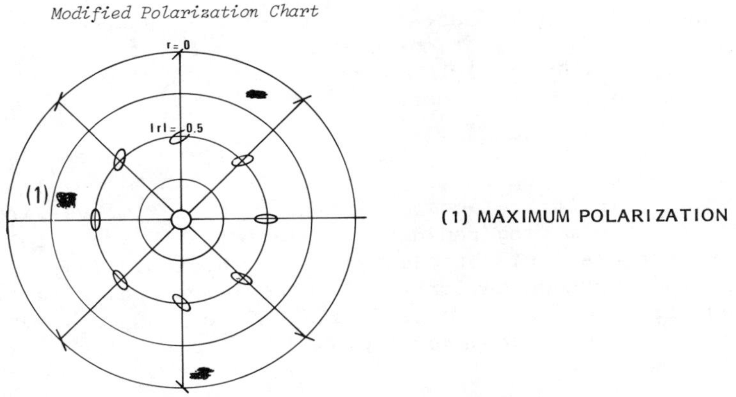

Fig. 27 Characteristic polarizations of isolated apartment building; 50 consecutive polarization forks, observation time 0.99 s (transmit mode HP◄►VP alternately, 100 Hz prf, 2.0 μs pulse duration, 3.15 km sampling range)

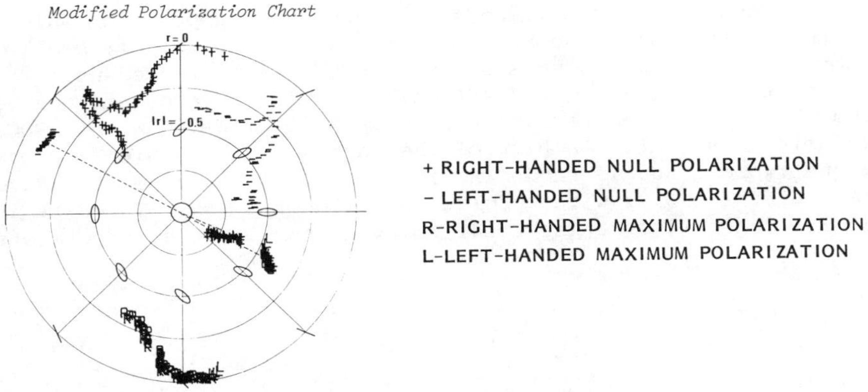

Fig. 28 Characteristic polarizations of fluctuating clutter; slow and fast moving foliage (short and long tracks respectively), observation time T = 32 ms (fast) and T = 48 ms (slow), 2.5 KHz prf, 0.4 μs pulse duration [18]

The results relating to the isolated building (Fig. 27) together with those for the boresight tower (Fig. 26) indicate that unwanted returns from an urban area dominated by fixed objects can be significantly suppressed by adaptive polarization. It should be noted that Fig. 27 also shows the mutual-coherent dual-channel receiver system to be very stable in operation, at least in the short-term sense, taking into account that the building represents a multiple-component body. The results for foliage clutter in Fig. 28 [18] indicate that single-cell clutter-suppression based on long-term averaged adaptive polarization would result in only a moderate improvement. However, if there was also spatial polarization-correlation a useful amount of suppression could probably be obtained via adaptive "window-averaged" polarization processing.

The general conclusion from these preliminary experiments is that man-made objects and ground clutter exhibit distinctive characteristic polarizations. In the case of time-varying clutter they are moving rapidly in polarization space, following apparent characteristic "trajectories". However, the exploitation of these phenomena for target discrimination, classification and identification seems to be in its infancy and much more data will have to be collected and analysed before an extensive range of algorithms can be derived and evaluated.

7. PERFORMANCE OF VECTOR-SIGNAL PROCESSING SCHEMES WITH EXPERIMENTALLY-DERIVED RAIN DATA

7.1 The Data Set

To study the effectiveness and degree of suppression which can be achieved with the signal processing schemes described in Chapter 5, many closed sets of data have been collected using the STC experimental X-band radar facility. These sets were derived from the backscatter radiation from rain illuminated with a RHCP transmitted signal. A plot of the incident field polarizations of one typical set is shown in Fig. 29(a). The transmitted signal had a prf of 200 Hz with a pulse duration of 2 μs. The data is for 255 sweeps (hence the total observation time was approximately

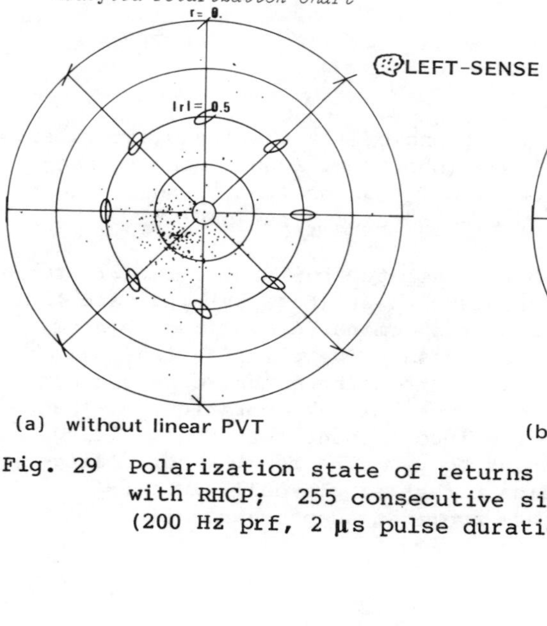

(a) without linear PVT

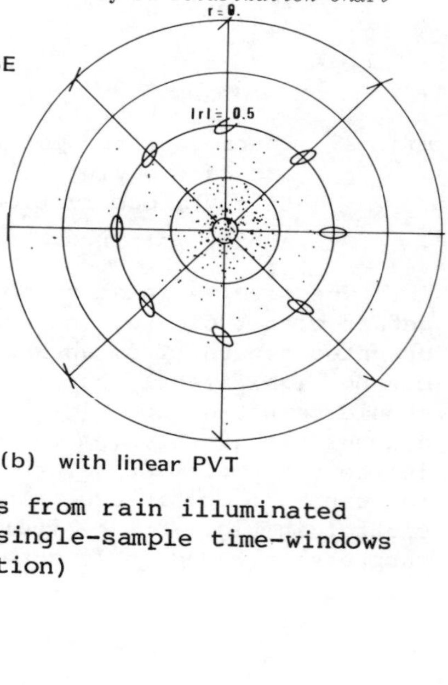

(b) with linear PVT

Fig. 29 Polarization state of returns from rain illuminated with RHCP; 255 consecutive single-sample time-windows (200 Hz prf, 2 μs pulse duration)

1.2 s) of a single range-cell located some 5 km from the radar, at an elevation angle of 17°.

The plots show that the polarization of the returns is predominantly left-handed (i.e., orthogonal to the polarization of the transmitted signal), but off-set from LHCP. Since this off-set is determined by environmental conditions and can be distributed anywhere around LHCP, the natural choice for the location of the centre of the suppression filter when attempting to suppress rain clutter is LHCP.

7.2 Configuration of Linear PVT plus MLP Filter

It is obvious from the plot of Fig. 29(a) that a sub-optimum value of suppression would be achieved if this data set were applied directly to only the MLP filter centred at LHCP, because of the circular layout of the notches and the off-set of the data from LHCP. The suppression can be increased by applying the linear PVT, whose controlling algorithm calculates the centroid of the set and translates this to LHCP with the data samples following accordingly. Figure 29(b) shows the centralizing effect of the linear PVT. The spread of the returns in polarization space is now more discernible and readily demonstrates the need for a filter with an area of deep suppression much wider than that of the antenna's response alone. It is also evident that for this data set the returns are approximately circularly distributed, with the majority contained in an area around LHCP and bounded by an ellipticity ratio of -0.7. Thus, "a priori", knowledge of the data set is in this case available, and so derivation of the filter parameters is easy.

Table 6 gives a summary of the performance which was achieved when processing this same data set with the linear PVT and a MLP filter of varying dimensions, based upon the typical design example of Section 5.2. The best obtainable improvement in suppression relative to the use of a single-notch filter alone at LHCP, was approximately 15 dB.

Table 6 Configuration of linear PVT and MLP filter; summary of results with the experimentally-derived data set from rain of Fig. 29(a)

Ellipticity ratio at edge of deep suppression of MLP filter (r_f)	Suppression due to use of linear PVT	Additional suppression due to the q-th ring					Increased suppression of 5-ring filter over single-notch filter at LHCP	Overall suppression relative to total received power
		q=1	q=2	q=3	q=4	q=5		
- 0.59	3.34 dB	3.20 dB	3.33 dB	2.26 dB	1.68 dB	1.34 dB	15.15 dB	32.69 dB

7.3 Configuration of the Linear PVT plus the Non-linear PVT and a Single-notch Filter

The data set referred to in Section 7.1 was applied to a system using all three processes described in Chapter 5, but with only a single-notch filter for suppression. The unprocessed data, having been centralized around LHCP by the linear PVT, were applied to the non-linear PVT with translation gradient parameter m of order 2 and different values of transition constraint parameter $\hat{r}$. Figures 30(a) and (b) show two typical plots of the polarization states of the processed data for $\hat{r} = -0.8$ and -0.2, respectively. The nature of the non-linear PVT is well-demonstrated by Fig. 30(a). Because many of the returns lie outside the circle of ellipticity ratio -0.8, processing by the non-linear PVT when $\hat{r} = -0.8$ translates them away from the single-notch filter; on the other hand, the returns inside this circle are compressed to LHCP. In fact the overall effect for this data set is to reduce the degree of suppression, as is evident from Table 7. However, as $\hat{r}$ is increased, more and more samples are translated towards LHCP, and therefore suppressed. Figure 30(b), for $\hat{r} = -0.2$, shows almost all the returns as being close to LHCP, and Table 7 gives the corresponding degrees of suppression.

For this particular data set, the linear PVT plus the non-linear PVT and the single-notch filter configuration gave an overall suppression, relative to that with a single-notch filter alone, of 16.8 dB. This was approximately 1.6 dB more than was obtained with a linear PVT plus MLP filter configuration processing the same data set.

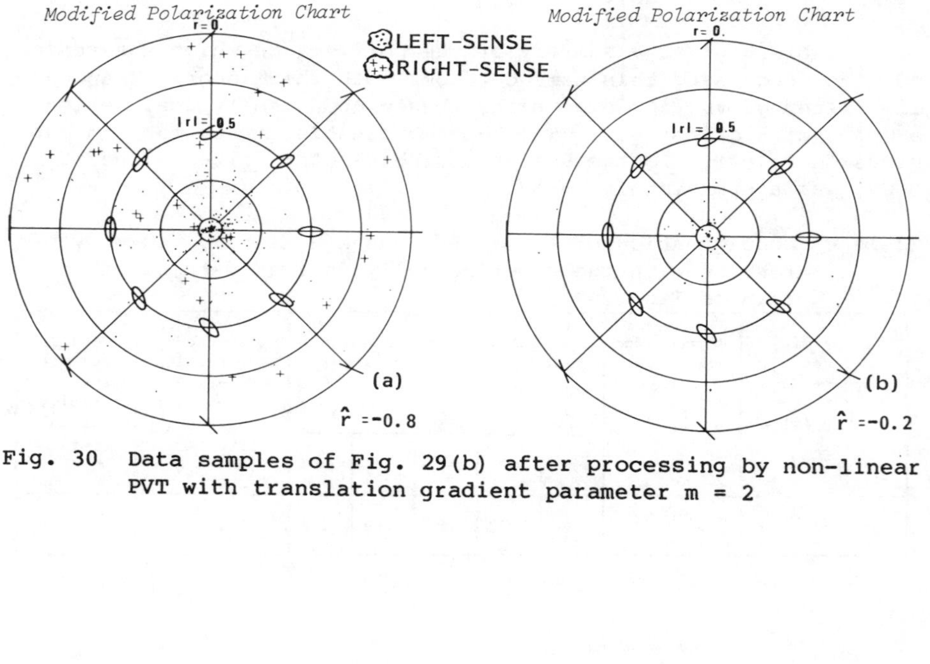

Fig. 30 Data samples of Fig. 29(b) after processing by non-linear PVT with translation gradient parameter m = 2

Table 7 Configuration of linear PVT plus non-linear PVT and a single-notch filter; summary of results with the experimentally derived data set from rain of Fig. 29(a)

Suppression due to use of linear PVT	Additional suppression due to use of non-linear PVT of order m = 2 for constraint values of $\hat{r}$ ranging from -0.8 to 0					Increased suppression over single-notch at LHCP; m = 2 and $\hat{r}$ = -0.2	Overall suppression relative to total received power; m = 2 and $\hat{r}$ = -0.2,
	-0.8	-0.6	-0.4	-0.2	0.0		
3.34 dB	-5.70 dB	2.09 dB	9.23 dB	13.43 dB	18.48 dB	16.77 dB	34.31 dB

7.4 Discussion of Processor Applications

It should be emphasised that in the application related to the detection of targets in an environment containing clutter and/or other unwanted signals, it is the ratio of average target-signal power to average unwanted-signal power which is important. If the polarizations of the wanted returns (i.e., those from a target) are close to those of the unwanted, the target signals will also be suppressed, and hence little improvement would result.

It was stated in Chapter 2 that approximately 3 dB loss in target detectability can be expected when using fixed circular polarization in a clear and quiet environment, relative to the use of fixed horizontal polarization. In conjunction with Fig. 1 this fact suggests that the returns from targets are, on average, linearly polarized. It follows also from Fig. 14 that target returns then suffer an additional loss when processed by the MLP filter centred about LHCP, and this loss increases with increasing number of rings, to reach about 2.3 dB for the 5-ring system of Section 5.2. Hence, the improvement-in-suppression figures in Table 6 should be reconsidered taking this additional loss into account; it follows then that if the target returns are, on the average, linearly polarized, an improvement of approximately 13 dB can be expected in target-to-rain clutter power ratio, relative to the use of a single-notch filter at LHCP.

In the non-linear PVT case the situation is more complex. The evaluation in [21] indicates that in a target-plus-rain environment where the signal-to-clutter ratio (S/C) > -5 dB, the effectiveness of a system using non-linear PVT characterized by

$\hat{r} = -0.2$ and $m = 2$ plus a single-notch filter is of the same order as that of the MLP filter, but incurring an extra "insertion loss" of approximately 5 dB. However, this performance can be improved by selecting $\hat{r} > -0.2$. Furthermore, in the less problematic environment where the target and the rain are spatially separated, the non-linear PVT incurs minimal "insertion loss", because the target returns are more distinguishable from the unwanted returns due to the separating action of the non-linear PVT, where the degree of separation is governed by $\hat{r}$ and m.

Since it is reasonable to expect a real-time adaptive system to react in a much shorter time than the observation periods discussed here (Fig. 29), the effective spread of the returns could well be less, and a larger overall improvement in suppression could be obtained with the processing schemes discussed in Chapter 5. Moreover, a smaller spread can mean reductions in the large amount of hardware and the lengthy processing times involved.

The realization of a 61-notch MLP filter designed to operate in real-time already involves a considerable amount of hardware, whereas the non-linear PVT approach requires less hardware and can achieve approximately the same performance as an equivalent MLP filter when (S/C) > -5 dB. However, the use of the non-linear PVT in conjunction with, say a 2-ring MLP filter, may result in an even better performance (see Section 5.5).

Encouraged by the results of these studies [19,20,21], preliminary investigations have been made regarding the realization of real-time practical processors based upon the MLP filter, linear PVT, and non-linear PVT.

8. FUTURE STUDIES

STC's future research work related to the utilization of polarization information in primary radar will be devoted to such of the following subjects as resources permit:

(a) The large scale measurement of complete scattering matrix data of aircraft targets-of-opportunity, ground clutter, and weather clutter; these measurements will be used to assess the statistics of the object's polarization signatures at fixed and multiple frequencies.

(b) The development of statistical models for target/clutter backscattering properties in terms of polarization, from measured data.

(c) Off-line analysis (using large-scale live data) of the effectiveness of the developed vector-signal processing algorithms when used for target enhancement in precipitation clutter.

(d) Theoretical assessment of the effectiveness of the developed vector-signal processing algorithms when used for target enhancement in a random dipole cloud.

(e) Feasibility studies related to the real-time implementation in primary radar systems of the vector-signal processing algorithms developed.

(f) Development of algorithms for the adaptation of the transmit polarization in a VPA processor, and suitable for real-time implementation.

(g) Assessment (using live data) of the compatibility of frequency agility and vector-signal processing in a weather clutter environment.

(h) Off-line analysis (using live precipitation data) of the applicability/effectiveness of combined frequency-domain and polarization-domain signal processing.

(i) Assessment (using live data) of the applicability of ortho-coherent MTI-like signal processing (see Fig. 31) for target enhancement [17,18].

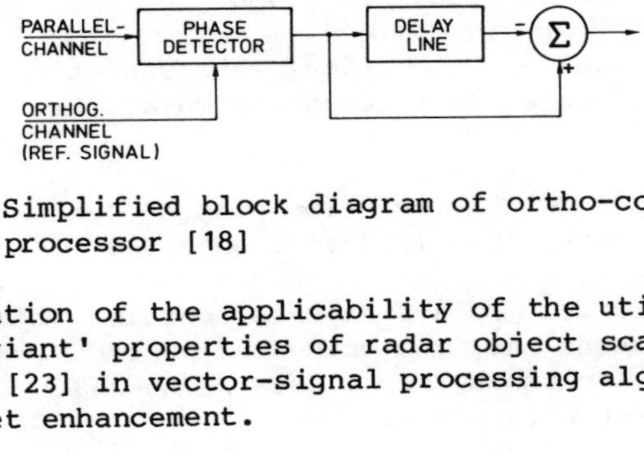

Fig. 31 Simplified block diagram of ortho-coherent MTI processor [18]

(j) Investigation of the applicability of the utilization of 'invariant' properties of radar object scattering matrices [23] in vector-signal processing algorithms for target enhancement.

9. ACKNOWLEDGEMENT

The authors would like to take this opportunity to acknowledge their colleagues who have been, and still are, involved with this project. These are: Mr. G.J. de Zwart, Mr. K.W.P. Krom, Mr. L. van Slingerland, Mr. A.J. Lipman, and Mr. J.P. van der Voort, all of whom provided the technical support, and the members of the Drawing Section and the Workshop Section. Further, thanks are expressed for the secretarial assistance and to colleagues in the Editorial Services Branch, in particular Mr. R. Cameron (who has been involved with the editing of the majority of the publications related to this project) and Miss R. Brown who typed the final version of this presentation.

10. REFERENCES

[1] BOERNER, W.M., EL-ARINI, M.B., CHAN, C.J., and MASTORIS, P.M.: 'Polarization dependence in electromagnetic inverse problems', IEEE Trans., 1981, AP-29, 2, pp. 262-271

[2] WHITE, C.D.: 'Circular radar cuts rain clutter', Electronics, 1954, 27, pp. 158-160

[3] SKOLNIK, M.I.: 'Radar Handbook', McGraw-Hill, 1970, Section 28.4

[4] SINCLAIR, G.: 'The transmission and reception of elliptically polarized waves', Proc. IRE, 1950, 38, pp. 148-151

[5] DESCHAMPS, G.A.: 'Geometrical representation of the polarization of a plane electromagnetic wave', Proc. IRE, 1951, 39 pp. 540-544

[6] BORN, M., and WOLF E.: 'Principles of optics', Pergamon Press, New York, 1965 (3rd edition), Section 10.8

[7] KENNAUGH, E.M.: 'Polarization properties of radar reflections', Antenna Lab, The Ohio State University Research Foundation, Columbus, Rept. 389-12 (AD 2494), March 1952, RADC Contract AF28(099)-90

[8] GRAVES, C.D.: 'Radar polarization power scattering matrix', Proc. IRE, 1956, 44, pp. 248-252

[9] HUYNEN, J.R.: 'A new approach to radar cross-section measurements', IRE Internat. Conf. Record, Pt. 5, 10, pp. 3-11.

[10] HUYNEN, J.R.: 'Measurement of the target-scattering matrix', Proc. IEEE, 1965, 53, pp. 936-946

[11] POELMAN, A.J.: 'Performance evaluation of two types of radar system having a circular-polarization facility', SHAPE Technical Centre, The Hague, The Netherlands, Technical Memorandum TM-276 (AD 883 960), April 1971 (NATO Unclassified, limited distribution)

[12] POELMAN, A.J. and TIMMERS, F.L.: 'Detection performance of an incoherent radar with two orthogonally-polarized receiving channels', SHAPE Technical Centre, The Hague, The Netherlands, Technical Memorandum TM-435 (AD 921 015), June 1974 (NATO Unclassified, limited distribution)

[13] POELMAN, A.J.: 'On using orthogonally-polarized receiving channels to detect target echoes in gaussian noise', IEEE Trans., 1975, AES-11, 4, pp. 660-663

[14] POELMAN, A.J.., and van der VOORT, J.I.: 'The polarization dependence of received back-scattered power from a random dipole cloud', SHAPE Technical Centre, The Hague, The Netherlands, Technical Memorandum TM-335 (AD 520 934), May 1972 (NATO Unclassified, limited distribution)

[15] POELMAN, A.J.: 'Cross-correlation of orthogonally-polarized backscatter components', IEEE Trans., 1976, AES-12, 6, pp. 674-682

[16] POELMAN, A.J.: 'Reconsideration of the target detection criteria based on adaptive antenna polarizations', AGARD Conf. Proc. No. 197 on new devices, techniques and systems in radar, February 1977

[17] POELMAN, A.J.: 'The applicability of controllable antenna polarizations to radar systems', Tijdschrift van het Nederlands Elektronica- en Radiogenootschap (The Netherlands), 1979, 44, pp. 93-106 (in English)

[18] POELMAN, A.J.: 'Virtual polarization adaptation, a method of increasing the detection capability of a radar system through polarization-vector processing', Proc. IEE, Pt. F, 1981, 128, 5, pp. 261-270. (The substance of this paper was first presented at the Military Microwave '80 Conference held in London in October 1980 under the title 'A study of controllable polarization applied to radar'.)

[19] POELMAN, A.J.: 'Polarization vector translation in radar systems', Proc. IEE, Pt. F. 1983, 130, 2, pp. 161-165

[20] POELMAN, A.J., and GUY, J.R.F.: 'Multinotch logic-product polarization suppression filters; a typical theoretical design example and its performance using live rain data', submitted to Proc. IEE, Pt. F, in March 1983; accepted September 1983

[21] POELMAN, A.J., and GUY, J.R.F.: 'Non-linear polarization vector translation in radar systems; a promising concept for real-time polarization-vector signal processing via a single-notch suppression filter', submitted to Proc. IEE, Pt. F, in April 1983; accepted September 1983

[22] POELMAN, A.J.: 'Complex radar objects and polarization dependence of received backscattered echoes', Elec. Lett. (UK), 1977, 13, 15, pp. 433-434

[23] BICKEL, S.H.: 'Some invariant properties of the polarization scattering matrix', Proc. IEEE, 1965, 53, 8, pp. 1070-1072

III.3 (MM.2)

POLARIZATION MEASUREMENTS IN RADIOASTRONOMY: DETERMINATION OF THE POLARIZATION - REFLECTION - MATRIX

Manfred A.F. Thiel

Angewandte Mathematik, Maschinenbauwesen
Fachhochschule Köln, Am Reitweg 1
D-5000 KOLN 1, FR Germany

Synopsis: The following represents a summary of the paper presented by Prof. Manfred Thiel during the Workshop (MM.2) which in great detail is contained in an excellent monograph (Thiel, 1970) containing the findings of his diploma thesis (1968).

Determination of the Polarization - Reflection - Matrix

When electromagnetic radiation is reflected by some device, its state of polarization is, in general, changed. In the quasi-monochromatic approximation, this can be described (Thiel, 1968) by a linear transformation, the operator of which is chatacteristic for the device and does not depend on the incident radiation.

Let us represent the state of polarization by the Stokes vector $\vec{s}$, unprimed for the incident state and primed for the transformation state. Then we have:

$$\vec{s}' = [T] \cdot \vec{s}$$

In order to determine the transformation matrix $[T]$, we have to

W.-M. Boerner et al. (eds.), Inverse Methods in Electromagnetic Imaging – Part 1, 573–575.

measure four (4) different test states (Thiel, 1970), which we shall choose to be linearly independent.

$$[\overset{1}{\vec{s}}{}', \overset{2}{\vec{s}}{}', \overset{3}{\vec{s}}{}', \overset{4}{\vec{s}}{}'] = [T] \cdot [\overset{1}{\vec{s}}, \overset{2}{\vec{s}}, \overset{3}{\vec{s}}, \overset{4}{\vec{s}}] \quad ,$$

the matrix $[T]$ being the same for each transformation.

Combining formally the different Stokes-vectors to one 4x4 matrix,

$$[s'] \equiv [\overset{1}{\vec{s}}{}', \overset{2}{\vec{s}}{}', \overset{3}{\vec{s}}{}', \overset{4}{\vec{s}}{}'] \text{ and } [s] \equiv [T] \cdot [\overset{1}{\vec{s}}, \overset{2}{\vec{s}}, \overset{3}{\vec{s}}, \overset{4}{\vec{s}}],$$

we have

$$[\vec{s}] = [T] \cdot [s] \quad .$$

Let us now compute the inverse matrix $[s]^{-1}$ of matrix $[s]$. This is possible, because its column-vectors are linearly independent, as we have chosen above.

Now, right-hand-multiplication with the inverse matrix $[s]^{-1}$ yields the transformation matrix:

$$[T] = [s'] \cdot [s]^{-1}$$

The transformation matrix $[T]$ is the matrix product of the matrix $[s']$, containing the transformed states, and of matrix $[s]^{-1}$, which is the inverse matrix $[s]$, containing the incident states as column-vectors.

Acknowledgement: The author wishes to express his gratitude to Prof. Wolfgang-M. Boerner for inviting him to this inspiring Workshop, for providing the opportunity of presenting his paper, and for accepting this summary for publication in the Proceedings.

References

M.A.F. Thiel, Ein Polarisationsmodulator und seine Anwendung in einem Radioastronomischen Polarimeter, Diplomarbeit, 1968, Max-Planck-Institut für Radioastronomie, Auf dem Hügel 69, D-53 Bonn 1, FR Germany.

M.A.F. Thiel, Darstellungs- und Transformationstheorie Quasi-Monochromatischer Strahlungsfelder, Beiträge zur Radioastronomie, Max-Planck-Institut für Radioastronomie, Bonn, Band I, Heft 5, März 1970.

M.A.F. Thiel, Error Calculation of Polarization Measurements, Max-Planck-Geselllschraft zür Förderung der Wissenschaften, e.V., MPIFR, Bonn, Sonderdruck Nr. 87, Ser.A, June 23, 1975 (also see: J. Opt. Soc. of Amer. Vol. 66(1), pp. 65-67, Jan. 1976.

III.4 (RS.4)

POLARIZATION-DEPENDENCE IN MICROWAVE RADIOMETRY

K. Grüner

Deutsche Forschungs- und Versuchsanstalt
für Luft- und Raumfahrt
Institut für Hochfrequenztechnik
8031 Oberpfaffenhofen, FRG

ABSTRACT

Polarization plays more and more an important role also in the field of microwave radiometry. It helps to find new signatures and to improve contrast or image quality. So ambiguity proplems can be solved in a better way. Disadvantages because of worse angle resolution in comparison to passive remote sensing methods in the IR- and optical region are partly compensated.

At first after some fundamentals a series of examples obtained by stationary measurements at 32 GHz and 90 GHz and airborne measurements at 32 GHz demonstrates the influence of polarization on natural microwave radiation. Various media or objects were observed at vertical and horizontal polarization for variing nadir angles. Then some remarks with respect to the complete state of polarization of the natural microwave radiation are made. It turns out, that also the circular polarization should be taken into consideration. Finally a proposal for a correlation measuring method is made which in connection with polarization switching could contribute to a better solution of the inverse radiation problem involved in microwave radiometry.

1. INTRODUCTION

The principle objetives of microwave radiometry are the detection of natural thermal microwave radiation of matter. It may therefore be considered as consistent development of passive remote sensing methods, successfully established in the optical and IR region, towards the cm- and mm wavelength region of the electromagnetic spectrum. By the transition to longer wavelengths disadvantages

W.-M. Boerner et al. (eds.), Inverse Methods in Electromagnetic Imaging – Part 1, 577–597.

and advantages arise. The main disadvantages may be the essential deterioration of the angular resolution and much smaller frequency bandwidths which to a high degree determine the achievable data rate for an assumed detection sensitivity or signal to noise ratio. But because of a minor weather sensitivity of microwave radiation, a higher penetration depth and an extraordinary variety of contrast in the microwave region, microwave radiometry nevertheless has gained great interest for many applications.

The variety of contrast is the consequence of strong variations of natural microwave radiation caused by strong variations of ground reflectivity and a background (sky) radiation decreasing very hard with decreasing zenith angle especially at lower microwave frequencies. Thereby the reflectivity mostly not only depends on the material but also on the look angle of the radiometer and the polarization of the received radiation. These latter characteristics, not so significant in the IR- and the optical region, appear because many surfaces are smooth in the sense of microwaves. By that way further independent informations with respect to an object can be gained, very important when contrast must be improved to solve ambiguity problems or new signatures are looked for.

In the following after some fundamentals a series of examples is given which shows the polarization dependence of different media for vertical and horizontal polarization.

2. FUNDAMENTALS AND EXAMPLES FOR THE POLARIZATION-DEPENDENCE

The radiation received by a radiometer from a given object is composed of a part of emitted thermal radiation and a part of reflected or scattered radiation originating from the surroundings. Usually a "brightness-temperature" (T_B) - more correct an "apparent brightness-temperature" - is attached to the received power.

For the most simple case of a homogeneous semiinfinite medium with an uniform temperature distribution bounded by a horizontal specular surface (Fig. 1) we have:

$$T_B(\Theta_n) = e(\Theta_r)\, T_{PHYS} + r(\Theta_i)\, T_{SKY}\,(\Theta_i) \qquad (1)$$

where:

T_B : brightness-temperature of the medium as a function of the angle of observation Θ_n (nadir angle)

T_{PHYS}: true temperature of the medium

T_{SKY} : brightness-temperature of the sky as a function of the angle of incidence $\Theta_i = \Theta_n$

$e(\Theta_r)$: emission coefficient of the medium as a function of the angle of reflection $\Theta_r = \Theta_i$

$r(\Theta_i)$: power reflection coefficient of the medium as a function of the angle of incidence.

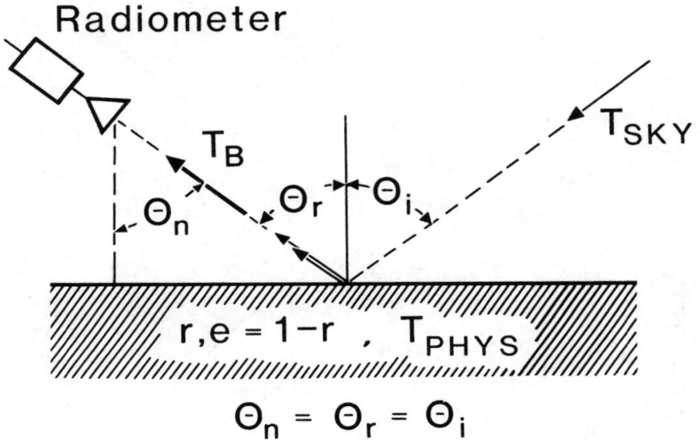

Figure 1. Sketch of the measuring configuration.

The first product at the right side of equation (1) corresponds to the equivalent blackbody temperature T_{EMIT} of the medium at a lookangle of $\Theta_n = \Theta_r$, and the second to a reflected temperature T_{REFL} which is the equivalent blackbody temperature of the sky at an angle of incidence Θ_i weighted by the power reflection factor for the same angle. By second law of thermodynamics a simple relation between the emission coefficient and the power reflection coefficient exists:

$$e(\Theta_r) = 1 - r(\Theta_i). \tag{2}$$

Thus equation (1) becomes:

$$T_B(\Theta_n) = T_{PHYS} + r(\Theta_i)\,(T_{SKY}(\Theta_i) - T_{PHYS}). \tag{3}$$

This formula is valid for either horizontal or vertical polarization whereby only $r(\Theta_i)$ and not $T_{SKY}(\Theta_i)$ shows normally a polarization-dependence. The radiation emitted by the atmosphere is said to be randomly polarized that means unpolarized.

In Fig. 2 the reflection coefficient for a water surface for horizontal and vertical polarization - electric field perpendicular and parallel respectively to the plane of incidence - is given as a function of the angle of incidence.

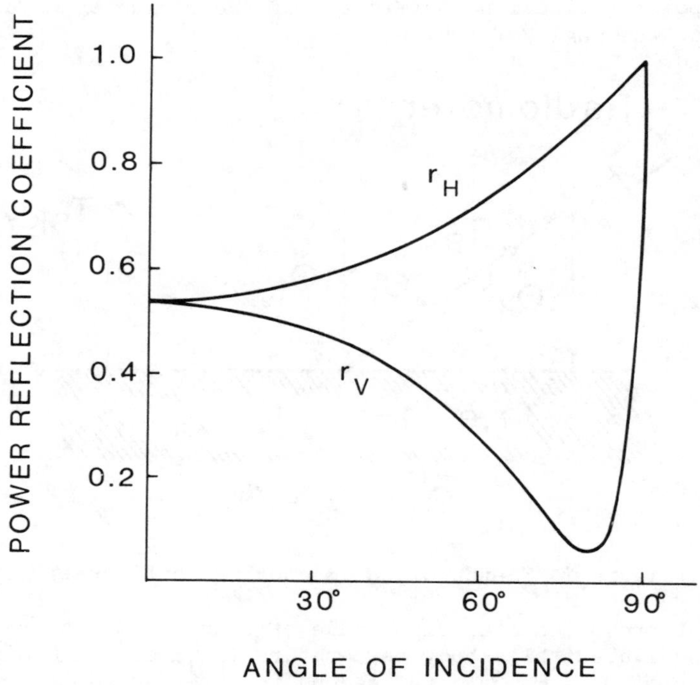

Figure 2. 32 GHz power reflection coefficient for a water surface as a function of the angle of incidence for horizontal and vertical polarization theoretical values, rel. permittivity 17-j27, water-temperature 10° C.

At 0^{0} a value of 0.54 is obtained for both polarizations. Increasing the angle of incidence the curve for horizontal polarization rises continuously until a value of 1 reached, while the curve for vertical polarization decreases at first. At the Brewster angle ($\approx 80^{0}$) a minima is obtained and then a steep ascent also to a value of 1 can be registrated.

The characteristics obtained by evaluation of the wellknown Fresnels equation (4a) and (4b), [1] are typical for homogeneous dielectric media (relative permittivity $\varepsilon = \varepsilon_r - j\,\varepsilon_i$)

$$r_v(\Theta_i) = \left| \frac{\varepsilon \cos \Theta_i - \sqrt{\varepsilon - \sin^2 \Theta_i}}{\varepsilon \cos \Theta_i + \sqrt{\varepsilon - \sin^2 \Theta_i}} \right|^2 \qquad (4a)$$

$$r_h(\Theta_i) = \left| \frac{\cos \Theta_i - \sqrt{\varepsilon - \sin^2 \Theta_i}}{\cos \Theta_i + \sqrt{\varepsilon - \sin^2 \Theta_i}} \right|^2 \qquad (4b)$$

The Brewster angle is influenced in the main by the real part of the relative permittivity of the medium and increases with an increasing value. No value below 45^0 is observed.

When we use the characteristics of Fig. 2 for the evaluation of eq. (1) curves are obtained for T_{EMIT} and T_{REFL} as shown in Fig. 3 (solid- and dashed lines). A clear atmosphere and T_{PHYS} equal 293 K are assumed.

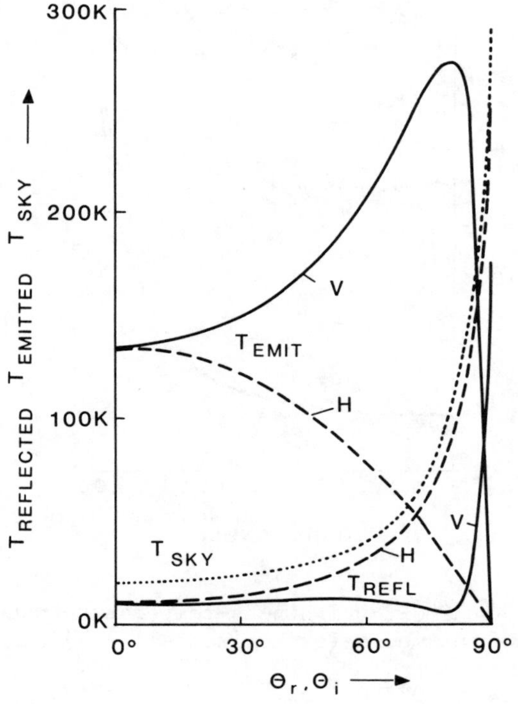

Figure 3. Brightness-temperatures at 32 GHz attached to the emitted- and the reflected radiation for a water surface and the sky radiation.

The brightness-temperature of the sky was measured at 500 m above sea level.

The polarization-dependences for both, the emitted and the reflected part of the radiation can clearly be detected, whereby T_{EMIT}, which shows reverse behaviour in comparison to Fig. 2 because of condition (2), dominates for angles up to 72^0 for horizontal and 88^0 for vertical polarization. This phenomenon depends on the background conditions however.

The radiometer measures the sum of T_{EMIT} and T_{REFL}. It is represented together with the (apparent) brightness-temperature characteristics of othder media in Figure 4.

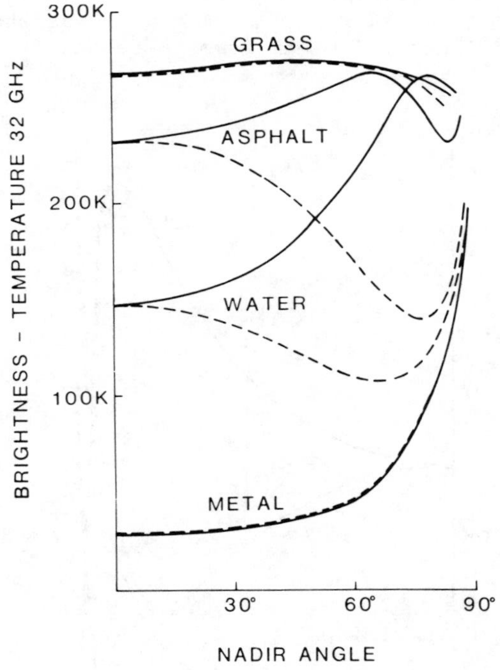

Figure 4. 32 GHz brightness-temperature characteristics of different media for horizontal (---) and vertical (——) polarization
measurements DFVLR 1980.

Discussing the characteristics for water first, we again observe the typical divergence of the characteristics for horizontal and vertical polarization, whereby the brightness-temperature for vertical polarization increases with increasing nadir angle, until near the Brewster angle a maximum is reached. On the other side the brightness-temperature for horizontal polarization decreases. The steep ascent above 70^0 is effected by the strong rise of the skytemperature near the horizon. The brightness-temperature behaviour of an ideal specular reflecting metal surface also represented in Fig. 4 corresponds to that of the sky and vice versa. No polarization effect occurs.

For an asphalt surface principally similiar curves as for water are obtained but at a higher temperaturelevel. The maximum for vertical polarization is already reached for 65^0.

The different behaviours for vertical and horizontal polarization disappear more and more by increasing heterogeneity or losses of a medium or increasing roughness of its surface, characteristics which must be partly seen relatively to wavelength yet as already stated above. The differences disappear in our case almost totally for a media as grass. For inhomogenious media or rough surfaces equation (1) must be replaced by a more complicate integral-expression [2].

Still four general comments should be given with respect to Fig. 4 important for ground mapping in the mm wavelength region.

- An increase of the dynamic range at the radiometer output for natural surfaces can be obtained by measuring the horizontally polarized component at higher nadir angles.

- The media specified in Figure 4 are typical representatives of four classes which are observed in the main. a: metallic surfaces; b: water, wet smooth surfaces; c: asphalt, concrete, gravelly soil, wet rough surfaces; d: vegetation, rough dry soil.

- At a nadir angle of about 50^0 the contrast between these classes is linearized for horizontal polarization by which the image quality can be improved as long as the ground resolution is still sufficient. For vertical polarization we obtain worse results.

- Ambiguity problems because of different background conditions and insufficient ground resolution might be solved by polarization switching at certain nadir angles.

In Fig. 5 passive 32 GHz terrain mapping at nadir angles of 45^0, 60^0 and 78^0 for vertical and horizontal polarization is demonstrated.

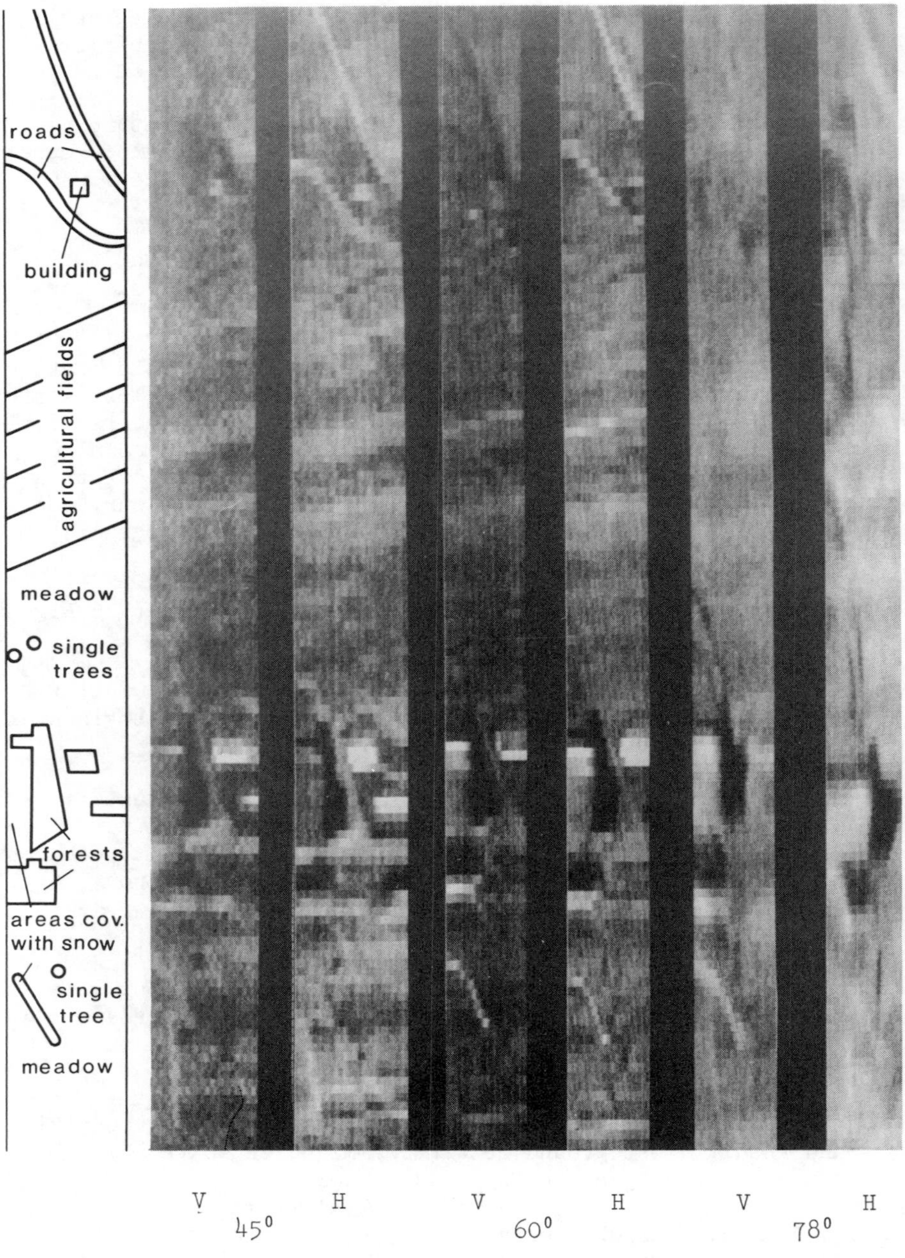

Figure 5. 32 GHz images of the same scene taken at different nadir angles and for vertical and horizontal polarization, sketch of the scene at the left-hand side.

The same scene was overflown with a linescanner at short intervals. The scene contains roads, fields, meadows, small forests, single trees, small areas with a snow cover and buildings (see sketch at the left-hand side). Black corresponds to brightness-temperatures near T_{PHYS}, white to lower brightness-temperatures (220 K at 45°, 210 K at 60° and 210 K/190 K at 78° for vertical/horizontal polarization). The statements given above are confirmed. The dynamic range increases with increasing nadir angle; a higher contrast is obtained for horizontal polarization; roads for instance can much better be detected at this polarization; small areas with snow cover are much better detected at 60° and 78° although the ground resolution became worse; at low depression angles for horizontal polarization distinct signatures for single trees are obtained (black lines).

This short discussion of Fig. 5 already proves the extraordinary importance of polarization in the field of passive microwave imaging.

We now return to further examples of stationary measurements. Fig.6 still shows an example where the typical characteristics obtained for water or asphalt respectively are no longer observed. The diagrams were obtained for a dry hard frozen snow layer of about 10 - 15 cm thickness on a grassy soil.

By reflection processes at the lower boundary interferences occur. Probably also (volume) scattering at the snow crystals happens, which would be an explanation for the low brightness temperatures at 90 GHz but no polarization dependences. This example gives a feeling for the complex radiation behaviour of already three layer dielectric media. In comparison to the Fresnels formulas much more complicate expressions are obtained. When we assume a layer of thickness d (rel. permittivity = ε_2) bounded by two semiinfinite media (rel. permittivity $\varepsilon_1 = 1$ (air), ε_3) then the formulas (5a) and (5b) are obtained for the effective power reflection coefficient at the boundary between media 1 and 2:

$$r_v(\Theta_i) = \left| - \frac{Z_{in,v} - \cos\Theta_i}{Z_{in,v} + \cos\Theta_i} \right|^2 \qquad (5a)$$

and

$$r_h(\Theta_i) = \left| \frac{Z_{in,h} - \sec\Theta_i}{Z_{in,h} + \sec\Theta_i} \right|^2 \qquad (5b)$$

with

$$Z_{in,v} = \frac{\varepsilon_2'}{\varepsilon_2} \quad \frac{1 - \dfrac{\varepsilon_2\,\varepsilon_3' - \varepsilon_3\,\varepsilon_2'}{\varepsilon_2\,\varepsilon_3' + \varepsilon_3\,\varepsilon_2'}\, e^{-j\,4\pi/\lambda_o\,\varepsilon_2/\varepsilon_2'\,d}}{1 + \dfrac{\varepsilon_2\,\varepsilon_3' - \varepsilon_3\,\varepsilon_2'}{\varepsilon_2\,\varepsilon_3' + \varepsilon_3\,\varepsilon_2'}\, e^{-j\,4\pi/\lambda_o\,\varepsilon_2/\varepsilon_2'\,d}}$$

$$Z_{in,h} = \frac{1}{\varepsilon_2'} \frac{1 + \dfrac{\varepsilon_2' - \varepsilon_3'}{\varepsilon_2' + \varepsilon_3'} e^{-j\, 4\pi/\lambda_o\, \varepsilon_2/\varepsilon_2'\, d}}{1 - \dfrac{\varepsilon_2' - \varepsilon_3'}{\varepsilon_2' + \varepsilon_3'} e^{-j\, 4\pi/\lambda_o\, \varepsilon_2/\varepsilon_2'\, d}}$$

where

$$\varepsilon_2' = \sqrt{\varepsilon_2 - \sin^2\Theta_i}\ , \quad \varepsilon_3' = \sqrt{\varepsilon_3 - \sin^2\Theta_i}\ .$$

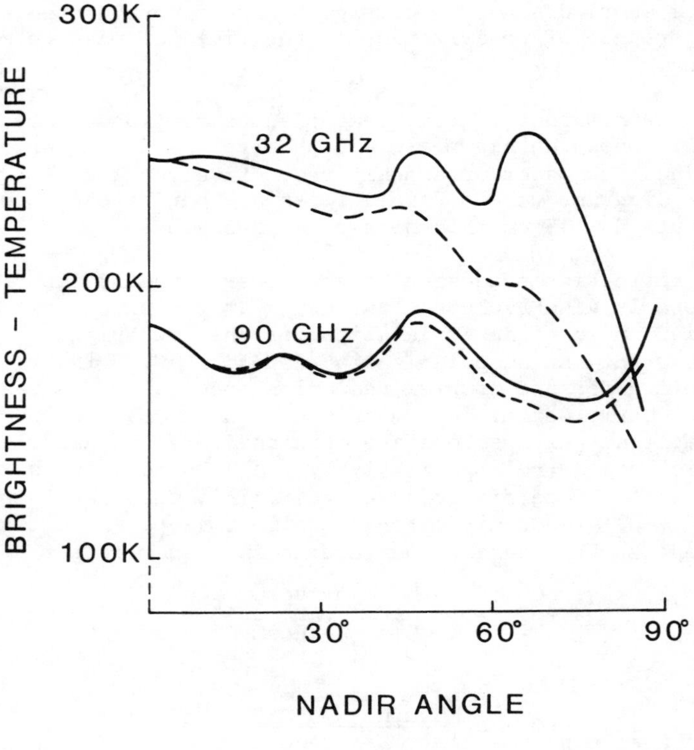

Figure 6. 32 GHz and 90 GHz brightness-temperature characteristics for a snow layer on a meadow for horizontal (---) and vertical (——) polarization
snow thickness 10-15 cm, snow hard frozen and dry, $T_{PHYS} = -6^o$ C, measurement DFVLR 1982.

The wavelength of the incident wave is given by λ_o. For the case of an uniform temperature profile the expressions (5a) and (5b) can immediately be used together with equation (3) for the computation of the brightness-temperatures.

Another method for the representation of measurements for horizontal and vertical polarization is chosen in Fig. 7.

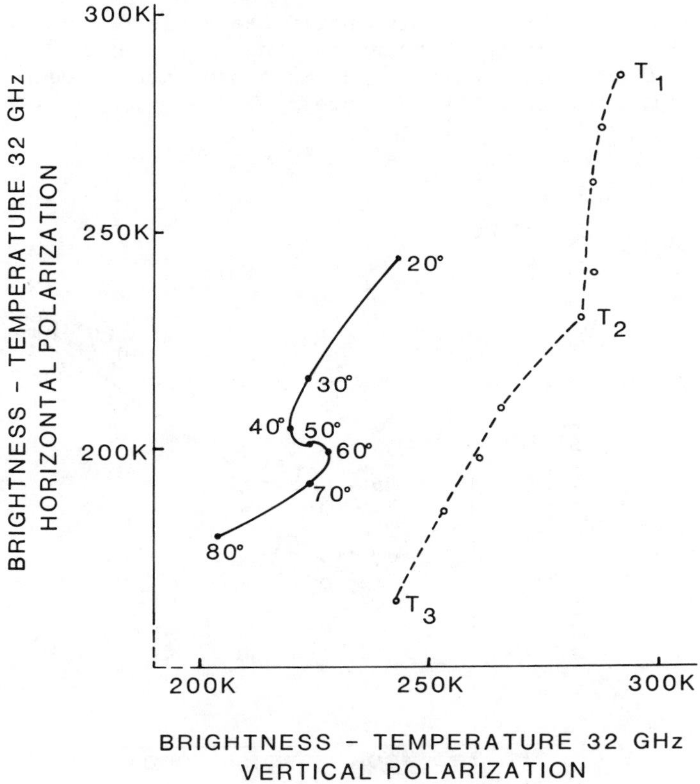

Figure 7. 32 GHz signatures of different objects by brightness-temperature measurements at horizontal and vertical polarization —— snow layer on a meadow, thickness ~10 cm, atmospheric temperature -3 °C, variation of nadir angle (20° - 80°), --- foam on a water surface, variation of time, multilayers of bubbles at T_1, single layer at T_2, pure water surface at T_3.

The results are drawn in a plane defined by the brightness-temperatures for the two orthogonal polarizations. The diagramm at the left-hand side was obtained looking again to a layer of dry but now fresh snow (thickness ~ 10 cm) on a meadow at different nadir angles. The characteristic at the right-hand side was obtained during a period $T_1 - T_2 - T_3$ looking at a constant nadir angle of 68° to a water surface with foam on it. At the time T_1 we had multilayers of air-bubbles. In the following, the bubbles bursted continuously until at the time T_2 only a single layer covered the water surface. Afterwards also this single layer disappeared. The pure water surface was observed at the time T_3.

Both examples show very distinctly the different behaviour for vertical and horizontal polarization. Characteristic signatures are obtained which confirm once more the extraordinary role of polarization for microwave radiometry.

A last example is finally given in Fig. 8.

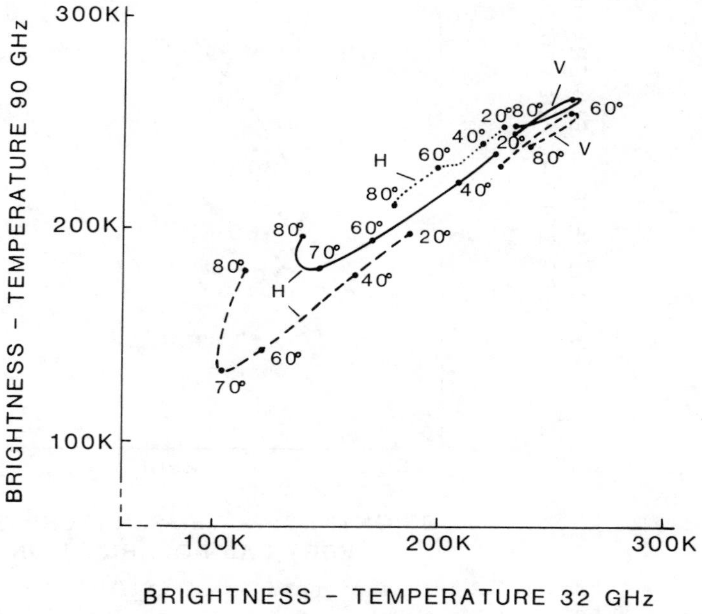

Figure 8. Signatures of asphalt for several surface conditions and horizontal and vertical polarization by simultaneous brightness-temperature measurements at 32 GHz and 90 GHz; surface dry (——), wet (---) and with a snow cover of 3.5 cm (...), result for snow and vertical polarization not drawn, measurement DFVLR, 1980.

Simultaneous measurements at 32 GHz and 90 GHz for horizontal and vertical polarization are represented once more, but now in a

32 GHz/90 GHz-plane. The curves were obtained, when looking to a asphalt surface at various nadir angles and for several surface conditions (dry ——, wet ---, covered with a 3.5 cm thick snow layer ...). The variations obtained for horizontal polarization are much more distinct than those for vertical polarization where it was renounced to draw the curve for snow cover.

The examples presented should give a first feeling what big variety of signatures can be obtained in the microwave region by using the informations contained in the natural microwave radiation when receiving the horizontally and vertically polarized components at various nadir angles and frequencies. At the end of this chapter, however, we must ask us the question whether we have extracted really all the polarization dependent informations hidden in the natural microwave radiation, when we measure the horizontally and vertically polarized components. By depolarization effects for instance there should be a (weak?) correlation between the horizontally polarized and vertically polarized electric-field components giving further independent informations. A short discussion is carried out by help of the Stokes parameter of a partially polarized wave.

2. SOME REMARKS WITH RESPECT TO THE COMPLETE STATE OF POLARIZATION

Natural microwave radiation generated by thermal processes of matter is completely randomly polarized or completely unpolarized at first. By reflection and refraction at the boundaries of different media a partially polarized wave results however within the frequency bands considered. In a certain sense, the boundaries act as a polarization-filter or -transformer.

A partially polarized wave can be regarded as the sum of a completely unpolarized wave and a completely polarized wave. To deal with partial polarization it is convenient to use the normalized Stokes parameters S_o, S_1, S_2, S_3 which completely describe the state of polarization of a wave [3] and which can be defined as a matrix of the column vector-type (Stokes vector $\underline{g}$).

The rules which are with Stokes parameters or vectors are very simple [3, 4]:

- To determine the Stokes parameters only power measurements and no phase measurements are necessary.
- If several independent waves propagating in the same direction are superimposed, the Stokes vector of the resultant wave is the sum of the Stokes vectors of the individual waves.
- The Stokes vector of a reflected or a refracted wave is related to the Stokes vector of an incident wave by a 4x4 real matrix (Müller matrix, Stokes reflection -/refraction matrix)

which characterizes the object and from which its scattering matrix can be derived e.g.:

$$g^r(\Theta_r) = [M(\Theta_r, \Theta_i)]\, g^i(\Theta_i) \ . \tag{6}$$

Following [3], the normalized Stokes parameter of a plane wave (frequency f), which can be completely described in a xyz-coordinate-system (wave traveling in positive z-direction, point of measurement z = 0) by the orthogonally polarized components of the electric field

$$\begin{aligned} e_x(t) &= E_x(t)\sin(2\pi\, ft - \delta_x(t)) \\ e_y(t) &= E_y(t)\sin(2\pi\, ft - \delta_y(t)) \end{aligned} \tag{7}$$

are:

$$S\begin{bmatrix} S_o \\ S_1 \\ S_2 \\ S_3 \end{bmatrix} = S\begin{bmatrix} 1 \\ (\langle E_x^2\rangle - \langle E_y^2\rangle)/ZS \\ 2\,\langle E_x\, E_y\, \cos\delta\rangle/ZS \\ 2\,\langle E_x\, E_y\, \sin\delta\rangle/ZS \end{bmatrix} \tag{8}$$

where

$\langle\ \rangle$ indicates the time average; the time variations of $E_x(t)$, $E_y(t)$, $\delta_x(t)$, $\delta_y(t)$ are assumed slow compared to frequency f which is only valid for a limited bandwidth,

δ phase difference $\delta_x - \delta_y$,

Z, S characteristic impedance of the medium and total power density(|Poynting vector|)of the wave at the point of measurement.

The coordinate-system could be selected in that way that the x- and y-directions coincide with the "horizontally" and "vertically" directed electric field components as defined in chapter 1.

The polarization state of a partially polarized wave is completely described by S_1, S_2 and S_3. S_1 is proportional to the difference of the powers received at horizontal and vertical polarization. S_2 and S_3 which are <u>independent</u> quantities are like different weighted cross-correlation functions of E_x and E_y. S_1, S_2 and S_3 are equal zero in the case of a totally unpolarized wave because $\langle E_x^2\rangle = \langle E_y^2\rangle$ and E_x and E_y are uncorrelated. S_1, S_2 and S_3 determine a (time averaged) polarization ellipse for the completely polarized part of the wave. Also a "degree of polarization" (d) which is the ratio of the power of the completely polarized wave to the power of the total wave can be derived.

$$0 \leq d = \sqrt{S_1^2 + S_2^2 + S_3^2} \leq 1 . \tag{9}$$

The determination of the Stokes parameter requires power measurements only, as stated above. Four independent measurements would be enough (S, S_1, S_2, S_3), but in order to simplify the following discussion six measurements are done [3]. We can carry out these by the help of a radiometer which has the same effective antenna aperture for all polarizations giving the responses: T_x, T_y, T'_x, T'_y, T_l, T_r (brightness-temperatures) where:

- T_x, T_y: obtained for linear x- and y directed polarizations
- T'_x, T'_y: obtained for the linear x- and y directed polarizations when the xyz coordinate system is turned -45° around the z-axis
- T_l, T_r: obtained for left- and right-hand circular polarization.

With these values, the following expressions are obtained for S_1, S_2 and S_3:

$$S_1 = \frac{T_x - T_y}{T_x + T_y} , \qquad S_2 = \frac{T'_x - T'_y}{T_x + T_y}$$
$$S_3 = \frac{T_l - T_r}{T_x + T_y} \qquad (S \sim T_x + T_y). \tag{10}$$

A short analysis shows, that the condition $S_1 \neq 0$ is valid for the examples presented in the last chapter for nadir angles > 0. $S_2 \neq 0$ is only true, when the surface or the medium observed are not longer smooth or homogeneous and certain unsymmetries are given. Finally $S_3 \neq 0$ would be valid, if the surface or the medium influences a left and a right circular polarized wave in a different way. A certain inhomogeneity in depth is necessary that this might be possible. Only small quantities with respect to S_2 and S_3 can be awaited because generally polarization-dependence decreases for rough surfaces or inhomogeneous media.

To verify this, brightness-temperatures of some natural media were measured for right- and left circular polarization. The results are drawn in Figure 9. Only very small differences between the two polarizations could be registrated, which are in the order of magnitude of the temperature resolution of the radiometer used and might be errors of measurements therefore.

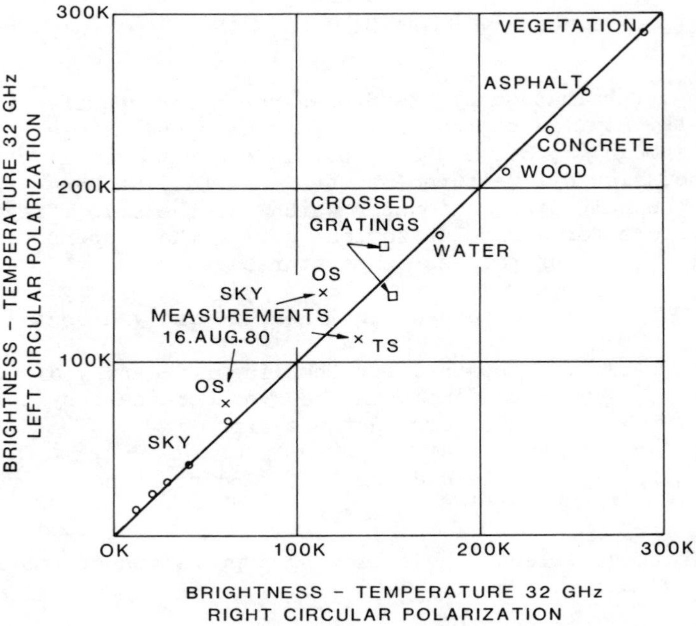

Figure 9. 32 GHz measurements for right- and left circular polarization.
DFVLR measurements 1980, 1981.

Also for the sky normally no differences were observed exept for one single case (X) several years ago when looking towards the sun (TS) or looking to the opposite side (OS). At time a complete physical explanation of this effect cannot be given. The results also could not be confirmed at a later time. Reproducable results were obtained on the other hand for artificial structures as gratings arranged crosswise at a certain distance (⌑).

In Fig. 10 a synopsis of all formulas important when dealing with partially polarized waves is given once more.

As we have seen, there is a physical basis for further signatures hidden in the natural microwave radiation. But the greatest variety of signatures are an useless phenomenon, as long as sensor inability does not allow the extraction of these informations. It is supposed, that in the case of natural objects accuracy of measurements must be improved very much to come to significant results (S_2, $S_3 \ll 1$). Low noise detectors are necessary and the properties of the antennas as the effective aperture (inclusive receiving characteristic and input impedance behaviour) must be known very exactly.

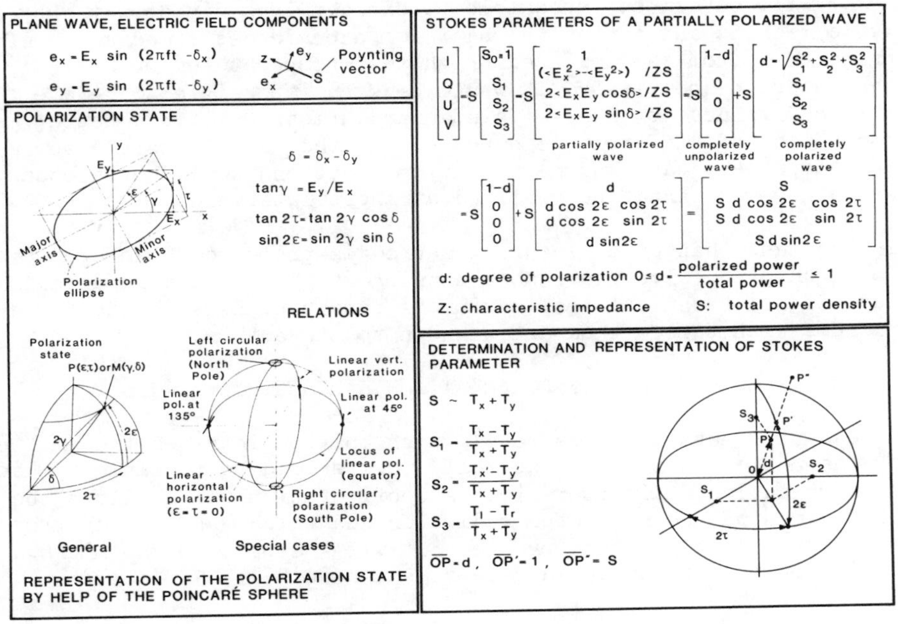

Figure 10. Polarization-relations.

If this could be realized we would have a very valuable instrument for sensing inhomogeneities transversal or in depth. Using the conventional measuring mehtod as shown in Fig. 1, the inner of the normalized Poincaré sphere (radius of 1), which is normally used to describe definitely an arbitrary polarization state of a completely polarized wave (d = 1) by a point on its surface, could immediately function as a normalized three dimensional cluster space for representation of S_1, S_2 and S_3 (see Fig. 10). When a normalization (S = 1) is not desired the true value of S could be a further criteria for discrimination (see distance OP" in Fig. 10) to reduce ambiguity problems.

It must be clear however, that this signature space must be filled at first by many values gained by aid of experimental and theoretical investigations before it can help to solve the inverse "scattering" problem of microwave radiometry.

Generally there is a great disadvantage with the measuring method outlined in Fig. 1. The radiometer normally receives at the same time emitted and reflected radiation. In the case of a complex terrain medium the power density is influenced by the physical temperature distribution of the ground as well as by the brightness-temperature distributions of the environment (weighting functions). So it will be almost impossible to extract in a direct way - by analytic methods and not on the basis of experiences - media specific properties from the measured brightness-temperatures or Stokes parameters respectively. It would be even impossible to relate the measured values to those of a standard atmosphere for a better comparison of different measurements on the same object. A series of independent (and known) weighting functions would be necessary which however cannot be realized at short intervals.

In the last chapter a measuring method is proposed therefore which could help to solve the remote sensing problem in a more direct way, if its technical realization is possible. Experimental experiences with this method are unknown to the author.

3. AN IMPROVED METHOD OF POLARIZATION RADIOMETRY

A measuring method is desirabel which only registrates the emitted <u>or</u> the reflected part. Principally this should be possible by use of correlation radiometers [5]. A configuration which allows the extraction of the reflected radiation and which was already proposed in [5] to solve the same problem but with respect to the microwave radiation from the moon surface is shown in Fig. 11.

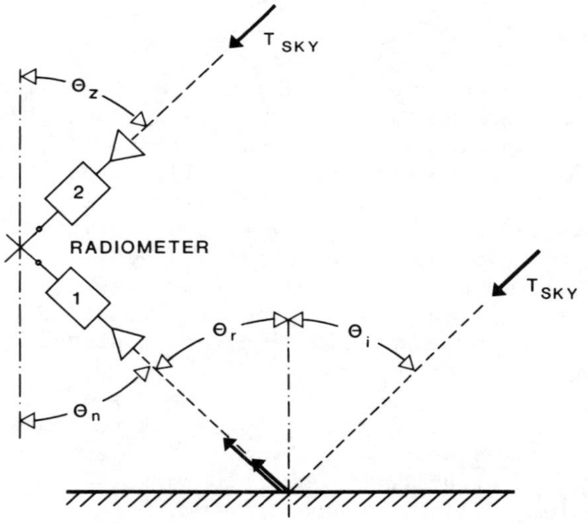

Figure 11. Correlation radiometer for the extraction of the reflected radiation.

A second radiometer looks to that region of the sky, which is the source of the reflected part. The polarization states of both systems should be the same for the present. In dependence of certain system parameters the correlation process would be carried out in the IF-regions of the radiometer receivers (IF type, the signals themselves are correlated) or after the square law envelope detection (Envelope type, only the amplitudes are correlated). The degree of correlation then immediately would be proportional to the power reflection coefficient $r(\Theta_i)$ at the ground ($\Theta_i = \Theta_z = \Theta_n$) for the selected polarization. Imaging the power reflection coefficient should lead to better results, because the relative contrast is higher than that for the (apparent) brightness-temperatures. In the case of a complex terrain medium the look angle (Θ_z) of the sky-radiometer had to be varied to obtain the power scattering characteristic $\sigma(\Theta_r, \Theta_i = \Theta_z)$ or the receiving characteristic of the sky-radiometer had to be changed.

Such a radiometer configuration probable could supply some more informations, by switching of different polarization states at both radiometer channels. In terms of Stokes parameters, switching the polarization state of the sky-radiometer means the extraction (filtering) of different incident waves with different Stokes vectors and switching the polarization state of the downward looking radiometer means the respective determination of the Stokes parameters of the reflected wave as already described in the last chapter. By a virtual manipulation of the input field a complete series of measurements at short intervals should make it possible to determine all independent elements of the Mueller matrix (bistatic case), which definitely characterizes the object. Thus <u>two</u> advantages make this method superior to the conventional method.

Principally a correlation method can also be realized for the emitted part of the natural radiation by using two radiometers the lines of sight of which cross at the object. In doing so, it would be of great advantage that the emitted radiation is mostly much stronger than the reflected radiation (see Fig. 3) by which a better signal to noise ratio is obtained. On the other hand the set of informations would be not so complete as in the upper case, because it is not possible to seperate and virtually manipulate the radiation of the source which generates the emitted part. Besides of that it turns out, that the signal to noise ratio when extracting the reflected part becomes better for bad weather conditions (see Fig. 12) or by shifting the frequency f to the absorption bands of atmosphere (22 GHz, 60 GHz, 120 GHz, 180 GHz ...), conditions which normally affect radiometric contrast ($T_{PHYS} - T_B$) very hard, because that profits from the difference($T_{PHYS} - T_{SKY}$) (compare equation (3)).

At any case, the realization of this method will require greatest efforts to be successfull not only for strong but also for weak reflecting media. A series of system parameters as the preferable frequency bands, optimum microwave-bandwidths, necessary temperature resolutions, the efficiency of the IF-type- and Envelope-type correlation methods, optimum antenna characteristics, calibration methods, possible flight profils and so on must be investigated.

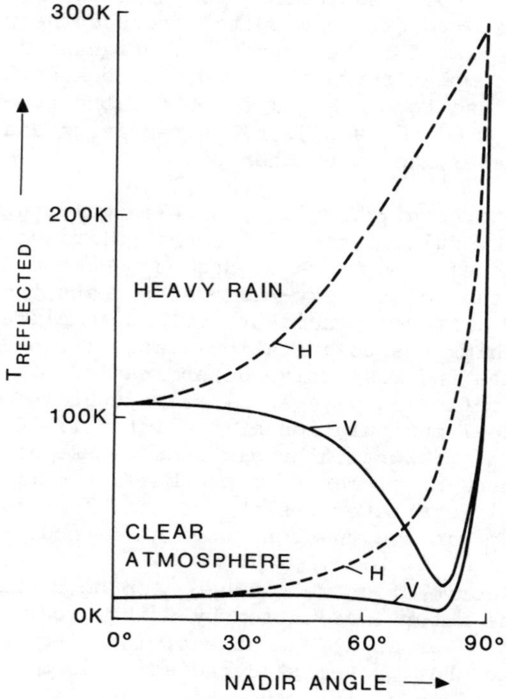

Figure 12. Example for the influence of the weather condition on the reflected part of the natural microwave radiation
water surface as above, horizontal (---) and vertical (——) polarization.

4. CONCLUSION

Some examples have shown, that polarization plays a very important role also in the field of microwave radiometry.

By discussion of the Stokes parameter of a partially polarized wave it was demonstrated that thermal microwave radiation after interaction with differen media should contain more informations, than obtained by only measuring the vertically and horizontally polarized components. This could lead to new signatures or new textures at least and microwave images could have a better quality when all independent informations are summarized.

Finally a correlation method was outlined, which should allow a better solution of the remote sensing problem of microwave radiometry by separation of the reflected sky radiation from the emitted ground radiation. At time, the whole future development with respect to this method cannot be told. A series of system parameters must be optimized.

Thus once more again the prerequisite of all these activities must be the improvement of the sensor technology, to realize an absolute accuracy of measurement, which is almost one order of magnitude better than today.

5. REFERENCES

[1] M. Born and E. Wolf,
"Principles of Optics", Pergamon Press, London, 1965.

[2] F.T. Ulaby, R.K. Moore and A.K. Fung,
"Microwave Remote Sensing", Volume 1, Addison-Wesley Publishing Company, London, 1981.

[3] J.D. Krans
"Radio Astronomy",
Mc Graw-Hill Book Company, New York, p. 108-130, 1966.

[4] W-M. Boerner, M.B. El-Arini et al.
"Polarization Utilization in Radar Target Reconstruction", Electromagnetic Imaging Division, Communications Laboratory, University of Illinois at Chicago, Technical (Interim) Report UICC, CL-EMID-NANRAR-81-01, 1981.

[5] K. Fujimoto,
"On the Correlation Radiometer Technique",
IEEE Trans. Microwave Theory Tech., Vol MTT-2,
pp 203-211, March 1964.

6. ACKNOWLEDGMENT

The author wishs to thank Professor Boerner, whose enthusiasm for all kinds of polarization phenomena induced me to write this paper. I further thank Mrs. Kappelmaier and Mrs. Malchow for typing this manuscript.

III.5 (SP.3)

MULTISTATIC VECTOR INVERSE SCATTERING

Gregory E. Heath

Massachusetts Institute of Technology
Lincoln Laboratory
Lexington, MA 02173-0073, U.S.A.

ABSTRACT. Two new vector (i.e., polarization dependent) methods for constructing a three-dimensional radar image of a target from transient or multi-frequency far field multistatic scattering data have been formulated. In both methods equivalent current image functions can be constructed directly from transient or multifrequency far field scattering data by using efficient inverse Radon or Fourier Transform computations, respectively. Analytic continuation of far field data into the near field is unnecessary.

I. INTRODUCTION

The first inverse scattering method to utilize complete polarization information in a logical and consistent manner was developed during 1968-1974 [1-5] for opaque scatterers satisfying impedance boundary conditions. However, successful target reconstruction required the computationally difficult and inefficient task of using analytic continuation to construct near field quantities from far field measurement data in order to implement inverse boundary conditions. Later multistatic investigations of perfectly conducting targets (1976 [6], 1979/80 [7-9]) avoided the need for analytic continuation by using various modifications of physical optics to express the target image as a computationally efficient inverse Fourier Transform of far field data. However, the role of polarization was not treated adequately.

Recently, two new vector (i.e., polarization dependent) methods for constructing a three-dimensional radar image of a target from transient or multi-frequency far field multistatic scattering data have been formulated. Multistatic vector physical optics far field inverse scattering (POFFIS) extends Bojarski's

W.-M. Boerner et al. (eds.), Inverse Methods in Electromagnetic Imaging - Part 1, 599–608.

polarization independent multi-monostatic technique for a perfectly conducting target [10] to the vector case of arbitrarily polarized multistatic scattering [11]. The inverse method of solenoidal equivalent currents (IMSEC) replaces the restriction of perfectly conducting targets and the use of physical optics with one, less restrictive, assumption: the scattering from an arbitrary (i.e., transparent or opaque) target can be represented by reradiation from equivalent magnetic and electric currents whose solenoidal (divergence-free) components depend on the incident field through a polarization dependent, linear, time-invariant, causal relationship that is spatially local and temporally isotropic (irrotational (curl-free) current components are ignored because they do not radiate) [12]. As a result of the IMSEC assumption, the solenoidal current components are required to vanish outside of the volume enclosed by the external target surface and, like physical optics equivalent currents, can be used to construct three-dimensional image functions that determine the location and shape of the target. In both methods equivalent current image functions can be constructed directly from transient or multi-frequency far field scattering data by using efficient inverse Radon and Fourier Transform computations, respectively. Analytic continuation of far field data into the near field region is not necessary.

The new POFFIS formulation obviates some of the objections to previous work by expressing equivalent magnetic and electric current image functions in terms of measurable, polarization dependent, far field quantities satisfying reciprocity and bistatic-monostatic equivalence. However, when the polarization reference plane for each radar is the propagation plane containing the transmitter and receiver lines of sight, reconstruction requires only principal polarization measurement data. In contrast, IMSEC yields exact reconstruction formulae without eliminating depolarization effects. Since actual induced currents may have solenoidal components that do not vanish in current-free regions (the non-zero solenoidal and irrotational components cancel) they are replaced by equivalent currents whose solenoidal components satisfy the IMSEC assumption. However, in order to account for arbitrarily polarized scattered fields, both magnetic and electric equivalent currents must be used. Although solenoidal components and the corresponding image functions can be constructed directly from far field scattering data, irrotational components cannot be determined without additional information regarding the nature of the fields or currents within the target volume.

II. EQUIVALENT CURRENT SCATTERING THEORY

The incident field is assumed to be a linearly polarized uniform plane wave of the form

$$[\vec{E}^i(\hat{k}_i,\vec{r},t),\ \eta\vec{H}^i(\hat{k}_i,\vec{r},t)] = [\hat{e},\ \hat{h}]f(t-\hat{k}_i\cdot\vec{r}/c) \qquad (1)$$

where $[\vec{E}^i, \vec{H}^i]$ are the incident electric and magnetic field vectors at position $\vec{r}$ and time t, η is the intrinsic impedance of the uniform medium surrounding the target, $\hat{k}_i$ is the unit incident wave normal, c is the speed of light and f(t) is an arbitrary time dependent incident waveform. The unit polarization vectors $[\hat{e}, \hat{h}]$ and $\hat{k}_i$ form the right handed orthogonal triplet $[\hat{e}, \hat{h}, \hat{k}_i]$ so that

$$[\vec{E}^i,\ \eta\vec{H}^i] = \hat{k}_i \times [-\eta\vec{H}^i,\ \vec{E}^i], \qquad \hat{k}_i\cdot [\vec{E}^i,\ \eta\vec{H}^i] = 0\ . \qquad (2)$$

The incident field induces electric currents on the target surface and within the target volume. The scattered fields can be interpreted as reradiation, in the presence of the target, from the induced electric current. According to the Equivalence Principle, the scattered fields outside of the target can also be represented by the radiation, in the absence of the target, of nonunique equivalent magnetic and electric currents.

A. POFFIS Currents

POFFIS currents are conveniently described in terms of the target surface describing function $a(\vec{r})$(< 0 outside of the target volume; = 0 on the target surface; > 0 inside of the target volume) and the target characteristic function $\gamma(\vec{r}) = u[a(\vec{r})]$ = (0 outside; 1/2 on the surface; 1 inside) where u is the Heaviside unit step function. Consequently

$$\nabla\gamma(\vec{r}) = \delta(a)\ \hat{n}_o(\hat{n}_o\cdot\nabla)a \qquad (3)$$

where δ is the Dirac delta distribution and $\hat{n}_o(\vec{r})$ is the unit outward surface normal. The target surface is decomposed into the illuminated surface, S_1 ($\hat{k}_i\cdot\hat{n}_o < 0$), the shadow boundary, C ($\hat{k}_i\cdot\hat{n}_o = 0$), and the shadowed surface, S_2 ($\hat{k}_i\cdot\hat{n}_o > 0$).

Equivalent volume current densities for a perfectly conducting target can be expressed in the form

$$[\eta\vec{J},\ \vec{M}] = [\eta\vec{J}_o(\hat{k}_i,\hat{e},\vec{r}),\ \vec{M}_o(\hat{k}_i,\hat{e},\vec{r})]f(t-\hat{k}_i\cdot\vec{r}/c) \qquad (4)$$

where

$$[\eta\vec{J}_o,\ \vec{M}_o] = [\mp\nabla\gamma \times \hat{h},\ \hat{e} \times \nabla\gamma] \qquad (5)$$

are non-zero only on the target surface because of $\delta(a)$. The upper and lower signs in (5) and (6) apply to S_1 and S_2, respectively. The equivalent surface current densities corresponding to (5) are given by

$$[\eta\vec{J}_s,\ \vec{M}_s] = [\pm\hat{n}_o \times \eta\vec{H}^i,\ -\vec{E}^i \times \hat{n}_o]\ \cdot \qquad (6)$$

All of the equivalent surface currents on S_2 can be replaced by equivalent surface currents on S_1 to obtain the conventional physical optics result $[\eta\vec{J}_s, \vec{M}_s] = [2\hat{n}_o \times \eta\vec{H}^i, 0]$ on S_1, and [0, 0] on S_2. The difference between the two surface current distributions is $[\hat{n}_o \times \eta\vec{H}^i, \vec{E}^i \times \hat{n}_o]$ which produces no field outside of the target surface because the incident field $[\vec{E}^i, \vec{H}^i]$ satisfies the sourceless Maxwell equations on and within the target surface. It should be obvious, on physical grounds, that the currents on S_2 in (5) can contain no information regarding the shape of either S_1 or S_2. Accordingly, the Maggi-Rubinowicz [13,11] transformation can be applied to replace the currents on S_2 by equivalent line currents on the shadow boundary C.

Only some of the equivalent surface currents on S_1 can be replaced by equivalent surface currents on S_2 or by equivalent line currents on C. The remaining currents on S_1 contain information on the illuminated surface shape. Exact (i.e., within the assumptions of the physical optics approximation) target reconstruction can be obtained from direct illumination of the complete surface by using two oppositely directed transmitters.

The three $(\hat{e}, \hat{h}, \hat{k}_i)$ components of $\nabla\gamma$ are suitable image functions because they have the characteristics of Dirac delta distributions located on the target surface and contain the same shape information as the components of $[\eta\vec{J}_o, \vec{M}_o]$.

B. IMSEC Currents

All equivalent currents can be uniquely decomposed into irrotational (curl-free) and solenoidal (divergence-free, rotational) parts. However, only the solenoidal part of the equivalent currents radiate. Therefore, neglecting creeping and traveling wave effects, target reconstruction using scattered field measurements will only yield images located in the support of the solenoidal part. Unfortunately, the solenoidal part may be non-zero in regions where the total equivalent current vanishes (the non-zero solenoidal part is canceled by the non-zero irrotational part). Therefore assumptions that guarantee vanishing solenoidal parts outside of the target volume are desirable.

The solenoidal part of IMSEC equivalent currents are assumed to be determined by the local incident field through the following polarization dependent, linear, time invariant, causal, temporally isotropic relationship:

$$\begin{bmatrix} \eta\vec{J}_R \\ \vec{M}_R \end{bmatrix} = \int_{-\infty}^{t} d\zeta \begin{bmatrix} w_e(t-\zeta)\overleftrightarrow{\Gamma}_e(\hat{k}_i,\vec{r}) \cdot \vec{E}^i(\hat{k}_i,\vec{r},\zeta) \\ w_m(t-\zeta)\overleftrightarrow{\Gamma}_m(\hat{k}_i,\vec{r}) \cdot \eta\vec{H}^i(\hat{k}_i,\vec{r},\zeta) \end{bmatrix}$$

$$= \begin{bmatrix} \eta \vec{J}_o(\hat{k}_i,\hat{e},\hat{r})\ g_e(t-\hat{k}_i \cdot \vec{r}/c) \\ \vec{M}_o(\hat{k}_i,\hat{e},\hat{r})\ g_m(t-\hat{k}_i \cdot \vec{r}/c) \end{bmatrix} \tag{7}$$

where

$$[\eta\vec{J}_o,\ \vec{M}_o] = [\overleftrightarrow{\Gamma}_e \cdot \hat{e},\ \overleftrightarrow{\Gamma}_m \cdot \hat{h}]\ , \qquad [\overleftrightarrow{\Gamma}_e,\ \overleftrightarrow{\Gamma}_m] \cdot \hat{k}_i = 0 \tag{8}$$

and $g_{e,m}(t)$ is the time convolution of $w_{e,m}(t)$ and $f(t)$. The requirement of solenoidality for <u>arbitrary</u> $f(t)$ results in the following auxiliary constraints:

$$\nabla \cdot [\overleftrightarrow{\Gamma}_e,\ \overleftrightarrow{\Gamma}_m] = 0, \quad \hat{k}_i \cdot [\overleftrightarrow{\Gamma}_e,\ \overleftrightarrow{\Gamma}_m] = 0 \quad . \tag{9}$$

In addition, the solenoidal currents must be curl-free in the current free region outside of the target. Therefore, outside of the target,

$$\nabla \times [\overleftrightarrow{\Gamma}_e,\ \overleftrightarrow{\Gamma}_m] = 0, \quad \hat{k}_i \times [\overleftrightarrow{\Gamma}_e,\ \overleftrightarrow{\Gamma}_m] = 0 \quad . \tag{10}$$

Combining (9) and (10) shows that $[\overleftrightarrow{\Gamma}_e,\ \overleftrightarrow{\Gamma}_m]$ vanishes outside of the target. Consequently, combinations of the cartesian components of $\overleftrightarrow{\Gamma}_e$ and $\overleftrightarrow{\Gamma}_m$ are suitable for image functions.

III. <u>SCATTERED FIELD POLARIZATION STRUCTURE</u>

The temporal Fourier component of the scattered electric field, $\vec{E}^s$, can be expressed in terms of the range normalized far field scattering dyadic, $\overleftrightarrow{S}$, through the relations

$$\vec{E}^s = 2\sqrt{\pi}\ F(jck)\ g(k,r)\ \overleftrightarrow{S}(k,\hat{k}_i,\hat{k}_s) \cdot \hat{e} \quad , \tag{11}$$

where F is the temporal Fourier transform of the incident pulse at the frequency $ck/2\pi$, and $g = \exp(-jkr)/4\pi r$ $(r=|\vec{r}|)$ is the scalar free space Green's function.

The linear and circular polarization representations of $\overleftrightarrow{S}$ are given by

$$\overleftrightarrow{S} = \hat{H}_s(a_{HH}\hat{H}_i + a_{HV}\hat{V}) + \hat{V}(a_{VH}\hat{H}_i + a_{VV}\hat{V})$$

$$= \hat{R}_s(a_{RR}\hat{R}^*_i + a_{RL}\hat{L}^*_i) + \hat{L}_s(a_{LR}\hat{R}^*_i + a_{LL}\hat{L}^*_i) \tag{12}$$

where the a's are the corresponding scattering matrix coefficients [14,15] and $\hat{V}$ is the vertical polarization unit vector in the $\hat{k}_s \times \hat{k}_i$ direction. The scattered and incident wave horizontal polarization unit vectors are defined by $\hat{H}_s = \hat{V} \times \hat{k}_s$ and $\hat{H}_i = \hat{k}_i \times \hat{V}$, respectively. The corresponding right (upper sign) and left (lower sign) circular polarization unit vectors are given by $\hat{R}_s, \hat{L}_s = (\hat{H}_s \mp j\hat{V})/\sqrt{2}$ and $\hat{R}_i, \hat{L}_i = (\hat{H}_i \pm j\hat{V})/\sqrt{2}$.

A. POFFIS Polarization

The physical optics scattering dyadic can be uniquely decomposed into reflected and diffracted parts [11], i.e.,

$$\overleftrightarrow{S} = \overleftrightarrow{S}_1 + \overleftrightarrow{S}_2 = \overleftrightarrow{S}_R + \overleftrightarrow{S}_D \tag{13}$$

where $\overleftrightarrow{S}_{1,2}$ are contributions from the surface currents on $S_{1,2}$, respectively. The diffracted component, $\overleftrightarrow{S}_D$, can always be interpreted as reradiation from shadow boundary line currents. It contains no information concerning the shape of S_1 and S_2 outside of their common boundary, C. Therefore, it is not used in POFFIS target reconstruction. Alternate forms of $\overleftrightarrow{S}_D$ in terms of $\vec{N}_1$, $\vec{N}_2$ and $\vec{L}$ are given in [11] where $\vec{N}_1(\vec{K}), \vec{N}_2(\vec{K})$ and $\vec{L}(\vec{K})$ $(\vec{K} \equiv k(\hat{k}_s - \hat{k}_i))$ are integrals of $\hat{n}_o(\vec{r})\exp(j\vec{K}\cdot\vec{r})$ over S_1, S_2 and C, respectively.

The reflected component, $\overleftrightarrow{S}_R(k, \hat{k}_i, \hat{k}_s)$ must be interpreted as reradiation from illuminated area surface currents. It is given by [11]

$$\overleftrightarrow{S}_R = a_R(\hat{H}_s\hat{H}_i + \hat{V}\hat{V}) = a_R(\hat{L}_s\hat{R}_i^* + \hat{R}_s\hat{L}_i^*) \tag{14}$$

where

$$\begin{aligned} a_R &= (-j/2\sqrt{\pi})\, \vec{K} \cdot \vec{N}_1(\vec{K}) \\ &= (a_{HH} + a_{VV})/2 = (a_{LR} + a_{RL})/2 \,. \end{aligned} \tag{15}$$

All eight linear and circular polarization scattering matrix coefficients are decomposed into reflected and diffracted parts in Table 1 of [11].

Although all of the physical optics scattering coefficients satisfy the principle of bistatic-monostatic equivalence [16], the equivalence is satisfied with a lower degree of error by the combinations in (15) that yield the reflected coefficient a_R [11].

B. IMSEC Polarization

The transverse nature of $\overleftrightarrow{\Gamma}_e$ and $\overleftrightarrow{\Gamma}_m$ indicated in (9) results in reradiated electric fields caused by electric and magnetic solenoidal equivalent currents that are polarized perpendicular and parallel, respectively, to the propagation plane containing the transmitter and receiver lines of sight [12]. Consequently

$$\overleftrightarrow{S} = \overleftrightarrow{S}_e + \overleftrightarrow{S}_m \tag{16}$$

where

$$\overleftrightarrow{S}_e = \hat{V}[a_{VH}\hat{H}_i + a_{VV}\hat{V}]$$

$$= (-jk/2\sqrt{\pi})w_e(jck)\int_V d^3\vec{r}'\,\overleftrightarrow{\Gamma}_e(\hat{k}_i,\vec{r}')\exp(j\vec{K}\cdot\vec{r}') \qquad (17)$$

and

$$\overleftrightarrow{S}_m = \hat{H}_s[a_{HH}\hat{H}_i + a_{HV}\hat{V}]$$

$$= (-jk/2\sqrt{\pi})w_m(jck)\int_V d^3\vec{r}'[-\hat{k}_s \times \overleftrightarrow{\Gamma}_m \times \hat{k}_i]\exp(j\vec{K}\cdot\vec{r}'). \qquad (18)$$

IV. RECONSTRUCTION ALGORITHMS

The results of the last chapter can be used to obtain inversion formulae in terms of spatial inverse Fourier Transforms $[\vec{K} \to \vec{r}]$ of multifrequency far field multistatic scattering data. The conventional relationship between Fourier and Radon Transforms [17] can then be used to obtain alternate inversion formulae in terms of inverse Radon transforms of transient scattering data. Filtered and prefiltered back-projections of both types of data can also be used [12,18,19].

A. POFFIS Inversion Formulae

The divergence theorem can be used to establish the following spatial Fourier Transform $(\vec{r} \leftrightarrow \vec{K})$ pairs [11]:

$$-\nabla\gamma \leftrightarrow (2j\sqrt{\pi}/K^2)\vec{K}[a_{HH}(k,\hat{k}_i,\hat{k}_s) + a^*_{HH}(k,-\hat{k}_i,-\hat{k}_s)] \qquad (19a)$$

$$\leftrightarrow (2j\sqrt{\pi}/K^2)\vec{K}[a_{VV}(k,\hat{k}_i,\hat{k}_s) + a^*_{VV}(k,-\hat{k}_i,-\hat{k}_s)] \qquad (19b)$$

$$\leftrightarrow (2j\sqrt{\pi}/K^2)\vec{K}[a_R(k,\hat{k}_i,\hat{k}_s) + a^*_R(k,-\hat{k}_i,-\hat{k}_s)] \qquad (19c)$$

where a_R can be obtained from linear or circular principal polarization measurements using (15). The three forms on the right hand side of (19) are equivalent only in the physical optics limit. Although each reduces to Bojarski's polarization independent monostatic results [10] when $\hat{k}_s = -\hat{k}_i$, only (19c) is consistent with the polarization dependent monostatic result of Boerner and Ho [17]. The forms in (19a) and (19b) also apply to scalar wave scattering with Neumann and Dirichlet boundary conditions, respectively. For two-dimensional scattering the right hand sides of (19) must be multiplied by $\sqrt{(jk/\pi)}$.

Physical optics scattering coefficients do not, in general, satisfy electromagnetic reciprocity. However, the electromagnetic reciprocal of the physical optics approximation is the image induction approximation [20] which also yields (19) [11]. Therefore (19) is consistent with reciprocity.

The ability to use only HH or VV scattering coefficients in (19a) and (19b) does not imply that multistatic POFFIS inversion should be attempted with a disregard for polarization effects. The simple principal polarization forms in (19) are valid only when the propagation plane is the polarization reference plane. Consequently, the $\hat{H}$ and $\hat{V}$ polarization directions indicated in (19) vary with receiver position. More practical polarization conventions require orthogonal polarization measurements. However, known target symmetries can be exploited to reduce the number of measurements [14,15,21].

B. IMSEC Inversion Formulae

IMSEC inversion formulae follow directly from (17) and (18), i.e.,

$$\overleftrightarrow{\Gamma}_e \leftrightarrow [2j\sqrt{\pi}/kw_e(jck)]\hat{V}[a_{VH}\hat{H}_s + a_{VV}\hat{V}] \tag{20}$$

and

$$\overleftrightarrow{\Gamma}_m \leftrightarrow [2j\sqrt{\pi}/kw_m(jck)]\hat{V}[a_{HV}\hat{H}_s - a_{HH}\hat{V}] \ . \tag{21}$$

V. EXAMPLES

Four IMSEC images of an infinitely long perfectly conducting circular (radius = a) cylinder in the $\hat{z}$ direction were obtained for the two-dimensional case of broadside incidence ($\hat{k}_i = -\hat{x}$) and reception ($\hat{k}_s = \cos\beta\ \hat{x} + \sin\beta\ \hat{y}$). The data were simulated from the exact eigenfunction series expansion and a 64 x 64 two-dimensional FFT (zero padded with 16 x 16 data points) was used to approximate the inverse Fourier Transform. Sample points in $\vec{K}$ space were chosen to yield a 4a ambiguity interval, i.e., $\Delta K_x = \Delta K_y = 2\pi/4a$. Spatial resolution was $\Delta x = \Delta y = 4a/16 = a/4$. The resulting range of bistatic angles and circumference to wavelength ratios was:

$$-133° \leq \beta = 2\tan^{-1}(K_y/K_x) \leq 127° \tag{22}$$

and

$$2.75 \leq ka = (1/2)[K_x + K_y^2/K_x]\ a \leq 17.2. \tag{23}$$

Therefore resonant region data was used for a specular point range of approximately ±65° about the transmitter line of sight.

The images in Figure 1 show little difference between the three principal polarization cases with image functions $|\vec{M}_{oz}|^2$ (HH), $|\eta\vec{J}_{oz}|^2$ (VV) and $(1/4)|\eta\vec{J}_{oz} + \vec{M}_{oz}|^2$ (LR). However, the computer printout indicated that background noise was three orders of magnitude lower in the circular polarization case. In all three cases the circular shape is recovered within ≈ ±45° of the transmitter line of sight without severe distortion. In contrast, the circular orthogonal polarization image function $(1/4)|\eta\vec{J}_{oz} - \vec{M}_{oz}|^2$ (RR) yields a distorted shape

IMSEC IMAGES OF A CIRCULAR CYLINDER

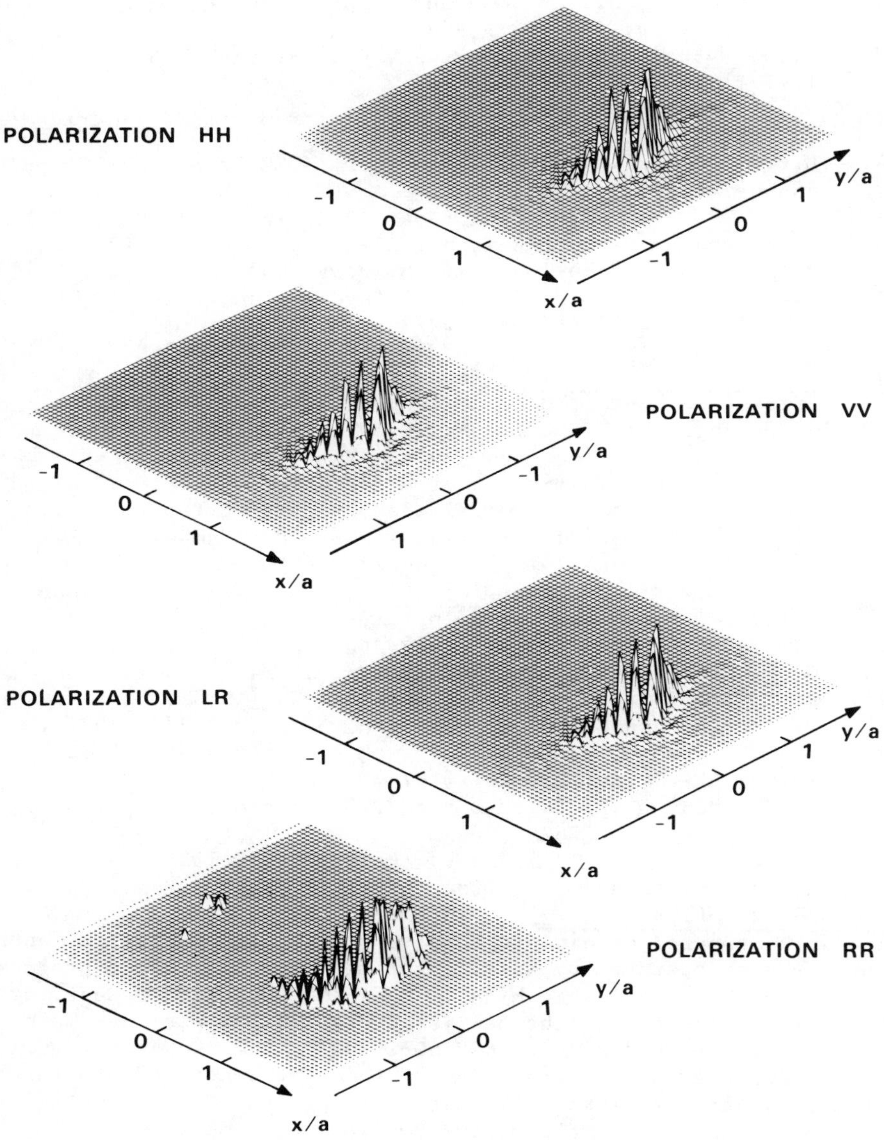

and observable creeping wave contamination. Experimentation with spectral weights ($w_e = w_m$) consisting of various powers of jk did not substantially improve the RR image. There was no observable difference between the principal polarization IMSEC images and the POFFIS images with the $(k,-\hat{k}_i,-\hat{k}_s)$ terms omitted (not shown) because of the simplicity of the target.

REFERENCES

1. V. H. Weston, J.J. Bowman and E. Ar, Arch. Rational Mech. Anal., 31, pp. 199-213 (1968).
2. V. H. Weston and W. M. Boerner, Can. J. Phys., 47, pp. 1177-1184 (1969).
3. W. M. Boerner and H.P.S. Ahluwalia, Can. J. Phys., 50, pp. 3023-3061 (1972).
4. H.P.S. Ahluwalia and W. M. Boerner, IEEE Trans. AP-21, pp. 663-672 (1973).
5. ______________ IEEE Trans. AP-22, pp. 673-682 (1974).
6. S. Rosenbaum-Raz, IEEE Trans. AP-24, pp. 66-70 (1976).
7. J. Detlefsen, "Abbildung mit Mikrowellen", Fortschrittberichte, der VDI-Zeitschriften, Reihe 10, Nr. 5. VDI-Verlag Dusseldorf (1979).
8. __________ IEEE Trans. AP-28, pp. 377-380 (1980).
9. ________________ 1980 International U.R.S.I. Symposium, Munich, Germany (August 26-29) pp. 322C/1.
10. N. N. Bojarski, IEEE Trans. AP-30, pp. 980-989 (1982).
11. G. E. Heath, "Direct and Inverse Multistatic Physical Optics: The Significance of Reciprocity, Polarization and Bistatic-Monostatic Equivalence", M.I.T./LL (to be publ.).
12. G. E. Heath, 1981 International IEEE/AP-S Symposium, Los Angeles, CA (June 19) pp. 620-623.
13. A. Rubinowicz, Progress in Optics, 5, E. Wolf, Ed. (John Wiley and Sons, Inc., 1965)pp. 199-240.
14. G. E. Heath, IEEE Trans. AP-29, pp. 523-525 (1981).
15. G. E. Heath, IEEE Trans. AP-29, pp. 429-434 (1981).
16. R. E. Kell, Proc. IEEE, 53, pp. 983-988 (1965).
17. W. Boerner and C. Ho, Wave Motion, 3, pp.311-333 (1981).
18. A. J. Rockmore, R. V. Denton and B. Friedlander, IEEE Trans. AP-27, pp. 239-241 (1979).
19. M. Y. Chiu, H. H. Barrett and R. G. Simpson, J. Opt. Soc. Am., 70, pp. 755-762 (1980).
20. R. F. Harrington, IRE Trans. AP-7, pp. 150-153 (1959).
21. M. Davidovitz and W. M. Boerner, IEEE Trans. AP-31, pp. 237-242 (1983).

The work reported herein was done at Lincoln Laboratory, a center for research operated by Mass. Inst. of Technology. The program is sponsored by Ballistic Missile Defense Prog.Office, Dept. of the Army; supported by the Ballistic Missile Defense Adv. Tech. Ctr. under A.F. Contract F19628-80-C-0002. The views and conclusions contained herein are those of the contractor and should not be interpreted as necessarily representing the official policies, either expressed or implied, of the U.S. Government.

III.6 (RP.2)

EXTENSION OF KENNAUGH'S OPTIMAL POLARIZATION CONCEPT TO THE ASYMMETRIC MATRIX CASE

Marat Davidovitz and Wolfgang-M. Boerner

Electromagnetic Imaging Division, Communications Lab.
Department of Electrical Engineering & Computer Science
University of Ill. at Chicago P.O. Box 4348, M/C 154
Chicago, Illinois 60680

Abstract

The polarization scattering matrix measured by a bistatic radar system generally will be asymmetric. Similarly, when the radar system is monostatic but the intervening propagation medium is non-reciprocal, the scattering matrix is asymmetric. In this paper we generalize Kennaugh's optimal polarization theory to the case of the asymmetric scattering matrix. The radar antenna polarizations to be used for maximum and zero power reception are defined and geometrically interpreted on the Poincaré sphere. These polarizations, termed optimal polarizations, may be used to enhance the level of target echo return, discriminate against undesired interference sources, or to classify radar targets.

I. Introduction

A monochromatic plane wave with a fixed polarization state impinging upon a radar target induces a surface current distribution, which in turn gives rise to a reradiated or scattered field. The polarization state of a wave scattered in the direction of observation will, in general, differ from that of the incident wave. The transformation of the polarization state upon reflection can be represented by a complex scattering matrix, characteristic of the target at a specific aspect and source frequency. The formula used to calculate the target radar cross section or echoing area involves the scattering matrix as well as the transmitting and receiving antenna polarizations. The problem facing the radar operator is the one of selecting the optimal radar polarizations which will, for example, maximize the level of echo signal or minimize the effect of undesired sources, such as clutter return or jamming signals.

W.-M. Boerner et al. (eds.), Inverse Methods in Electromagnetic Imaging - Part 1, 609–628.

Utilizing the principle of electromagnetic reciprocity, one can show that when the radar system is monostatic and the intervening propagation medium is reciprocal, the scattering matrix is symmetric. The optimal polarizations associated with a symmetric scattering matrix were first defined by Kennaugh [1] in 1950. In particular, he found that a co-polarized transmitting and receiving radar antenna can be used to obtain the maximum echo return possible and that there exists only one polarization for which this maximum is achieved. Kennaugh also showed that there are two polarizations for which the back-scattered wave is orthogonal to the co-polarized radar antenna, thus rendering the radar "blind" to the incoming wave.

In a situation where the radar system is bistatic or one in which the radar system is monostatic but the medium of propagation is non-reciprocal, the scattering matrix generally is asymmetric. In this paper we attempt to generalize the optimal polarizations theory to the case of the asymmetric scattering matrix. In particular, we show that the maximum echo return can no longer be obtained with an identically polarized transmitting and receiving antenna. The optimal transmitting and receiving antenna polarizations are defined in terms of the scattering matrix elements. It is subsequently proven that for the case of the asymmetric scattering matrix, the co-polarized echo return will always be less than the absolute maximum obtained with separately polarized transmitting and receiving antennas. Finally, we represent the optimal polarizations on the Poincaré sphere and prove certain geometrical properties of the resulting configuration.

II. Overview and Statement of the Problem

Under the condition that the radar antennas are located sufficiently far from the target, we may consider the wave incident on the target and the scattered wave observed at the receiving antenna to be quasi-planar. In such case the electric field vectors $\underline{E}^i$ and $\underline{E}^s$ of the incident and the scattered wave, respectively, will lie entirely in the planes perpendicular to directions of incidence and observation. The standard two-dimensional orthonormal polarization bases may then be used to represent the electric field vectors [2]. Due to the linearity of Maxwell's equations, the components of the vectors $\underline{E}^i$ and $\underline{E}^s$ can be related by a 2x2 matrix transformation S as follows:

$$\underline{E}^s = S \cdot \underline{E}^i \tag{1}$$

The scattering matrix S serves as a discriptor of the polarizing properties of the target.

The measurable quantity of primary interest to the radar operator is the target radar cross section or echoing area. For a given incident wave $\underline{E}^i$, the echoing area A is a measure of the power contained in the wave scattered toward the receiver. The echoing area A is calculated according to the following standard formula

$$A = 4\pi R^2 \frac{\| S \cdot \underline{E}^i \|}{\| \underline{E}^i \|} \tag{2}$$

where $\| \cdot \|$ denotes the norm of the enclosed vector quantity. The norm is defined as $\| \underline{E} \| = \tilde{\underline{E}}^* \cdot \underline{E}$, where "~" and "*" denote transposition and conjugation, respectively.

However, when there is a mismatch between the polarization state of the scattered wave and that of the receiving antenna, only a fraction of the reflected wave's power is sampled. The formula for the resulting effective echoing area must be appropriately redefined to account for the dependence on the receiving antenna polarization. The following bilinear matrix form defines the effective echoing area A_e

$$A_e = |\tilde{\underline{h}}^r \cdot S \cdot \underline{h}^t|^2 \tag{3}$$

where $\underline{h}^t$ and $\underline{h}^r$ are the normalized ($\| \underline{h}^r \| = 1 = \| \underline{h}^t \|$) vector heights of the transmitting and receiving antenna, respectively [3]. Note, the quantity $A_e = A_e(\underline{h}^t, \underline{h}^r)$ is a function of two independent variables, namely $\underline{h}^t$ and $\underline{h}^r$. The problem which we pose in this paper is the one of finding the maxima and minima of A_e and the corresponding antenna polarizations for which the optimum values of A_e are obtained. Such polarizations are termed optimal.

We first consider the somewhat simplified case of expression (3), namely one in which the transmitting and the receiving antenna are identically polarized, i.e. $\underline{h}^t = \underline{h}^r = \underline{h}$. At this point it is proper to raise the question of just what do we mean when we say identically polarized antennas, since in the general case of a bistatic radar system, the transmitting and the receiving antenna polarizations are referred to two separate local coordinate systems. Evidently there is no unique answer to the posed question, since several definitions of co- and cross-polarizations have been offered in the past [4,5]. However, it is important to note that a unique definition of co-polarization is not crucial to the theory presented in this paper. The main reason is that the optimal polarizations are physically defined by the radar target itself, and are, therefore, invariant to the various definitions of coordinate systems. In other words, even though the mathematical expressions representing the optimal polarizations may change depending on the particular choice of the reference coordinate systems, the physical attributes of the optimal polarizations, such as ellipticity, sense of rotation, and the orientation angle, referred to the plane defined by the directions of incidence and observation, remain invariant. We may, therefore, without any loss of generality, consider the plane defined by the directions of incidence and observation to be the polarization reference plane. We then proceed to construct right-handed coordinate frames at the transmitter and receiver sights, letting the y-axis lie in the reference plane and x-axis be perpendicular to it, while the z-axis will be directed at the target [3].

Returning to expression (3) and setting $\underline{h}^t = \underline{h}^r = \underline{h}$, we obtain the following quadratic matrix form

$$A_e = |\tilde{\underline{h}}\cdot S\cdot\underline{h}|^2 \tag{4}$$

The scattering matrix S appearing in (4) will, in general, be asymmetric. From elementary matrix theory we know that any asymmetric matrix can be decomposed into a symmetric (S_s) and a skew- or anti-symmetric (S_a) component as follows:

$$S_s = (S+\tilde{S})/2 \tag{5a}$$

$$S_a = (S-\tilde{S})/2 \tag{5b}$$

The value of the quadratic matrix form defined by (4) remains unchanged if the scattering matrix S is replaced by its symmetric component S_s.

Taking into consideration the discussion of the previous paragraph, we may subdivide the general problem into two parts. In the first part we will investigate the properties of symmetric scattering matrices under the restriction of identically polarized transmitting and receiving antennas. In the second part this restriction will be removed, thus forcing us to consider a more general problem of the asymmetric scattering matrix and separately polarized radar antennas.

III. Polarization Basis Transformations

Transformation of polarization vectors from one orthonormal basis to another is accomplished through the use of unitary matrices.

Unitary change-of-basis transformations have the important property of preserving the total power in the wave [2], i.e. the norm of the electric field vector $\|\underline{E}\|$ is invariant under unitary transformations. A unitary change-of-basis matrix T must satisfy the following

$$|\det\{t\}| = 1 \tag{6a}$$

$$T^{-1} = \tilde{T}^{*} \tag{6b}$$

The most general form of a unitary matrix, satisfying the above requirements, is given by

$$T = \begin{bmatrix} e^{j\psi_1}\cos(\gamma/2) & -e^{j\psi_2}\sin(\gamma/2) \\ e^{j\psi_3}\sin(\gamma/2) & e^{j\psi_4}\cos(\gamma/2) \end{bmatrix} \tag{7}$$

where, because of condition (6a), the phases of the matrix are interrelated as follows: $\psi_1 + \psi_4 = \psi_2 + \psi_3$. Without any loss of generality we may choose to set $\psi_1 = \psi_4 = 0$ and let $\psi_3 = -\psi_2 = \delta$, in which case the matrix T takes on the following form [6,7]

$$T = \begin{bmatrix} \cos(\gamma/2) & -e^{-j\delta}\sin(\gamma/2) \\ e^{j\delta}\sin(\gamma/2) & \cos(\gamma/2) \end{bmatrix} = \frac{1}{(1+\rho\rho^*)^{1/2}} \begin{bmatrix} 1 & -\rho^* \\ \rho & 1 \end{bmatrix} \tag{8}$$

where $\rho = \tan(\gamma/2)e^{j\delta}$.

The particular explicit form of T given in (8) was chosen because of its intimate connection to the geometrical representation of unitary transformations on the Poincaré sphere. In fact, when the Poincaré sphere is used to represent two orthonormal bases, namely $[\underline{e}_A, \underline{e}_{A'}]$ and $[\underline{e}_B, \underline{e}_{B'}]$, the geometrical transformation analogous to (8) is accomplished using the angles γ and δ as shown in Fig. 1.

Utilizing the change-of-basis matrix defined above, we proceed to transform the vectors $\underline{E}^i$, $\underline{E}^s$ from the $[\underline{e}_A, \underline{e}_{A'}]$ basis into the $[\underline{e}_B, \underline{e}_{B'}]$ basis in the following manner

$$\underline{E}^i(A,A') = T \cdot \underline{E}^i(B,B') \tag{9a}$$

$$\underline{E}^s(A,A') = T^* \cdot \underline{E}^s(B,B') \tag{9b}$$

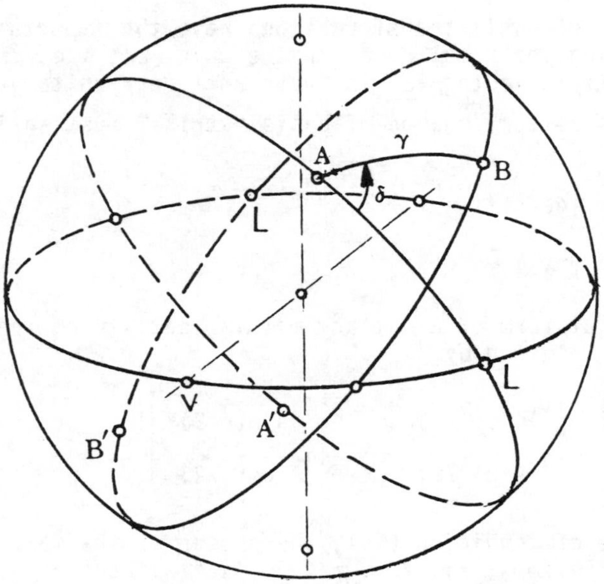

FIGURE 1: POLARIZATION BASIS TRANSFORMATION ON THE POINCARÉ SPHERE

Note that the incident and the scattered wave are propagating in opposite directions. Consequently, in order for the transformation to have the same effect on the polarization state of $\underline{E}^i$ and $\underline{E}^s$, the change-of-basis matrices used to transform $\underline{E}^i$ and $\underline{E}^s$ must be related through conjugation [2,6]. The latter fact is reflected in relations (9a,b).

Recalling that the relation between $\underline{E}^i(A,A')$ and $\underline{E}^s(A,A')$ is given by

$$\underline{E}^s(A,A') = S(A,A') \cdot \underline{E}^i(A,A') \tag{10}$$

we can transform the scattering matrix $S(A,A')$ into the $[\underline{e}_B, \underline{e}_{B'}]$ basis. Substituting the relations in (9a,b) into (10), we obtain the scattering matrix representation in the $[\underline{e}_B, \underline{e}_{B'}]$ basis. The resulting expression is given by

$$S(B,B') = \tilde{T} \cdot S(A, A') \cdot T \tag{11}$$

The general situation is conveniently summarized in the following mneumonic diagram

$$\begin{array}{ccc} \underline{E}^{i}(B,B') & \xrightarrow{\quad S(B,B') \quad} & \underline{E}^{s}(B,B') \\ T \downarrow & & \downarrow T^{*} \\ \underline{E}^{i}(A,A') & \xrightarrow[\quad S(A,A') \quad]{} & \underline{E}^{s}(A,A') \end{array}$$

Finally, we prove that the type of transformation prescribed by equation (11) preserves the (skew-) symmetry of a matrix. Starting with a symmetric matrix S_s or a skew-symmetric matrix S_a and applying the transformation yields

$$\tilde{T}\cdot S_{s,a}\cdot T = S'_{s,a} \tag{12}$$

Taking note of the fact that $\tilde{S}_{s,a} = \pm S_{s,a}$, leads to following equation

$$\tilde{S}'_{s,a} = \widetilde{(\tilde{T}\cdot S_{s,a}\cdot T)} = \tilde{T}\cdot\tilde{S}_{s,a}\cdot T = \pm(\tilde{T}\cdot S_{s,a}\cdot T) = \pm S'_{s,a} \tag{13}$$

where the lower sign corresponds to the lower subscript.

Moreover, it is a well-known fact that the transformation of equation (11) can take a complex symmetric matrix into the complex diagonal form [2]. This important fact is used in the next section, where the optimal polarization properties of symmetric scattering matrices are considered.

IV. Optimal Polarizations Associated with a Symmetric Scattering Matrix

Consider a situation in which the scattering matrix, obtained either through measurement or theoretical computation, is represented in an orthonormal basis $[\underline{e}_A, \underline{e}_{A'}]$ by a complex asymmetric matrix $S(A,A')$. The effective echoing area obtained by transmitting and receiving with the same polarization $\underline{h}(A, A')$ ($\|\underline{h}(A, A')\| = 1$) is given by

$$A_e = |\tilde{\underline{h}}(A,A')\cdot S(A,A')\cdot\underline{h}(A,A')|^2 = |\tilde{\underline{h}}(A,A')\cdot S_s(A,A')\cdot\underline{h}(A,A')|^2 \tag{14}$$

where we have replaced the matrix $S(A,A')$ by its symmetric component $S_s(A,A')$. Note, no loss in generality occurs if the matrix $S(A,A')$ is symmetric from the outset.

The mathematical computations involved in optimization of the quadratic form for A_e can be simplified by finding a basis in which the scattering matrix is diagonal. Recalling that the type of transformation described in the preceeding section can diagonalize a complex symmetric matrix, we pose the problem of finding the change-of-basis matrix T_M such that

$$\tilde{T}_M \cdot S(A,A') \cdot T_M = S_d(M,M') \tag{15}$$

where $S_d(M,M')$ is a complex diagonal matrix. Conjugating and transposing (15) and multiplying both sides of the original equation by the corresponding sides of the new equation, we obtain an equivalent formulation of the problem. The resulting equation is given by

$$\tilde{T}_M^{*} \cdot P(A,A') \cdot T_M = T_M^{-1} \cdot P(A,A') \cdot T_M = P_d(M,M') \tag{16}$$

where $P(A,A') = S_s^{*}(A,A') \cdot S_s(A,A')$ and $P_d(M,M') = S_d^{*}(M,M') \cdot S_d(M,M')$. We recognize that $P(A,A')$, $P_d(M,M')$ are the so-called power scattering matrices [2]. We also note the fact that the diagonal elements of $P_d(M,M')$ are the eigenvalues of $P(A,A')$ and are equal to the squared magnitudes of the respective elements of the matrix $S_d(M,M')$. The basis $[\underline{h}_M, \underline{h}_{M'}]$ consists of the eigenvectors of $P(A,A')$ and can, therefore, be termed the "characteristic" basis. It can easily be proven that such a basis is orthogonal [8]. Moreover, Kennaugh showed [1] that the maximum value of A_e is equal to the larger eigenvalue of $P(A,A')$ and the maximum polarization is the corresponding eigenvector.

The eigenvalues of $P(A,A')$ are found in the usual manner, i.e., by solving the equation $\det\{P(A,A') - |\lambda|^2 I\} = 0$. The solutions satisfying this equation are

$$|\lambda_{1,2}|^2 = \frac{tr\{P(A,A')\} \pm \sqrt{tr^2\{P(A,A')\} - 4 \cdot \det\{P(A,A')\}}}{2} \tag{17}$$

where $\lambda_{1,2}$ are the complex diagonal elements of $S_d(M,M')$. It is interesting to note that $tr\{P(A,A')\} = tr\{P_d(M,M')\}$ and $\det\{P(A,A')\} = \det\{P_d(M,M')\}$, i.e. the trace and the determinant are invariant under the change-of-basis transformation. Furthermore, we can show that

$$tr\{P(A,A')\} = span\{S_s(A,A')\} \tag{18a}$$

and

$$\det\{P(A,A')\} = |\det\{S_s(A,A')\}|^2 \tag{18b}$$

where the span of a complex matrix is defined as the sum of the magnitudes squared of its elements.

To find the change-of-basis matrix T_M and the eigenvectors corresponding to $|\lambda_{1,2}|^2$, we rewrite equation (16) as follows:

$$P(A,A')\cdot T_M = T_M \cdot P_d(M,M') \tag{19}$$

The matrix equation (19) corresponds to four scalar equations. If we use the explicit form of T_M defined in Section III, any one of the four scalar equations yields the complex parameter ρ_M

$$\rho_M = \frac{|\lambda_1|^2 - P_{AA}}{P_{AA'}} \tag{20}$$

where P_{AA}, P'_{AA} are the elements of the power scattering matrix $P(A,A')$ and $|\lambda_1|^2$ is given by (17). The eigenvectors $\underline{h}_M$ and $\underline{h}_{M'}$ and the change-of-basis matrix T_M can be defined in terms of ρ_M as follows:

$$\underline{h}_M = C_M \begin{bmatrix} 1 \\ \rho_M \end{bmatrix}, \quad \underline{h}_{M'} = C_M \begin{bmatrix} -\rho_M^* \\ 1 \end{bmatrix}, \quad C_M = \frac{1}{\sqrt{1+\rho_M \rho^*{}_M}} \tag{21}$$

and

$$T_M = [\underline{h}_M, \underline{h}_M] = C_M \begin{bmatrix} 1 & -\rho_M^* \\ \rho_M & 1 \end{bmatrix} \tag{22}$$

From equation (17) it is clear that $|\lambda_1|^2 > |\lambda_2|^2$ and we therefore conclude that $A_{e,max} = |\lambda_1|^2$ and the maximum polarization is specified by $\underline{h}_M$.

We may use the following formula to calculate the effective echoing area A_e

$$A_e = |\tilde{\underline{h}}\cdot(M,M')\cdot S_d(M,M')\cdot \underline{h}(M,M')|^2 \tag{23}$$

The matrix $S_d(M,M')$ has been determined by straight-forward matrix multiplication in accordance with equation (15) and using the matrix T_M displayed in (22).

We now turn to the problem of finding the so-called "null" polarization in the characteristic basis $[\underline{h}_M, \underline{h}_{M'}]$. The "null", $\underline{h}_o$, is defined by the condition $A_e = 0$ [9], which in turn leads to the following equation

$$\tilde{\underline{h}}_o(M,M') \cdot S_d(M,M') \cdot \underline{h}_o(M,M') = 0 \tag{24}$$

Two "null" polarizations generally will be found by expanding equation (24). The polarization ratios defining the "null" polarizations are given by [1]

$$\rho_{1,2} = \pm j \sqrt{\frac{S_{MM}}{S_{M'M'}}} = \sqrt{\frac{|S_{MM}|}{|S_{M'M'}|}} \exp[j\frac{(\phi_{MM}-\phi_{M'M'})}{2} \pm j(\pi/2)] =$$

$$= \tan(\gamma_{1,2}/2)e^{j\delta_{1,2}} \tag{25}$$

In the characteristic basis, the magnitudes of the polarization ratios $\rho_{1,2}$ are equal, i.e. $\gamma_1 = \gamma_2 = \gamma$, and the phase difference $\delta_1 - \delta_2$ is equal to π. There is an interesting geometrical interpretation of these facts.

We first represent the orthonormal basis $[\underline{h}_M, \underline{h}_{M'}]$ on the Poincaré sphere by antipodal points M,M', as shown in Fig. 2. In accordance with the Poincaré sphere rules [9,10,11], all polarizations with equivalued polarization ratio magnitude lie on a small circle, centered on the diameter joining M and M'. Change in the phase of a polarization ratio, on the other hand, rotates the point representing the polarization about the diameter MM'. Using these rules to map the "null" polarizations $N_{1,2}$ on the Poincaré sphere we come to the following conclusions; since the difference in the rotations of the points N_1, N_2 about MM' is equal to π, they are located on the same great circle. Moreover, the fact that $\gamma_1 = \gamma_2 = \gamma$ implies that the angle N_1ON_2 (O is the center of the sphere) is bisected by MM'. The conclusions are illustrated in Fig. 2.

V. Optimal Polarizations Associated with the Asymmetric Scattering Matrix

In the preceeding section we have constrained our radar system to transmit and receive with the same polarization. In this section we remove this restriction and attempt to find distinct transmitting and receiving antenna polarizations, such that the effective echoing area is maximized.

We will start by transforming the general asymmetric scattering

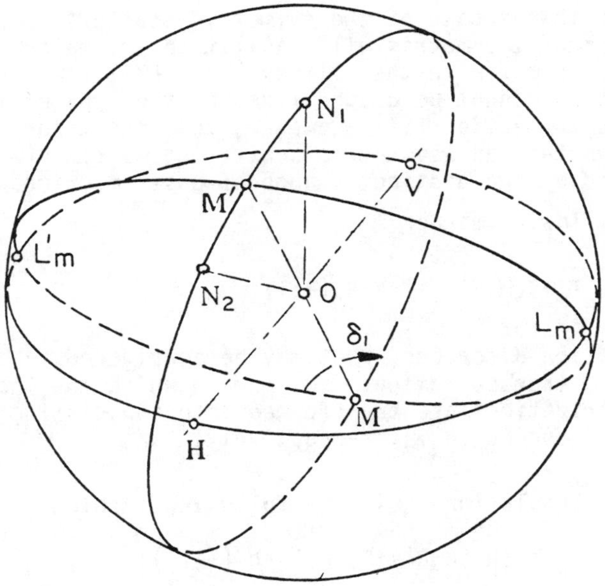

FIGURE 2: OPTIMAL POLARIZATION FOR CO-POLARIZED TRANSITORY AND RECEIVING ANTENNA

matrix S(A,A') into the characteristic basis $[\underline{h}_M, \underline{h}_{M'}]$. As we have seen, this transformation diagonalizes the symmetric component of S(A,A'). We also recall that this type of transformation preserves the skew-symmetry of the anti-symmetric component ($S_a(A,A')$) of S'(A,A'). Therefore, the transformed matrix S(M,M') can be ex pressed as the sum of a diagonal and a skew-symmetric matrix, i.e.

$$S(M,M') = \tilde{T}_M \cdot S(A,A') \cdot T_M = \tilde{T}_M \cdot S_s(A,A') \cdot T_M + \tilde{T}_M \cdot S_a(A,A') \cdot T_M =$$

$$= S_d(M,M') + S_a(M,M') = \begin{bmatrix} S_{MM} & S_{MM'} \\ -S_{MM'} & S_{M'M'} \end{bmatrix} \qquad (26)$$

We next turn to the problem of finding the polarizations $\underline{h}^{t,r}(M,M')$ ($\|\underline{h}^{t,r}(M,M')\|=1$) which will maximize the effective echoing area A_e, computed according to the formula

$$A_e = |\tilde{\underline{h}}^r(M,M') \cdot S(M,M') \cdot \underline{h}^t(M,M')|^2 \qquad (27)$$

Again, just as in the case of the symmetric scattering matrix, diagonalization of the matrix will facilitate the mathematical procedure used to maximize the bilinear form (27). As we know, an asymmetric matrix cannot be diagonalized by the type of transformation defined in Section III. However, from the theory of matrices it is known that an asymmetric scattering matrix S(M,M') can be diagonalized by two distinct change-of-basis matrices T_I and T_R used in the following manner

$$\tilde{T}_R \cdot S(M,M') \cdot T_1 = S_d(R;I) \tag{28}$$

The above equation discribes, what may be considered, a mixed change-of-basis transformation, i.e. one in which the incident and scattered polarizations are transformed into two distinct bases, namely $[\underline{e}_I, \underline{e}_{I'}]$ and $[\underline{e}_R, \underline{e}_{R'}]$, respectively.

Rewriting the formula for A_e in the following fashion

$$A_e = |\tilde{h}^r(R,R') \cdot S_d(R;I) \cdot \underline{h}^t(I,I')|^2 \tag{29}$$

We can show that the maximum value of A_e is equal to the squared magnitude of the larger non-zero element of $S_d(R,I)$. The polarizations for which this maximum is achieved are given by either $\underline{h}^r = \underline{e}_R$, $\underline{h}^t = \underline{e}_I$ or $\underline{h}^r = \underline{e}_{R'}$, $\underline{h}^t = \underline{e}_I$, depending on which matrix element has a greater magnitude. Therefore, our next objective is to find the change-of-basis matrices T_I and T_R, the matrix $S_d(R;I)$ and express them in terms of the elements of S(M,M'). Conjugating and transposing equation (28) we get

$$\tilde{T}_I^* \cdot \tilde{S}^*(M,M') \cdot T_R^* = \tilde{S}_d^*(R;I) \tag{30}$$

Now, pre- and post- multiplying both sides of the equation (28) by the corresponding sides of equation (30), we reduce the more difficult original problem to two problems whose solutions are derived in Section IV. The resulting equations are given by

$$\tilde{T}_I^* \cdot P(M,M') \cdot T_I = P_d(R;I) \tag{31a}$$

$$\tilde{T}_R \cdot P'(M,M') \cdot T_R^* = P_d(R;I) \tag{31b}$$

where $P(M,M') = \tilde{S}^*(M,M') \cdot S(M,M')$; $P'(M,M') = S(M,M') \cdot \tilde{S}^*(M,M')$;

$$P_d(R;I) = \tilde{S}^*_d(R;I) \cdot S_d(R;I)$$

As we immediately observed, the type of problem posed by equations (31a,b) was solved in the preceeding section. The non-zero elements of $P_d(R;I)$ are the eigenvalues of both $P(M,M')$ and $P'(M,M')$, and are equal to the squared magnitudes of the respective elements of $S_d(R;I)$. The bases $[\underline{e}_I, \underline{e}_{I'}]$ and $[\underline{e}_R, \underline{e}_{R'}]$ consist of the eigenvectors of P and P', respectively. Following the procedure used in the preceeding section, but omitting some of the details, we obtain the following results:

$$|\mu_{1,2}| = \frac{tr\{P\} \pm \sqrt{tr^2\{P\} - 4\det\{P\}}}{2} \quad (32)$$

where $\mu_{1,2}$ are the diagonal elements of $S_d(R;I)$ and P stands for either $P(M,M')$ or $P'(M,M')$;

$$\rho_I = \frac{|\mu_1|^2 - P_{MM}}{P_{MM'}} \; ; \; \rho_R = \left(\frac{|\mu_1|^2 - P'_{MM}}{P'_{MM'}}\right)^* \quad (33a,b)$$

where P_{MM}, $P_{MM'}$ are the elements of $P(M,M')$ and P'_{MM}, $P'_{MM'}$ are the elements of $P'(M,M')$;

$$\underline{e}_I = C_I \begin{bmatrix} 1 \\ \rho_I \end{bmatrix}; \; \underline{e}_{I'} = C_I \begin{bmatrix} -\rho_I^* \\ 1 \end{bmatrix}; \; T_I = [\underline{e}_I \underline{e}_{I'}] = C_I \begin{bmatrix} 1 & -\rho_I^* \\ \rho_I & 1 \end{bmatrix} \quad (34)$$

$$\underline{e}_R = C_R \begin{bmatrix} 1 \\ \rho_R \end{bmatrix}; \; \underline{e}_{R'} = C_{R'} \begin{bmatrix} -\rho_R^* \\ 1 \end{bmatrix}; \; T_R = [\underline{e}_R \underline{e}_{R'}] = C_R \begin{bmatrix} 1 & -\rho_R^* \\ \rho_R & 1 \end{bmatrix} \quad (35)$$

and $$C_{I,R} = \frac{1}{\sqrt{1 + \rho_{I,R} \cdot \rho^*_{I,R}}}$$

From equation (32) it is clear that $|\mu_1|^2 \geq |\mu_2|^2$ and we, therefore, conclude that the maximum value of A_e is given by $|\mu_1|^2$ and the polarizations which are used to obtain $A_{e,max} = |\mu_1|^2$ are, in fact, $\underline{h}^t = \underline{e}_I$, $\underline{h}^r = \underline{e}_R$.

In Section IV we have demonstrated that when the radar was under constraint to transmit and receive with the same polarization, the maximum echoing area obtainable was given by $A_{e,max} = |S_{MM}|^2 = |\lambda_1|^2$. In this section we saw that by using distinctly polarized transmitting and receiving antennas we obtain the maximum echoing area $A_{e,max} = |\mu_1|^2$. In what follows, it is proven that $|\mu_1|^2$ is the absolute maximum of the function A_e, exceeding the value of $|\lambda_1|^2$. The proof utilizes the fact that the trace and the determinant of the power cattering matrix are invariant under all unitary change-of-basis transformations, i.e.

$$\mathrm{tr}\{P_d(R;I)\} = |\mu_1|^2 + |\mu_2|^2 = \mathrm{tr}\{P(M,M')\} = |\lambda_1|^2 + |\lambda_2|^2 + 2|S_{MM'}|^2 \tag{36}$$

and

$$\det\{P_d(R;I)\} = |\mu_1|^2 \cdot |\mu_2|^2 = \det\{P(M,M')\} = = |\lambda_1|^2|\lambda_2|^2 + |S_{MM'}|^4 + 2|\lambda_1||\lambda_2|^2|S_{MM'}|^2 \cdot \cos\alpha \tag{37}$$

where

$$\lambda_1 = S_{MM},\ \lambda_2 = S_{M'M'},\ \text{and } \alpha = \phi_{MM} + \phi_{M'M'} - 2\phi_{MM'}\ .$$

From equation (36,37) it follows that

$$(|\mu_1|^2 - |\mu_2|^2)^2 = (|\lambda_1 - |\lambda_2|^2)^2 + +4|S_{MM'}|^2 (|\lambda_1|^2 + |\lambda_2|^2 - 2|\lambda_1||\lambda_2|\cos\alpha) \tag{38}$$

and consequently

$$(|\mu_1|^2 - |\mu_2|^2)^2 > (|\lambda_1|^2 - |\lambda_2|^2)^2 \tag{39}$$

From equations (36) and (39) we obtain the following set of inequalities

$$\left.\begin{array}{l} |\mu_1|^2 + |\mu_2|^2 > |\lambda_1|^2 + |\lambda_2|^2 \\ |\mu_1|^2 - |\mu_2|^2 > |\lambda_1|^2 - |\lambda_2|^2 \end{array}\right\} |\mu_1|^2 > |\lambda_1|^2 \tag{40}$$

which yield the sought for proof of the fact that $|\mu_1|^2 > |\lambda_1|^2$.

Finally, we represent the maximum polarizations $\underline{h}^t = \underline{e}_I$ and $\underline{h}^r = \underline{e}_R$ on the Poincaré sphere. The optimal polarizations $h^{t,r}$ may be specified by their complex polarization ratios, as follows:

$$\rho^t = \frac{h^t_{M'}}{h^t_{M'}} = \rho_I = \tan(\frac{\xi_I}{2})e^{j\delta_I} \; ; \text{ and } \rho^r = \frac{h^r_{M'}}{h^r_{M'}} = \rho_R = \tan(\frac{\xi_R}{2})e^{j\delta_R} \qquad (41a,b)$$

Taking note of the fact that $P'_{MM} = P_{MM}$ and $P'_{MM'} = -P^*_{MM'}$ and using equations (33a,b), we obtain the following result

$$\frac{\rho^t}{\rho^r} = \frac{|\mu_1|^2 - P_{MM}}{P_{MM'}} \bullet \frac{-P_{MM'}}{|\mu_1|^2 - P_{MM}} = -1 \qquad (42)$$

The implications of the above result are that $\xi_R = \xi_I = \xi$ and that $\delta_R - \delta_I = \pi$. In accordance with the Poincaré sphere rules expounded at the end of the preceeding section we see that the optimal polarizations $\underline{h}^{t,r}$, represented by points I,S respectively, are located on the same great circle and the angle IOS is bisected by the diameter MM'. Furthermore, we can show that the angle β included between the planes of the great circles containing N_1, N_2 and I,S is given by

$$\beta = \text{arccot}\{\frac{|\lambda_1|+|\lambda_2|}{|\lambda_1|-|\lambda_2|} \tan\nu\} \qquad (43)$$

where $\nu = \phi_{MM'} - 1/2(\phi_{MM} + \phi_{M'M'})$

The complete optimal polarizations configuration on the Poincaré sphere is illustrated in Fig. 3.

V. Effect of Faraday Rotation on the Optimal Polarizations

Faraday rotation is the anisotropic effect exhibited by magneto-ionic media, such as the ionosphere. As a result of this effect, the polarization plane of a linearly polarized wave rotates continuously along the direction of propagation.

The purpose of this section is to investigate the effects of Faraday rotation on the optimal polarizations defined in the preceeding sections. In order to keep the exposition intuitively

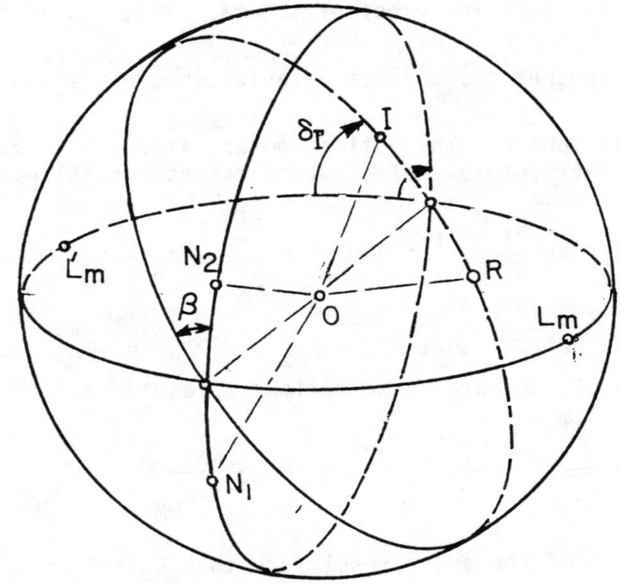

FIGURE 3: OPTIMAL POLARIZATION CONFIGURATION FOR AN ASYMMETRIC SCATTERING MATRIX

clear, the example considered here will be simple, although not completely devoid of practical value.

Consider a body of revolution or any other plane symmetric scatterer illuminated by a monostatic radar with linearly polarized transmitting and receiving antenna. It is a well-known fact that if the radar's line of sight and the vibration plane of the linearly polarized antenna lie in the target's symmetry plane, no cross-polarization is observed, i.e. the scattering matrix is diagonal [1]. Denoting the linear polarizations by x and y, the scattering matrix for such a target can be written as follows:

$$S_d(x,y) = \begin{bmatrix} S_{xx} & 0 \\ 0 & S_{yy} \end{bmatrix} \tag{44}$$

As the term Faraday rotation implies, the influence of the intervening magneto-ionic propagation medium on the incident linear polarization can be accounted for by a simple rotation matrix, given by [12]

$$R(\phi) = \begin{bmatrix} \cos\phi & \sin\phi \\ -\sin\phi & \cos\phi \end{bmatrix} \tag{45}$$

The preceeding example served to illustrate a potential application of the optimal polarization theory. As the results indicate, under special circumstances the optimal polarizations are intimately related to a physical parameter of the total scattering problem. In particular we see that the polarization ratios of the maximum polarizations (equations 47a,b) are simply related to the physical parameter of the propagation medium, namely the Faraday rotation angle ϕ.

VI. Conclusion

The optimal polarizations of the scattering matrix are derived for the general asymmetric case. In particular, it is found that unlike in the symmetric scattering matrix case, the maximum return is no longer obtained utilizing identically polarized transmit-receive antennas. Therefore, in addition to the optimal polarizations found for the symmetric matrix, there are two more optimal polarizations associated with the asymmetric scattering matrix. When these polarizations are used by the transmit and receive antennas maximum target echo is observed.

It is also shown that when the optimal polarizations are mapped onto the surface of the Poincaré sphere, certain voltage and power calculations associated with the scattering matrix can be very conveniently performed by geometrical methods. For example, it is shown that the product of the chord lengths extending from point A, representing the transmit-receive antenna polarization to the null polarizations N_1 and N_2, is proportional to the magnitude of the voltage recorded at the receiver. Similarly, a certain combination of the three chord lengths associated with the chordal triangle N_1AN_2 directly determines the phase of the received voltage.

Finally, it is shown that the amount of scattering matrix measurements required to characterize completely bistatic scattering by targets having inversion symmetry can be significantly reduced. The necessary relations are derived upon simple geometrical arguments. The relations are subsequently numerically verified for specific geometries.

It should be noted that the theory presented here, and developed in greater detail in the M.Sc. thesis of Davidovitz [12], namely the derivation of optimal polarizations for the asymmetric scattering matrix, constitutes a direct extension of the concepts first developed by Kennaugh [1], and later expounded by Huynen [8], for the symmetric scattering matrix. In fact, the generality of the present theory is verified by showing that Kennaugh's results can be obtained therefrom in a straight-forward manner.

The Poincaré sphere results obtained here, are derived in part

The angle ϕ is the total amount by which the plane of polarization is rotated while traveling in the anisotropic medium. Since the medium is non-reciprocal, the rotation of the polarization plane resulting from back-propagation to the antenna will be in the same direction and will, therefore, be represented by the same rotation matrix $R(\phi)$. The overall scattering matrix $S'(x,y)$, including the effect of the target as well as that of the magneto-ionic propaga-tion medium can be constructed as follows:

$$S'(x,y) = R(\phi)\cdot S_d(x,y)\cdot R(\phi) =$$

$$= \begin{bmatrix} S_{xx}\cos^2\phi - S_{yy}\sin^2\phi & (S_{xx} + S_{yy})\cos\phi\sin\phi \\ -(S_{xx} + S_{yy})\cos\phi\sin\phi & -S_{xx}\sin^2\phi + S_{yy}\cos^2\phi \end{bmatrix} \tag{46}$$

The resulting matrix is of the form shown in (26). We, therefore, conclude that the basis $[\underline{h}_x, \underline{h}_y]$ is the so-called characteristic basis. This in turn implies that $|S'_{xx}|^2$ is the maximum echoing area obtainable with a co-polarized antenna. The maximum polariza-tion is, of course, specified by $\underline{h}_x$.

However, it is intuitively clear that, in fact, $|S_{xx}|^2$ and not $|S'_{xx}|^2$ is the absolute maximum echo return obtainable from that target. Solution of equation (32) shows that the larger eigenvalue of $P'(x,y) = \tilde{S}'^*(x,y)\cdot S'(x,y)$ is indeed equal to $|S_{xx}|^2$. Further more, the polarization ratios of the maximum polarizations, speci-fied by equations (33a,b), are found to be

$$\rho_I = \tan\phi \ ; \quad \rho_R = -\tan\phi \tag{47a,b}$$

The "null" polarizations associated with the scattering matrix $S'(x,y)$ are given by $\rho'_{1,2}$, where

$$(\rho'_{1,2}) = \frac{\tan^2\phi + \rho^2_{1,2}}{1+\rho^2_{1,2}\tan^2\phi} \tag{48}$$

and $\rho_{1,2} = \pm\sqrt{-\frac{S_{xx}}{S_{yy}}}$ are the "null" polarizations of the scatter-ing matrix $S_d(x,y)$.

from the work of Kennaugh [1], as well as from the papers on the theory of interference of polarized light of Pancharatnam [11].

In light of the present results, several recommendations can be made:

1) Phenomenological approach to target analysis utilized by Huynen in his thesis [8] should be applied to the bistatic scattering matrix case, in order to establish what useful information, such as target shape, compositon, etc. can be extracted from the scattering matrix parameters.

2) The Poincaré sphere theory should be extended further by devising a method of representing complex bilinear transformations on the surface of the sphere. This will provide a geometrical means of determining polarization of the scattered wave, given the incident polarization and the Poincaré sphere representation of the scattering matrix.

   A great wealth of the Poincaré sphere theory in both the fully, as well as the partially-polarized wave cases can be found in the papers of Pancharatnam referenced in [11].

3) Finallly, it is proposed, that the theory of matrices be applied to the problem of scattering matrix analysis in order to justify the already existing results by rigorous mathematical methods.

VII. Acknowledgements

The research reported in this paper was supported under Contract No. ONR N00001480-C-0773, and we thank Dr. Richard G. Brandt for the continued interest he has shown in our research. We also thank Mrs. Deborah Foster and Mr. Richard Foster for skillfully typing this manuscript and for preparing the illustrations.

VIII. References

[1] E.M. Kennaugh, "Polarization Properties of Radar Reflections", M. Sc. thesis, Dept. Elec. Engr., Ohio State Univ., Columbus, OH., 1952 (Antenna Laboratory, Rept. 389-12, March 1, 1952).

[2] C.D. Graves, "Radar Polarization Power Scattering Matrix", Proc. IRE, Vol. 44, pp. 248-252, Feb. 1956.

[3] E.M. Kennaugh, "Contributions to the Polarization Properties of Radar Targets", Commemorative Collection of Unpublished Notes and Reports in Four Volumes", ElectroScience Lab., Dept. of Electrical Engineering, Ohio State University, Columbus, OH 43212, June, 1984.

[4] A.C. Ludwig, "The Definition of Cross-Polarization", IEEE, Trans A&P-1, (Comm.) Jan. 1973,

[5] G.A. Deschamps, "Polarization Descriptors - Utilization of Symmetries", These Proceedings, Part 1, Topic III.1, pp. 484-511, 1984.

[6] A.L. Maffet, "Scattering Matrices in Methods of Radar Cross Section Analysis", Ed. by J.W. Crispin, K.M. Siegal, Academic Press, NY, 1968.

[7] W-M. Boerner, "Use of Polarization in Electromagnetic Inverse Scattering", Radio Sci., Vol. 16(6), (Special Issue including Papers: 1980 Munich Symposium on Electromagnetic Waves), pp. 1037-1045, Nov./Dec. 1981.

[8] J.R. Huynen, "Phenomenological Theory of Radar Targets", Ph.D. Dissertation, Technical University of Delft, The Netherlands, 1970.

[9] D.B. Kanareykin, N.F. Pavlov, and V.A. Potekhin, The Polarization of Radar Signals. Moscow: Sovyetskoye Radio, (Russian Version) 1966.

[10] G.A. Deschamps, "Geometrical Representation of the Polarization of a Plane Electromagnetic Wave", Proc. IRE, Vol. 39(5), pp. 543-548, 1951.

[11] S. Pancharatnam, "Generalized Theory of Interference and its Applications", Proc. Ind. Acad. Sci., A, Vol.44, 1956, pp.247-262.

[12] M. Davidovitz, "Analysis of Certain Characteristic Properties of the Bistatic, Asymmetric Scattering Matrix, M.Sc. thesis, Univ. of Ill. at Chgo., Electr. Engr. & Comp. Sci. Dept., Communications Lab. Rept. No. 83-04-15, April 15, 1983.

III.7 (RP.3)

ON THE NULL REPRESENTATION OF POLARIZATION PARAMETERS

Bob Raven

Westinghouse DEC, Baltimore, MD

Abstract-A sampling of simulated, complex RCS scattering matrix data are represented as null pairs and plotted on an equal-area projection of the Poincaré sphere. Examples include randomly oriented collections of conducting thin wires, dihedrals, and trihedrals and rotating composite dihedrals.

I. INTRODUCTION

This paper is concerned with the graphical display of polarization parameters of monostatic RCS scattering matrices, specifically, with the display of the polarization nulls of simulated, complex reflectors. Graphical displays of empirical data are desireable to exhibit relationships and for preliminary data analysis. For scattering matrix parameters, the problem is the common one of showing an object of more than three dimensions in a two-dimensional format. A monostatic scattering matrix is six-dimensional (three complex numbers). Typically, one or two of these, representing the distance to the target in wavelengths and a measure of the reflected signal magnitude are suppressed to reduce the dimensionality. In the following discussion, both of these parameters are suppressed leaving the "pure" polarization matrix as a four-dimensional object to be displayed. Several canonical forms have been employed: (1) standard transmit and receive polarizations usually horizontal and vertical or right and left circular, (2) the eigenpolarization and values, and (3) the null polarizations of E. M. Kennaugh [1]. For present purposes, the last form was selected because it lends itself to a representation as point pairs on the surface of the Poincaré sphere which is suitable for a finite, equal-area projection. Null pair diagrams have beem previously used

W.-M. Boerner et al. (eds.), Inverse Methods in Electromagnetic Imaging - Part 1, 629–641.

for assessment of polarimetric data [2],[3],[4] but with orthographic or rectangular coordinates which produce severe distortion on the equator or at the poles. An equal-area projection preserves areas while introducing modest shape distortion at the edges. It is useful for preliminary analysis of null distribution and clustering. A number of equal-area projections have been used by cartographers. For this study, the projection selected is a modification of the Lambert meridional projection applied to the whole sphere by Aitoff [5].

II. COMPUTATIONS

A. General Description

Two groups of reflector models were examined: (1) combinations of multiple conducting thin wires, dihedrals,and trihedrals with random amplitudes, phases, and orientations, and (2) specular reflections from single rotating composite dihedrals. In the first category, scattering matrices, S, were constructed as sums of reflector matrices, S_{rk}, multiplied by a complex Gaussian variate, Z_{rk}, plus, in some cases, a complex Gaussian noise matrix, S_n.

$$S = \sum_{k=1}^{N} Z_{rk} S_{rk} + S_n \tag{1}$$

Nulls were then calculated for each S and their coordinates plotted on a projection of the surface of the Poincaré sphere. Sixteen such cases are displayed in following figures.

Composite material dihedrals illustrate the marked deviation from the case of conducting surfaces when a finite Brewster angle is present. The five material combinations listed in table 1 were considered:

Table 1.
Dihedral Material Combinations

Steel/Sea water	Wood/Sea water
Steel/Dry ground	Wood/Moist ground
Steel/Moist ground	

Large smooth surfaces giving specular reflections with no diffraction or surface roughness effects were assumed with Fresnel's expressions, given below, used to determine reflected polarization as a function of incidence angle.

$$R_H = \frac{\sin\psi - \sqrt{e_c - \cos^2\psi}}{\sin\psi + \sqrt{e_c - \cos^2\psi}} \tag{2}$$

$$R_V = \frac{e_c\sin\psi - \sqrt{e_c - \cos^2\psi}}{e_c\sin\psi + \sqrt{e_c - \cos^2\psi}} \tag{3}$$

$$e_c = e_1 - i60\lambda\sigma \quad (\lambda m \ \& \ \sigma mhos) \tag{4}$$

A K_a-band frequency was assumed. Published E-M material properties at this frequency are fragmentary. The following data were adapted from several sources for this study [6],[7].

Table 2
E-M Properties Assumed at K_a-band

	e_1	σ-mhos/m
Steel	1.0	10^7
Sea water	65.0	4.3
Moist grnd	20.0	.5
Dry ground	3.0	.01
Wood	2.0	.2

Variation of $|R_V|$ and $|R_H|$ with incidence angle is plotted in fig. 1.

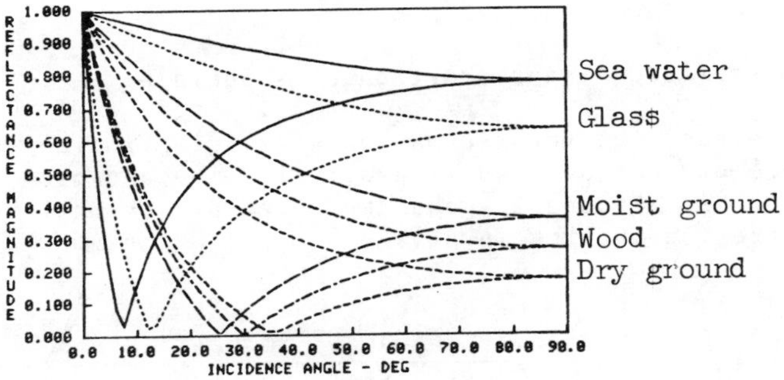

Figure 1. Reflection Coefficient Magnitudes at K_a-band

For the composite dihedrals, the dihedral was horizontal and normal to the incident direction and rotated about the line of intersection such that the dielectric surface traversed incidence angles of from 2.5° to 85°. The null trajectories were then plotted to show the marked change in reflection characteristics as the incidence angle traversed the Brewster angle of the dielectric surface. A polarization transformation subroutine which computed the reflected polarization from arbitrarily oriented surfaces with arbitrary incidence, polarization, and complex dielectric constant was used to track the direction of propagation and polarization from surface to surface.

B. The Aitoff Projection

The Aitoff projection used to display nulls on the Poincaré sphere is a Lambert meridional projection stretched by a factor of two in the horizontal direction. It is implemented as follows:

A_S = azimuth angle
E_S = elevation angle
B = chord angle
C = chord length
D = polar angle from the equator to the chord plane
K = scale factor
X,Y = rectangular coordinates

1. Transform spherical angles to hemispherical angles by halving the azimuth angle.

$$A_H = .5A_S$$
$$E_H = E_S \qquad (5)$$

2. Compute the chord angle, chord, and polar angle from the equatorial point $A_H, E_H = 0,0$ to A_H, E_H.

$$B = \cos^{-1}(\cos .5A_S \cos E_S)$$
$$C = 2^{1/2} \sin .5B$$
$$D = \sin^{-1}(\sin E_S / \sin B) \qquad (6)$$

3. Transform to rectangular coordinates doubling the X component.

$$X = 2KC\cos D$$
$$Y = KC\sin D \qquad (7)$$

C. Polarization Transformation by a Surface

Because the transformation of a polarization vector by a surface may not be readily available, the approach used is summarized below. Figure 2 shows the seven directional vectors and two polarization vectors involved.

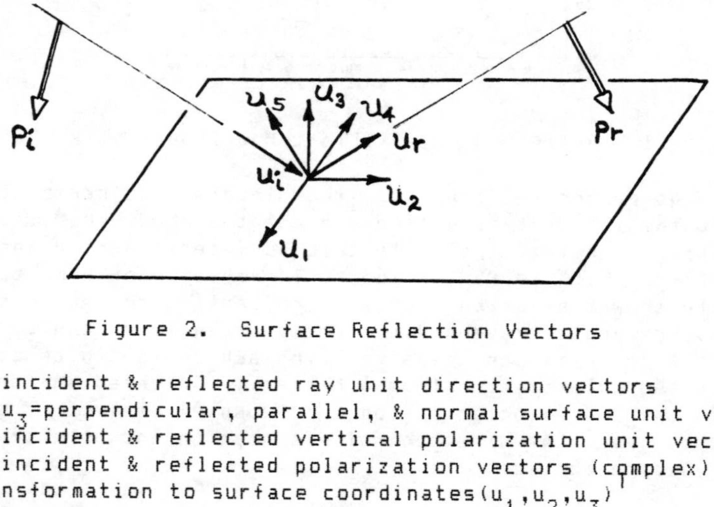

Figure 2. Surface Reflection Vectors

u_i, u_r =incident & reflected ray unit direction vectors
u_1, u_2, u_3 =perpendicular, parallel, & normal surface unit vectors
u_4, u_5 =incident & reflected vertical polarization unit vectors
p_i, p_r =incident & reflected polarization vectors (complex)
T =transformation to surface coordinates $(u_1, u_2, u_3)^T$
S_i =transformation to incident polarization coordinates $(u_1, u_4, u_i)^T$
S_r =transformation to reflected polarization coordinates $(u_1, u_5, u_r)^T$
R =polarization reflectance matrix $\begin{bmatrix} R_V & 0 \\ 0 & R_H \end{bmatrix}$

J =ray reflection matrix $\begin{bmatrix} 1 & & 0 \\ & 1 & \\ 0 & & -1 \end{bmatrix}$

Given: u_i, u_3, p_i

$$u_1 = \frac{u_i \times u_3}{|u_i \times u_3|} \qquad u_2 = u_3 \times u_1 \qquad u_4 = u_1 \times u_i \qquad u_5 = u_1 \times u_r \tag{8}$$

With these definitions, transformation of the ray and the polarization can be expressed compactly. The ray is transformed to surface coordinates by T, reflected by J, and transformed back to the initial coordinate system by T^T.

$$u_r = T^T J T u_i \tag{9}$$

Similarly, the polarization is transformed to incident polarization coordinates by S_i, multiplied by the complex reflectance matrix R, and transformed back to the initial coordinate system by S_r^T.

$$p_r = S_r^T R S_i p_i \tag{10}$$

III. RANDOM ELEMENT COLLECTIONS

Sixteen plots of random collections of elements are shown in figs. 3-18. These examples are not intended to model real clutter but to illustrate phenomena which can occur in idealized scattering models. Figure 3 shows the distribution of nulls for random noise matrices. The nulls appear to be uniformly distributed. This is an important property if the representation is to be useful and suggestive. The author is unaware of a mathematical demonstration of this property. In fig. 4, the null pairs in fig. 4 are connected. Fig. 5 shows the nulls of single conducting thin wires or dipoles with random orientations, amplitudes, and phases. The nulls of such reflectors are coincident and on the equator with their azimuth equal to twice the rotation angle. In figs. 6 and 7, collections of two and five wires with random orientations, amplitudes, and phases are displayed. Only two elements are sufficient mask the linear element characteristic of fig. 5, while five elements already seem an adequate model for random noise.

In figs. 8-18, combinations of conducting dihedrals with random orientations, amplitudes, and phases in combination with noise and trihedrals are shown. Figure 8 shows only single dihedrals. The nulls plot as equatorial pairs spaced by 180° with their azimuth given by twice the rotation angle. In figs. 9 and 10, collections of two and five such elements are plotted. A dihedral characteristic is preserved in that all null pairs occur at the same elevation. Only two elements, however, are sufficient to disturb the nulls substantially from the equator. Further, there is little to distinguish the two and five element cases. In fig. 11, the effect of a signal--to-noise ratio of 10dB on the case in fig. 10 is shown. Signal-to-noise is defined here as the ratio of matrix components. In fig. 12, the rotation angle of the dihedrals were limited to 30° to show the sensitivity to this variable. Figures 13, 14, and 15 show the effect of various mixtures of dihedrals and trihedrals. Not surprisingly, an equal power ratio has considerable resemblance to random noise. Power ratios in these cases were defined in terms of the total power of all matrix components for the two constituents. The last three plots, figs. 16, 17, and 18, show single trihedrals and dihedrals for three signal-to-noise ratios. Significant deteriation of polarization features is apparent at a 10dB signal-to-noise ratio.

VI. SPECULAR REFLECTIONS FROM COMPOSITE DIHEDRALS

Figures 19-23 show five examples of specular reflections from composite dihedrals. In these examples, horizontal dihedrals were rotated about their line of intersection such that the angle of incidence on the first plane varied from 2.5^o to 85° and the null migration plotted. In the first three cases, one surface is conducting and the other is dielectric modeling, for instance, the side of a ship and the sea or a van and the ground. In these examples, the null migration is such that the object first appears to be a conducting dihedral, then a wire, and lastly a trihedral or plate as the Brewster angle of the dielectric surface is traversed. Figure 21 is noteworthy in that, by chance, one of the sample incidence angles was very close to the Brewster angle for the dielectric surface, and the reflection is almost identical to that from a thin horizontal wire. Figures 22 and 23 show null migrations for a similar geometry where both surfaces are dielectrics, and there are two Brewster angles. In these cases, the nulls perform loops first aping trihedrals, then dihedrals, and lastly back to trihedrals as the two Brewster angles are traversed.

IV REFERENCES

[1] E.M. Kennaugh, "Effects of type of polarization on echo characteristics," Antenna Lab Rpts. 389-1/24, Ohio State Univ., 1949-1954.
[2] J. R. Huynen, "Phenomenological theory of radar targets," Dissertation Univ. of Delft, Drukkerij BronderOffset, Rotterdam, 1970.
[3] L. A. Morgan & S. Weisbrod, "RCS matrix studies of sea clutter," Workshop on Polarization Technology, Vol. 1, GACIAC, IIT Res. Inst., Chicago, IL, 1980, pp. 103-120.
[4] J. C. Daley, "Radar target discrimination based on polarization effects," Workshop on Polarization Technology, Vol. 1, GACIAC, IIT Res. Inst., Chicago, IL, 1980, pp. 185-213.
[5] C. H. Deetz & O. S. Adams, "Elements of map projection," Spec. Pub. No. 68, U.S. Gov't Printing Off., 1945, pp. 160-163.
[6] D. J. Kerr, Ed., "Propagation of short radio waves," Vol. 13, MIT Rad Lab Series, McGraw Hill, 1951, p. 398.
[7] H. Jasik, Ed., "Antenna engineering handbook," McGraw Hill, 1961

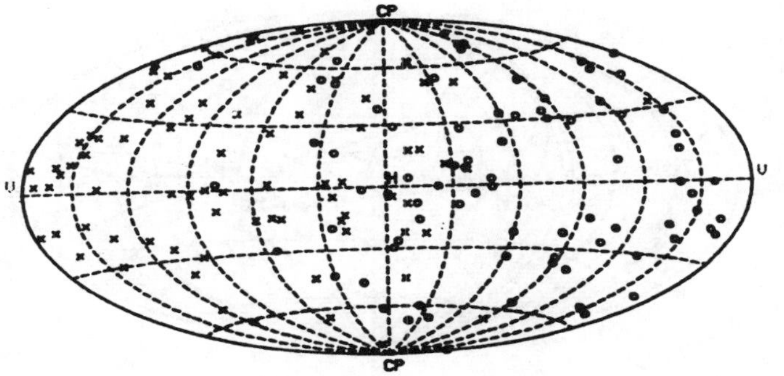

Figure 3. Aitoff Equal-area Projection of the Poincaré Sphere. Co-pol Nulls of 80 Random Noise Samples.

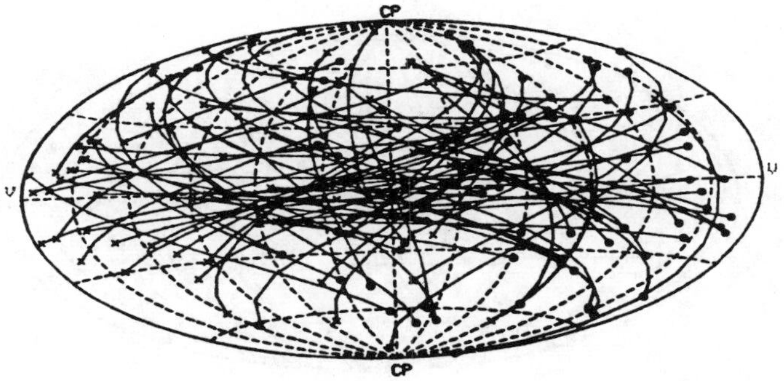

Figure 4. Aitoff Equal-area Projection of the Poincaré Sphere. Co-pol Nulls of 80 Random Noise Samples (connected).

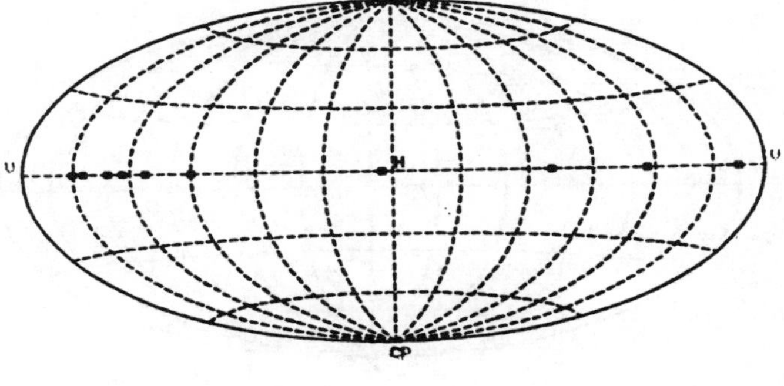

Figure 5. Aitoff Equal-area Projection of the Poincaré Sphere. Co-pol Nulls of 10 Wire Reflectors. Single wires with Random Amplitudes, Phases, & Attitudes.

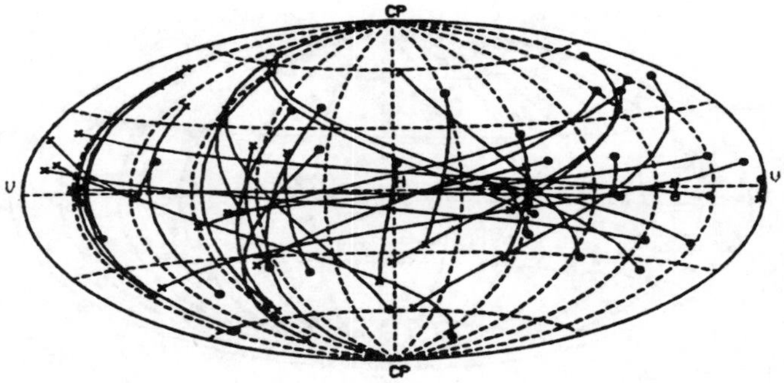

Figure 6. Aitoff Equal-area Projection of the Poincaré Sphere. Co-pol Nulls of 40 Wire Reflectors. Two Wire Elements with Random Amplitudes, Phases, & Attitudes.

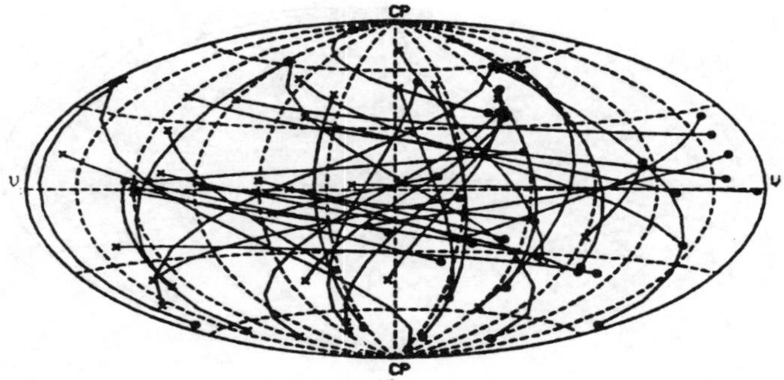

Figure 7. Aitoff Equal-area Projection of the Poincaré Sphere. Co-pol Nulls of 40 Wire Reflectors. Five Wire Elements with Random Amplitudes, Phases, & Attitudes.

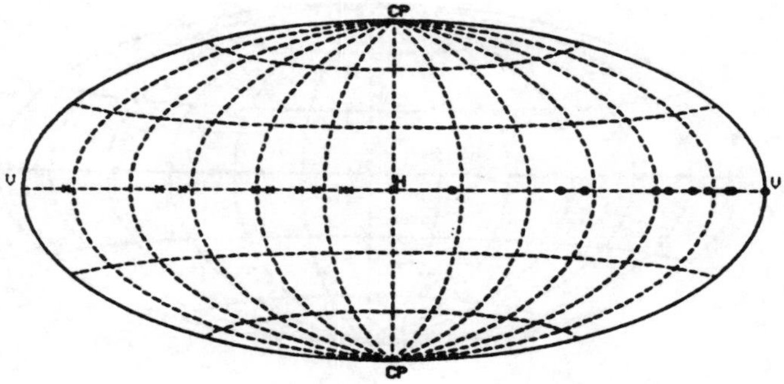

Figure 8. Aitoff Equal-area Projection of the Poincaré Sphere. Co-pol Nulls of 40 Dihedral Reflectors. Single Dihedral Elements with Random Amplitudes, Phases, & Attitudes.

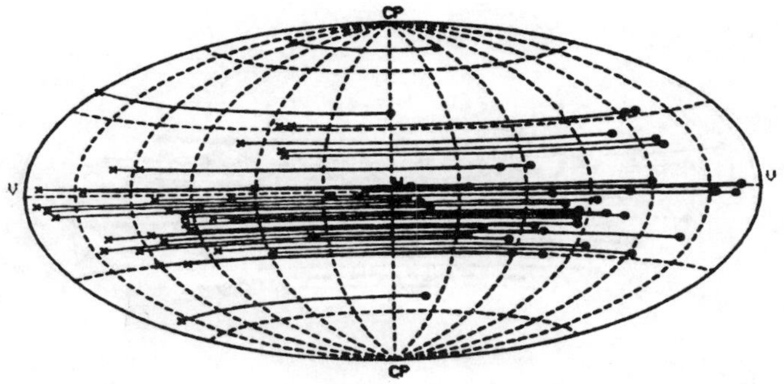

Figure 9. Aitoff Equal-area Projection of the Poincaré Sphere. Co-pol Nulls of 40 Dihedral Reflectors. Two Dihedral Elements with Random Amplitudes, Phases, & Attitudes.

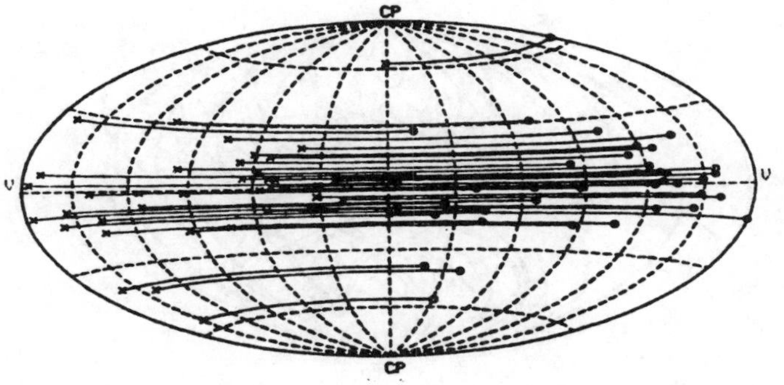

Figure 10. Aitoff Equal-area Projection of the Poincaré Sphere. Co-pol Nulls of 40 Dihedral Reflectors. Five Dihedral Elements with Random Amplitudes, Phases, & Attitudes.

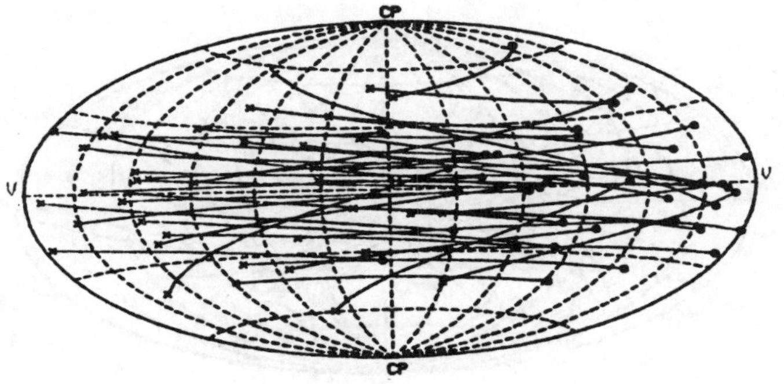

Figure 11. Aitoff Equal-area Projection of the Poincaré Sphere. Co-pol Nulls of 40 Dihedral Reflectors. Five Dihedral Elements with Random Amplitudes, Phases, & Attitudes plus Noise. SNR=10dB.

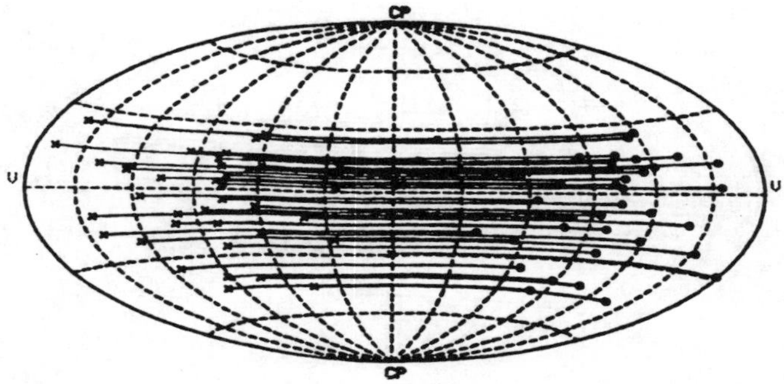

Figure 12. Aitoff Equal-area Projection of the Poincaré Sphere. Co-pol Nulls of 40 Dihedral Reflectors. Five Dihedral Elements with Random Amplitudes, Phases, & Attitudes. $|\Theta|<30^{\circ}$.

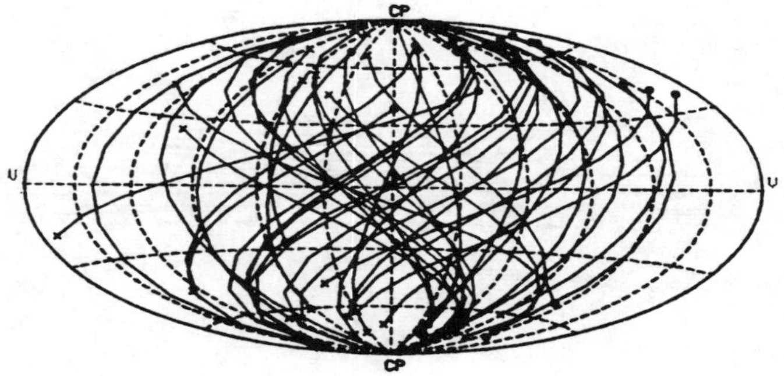

Figure 13. Aitoff Equal-area Projection of the Poincaré Sphere. Co-pol Nulls of 40 Dihedral/Trihedral Reflectors. Five Dihedrals & One Trihedral with Random Amplitudes, Phases, & Attitudes. TDR=10dB.

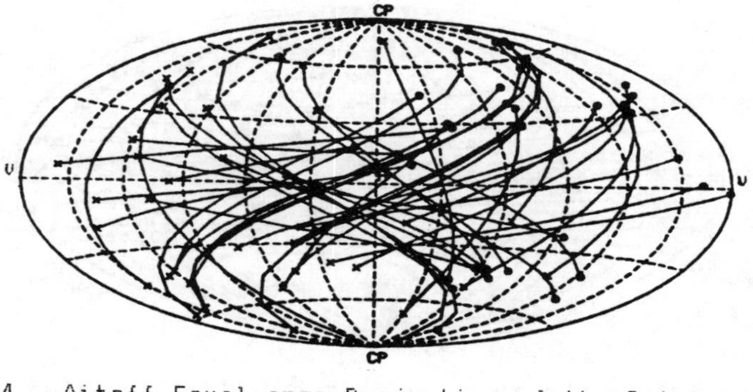

Figure 14. Aitoff Equal-area Projection of the Poincaré Sphere. Co-pol Nulls of 40 Dihedral/Trihedral Reflectors. Five Dihedrals & One Trihedral with Random Amplitudes, Phases, & Attitudes. TDR=0dB.

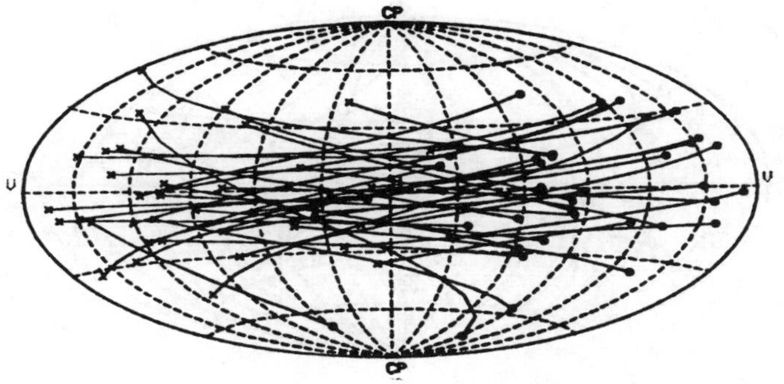

Figure 15. Aitoff Equal-area Projection of the Poincaré Sphere. Co-pol Nulls of 40 Dihedral/Trihedral Reflectors. Five Dihedrals & One Trihedral with Random Amplitudes, Phases, & Attitudes. TDR=-10dB.

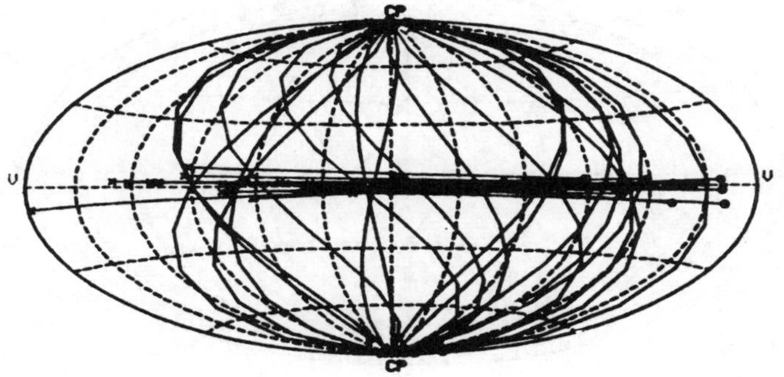

Figure 16. Aitoff Equal-area Projection of the Poincaré Sphere. Co-pol Nulls of 40 Dihedral & Trihedral Reflectors. Single Elements with Random Amplitudes, Phases, & Attitudes plus Noise. SNR=30dB.

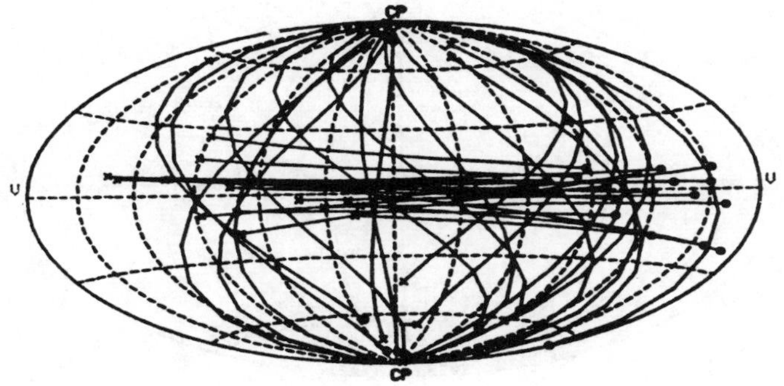

Figure 17. Aitoff Equal-area Projection of the Poincaré Sphere. Co-pol Nulls of 40 Dihedral & Trihedral Reflectors. Single Elements with Random Amplitudes, Phases, & Attitudes plus Noise. SNR=20dB.

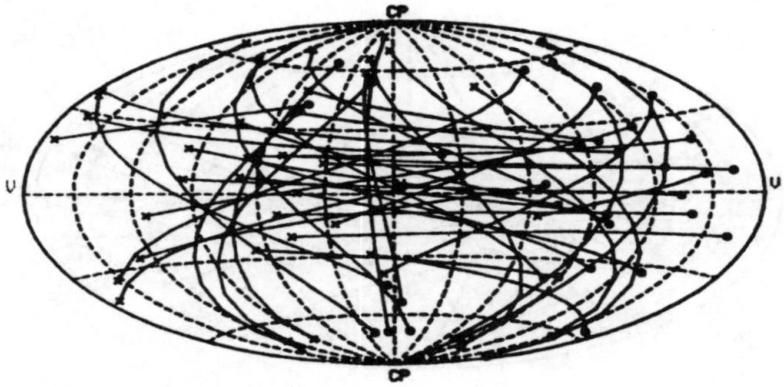

Figure 18. Aitoff Equal-area Projection of the Poincaré Sphere. Co-pol Nulls of 40 Dihedral & Trihedral Reflectors. Single Elements with Random Amplitudes, Phases, & Attitudes plus Noise. SNR=10dB.

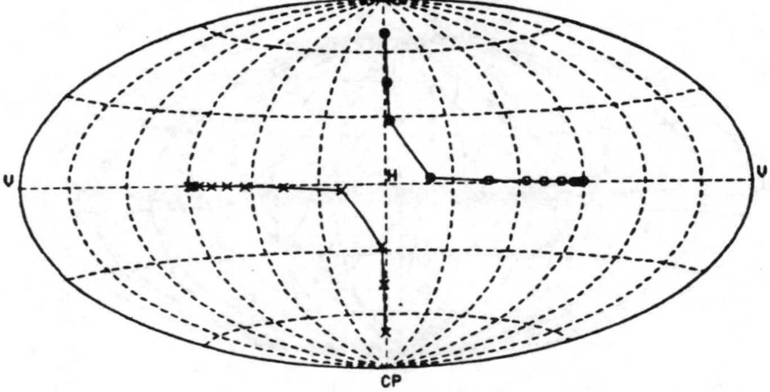

Figure 19. Aitoff Equal-area Projection of the Poincaré Sphere. Co-pol Null Migrations of Composite Dihedral with Incidence from 2.5° to 85°. Steel Cond./Sea Water. K_a-band.

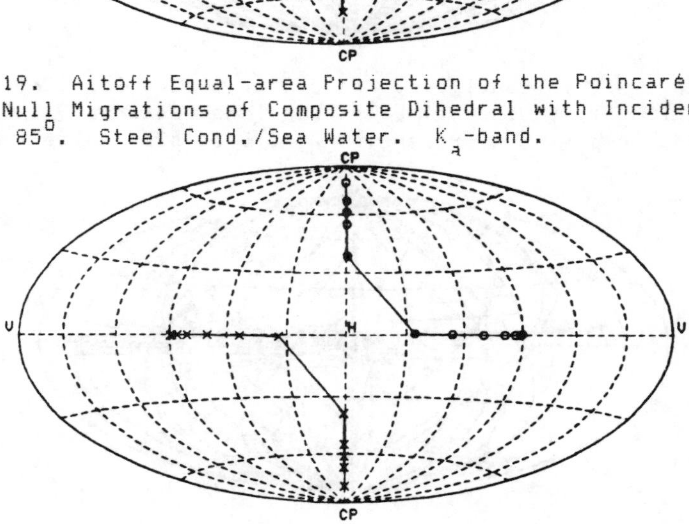

Figure 20. Aitoff Equal-area Projection of the Poincaré Sphere. Co-pol Null Migrations of Composite Dihedral with Incidence from 2.5° to 85°. Steel Cond./Moist Ground. K_a-band.

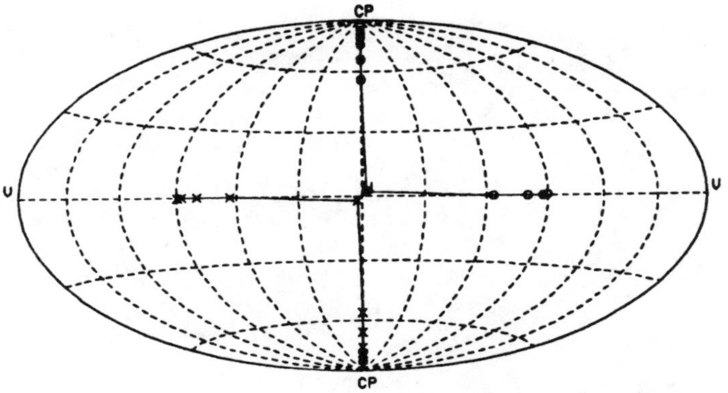

Figure 21. Aitoff Equal-area Projection of the Poincaré Sphere. Co-pol Null Migrations of Composite Dihedral with Incidence from 2.5^o to 85°. Steel Cond./Dry Ground. K_a-band.

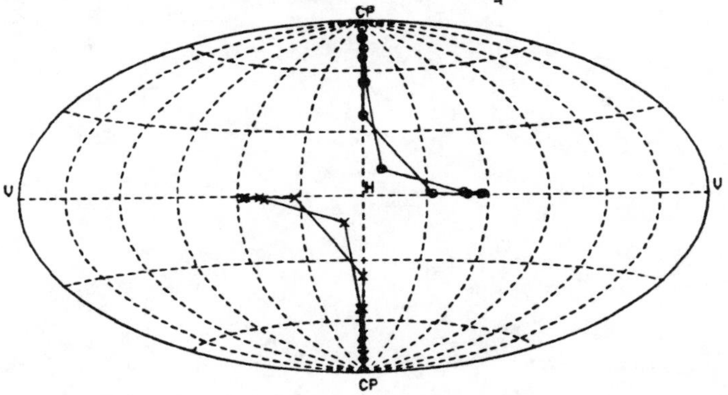

Figure 22. Aitoff Equal-area Projection of the Poincaré Sphere. Co-pol Null Migrations of Composite Dihedral with Incidence from 2.5^o to 85°. Wood/Sea Water. K_a-band.

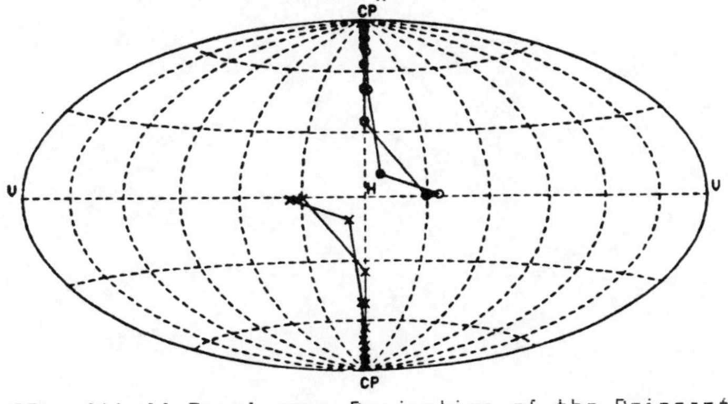

Figure 23. Aitoff Equal-area Projection of the Poincaré Sphere. Co-pol Null Migrations of Composite Dihedral with Incidence from 2.5^o to 85°. Wood/Moist Ground. K_a-band.

III.8 (MM.3)

THEORY AND DESIGN OF A DUAL POLARIZATION RADAR FOR CLUTTER ANALYSIS

Jerald D. Nespor, Amit P. Agrawal, Wolfgang-M. Boerner

Communications Laboratory, Polarization Radar Data Processing Division, Department of Electrical Engineering & Computer Science, Mail Stop M/C 154, 851 S. Morgan St., P.O. Box 4348, 1141-SEO, 4210-11-SEL, Chicago, IL 60680, USA

ABSTRACT

Solutions of the inverse problem, that of extracting orientation and shape information from polarization measurements, provide useful meteorological data.

Currently, such information is obtained from circular and linear dual-polarization measurables derived from auto and cross correlations of the main and orthogonal channels.

In this paper, an alternative integrated approach is used to deduce meteorological information from measurements of the relative backscattering matrix derived from polarimetric techniques of Kennaugh. For meteorological scatterers, such measurements would be gathered in a time series.

The time elements gathered from a single measurement of the relative backscattering matrix is used to determine the co and cross pol nulls on the Poincaré sphere or polarization chart. The optimal polarizations are then given in terms of elliptical and/or circular dual polarization measurables where the equivalency of the coherency and Stokes parameter measuring radar is given. Finally, the statistics of the co-pol null distributions on the Poincaré sphere are given.

1. Introduction

Polarization, along with amplitude, frequency, phase and doppler, are the five parameters that completely describe an electromagnetic wave. In propagation through a precipitation medium, the polarization state can be the parameter most significantly changed. Due to engineering limitations prior to the last decade, such as

W.-M. Boerner et al. (eds.), Inverse Methods in Electromagnetic Imaging – Part 1, 643–659.

achieving high polarization purity within an antenna beam, coupled with the lack of theoretical development, polarization diversity techniques applied to radar meteorology have had a low level of research effort.

There has, however, been substantial progress within the last decade with the development of theory for predicting and interpreting backscatter and propagation measurements with the construction of antennas with good polarization characteristics, as well as, fast polarization switching techniques.

As the scope of research relating to polarization has increased, new ideas for meteorological applications have been proposed. It is the objective of this paper to suggest an integrated approach to polarization diversity techniques which have the possibilities of adding greater insight and reliability to meteorological identification and classification of different precipitation states.

2. The Scattering Matrix

It was first shown by Sinclair [1950] that a radar target acts like a polarization transformer which is described by its associated scattering matrix [S]. At the target scatterer, then, the target scattering matrix characterizes the scattering properties of the target such that the scattered electric field can be related to the incident electric field by:

$$\begin{bmatrix} E_{s1} \\ E_{s2} \end{bmatrix} = \frac{\exp(-j\, k_o r)}{r} \begin{bmatrix} S_{11} & S_{12} \\ S_{21} & S_{22} \end{bmatrix} \begin{bmatrix} E_{i1} \\ E_{i2} \end{bmatrix} . \tag{1a}$$

where the subscripts s and i refer to the scattered and incident electric fields, respectively. The subscripts 1 and 2 refer to any two general orthogonal polarizations.

In vector notation, equation (1a) can be written as:

$$\underline{E}_s = \frac{\exp(-jk_o r)}{r} [S]\, \underline{E}_i \tag{1b}$$

Assuming that the propagation space is reciprocal, one can now represent the scattering matrix in terms of amplitudes and phases where for a monostatic radar the relative scattering matrix becomes symmetric and is given by:

$$[S]_{SMR} = \begin{bmatrix} |S_{11}|\, e^{j(\phi_{11} - \phi_{12})} & |S_{12}| \\ |S_{21}| & |S_{22}|\, e^{j(\phi_{22} - \phi_{12})} \end{bmatrix} \tag{2}$$

where the || brackets denote absolute value.

For the monostatic reciprocal radar case, it can be shown from the Reciprocity Theorem that $|S_{21}| = |S_{12}|$ and $\phi_{21} = \phi_{12}$. Throughout the remaining discussion, it will be assumed that the [S] implies the monostatic $[S]_{SMR}$ case unless specifically stated otherwise.

The radar cross section of a target is related to the scattering matrix by:

$$\sigma_{ij} = |S_{ij}|^2 \tag{3}$$

where i, j = 1 or 2

The radar cross section of a random array of objects (such as precipitation particles) fluctuates with pulse and range time. The fluctuations are not caused by changes in the radar cross sections of the individual objects; but rather they are due almost entirely to phase effects. Each object in the contributing region gives rise to a scattered wavelet with a certain phase, and the total scattered electric field is the sum of the interference effects of the individual wavelets. Since the relative phases of the wavelets change as the objects move about during the interval between radar pulses, the total echo power varies with time. The relative phases also vary with time in a random fashion, so the fluctuations are called "random phase" fluctuations. The amplitudes of the wavelets (or equivalently, the random cross sections of individual objects) sometimes vary as well. However, any echo fluctuations due to such amplitude changes will be superimposed or masked out by the "random phase" fluctuations.

Random-phase fluctuations are a characteristic feature of precipitation echoes. Because of the random-phase fluctuations, the radar cross-section for a distributed target can be expressed as:

$$\overline{\sigma}_{ij} = \overline{|S_{ij}|^2} \tag{4}$$

where i, j = 1 or 2 and the bar above the elements denote averaging for an ensemble of particles within the radar pulse.

Random-phase fluctuations also occur in range-time as the contributing region moves out from the radar; thus, the contributing region contains a different set of hydrometeors each time it moves out by its own length. Fluctuations of a similar nature can also occur as the antenna is rotated because the contributing region moves with the antenna beam.

The echo power $[P_R]$ received from a contributing region at range r for any one radar pulse is not, in general, equal to the average power $<[P_R]>$ because of the random phase fluctuations. Thus, the power, $[P_R]$, may be higher or lower than $<[P_R]>$ by an amount that can only be predicted statistically. Measuring the average echo power $<[P_R]>$ from any contributing region exactly would require averaging over an infinite time interval. In practice, however, it is therefore necessary to estimate the average echo power $<[P_R]>$ by observing only a limited number of echoes from the contributing region.

From Figure 1, the following equation for the ensemble average radar cross-section is:

$$\overline{\sigma} = \sum_{i=1}^{N} \overline{|S_i|^2} \quad (5)$$

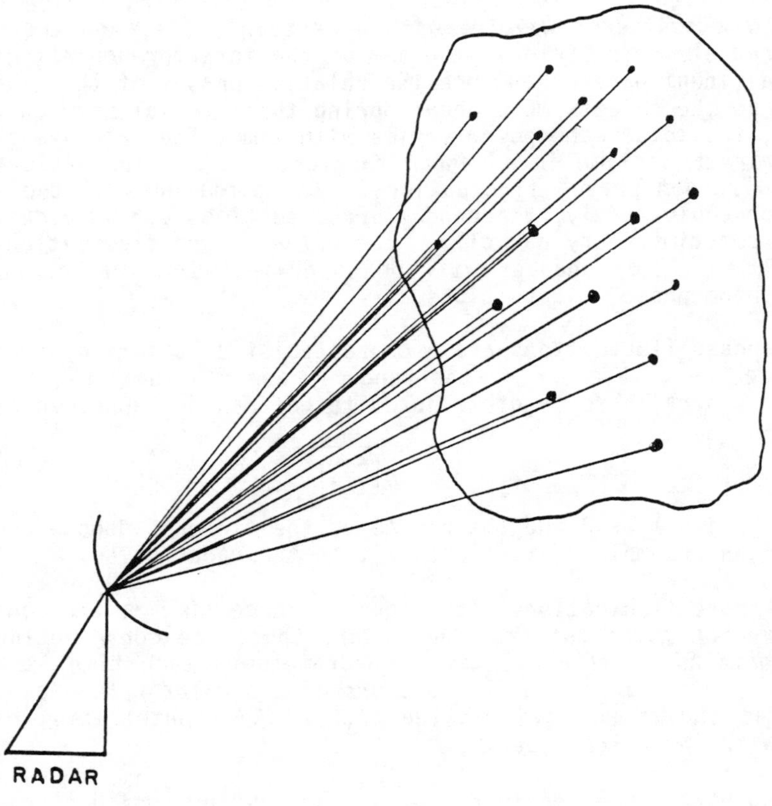

FIGURE 1: AMPLITUDE STATISTICS OF THE RADAR CROSS SECTION

Thus, by the Central Limit Theorem which states that the mean of N samples follows a Gaussian distribution whatever the distribution of the individual measurements. It can then be assumed that the scattering matrix elements are zero-mean Gaussian and statistically independent when N is large. Consequently, the radar cross-section is Rayleigh distributed, because it is the product of two Gaussian distributions. It can be seen that the above assumptions are in total agreement with the earlier work of Marshall and Hitschfeld [1953].

3. Transformations From Linear To Circular And Elliptical Polarizations

Thus far, no stipulation has been made about a specific polarization basis in the above sections. Up to this point, the only requirement has been that the transmitted polarizations be orthogonal. Through the appropriate congruent transformations, it has been shown in Crispin and Siegal [1969] that the linear scattering matrix (LSM), $[S]_\ell$, can be transformed to a circular scattering matrix (CSM), $[S]_c$, or, in general, to an elliptical scattering matrix (ESM), $[S]_e$.

To transform [S] to different polarization states, the following matrix transformation can be used:

$$[S']_i = [T][S][T]^T \qquad (6)$$

where the subscript $i \to \ell$ represents the scattering matrix in a linear polarization basis where ℓ is usually implied when no subscript is present. For the scattering matrix in circular or elliptical polarization basis, $I \to c$ or e, respectively. The superscript T denotes the transpose of the transformation matrix [T].

The most general form of [T] can be found to be:

$$[T] = \begin{bmatrix} e^{j\phi_1}\cos\beta & -e^{j\phi_2}\sin\beta \\ e^{j\phi_3}\sin\beta & e^{j\phi_4}\cos\beta \end{bmatrix} \qquad (7)$$

where $\phi_2 - \phi_1 = \phi_4 - \phi_3$. It should be noted that the most general basis is elliptical; but when all the phases are set to zero, [T] reduces to an ordinary rotational matrix which rotates a particular linear polarization base by an angle β. There exist numerous transformation matrices for transforming between unit vectors of any polarization basis depending on one's choice of ϕ_1, ϕ_2, ϕ_3, ϕ_4 and β.

In this work, the transformation matrix will be taken as:

$$[T] = \frac{1}{\sqrt{2}}\begin{bmatrix} 1 & -j \\ 1 & +j \end{bmatrix} \tag{8}$$

to transform between the linear and the circular scattering matrices. Hence, from equation (6),

$$[S]_c = \frac{1}{2}\begin{bmatrix} 1 & -j \\ 1 & +j \end{bmatrix}\begin{bmatrix} S_{11} & S_{12} \\ S_{12} & S_{22} \end{bmatrix}\begin{bmatrix} 1 & 1 \\ -j & +j \end{bmatrix} \tag{9}$$

Finally, the scattering matrix in circular polarization basis becomes:

$$[S]_c = \frac{1}{2}\begin{bmatrix} S_{11} - S_{22} - j2S_{12} & S_{11} + S_{22} \\ S_{11} + S_{22} & S_{11} - S_{22} + j2S_{12} \end{bmatrix} \tag{10}$$

The transformation given in equation (7) is necessary if the non-ideal nature of the radar antenna is to be taken into account. Whether one desires to use linear or circular polarizations, in practice, the received polarizations will be elliptical. For any practical analysis for a dual-polarization system, antenna effects should be included by using the rotational and elliptical transformations presented above.

4. The Backscattered Coherency Matrix

For a radar volume filled with scatterers, the received power is due to the backscattered components of all N scatterers within the radar volume. Thus, the partially polarized signals can be represented using the coherency matrix [J] as:

$$[J]_s = \langle \underline{E}_s \underline{E}^{*T}_s \rangle = \begin{bmatrix} \langle E_{s1}E_{s1}^* \rangle & \langle E_{s1}E_{s2}^* \rangle \\ \langle E_{s2}E_{s1}^* \rangle & \langle E_{s2}E_{s2}^* \rangle \end{bmatrix} \tag{11}$$

$$= \langle \frac{1}{r^2} [S] \underline{E}_i \underline{E}_i^{*T} [S]^{*T} \rangle \tag{12}$$

where * and T denote the complex conjugate and the transpose, respectively and the subscripts s and i denote the scattered and incident coherency matrix, respectively.

Expanding the above equation (12) for the monostatic reciprocal case ($S_{12} = S_{21}$) yields:

$$J_{s11} = \langle |S_{11}|^2 |E_{i1}|^2 + 2\mathrm{Re}[S_{11}S_{12}^* E_{i1}E_{i2}^*] + |S_{12}|^2 |E_{i2}|^2 \rangle$$

$$J_{s12} = \langle S_{11}S_{12}^* \, |E_{i1}|^2 + S_{11}S_{22}^* \, E_{i1}E_{12}^* + |S_{12}|^2 \, E_{i2}E_{i1}^* + \\ + S_{22}^*S_{12} \, |E_{i2}|^2 \rangle \tag{13}$$

$$J_{s21} = J_{s12}^*$$

$$J_{s22} = \langle \, |S_{22}|^2 \, |E_{i2}|^2 + 2\mathrm{Re}[S_{22}S_{12}^* \, E_{i2}E_{i1}^*] + |S_{12}|^2 |E_{i1}|^2 \rangle$$

5. The Stokes Vector And The Average Mueller Matrix

up to this point, the polarization properties of distributed targets such as precipitation have been represented in terms of the incident and scattered coherency matrices and the 2x2 scattering matrix. The partially polarized wave backscattered from the distributed target can also be represented in terms of the 4x1 Stokes vector and the average Mueller matrix $[\bar{M}]$ can be defined (Kennaugh 1949-1954) as:

$$\underline{g}_s = [\bar{M}]\underline{g}_i \tag{14}$$

where $\underline{g}_s$ and $\underline{g}_i$ are, respectively, the scattered (partially polarized) and incident (completely polarized) waves, expressed as Stokes vectors and $[\bar{M}]$ is the 4x4 real and symmetric average Mueller matrix used to represent partially polarized signals. It has been shown by Chan [1981] that a transformation exists between the scattering matrix [S] and the coherent Mueller matrix [M]. Equation (14) can then be expressed as:

$$\begin{bmatrix} g_0 \\ g_1 \\ g_2 \\ g_3 \end{bmatrix}_s = \begin{bmatrix} \bar{m}_{11} & \bar{m}_{12} & \bar{m}_{13} & \bar{m}_{14} \\ \bar{m}_{21} & \bar{m}_{22} & \bar{m}_{23} & \bar{m}_{24} \\ \bar{m}_{31} & \bar{m}_{32} & \bar{m}_{33} & \bar{m}_{34} \\ \bar{m}_{41} & \bar{m}_{42} & \bar{m}_{43} & \bar{m}_{44} \end{bmatrix} \begin{bmatrix} g_0 \\ g_1 \\ g_2 \\ g_3 \end{bmatrix}_i \tag{15}$$

The Stokes parameters and the elements of the coherency matrix are related by:

$$g_0 = J_{11} + J_{22}$$

$$g_1 = J_{11} - J_{22}$$

$$g_2 = J_{12} + J_{21}$$

$$g_3 = j(J_{12} - J_{21}) \tag{16}$$

6. Optimal Polarizations Concept For Dual Polarization Radar

Optimal polarizations are those which produce nulls in the co-polarized and the cross-polarized components of backscattered radiation. The nulls in the co-polarized and the cross-polarized channels have been termed as 'co-pol' and 'x-pol' nulls, respectively, in this paper. These are the polarizations where minimum and maximum voltages are recieved across the antenna terminals of a radar.

In order to obtain all the backscatter properties of an ensemble of scatterers, it is necessary to make a complete measurement of the polarization, amplitude and relative phase characteristics of the backscattered signals. This can be done, for example, by means of measuring either the Stokes parameters or the coherency matrix. The two methods have been seen to be equivalent in the previous section. Such measurements for a dual polarization radar require two separate sequential orthogonal transmissions. Part of the information contained in this system of measurements can also be obtained from a knowledge of the two optimal polarizations, say one co-pol and one cross-pol null. Optimal polarizations do not in themselves constitute a complete backscatter description, but they can be made complete by a measurement of the cross-section and the degree of polarization. Alternatively, the measurement can be completed by obtaining the power in the co-pol channel and the orthogonal channel [G.C. McCormick, 1982].

7. Optimal Polarizations For Completely Polarized Signals

The backscattered signal from a single particle is completely polarized. First of all, the optimal polarizations for a single particle will be formulated and then these formulations will be extended for an ensemble of particles. Making the appropriate substitutions, equation (7) yields:

$$[T] = \frac{1}{(1+\rho\rho^*)^{1/2}} \begin{bmatrix} 1 & -\rho \\ \rho^* & 1 \end{bmatrix} \tag{17}$$

where $\rho = e^{j\alpha}\tan {}^{\gamma}/2$. It should be noted that the transformation matrix [T] is also uniitary.

Recalling from equation (1b) that the incident electric field was represented in matrix notation by $\underline{E}_i$, and the scattered electric field by $\underline{E}_s$; then the transformations of the electric field to the new optimal polarization basis are given by:

$$\underline{E}_i' = [T]\, \underline{E}_i \tag{18}$$

$$\underline{E}_s' = [T^*]\, \underline{E}_s \tag{19}$$

where the prime denotes the new change of basis and the * denotes the complex conjugate.

When equations (18) and (19) are substituted into equation (1a), the electric fields can be expressed in the new optimal basis as:

$$\begin{bmatrix} E'_{s1} \\ E'_{s2} \end{bmatrix} = \frac{\exp(-jk_o r)}{r} [T^*] \begin{bmatrix} S_{11} & S_{12} \\ S_{12} & S_{22} \end{bmatrix} [T^*]^T \begin{bmatrix} E'_{i1} \\ E'_{i1} \end{bmatrix} \tag{20}$$

Alternatively, in matrix notation equation (1b) becomes:

$$\underline{E}'_s = \frac{\exp(-jk_o r)}{r} [S'] \underline{E}'_i \tag{21}$$

where

$$[S'] = [T^*][S][T^*]^T \tag{22}$$

and is the scattering matrix in the optimal polarization base.

Expanding equation (22), the elements of [S'] become:

$$S'_{11} = (1 + \rho\rho^*)^{-1} (\rho^2 S_{22} + 2\rho S_{12} + S_{11}) \tag{23a}$$

$$S'_{12} = (1 + \rho\rho^*)^{-1} \left(\rho^2 S_{22} + (1 - \rho\rho^*) S_{12} - \rho^* S_{11}\right) \tag{23b}$$

$$S'_{21} = S'_{12} \tag{23c}$$

$$S'_{22} = (1 + \rho\rho^*)^{-1} (\rho^{*2} S_{11} - 2\rho^* S_{12} + S_{22}) \tag{23d}$$

Expressions for the co and cross pol nulls can be calculated by assuming the first normalized sequential pulse (polarization 1) is transmitted. Equation (20) gives:

$$\begin{bmatrix} E'_{s1} \\ E'_{s2} \end{bmatrix} = \frac{\exp(-jk_o r)}{r} [S'] \begin{bmatrix} 1 \\ 0 \end{bmatrix} \tag{24}$$

Setting $E'_{s1} = 0$ in equation (20) yields the following expression for the co-pol nulls:

$$\rho_{1co} = \frac{-S_{12} \pm (S_{12}^2 - S_{11}S_{22})^{1/2}}{S_{22}} \tag{25}$$

Setting $E'_{s2} = 0$ in equation (20) yields the following expression for the cross-pol nulls:

$$\rho_{1x} = \frac{-B \pm (B^2 - 4AC)^{1/2}}{2A} \tag{26}$$

where

$$A = S_{22}S^*_{12} + S^*_{11}S_{12}$$

$$B = |S_{11}|^2 - |S_{22}|^2$$

$$C = -A^*$$

Alternatively, one can obtain expressions for the co and cross-pol nulls in polarization 2 by assuming that the normalized second sequential pulse is transmitted:

$$\begin{bmatrix} E'_{s1} \\ E'_{s2} \end{bmatrix} = \frac{\exp(-jk_o r)}{r} [S'] \begin{bmatrix} 0 \\ 1 \end{bmatrix} \tag{27}$$

8. Optimal Polarizations for Completely Polarized Signals and Their Relationship to the Elliptical Depolarization Ratio

In general, the backscattered signal from a target will be ellipticlaly polarized, and it may be desirable for an adaptive dual-polarization radar to receive an elliptical mode. An elliptical depolarization ratio (EDR) may be defined where CDR is just a special case.

Recalling from Section 3 that the scattering m matrix in an transformed elliptical polarization base can be expressed as:

$$[S']_e = \begin{bmatrix} S'_{e11} & S'_{e12} \\ S'_{e12} & S'_{e22} \end{bmatrix} \tag{28}$$

or

$$[S']_e = S'_{e12} = \begin{bmatrix} \eta'_{e11} & 1 \\ 1 & \eta'_{e22} \end{bmatrix} \tag{29}$$

Following the same procedure as in the previous section, optimal polarizations can be expressed in terms of the non-logarithmic parameters used to calculate EDR as:

$$\rho_{1co} = \frac{-1 \pm \sqrt{1 - \eta'_{e11}\eta'_{e22}}}{\eta'_{e11}} \tag{30}$$

$$\rho_{1x} = \frac{1}{2} \frac{|\eta'_{e11}|^2 - |\eta'_{e22}|^2}{|\eta'^*_{e11} + \eta'_{e22}|} \pm \sqrt{1 + \frac{\left(|\eta'_{e11}|^2 - |\eta'_{e22}|^2\right)^2}{4|\eta'^*_{e11} + \eta'_{e22}|^2}} \quad (31)$$

9. Optimal Polarizations For Partially Polarized Signals

The backscattered signal from an ensemble of particles (e.g. precipitation) is always partially polarized. Thus, expressions for the optimal polarizations for the partially polarized signals will be developed in this section.

The backscattered coherency matrix can now be expressed in terms of the optimal scattering matrix elements, as given in equations (23 a-d). When polarization 1 is transmitted, the backscattered coherency matrix elements in the optimal polarization base simplify to:

$$J'_{s11} = \langle |S'_{11}|^2 \; |E'_{i1}|^2 \rangle$$

$$J'_{s12} = \langle S'_{11} S'^*_{12} \; |E'_{i1}|^2 \rangle \quad (32)$$

$$J'_{s21} = J'^*_{s12}$$

$$J'_{s22} = \langle |S'_{12}|^2 \; |E'_{i1}|^2 \rangle$$

For either J'_{s11} or J'_{s22} a minimum, the optimal polarizations can be seen to be formally the same as for the completely polarized case presented in Section 7 with S_{11}, S_{12} and S_{22} being replaced by their ensemble averages:

$$\rho_{1co} = \frac{-\overline{S_{12}} \pm \sqrt{\overline{S_{12}}^2 - \overline{S_{11}}\;\overline{S_{22}}}}{\overline{S_{22}}} \quad (33)$$

$$\rho_{1x} = \frac{-\overline{B} \pm \sqrt{\overline{B}^2 - 4\overline{A}\;\overline{C}}}{2\overline{A}} \quad (34)$$

where

$$\overline{A} = \overline{S_{22}\; S_{12}^*} + \overline{S_{11}^* S_{12}}$$

$$\overline{B} = |\overline{S}_{11}|^2 - |\overline{S}_{22}|^2$$

$$\overline{C} = \overline{S_{11}\; S_{12}^*} - \overline{S_{22}^*\; S_{12}}$$

where the bar denotes ensemble averaging.

Similarly, expressions for the optimal polarizations can be obtained for polarization 2.

10. Optimal Polarizations For Partially Polarized Signals And Their Relationship To The Elliptical Depolarization Ratio

It can be seen that the optimal polarizations for partially polarized signals in terms of the non-logarithmic parameter used to calculate EDR are:

$$\rho_{1co} = \frac{-1 \pm \sqrt{1 - \overline{\eta}'_{e11}\,\overline{\eta}'_{e22}}}{\overline{\eta}'_{e11}} \qquad (35)$$

$$\rho_{1x} = \frac{1}{2}\quad \frac{|\overline{\eta}'_{e11}|^2 - |\overline{\eta}'_{e22}|^2}{|\overline{\eta}'^{*}_{e11} + \overline{\eta}'_{e22}|^2} \pm \sqrt{1 + \frac{(|\overline{\eta}'_{e11}|^2 - |\overline{\eta}'_{e22}|^2)^2}{4|\overline{\eta}'^{*}_{e11} + \overline{\eta}'_{e22}|^2}} \qquad (36)$$

11. The Degree of Polarization

The degree of polarization, P, is defined as the ratio of the intensity of the polarized portion of the wave to the total intensity of the wave and is given by:

$$P = \sqrt{1 - \frac{4\det[J]}{(Tr[J])^2}} \qquad (37)$$

When polarization 1 is transmitted, the $\det[J_s']$ and $Tr[J_s']$ can be calculated from equation (32) as:

$$\det[J_s'] = <|S_{11}'|^2\ |E_{i1}'|^2> <|S_{12}'|^2\ |E_{i1}'|^2> - $$
$$ - <S_{11}'S_{12}'^{*}\ |E_{i1}|^2> <S_{11}'^{*}S_{12}'\ |E_{i1}|^2> \qquad (38)$$

$$Tr[J_s'] = <|S_{11}'|^2\ |E_{i1}'|^2> + <|S_{12}'|^2\ <|E_{i1}'|^2> \qquad (39)$$

The degree of polarization, therefore, is obtained by substituting equations (38) and (39) into the above expression for P, yielding:

$$P_1 = \sqrt{1 - \frac{4[\overline{|S_{11}|^2}\ \overline{|S_{12}|^2} - \overline{|S_{11}S_{12}|^2}]}{(\overline{|S_{11}|^2} + \overline{|S_{12}|^2})^2}} \qquad (40)$$

Similarly, P_2 can be found to be:

$$P_2 = \sqrt{1 - \frac{4[\overline{|S_{22}|^2}\ \overline{|S_{12}|^2} - \overline{|S_{22}\ S_{12}|^2}]}{(\overline{|S_{22}|^2} + \overline{|S_{12}|^2})^2}} \tag{41}$$

12. The Degree Of Coherence (Orientation)

The degree of coherence, μ, is the complex correlation factor between intensities in two recieving channels and it can be expressed as:

$$\mu = \frac{J_{12}}{\sqrt{J_{11}J_{22}}} \tag{42}$$

Thus, when polarization 1 is transmitted the degree of coherence is given by:

$$\mu_1 = \frac{\overline{S_{11}S_{12}^*}}{[\overline{|S_{11}|^2} \cdot \overline{|S_{12}|^2}]^{1/2}} \tag{43}$$

Similarly, when polarization 2 is transmitted, the degree of coherence becomes:

$$\mu_2 = \frac{\overline{S_{22}\ S_{12}^*}}{[\overline{|S_{12}|^2} \cdot \overline{|S_{22}|^2}]^{1/2}} \tag{44}$$

13. Mean And Spread Of The Optimal Polarizations For Partially Polarized Signals

When optimal polarizations are calculated as in Section 9, a distribution of co-pol and cross-pol nulls will then result from fluctuations due to the motion of the scatterers in the distributed target.

The mean values of the optimal polarizations are σ_{co} and σ_x, as calculated in Section 9. The variances of the clustering about the mean values are derived below.

In Section 2, it was assumed that S_{11}, S_{12} and S_{22} were Gaussian. Thus, η'_{11} is also Gaussian.

Now, from equation (32), it can be seen for the polarization 1 being transmitted, that the mean of the square of EDR becomes:

$$<|\eta'_{11}|^2> = \frac{<|S'_{e11}|^2>}{<|S'_{e12}|^2>} = \frac{J'_{11min}}{J'_{22_{max}}} \simeq \frac{\det[J']}{(Tr[J'])^2} \tag{45}$$

Since $<|\eta'_{11}|^2>$ is assumed to be a zero-mean Rayleigh distribution (i.e., $\eta'_{11} = 0$), $<|\eta'_{11}|^2>$ becomes the mean square deviation, and the root mean square deviation (spread) of the co-pol nulls distributed on the Poincaré Sphere is given by:

$$\sigma_{co} \simeq \sqrt{\frac{\det[J']}{(Tr[J'])^2}} \simeq \frac{1}{2}\sqrt{1 - P^2} \tag{46}$$

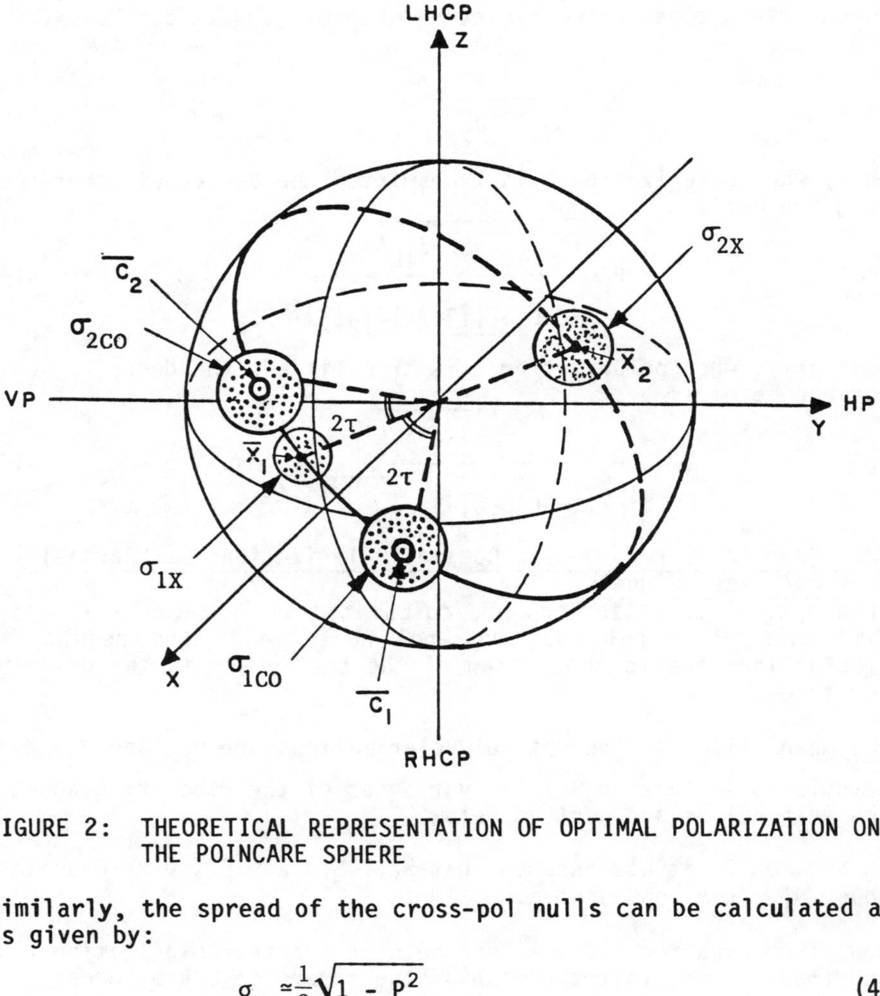

FIGURE 2: THEORETICAL REPRESENTATION OF OPTIMAL POLARIZATION ON THE POINCARE SPHERE

Similarly, the spread of the cross-pol nulls can be calculated and is given by:

$$\sigma_X \simeq \frac{1}{2}\sqrt{1 - P^2} \tag{47}$$

The spread has general validity because all optimal polarizations represented on the Poincaré Sphere are unique.

14. Graphical Representation Of The Optimal Polarizations

The optimal polarizations for completely and partially polarized backscattered signals can be represented on the Poincaré Sphere.

For ensemble of particles (precipitation), the optimal polarizations will distribute themselves on the sphere because of the fluctuations due to the motion of the scatterers as described in Section 2. This distribution of optimal polarizations represented on the Poincaré Sphere along with mean and spread is demonstrated in Figure 2.

In Figure 3, plots of the optimal polarizations are given for rain. These plots for rain on the polarization chart were obtained from Poelman [1983] of the SHAPE-TC. The prescribed clustering distribution is again evident. The theory developed above can be applied for classifying and identifying the types of the clutter.

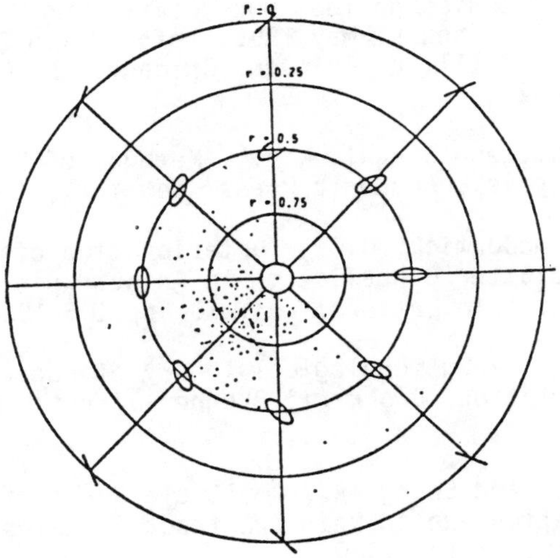

FIGURE 3: OPTIMAL POLARIZATIONS REPRESENTED ON THE POLARIZATION CHART FOR RAIN TRANSMITTING RHCP (Poelman, 1983)

16. Acknowledgements

This research was supported, in part, by Army Research Office contract No. DAAG-20-80-K0027, and Office of Naval Research Grant No. N00001480-C-07730, and we wish to express our sincere gratitude to Drs. James W. Mink, Walter Flood, and Richard J. Brandt for their continued interest in our research. The authors wish to express special thanks to Mr. Richard and Mrs. Deborah Foster for the efficient typing and the supplying of the figures used in this manuscript.

19. References

[1] Boerner, W-M., El-Arini, M.B., Chan, C-Y., and Mastoris, P.M., (1981), "Polarization Dependence in Electromagnetic Inverse Problems", IEEE Trans. A & P, Vol. AP-29, No. 2, pp. 262-271, March, 1981.

[2] Boerner, W-M., and El-Arini, M.B., "Utilization of the Optimal Polarization Concept in Radar Meteorology", 20th Conf. on Radar Meteorology, Boston, pp. 656-665, 1981.

[3] Bolinder, E.F., "Geometric Analysis of Partially Polarized Electromagnetic Waves", IEEE, Trans A&P, Vol. 15(1), pp. 37-40, 1967.

[4] Born, E., and Wolf, E., "Principles of Optics", 2nd Edition, pp. 552. Pergamon, NY, 1964.

[5] Chan, C-W., "Studies on the Power Scattering Matrix of Radar Targets", M.Sc. Thesis, May 1981, Information Engineering Dept., Univ. of Ill. at Chicago, Chicago, IL 60680, #CL-EMID-81-02.

[6] Crispin, J.W., and Siegal, K.M., "Methods of Radar Cross Section Analysis", Acadenic Press, New York, 1969.

[7] Hendry, A., McCormick, G.C., "Deterioration of Circular Polarization Clutter Cancellation in Anisotropic Precipitation Media", Electronic Letters, Vol. 10, pp.165-166, 1974.

[8] Huynen, J-R., "Phenomenological Theory of Radar Targets", Ph.D. Dissertation, Drukkerij Bronder-Offset N.V., Rotterdam, 1970.

[9] Ishimaru, A., and Chung, R., "Multiple Scattering Effects on Wave Propagation Due to Rain", Extract Annuales DesTelecommunications, pp. 11-12, 1980.

[10] Jameson, A.R., "Microphysical Interpretation of Multi-Parameter Radar Measurements in Rain", Part 1, Interpretation of Polarization Measurements and Estimation of Raindrop Shapes, Submitted to the Jornal of Atmos. Sci., 1983.

[11] Jameson, A.R., and Beard, K.V., "Raindrop Axial Ratios", J. Applied Meteorology, Vol. 21, pp. 257-259, 1982.

[12] Kennaugh, E.M., "Effects of Type of Polarization on Echo Characteristics", Reports 389-1 to 389-24, Antenna Lab., Ohio State Univeristy, Columbus OH, 1949 to 1954.

[13] Marshall, J.S., and Hitschfeld, W., "Interpretation of the Fluctuating Echo from Randomly Distributed Scatterers", Canadian J. Physics, Part 1, pp. 962-995, 1953.

[14] McCormick, G.C. "Private Communications", May, 1982.

[15] McCormick, G.C. and Hendry, A., "Polarization-Related Parameters for Rain: Measurements Obtained by Radar", Radio Sci., Vol. 11, pp. 731-740, 1976.

[16] McCormick, G.C. and Hendry, A., "Principles for the Radar Determination of the Polarization Properties of Precipitation", Radio Science, Vol. 10(4), pp. 421-434, 1975.

[17] McCormick, G.C. and Hendry, A., and Barge, B.L., "The Anisotropy of Precipitation Media", Nature, 238, pp. 214-216, 1972.

[18] Nespor, J.D., "Theory and Design of a Dual Polarized Radar for Meteorological Studies", M.Sc. Thesis, University of Illinois at Chicago, Chicago, IL 60680, July, 1983.

[19] Nespor, J.D., Jameson, A.R., Boerner, W-M., "Sensitivity of Optimal Dual Polarization Measurements to the Estimation of Raindrop Shapes", Proc. 21st Conf. on Radar Meteor., American Meteor. Soc., Edmonton, Alberta, Canada, Sept. 1983.

[20] Poelman, A.J., "Multi-Notch Logic-Product Polarization Suppression Filters", Proc. of the 2nd Workshop on Polarimetric Radar Technology, May, 1983.

[21] Poelman, A.J.. "Cross Correlation of Orthogonolly Polarized Backscatter Components", IEEE Trans AES, Vol. 12(6), pp. 674-682, 1976.

[22] Rosien, R., Hammers, D., Ioannidis, G., Bell, J., and Nemit, J., "Implementation Techniques for Polarization Control for ECCM", RADC-TR-79-4, Final Tech. Rept., Feb., 1979.

[23] Seliga, T.A., and Bringi, V.N., "Potential Use of Radar Differential Reflectivity Measurements at Orthogonal Polarizations for Measuring Precipitation", J. Applied Meteor., Vol. 15, pp. 69-76, 1976.

[24] Sinclair, G., "The Transformation and Reception of Ellipitically Polarized Waves", Proc. IRE, Vol. 38, pp. 148-151, Feb., 1950.

III.9 (SR.3*)

RADAR TARGET HANDLING IN CLUTTER, CONSIDERATION OF THE MEASUREMENT SYSTEM.

D. J. R. Stock

AEG-Telefunken

Introduction

Some ideas of Dunn and Howard (1) about glint can be extended to polarisation-glint as concept.
When an antenna illuminates a fixed target in vacuum, the waves reflected from the target combine on the aperture as a vectorsum. Since the target-antenna distance is fixed, no statistical amplitude-changes are detected as a function of time (excluding noise).

If the target is moving, the waves reflected from various parts of the target show distance variations with time which are very small compared to the distance between target and antenna, but may be multiples of $\lambda/4$. The vectorsum of the waves on the antenna-aperture shows a "glint" component.
Poelman (2) based on Born + Wolf (3) gives a similar argument for what may be called polarisation-glint. It is the purpose of this paper to consider this problem with special emphasis on the antenna system rather than on special properties of the target.

We assume following Poelman that a transmitting antenna radiates a wave given by the vector.[1)]

$$e_T = \begin{bmatrix} \cos \varepsilon_T \\ \sin \varepsilon_T \; e^{j\delta_T} \end{bmatrix} \qquad /1/$$

For convenience we suppose that there are two receiving antennas, orthogonally polarized to each other.

W.-M. Boerner et al. (eds.), Inverse Methods in Electromagnetic Imaging - Part 1, 661–671.

$$e_R = \begin{bmatrix} \cos \varepsilon_R \\ \sin \varepsilon_R \, e^{j\delta_R} \end{bmatrix} \quad e_{R\perp} = \begin{bmatrix} \sin \varepsilon_R \, e^{j\delta_R} \\ - \cos \varepsilon_R \end{bmatrix} \qquad /2/$$

If the target be represented by a 2x2 scattering Matrix T, then the backscatter from the target is:

$$T \, e_T \qquad /3/$$

The average backscatter power is then:

$$P_T = \langle (T \, e_T)^* \, T \, e_T \rangle = \langle \widetilde{e_T}^* \, \widetilde{T}^* \, T \, e_T \rangle \qquad /4/$$

a hermitian form.[2]
In addition to the matrix

$$\widetilde{T}^* \, T$$

called "the power polarization matrix of Graves " in the literature, one can introduce the correlation matrix of (2), (3):

$$J = \begin{bmatrix} J_{yy} & J_{xy} \\ J_{yx} & J_{xx} \end{bmatrix} \qquad /5/$$

Both matrices appear in quadratic forms which lead to the following polynomials:

$$\det \, (\widetilde{T}^* \, T - k_0^2 \,) = 0$$
$$\det \, (J - k^2 \,) = 0 \qquad /6/$$

The solutions are easily found by expansion:

$$2 \, k^2 = \mathrm{Tr} \, J \, \left(1 \pm \sqrt{1 - \frac{4 \det J}{(\mathrm{Tr} \, J)^2}} \right) \qquad /7/$$

The polarizationfactor /2/, /3/ is defined as:[3]

$$p = \sqrt{1 - \frac{4 \det J}{(\mathrm{Tr} \, J)^2}} \qquad /8/$$

It may be easily shown that:

$$\langle \widetilde{e_T}^* \, \widetilde{T}^* \, T \, e_T \rangle = \mathrm{Tr} \, J \qquad /10/$$

and when $p = 1$

$$\mathrm{Tr}\ J = k^2{}_{max}$$

A cross-correlation function (2), (3) in form

$$\mu = \frac{J_{xy}}{\sqrt{J_{xx}\ J_{yy}}}$$

plays an important part in the work that follows, since the various receiver systems discussed all make use of two (othogonal) channels.
The backscatter vector sums of waves falling on the two receive-apertures give rise to the bilinear forms:

$$V_R = (\widetilde{T\ e_T})\ e_R$$

$$V_R = (\widetilde{T\ e_T})\ e_{R\perp}$$

with associated powers:

$$P_R = V_R\ V_R^*$$

$$P_{R\perp} = V_{R\perp} V_{R\perp}^*$$

Discussion of the Radarsystem

As discussed above and shown in Fig. 1, the antenna system consists of three antennas, a configuration which may be easily realized for multistatic radar. In pratice today, however, the weather radars for example (4) have only one antenna, so that such systems will be preferred here.[4]
Such a system is switchable:

1. The polarization e_T is radiated, and the copolar polarization e_R received.
2. The crosspolarization $e_{T\perp}$ is radiated and $e_{R\perp}$ received.

superposition is assumed.

The measurement Quantities

As noted above, the backscatter power is:

$$P_T = \langle \widetilde{e_T{}^*}\ \widetilde{T^*}\ T\ e_T \rangle$$

since $\widetilde{T^*}\ T$ is hermitian, P_T is a hermitian form.
Since $P_T = 0$ not only for the trivial case[5] $e_T = 0$,

it is positive semi-definite (PSD).
The received voltages give rise to the power forms:

$$P_R = \widetilde{e_T}\, \widetilde{T}\, e_R \cdot \widetilde{e_T}^*\, \widetilde{T^*}\, e_R^*$$
$$P_{R\perp} = \widetilde{e_T}\, \widetilde{T}\, e_{R\perp} \cdot \widetilde{e_T}^*\, \widetilde{T^*}\, e_{R\perp}^* \qquad /11/$$
$$V_R V_{R\perp}^* = \widetilde{e_T}\, \widetilde{T}\, e_R \cdot \widetilde{e_T}^*\, \widetilde{T^*}\, e_{R\perp}^*$$

We rewrite $e_{R\perp}$ as:

$$e_{R\perp} = e^{-j\delta} \begin{bmatrix} e^{-j\delta_R} & 0 \\ 0 & e^{j\delta_R} \end{bmatrix} \begin{bmatrix} 0 & 1 \\ -1 & 0 \end{bmatrix}$$
$$= e^{-j\delta_R} \Lambda \sigma_1 e_R \qquad /12/$$

where

$$\widetilde{\Lambda}^* \Lambda = 1$$
$$\widetilde{e}_R\, \widetilde{\sigma}_1\, e_{R\perp} = 1$$

so that

$$P_{R\perp} = \widetilde{e_T}\, \widetilde{T}\, \sigma_1\, e_R \cdot \widetilde{e_T}^*\, \widetilde{T^*}\, \sigma_1\, e_R^*$$
$$V_R V_{R\perp}^* = (\widetilde{e_T}\, T\, e_R \cdot \widetilde{e_T}^*\, \widetilde{T^*}\, \Lambda^*\, \sigma_1\, e_R^*)\, e^{j\delta_R} \qquad /13/$$

from /6/ one obtains

$$2\, k_0^2 = \mathrm{Tr}\, (\widetilde{T^*}\, T)\, (1 \pm p_0)$$

where

$$p_0 = \sqrt{1 - \frac{4 \det \widetilde{T^*}\, T}{(\mathrm{Tr}\, \widetilde{T^*}\, T)^2}}$$

Bickel (5) shows that Tr $\widetilde{T^*}\, T$ is invariant for polarizations-rotation of the receiving antenna , and Graves (6) shows the invariance of det $T^*\, T$ under this condition.
According to Mirsky (7)

$$k_{0\,min}^2\, \widetilde{e_T}^*\, e_T \le \widetilde{e_T}^*\, \widetilde{T^*}\, T\, e_T \le k_{0\,max}^2\, \widetilde{e_T}^*\, e_T$$

since

$$\tilde{e}_T^* \, e_T = 1$$

$$k_{0\,min}^2 \le \tilde{e}_T^* \, \tilde{T}^* \, T \, e_T \le k_{0\,max}^2 \qquad /14/$$

Then $k_{0\,max}^2$ must be the entire radiated power. Equality right in the inqualitiy results when the target reflects the entire power, i.e., it is lossless.

The"orthogonal" target

To show the simplicity of the measurement systems, we define an orthogonal target:

$$T_\perp$$

using:

$$P_{R\perp} = \tilde{e}_T \, T_\perp \, e_R \cdot \tilde{e}_T^* \, T_\perp^* \, e_R$$

where

$$T_\perp = \sigma_1 \Lambda_T \, T \, \Lambda \sigma_1 \qquad \text{and } \Lambda_T = \begin{bmatrix} e^{-j\delta_T} & 0 \\ 0 & e^{j\delta_T} \end{bmatrix}$$

is the backscattermatrix of the orthogonal target. Noting that

$$P_{T\perp} = \tilde{e}_T^* \, \tilde{\sigma}_1 \, \Lambda^*_T \, \tilde{T}^* \, \Lambda^* \, \sigma_1 \, \tilde{\sigma}_1 \, \Lambda_T \, \tilde{T} \, \Lambda \, \sigma_1 \, e_T$$

and that

$$\Lambda^*_T \, \sigma_1 \, \tilde{\sigma}_1 \, \Lambda = 1 \qquad \sigma_T = \sigma_R$$

we have

$$P_{T\perp} = \tilde{e}_T^* \, T_\perp^* \, T_\perp \, e_T$$

analogous to /4/.
Following Bickel and Graves one can also show that:

$$\det \tilde{T}_\perp^* \; T_\perp \quad \text{and} \quad \mathrm{Tr}\, \tilde{T}_\perp^* \; T_\perp$$

are invariant under rotation.
McCormick and Hendry (8) show a system in which a network of couplers and phase shifters behind the antenna separate the signal into two orthogonal components. Each component feeds a separate receiver, and the receiver outputs are fed to a correlator. They compare this method (9) with a method of Seliga and Bringi (10) where the polarizations are

switched.
The two methods are really the same when one uses the concept of orthogonal targets. As shown above, the orthogonal splitting can take place in the target plane, or in the network as shown by McCormick and Hendry.

Antenna adjustment for measurement

McCormick and Hendry (11) show that when the antenna polarization is adjusted so that a minimum exists in one of the orthogonal channels, one obtains:

$$\frac{k^2_-}{k^2_+} = \frac{\text{Tr } J - \sqrt{\text{Tr } J - 4 \det J}}{\text{Tr } J + \sqrt{\text{Tr } J - 4 \det J}} \approx \frac{\det J}{(\text{Tr } J)^2}$$

(our notation)
They further note that a non-zero value in one channel ($k^2_- \neq 0$) results from a non-coherent generator[6)].
The concept of orthogonal targets allows this antenna adjustment to be made, although from the following considerations this may not be necessary.
One defines the orthogonal matrix J by a derivation using $e_{T\perp}$ instead of e_T.
It is then easy to show in general without specification of the target that although Tr J is not an invariant under polarisation-rotation,

$$\text{Tr } (J + J_\perp)$$

is independent of ε_T, δ_T. That is, Tr ($J + J_\perp$) may be measured with two arbitrary orthogonal polarizations.
Further this target invariant is related to Graves' invariant:

$$\text{Tr } (\widetilde{T^*}\, T) = \text{Tr } (J + J_\perp)$$

An alternative statement of the McCormick and Hendry-method is: using /14/ the extreme value of $\widetilde{T^*}\, T$ can be obtained by antenna adjustment.

$$k_{0\,-}^{2} \leq \text{Tr } J \leq k_{0\,+}^{2}$$

$$k_{0\,-\perp}^{2} \leq \text{Tr } J_\perp \leq k_{0\,+\perp}^{2}$$

and the constant values are:

$$\mathrm{Tr}\ (J + J_{\perp}) = \mathrm{Tr}\ \widetilde{T}^{*}\ T = k_{O+}^{2} + k_{O-}^{2}$$

of special interest here is the result that the properties of the polarization-switchable elements of a phased-array antenna can be used (12). The measurement steps are:

1. The arbitrary polarization e_T is radiated.
2. The copolarization e_R is received.
3. The antenna is switched to the orthogonal polarization.
4. The polarization $e_{T\perp}$ is radiated.
5. The "cross"-polarization $e_{R\perp}$ is received.

see Fig. 2 and 3.

Summary

The problem of target glint as expressed by the J-matrix was considered here more from the standpoint of the measuring system than from target properties. Only pairs of orthogonal targets were considered in order to throw emphasis on the antenna-receiver system. It is then apparent that the various systems in the literature are conceptually identical.

Acknowledgement

Discussion with H. Trogus, AEG-TELEFUNKEN concerning the KR-75 coastel Radar with phased-array antenna is gratefully acknowledged.

Index:

1.) This form is chosen to emphasize the polarisation properties. The wave is described fully by:

$$\vec{E} = \vec{E_T}\ e^{j(\omega t - \beta z)}$$

where $\vec{E_T} = E_O\ \vec{e_T}$

and $\vec{e_T} = \vec{1}_x \cos \varepsilon_T + \vec{1}_y \sin \varepsilon_T\ e^{j\delta_T}$

2.) If T is a n x n-matrix, $\widetilde{T}$ is its transpose, T* its complex conjugate, T^{-1} the inverse, det T the determinant, and Tr (T) the trace of T. T is hermitian, when

FIG. 1

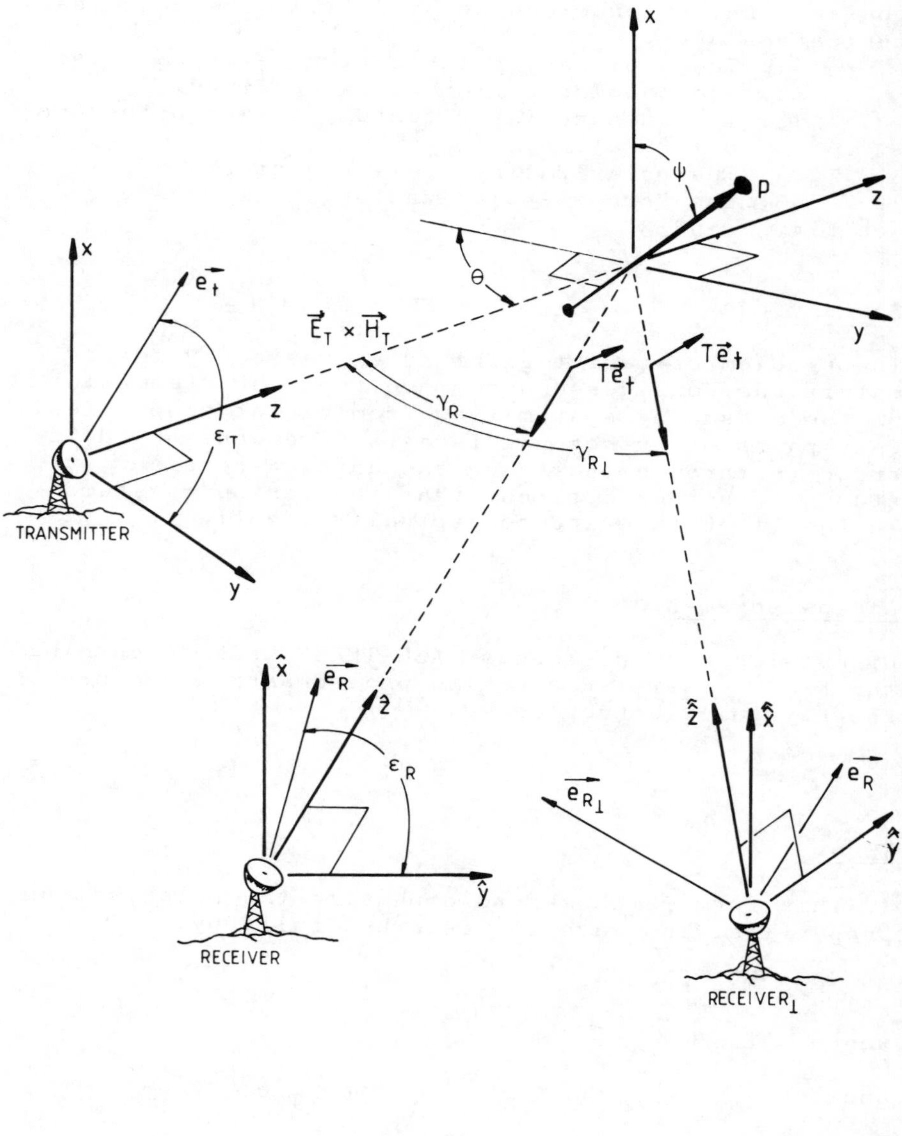
x
ψ
p
z
Θ
y
x
$\vec{e_t}$
$\vec{E}_T \times \vec{H}_T$
$T\vec{e}_t$
$T\vec{e}_t$
z
γ_R
ε_T
$\gamma_{R\perp}$
TRANSMITTER
y
$\hat{x}$
$\vec{e_R}$
$\hat{z}$
ε_R
$\hat{y}$
RECEIVER
$\hat{\hat{z}}$
$\hat{\hat{x}}$
$\vec{e_R}$
$\vec{e_{R\perp}}$
$\hat{\hat{y}}$
RECEIVER$_\perp$

FIG. 2

Receiving antenna

orthogonal polarized receiving antenna

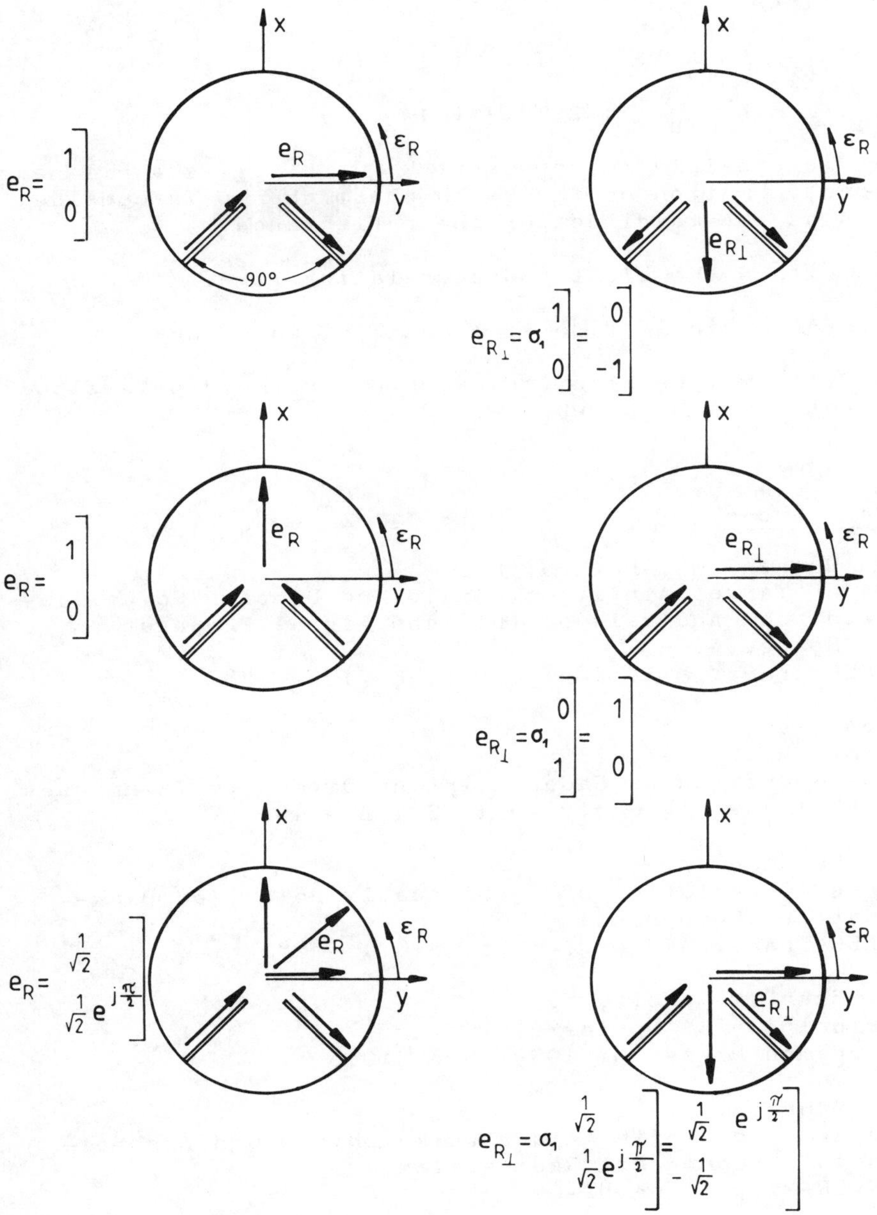

$T^* = \tilde{T}$, T is unitary when $\tilde{T}^* \cdot T = 1$. For a complex matrix T, $\tilde{T}^* \cdot T$ is hermitian.

3.) Since

$$k^2_{max} = 1/2 \; Tr \; J \; (1 + p) \qquad /9/$$
$$k^2_{min} = 1/2 \; Tr \; J \; (1 - p)$$

it is possible to represent k^2_{max}/k^2_{min} as a bilinear transformation on the smith chart when p is considered as the magnitude of the reflectance.

4.) within the clutter decorrelation time.

5.) for a single dipole, T and $\tilde{T}^* \cdot T$ have rank 1.

6.) this may be seen for example for the dipole cloud of Poelman (2) when $\Theta_0 = 0$.

References:

1. Dunn, J.H., Howard, D.D.
Radar Target Amplitude, Angle and Doppler Scintillation from Analysis of the Echo Signal Propagation in Space.
IEEE Trans. MTT, Sept. 1968, S 715 - 728

see also
Sims, R.J. Graf, E.R.
The Reduction of Radar Glint by Diversity Techniques.
IEEE Trans. AP, July 1971, S 462 - 468

2. Poelman, A.J.
Cross Correlation of Orthogonally Polarized Backscattern Components.
IEEE Trans. AES Nov. 1976, S 674 - 682

3. Born, M., Wolf, E.
Principles of Optics
Pergamon Press, NY 1959, Sec. 10.8

4. Schroth, A.
Radarmeteorologie in der Wolkenphysik und Ausbreitungsforschung mit Radiowellen.
DFVLR-Mitteilung 82-09

5. Bickel, S.H.
Some Invariant Properties of the Polarization Scattering Matrix
Proc. IEEE Aug. 1965, S 1070 - 1072

6. Graves, C.D.
Radar Polarization Power Scattering Matrix
Proc. IRE Feb. 1956, S 246 - 252

7. Mirsky, L.
An Introduction to Linear Algebra
Clarendon Press, Oxford 1961, S 388

8. McCormick, G.C., Hendry, A.
Principles for the Radar Determination of the Polarization Properties of Precipitation
Radio Science, April 1975, S 421 - 434

9. McCormick, G.C.
Relationship of differential Reflectivity to Correlation in Dual-Polarization Radar.
Electronic Letters, 10 May 1979, S 265 - 266

10. Seliga, T.A., Bringi, V.N.
Potential Use of Radar Differential Reflectivity Measurements of Orthogonal Polarization for Measuring Precipitation
J. Appl. Meteorol. 1976, S 69 - 76

11. McCormick, G.C. Hendry, A.
Techniques for the Determination of the Polarization Properties of Precipitation
Radio Science, Nov. - Dec. 1979, S 1027 - 1040

12. Trogus, H.
Costal Radar KR-75 a Contribution Towards the Prevention of Collisions.
AEG-TELEFUNKEN Progress 1980, S 23 - 30